One Hundred Years of General Relativity

From Genesis and Empirical Foundations to Gravitational Waves, Cosmology and Quantum Gravity

Volume 1

One Hundred Years of General Relativity

From Genesis and Empirical Foundations to Gravitational Waves, Cosmology and Quantum Gravity

Volume 1

Editor

Wei-Tou Ni

National Tsing Hua University, Hsinchu

World Scientific

NEW JERSEY • LONDON • SINGAPORE • BEIJING • SHANGHAI • HONG KONG • TAIPEI • CHENNAI • TOKYO

Published by

World Scientific Publishing Co. Pte. Ltd.
5 Toh Tuck Link, Singapore 596224
USA office: 27 Warren Street, Suite 401-402, Hackensack, NJ 07601
UK office: 57 Shelton Street, Covent Garden, London WC2H 9HE

Library of Congress Cataloging-in-Publication Data
Names: Ni, Wei-Tou, 1944– editor.
Title: One hundred years of general relativity : from genesis and empirical foundations to gravitational waves, cosmology and quantum gravity / editor, Wei-Tou Ni, National Tsing Hua University, Hsinchu.
Description: Singapore ; Hackensack, NJ : World Scientific, [2015] | Includes bibliographical references.
Identifiers: LCCN 2015032705| ISBN 9789814635127 (set : alk. paper) | ISBN 981463512X (set : alk. paper) | ISBN 9789814678483 (v.1 : alk. paper) | ISBN 9814678481 (v.1 : alk. paper) | ISBN 9789814678490 (v.2 : alk. paper) | ISBN 981467849X (v.2 : alk. paper)
Subjects: LCSH: General relativity (Physics)--History. | Gravitational waves. | Cosmology. | Quantum gravity.
Classification: LCC QC173.6 .O54 2015 | DDC 530.11--dc23
LC record available at http://lccn.loc.gov/2015032705

British Library Cataloguing-in-Publication Data
A catalogue record for this book is available from the British Library.

The cover figure is adapted from a figure by David J. Champion (MPI for Radioastronomy).

The articles in this two-volume set were previously published in various issues of *International Journal of Modern Physics D*.

Desk Editor: Ng Kah Fee

Typeset by Stallion Press
Email: enquiries@stallionpress.com

Foreword

General Relativity (GR) is founded on the observation of Mercury perihelion precession anomaly discovered by Le Verrier and improved by Newcomb, the Michelson–Morley experiment, and the precision Eötvös experiment. Its theoretical basis is based on Special Relativity (previously called the restricted theory of relativity), Einstein Equivalence Principle (EEP) and the realization that the metric is the dynamical quantity of gravity together with the Principle of General Covariance and the absence of other dynamical quantities. The establishment of GR in 1915 is a community effort with Albert Einstein clearly playing the dominant role. For one hundred years, its applicability through solar system to cosmology is prevailing. If one includes the cosmological constant (proposed in 1917 by Einstein) in GR, there have not been any fully established non-applicable places. The only possible potential exception is the missing mass (dark matter)-deficient acceleration issue. Dark energy and quantum gravity are needed in the present theoretical foundation of physics; however, more experimental clues are needed. The framework applicability of GR is already demonstrated in theoretical inflation models with quantum fluctuations leading to structure formation with experimentally observed spectrum.

To celebrate the GR centennial, we solicit the writing of 23 chapters in these two volumes consisting of five parts:

Part I. Genesis, Solutions and Energy.
Part II. Empirical Foundations.
Part III. Gravitational Waves.
Part IV. Cosmology.
Part V. Quantum Gravity.

Volume 1 consists of Part I, Part II and Part III; Volume 2 consists of Part IV and Part V.

In Part I, Valerie Messager and Christophe Letellier start in Chapter 1 with a genesis of special relativity to set the stage. They rely on the original literature to make the development clear and connected. Thanks to many thorough researches in the last 50 years, the path to general relativity is clear. A concise exposition of the path is presented in Chapter 2. In Chapter 3, Christian Heinicke and Friedrich Hehl present the historical development and detailed properties of the basic and fundamental spherical Schwarzschild and axisymmetric Kerr solutions. In Chapter 4, Chiang-Mei Chen, James Nester and Roh-Sung Tung expound the important and useful concept of energy with its many facets and various applications.

In Part II, the empirical foundations of GR are examined. First, the cornerstone Einstein Equivalence Principle (EEP) is explored. Ever since about 100 ps (the time of electroweak phase transition or the equivalent/substitute) at the quasi-equilibrium Higgs/intermediate boson energy scale from the Big Bang (or equivalent/substitute), photons and charged particles are abundant. With the premetric formulation of electrodynamics, we examine the tests of EEP via the metric-induced spacetime constitutive tensor density. The non-birefringence of the cosmic electromagnetic wave propagation in spacetime is observed to ultrahigh precision. This constrains the spacetime constitutive tensor density to Maxwell–Lorentz (metric) form plus a scalar (dilaton) degree of freedom and a pseudoscalar (axion) degree of freedom to high precision. The accurate agreement of cosmic microwave background spectrum with the Planck spectrum constrains the fractional change of the cosmic dilaton to be less than 8×10^{-4}. The Galileo weak equivalence principle (WEP) experiments (Eötvös-type experiments) constrain the fractional dilatonic change in the solar system to be less than 10^{-10}. Accompanying the axion degree of freedom is the rotation of linear polarization in the cosmic propagation of electromagnetic waves called cosmic polarization rotation (CPR). Sperello di Serego Alighieri reviews the constraints from radio galaxy observations and CMB polarization observations to give a general constraint of 0.02 rad for the mean (uniform) CPR and also a constraint of 0.02 rad for the CPR fluctuations. In many inflation models dilatons and axions play important roles; these investigations are crucial to give clues or constraints on the models. Frequency and time are the most precise metrological quantities. Their uses in gravity experiments are unavoidable. The use of GR in time synchronization and in GPS, GLONNESS, Galileo and Beidou becomes a folk talk. There are two good ways to compare precision clocks: (i) fiber links; (ii) space optical links using laser ranging. Étienne Samain expounds the space optical link approach and addresses the laser ranging missions T2L2 (Time transfer by Laser link), LRO (Lunar Reconnaissance Orbiter) and LTT (Laser Time Transfer) together with future space mission proposals for fundamental physics, solar system science/navigation in which laser links are of prime importance. Solar-system observation provides the original impetus and the first confirmation of GR. Chapter 8 summarizes the progress of classical solar system tests and explores its potential in the future. Improvement of three or more orders of magnitude is still possible.

Perhaps the most dramatic development in testing relativistic gravity and in improving the dynamical foundations of general relativity is the discovery and observation of pulsars, binary pulsars, millisecond pulsars and double pulsars since 1967, 1974, 1982 and 2003 respectively. Richard Manchester reviews the pulsar observation in its relation with gravity in Chapter 9 with a brief introduction to basic pulsar properties and pulsar timing. He presents a rather thorough account of dynamical tests of GR and the strong equivalence principle together with a lucid but in-depth account of GW detection using pulsar timing arrays (PTAs). See front cover for an illustrative schematic of a PTA.

In 1916 Einstein predicted gravitational waves (GWs) in GR almost immediately after his founding of it. The existence of gravitational waves is the direct consequence of general relativity and unavoidable consequences of all relativistic gravity theories with finite velocity of propagation. Their importance in GR is like that electromagnetic waves in Maxwell–Lorentz theory of electromagnetism. Einstein's general relativity and relativistic gravity theories predict the existence of gravitational waves. Gravitational waves propagate in spacetime forming ripples of spacetime geometry. In the introductory chapter of Part III, Kazuaki Kuroda, Wei-Ping Pan and I review and summarize the complete GW spectrum, the methods of detection, and the detection sensitivities in various frequency bands with a brief introduction to GW sources. At the time Einstein predicted GWs in GR, he estimated that GWs were experimentally not detectable due to feeble strengths. However, thanks to one hundred years of development of experimental methods and technology together with the discovery of various astrophysical compact objects and cosmological sources, GWs are now on the verge of detection in three frequency bands. The very low frequency band (10 fHz–300 pHz) GWs are on the verge of detection by the PTAs; Richard Manchester covers this part in his chapter on pulsars and gravity in Part II. As mentioned in Chapter 10, the observation of PTAs has already constrained the isotropic GW background to a level excluding most current models of supermassive black hole formation. This is a strong signal that PTA observation is on the verge of detecting GWs. The high frequency band (10 Hz–100 kHz) GWs are on the verge of detection by ground-based interferometers; Kazuaki Kuroda addresses the detection methods and the sources in the second chapter of Part III. The extremely low (Hubble) frequency band (1 aHz–10 fHz) GWs may also be on the verge of detection by CMB polarization observations; the present status is briefly reviewed in the introduction chapter of Part III. The low frequency band (100 nHz–100 mHz) and the middle frequency band detections will have the greatest S/N ratios according to the present expectation. We review the sources, goal sensitivities, various mission proposals together with the current supporting activities in the third chapter of Part III. The GW quadrupole radiation formula has already been verified by the binary pulsar observations. In the next hundred years we will see great discoveries and immense focused activities toward the establishment and flourish of GW astronomy and GW cosmology. GW physics and GW astronomy will become a precision discipline in the coming century.

The development of cosmology is most dramatic during the last hundred years. From Kapteyn universe in 1915 of observed disk star system of 10 kpc diameter and 2 kpc thickness with the Sun near its center to full-fledged precision cosmology now is monumental in the human history. It is fortunate that the development of observational cosmology has GR theory as a theoretical basis and goes hand-in-hand with the development of general relativity. This is fortunate both for observational cosmology and for GR. Using the Cosmological Principle Einstein looked into cosmological solutions in GR in 1917. The fast development of observational distance ladder around that time soon extends the reach of astronomy to modern cosmos.

Studies in the fundamental issues on the origins of cosmos lead to anthropological principle, cosmic inflation, and cosmic landscape scenarios. The cosmos is believed to be open in (extended beyond) the Hubble distance scale. Part III consists of seven chapters: Martin Bucher and I present some introductory remarks with a discussion of missing mass-deficient acceleration issue in the first chapter; Marc Davis reviews the observation and evolution of cosmic structure; Martin Bucher give a rather comprehensive exposition of the physics (almost on every aspect of cosmology) of CMB; Xiangcun Meng, Yan Gao and Zhanwen Han review the SNIa as a standardizable distance candle, its nature, its progenitors and its role in the cosmology together with related current issues; Toshifumi Futamase on the gravitational lensing in cosmology; K. Sato and Juni'ichi on cosmic inflation with a brief historical exposition on the development in Japan and Russia; David Chernoff and Henry Tye on inflation and cosmic strings from the point of view of string theory.

The quest for a satisfactory quantum description of gravity began very early. Einstein thought that quantum effects must modify general relativity in his first paper on GWs in 1916. Klein argued that the quantum theory must ultimately modify the role of spatiotemporal concepts in fundamental physics in 1927. Part V on Quantum Gravity consists of 4 chapters. Chapter 20 gives a bird's-eye survey on the development of fundamental ideas of quantum gravity together with possible observations of quantum gravitational effects in the foreseeable future. The classical age (1958–1969; according to the chronological classification of Rovelli) started with ADM canonical formalism and concluded with DeWitt–Wheeler equation and DeWitt's derivation of Feynman rules for perturbative GR. In the middle ages (1970–1983), the discovery of black hole thermodynamics and Hawking's derivation of black hole radiation radically affected our understanding of general relativity. In the renaissance period (1984–1994), there are two influential developments. From the covariant approach, attempts to get rid of infinities merge into string theory. The use of strings and branes extends the theoretical framework of quantum field theory. From the canonical approach, background-independent loop quantum gravity emerged 20 years after DeWitt–Wheeler equation. In Chapter 21, Richard Woodard starts with experiences of two personal academic careers through the classical and middle ages, advocates that the cosmological data from the epoch of primordial inflation is catalyzing the maturation of quantum gravity from speculation into a hard science, explains why quantum gravitational effects from primordial inflation are observable, reviews what has been done in perturbative quantum gravity, tells us what the future holds both theoretically and observationally, and discusses what this tells us about quantum gravity. In Chapter 22, Steven Carlip reviews the discovery of black hole thermodynamics and summarizes the many independent ways of obtaining the thermodynamic and statistical mechanical properties of black holes. This has offered us some early hints about the nature of quantum gravity. Steven then describes some of the remaining puzzles, including the nature of the quantum microstates, the problem of universality, and the information loss paradox. In the last chapter, Dah-Wei Chiou gives us a rather self-contained introductory review on loop quantum gravity — a background-independent nonperturbative approach to a consistent quantum theory of gravity placing emphasis on the fundamental ideas and their significance. The review presents the canonical formulation of loop quantum gravity as the central topic and covers briefly the spin foam theory, the relation to black hole thermodynamics and the loop quantum cosmology with current directions and open issues summarized.

Although we do not yet have a consistent calculable quantum gravity theory which has a good degree of completeness like quantum electrodynamics or quantum chromodynamics, the efforts to find one already led to the consistent renormalization of the gauge theory in 1960's. The new development since 1980's together with more understanding and further development of perturbation theory may give clues to a consistent theory. During these endeavors, the quest for a well-developed quantum gravity phenomenology including the quest to find a correct inflationary (or non-inflationary) scenario may play a significant role.

The hope is that we will have one within a generation. This book is written and assembled for graduate students and general scientific-oriented readers alike. Each chapter is basically a review article. The five Parts are interconnected. Different combinations can be designed for special topics for graduate students and advanced undergraduates. For example, following combinations are suitable for each topic named:

(i) Basics (Selected Topics in GR): Part I, Chapters 8, 9, 10, 13, 20;
(ii) Empirical Foundations (Empirical Foundations of Relativistic Gravity): Chapter 2, Part II, Chapters 10, 11, 13, 14, 16, 20;
(iii) Gravitational Waves: Chapters 2, 9, Part III, Chapters 15, 18, 19;
(iv) Cosmology: Chapters 5, 6, 10, 12, Part IV, Chapter 20;
(v) Quantum Gravity: Chapters 3, 4, 10, 18, 19, Part V.

There can be various other combinations too.

We are grateful to all contributors for agreeing to write comprehensive reviews to make this publication possible. We would also like to thank all the referees for their valuable comments and suggestions: Martin Bucher, Stephen Carlip, Dah-Wei Chiou, Sperello di Serego Alighieri, Angela Di Virgilio, John Eldridge, Jeremy Gray, Friedrich Hehl, Jim Hough, Ekaterina Koptelova, Ettore Majorana, James Nester, Ulrich Schreiber, Alexei Starobinsky, David Tanner, Richard Woodard, An-Ming Wu, Masahide Yamaguchi. We thank the World Scientific staff, especially Dr. K. K. Phua and Kah Fee Ng for their generous support in completing the book.

We dedicate this two-volume GR centennial book to the founders of GR and various communities who have contributed to this dramatic century of development and applications of GR.

Wei-Tou Ni
November, 2015

Note Added in Proof

After the foreword was written, LIGO Scientific and Virgo Collaborations announced in February 2016 and in June 2016 the first direct detections of gravitational waves (GWs) by LIGO Hanford and LIGO Livingston detectors in September 2015 and in December 2015. With the LIGO discovery announcements, two important things are verified: (i) GWs are directly detected in the solar-system; (ii) Black holes (BHs), binary BHs and BH coalescences are discovered and measured experimentally and directly with the distances reached more than 1 billion light years. These discoveries constitute the best celebration of the centennial of the genesis of general relativity. We refer the readers to Refs. 1 and 2 for the discovery and Refs. 3 and 4 for a brief history of gravitational wave research.

A web page will be set up for updates of the reviews of these two volumes. Please see http://astrod.wikispaces.com/ for announcement.

References

1. B. P. Abbott *et al.* (LIGO Scientific and Virgo Collaborations), Observation of gravitational waves from a binary black hole merger, *Phys. Rev. Lett.* **116** (2016) 061102.
2. B. P. Abbott *et al.* (LIGO Scientific and Virgo Collaborations), GW151226: Observation of gravitational waves from a 22-solar-mass binary black hole coalescence, *Phys. Rev. Lett.* **116** (2016) 241103.
3. J. L. Cervantes-Cota, S. Galindo-Uribarri and G.F. Smoot, A brief history of gravitational waves, *Universe* **2**(22) (2016) 09400.
4. C.-M. Chen, J. M. Nester, W.-T. Ni, A brief history of gravitational wave research, *Chinese Journal of Physics* **55** (2017) 142–169.

Contents

Volume 1

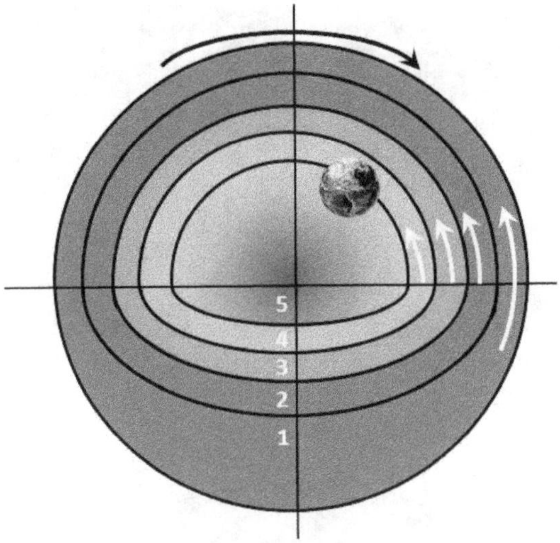

Chapter 1, Fig. 1. System of five concentric spheres representing the motion of a planet (here the Earth) in Eudoxus' mathematical representation.

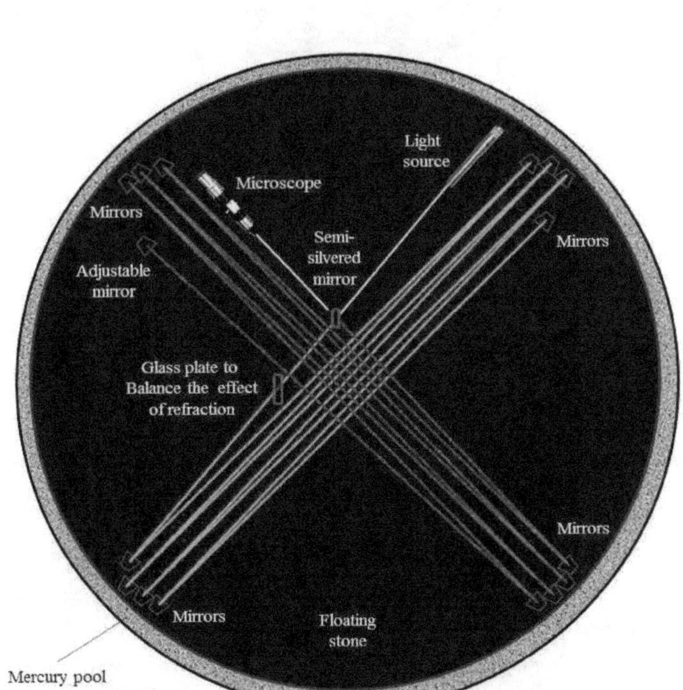

Chapter 1, Fig. A.2. Sketch of the experimental device conducted by Michelson and Morley in 1887.

Chapter 3, Fig. 7. Not quite seriously: “Schwarzschild” (left) versus “Kerr” (right).

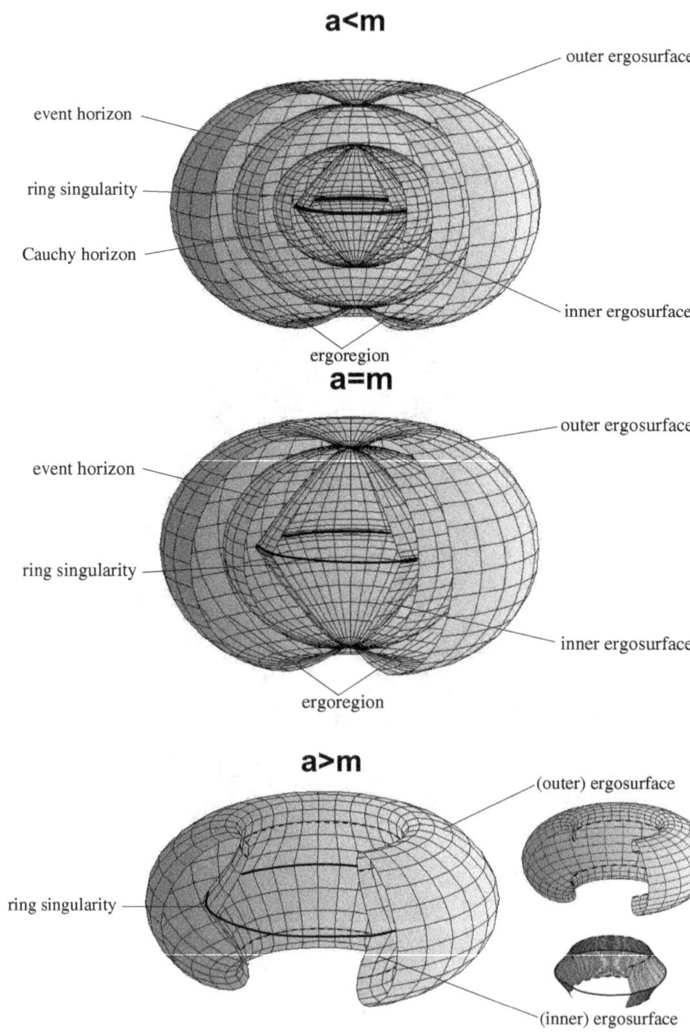

Chapter 3, Fig. 12. Ergosurfaces, horizons, and singularity for slow, extremal (“critical”), and fast Kerr black holes.

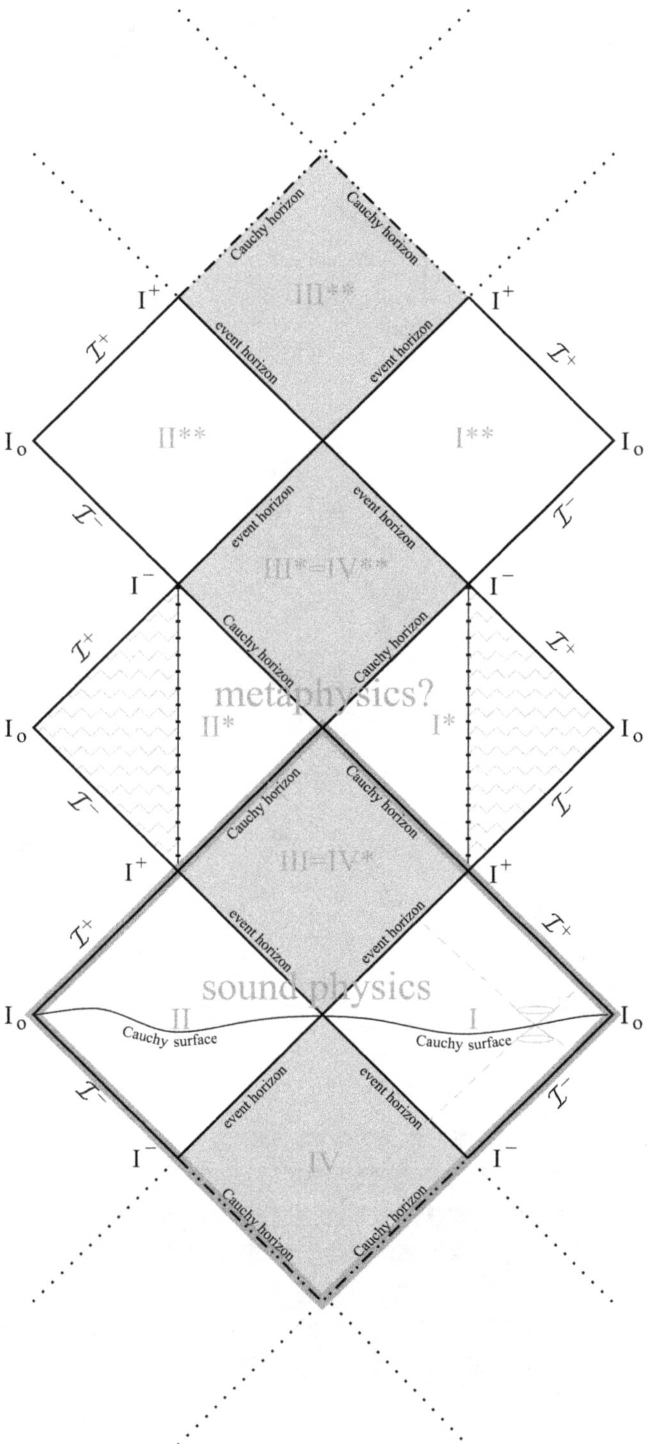

Chapter 3, Fig. 15. Maximal analytic extension of the Kerr spacetime.

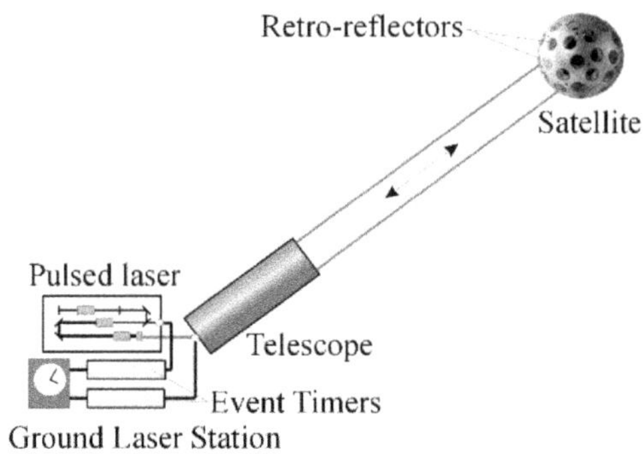

Chapter 7, Fig. 1.

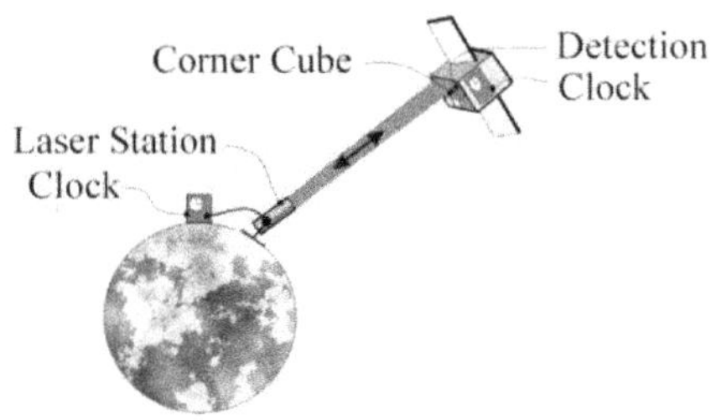

Chapter 7, Fig. 2.

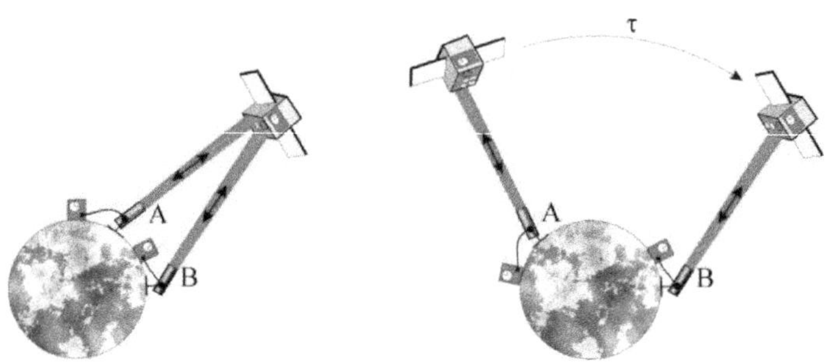

Chapter 7, Fig. 3.

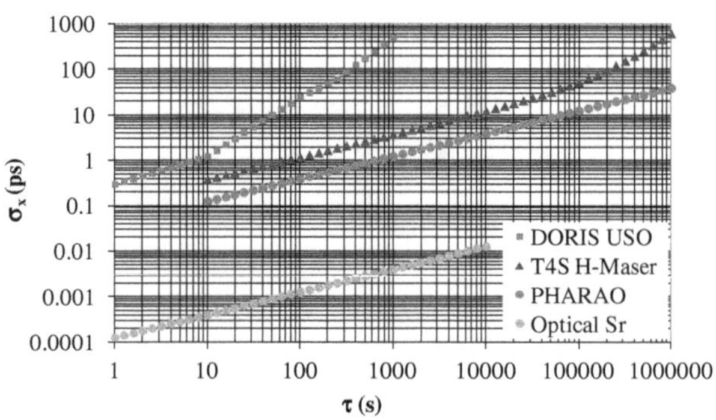

Chapter 7, Fig. 5.

Chapter 7, Fig. 10. Photography of the MeO laser station at Grasse (France) built by the end of the seventies for lunar laser ranging and redesigned for satellite and time transfer in 2005.

Chapter 7, Fig. 12. Photography of units A (right) and B (left) of the T2L2 space instrument. The cylinders on the right are the detection modules (linear and nonlinear). The LRA module is not integrated into the photo.

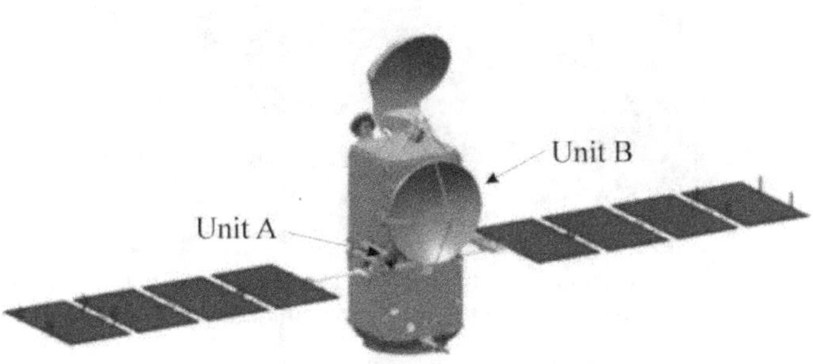

Chapter 7, Fig. 13. CAO view of the whole Jason-2 satellite. T2L2 instrumentation is shared into two units A and B respectively outside and inside the satellite.

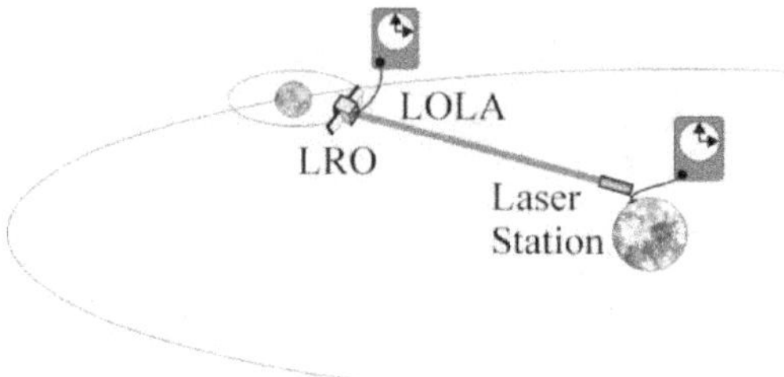

Chapter 7, Fig. 20. Earth to Moon one-way laser ranging with LRO through the LOLA of the spacecraft and some laser stations on ground.

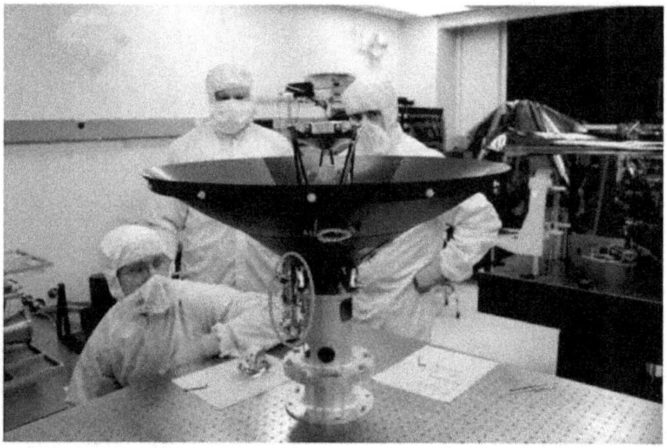

Chapter 7, Fig. 21. RF antenna. The Laser ranging receiver telescope is on the left from the center of the main RF antenna (red ellipse). LOLA is coupled with the telescope through an optical fiber. (Courtesy: NASA Goddard Space Flight Center).

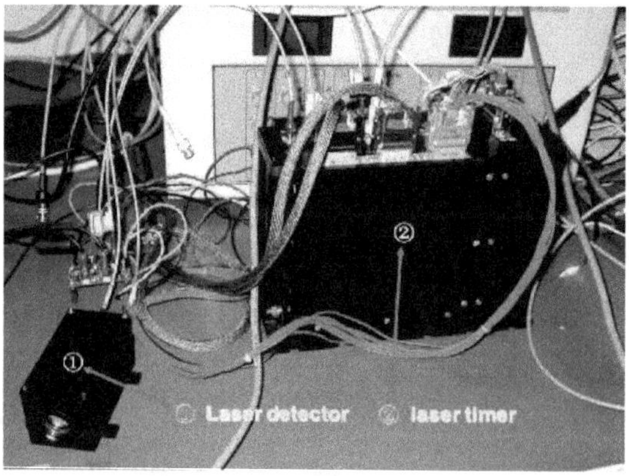

Chapter 7, Fig. 23. LTT equipment. On the left is the detector; in the middle lies the main electronic package including the event timer. (Courtesy: Shanghai Astronomical observatory).

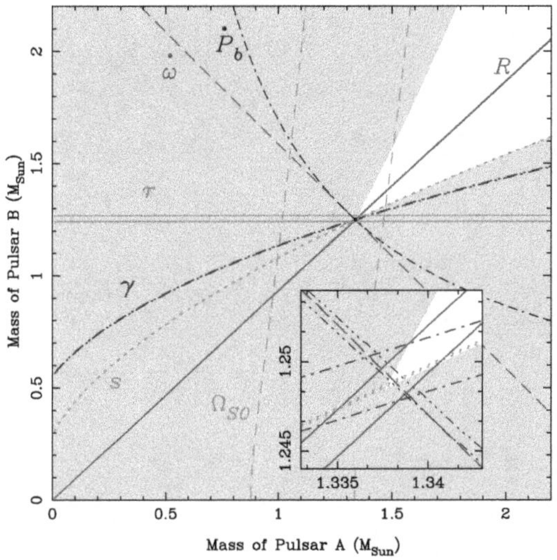

Chapter 9, Fig. 8. Plot of companion mass (m_2) versus pulsar mass (m_1) for PSR J0737−3039 with observed constraints interpreted in the framework of GR. The inset shows the central region at an expanded scale, illustrating that GR is consistent with all constraints (M. Kramer, private communication).

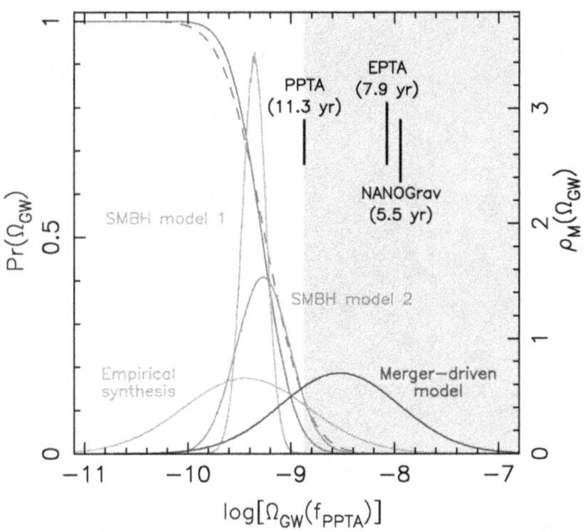

Chapter 9, Fig. 20. Limits on the relative energy density of the GWB, Ω_{GW} at a GW frequency of 2.8 nHz based on the *PPTA* data sets, together with predictions for Ω_{GW} based on several different models for the GWB.[119] The solid and dashed lines that are asymptotic to 1.0 at low Ω_{GW} show the probability *Pr* that a GWB signal of energy density Ω_{GW} can exist in the *PPTA* data sets, based on Gaussian and non-Gaussian GWB statistics respectively. The shaded region is ruled out with 95% confidence by the *PPTA* data. Corresponding limits from analysis of *EPTA*[138] and *NANOGrav*[40] data sets, scaled to $f_{GW} = 2.8\,\text{nHz}$, are also shown. The Gaussian curves show the probability density functions ρ_M for the existence of a GWB with energy density Ω_{GW} based on a merger-driven model for growth of SMBHs in galaxies,[89] an empirical synthesis of observational constraints on SMBHs in galaxies,[113] and based on the Millennium dark matter simulations[18] together with semi-analytic models for growth of SMBHs in galaxies (see Ref. 119 for more details).

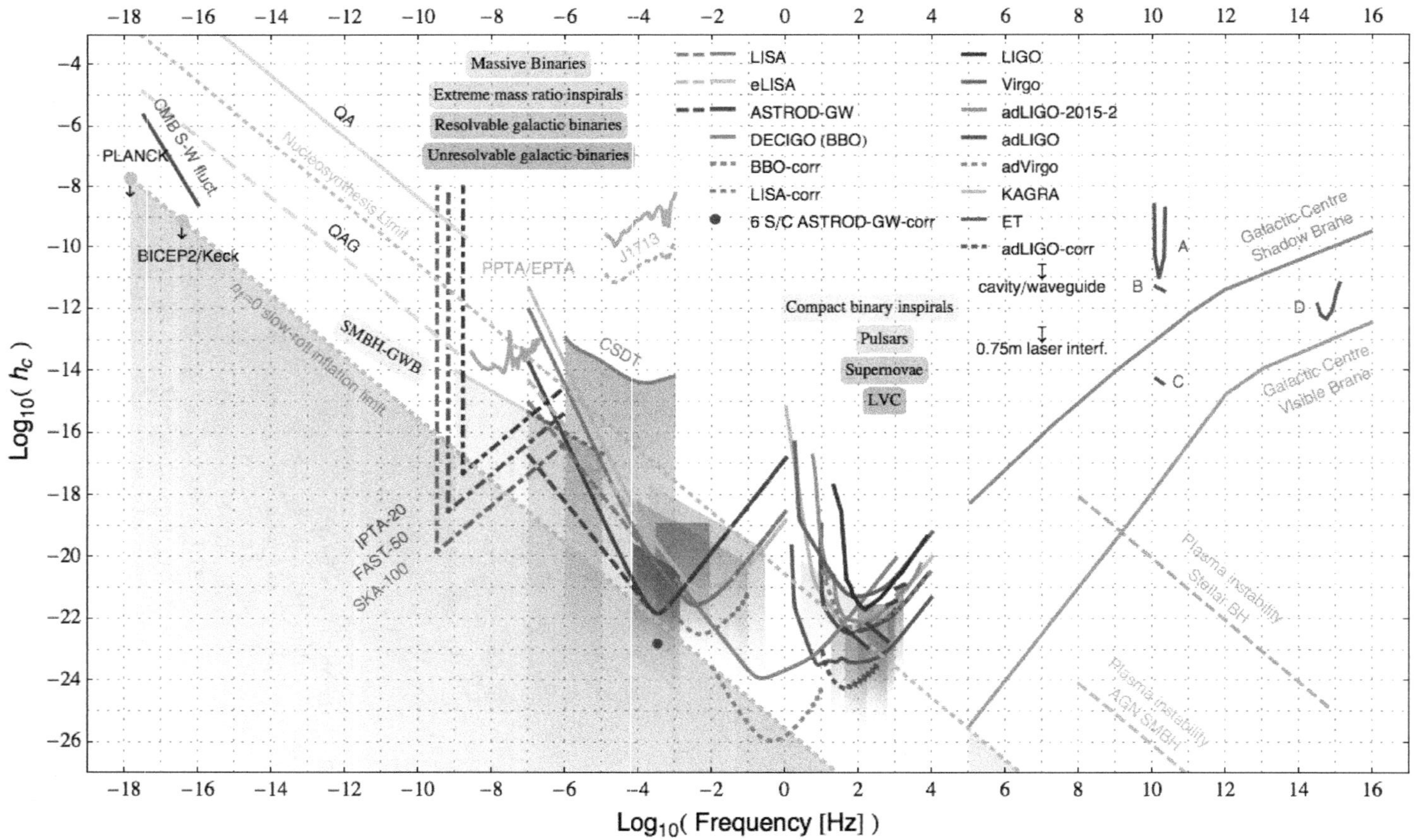

Chapter 10, Fig. 2. Characteristic strain h_c versus frequency for various GW detectors and sources. [QA: Quasar Astrometry; QAG: Quasar Astrometry Goal; LVC: LIGO-Virgo Constraints; CSDT: Cassini Spacecraft Doppler Tracking; SMBH-GWB: Supermassive Black Hole-GW Background.]

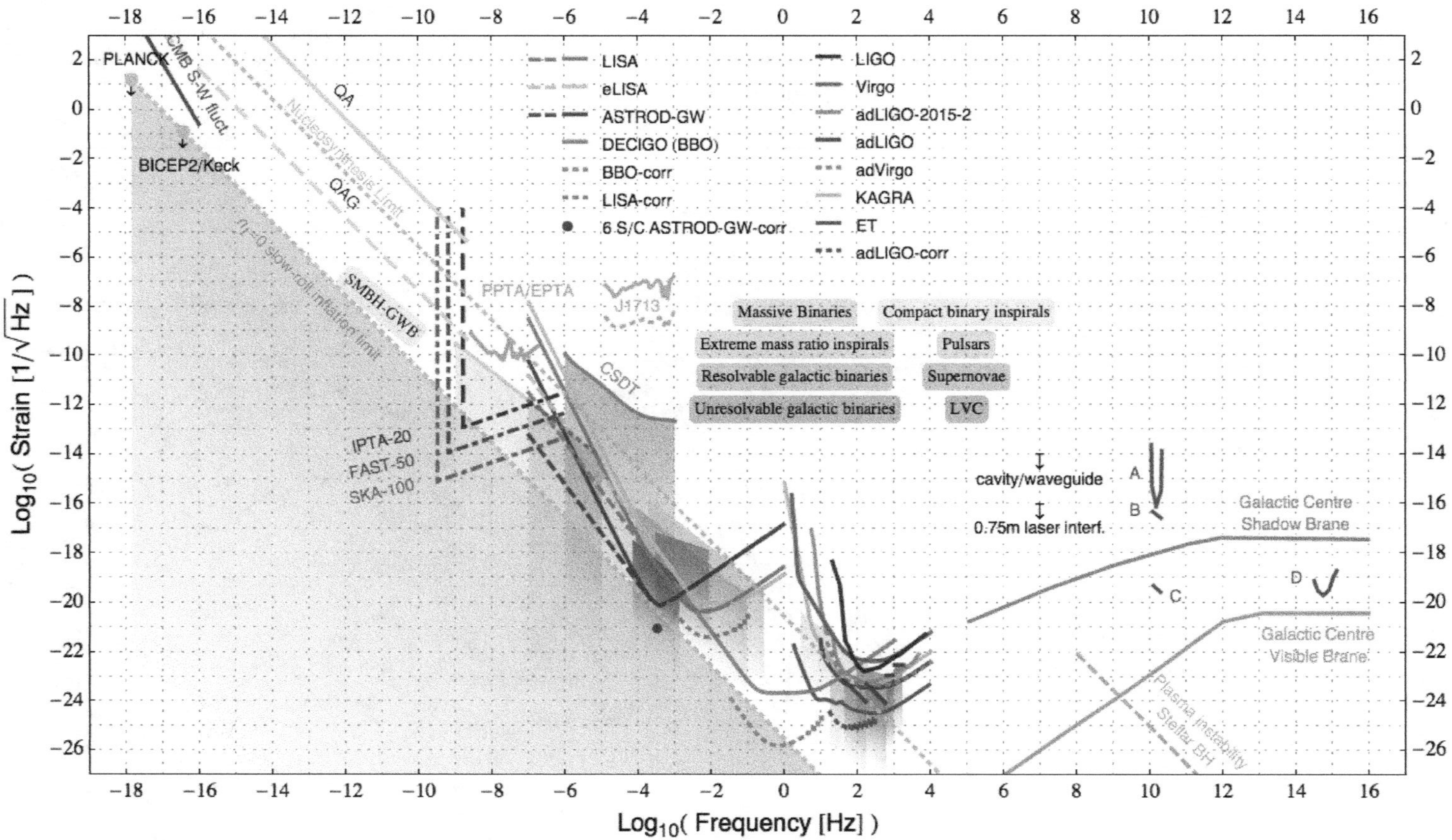

Chapter 10, Fig. 3. Strain psd amplitude versus frequency for various GW detectors and GW sources. See Fig. 2 caption for the meaning of various acronyms.

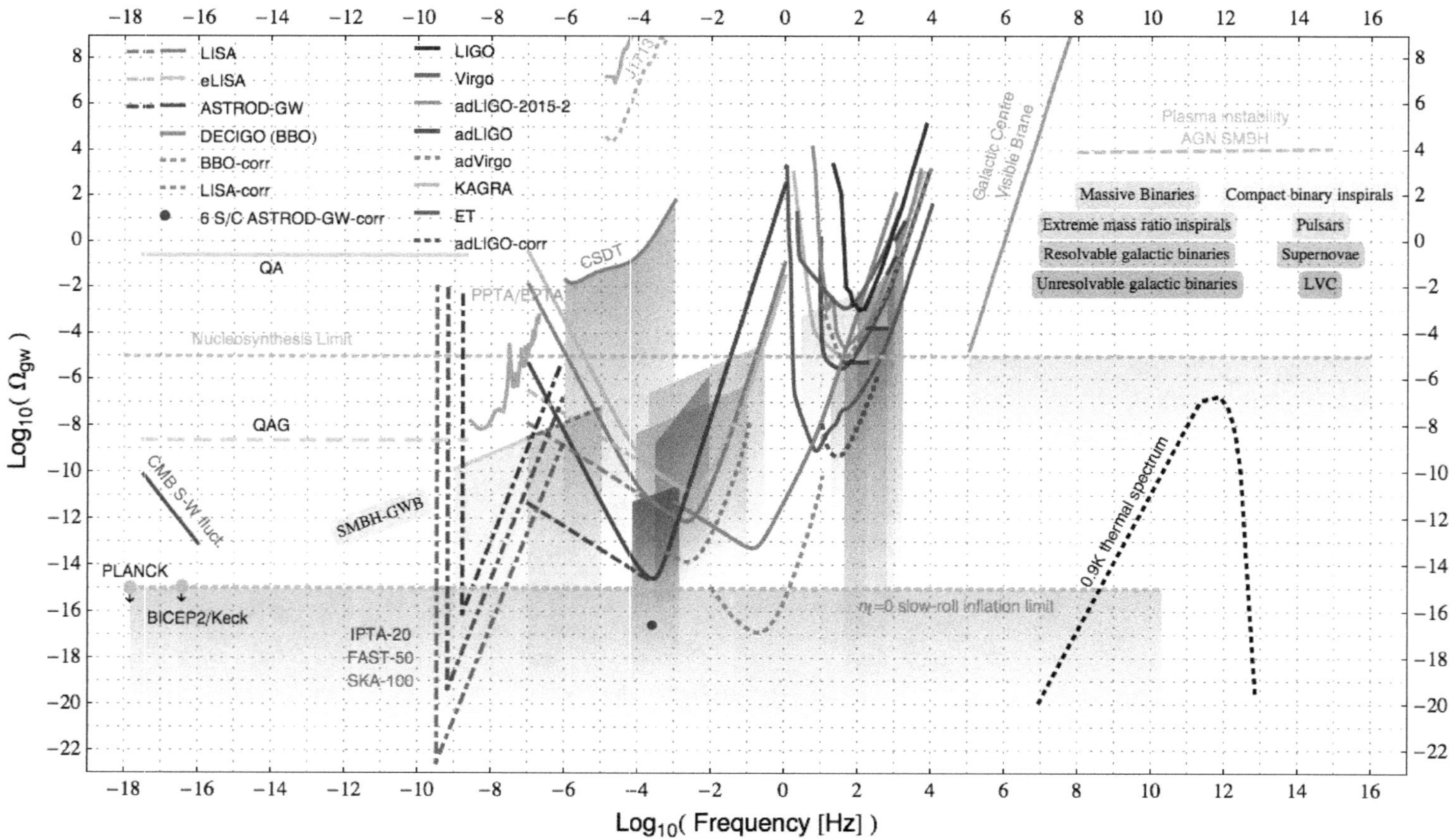

Chapter 10, Fig. 4. Normalized GW spectral energy density $\Omega_{\rm gw}$ versus frequency for GW detector sensitivities and GW sources. See Fig. 2 caption for the meaning of various acronyms.

Chapter 11, Fig. 8. Explorer resonant antenna in Rome. *Source*: Photo is reprinted from http://www.roma1.infn.it/rog/explorer/.

Chapter 11, Fig. 11. AURIGA cryogenic resonant antenna is the twin of NAUTILUS, which is placed at Legnaro in Padova. *Source*: Photo is reprinted from http://www.auriga.lnl.infn.it.

Chapter 11, Fig. 21. LIGO project started in 1994 to construct a pair of 4 km baseline length scale facilities for laser interferometers separated by 3030 km, which were in Livingston, Louisiana and in Hanford, Washington. *Source*: These pictures are taken from Ref. 72.

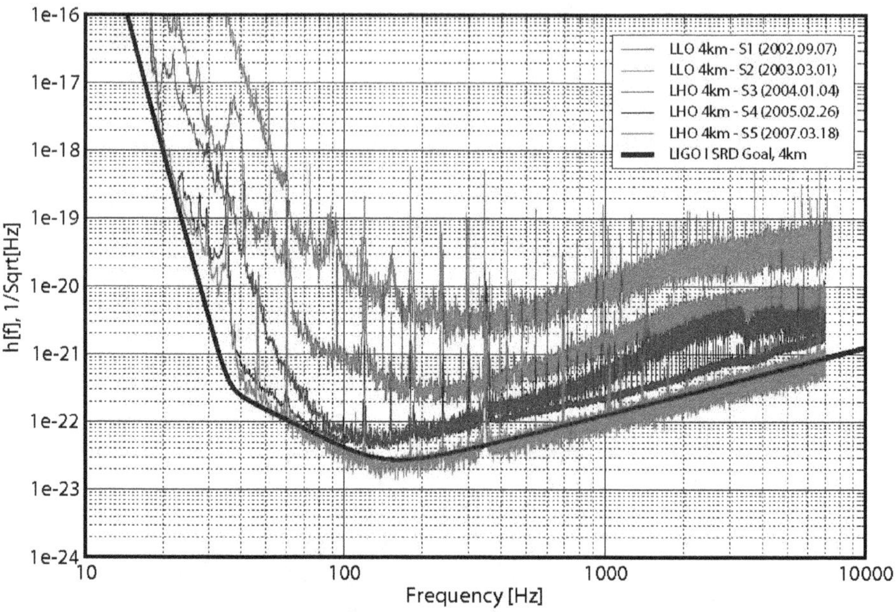

Chapter 11, Fig. 22. During the initial LIGO project, it took about five years to attain the target design sensitivity after the installation. However, much more short period is expected to achieve the sensitivity of the advanced LIGO, the installation of which is finished in 2015.

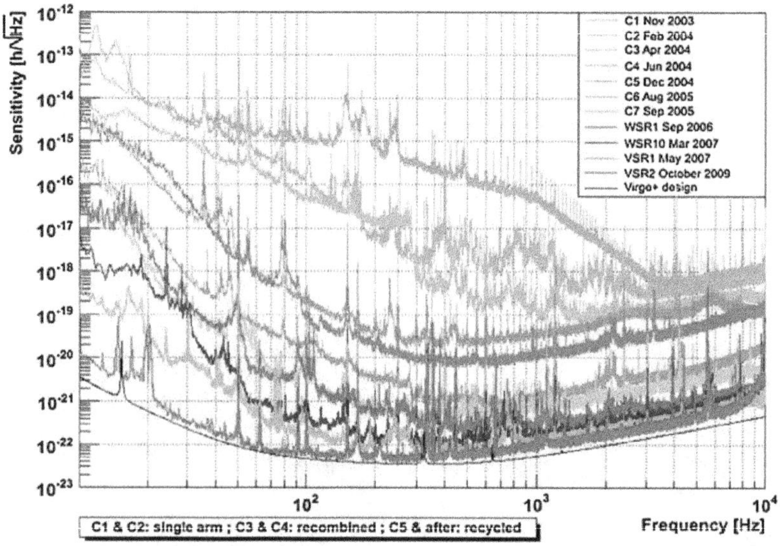

Chapter 11, Fig. 26. Sensitivity improvement of Virgo. The sensitivity is inferior to that of LIGO at around mid frequencies. However, it is much better than LIGO at lower frequencies. *Source*: The figure is taken from a paper after VSR2 (see Ref. 153).

Chapter 11, Fig. 35. An end cryostat of CLIO placed underground at Kamioka mine. Thermal noise at cryogenic temperature, 10 K, was achieved in 2009.

Chapter 11, Fig. 37. KAGRA is a 3 km baseline length power-recycled Fabry–Perot Michelson interferometer having RSE configuration with cryogenic mirrors, and is placed underground at Kamioka in Gifu prefecture.

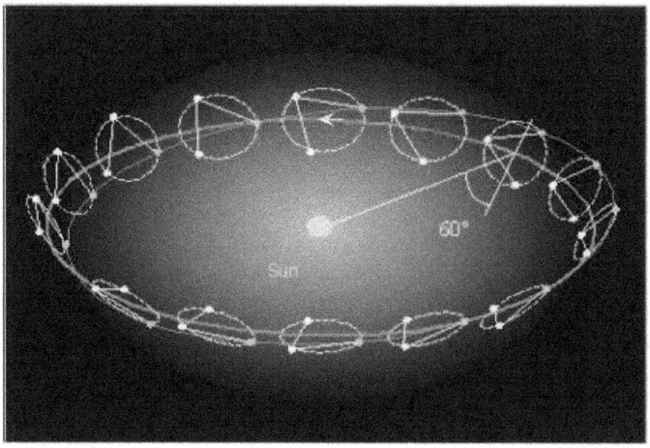

Chapter 12, Fig. 1. Schematic of LISA-type orbit configuration in Earthlike solar orbit.

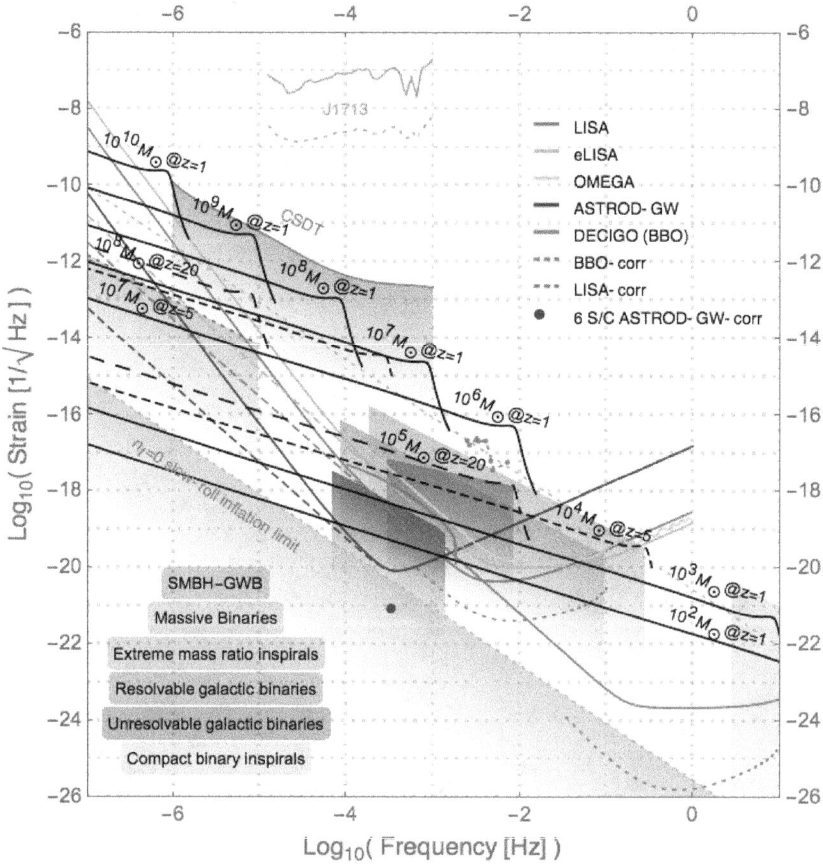

Chapter 12, Fig. 4. Strain PSD amplitude versus frequency for various GW detectors and GW sources. The black lines show the inspiral, coalescence and oscillation phases of GW emission from various equal-mass black-hole binary mergers in circular orbits at various redshift: solid line, $z = 1$; dashed line, $z = 5$; long-dashed line $z = 20$. See text for more explanation. [Cassini Spacecraft Doppler Tracking (CSDT); Supermassive Black Hole-GW Background (SMBH-GWB).]

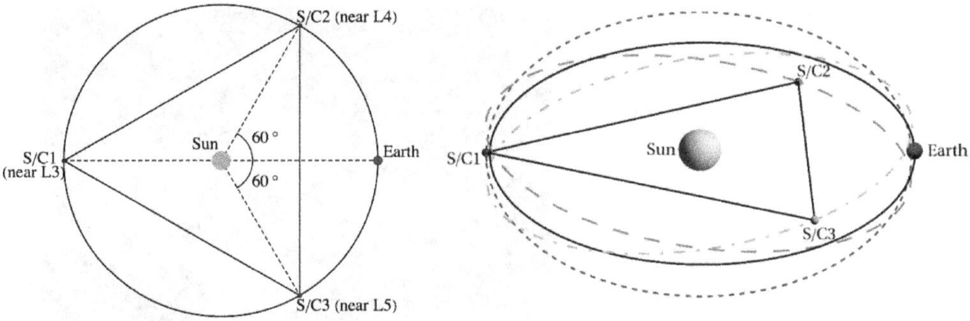

Chapter 12, Fig. 2. Schematic of ASTROD-GW orbit configuration with inclination. Left, projection on the ecliptic plane; Right, 3D view with the scale of vertical axis multiplied tenfold.[3,41]

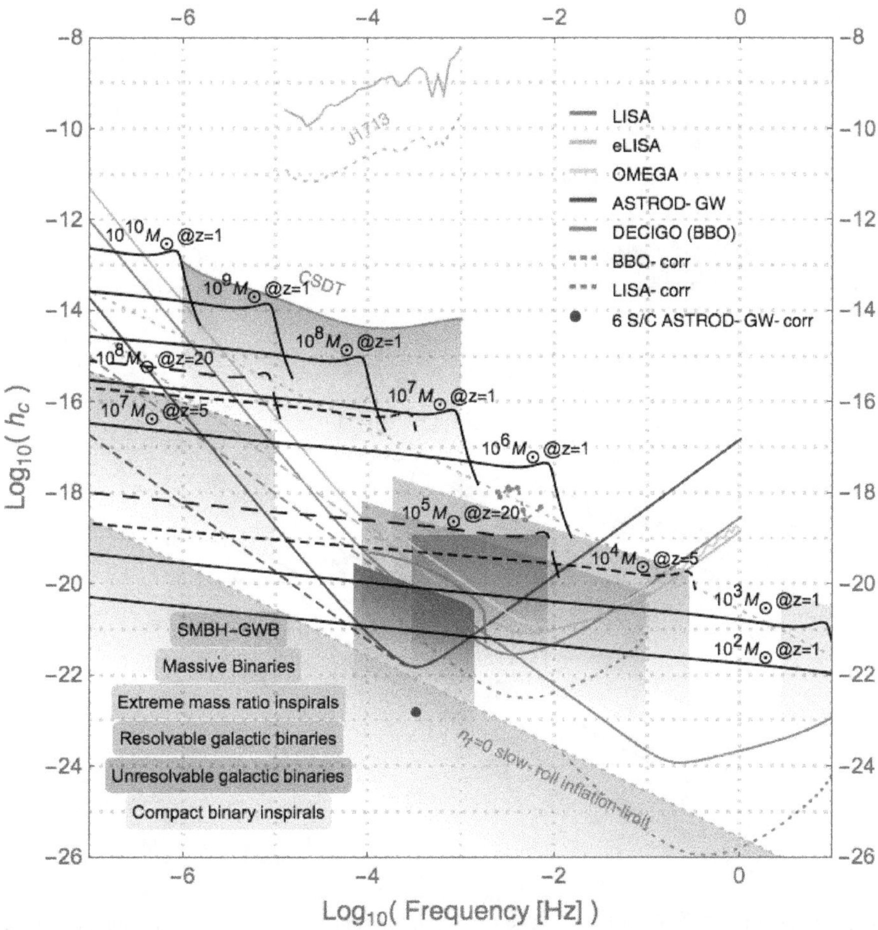

Chapter 12, Fig. 5. Characteristic strain h_c versus frequency for various GW detectors and sources. The black lines show the inspiral, coalescence and oscillation phases of GW emission from various equal-mass black-hole binary mergers in circular orbits at various redshift: solid line, $z = 1$; dashed line, $z = 5$; long-dashed line $z = 20$. See text for more explanation. [Cassini Spacecraft Doppler Tracking (CSDT); Supermassive Black Hole-GW Background (SMBH-GWB).]

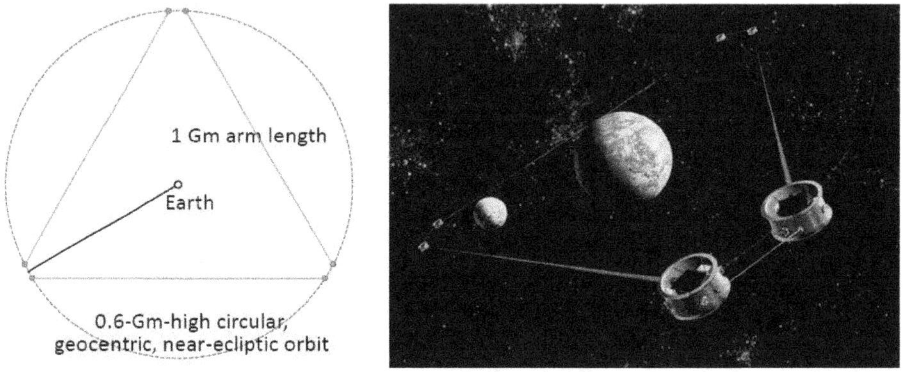

Chapter 12, Fig. 3. Schematic (left) and artist's conception (right) of the OMEGA mission configuration.[55]

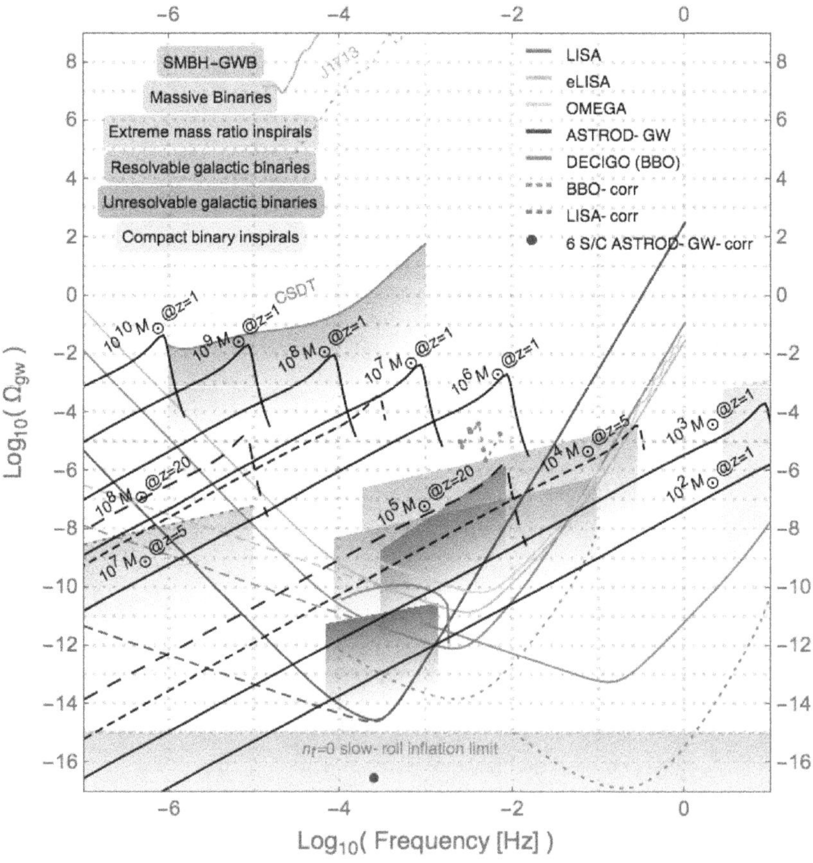

Chapter 12, page I-587. Normalized GW spectral energy density Ω_{gw} versus frequency for various GW detectors and GW sources. The black lines show the inspiral, coalescence and oscillation phases of GW emission from various equal-mass black-hole binary mergers in circular orbits at various redshift: solid line, $z = 1$; dashed line, $z = 5$; long-dashed line $z = 20$. See text for more explanation. [Cassini Spacecraft Doppler Tracking (CSDT); Supermassive Black Hole-GW Background (SMBH-GWB).]

Part I. Genesis, Solutions and Energy

Chapter 1

A genesis of special relativity

Valérie Messager* and Christophe Letellier†

CORIA – Normandie Université,
CNRS – Université et INSA de Rouen,
Campus Universitaire du Madrillet,
F-76801 Saint-Etienne du Rouvray, France
**valerie.messager@coria.fr*
†*christophe.letellier@coria.fr*

The genesis of special relativity is intimately related to the development of the theory of light propagation. When optical phenomena were described, there are typically two kinds of theories: (i) One based on light rays and light particles and (ii) one considering the light as waves. When diffraction and refraction were experimentally discovered, light propagation became more often described in terms of waves. Nevertheless, when attempts were made to explain how light was propagated, it was nearly always in terms of a corpuscular theory combined with an ether, a subtle medium supporting the waves. Consequently, most of the theories from Newton's to those developed in the 19th century were dual and required the existence of an ether. We therefore used the ether as our Ariadne thread for explaining how the principle of relativity became generalized to the so-called Maxwell equations around the 1900's. Our aim is more to describe how the successive ideas were developed and interconnected than framing the context in which these ideas arose.

Keywords: Special relativity; ether; light propagation; electrodynamics; Maxwell equations.

"*Comprendre la genèse de la science* [. . .] *est indispensable pour l'intelligence complète de la science elle-même*"

Henri Poincaré, *La Science et l'Hypothèse*, p. 163, 1902.

1. Introduction

One of the very first contributions to the principle of relativity is due to Nicola Cusano (1401–1464), who addressed the question: "*How would a person know that a ship was in movement, if, from the ship in the middle of the river, the banks were invisible to him and he was ignorant of the fact that water flows?*"[1] He was followed by Giordano Bruno (1548–1600), an Italian Dominican friar who was condemned for his theological positions which were, for instance, that "*Christ was not God but merely an unusually skillful magician, that the Holy Ghost is the soul of the world, that the Devil will be saved, etc.*"[2] In one of his books,[3] he considered the problem of someone dropping a stone from the top of a mast of a ship moving at a constant velocity. Bruno thus stated that a difference exists between the motion of a ship and the motion of what it contains, otherwise, "*one could never draw something along a straight line from one of its corners to the other, and that it would not be possible for one to make a jump and return with his feet to the point from where he took off*". Consequently [Ref. 3, p. 85],

> if someone was placed high on the mast of that ship, move as it may however fast, he would not miss his target at all, so that the stone or some other heavy thing thrown downward would not come along a straight line from the point E which is at the top of the mast, or cage, to the point D which is at the bottom of the mast, or at some point in the bowels and body of the ship. Thus, if from the point D to the point E someone who is inside the ship would throw a stone straight [up], it would return to the bottom along the same line however far the ship moved, provided it was not subject to any pitch and roll.

Bruno was arrested by the Inquisition in Venice where he was up to May 22, 1592. Galilei Galileo (1564–1642) arrived on March 23, 1592 at Padua University (The University of the Republic of Venice). What did Galileo actually learn from Bruno is not documented, neither whether they met each other, most likely due to Bruno's condemnation, but similar ideas were then developed by Galileo through the numerous subjects he investigated during his career, and which can be considered as premises to relativity as it was defined in 1905.

While mostly associated with mechanics, the developments of the so-called special relativity were also performed in various scientific domains such as optics and electrodynamics in not always straightforward connections: It is therefore a little bit delicate to establish the genesis of this theory. However, when the evolution of ideas in electrodynamics and optics is redrawn to lead to the structural foundations of special relativity, a fundamental concept always occurred in a recurrent way and cannot be avoided: The ether. Always associated with propagative phenomena of any origin such as light propagation in optics as well as the flow of electrical charges in electrodynamics, the ether — this is the name used for designating the subtle medium in which propagation (of any kind) is described — is used in all theories developed during 200 years until 1905 with no exception.

Consequently, the genesis of special relativity cannot be provided without discussing the concept of ether; we choose to develop it from its origins. What is it? Where does it come from? In what and how does it intervene in wave propagation? All these questions were so often discussed during the early developments of special relativity that it would not be possible to evidence a certain consistency to the succession of ideas and discoveries without it. The ether was thus considered by us as an Ariadne thread. By investigating its fundamental role in optics and electrodynamics, we were able to track how these theories fused and from which special relativity became the outcome.

The subsequent parts of this chapter are organized as follows. Section 2 is devoted to the origin of the ether and its first relationships with light. Section 3 discusses Galileo's composition law for velocities. We addressed the nature of light in Sec. 4 and we discussed how optical phenomena were linked with electromagnetic fields in Sec. 5. Section 6 is devoted to the invariance of the equations describing light propagation under a coordinate transformation between two frames with a uniform translation between each other. Section 7 discusses Poincaré's contribution (most of this section was provided in Ref. 4) and Sec. 8 briefly describes Einstein's 1905 contribution. Section 9 gives a short conclusion.

2. The Ether: From Celestial Body Motion to Light Propagation

2.1. *Its origin*

The concept of an ether — an elastic medium with subtle properties — was first introduced by Aristotle (−384 to −322)[5] for explaining the motion of planets and to propose a mechanical model based on the mathematical theory describing the motion of celestial bodies proposed by Eudoxus of Cnidus (−405 to −355). According to the latter, a simple mathematical solution should exist to describe the complex motions of planets by combining simple circular motions, including their apparent irregularities (Ref. 6, p. 106 and Ref. 7, pp. 111–122).

These complex motions result from the rotation of concentric spheres, each with a different velocity, the Earth being motionless in the center of that system. In order to reproduce the complexity of these motions, Eudoxus distinguished the uniform circular motion of the sphere holding the stars, which only requires a single circular motion, from the motions of the Sun, the Moon and the five planets which mostly take place in the ecliptic plane: The latter rotations are in the opposite direction (compared to the rotation of the stars).

In this model, the solar and lunar motions are explained by combining the circular motion of three concentric spheres, and those corresponding to the five planets (Venus, Mercury, Mars, Jupiter and Saturn) by combining the circular motion of five concentric spheres. For each system of concentric spheres, the most external sphere motion was induced by the movement of fixed stars, the other internal spheres being animated by contra-rotative motions. The most internal sphere of each system was supporting the corresponding celestial body (Fig. 1).

For the planets, the three concentric spheres between the most external and the most internal spheres were introduced for reproducing visible movements (sphere 2) as well as stations and downgradings of planets (spheres 3 and 4) (Fig. 1). Since solar and lunar motions do not present retrograde phenomena, spheres 3 and 4 are useless and, consequently, their motions were only described by three concentric spheres.

In order to take into account the seasonal variability, Callippus (born at Cyzicus) (−370 to −300) added two circular motions for describing the solar and lunar motions, and one for reproducing those of Mars, Venus and Mercury (Ref. 6, p. 111).

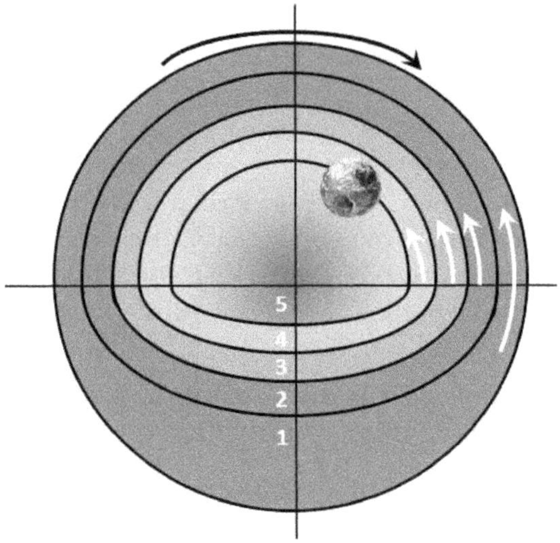

Fig. 1. System of five concentric spheres representing the motion of a planet (here the Earth) in Eudoxus' mathematical representation. (For color version, see page I-CP1.)

The resulting mechanical model remains in agreement with Aristotle's mechanics for explaining how the rotation of the spheres propagates from the stars up to the sublunar realm in the center of which is the Earth. In Aristotle's world,[5] there are two subworlds, the supralunar realm where are located the spheres associated with stars, planets, the Sun and the Moon, and the sublunar realm bounded by the sphere holding the Moon. The origin of motion is thus divine by nature since heaven is commonly located at "*the extremity or upper region, which we take to be the seat of all that is divine*" (Ref. 5, Book I, Sec. 9). It is thus propagated by friction from one sphere to the other. For Aristotle, motion is necessarily associated with body (Ref. 5, Book 1, Sec. 15) and, more generally, to matter (Ref. 5, Book I, Sec. 2):

> All natural bodies and magnitudes we hold to be, as such, capable of locomotion; for nature, we say, is their principle of movement. But all movement that is in place, all locomotion, as we term it, is either straight or circular or a combination of these two, which are the only simple movements.

In addition to that, the four elements (earth, air, water and fire) from which everything belonging to the sublunar realm including the Earth is made, are animated of straight motion, either up or down with respect to the center of the Earth; since these motions cannot be sustained forever, the sublunar realm is subject to change, it is "*corruptible*". In contrast to this, the supralunar realm, close to the divine entity, must be ungenerated, not corruptible and immutable: Celestial bodies are thus necessarily animated of circular motion, the simplest one to be periodic, that is, to be repeated equal to itself forever. Such a perfect motion is, by essence, natural (nonforced) (Ref. 5, Book I, Sec. 2):

> By simple bodies I mean those which possess a principle of movement in their own nature [...]. Supposing, then, that there is such a thing as simple movement, and that circular movement is an instance of it, and that both movement of a simple body is simple and simple movement is of a simple body [...] then there must necessarily be some simple body which revolves naturally and in virtue of its own nature with a circular movement.

By definition, a body which naturally has a circular motion cannot be one of the four sublunar elements, nor made of them. Consequently, another substance must exist, a fifth element, which is naturally and eternally animated of a circular motion. Moreover, the (Ref. 5, Book I, Sec. 2)

> circular motion is necessarily primary. For the perfect is naturally prior to the imperfect, and the circle is a perfect thing. This cannot be said of any straight line [...]. These premises clearly give the conclusion that there is in nature some bodily substance other than the formations we know, prior to them all and more divine than them.

According to this principle stating that any motion is matter, vacuum cannot exist: Otherwise, for Aristotle who believed that velocity was inversely proportional to the medium density, the velocity could be infinitely large, a feature which was not conceivable. The fifth (divine) substance, associated with the circular motion of celestial bodies, can only be as the supralunar realm looks like, that is, ungenerated, not corruptible, "*exempt from increase and alteration*" (Ref. 5, Book I, Sec. 3):

> For all men [who] have some conception of the nature of the gods, and [...] who believe in the existence of gods at all, whether Barbarian or Greek, agree in allotting the highest place to the deity, surely because they suppose that immortal is linked with immortal and regard any other supposition as inconceivable. [...] so, implying that the primary body is something else different from earth, fire, air and water, the Older gave the highest place a name of its own, aether, derived from the fact that it "runs always" for an eternity of time.

The ether, a simple natural body, was thus associated with circular motion of celestial body: It was constituting the supralunar realm up to the heavens. It has a divine character and very particular properties (Ref. 5, Book I, Part 3):

> The body [...] which moves in a circle cannot possibly possess either heaviness or lightness. For neither naturally nor unnaturally can it move either towards or away from the center.

Aristotle wants not only to explain what produces the original motion, but also how it propagates through the universe. He thus explains that the motion of the primary sphere is responsible for all the others (Ref. 5, Book II, Part 12):

> In thinking of the life and moving principle of the several heavens one must regard the first as far superior to the others. Such a superiority would be reasonable. For this single first motion has to move many of the divine bodies, while the numerous other motions move only one each, since each single planet moves with a variety of motions.

Aristotle does not leave the center of circular motion without any associated body (Ref. 5, Book II, Part 3):

> Earth then has to exist; for it is Earth which is at rest at the center [...]. Earth is required because eternal movement in one body necessitates eternal rest in another.

This geocentric system, proposed by Aristotle and based on mathematical laws proposed by Eudoxus of Cnidus, remained well accepted for a long time, with very few exceptions as discussed, for instance, by Aristarchus of Samos (−310 to −230).

2.2. *The luminiferous ether*

For Aristotle, the ether was thus a medium propagating the circular motions that animate all celestial bodies. It became the medium for light propagation with the Aristotelian Bishop of Lincoln, Robert Grosseteste (1175–1253). For him, light is the "*first corporeal form*" (Ref. 8, p. 10), the "*bodily spirit*" (Ref. 8, p. 13), whose lack of determined properties allows it to be transformed into any substance, whatever its nature, and, consequently, into the four sublunar elements.[8] Being able to be diffused at infinity, light is "*inseparable from matter*" (Ref. 8, p. 13) which can thus propagate as sound or heat: Sound and heat therefore contain light.

Consequently, the universe was created from light which (Ref. 8, p. 13)

> by extending first matter into the form of a sphere and by rarefying its outermost parts to the highest degree, actualized completely in the outermost sphere the potentiality of matter, and left this matter without any potency to further impression. And thus the first body in the outermost part of the sphere, the body which is called the firmament, is perfect, because it has nothing in its composition but first matter and first form [...]. When the first body, which is firmament, has in this way been completely actualized, it diffuses its light (*lumen*) from every part of itself to the center of the universe.

Thus light, which is considered as "*the perfection of the first body*" (Ref. 8, p. 13) expanded and brought together from the first body [firmament] toward the center of the universe, gathered together the mass existing below the first body. Moreover (Ref. 8, p. 13)

> since this light (*lux*) is a form entirely inseparable from matter in its diffusion from the first body, it extends along with itself the spirituality of the matter of the first body.

As a consequence, the firmament is the place where light is the simplest (less dense), becoming denser and denser while approaching the center of the universe from the primary sphere to the "*ninth and lowest sphere, the dense mass which constitutes the matter of the four elements*" (Ref. 8, p. 14). Since vacuum does not exist, the firmament, the boundary of the world, must be finite. Matter can thus be expanded from earth to fire through water and air, light being what remains beyond the realm of fire. In Grosseteste's world, the nine heavenly spheres are not subject to change, since already in the form of light, contrary to this, the sublunar realm can change, that is, an element can be transformed into another one by expansion (rarefaction) or by compaction (condensation). According to Grosseteste, matter can be viewed as a high compaction of light, matter being made from light. There is a kind of equivalence between matter (mass) and light. Higher (lighter) bodies are more spiritual, and lower (heavier) bodies are more corporeal. Since the supralunar realm is eternal, there is no rarefaction nor condensation (of light) and, consequently, the sole possible motion is circular motion. It is interesting to note that the constraint on light motion determines the motion of the heavenly spheres.

Grosseteste thus proposed a mechanical explanation for the motion of the 13 spheres of the world (Ref. 8, p. 16):

> The higher body receives its motion from the same incorporeal moving power by which the higher body is moved. For this reason the incorporeal power of intelligence or soul, which moves the first and highest sphere with a diurnal motion, moves all the lower heavenly spheres with this same diurnal motion. But in proportion as these spheres are lower they receive this motion in a more weakened state, because in proportion as a sphere is lower the purity and strength of the first corporeal light is lessened in it.

For Grosseteste, the first body light is nothing else than "*unchangeable*" (Ref. 8, p. 14), and "*rarefied to the highest degree*" (Ref. 8, p. 13): It is of divine spirit and responsible for the motions of planets. It has exactly the same identity as Aristotle's ether, leading to a unique entity named the luminiferous ether. According to Grosseteste's conceptions, light is present inside matter. The fact that light was "*filling*" matter was later explained by René Descartes (1596–1650) (Ref. 9, p. 5):

> I first suppose that water, earth, air and fire and any other body of our environment are made of many small parts of different shapes and sizes, which are never so well arranged nor so accurately joined together that it

> remains always some intervals around them, and that these intervals are not empty but filled of this so subtle matter, by the means of which is propagated the action of light. Moreover, one must think that the subtle matter filling the intervals between parts of these bodies is of such a nature that it never ceases to move very rapidly but not exactly with the same speed in any place nor in any time, but it commonly moves slightly faster toward the Earth surface than it does towards the heavens and faster towards places close to the equator than towards poles, and at the same place, faster during summer than during winter, and during day than night.

For Descartes, light is still made of bodies but is now responsible for the motion of the ether (Ref. 9, p. 6):

> Light is nothing else than a certain motion or an action whose luminous bodies push this subtle matter in a straight line in any direction around them.

Descartes's ether is required for propagating light as well as any other propagative phenomenon like, for instance, heat (Ref. 9, p. 7):

> Marbles and metals seem colder than woods, [and] one must think that their pores do not receive so easily subtle parts of this matter [ether].

Vibrations were mentioned by Nicholas Malebranche (1638–1715) who proposed to describe interactions between light and ether by considering ether as a fluid governed by vortices (as in Descartes's work) (Ref. 10, p. 377):

> Since reflection and refraction of rays are not produced by the action of air nor glass during transition from one to the other, it is thus necessary that the cause comes from the action of the subtle matter, since there is here only air, glass and subtle matter. In order to explain the manner in which it happens, it must be remarked that any part of ether, or all small whirls from which I believe to have shown that it is made of, are also pressed on and at equilibrium each other or always tend toward to be so [...]. Let us assume that all these small whirls of ether are equally and as infinitely pressed on, and that they counterbalance each other by their centrifugal forces, as soon as the small parts of a luminous body squeeze the small whirls that are encountered, their pressure is communicated to all the others up to us, and doing so in an instant, because there is no vacuum. These small parts of a luminous body, by their various motions squeezing by shaking the whirls which are resisting, induce in them vibrations of pressure. And all these vibrations of pressure are made in a straight line, until they are in ether [...]. These rays cannot change in direction, but when they meet obliquely a glass surface, they are subject to a refraction and deviate towards the perpendicular line to this surface; this refraction is as large as the bodies

> in which they enter are more weighted and denser than those from which they are issued.

Light propagation thus results from the interactions between luminous bodies constituting the light and the ether considered as a fluid whose whirls are at the origin of vibrating properties. Ether as light particles are in motion but only light is propagated at long distances.

3. Galileo's Composition Law for Velocities

The very first to support the idea of an heliocentric system — following for instance Aristarchus of Samos — was Mikolaj Kopernik (1473–1543). He did not want to introduce a rupture in Aristotelian conceptions but rather to better match with Aristotle's first principle according which celestial bodies are subject to circular motions. In fact, Kopernik was looking for a system where less numerous combinations of circular motions would have been required[11]: In particular, he wanted to remove from his heliocentric system the equant that he considered as against Artistotle's principles. In a certain sense, Kopernik was more a conservative than a revolutionary! In contrast to this, Galileo, when he compared the "*two main chief world systems*",[12] was more motivated by evidencing the weaknesses of Aristotle's rationale than by proving the correctness of the heliocentric system. In doing this, Galileo started to discuss some problems related to what is now called the principle of relativity.

Thus, in the second day of his discourses, Galileo remarked that one is unable to detect relative motions[12]:

> Whatever motion comes to be attributed to the Earth must necessarily remain imperceptible to us and as if nonexistent, so long as we look only at terrestrial objects; for as inhabitants of the Earth, we consequently participate in the same motion [...].
>
> Motion, in so far as it is and acts as motion, to that extent exists relatively to things that lack it; and among things which all share equally in any motion, it does not act, and is as if it did not exist.
>
> It is obvious, then, that motion which is common to many moving things is idle and inconsequential to the relation of these movables among themselves, nothing being changed among them, and that it is operative only in the relation that they have with other bodies lacking that motion, among which their location is changed.

Galileo thus proposed a procedure to evidence the motion of Earth:

> The true method of investigating whether any motion can be attributed to the Earth, and if so what it may be, is to observe and consider whether bodies separated from the Earth exhibit some appearance of motion which belongs equally to all. For a motion which is perceived only, for example,

> in the Moon, and which does not affect Venus or Jupiter or the other stars, cannot in any way be the Earth's or anything but the Moon's.
>
> Now there is one motion which is most general and supreme over all, and it is that by which the Sun, Moon and all other planets and fixed stars — in a word, the whole universe, the Earth alone excepted — appear to be moved as a unit from east to west in the space of twenty-four hours. This, in so far as first appearances are concerned, may just as logically belong to the Earth alone as to the rest of the universe, since the same appearances would prevail as much in the one situation as in the other.

He then clearly addressed the problem of describing a motion in a frame at rest and compared it to its description in a frame animated with a uniform translation[12]:

> If a stone is dropped from the top of a mast with a large velocity falling down exactly at the same place of the ship as if it would have been at rest, how can this fall serve you to decide whether the ship is at rest or in motion?
>
> [...] the same argument being valid for the ship as for the Earth, one cannot be conclusive about the motion or the rest of the Earth.
>
> [...] with respect to the Earth, the tower and to us, that all are moving with the daily motion, simultaneously to the stone, the daily motion is as it was nothing, it remains indifferent, not perceptible, and has no action; only observable to us is the motion that we are lacking, the motion of the stone which skims along the tower while falling.

Thus, nobody can perceive a motion when it is not related to a reference frame which is not animated by this motion. Many other examples of relative motions were developed by Galileo to support his explanations. As an ultimate experiment, he proposed[12]:

> Let be with a friend in the largest cabin under the deck of a large ship and take with you some flies, butterflies and other small animals that are flying, take also a large vessel filled with water with small fishes, hang on also a small bucket from which water is flowing droplet per droplet in another vase with a small aperture in its bottom. When the ship is at rest, observe carefully how these small flying animals move with the same velocity in any direction of the cabin, how fishes equally swim in any direction [...]; if you bunny hop, as we say, you will move by equal distances in any direction.
>
> When you will have carefully observed this, although that there is no doubt that it must occur as the ship was at rest, let the ship move at the velocity you want — until the motion is uniform without pitch or roll, you will not remark the least change in all the effects that we just described. None will allow you to detect whether the ship is moving or is at rest [...]. By jumping, you will move on the floor by the same distances as before, and you will not jump further towards the bow or the stern because the ship

> will move very fastly; however, during the time you were in the air, the floor below you runs in the direction opposite to your jump [...], butterflies and flies will continue to fly equally in all directions, you will never see them taking refuge towards the wall in the stern as they were tired to follow the fast move of the ship [...]. If these effects match each other, it results from the fact that the ship motion is shared with everything it contains as well as to the air; this is why I asked you to be under the deck.
>
> If you have be on the deck, the air would not follow the ship run and we would have observed more or less noticeable differences.

Galileo thus established that it is impossible to assert whether a body with whom we shared a common uniform motion is moving or not. He also argued against Aristotle for the nonexistence of vacuum. This can be seen as a first attempt to remove the ether but this is not performed in the context of propagative phenomena: Galileo therefore had no need to consider an ether! While investigating falling bodies in medium, he understood that only the density of the medium is responsible for various velocities. Extrapolating such a result in medium with a null density, he came to the conclusion that "*if we had fully removed the medium resistance, all bodies would fall at the same velocity*". Vacuum thus became for Galileo an ideal frictionless medium for investigating motion.

His repeated studies on mechanical laws governing falling bodies led him, in the early 1630's, to investigate the motion of a ball rolling on and falling from a table. He thus understood that the parabola described by the ball once it had left the table was the result of a combination of two motions as follows: "*A projectile which is carried by a uniform horizontal motion compounded with a naturally accelerated vertical motion describes a path which is a semiparabola*".[13]

From these observations, the composition law for velocities was later formulated and became designated as the principle of Galilean relativity. Nevertheless, if it is common in mechanics, the Galilean composition law was quickly questioned when applied to the propagation of light. Light propagation was considered as infinitely fast, if not instantaneous. Galileo seems to be one of the first to imagine an experiment for measuring the light velocity[13]:

> Let each of two persons take a light contained in a lantern, or other receptacle, such that by the interposition of the hand, the one can shut off or admit the light to the vision of the other. Next, let them stand opposite each other at a distance of a few cubits and practice until they acquire such skill in uncovering and occulting their lights that the instant one sees the light of his companion he will uncover his own. After a few trials the response will be so prompt that without sensible error [svario] the uncovering of one light is immediately followed by the uncovering of the other, so that as soon as one exposes his light he will instantly see that of the other. Having acquired skill at this short distance let the two experimenters, equipped as before, take up positions separated by a distance of two or three miles and let

> them perform the same experiment at night, noting carefully whether the exposures and occultations occur in the same manner as at short distances; if they do, we may safely conclude that the propagation of light is instantaneous; but if time is required at a distance of three miles which, considering the going of one light and the coming of the other, really amounts to six, then the delay ought to be easily observable.

Galileo thus concluded that if the light propagation is not instantaneous, "*it is at least extremely fast, nearly immediate*".[13] But Galileo did not mention any value for the light velocity.

The first evaluation of light velocity was provided by Ole Roemer (1644–1710) who used the eclipses of Jupiter's first satellite.[14] In order to assess the light velocity, Roemer supposed that, according to the additive law for velocities, when the Earth moves toward Jupiter during its revolution around the Sun, the light should take less time for traveling from Jupiter to the Earth than when it moves in the opposite direction (six months later) (Fig. 2). The results of Roemer's experiments were equivalent to a light velocity equal to $2 \cdot 10^8\,\mathrm{m \cdot s^{-1}}$.

Later, James Bradley (1693–1762) showed that the phenomenon investigated by Roemer, that he named *aberration*, was an annual motion shown by all stars and which was resulting from the combination of the Earth's velocity around the Sun — which is also the velocity of the observer — with the light velocity. Since a single apparent motion was shown by all stars, the light velocity should be the same for every star and should be uniform over any distance between the celestial body from which it is issued and the Earth.[15] The aberration was thus correctly explained with

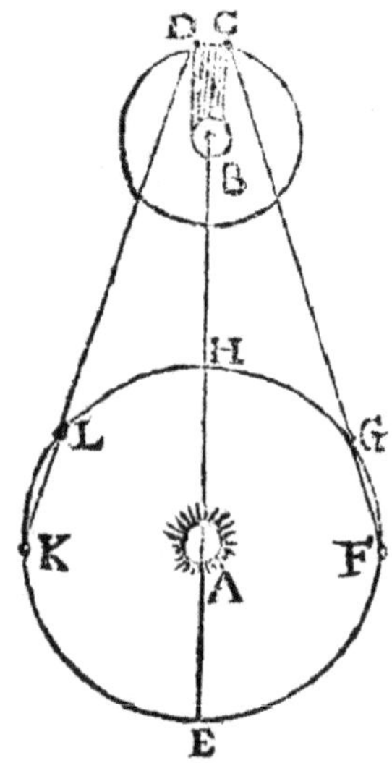

Fig. 2. Light velocity measurement from the Jupiter's first satellite, with A the Sun, B Jupiter and C the first satellite which enters the shade of Jupiter to go out of it at D, and EFGHKL the Earth placed at different distances from Jupiter (Ref. 14, p. 234).

a single light velocity — the same for light issued from every star — and with the classical composition law for velocities. Such a conclusion was not widely accepted, in particular by those who were considering light as made of corpuscles such as John Michell or François Arago as we will see later; for them, light velocity should depend on stellar size which was directly related to their distance from Earth.

4. Questioning the Nature of Light: Waves or Corpuscles?

By the end of the 17th century, two main theories concerning the nature of light were discussed. For both of them, the existence of a luminous ether was not questioned and light (regardless of wave or particle) must interact with the ether to be propagated. One of the two main theories was developed by Christiaan Huygens (1629–1695)[16] and was based on Grosseteste's ideas: Mostly light was considered as a power propagating through an ether. It led to the wave theory that Thomas Young (1773–1829) used for explaining diffraction and interference patterns.[17,18] The second theory was pushed by Isaac Newton (1643–1727) who considered light as made of small weighted corpuscles: This is the corpuscular theory also named the emission theory. Nevertheless, as we will see, these two theories are both mixed in the sense that they are not fully based on waves nor corpuscles, but rather combined the two conceptions.

The corpuscular theory of light was mainly developed in the queries written by Newton at the end of his book *Optiks*[19] which is mostly devoted to the reflection and refraction of light rays as well as to a color theory. Newton only addressed the nature of light when he questioned the production of light as follows (Ref. 19, Query 8):

> Do not all fix'd Bodies, when heated beyond a certain degree, emit Light and shine; and is not this *Emission* perform'd by the vibrating motions of their parts?

Heat and light are here clearly related and associated with vibrations of what we would call today atoms or molecules. Indeed, Newton questioned interactions between light and matter and not only between light and ether (Ref. 19, Query 5):

> Do not Bodies and Light act mutually upon one another; that is to say, Bodies upon Light in emitting, reflecting, refracting and inflecting it, and Light upon Bodies for heating them, and putting their parts into a vibration motion herein heat consists?

Nevertheless, Newton suggested that light — when rays are considered — would be made of small corpuscles (Ref. 19, Query 29):

> Are not the Rays of Light very small Bodies emitted from shining Substances? For such Bodies will pass through uniform Mediums in right Lines without bending into the Shadow, which is the Nature of the Rays of Light.

> They will also be capable of several properties, and be able to conserve their Properties unchanged in passing through several Mediums, which is another Condition of the Rays of Light.

Light rays propagate in straight lines through media but an ether is not specified here. Some periodic vibrations are also associated to light propagation, today one would say waves, as clearly suggested by the analogy with a falling stone in water (Ref. 19, Query 17):

> If a stone be thrown into stagnating Water, the Waves excited thereby continue some time to arise in the place where the Stone fell into the Water, and are propagated from thence in concentrick Circles upon the Surface of the Water to great distance. And the Vibrations or Tremors excited in the Air by percussion, continue a little time to move from the place of percussion in concentrick Spheres to great distances. And in like manner, when a Ray of Light falls upon the Surface of any pellucid Body, and is there refracted or reflected, may not Waves of Vibrations, or Tremors, be thereby excited in the refracting or reflecting Medium at the point of Incidence, and continue to arise there, and to be propagated from thence as long as they continue to arise and be propagated, when they are excited in the bottom of the Eye by the Pressure or Motion of the Finger, or by the Light which comes from the Coal of Fire in the Experiments abovemention'd? And are not these Vibrations propagated from the point of Incidence to great distance? And do they not overtake the Rays of Light, and by overtaking them successively, do they not put them into the Fits of easy Reflection and easy Transmission described above? For if the Rays endeavour to recede from the densest part of the Vibration, they may be alternately accelerated and retarded by Vibrations overtaking them.

Newton's theory of light propagation uses the corpuscular as well as the undulatory nature of light depending on the medium in which it propagates. Interactions between light and matter (here seen as a refracting or reflecting medium) is thus associated with waves that can propagate over great distances. Moreover, vibrations (waves) are possible for any deflection from the straight line as specified in Query 29[19]:

> Nothing more is requisite for putting the Rays of Light into Fits of easy Reflection and easy Transmission, than that they be small Bodies which by their attractive Powers, or some other Force, stir up Vibrations in what they act upon, which Vibrations being swifter than the Rays, overtake them successively, and agitate them so as by turns to increase and decrease their Velocities, and thereby put them into those Fits.

Moreover there is a force, not specified, responsible for the change in light propagation.

For a mechanical explanation of wave propagation, a medium, different from the air, is required: Such a medium, the ether, can be found in matter (light propagates through glass) as well as in the heavens (Ref. 19, Query 18):

> Is not the Heat of the warm Room convey'd through the *Vacuum* by the Vibrations of a much subtler Medium than Air, which after the Air was drawn out remained in the *Vacuum*? And is not this Medium the same with that Medium by which Light is refracted and reflected, and by whose Vibrations Light comunicates Heat to Bodies, and is put into Fits of easy Reflexion and easy Transmission? [...] And is not this Medium exceeding more rare and subtile Air, and exceeding more elastik and active? And doth it not readily pervade all Bodies? And is it not (by its elastik force) expanded throughout all the Heavens?

Ether would be rarer in dense matter (light propagates less easily in those bodies) than in the empty celestial space. Newton stated that "*light moves from the Sun to us in about seven or eight minutes of time*" (Ref. 19, Query 21) as "*this was observed first by Roemer, and then by others, by means of the Eclipses of the Satellites of Jupiter* (Ref. 19, Prop. XI), the actual value being 8.3 min.

Ether was also made of particles which were smaller than those of light, being in turn smaller than those of air; but Newton did not know what these ethereal particles could be. He thus proposed two properties attributed to the ether: One is resulting from the surface forces, thus inducing that the ether is "*less able to resist the motions of Projectiles*" (Ref. 19, Query 21) and one due to the volume forces, leading to the fact that the ether is "*exceedingly more able to press upon gross Bodies, by endeavouring to expand itself*" (Ref. 19, Query 21). In Query 31, Newton suggested that interactions between "*small particles of bodies*" could be in a certain way similar to those between light rays: He even suggested that "*a more attractive power*" than gravity, magnetism and electricity could exist. Light might be not a special phenomenon but one of the phenomena in nature governed by similar laws as matter. Thus, in his 29th query, he discussed the double refraction investigated by Huygens[16] in Island Crystal[a] and proposed to attribute "*some kind of attractive virtue lodged in certain sides both of the rays and of the particles of the Crystal*" (Ref. 19, Query 29), two sides — acting "*as the Poles of two Magnets answer to one another*" (Ref. 19, Query 29) — explaining the "*usual refraction*", and two other sides responsible for the "*unusual refraction*". According to Newton, the double refraction observed in Island Crystal could not be explained with a wave theory (Ref. 19, Query 28):

> Are not all Hypotheses erroneous, in which Light is supposed to consist in Pression or Motion, propagated through a fluid Medium? [...] To explain

[a]We choose to use the term "Island" as used by Newton, Malus or Huygens and not "Iceland" as sometimes encountered.

> the unusual Refraction of Island Crystal by Pression or Motion propagated, has not hitherto been attempted (to my knowledge) except by *Huygens*, who for that end supposed two several vibrating Mediums within that Crystal. But when he tried the Refractions in two successive pieces of that Crystal, and found them such as is mention'd above; he confessed himself at a loss for explaining them. [...] To me, at least, this seems inexplicable, if Light be nothing else than Pression or Motion propagated through *Aether*.

According to Newton, light is made of bright corpuscles of very small size, possessing properties similar to magnetism, and whose propagation is induced by shaking the ether. The vibrations of the ether so produced allow, by the action of the forces of attraction and repulsion between the bright corpuscles and etherous molecules, to carry the light corpuscles from a point to another one. Newton's ether was thus a medium with the same properties as a fluid could have: (i) Seen as a continuous medium allowing wave propagation and (ii) made of particles allowing him to describe its behavior using forces in a mechanical description according to his own mechanical laws. Such a mechanical description was the ultimate aim as evidenced by Jean-Baptiste Biot (1774–1862) while discussing the double refraction (Ref. 20, p. XV):

> You might thus [...] examine the phenomena under all its sides [...]. The law for this phenomenon is still only established in an experimental manner; it can only be viewed as exact in the limit of accuracy of the experiments. In order to make it fully sure and rigorous, it must be brought back to the general laws of mechanics, that is, to obtain it from general conditions for motion and equilibrium as required by these laws. Because, when such a reduction can be fully performed, it necessarily evidences the character of forces by which these phenomena are produced, that is the last end where science can go.

Working according to this paradigm, François Arago (1786–1853), supporting a corpuscular theory of light, was investigating since 1809 the effect of light velocity on refraction, one of the most studied optical phenomena. Arago was also aware of a paper published in 1783 by John Michell (1724–1793) in which light was considered as being made of small particles. When emitted by a star, these light particles would have their velocity reduced by the gravitational field as any other material particle and, consequently, the light emitted by stars should depend on their sizes.[21] Bradley's results showing the same aberration for all the stars was therefore only due to a lack of accuracy in his observations. Arago wanted to investigate in a deeper way Bradley's aberration by using prisms that he considered (with Michell) as more adequate than direct observation due to their sensitiveness "*to slight equalities*". Arago hoped to be able to determine whether the size of stars — "*a circumstance which must produce significant differences in* [light] *velocities emanating from these various bodies*" — affects the light velocity or not (Ref. 22, p. 40):

> Since the deviation of the light rays penetrating sidewise in diaphanous bodies is a given function of their initial velocity, we will see that the observation of their total deviation while passing through a prism provides a natural measure of their velocities.

In other words, the deflection of light rays in a refringent medium as a prism, should depend on their velocity in it and, consequently, on their initial velocities. These experiments led Arago to conclude that (Ref. 22, p. 40):

> Light moves with the same velocity not depending on the bodies from which it is issued or, at least, if there exists some differences, they cannot, in any way, alter the exactness of astronomical observations.

Arago tried to show, "*by direct experiments, that there is an increase in the velocity required by light rays during a switch from a rare medium into a dense medium*" (Ref. 22, p. 41). He used for doing this the property that "*an inequality in velocity produces an inequality in deflection, a fact which directly results from Newton's explanation for refraction*" (Ref. 22, p. 41). He also considered how refraction was affected by the velocity of the body through which the light was passing (Ref. 22, p. 42):

> The difficulty presented [...] by the verification of Newton's theory results from the principle which is a consequence of it follows: Light velocity, in any diaphane medium, must be the same, for any kind and number of media previously traversed. One can however remark that, when refringent bodies are in motion, refraction induced by a body must no longer be computed with the [light] absolute value, but with this same velocity augmented or reduced by the velocity of the body, that is, with relative velocity of the ray; motions that we can give to bodies on the Earth being far too small to sensitively influence light refraction, one might search in much faster planet motions, some circumstances appropriate for making significant these inequalities in refraction.

Arago thus considered that Earth's motion combined with his experiment could lead to sensitive enough differences.

In order to obtain a sensitive measurement of the deflection of light rays according to their velocity, Arago performed an experiment using an achromatic prism for avoiding chromatic aberration and, consequently, which should allow a better spectral separation of refracted rays without diffusion among them. By measuring "*distances at zenith of a large number of stars*" (Ref. 22, p. 44), thus using light rays with supposed different velocities due to various sizes of the observed stars, significant deflections through the prism were expected. In spite of this, measured deflections in these experiments were of the same order of magnitude as in experiments using direct observations. Arago expected differences by 1/10,000 induced by the Earth rotation. Contrary to this, he observed no difference in his observations, "*rays from all stars being affected by the same deflections*" (Ref. 22, p. 46). Arago thus concluded (Ref. 22, p. 47):

> This result seems to be, at first, in a clear contradiction with the Newtonian theory of refraction because an actual inequality in the velocity of the rays does not imply an inequality in deflections that are induced.

Not ready to leave Newton's theory, Arago justified his negative result by arguing that (Ref. 22, p. 47)

> one can only explain [these results] by assuming that luminous bodies emit rays with any kind of velocities, provided that one also admits that these rays are visible only when their velocities are between given limits: According to this hypothesis, indeed, the visibility of rays will depend on their relative velocities and, since their equal velocities determine the quantity of refraction, visible rays are equally refracted.

In spite of this, Arago remained unable to provide a convincing explanation. He then asked to his friend, Augustin Fresnel (1788–1827) to address this problem using the wave theory that he defended. Fresnel thus provided an explanation considering "*light as in the vibrations of a universal medium*" (Ref. 23, p. 628), the so-called ether. For Fresnel, "*the velocity with which waves are propagating is independent from the motion of the body from which they are emitted*" (Ref. 23, p. 628). But he was not able to explain the stellar aberration with the theory according to which the Earth "*induces its motion to the surrounding the ether*" (Ref. 23, p. 628) and that would easily lead one to conceive "*why the same prism always refracts light in the same manner, for any side from where it comes*" (Ref. 23, p. 628). Fresnel only obtained an explanation of the phenomena by assuming that the ether goes freely through the globe, and that "*the velocity communicated to this subtle fluid is only a small part of the velocity of the Earth*" (Ref. 23, p. 628), typically less than one-hundredth. This line of reasoning led to the assumption for a partial driving by the motion of the Earth of the ether contained in transparent medium, an explanation in agreement to "*the extreme porosity of bodies*" (Ref. 23, p. 628), then supported by the most important scientists.

Newton's theory was, in his main concepts, a dual theory combining waves and particles. But as evidenced by Young, a wave theory had some indefectible advantages for explaining some optical phenomena. That is what Fresnel summed up as "*wave theory is more adaptable to all phenomena than Newton's theory*" (Ref. 24, p. 12, Sec. 7); indeed, Newton had to multiply the assumptions to describe optical phenomena using particles (Ref. 24, p. 12, Sec. 5):

> Situations with easy reflexion and easy transmission are nearly non explainable in Newton's system. Thus, he presents them as new properties of light, and does not try to link them to the basis of his theory. It seems to me that these periodic variations in light disposals would be easier to conceive by

> considering light as produced by vibrations of calorific medium, because, in the same wave, it would have successively different velocities, different degrees of pressure and would repeat itself in the next undulations.

Considering Newton's explanation on the double refraction observed in the Island Crystal, Fresnel added (Ref. 24, p. 12, Sec. 6):

> The double refraction forced Newton to make again a new assumption, which is quite extraordinary; luminiferous molecules have poles, and the Island Crystal turns in the same direction poles of the same kind. Malus proved, by his beautiful experiments on the polarization of light, that it was modified in the same way as it is reflected with a certain angle by a non-tined mirror. It is necessary to admit poles in luminiferous molecules to conceive this phenomenon, and might it not be possible to assume that the mirror impose vibrations to the light, along the reflexion plane, a particular modification, which makes that it is more able to be reflected in a direction than in the other?

Based on his objections against the Newtonian theory of light, in particular for the explanation of reflection and refraction, Fresnel began a series of meticulous experiments on diffraction whose principle was similar to those conducted by Newton, and which allowed him to develop a first theory for diffraction using light considered as waves.[24] Some difficulties were still remaining as those encountered for explaining reflexion and diffraction — which were only explained using a corpuscular theory — as well as the polarization of light. Fresnel wanted to overcome this difficulty and to explain all optical phenomena (Ref. 25, p. 4):

> New phenomena, compared to those previously known, daily increase the probabilities in favor of a system of undulations. Although neglected for a long time, and more difficult to follow on its mechanical consequences than the emission theory, [a wave theory] already provides larger computations abilities [...]. It is interesting for improvements in optics and everything related, that is, the whole physics and chemistry, to know whether luminous molecules are launched from bodies which are sending light on us up to our eyes, or whether light is propagated by vibrations of an intermediary fluid to which particles of these bodies transfer their oscillations.

The wave theory was very rarely used at that time but had received "*a striking confirmation*" by "*curious experiments*" (Ref. 25, p. 4) conducted by Thomas Young. Comparing the advantages and disadvantages of a corpuscular theory to those provided by a wave theory, Fresnel thus evidenced the difficulties encountered in explaining Young's experiments and concluded: "*Diffraction phenomena are not explainable with the emission theory*" (Ref. 25, p. 33). Consequently, light must be considered as a wave. During its propagation in an elastic medium, such a wave gives to the molecules of the etherous fluid an oscillatory motion to which it is possible to apply laws from mechanics and to explain the manner in which interfringes are produced (Ref. 25, p. 36):

> It is sufficient that these movements are oscillatory, that is, carry molecules alternatively in two opposite directions, to have the effect of a wave series destroyed by the effect of another series with the same intensity [...]. For instance, in waves formed at the surface of a liquid, oscillations are vertical, and propagation is horizontal and, consequently, along a direction perpendicular to the first one.

Considering the etherous fluid being homogeneous and isotropic, light waves propagate with a constant velocity because "*the velocity for propagation* (*which should not be mistaken with the absolute velocity of molecules*) *only depends on the density and the elasticity of the fluid*" (Ref. 25, p. 37). Light intensity therefore depends on the vibration intensity of the ether "*and color will depend on the duration of each oscillation or on the wave-length since one is proportional to the other*" (Ref. 25, p. 43). Fresnel completed "*the foundations to build the general theory of diffraction*" (Ref. 25, p. 59) by investigating Huygens' principle according to which (Ref. 25, pp. 59–60):

> Vibrations of a light wave in each of its points can be seen as the result of elementary movements sent at the same time, acting independently, by all parts of this wave considered in any one of its previous positions;

and that Fresnel considered as "*a rigorous consequence*" of the system of the undulations. These elementary movements, also called *shakings* (*ébranlements*) by Fresnel, have certain properties as follows (Ref. 25, p. 60):

> I will assume that these shakings, in an infinite number, are all of the same kind, occur simultaneously, are adjoining and located in the same plane or in the same spherical surface. I will make yet an assumption related to the nature of these shakings: I will assume that the velocities given to molecules are all oriented in the same direction, perpendicular to the spherical surface, and are however proportional to condensations, in a ratio such that molecules cannot have retrograde movements. I will have thus reconstructed a wave from its partial shakings.

The relevant aspect of Huygens' principle is that any wave can be decomposed in "elementary" or "partial" movements. This is the key concept which allows Fresnel to understand that double refraction is the signature of the decomposition of a wave into two different components with different velocities (Ref. 25, p. 93):

> The difference between the squares of the propagation velocities of ordinary and extraordinary rays is proportional to the square of the sine of the angle that there is between each their direction and the [crystal] axis. It results from these facts that the two beams produced by the double refraction do

> not have the same optical properties around their direction, since they are affected either by the ordinary refraction or by the extraordinary refraction, depending on whether the main section of the second crystal is oriented along a certain plane or perpendicular to this plane. If we draw straight lines perpendicular to rays along these planes, and if we conceive them as carried by the set of waves in its course, they will indicate the two directions in which [this crystal] presents opposite optical properties.

Fresnel then explained how there could be different properties of the waves in the two beams made of (Ref. 25, p. 96)

> *transverse* movements (I call transverse movements the oscillations of etherous molecules that would occur perpendicular to the direction of rays) which could not be the same in the two directions [...]. This is not only by its path through a crystal which splits it into two distinct beams that light receives this particular modification: It can be also polarized by a simple reflexion on a surface of a transparent surface, as Malus was the first to observe.

Fresnel thus built a mechanical description of light propagation by treating the velocities of oscillations as they are treated in classical problem, that is, a velocity has some components which can be isolated or determined. Nevertheless, in mechanics forces are responsible for motion. In the case of light, since there is no force yet identified, these velocities associated with the "*elementary movements*" seem for Fresnel to play the role of forces. Such a status thus explains why it was so difficult for him to establish that these oscillations, then considered as responsible for the motion (propagation) of light, were perpendicular to the direction of propagation. These directions are thus needed for describing light propagation: The main one corresponds to the axis along which the propagation occurs; perpendicular to it, there is the plane in which there are the two components evidenced by using the double refraction. These "*elementary movements*" are perpendicular to light rays (Ref. 25, p. 101):

> the two components of the velocity are also proportional to $\sin i$ and $\cos i$, according to the principle of composition and decomposition of small motions in fluid, which must be [decomposed] as forces in statics. Malus' law seems thus to indicate that oscillatory movements of etherous molecules arise perpendicularly to rays.

Fresnel was thus able to show that light was associated with wave propagation made of transverse oscillations which can be decomposed (polarized) into two components. It is interesting to note (i) that the term "*polarization*" — first introduced by Malus to designate this light property — comes from the first attempt to explain the double refraction using the "*poles*" of magnets and (ii) that Fresnel pushed a little bit further the wave theory by understanding that the two components were not poles but particular directions along which elementary oscillatory movements can be decomposed. Nevertheless, Fresnel attributed these oscillations to an universal fluid — the ether — as all his predecessors and understood that its oscillations more than its molecules were taking part in light–matter interactions (Ref. 25, p. 141):

> If light is only a certain type of vibrations of an universal fluid, as diffraction phenomena show it, one must no longer assume that its chemical action on bodies consists in a combination of its molecules with their own ones, but in a mechanical action that vibrations of this fluid are exerted on ponderable particles, and that force them to new arrangements, to a new more stable equilibrium system, for the type or the energy of vibrations to which they are expressed.

Thus, Fresnel's works on diffraction extended to bright phenomena led to the establishment of a rigorous theory of light which was considered as a transverse polarized wave propagating through oscillations of etherous molecules according to laws of mechanics.

5. From Electrodynamics to Light

5.1. *Ampère's law*

On the one hand, Fresnel's contribution widely convinced that wave theory had serious advantages in explaining optical phenomena. On the other hand, magnetic and elastic forces were briefly suggested by Newton to explain some optical phenomena but no conclusive explanations were provided. Nevertheless, in both cases, laws governing optical phenomena were not yet established, nor were those for electrical and magnetical phenomena. One noticeable contribution for the latter was provided by André-Marie Ampère (1775–1836) who proposed an empirical law for describing the reciprocal action between two current elements. Following Newton's methodology, Ampère's approach consists in (Ref. 26, p. 176):

> First observing the facts, varying their circumstances as much as possible, joining to this first work accurate measurements in order to conclude with general laws based on experiments, and deducing from the so-obtained laws, independently from any assumption on the nature of the forces producing the phenomena, the mathematical value of these forces, that is, the mathematical formula that describe them.

In spite of this, Ampère was forced to add a few assumptions when he considered the mutual action of two elements of current, mainly because he was unable to conduct experiments with infinitely small parts of voltaic circuits.

By restricting himself to the observations of balance between two photovoltaic elements, he thus assumed that the action between two elements of current is according to a force along the straight line that joins them. This is one of the simplest hypothesis used by Newton for describing any force acting between particles: Such a force is always acting (Ref. 26, p. 178)

> along the straight line which joins them, in such a way that the action produced by one onto the other is equal and opposed to the force simultaneously produced by the latter onto the former.

In such a case, when these two particles are permanently linked to each other, there is no motion resulting from their mutual action. Applying this to two current elements placed at a given distance one from the other, Ampère found that "*their mutual action depends on the lengths, the intensities of currents, and on their respective positions*" (Ref. 26, p. 200). The mathematical expression corresponding to this feature was then (Ref. 26, p. 204)

$$\frac{ii'}{r^n} ds\, ds' (\cos\epsilon + h \cos\theta \cos\theta'), \tag{1}$$

for any two current elements of lengths ds and ds', having for intensities i and i', respectively, and where ϵ is the angle between the two current elements, θ and θ' the angles that present these elements with respect to the direction of their currents with the length r of the line joining their centers, and the constant $h = k - 1$ where k represents the action of one element onto the other. He then rewrote Eq. (1) as (Ref. 26, p. 207)

$$-\frac{ii' ds\, ds'}{r^n}\left(r \frac{dr^2}{ds\, ds'} + k \frac{dr}{ds}\frac{dr}{ds'}\right) \tag{2}$$

and showed that the action is equal to the reaction only when $n = 2$ (Ref. 26, p. 232).

Ampère mainly investigated the phenomena produced by electric current. He clearly distinguished these phenomena that he designated as "*electrodynamic*" phenomena from those produced by the interaction between a magnet and an electric current which were commonly designated as "*electromagnetic*" phenomena (Ref. 26, p. 298). Nevertheless, by these times, the term "*electromagnetic*" was already used for designating phenomena produced by two current elements, Ampère's terminology was therefore not retained.

Ampère conceded that more difficult researches should be conducted for investigating whether "*electrodynamic phenomena*" explained in terms of movement of the ether could also lead to the same formula. Such a question left open by Ampère already suggests a possible analogy, if not more, between, on the one hand, electromagnetic and electrodynamic phenomena and, on the other hand, light propagation (Ref. 26, p. 301):

> If it were possible to prove on the basis of this consideration, that the reciprocal action of two elements was in fact proportional according to the formula that I have described it, then this account of the fundamental fact of the entire theory of electrodynamic phenomena would obviously have

> to be preferred to every other theory; it would, however, require investigations with which I have had no time to perform myself, neither the still more difficult investigations which one would have to undertake in order to ascertain whether the opposing explanation, whereby one attributes electrodynamic phenomena to motions imparted by the electrical currents of the ether, could lead to the same formula.

Unfortunately, Ampère's law (2) does not explain a certain class of electrodynamic phenomena such as the Volta induction discovered by Michael Faraday (1791–1867)[27] and corresponding to the reciprocal action produced by (i) two electric charges, one moving with respect to the other or, (ii) two current elements, one current varying with respect to the other. Looking for a generalized formula explaining these latter phenomena, Wilhelm Weber (1804–1891) expected a single law for electrostatics as well as for electrodynamics.[28] In fact, Ampère expected such a single formula too but without taking into account the possibility for the two current elements to be in a relative movement, nor considering varying currents. From Weber's point of view (Ref. 28, p. 82):

> [Ampère's] law holds only as a particular law and still requires a definitive law with truly general validity applicable to all electrodynamic phenomena to replace it.

In order to do that, Weber considered electric current as made of electric masses: "*The electrical fluids in the two current elements themselves have in them like amounts of positive and negative electricity, which, in each element, are in motion in an opposite fashion*" (Ref. 28, p. 83). The mutual action between two current elements thus results from (Ref. 28, p. 83)

> four reciprocal actions of electrical masses to consider *two repulsive* between the two positive and between the two negative masses in the current element, and *two attractive*, between the positive mass in the first and the negative mass in the second, and between the negative mass in the first and the positive mass in the second.

Weber also investigated situations where the two current elements had various orientations, relative velocities and relative accelerations in order to get a general formula also describing induction phenomena evidenced by Faraday. Using the electrostatic system of units, he thus obtained the force with which two charged masses act upon another expressed as (Ref. 28, p. 89)

$$\frac{ee'}{r^2}\left(1 - a^2\frac{dr^2}{dt^2} + 2a^2 r\frac{d^2r}{dt^2}\right), \tag{3}$$

where e and e' are isolated electrical charges with positive or negative values, r is the distance between them, $\frac{dr^2}{dt^2}$ is the square of the relative velocity between the two masses and $\frac{d^2r}{dt^2}$ their relative acceleration. The constant a^2 remained to be determined. The relative velocity $\frac{dr}{dt}$ between two electrical masses could be positive or negative, depending on whether the two masses are moving away from or approaching one another. When the two electrical charges are at rest ($\frac{dr}{dt} = 0$) this law reduces to Coulomb's electrostatic law.

It was later demonstrated by Henri Poincaré (1854–1912) that in the electromagnetic system of units, this force between two electric charges is expressed as (Ref. 29, p. 34):

$$+\frac{ii'ds\,ds'}{r^2}\left(2r\frac{dr^2}{ds\,ds'} - \frac{dr}{ds}\frac{dr}{ds'}\right), \tag{4}$$

which is in complete agreement with Ampère's law. Thus, the reciprocal action is directly proportional to the current intensities in the two current elements and inversely proportional to the square of the distance between them; moreover it can distinguish repulsive from attractive forces (Ref. 28, p. 92):

> The force [not only depends on] the magnitude of the masses and their distance from one another, but also on their relative velocity and relative acceleration.

Weber finally got "*the general law*" (Ref. 28, p. 98)

$$\frac{ee'}{r^2}\left(1 - \frac{a^2}{16}\frac{dr^2}{dt^2} + \frac{a^2}{8}r\frac{dr^2}{dt^2}\right), \tag{5}$$

for the force between two ponderable charges. He was thus able to conclude that this force did not only depend on the two ponderable charges but also "*on the presence of a third body*" (Ref. 28, p. 141). For Weber, there is thus a medium transmitting the force between the ponderable charges. Moreover, while discussing Faraday's experiments on the influence of electrical currents on light,[30] Weber explained that it is (Ref. 28, p. 142)

> not improbable that the all-pervasive neutral electrical medium is itself that all-pervasive ether, which creates and propagates light vibrations, or that at least the two are so intimately interconnected that observations of light vibrations may be able to explain the behavior of the neutral electrical medium.

In 1852, Weber replaced the constant $\frac{a^2}{16}$ in formula (5) by $\frac{1}{\tilde{c}^2}$ and obtained the new formulation

$$\frac{ee'}{r^2}\left(1 - \frac{1}{\tilde{c}^2}\frac{dr^2}{dt^2} + \frac{2r}{\tilde{c}^2}\frac{d^2r}{dt^2}\right), \tag{6}$$

where $\tilde{c}$ designates Weber's constant[b] and is different from the constant c used for expressing the light velocity in vacuum. The constant $\tilde{c}$ represents the number

[b]In fact, Weber designated his constant as c but this constant had a different value than the one known as the light velocity, we changed the notation to make a distinction between Weber's constant c and c the velocity of light.

of electrostatic units in one electromagnetic unit of electricity and its value was determined ten years later by Weber and Rudolf Kohlrausch (1809–1858)[31]; they obtained

$$\frac{E}{B} = \tilde{c} = 4.39 \cdot 10^8 \, \mathrm{m} \cdot \mathrm{s}^{-1}. \tag{7}$$

Then, taking into account that for having the same effect, units for electromagnetic current i are greater by a factor $\sqrt{2}$ than those used for electrodynamic current j ($j = \sqrt{2}i$) (Ref. 28, p. 18), it came in fact that

$$c = \frac{E}{B} = \frac{\tilde{c}}{\sqrt{2}} = 3.10741 \cdot 10^8 \, \mathrm{m} \cdot \mathrm{s}^{-1}. \tag{8}$$

For Weber, the physical meaning of this constant $\tilde{c}$ was the relative velocity between particles with charges e and e', respectively, to have and to keep a null action to each other.

In 1857, Gustav Kirchhoff (1824–1887) used Weber's theory to develop a theory of electric current in conductors of any form, consisting in fact of a generalization of his theory for electric current in linear conductors that he developed in earlier papers.[32,33] As in the case of linear conductors, Kirchhoff proved[34]

> that [...] the electricity in the wire progresses like a wave in a taught string with the velocity of light in empty space.

Stimulated by Kirchhoff's work, Weber came to the conclusion that[35] "$\frac{\tilde{c}}{\sqrt{2}}$ *is the limit towards which converges all propagation velocities and* [...] *this limit has for value* $\frac{\tilde{c}}{\sqrt{2}} = 310.740 \cdot 10^6 \, \mathrm{m} \cdot \mathrm{s}^{-1}$" (Ref. 35, p. 622), thus confirming Kirchhoff's results. He then suggested a possible connection between light and electromagnetic phenomena:

> If this approximated agreement between the propagation speed of electrical waves and the light speed could be seen as an indication of a close relation between these two doctrines, it would deserve a great interest, because the research for such a relation is of a great importance.

5.2. *Maxwell's electromagnetic waves as light*

The breakthrough was provided by James Clerk Maxwell (1831–1879) who published the first electromagnetic theory[36] in 1865, claiming that light could be an electromagnetic wave. He started with Weber's theory which was the most complete by these times, but he encountered some difficulties in providing a mechanical explanation of its functioning due to the assumption that particles were acting at a distance with forces depending on their velocities: These mechanical difficulties prevented him from considering Weber's theory "*as an ultimate one*". For Maxwell, phenomena must be explained (Ref. 36, p. 460)

> by supposing them to be produced by actions which go on in the surrounding medium as well as the excited bodies, and endeavouring to explain

> the action between distant bodies without assuming the existence of forces capable of action directly at sensible distances.

In fact, Maxwell replaced any action at a distance by the concept of the electromagnetic field as suggested by Faraday. The electromagnetic field was thus "*a part of space which contains and surrounds bodies in electric or magnetic conditions*" (Ref. 36, p. 460). It was clear for Maxwell that "*in this space, there is matter in motion, by which the observed electromagnetic phenomena are produced*" (Ref. 36, p. 460); this was therefore a "*dynamical theory*".

Following Ampère's and Weber's assumptions, Maxwell had (Ref. 36, p. 460)

> therefore some reason to believe from the phenomena of light and heat, that there is an aethereal medium filling space and permeating bodies, capable of being set in motion and of transmitting that motion from one part to another and affects it in various ways.

Maxwell thus avoided to consider action at a distance by using the ether "*filling space and permeating bodies*" (Ref. 36, p. 460) with the properties which were already encountered in the works by Newton and Fresnel; for instance, the ether has "*small but real density,* [is] *capable of being set in motion, and of transmitting motion from one part to another with great, but not infinite velocity*; [it has] *certain kind of elasticity yielding*" (Ref. 36, p. 460). Using the electromagnetic field, Maxwell was thus able to explain how two current elements were interacting at a distance (Ref. 36, p. 464):

> When an electric current is established in a conducting circuit, the neighboring part of the field is characterized by certain magnetic properties, and that if two circuits are in the field, the magnetic properties of the field due to the two currents are combined. Thus each part of the field is in connexion with both currents, and the two currents are put into connexion with each other in virtue of their connexion with the magnetization of the field.

As in Fresnel's essay, double refraction was a key phenomenon to be explained; Maxwell did it by using the electromagnetic field to which it became relevant to link polarization. He thus explained Faraday's experiments in which light polarization was affected by a magnetic field as follows (Ref. 36, p. 461):

> The luminiferous medium is in certain cases acted on by magnetism, [as evidenced by Faraday's experiments[30] showing that] when a plane polarized ray traverses a transparent diamagnetic medium in the direction of the lines of magnetic field produced by magnets or currents in the neighborhood, the plane of polarization is caused to rotate.

The ether has therefore magnetic properties. It is subject to motion as well as to vibrations: Electromagnetic phenomena were more related to motion and light to vibrations. Maxwell was able to draw some analogy between these two types of phenomena (Ref. 36, p. 464):

> It appears therefore that certain phenomena in electricity and magnetism lead to the same conclusion as those of optics, namely, that there is an aethereal medium pervading all bodies, and modified only in degree by their presence; that the parts of this medium are capable of being set in motion by electric currents and magnets, that this motion is communicated from one part of the medium to another by forces arising from the connexions of those parts; that under the action of these forces there is a certain yielding depending on the elasticity of these connexions, and that therefore energy in two different forms may exist in the medium, the one form being the actual energy of motion of its parts, and the other being the potential energy stored up in the connexions, in virtue of their elasticity.

Maxwell summed up all existing electromagnetic phenomena into 20 differential equations; in doing so, he used a local electromagnetic field in the ether filling the space surrounding electric and magnetic bodies. He thus described the mechanical actions applied to these bodies.

Expressed in the electrostatic system of units, the units used for describing mechanical actions between electrified bodies must be corrected by a coefficient k taking into account the mechanical action between currents in the electromagnetic system of units. For Maxwell, the coefficient k represents "*the coefficient of electric elasticity in the medium in which the experiments are made, i.e. common air*" (Ref. 36, pp. 491–492); it is related to v, the number of electrostatic units in one electromagnetic unit, by the relation

$$k = 4\pi v^2, \tag{9}$$

where v is in fact Weber's constant $c = \frac{\tilde{c}}{\sqrt{2}}$ that is not so different from the value of light velocity. A natural task, for Maxwell, once the ether was required to explain light propagation as well as electromagnetic phenomena, was to question (Ref. 36, p. 497)

> whether these properties of that which constitutes the electromagnetic field, deduced from electromagnetic phenomena alone, are sufficient to explain the propagation of light through the same substance.

Applying his equations governing the electromagnetic field to the propagation of a plane wave, Maxwell showed "*that the wave is propagated in either direction with a velocity*" (Ref. 36, p. 498)

$$V = \pm\sqrt{\frac{k}{4\pi\mu}}, \tag{10}$$

where μ is the coefficient of magnetic induction which depends "*on the nature of the medium, its temperature and the amount of magnetization already produced*" (Ref. 36, p. 482). Since experiments to determine the value of k were only made in air in which $\mu = 1$, the velocity of light V in air is equal to

$$V = v = \frac{\tilde{c}}{\sqrt{2}} = 3.1074 \cdot 10^8 \, \mathrm{m} \cdot \mathrm{s}^{-1}. \tag{11}$$

This value was only slightly different from the light velocity experimentally measured by Fizeau ($V = 314.858.000 \, \mathrm{m} \cdot \mathrm{s}^{-1}$)[37] or the value deduced from the coefficient for light aberration ($V = 308.000.000 \, \mathrm{m} \cdot \mathrm{s}^{-1}$). Since these two measurements did not involve electricity or magnetism, Maxwell was led to conclude (Ref. 36, p. 499):

> The agreement of the results seems to show that light and magnetism are affections of the same substance, and that light is an electromagnetic disturbance propagated through the field according to electromagnetic laws.

He also showed with his set of electromagnetic field equations that only transversal vibrations could be propagated through the medium and, consequently, that (Ref. 36, p. 499)

> this wave consists entirely of magnetic disturbances, the direction of magnetization being in the plane of the wave. No magnetic disturbance whose direction of magnetization is not in the plane of the wave can be propagated as a plane wave at all.
>
> Hence magnetic disturbances propagated through the electromagnetic field agree with light in this, that the disturbance at any point is transverse to the direction of propagation, and such waves may have all the properties of polarized light.

Fresnel's result for light was thus confirmed here. Hence, he concluded that (Ref. 36, p. 501)

> electromagnetic science leads to exactly the same conclusions as optical science with respect to the direction of the disturbances which can be propagated through the field, both affirm the propagation of transverse vibrations and both give the same velocity of propagation. On the other hand, both sciences are at a loss when called on to affirm or deny the existence of normal vibrations.

The velocity c of light — which emerges from all these theories as a constant — was presented by George Stoney (1826–1911) at the Belfast meeting (1874) of the British Association for the Advancement of the Science[38] as one of three absolute quantities, the "*velocity of Maxwell*" that is also " *the maximum of the velocity of light*", the constant of gravitation and the unit quantity of electricity. For Stoney, using the constant c as the unit velocity would provide "*an immense simplification* [...] *in our treatment of the whole range of electric phenomena, and probably into our study of light and heat*".[38]

5.3. *Helmholtz's theory*

When Helmholtz turned his attention in electrodynamics in 1870,[39] the main three theories were the potential law of Frantz Neumann (1798–1895),[40] the reciprocal action between two moving ponderable charges by Weber[35] and the theory of the electromagnetic field in the ether by Maxwell.[36] These theories were not so widely accepted for various reasons. Maxwell's ideas were criticized for their lack of rigor and experimental evidences; Weber's formula did not obey to the principle of energy conservation due to its dependence on the relative velocity and on the relative acceleration between ponderable charges, a principle that Helmholtz showed to be universal.[41]

In order to overcome such a weakness, Helmholtz also proposed an electrodynamic theory including the previously existing theories and describing the propagation of light as Maxwell did, but in a more comprehensive form. The three theories successively developed by Weber, Neumann and Maxwell were then unified in a single formula by means of an arbitrary parameter k whose values were -1, $+1$ and 0, respectively. Like the previous ones, Helmholtz's theory was based on the existence of an ether.

From his general law, Helmholtz obtained the differential equations describing the motion of electricity. Then, by investigating the nature of his differential equations for the three values of the constant k, he concluded that for $k = -1$ corresponding to Weber's law, the motion of electric charges was unstable, contrary to what he obtained for the two other values of k. He also showed experimentally that motions of electricity in conducting bodies were nearly the same as he obtained with his equations for $k = +1$ and $k = 0$.

Nevertheless, comparing his theory to Maxwell's, he concluded that[42]

> the two theories are opposed to each other in a certain sense since according to the theory of magnetic induction originating with Poisson, which can be carried through in a fully corresponding way for the theory of dielectric, polarization of insulators, the action at a distance is diminished by the polarization, while according to Maxwell's theory on the other hand, the action at a distance is exactly replaced by the polarization [...]. It follows [...] from these investigations that the remarkable analogy between the motion of electricity in a dielectric and that of the light aether does not depend on the particular form of Maxwell's hypotheses, but results also on a basically similar fashion if one maintains the older view point about electrical action at a distance.

Although Maxwell's theory was later published in a slightly different form,[43] but without removing its internal contradiction, it was preferred to Helmholtz's theory as history has shown.

5.4. *Hertz's experiments for validating Maxwell's theory*

Maxwell left his theory without any experimental proof. Such an experimental evidence was provided by Heinrich Hertz (1857–1894)[44–46] while addressing the question related to the "Berlin Academy of Science Prize" proposed by Helmholtz for the year of 1879: "*To establish experimentally any relation between electromagnetic forces and the dielectric polarization of insulators*" (Ref. 47, p. 1). Hertz's researches were thus guided by the connection propounded by Helmholtz, according to which (Ref. 47, p. 6):

> If we start from the electromagnetic laws which in 1879 enjoyed universal recognition, and make certain further assumptions, we arrive at the equations of Maxwell's theory which at that time (in Germany) were by no means universally recognized.

For Hertz, the question was to experimentally investigate the validity of Maxwell's conclusion that light could be an electromagnetic wave. Hertz was initially convinced that such a claim was unacceptable (Ref. 47, p. 8):

> I reflected that it would be quite as important to find out that electric force was propagated with an infinite velocity, and that Maxwell's theory was false, as it would be, on the other hand, to prove that this theory was correct, provided only that the result arrived at should be definite and certain. [...]
>
> I have entered into these details here in order that the reader may be convinced that my desire has not been simply to establish a preconceived idea in the most convenient way by a suitable interpretation of the experiments. On the contrary, I have carried out with the greatest possible care these experiments (by no means easy ones) although they were in opposition to my preconceived views.

Hertz's experiments were as follows (Ref. 45, p. 107):

> In the first place, regular progressive waves were to be produced in a straight, stretched wire by means of corresponding rapid oscillations of a primary conductor. Next, a secondary conductor was to be exposed simultaneously to the influence of the waves propagated throughout the wire and to the direct action of the primary conductor propagated through the air; and thus both actions were to be made to interfere. Finally, such interferences were to be produced at different distances from the primary circuit,

> so as to find out whether the oscillations of the electric force at great distances would or would not exhibit a retardation of phase, as compared with the oscillations in the neighborhood of the primary circuit.

Hertz was thus investigating the influence of primary oscillations on a secondary circuit, taking into account their relative positions, a possibly relevant aspect for testing the electromagnetic effect independently of the electrostatic effect. From his experiments, he found "*that the waves in the wire have the same periodic time as the primary oscillations*" (Ref. 45, p. 111). Using interference phenomena resulting from the interaction between the waves propagated in air with the waves propagated in a wire, he observed that (Ref. 45, p. 117):

> Very little consideration will show that, if the action is propagated through the air with infinite velocity, it must interfere with the waves in the wire in opposite senses at distances of half a wave-length (i.e. 2.8 meters) along the wire. Again, if the action is propagated through the air with the same velocity as that of the waves in the wire, the two will interfere in the same way at all distances. Lastly, if the action is propagated through the air with a velocity which is finite, but different from that of the waves in the wire, the nature of the interference will alternate, but at distance which are farther than 2.8 meters apart.

Hertz was finally able to assess the velocity of electromagnetic waves (Ref. 45, p. 107):

> The experiments carried out in accordance with it have shown that the inductive action is undoubtedly propagated with a finite velocity. This velocity is greater than the velocity of propagation of electric waves in wires. According to the experiments made up to the present time, the ratio of these velocities is about 45:28. From this it follows that the absolute value of the first of these is of the same order as the velocity of light.

and concluded that "*the absolute velocity of propagation in air is* 320.000 *km per second*" (Ref. 45, p. 121).

Although he had experimentally proved that electromagnetic actions are propagated through air with a finite velocity, Hertz wanted to propose an experiment which could "*exhibit the propagation of induction through the air by wave-motion in a visible and almost tangible form*" that would permit "*a direct measurement of the wave-length in air*" (Ref. 46, p. 124). This new experiment was based on the interference between direct waves propagating in air and waves which were reflected by a wall, producing stationary waves. By repeating this interference experiment between two waves for various distances from the reflecting wall, Hertz showed that the velocity had the same order of magnitude as the velocity associated with light propagation. He thus concluded that (Ref. 46, p. 136)

> the experiments amount to so many reasons in favor of that theory of electromagnetic phenomena which was first developed by Maxwell from Faraday's views. It also appears to me that the hypothesis as to the nature of light which is connected with that theory now forces itself upon the mind with still stronger reason than heretofore. Certainly it is a fascinating idea that the processes in air which we have been investigating represent to us on a million-fold larger scale the same processes which go on in the neighborhood of a Fresnel mirror or between the glass plates used for exhibiting Newton's rings.

The result concerning the period of oscillations of electromagnetic waves in air obtained by Hertz presented a computational error by $\sqrt{2} : 1$, Poincaré first drew Hertz's attention to this.[47–49]

Hertz interpreted all the results provided by his rapid oscillations experiments from the standpoint associated with Helmholtz's theory, that is, from the fact that there are "*two forms of electric force — the electromagnetic and the electrostatic — to which* [...] *two different velocities are attributed*" (Ref. 47, p. 15). Moreover, in a special limiting case, Helmholtz's equations became "*the same as those of Maxwell's theory*: *only one form of the force remains, and this is propagated with the velocity of light*" (Ref. 47, p. 15). Hertz's experimental results conferred "*upon Maxwell's theory a position of superiority to all others*" (Ref. 50, p. 137). Hertz was thus able "*to show that the phenomena can be explained in terms of Maxwell's theory without introducing*" (Ref. 50, p. 137) a distinction between electromagnetic and electrostatic forces, as was done by Helmholtz.

In order to describe electric and magnetic forces acting in the ether at points $\mathbf{x}$ which, according to Maxwell, are such that "*the time-rate of change of the forces is independent upon their distribution in space*" (Ref. 50, p. 138), Hertz[c] used the equations (Ref. 50, p. 138)

$$\frac{1}{c}\frac{d\mathbf{H}}{dt} = \boldsymbol{\nabla} \wedge \mathbf{E} \tag{12}$$

and

$$\frac{1}{c}\frac{d\mathbf{E}}{dt} = -(\boldsymbol{\nabla} \wedge \mathbf{H}), \tag{13}$$

[c]We choose to translate the equations in the modern vectorial notations — introduced by Oliver Heaviside[51] — rather than leaving the quite cumbersome use of different letters for the components of a given vector. Apart from that, we left unchanged the way in which the equations were written in the original paper. For instance, Hertz introduce a minus sign in Eq. (13) and not in Eq. (12) as they are now written. These equations, describing the electric and magnetic fields in the ether, were rewritten with the sign as used today by Poincaré (Ref. 71, p. 373) that is, with a minus sign in Eq. (12) (and not in Eq. (13)) by Poincaré (Ref. 29, pp. 115–116). The corresponding equations for electromagnetic phenomena in bodies in motion were also rewritten with the corrected sign by Lorentz (Ref. 52, pp. 55 and 58).

where $\mathbf{E}$ is the electric force (whose components were designated by (X, Y, Z) by Hertz), $\mathbf{H}$ the magnetic force (whose components were designated by (L, M, N) by Hertz); t is the time and $\frac{1}{c}$ is the reciprocal of the light velocity.

For Hertz, $\frac{1}{c}$ is "*in reality an intrinsic constant of the ether; in saying this we assert that its magnitude is independent of the presence of any other body, or of any arbitrary stipulation on our part*" (Ref. 53, p. 202). To these equations, Hertz added two specific conditions (Ref. 50, p. 138)

$$\boldsymbol{\nabla} \cdot \mathbf{E} = 0 \tag{14}$$

and

$$\boldsymbol{\nabla} \cdot \mathbf{H} = 0, \tag{15}$$

which must be satisfied and which are required to distinguish the ether from the ponderable matter.

The total energy contained in a volume-element τ of the ether was the sum of the electric energy (Ref. 50, p. 138)

$$\frac{1}{8\pi} \int E^2 d\tau \tag{16}$$

and the magnetic energy

$$\frac{1}{8\pi} \int H^2 d\tau, \tag{17}$$

contained in that volume-element. When electric and magnetic forces act on the ether, Hertz imposed $\mu = 1$ and $\epsilon = 1$, as he specified: "*The specific inductive capacity* (*Dielecktricitätsconstante*) and the magnetic permeability (*Magnetisirungsconstante*) *of a susbstance* [are] *equal to unity for the ether; but this does not state any fact derived from experience; it is only an arbitrary stipulation on our parts.*" (Ref. 53, p. 200)

Concerning all these equations, Hertz wrote that (Ref. 50, p. 138):

> These statements form, as far as the ether is concerned, the essential parts of Maxwell's theory. Maxwell arrived at them by starting with the idea of action-at-a-distance and attributing to the ether the properties of a highly polarisable dielectric medium.

Hertz assumed that it was possible to reach the same statements by using other ways "*but in no way can a direct proof of these equations be deduced from experience*" (Ref. 50, p. 138). He was thus able to show that his experiments were in agreement with "*these much simpler assumptions of Maxwell's theory*" (Ref. 50, p. 159):

> In our endeavour to explain the observations by means of Maxwell's theory, we have not succeeded in removing all difficulties. Nevertheless, the theory

> had been found to account most satisfactorily for the majority of the phenomena; and it will be acknowledged that this is no mean performance. But if we try to adapt any of the older theories to the phenomena, we meet with inconsistencies from the very start, unless we reconcile these theories with Maxwell's by introducing the ether as dielectric on the manner indicated by Helmholtz.

Nevertheless, as many other scientists, Hertz had "*been compelled to abandon the hope of forming for [himself] an altogether consistent conception of Maxwell's ideas*" (Ref. 47, p. 29), thus justifying why his experiments were guided by Helmholtz's theory and not by Maxwell's. Unfortunately, his experiments led him to promote Maxwell's theory, leaving him with a lack of consistency in his methodology. Consequently, Hertz wanted "*to form for [himself] in a consistent manner the necessary physical conceptions starting from Maxwell's equations, but otherwise simplifying Maxwell's theory as far as possible*" (Ref. 47, p. 21). In fact, Hertz considered that "*Maxwell's theory is Maxwell's system of equations*" and added that (Ref. 47, p. 21)

> every theory which leads to the same system of equations, and therefore comprises the same possible phenomena, I would consider as being a form or special case of Maxwell's theory; every theory which leads to different equations, and therefore to different possible phenomena, is a different theory.

6. Invariance of the Field Equations from a Frame to Another One

6.1. *Hertz's electrodynamic theory*

The electromagnetic theory developed by Hertz was published in two papers, one for bodies at rest (Ref. 53, p. 195) and one for bodies in motion (Ref. 54, p. 241). From Maxwell's theory, Hertz kept only the equations he corrected according to his experiments and wanted to build a rigorous theory leading to them.

Although informed about similar work[55] led by Oliver Heaviside (1850–1925) whose results are identical to those obtained by Hertz, the latter remained convinced that his theory had to be preferred to all the others because it was describing accurately more phenomena. Hertz was working from the experimental fact to the equations describing them, using a methodology claimed by Newton or Ampère: "*The statement will be rather given as facts derived from experience, and experience must be regarded as their proof*" (Ref. 53, p. 197). For determining electric and magnetic forces as well as the total energy at a given point of the space, Hertz made "*an essential and important hypothesis*" which was (Ref. 53, p. 198)

> that the specification of a single directed magnitude is sufficient to determine completely the change of state under consideration. Certain phenomena, e.g, those of permanent magnetism, dispersion, etc., are not intelligible

> from this standpoint; they require that the electric or magnetic conditions of any point should be represented by more than one variable.

Considering his fundamental Eqs. (12)–(15) that he introduced in one of his previous papers,[50] Hertz proposed the equations governing electromagnetics phenomena — according to his experiments — in bodies of various characteristics (good or bad conductor, isotropic or crystalline medium), considering that phenomena quantitatively differ from those observed in the free ether "*in two respects*: *In the first place, the intrinsic constant has a value different from what it has in the ether*; *and in the second place, the expression for the energy per unit volume contains, as already explained, the constants* ϵ *and* μ" (Ref. 53, p. 202). As an example, the set of equations describing electromagnetic phenomena in a conducting isotropic and homogeneous medium are (Ref. 53, p. 205)

$$\frac{1}{c}\mu\frac{d\mathbf{H}}{dt} = \mathbf{\nabla} \wedge \mathbf{E} \tag{18}$$

and

$$\frac{1}{c}\epsilon\frac{d\mathbf{E}}{dt} = -\mathbf{\nabla} \wedge \mathbf{H} - \frac{4\pi\lambda}{c}\mathbf{E}, \tag{19}$$

where μ is the magnetic permeability of the medium; ϵ its specific inductive capacity and λ is the specific conductibility of the body. Consequently, the total energy of the electromagnetic field per unit of volume is (Ref. 53, p. 199)

$$\frac{\epsilon}{8\pi}E^2 + \frac{\mu}{8\pi}H^2. \tag{20}$$

As it must be, the set of equations proposed by Hertz and describing electromagnetic phenomena in bodies at rest is consistent with the principle of the conservation of energy. All electromagnetic phenomena as well as some optical phenomena such as reflection or refraction are well described by these equations. Nevertheless, they cannot provide an explanation of the dispersion phenomena which need more than one oriented magnitude as used in this theory.

Developing further his first paper for bodies at rest, Hertz extended his theory "*to embrace the course of electromagnetic phenomena in bodies which are in motion*" (Ref. 54, p. 241). By bodies in motion, Hertz considered ponderable matter and "*the disturbances of the ether which simultaneously arise* [*and*] *cannot be without effect*; *and of these we have no knowledge*" (Ref. 54, p. 241). In order to do that, he needed to add an assumption on the motion of the ether: In a ponderable body, there is a total driving of the ether. Hertz thus assumed that (Ref. 54, p. 242):

> Electric and magnetic phenomena must be compatible with the view that no [partial driving] occurs, but that the ether which is hypothetically assumed to exist in the interior of ponderable matter only moves with

> it. This view includes the possibility of taking into consideration at every point in space the condition of only one medium filling the space.

Here, Hertz proposed to fuse into a single medium the transparent body traversed by light and the ether that should exist in it; at least in a ponderable body, the ether could thus be omitted for explaining light propagation. Such an assumption was mainly motivated by the fact that, according to Hertz, he has no experimental data related to a partial driving.

He admitted that the theory he developed (Ref. 54, pp. 242–243)

> on such a foundation will not possess the advantage of giving to every question that may be raised the correct answer, or even of giving only one definite answer; but it at least gives a possible answer to every question that may be propounded, i.e. answers which are not inconsistent with the observed phenomena nor yet with the views which we have obtained as to bodies at rest.

Hertz then considered that each point of a body is characterized by the electric force $\mathbf{E}$ and the magnetic force $\mathbf{H}$, the electric polarization $\mathbf{P}$ and the magnetic polarization $\mathbf{M}$, the electric current $\mathbf{I}$, the electromotive force $\mathbf{E}'$, the magnetic and dielectric constants μ and ϵ, respectively, and the conductibility constant λ.

In his previous essay,[53] Hertz had obtained the following system of equations to describe the electromagnetic phenomena in bodies at rest according to electric and magnetic polarizations (Ref. 53, p. 211)

$$\frac{1}{c}\frac{d\mathbf{M}}{dt} = \boldsymbol{\nabla} \wedge \mathbf{E} \tag{21}$$

and

$$\frac{1}{c}\frac{d\mathbf{P}}{dt} = -\boldsymbol{\nabla} \wedge \mathbf{H} - \frac{4\pi}{c}\mathbf{I}. \tag{22}$$

By introducing the polarizations, the electromagnetic energy per unit volume of any body takes the form:

$$\frac{1}{8\pi}\mathbf{P} \cdot \mathbf{E} + \frac{1}{8\pi}\mathbf{M} \cdot \mathbf{H}. \tag{23}$$

> In these expressions there no longer appear any quantities which refer to any particular body. The statement that these equations must be satisfied at all points of infinite space, embraces all problems of electromagnetism; and the infinite multiplicity of these problems only arises through the fact that the constants ϵ, μ, λ, $\mathbf{E}'$ of the linear relations, may be functions of the space in a multiplicity of ways, varying partly continuously, and partly discontinuously, from point to point. (Ref. 53, p. 211)

As in his first paper, electric and magnetic polarizations must be "*regarded as a second and equivalent means of indicating the same conditions*" (Ref. 54, p. 243) as electric and magnetic forces.

To transpose his fundamental equations for bodies at rest to the case of bodies in motion, Hertz remarked that (Ref. 54, pp. 243–244)

> At any point of a body at rest the time-variation of the magnetic state is determined simply by the distribution of the electric force in the neighbourhood of the point. In the case of a body in motion there is, in addition to this, a second variation which at every instant is superposed upon the first and which arises from the distortion which the neighbourhood of the point under consideration experiences through the motion.

Hertz also introduced a relevant hypothesis according which "*the influence of the motion is of such a kind that, if it alone were at work, it would carry the magnetic lines of force with the matter*" (Ref. 54, p. 244).

> The corresponding statement holds good for the variation which the electric polarization experiences through the motion. These statements suffice for extending to moving bodies the theory already developed for bodies at rest; they clearly satisfy the conditions which our system of itself requires, and it will be shown that they embrace all the observed facts.

Under these considerations, Hertz obtained his fundamental electromagnetic equations for bodies in motion (Ref. 54, p. 245):

$$\frac{1}{c}\left[\frac{d\mathbf{M}}{dt} + \boldsymbol{\nabla}\wedge(\mathbf{M}\wedge\mathbf{V}) + \mathbf{V}\cdot(\boldsymbol{\nabla}\cdot\mathbf{M})\right] = \boldsymbol{\nabla}\wedge\mathbf{E}, \tag{24}$$

$$\frac{1}{c}\left[\frac{d\mathbf{P}}{dt} + \boldsymbol{\nabla}\wedge(\mathbf{P}\wedge\mathbf{V}) + \mathbf{V}\cdot(\boldsymbol{\nabla}\cdot\mathbf{P})\right] = -\boldsymbol{\nabla}\wedge\mathbf{H} - \frac{4\pi}{c}\mathbf{I}, \tag{25}$$

where $\mathbf{V}$ is the velocity with which the surface element was moving. These equations "*are completed by linear relations which connect the polarizations and the current-components with the forces. The constants of these relations are to be regarded as functions of the varying conditions of the moving matter, and to this extent as functions of the time as well*" (Ref. 54, p. 246). Hertz also provided some details about the reference frame used for writing these equations:

> Our method of deducing the [previous] equations does not require that the system of co-ordinates used should remain absolutely fixed in space. We can, therefore, without change of form, transform our equations from the system of co-ordinates first chosen to a system of co-ordinates moving in any manner through space, by taking V_x, V_y, V_z to represent the velocity-components with reference to the new system of co-ordinates, and referring

> the constants ϵ, μ, λ, $\mathbf{E}'$, which depend upon direction, at every instant to these. From this it follows that the absolute motion of a rigid system of bodies has no effect upon any internal electromagnetic processes whatever in it, provided that *all* the bodies under consideration, including the ether as well, actually share the motion. It further follows from this consideration that even if only a single part of a moving system moves as a rigid body, the processes which occur in this part follow exactly the same course as in bodies at rest.
>
> [The previous] equations tell us the future value of the polarizations at every fixed point in space or, if we prefer it, in each element of the moving matter, as a definite and determinate consequence of the present electromagnetic state and the present motion in the neighbourhood of the point under consideration (Ref. 54, p. 247).

Although Hertz asserted the invariance of his equations from a system of coordinates at rest to a system of coordinates in motion (with no restriction on the type of relative motion), he did not propose the coordinate transformation allowing such invariance which was not therefore proved.

As he did for bodies at rest, Hertz showed that the fundamental electromagnetic equations for bodies in motion are consistent with the principle of the conservation of energy as well as the principle of action and reaction, as they must be. He then concluded (Ref. 54, p. 268):

> I only attach value to the theory of electromagnetic forces in moving bodies here proposed from the point of view of systematic arrangement. The theory shows how we can treat completely the electromagnetic phenomena in moving bodies, under certain restrictions which we arbitrarily impose. It is scarcely probable that these restrictions correspond to the actual facts of the case. The correct theory should rather distinguish between the conditions of the ether at every point, and those of the embedded matter. But it seems to me that, in order to propound a theory in accordance with this view at present, we should require to make more numerous and arbitrary hypotheses than those of the theory here set forth.

Thus, even if the theory developed by Hertz for bodies in motion is consistent with the principles of action and reaction and of the conservation of the energy, it cannot explain certain optical phenomena. This lack of generality was addressed by Hendrick A. Lorentz (1853–1928) in 1892 when he investigated the electrodynamics of bodies in motion.[52]

6.2. *Voigt's wave equation*

The first coordinate transformation leaving invariant the system of electromagnetic equations was proposed by Woldemar Voigt (1850–1919). In order to explain the so-called Doppler effect, Voigt started to investigate the differential equation governing the propagation of a plane wave in an elastic medium, that is, a phenomenon equivalent to light propagation in an ether.[56,57] Assuming that plane waves were moving with a velocity c and a constant amplitude, Voigt was looking for a change of coordinates that would leave unchanged his wave equation when he switched from one illuminating surface to another one in a uniform translation by a velocity V with respect to the first one. Voigt had to introduce the coordinate transformation (Ref. 56, p. 50)

$$\left| \begin{aligned} & x' = x - Vt \\ & y' = \gamma y \\ & z' = \gamma z \\ & t' = t - \frac{V \cdot x}{c^2} \end{aligned} \right. , \tag{26}$$

where V (designated by χ by Voigt) is the uniform velocity of the moving axes (x', y', z') along the x-axis, $\gamma = \sqrt{1 - \frac{V^2}{c^2}}$ (γ was designated as q by Voigt) and c (ω in Voigt's notation) is the propagation velocity of the oscillations (or the plane waves). Voigt thus needed to introduce a local time t' to obtain the invariance of his wave equation when terms $\frac{V^2}{c^2}$ were neglected (Ref. 56, p. 50):

> [The two illuminating surfaces, one at rest and one in motion at velocity V] have identical forms only if $\gamma = 1$, i.e. V is small against c, that V^2 can be neglected with respect to c^2.

While considering a small illuminated sphere of radius r, he thus concluded that (Ref. 56, p. 50)

> a stationary observer, since the perpendicular to the wave surface through the location of observation gives the direction in which the light source is to be perceived, would see the illuminating point at the location where it was at time $\frac{r}{c}$; in other words, he would observe, if his radius vector r includes the angle ϕ with the direction of motion, an "aberration" of the size $\frac{r}{c} \sin \phi$ in the direction opposite to the motion of the point.

The aberration could be seen as a length contraction as it was later introduced by FitzGerald and Lorentz (see next section).

6.3. *Lorentz's electrodynamical theory*

When Lorentz started to investigate electrodynamical theories, he was mainly focused on the foundations used by Hertz for developing his electromagnetic theories for bodies at rest and in motion. Lorentz considered that, to establish his fundamental equations, Hertz "*doesn't take care of a link between the electromagnetics actions and the laws of the wave mechanic*" (Ref. 52, p. 6). He clarified his approach as follows (Ref. 52, p. 6):

> We are always tempted to return to the mechanical explanations. That's why it seemed to me useful to apply directly to the most general case the method of which Maxwell gave the example in his study of the linear circuits.
>
> I had another reason to endeavor these researches. In the report where Hertz treated bodies in motion, he admits that the ether which they contain moves with them. But, optical phenomena demonstrated for a long time that it is not there still so. I thus wished to know the laws which govern the electric movements in bodies which cross the ether without entraining it, and it seemed to me difficult to reach the purpose without having for guide a theoretical idea. The sights of Maxwell can be of use for the foundation of the theory we are looking for.

The optical phenomena not considered by Hertz's assumption were the experiments conducted by Fizeau[58] (see Appendix A.1) and those by Michelson and Morley[59] (see Appendix A.2). When he developed his theory, Lorentz had already in mind the latter experiment. He thus developed a "*theory of electromagnetic phenomena based on the idea that there is a ponderable matter perfectly permeable to the ether and able to move without transferring to this latter the smallest movement*" (Ref. 52, p. 70).

Starting from what we call now the Maxwell equations to describe the motion of an electrified body in an ether partly entrained and being the source of an electric current and a dielectric phenomenon, Lorentz added a few hypotheses:

(i) The electrified body and the resting ether are independent, although the ether can freely go across the body.
(ii) The electrified body is considered as a rigid body whose movements are limited to a translation and a rotation.
(iii) The position of any point taking part in the electromagnetic motion is determined as soon as one knows the position of all the charged particles of the system and the components of the dielectric displacement in the ether at every point.

To develop his theory, Lorentz needed to use two sets of equations, one related to the state of the ether and the second related to the reaction of this medium on the electrified particles.

The set of equations related to a system of charged particles moving in the ether but without entraining it was (Ref. 52, pp. 89–90)

$$\nabla \cdot \mathbf{D} = \rho, \tag{27}$$

where $\mathbf{D}$ is the dielectric displacement in the ether and ρ the density of the electric charges. We used here the vector notation as introduced by Heaviside[51] (Lorentz only used the operators $\boldsymbol{\nabla}$ and $\Box$). A similar equation was provided for the magnetic force $\mathbf{H}$ in the ether, that is

$$\boldsymbol{\nabla} \cdot \mathbf{H} = 0. \tag{28}$$

The last two equations were

$$\boldsymbol{\nabla} \wedge \mathbf{H} = 4\pi \mathbf{I} = 4\pi \left(\rho \mathbf{v} + \frac{\partial \mathbf{D}}{\partial t} \right), \tag{29}$$

where $\mathbf{I}$ the electric current and $\mathbf{v}$ the velocity of a point of a charged particle;

$$-4\pi c^2 (\boldsymbol{\nabla} \wedge \mathbf{D}) = \frac{\partial \mathbf{H}}{\partial t}, \tag{30}$$

where c is the light velocity in the ether. These equations correspond to a slightly modified form of Eqs. (12)–(15) proposed by Hertz. Their use was justified by Lorentz as follows (Ref. 52, pp. 90–91):

> As far as, in the field considered, there is no charged body, the formula given by Hertz in his first memoir are the simplest that one can admit to express the state of the ether.

For Lorentz, light propagation consists in (Ref. 52, p. 112)

> oscillatory movements that the charged particles could perform in the molecules of a dielectric. Accompanied with periodic changes in the state of the ether, these vibrations will constitute a beam of light, the propagation of which I suggest to study.

The previously established equations "*will serve to determine the state of the ether which is compatible with the movement of particles*" (Ref. 52, p. 112).

To establish the fundamental equations describing the propagation of light in bodies in motion, Lorentz needs to introduce two different sets of coordinates. Supposing that (Ref. 52, p. 136)

> all molecules of the dielectric are animated with the same velocity of translation parallel to the x-axis and independent of the time. I will designate by V this velocity and, keeping for now motionless axes O_x, O_y and O_z, I will introduce new axes which are fixedly linked to the ponderable matter. The first of these axes will match with O_x; the two others will be parallel to O_y and O_z and will match with these axes at time $t = 0$.

The coordinate transformation from a set of motionless axes to a set of axes in uniform translation at a velocity of V is then given by (Ref. 52, p. 136):

$$\left| \begin{aligned} x' &= x - Vt \\ y' &= y \\ z' &= z \end{aligned} \right. , \tag{31}$$

where (x', y', z') are the coordinates in uniform translation and (x, y, z) are the coordinates at rest. This transformation is completed by the partial derivatives expressed as (Ref. 52, p. 136)

$$\left| \begin{array}{l} \boldsymbol{\nabla} = \boldsymbol{\nabla}' \\ \\ \dfrac{\partial}{\partial t} = \dfrac{\partial}{\partial t'} - V\dfrac{\partial}{\partial x'}, \end{array} \right. \tag{32}$$

allowing Lorentz to rewrite his fundamental equations in the form (Ref. 52, p. 137)

$$\boldsymbol{\nabla} \cdot \mathbf{D} = \rho, \tag{33}$$

$$\boldsymbol{\nabla} \cdot \mathbf{H} = 0, \tag{34}$$

$$\boldsymbol{\nabla} \wedge \mathbf{H} = 4\pi \left[\rho(\mathbf{v} + \mathbf{V}) + \left(\frac{\partial}{\partial t} - \mathbf{V} \cdot \boldsymbol{\nabla} \right) \mathbf{D} \right], \tag{35}$$

where $\mathbf{V} = (V, 0, 0)$ "*does no longer designate the absolute velocity of a charged particle, but is the velocity with respect to the ponderable matter, so that the absolute velocity becomes* $(v_x + V, v_y, v_z)$" (Ref. 52, p. 137). The last equation is

$$-4\pi c^2 \boldsymbol{\nabla} \wedge \mathbf{D} = \left(\frac{\partial}{\partial t} - \mathbf{V} \cdot \boldsymbol{\nabla} \right) \mathbf{H}. \tag{36}$$

> As long as the charged particles have no other motion than the common velocity of the ponderable matter, we will have $\mathbf{v} = \mathbf{0}$ and the density at a point $\mathbf{x}$ will be independent of t. This will be no longer like this when molecules are the place of electric vibrations. (Ref. 52, p. 137)

With his new electromagnetic theory of light, Lorentz searched to explain the negative result provided by Michelson and Morley's experiment (see Appendix A.2), trusting Fresnel's theory for a partly driven ether. Lorentz computed, using the classical composition law for velocities and Fresnel's partial driving, that for this experiment "*the time required by light to travel forth and back between [the] two points regarded as fixed to Earth*" (Ref. 60, p. 1) should give a fringe shift of $\frac{lV^2}{c^2}$ where l is the distance between the two points, V is the velocity of the Earth and c is the light velocity. Unfortunately Michelson and Morley's experiment did not show any displacement of the fringes. Lorentz finally reached independently the same conclusion as FitzGerald (Ref. 60, p. 2) (see Appendix A.2):

> I found only one way to reconcile [Michelson's] result with Fresnel's theory. It consists in the assumption, that the line joining two points of a solid body does not conserve its length, when it is once in motion parallel to the direction of the Earth motion, and afterwards it is brought normal to it.

Lorentz quantified this contraction according to the coefficient

$$\alpha = \frac{1}{2}\frac{V^2}{c^2}. \tag{37}$$

He physically justified such a contraction as follows (Ref. 60, p. 2):

> What determines the size and shape of a solid body? Apparently the intensity of molecular forces; any cause that could modify it, could modify the shape and size as well. Now we can assume at present, that electric and magnetic forces act by intervention of the aether. It is not unnatural to assume the same for molecular forces, but then it can make a difference, whether the connecting line of two particles, which move together through the ether, is moving parallel to the direction of motion or perpendicular to it. One can easily see, that an effect of order $\frac{V}{c}$ is not expected, but an effect of order $\frac{V^2}{c^2}$ is not excluded and that is exactly what we need.

In 1895, he improved his 1892 theory.[61] Keeping the two different sets of coordinates introduced in his first theory — one set of fixed coordinates and one moving set of coordinates "*rigidly connected with ponderable matter and therefore its displacement*" (Ref. 61, Sec. 19) — he then proposed the new coordinate transformation (Ref. 61, Sec. 23)

$$\left| \begin{aligned} x' &= kx \\ y' &= y \\ z' &= z \end{aligned} \right. , \tag{38}$$

where $k = \frac{1}{\sqrt{1-\beta^2}}$, with $\beta = \frac{V}{c}$ as the contraction coefficient. The location of a point with respect to the fixed system was designated by (x, y, z) and the location of that point in the relative coordinates was designated by (x', y', z'). Lorentz was thus able to show that the new electric density was (Ref. 61, Sec. 23)

$$\rho' = \rho\sqrt{1-\beta^2}. \tag{39}$$

When he tried to express the magnetic force $\mathbf{H}$ in the resting frame, the resulting expression suggested to him the introduction of a new time (Ref. 61, Sec. 31)

$$t' = t - \frac{V_x x + V_y y + V_z z}{c^2}, \tag{40}$$

which can be regarded as a "*local time*" (Ref. 61, Sec. 31), in contrast to the "*general time*". Lorentz was looking for an invariance of the electric force $\mathbf{E}$, the magnetic force $\mathbf{H}$ and the dielectric polarization $\mathbf{P}$, that he formulated as follows (Ref. 61, Sec. 59):

> Namely, if a state of motion for a system of stationary bodies is known, where $\mathbf{P}$, $\mathbf{E}$, $\mathbf{H}$ are certain functions of x, y, z and t, then in the same system, if it is displaced by the velocity V, there can exist a state motion, where $\mathbf{P}'$, $\mathbf{E}'$, $\mathbf{H}'$ are exactly the same functions of x', y', z' and t',

where t' is the local time given by Eq. (40).

Using the coordinate transformations (38)–(40), Lorentz was able to conclude that (Ref. 61, Sec. 64)

> in general, the motion of Earth will never have an influence of the first order on experiments with terrestrial light sources.

In 1899, Lorentz modified the coordinate transformations (38) and (40) into (Ref. 62, p. 429)

$$\left| \begin{aligned} t' &= t - \frac{\beta x}{c(1-\beta^2)} \\ x' &= \frac{x}{\sqrt{1-\beta^2}} \\ y' &= y \\ z' &= z. \end{aligned} \right. \tag{41}$$

But this was not yet satisfactory and Lorentz obtained the invariance of the fields only when he neglected the terms depending on $(\frac{V}{c})^2 = \beta^2$. The imperfection of Lorentz's results were pointed out by Poincaré when he wrote for the celebration of the 25th anniversary of Lorentz's Ph.D. thesis.[63]

Stimulated by Poincaré's criticisms, Lorentz developed a third theory for attempting to overcome the difficulties so pointed out (Ref. 64, p. 18):

> Poincaré has objected to the existing theory of electric and optical phenomena in moving bodies that, in order to explain Michelson's negative result, the introduction of a new hypothesis has been required, and that the same necessity may occur each time new facts will be brought to light. Surely, this course of inventing special hypotheses for each new experimental result is somewhat artificial. It would be more satisfactory if it were possible to show, by means of certain fundamental assumptions, and without neglecting terms of one order of magnitude or another, that many electromagnetic actions are entirely independent of the motion of the system. Some years ago, I have already sought to frame a theory of this kind (Ref. 62, p. 507). I believe now to be able to treat the subject with a better result. The only restriction as regards the velocity will be that it be less than that of light.

Lorentz therefore proposed an additional correction to the change of coordinates (41) allowing for switching from a resting frame to a frame moving with a constant velocity V along the x-axis; he thus proposed (Ref. 64, p. 812):

$$\left| \begin{aligned} t' &= \frac{l}{k}t - kl\frac{V}{c^2}x \\ x' &= klx \\ y' &= ly \\ z' &= lz \end{aligned} \right. , \tag{42}$$

where t' is the local time in the moving frame, l is a numerical quantity to determine, $k = \frac{1}{\sqrt{1-\beta^2}}$ and $\beta = \frac{V}{c}$ is the contraction coefficient along the direction of the motion. With this new transformation, the electric density became (Ref. 64, p. 813)

$$\rho' = \frac{1}{kl^3}\rho. \tag{43}$$

Applying this new transformation to the electrons, Lorentz supposed that (Ref. 64, p. 818)

> the electrons, which I take to be spheres of radius R in the state of rest, have their dimensions changed by the effect of a translation, the dimensions in the direction of motion becoming kl times and thus in perpendicular directions l times smaller. In this deformation, which may be represented by $\left(\frac{1}{kl}, \frac{1}{l}, \frac{1}{l}\right)$, each element of volume is understood to preserve its charge.

Concerning the value of l, the single condition which must be taken into account is (Ref. 64, p. 823)

$$\frac{d(klV)}{dV} = k^3 l. \tag{44}$$

Since

$$\frac{d(kV)}{dV} = k^3, \tag{45}$$

this condition can be rewritten as

$$\frac{dl}{dV} = 0 \tag{46}$$

and, consequently, l must be a constant. Lorentz then concluded that "*the value of the constant must be unity, because we know already that, for* $V = 0$, $l = 1$" (Ref. 64, p. 824). Nevertheless, the coordinate transformation (42) was not yet fully correct as expressed by Poincaré who sent in two letters to Lorentz the correct form (May 1905, see Refs. 65 and 66)

$$\left| \begin{aligned} t' &= kl(t + \epsilon x) \\ x' &= kl(x + \epsilon t) \\ y' &= ly \\ z' &= lz \end{aligned} \right. , \tag{47}$$

where $-\epsilon$ is the velocity of the translation, the light velocity being taken equal to the unity. As we will see in Sec. 7, Poincaré proposed a more rigorous way to show that $l = 1$.[66]

The second theory of Lorentz is based on a few fundamental hypotheses; among them we have:

- There is a contraction of dimensions in the direction of motion by a coefficient $k\,l$ (when $l = 1$, this coefficient is in agreement with Kaufmann's experimental results).

- The forces observed between uncharged particles as well as between electrified particles are influenced by a translation in quite the same way as the electric force in an electrostatic system.

One of the very important conclusion by Lorentz is that[64]:

> It will be impossible to detect an influence of the Earth's motion in any optical experiments, made with a terrestrial source of light [...]. Many experiments on interference and diffraction belong to this class.

The Michelson and Morley experiment was thus explained (at a given order) with this theory. Lorentz was not able to prove at any order the invariance of Maxwell's equations. Nevertheless, he was also able to express the function $\Psi(\beta)$ introduced by Max Abraham (1875–1922) who proposed an equation expressing the dependence of the mass m on the velocity V,[67] that is,

$$\frac{e}{m} = \frac{e}{m_0}\frac{4}{3}\frac{1}{\Psi(\beta)}, \quad \text{where } \beta = \frac{V}{c} \tag{48}$$

and m_0 is the mass at rest. At that time, the electric charge e of the electron was not yet determined with a great accuracy but, according to Stoney,[38] it was mostly considered as being constant. This equation was experimentally tested by Walter Kaufmann (1871–1947).[68] Lorentz proposed to use

$$\Psi(\beta) = \frac{4}{3}\frac{1}{\sqrt{1-\beta^2}}. \tag{49}$$

When introduced in Eq. (48), one gets

$$\frac{e}{m} = \frac{e}{m_0}\frac{4}{3}\frac{1}{\frac{4}{3}\sqrt{1-\beta^2}} = \frac{e}{m_0}\sqrt{1-\beta^2}, \tag{50}$$

that is,

$$m = \frac{m_0}{\sqrt{1-\beta^2}}, \tag{51}$$

when the electric charge e is kept constant. Although he stated that the non-electromagnetic forces should be affected in the same manner as the electromagnetic forces, it was not clear for Lorentz whether the mass, in the "*classical*" sense, would vary as observed in Kaufmann's experiments. For this reason, Lorentz introduced the concept of "*electromagnetic mass*", with a "*longitudinal*" and a "*transverse*" components. He thus supposed that "*the masses of all particles are influenced by a translation to the same degree as the electromagnetic masses of the electrons*". Lorentz was not so conclusive because[64]:

> What we know about the nature of electrons is very little and the only means of pushing our way farther will be to test such hypotheses as I have here made.

6.4. *Larmor's theory*

Another theory was proposed by Joseph Larmor (1857–1942). It was based on Neumann's theory according to which the velocity of the ether is provided in amplitude and direction by the magnetic force rather than the electric force as in Fresnel's theory.[69] Investigating the propagation of a radiation in a material medium — the free ether — moving by a velocity V along the x-axis, "*whose dynamical equations have been definitely ascertained in quite independent ways from consideration of both the optical side and the electrodynamic side of its activity*" (Ref. 70, p. 161), Larmor introduced in 1897 a change of coordinates (Ref. 69, p. 299) that he modified three years later in (Ref. 70, p. 174)

$$\left| \begin{array}{l} t' = \dfrac{t}{\sqrt{1-\dfrac{V^2}{c^2}}} + \dfrac{1}{c^2}\dfrac{Vx}{\sqrt{1-\dfrac{V^2}{c^2}}} \\ x' = \dfrac{x}{\sqrt{1-\dfrac{V^2}{c^2}}} \\ y' = y \\ z' = z \end{array} \right. , \tag{52}$$

where t' is a local time. He used an assumption for length contraction as FitzGerald and Lorentz did. His change of coordinates was close to what Lorentz's introduced in 1899 (therefore two years after Larmor).

Larmor thus obtained (Ref. 70, p. 176)

> the result, correct to the second order, that if the internal forces of a material system arise wholly from electrodynamic actions between the systems of electrons which constitute the atoms, then an effect of imparting to a steady material system a uniform velocity of translation is to produce a uniform contraction of the system in the direction of the motion, of amount
>
> $$\frac{1}{\sqrt{1-\dfrac{V^2}{c^2}}}. \tag{53}$$

We saw few variants of the electrodynamical theory. None of them had clearly reached a preferential status, each of them having its own particularities leading them to be as probable as the others. Only a rigorous mathematical and physical analysis, using a uniform notation (as we did in this paper) for making possible comparisons between them could allow to discriminate them. Such analysis was performed by Poincaré who wrote (Ref. 71, Introduction)

> Although none of these theories seems to me fully satisfactory, each one contains without any doubt a part of the truth and comparing them may

be instructive. From all of them, Lorentz theory seems to me the one which describes in the better way the facts.

7. Poincaré's Contribution

Henri Poincaré started to work on light and optical phenomena when he got the chair previously occupied by Gabriel Lippman at the Faculté des Sciences de Paris in 1886. In his lectures, taught at La Sorbonne (Paris) in 1887–1888 (see Ref. 72) and published three years later, Poincaré made an overview of all the existing optical theories with a rigorous and quite extensive mathematical analysis. His main task was to determine which theory was explaining the largest number of phenomena. For Poincaré, scientific law must be general to be of interest and for one to be confident enough in it. This was justified as follows (Ref. 73, p. 306):

> These principles result from highly generalized experiments; but they seem to borrow from this generality a huge degree of certainty. The more general they are, the more often they can be checked, and these verifications, in becoming more numerous, in taking the most various and unexpected forms, lead by not leaving any doubt.

But Poincaré did not forget his philosophy and thus added (Ref. 72, p. I)

> Mathematical theories do not aim to reveal the true nature of things; this could be an unreasonable pretention. Their unique aim is to coordinate physical laws that experience taught to us, but that, without the help of mathematics, we could not even state.

To complete Poincaré's views on the status of theories, it should be stated that for him, "*any generalization is an hypothesis*": From that point of view, a theory is nothing else than a possible explanation. It cannot be the truth, not even a partial truth. Such an approach opened the way followed by Karl Popper (1902–1994) who pushed the idea that a theory is the expression of the most advanced knowledge at a given time.[74] Poincaré also wrote that[75]

> experiment [...] alone can teach us something new; it alone can give us certainty. It is not sufficient merely to observe: We must use our observations, and for that purpose we must generalize. Mathematical physics [...] must direct the generalization, so as to increase [...] the output of Science.

It is therefore understood why Poincaré always used the conditional form when he presented his results; he only considered them as a possible representation of the phenomena under study.

In all pre-existing theories, an ether was required from the undulations of which the light and/or electromagnetic field can be propagated. However Poincaré, according to his philosophical approach to theories, thus stated that (Ref. 72, p. I)

> it does not matter that the ether actually exists; this is metaphysicians' business; what is relevant for us is that it is as if it would exist and that this assumption is convenient to explain phenomena. After all, do we have another reason to believe in the existence of material objects? This is only a convenient assumption; and it will be always like that, while someday will come for sure where the ether will be rejected, being useless. But this day, laws for optics and equations which analytically described them will remain true, at least as a first approximation. It will be always useful to investigate a doctrine linking all together these equations.

From his mathematical point of view, introducing an ether is reduced to a simplification of the mathematical description of the electromagnetics phenomena (Ref. 75, p. 1171)

> Does our ether actually exist? One knows from where is coming the belief in the ether? If the light arises from a distant star, during many years, it is no longer on the star and it is not yet on the Earth; it is necessary that it is somewhere and, supported, so to speak, by some material support. One may express the same idea under a more mathematical and a more abstract form. What we observe, there are the changes affecting material molecules; [...]. In the classical mechanics, the state of the system under study only depends on its state at an immediately preceding time; the system thus obey to some differential equations. Contrary to this, if we do not believe in the ether, the state of the material universe would depend not only on the immediately preceding state, but also on much older states; the system would obey to some finite-difference equations. This is for escaping to this exemption to the general laws of mechanics that we invented the ether.

Poincaré reminded us that the ether was supposed to fill transparent material medium as well as the interplanetary space, just because (Ref. 72, p. 379)

> it is not possible to conceive light propagation from the Sun to the Earth without the existence of an elastic medium. Contrary to this, it can be unnecessary and perhaps philosophical to assume the existence of an ether in material media. Nevertheless, the phenomenon of astronomical aberration that evidences the relative motion of the ether and the ponderable medium that it penetrates seems to be absolutely opposed to the removal of this hypothesis; or at least, if this hypothesis is rejected, the explanation of the astronomical aberration would present so many difficulties that its maintenance is desirable.

This comment mainly results from some experiments which were performed with an astronomical telescope filled with air or with water: No difference was found.

This is therefore an experiment more or less similar to that of Fizeau which is here discussed by Poincaré under the name *aberration.*

Thus, in order to compare the main optical theories, Poincaré had on the one hand to propose a rigorous theory of the ether which was part of all existing theories and, on the other hand, to unify their mathematical expressions. For investigating the light propagation according to the wave theory which was then the most complete, Poincaré assumed that light waves are based on a molecular hypothesis. He thus devoted a complete study to the small movements of the distinct molecules constituting the elastic medium; he thus considered a discontinuous matter. He reproduced the most widely accepted description of what should be the ether, that is, "*formed of molecules distant one from the other*" (Ref. 72, p. 1) which are governed by some forces pushing them out of their equilibrium states and, left to themselves, "*oscillate by very small motions around their equilibrium state*" (Ref. 72, p. 1). The medium is considered as isotropic, meaning that any plane is a plane of symmetry and that the equations are left invariant when two or three axes are permuted or when x is replaced with $-x$. In other words, the equations do not depend on a particular choice for axes (Ref. 72, p. 23):

> There exists certain functions of the coefficients of these equations known under the name of invariants, which are not dependent on the choice for the axes; they must be isotropic functions. One of the invariants is the sum of the squared coefficients of the squared terms.

Poincaré retained Fresnel's results for wave propagation according to which "*experiments show that vibrations of ether are always transverse*" (Ref. 72, p. 53). He specified that (Ref. 72, p. 65)

> in experimental studies in optics, it is not possible to directly determine the direction for the vibrations of the ether propagating rectilinearly polarized light; what one might observe is that phenomena depend on the position of a certain plane, the so-called plane of polarization. By symmetry properties, the direction of vibrations must be either in the polarization plane or perpendicular to this plane. Fresnel admits that it is perpendicular to it, other scientists preferred the opposite hypothesis.

He explained that the quantities occurring in the equations describing the transverse movements "*can be considered as constant with respect to the duration of a certain number of vibrations*" (Ref. 72, p. 66), the velocity c of light propagation being considered as an absolute value in a homogeneous medium.

Poincaré used these mathematical elements characterizing the ether to establish a detailed mathematical analysis of various optical phenomena commonly considered by existing optical theories: Reflection, refraction, diffraction, dispersion, double refraction, aberration.... Poincaré choose to investigate aberration by using a wave theory, the ether being considered as the support to light propagation

(Ref. 72, p. 379):

> The phenomenon of astronomical aberration, which evidences the relative motion between the ether and the ponderable medium that it penetrates, seems to argue against the removal of this assumption; or at least, if this hypothesis was rejected, the explanation of the astronomical aberration would encounter such difficulties that its upholding is preferable.

Taking into account Fresnel's results as well as Michelson and Morley's experiment, both related to an ether partly driven, Poincaré applied the composition law for velocities for evaluating the light velocity c with respect to motionless axes in space, that is (Ref. 72, p. 387),

$$c = c' + V\left(1 - \frac{1}{n^2}\right)\cos\phi, \tag{54}$$

where c' would be the absolute value of light velocity, $V\left(1-\frac{1}{n^2}\right)$ the absolute driving velocity of the ether and ϕ the angle between these two velocity vectors.

When the moving axes are permanently linked to the medium in motion, the velocity c of the light becomes (Ref. 72, p. 388)

$$c = c' - V'\cos\phi, \tag{55}$$

where V' is the relative driving velocity of the ether. It is this last expression that Poincaré used to express the duration spent by light to travel the distance between a point A_0 and a point A_n in a moving medium, that is (Ref. 72, p. 389),

$$\sum \frac{A_1A_2}{c} + \frac{V}{c'^2}\mathrm{A}'_0\mathrm{A}'_n, \tag{56}$$

where c' is the absolute light velocity, V is the displacement velocity of the medium, $c = \frac{c'}{n}$ and $\mathrm{A}'_0\mathrm{A}'_n$ is the projection of $\mathrm{A}_0\mathrm{A}_n$ on the x-axis.

> The first term $\sum \frac{\mathrm{A}_1\mathrm{A}_2}{c}$ represents the duration that light would spend when the medium is at rest, the second only depends on the location of the extreme points and by no means of the path traveled by light for going from one point to the other (Ref. 72, p. 389).
>
> An important consequence of the preceding formula is that reflection and refraction laws, interference phenomena are not affected by the Earth motion. (Ref. 72, p. 389)

If one forgets to look for a mechanical explanation of light propagation and if one does not consider valid the assumption that light emitted by stars depends on their size as supported by Michell and Arago, one would have no difficulty to describe the astronomical aberration as Bradley did. All the problem arises when one wants to provide an explanation of light propagation (we clearly distinguish "describing" from "explaining"). They were considering (Poincaré included) the relative motion with the implicit aspect of the composition law for velocities, but never clearly addressed the problem in terms of the distance traveled by light. This is well evident from the statement of Poincaré that (Ref. 72, p. 391)

> optical phenomena only evidence relative motions with respect to the observations of the light source and of ponderable matter. This is what happens in aberration where the observer and the observed star are not animated with the same motion; this is what happens in Fizeau's experiments where the water contained in the tubes has a relative motion with respect to the observer. A single fact would not correspond to these conclusions, this is the variation of the polarization plane of reflected light.

But Poincaré, as his predecessors, was deeply concerned by "explanation", and was thus forced to remind us of the weak status of any theory (Ref. 72, p. 398):

> It is quite difficult to explain these aberration phenomena, and there is no satisfactory theory. There is no sufficient reason to choose a theory [...]. Indeed, we cannot complain to be in the impossibility to make a choice. This impossibility shows us that mathematical theories for physical phenomena must be only considered as research tools; very precious tools, this is true, but from which we must not remain as slaves and that we must reject as soon as they are in an actual contradiction with experiments.

The second monograph resulting from Poincaré's lectures is mostly devoted to Maxwell's electromagnetic theory. For Poincaré, Maxwell's theory is more a theory for electromagnetic phenomena than for light propagation and therefore deserved a specific monograph. According to Poincaré, Maxwell's aim was (Ref. 76, p. 192)

> to find an explanation for electric and electromagnetic phenomena, commonly attributed to a force acting at a distance, by means of an hypothetic fluid filling the space.

Poincaré summed up Maxwell's approach as follows (Ref. 76, pp. XIV–XV):

> In order to demonstrate the possibility for a mechanical explanation of electricity, we do not have to worry for finding an explanation in itself, it is sufficient for us to know the expressions of the two functions T and U [describing the kinetic energy and the potential energy, respectively] which are the two parts of the energy, to construct with these two functions Lagrange's equations, and to compare these equations with experimental laws.

He was clearly motivated for investigating Maxwell's theory by its unifying character between, on the one hand, the ether which propagates light according to Fresnel and, on the other hand, Maxwell's fluid used for explaining electromagnetic phenomena: Both fluids have the same properties in Maxwell's works. Such a correspondence was already pointed out by Maxwell (Ref. 43, Vol. II, p. 431) as quoted by Poincaré himself:

> To fill all the space with a new medium whenever any new phenomenon is to be explained is by no means philosophical, but if the study of two different branches of science has independently suggested the idea of a medium, and if the properties which must be attributed to the medium in order to account for electromagnetic phenomena are of the same kind as those that we attribute to the luminiferous medium in order to account for the phenomena of light, the evidence for the physical existence of the medium will be considerably strengthened.

Poincaré then wrote (Ref. 76, pp. 193–194):

> The ether and Maxwell's fluid having the same properties, light must be considered as an electromagnetic phenomenon and the vibratory movement which produces on our retina the impression of a light intensity must result from periodic perturbations of a magnetic field. If this is so, from general equations for this field must be deduced the explanation of light phenomena.

Poincaré saw in each case of experimental evidence where the optical and electrical constants of a given body are found with nearly (if not) equal values as new "*indirect but convincing*" (Ref. 76, p. 194) validations of the electromagnetic theory of light. He considered as one of the best validations the values for the velocity of light propagation found by Foucault, Fizeau and Cornu (for instance, Alfred Cornu (1842–1902) found $3.0004 \cdot 10^8\,\mathrm{m} \cdot \mathrm{s}^{-1}$ in 1876 (see Ref. 77)) and the value deduced from the electromagnetic theory. In fact, Poincaré remarked that the equations governing the propagation of an electromagnetic perturbation were similar to those governing the movements of an ether molecule, a similarity that he considered as a "*confirmation of the assumption concerning the electromagnetic nature of luminous vibrations*" (Ref. 76, p. 197). Moreover, these former equations lead to transverse periodic electromagnetic perturbations (as Maxwell showed) propagating with the velocity

$$c = \frac{1}{\sqrt{K\mu}}, \tag{57}$$

which reduces to $c = \frac{1}{\sqrt{K}}$ in the vacuum since μ is the permeability coefficient equal to 1 in the electromagnetic system of units and K is the permittivity. He found, as Maxwell, that c is also the quantity of electricity (in the electromagnetic system of units) in one unit of electromagnetism.

His study of Hertz's experiments where the velocity of electromagnetic waves was found with the same order of magnitude than light velocity led him to conclude that "*this is again a very satisfying validation of the electromagnetic theory of light, if one takes into account the difficulties in measuring the quantities involved in Hertz's computations*" (Ref. 76, p. 203). The double refraction, which is one of the most complex optical phenomena to explain, was fully explained using the electromagnetic theory of light as shown by Poincaré, proving the validity of this theory.

After the optical theories, Poincaré paid his attention to the electrodynamic theories leading to a new monograph published in 1891.[29] In his lectures, he reviewed the main electrodynamic theories of those times which were those of Ampère, Weber, Helmholtz and Maxwell. He paid particular attention to Helmholtz's aims to have a more general theory than the others and comparable to Maxwell's. He also investigated how Hertz's experiments could allow one to determine which theory could be the best one. In doing this, Poincaré thus showed that Ampère's theory (based on a force acting at a distance) is not a particular case of Helmholtz's (based on the action of a potential) and is, in fact, the single one which is able to explain the facts by action between two elements reduced to one force along the straight line joining them. As soon as one admits that Ampère's law results from a potential as assumed by Helmholtz's theory, and since the potential depends on the orientation of the current elements, "*derivatives [of the potential] with respect to the angles defining its orientation are not identically null*" (Ref. 29, p. 51). Applying the principles of mechanics to Helmholtz's theory, Poincaré showed that it provides an unstable equilibrium propagation when the constant k is negative (corresponding to Weber's theory), and that it must be rejected in this case. For $k = 0$ (Maxwell's theory) and $k = +1$ (Neumann's theory), Poincaré showed that it is possible to switch from Helmholtz's theory to Maxwell's by reducing the parameter λ — being equal to "1 *in the system of electrostatic units and is the square of the light velocity in the electromagnetic system*" (Ref. 29, p. 56) — to an infinitely small value: Maxwell's theory is thus a limiting case of Helmholtz's theory. He thus concluded that (Ref. 29, p. 110)

> in Maxwell's theory, there are only transverse vibrations and their propagation velocity V_2 is equal to the velocity c of light [...]. If one sets λ to a positive value different from 0, one has for V_2 a velocity greater than the velocity of light [which is not possible. Consequently,] Maxwell's theory can thus [only] be deduced from Helmholtz's by setting $\lambda = 0$.

Poincaré also preferred Maxwell's theory because, on the one hand, "*the ratio c among units is equal to the light velocity and is very well explained in it*" (Ref. 29, p. 114) and, on the other hand, the governing equations were written in a "*very elegant form*" by Hertz. In particular, Poincaré showed that the equations governing the electric displacement and those governing the current can be written as a function of the magnetic induction, thus exhibiting the correspondence between the electric displacement and the magnetic force (Ref. 29, pp. 115–116).

For choosing the best electromagnetic theory, Poincaré investigated whether the so-called "*principle of the unity of the electric force*" (Ref. 29, p. 122) is in agreement with all possible theories or not. This principle, introduced by Hertz, is described by Poincaré as follows (Ref. 29, p. 123):

> We will admit for electricity a principle analogous to the principle that everybody admits for magnetism. A magnet in a ring shape and whose magnetism varies or, what is equivalent, a closed solenoid traversed by a variable current, is equivalent to an electric layer of a convenient power, from the electric field point of view that it produces. It will act as this layer onto another electric layer; and, according to the principle of action and reaction, will receive from this second layer a reaction equal and opposite to the applied action. Thus, a variable closed solenoid in an electric field receives a mechanical action; and, as such a solenoid produces an electric field, two variable closed solenoids apply one onto the other a mechanical action identical to the action produced by two equivalent electric layers. This is the principle of the unity of the electric force.

Maxwell's theory was the single one found by Poincaré to obey to this principle: This was yet another argument in favor of this theory.

The last of Poincaré's monographs on electrodynamic theories was published in 1901 and was devoted to the electrodynamic theories developed by Hertz, Lorentz and Larmor, respectively. As done in his previous studies, he mathematically investigated these theories and confronted them with the principles of mechanics. The detailed analysis of Hertz's theory for bodies at rest and his comparison to Maxwell's allowed Poincaré to show that the two theories are in agreement in every point with the exception of "*the expression of the magnetic energy of Hertz [which] is thus the single acceptable one*" (Ref. 71, p. 362). A similar evaluation of Hertz's and Maxwell's theories for bodies in motion allowed Poincaré to determine the equivalence between these two theories and to state about their conformity with the principles of mechanics. He also demonstrated that (Ref. 71, p. 388)

> Hertz's equations keep the same form when we adopt motionless axes as well as when we adopt moving axes; in other terms, Hertz's equations keep the same form in relative motion as well as in absolute motion.

Such a demonstration led him to the following observation (Ref. 71, p. 389):

> It thus results from the theory that the derivative $\frac{\partial}{\partial t}$ plays with respect to relative motion, the same role as played by the derivative $\frac{\partial}{\partial t}$ with respect to absolute motion.
>
> This last remark induces two consequences: One is fortunate, the other is annoying. The fortunate consequence is that Hertz's equations are consistent with the principle of action and reaction; the annoying consequence is that these equations cannot describe certain optical phenomena.

Hertz was already aware of such a weakness in his theory and Lorentz tried to overcome it. Consequently, Hertz's theory was valid for describing electrical phenomena and consistent with the principle of mechanics but was unable to describe optical phenomena in motion (Ref. 71, p. 422).

In contrast to this, as Poincaré also showed that, Lorentz's theory "*quite well explains optical phenomena which were not explained by Hertz's theory but, unfortunately, it is not consistent with the principle of action and reaction*" (Ref. 71, p. 422). The main difference between these two theories, as evidenced by Poincaré, is related to the assumptions added by Lorentz, that is, there is "*neither magnetism, nor dielectric other than the vacuum*" (Ref. 71, p. 435). Moreover, in order to explain optical phenomena by Lorentz's theory, it must be admitted that (Ref. 71, p. 518):

> If one wants that optical phenomena are not affected by the Earth motion, one must neglect in the formulas terms of the order of the squared aberration [...]. In almost all experiments, these terms are indeed negligible; there is however one exception for Michelson's experiment which shows that the Earth motion has no influence on optical phenomena observed at the terrestrial surface and where it is found that terms of the order of aberration are no longer negligible.

Then description and comparison of the main existing theories allowed Poincaré to evidence what they have in common. He thus listed the conditions that any electrodynamical theory for a moving body should obey (see Ref. 78 and p. 602 of Ref. 71):

> (i) It should explain Fizeau's experiments, that is, the partial driving of light waves or, which is equivalent, of transverse electromagnetic waves.
> (ii) It must verify the principle for the conservation of electricity and magnetism.
> (iii) It should be compatible with the principle of the equality between action and reaction.

These conditions are imposed by known experiments. They are necessary conditions, matching with Poincaré's metaphysics according to which "*experience is the sole source of certainty.*" Concerning the theories successively developed by Hertz, Helmholtz and Lorentz, Poincaré established that none of them simultaneously fulfilled to these three conditions:

- The theory of Hertz[54] satisfies the conditions (ii) and (iii);
- the theory of Helmholtz[79] satisfies the conditions (i) and (iii);
- the theory of Lorentz[52,61] satisfies the conditions (i) and (ii).

He thus concluded that (Ref. 71, p. 611):

> One may ask whether this is due to the fact that these theories are not completed or whether these three conditions are actually compatible or would become so only by a deep modification of the admitted assumptions

> [...]. It is thus needed to give up on developing a perfectly satisfying theory and to temporarily retain the less flawed one which seems to be Lorentz's theory.

To overcome these problems, the next important step was related to the measure of time for describing physical phenomena which is a key point for writing the coordinate transformation for switching from a frame at rest to a moving frame.[80] The relevant question asked by Poincaré and related to these problems was: "*Can we reduce to a single measure facts which occur in different worlds*?" (Ref. 80, p. 2). A related question was "*can we transform a psychological time, which is qualitative, into a quantitative time*?" (Ref. 80, p. 2). Poincaré already answered this last question, asserting that "*we have no direct intuition of the equality of two intervals of time. The persons who believe they possess this intuition are deceived by an illusion*" (Ref. 80, p. 2). He thus explained (Ref. 80, p. 3):

> When I say, from noon to one the same time passes as from two to three, what meaning has this affirmation? The least reflection shows that by itself it has none at all. It will only have the one which I choose to give it, by a definition which will certainly possess a certain degree of arbitrariness. To measure time [physicists and astronomers] use the pendulum and they suppose by definition that all the beats of this pendulum are of equal duration. But this is only a first approximation; temperature, resistance of the air, barometric pressure, make the rate of the pendulum vary.

This is nothing else than a physical justification for the local character of the time since these physical conditions (temperature, pressure, etc.) depend not only on time but also on location. This is here a first, perhaps a little bit "naive", consideration of what will be became the concept of a local time.

Poincaré then considered how the unit of time is defined, leading to the conclusion that there is no rigor in its definition: "*When we use the pendulum to measure the time [we implicitly admit that] the duration of two identical phenomena is the same or, if we prefer, that the same causes take the same time to produce the same effects*" (Ref. 80, pp. 3–4). In order to do that, physicists and astronomers use the conservation of energy and Newton's laws which are only approximations since they are deduced from experiment. Consequently, Poincaré cannot avoid to conclude that "*there is no manner to measure time which is more 'true' than another; the one which is commonly adopted is only the most convenient.*" (Ref. 80, p. 6).

Considering the problem of how to determine the simultaneity of two phenomena, he pointed out the singular role played by light in this question (Ref. 80, p. 11):

> When an astronomer tells me that some stellar phenomenon, that his telescope reveals to him at this moment, happened, nevertheless, fifty years

> ago, I seek his meaning, and to that end I shall ask him first how he knows it, that is, how he has measured the velocity of light.
>
> He has begun by "supposing" that light has a constant velocity, and in particular that its velocity is the same in all directions. This is a postulate without which no measurement of this velocity could be attempted.

Poincaré thus insisted on the fact that when optical measurements are performed, it is always implicitly assumed that the light velocity is constant in a space which was considered as isotropic and homogeneous. More explicitly, a theory is required to justify the choice of the quantity measured and how it is related to the assumption tested. In the case considered, Poincaré insisted on the fact that the assumptions (the light velocity) and the theory are not independent: There is an inner consistency but the explanation provided is not unique. Once again, the retained solution was selected according to its simplicity. More importantly, Poincaré pointed out that it is impossible to dissociate the concept of simultaneity from the measure of time. This problem is necessarily associated with any change of reference frame.

Poincaré then revisited the principles of mechanics,[81] distinguishing what is learnt from experiment from what is obtained by mathematical reasoning, the latter only being a convention or an assumption. He started with a severe criticism of the foundations on which Newton's *Principia Mathematica* were constructed (Ref. 81, p. 458):

> (i) There is no absolute space, and we only conceive of relative motion [...].
> (ii) There is no absolute time [...].
> (iii) Not only we have no direct intuition of the equality of two periods, but we have not even direct intuition of the simultaneity of two events occurring in two different places.
> (iv) Finally, is not our Euclidean geometry in itself only a kind of convention of language? Mechanical facts might be enunciated with reference to a non-Euclidean space.
>
> [Consequently], we might endeavour to enunciate the fundamental laws of mechanics in a language independent of all these conventions [...]. No doubt that the enunciation of these laws would become much more complicated, because all these conventions have been adopted for the very purpose of abbreviating and simplifying the enunciation. (Ref. 81, p. 459)

All these considerations led Poincaré to introduce a local time and a non-Euclidean space.

In the same year (1900), Poincaré wrote some comments on the theory of Lorentz[75] in a compendium to celebrate the 25th anniversary of Lorentz's Ph.D. thesis. This theory was considered by Poincaré as the least bad theory, mainly because "*the principle of relativity of motion has been verified only imperfectly.*" In fact, this principle is only verified when $\frac{V^2}{c^2}$ is neglected. The main criticism provided by Poincaré concerned the principle of action and reaction which was no longer satisfied when it was applied to the matter alone, that is, when no momentum was attributed to light (Ref. 82, p. 269):

> In order to show experimentally that the principle of reaction is broken in reality as it is in Lorentz's theory, it was not sufficient to show that the devices producing energy undergo a recoil, which would be already quite difficult to do, it should be also shown that this recoil is not balanced by the movement of dielectrics and, in particular, by the movement of air crossed by electromagnetic waves, which would be obviously much more difficult to prove.

To argue that this objection is related to relative motion and not to absolute motion, Poincaré developed the example as follows (Ref. 82, pp. 270–271):

> Let A and B be two bodies, acting on each other, but not under any external action; if the action of one of each was not equal to the reaction of the other, one could attach one to the other with a rod of constant length in such a manner that they behave as a *single* solid body. The forces applied to this body being not at the equilibrium, the system would be in movement and this motion would go with a constant acceleration but at one condition, that is, the mutual action of the two bodies only depends on their relative positions and their relative velocities, but is independent on their absolute positions and their absolute velocities.
>
> The principle of reaction thus appears as a consequence of the principles of the [conservation of] energy and of the relative motion.

For Poincaré, Lorentz's theory can only be in agreement with experimental facts if "*phenomena are related, not to the true time t but to a certain local time t'*" (Ref. 82, p. 272), that he defined as follows (Ref. 82, pp. 272–273):

> I suppose that some observers placed at various points, synchronize their clocks using light signals. They attempt to correct these signals from the transmission duration, but they are not aware about the motion of translation with which they are animated and thus, believing that these signals travel equally fast in both directions, they limit themselves to cross the observations, sending one signal from A to B, followed by another one from B to A.
>
> The local time t is the time indicated by the clocks which are so adjusted.
>
> If c is the speed of light, and V is the speed of the Earth that we suppose to be parallel to the x-axis in the positive direction, one will have
>
> $$t' = t - \frac{Vx}{c^2}. \tag{58}$$

> The apparent energy propagates in a relative motion according to the same laws as the actual energy in absolute movement, but the apparent energy is not exactly equal to the actual corresponding energy.
>
> In relative movement, the bodies producing the electromagnetic energy are subject to an apparent complementary force which does not exist in absolute movement.

It is important to note here that due to the existence of a local time, some effects, as those measured in a "*resting frame*" are only apparent, due to a change of reference coordinates. This is a very important aspect which will be rediscussed a little further on.

From this argument, Poincaré concluded that, according to Lorentz's theory, the principle of reaction cannot be applied only to ponderable matter and should be applied to light. This is what should be understood when Poincaré endowed a momentum to the matter and to a "*fictional fluid*".[82] He made this quite explicit when he wrote that "*the electromagnetic energy behaves as a fluid which has inertia, we must conclude that, if any sort of device produces electromagnetic energy and radiates it in a particular direction, that device must recoil just as a cannon does when it fixes a projectile*" (Ref. 82, p. 260). Poincaré then provided an explicit example with an "*Hertzian exciter placed at the focus of a parabolic mirror*" (Ref. 82, p. 260):

> It is easy to evaluate the recoil quantitatively. If the device has a mass of 1 kg and if it emits three million joules in one direction with the velocity of light, the speed of recoil is $1\,\mathrm{cm}\cdot\mathrm{s}^{-1}$.

Due to the context in which this example occurred, it is very likely that Poincaré used a formula for the momentum like $p = \frac{E}{c}$, where p is the momentum and E the energy.[d] Consequently, the principle of relative motion should be applied to ponderable matter and to electromagnetic fields. Nevertheless, Fizeau's experiment was still in contradiction with the principle of reaction. Such a feature led Poincaré to conclude that, if (Ref. 82, p. 278)

> the driving of waves is only partial, this is due to the relative propagation of waves in a moving medium which does not obey the same law as the propagation in a resting medium, that is, the principle of a relative movement does not only apply to matter and it is required to apply a correction [...] which consists in the introduction of a local time. If this correction is not

[d]Such a relation has to be used for getting the numerical value for the speed of recoil. It was correctly introduced by Einstein in 1917.[83] There is another possibility to obtain this result by stating that the variation of the mass is related to the variation of the energy according to $dm = \frac{dE}{c^2}$ which was the form in which Einstein published the equivalence mass–energy in 1905[84]: "*The mass of a body is a measure of its energy content; if the energy changes by* dE*, the mass changes in the same sense by* $\frac{dE}{9\cdot 10^{20}}$*, if the energy is measured in ergs and the mass in grams.*"

> balanced by some others, we shall conclude that the principle of reaction is not true for the matter.
>
> Thus all theories obeying this principle would be condemned, unless we would consent to change all our ideas on electromagnetism.

In 1902, Poincaré published a book entitled *Science and Hypothesis.*[85] Often considered as written for a broad audience, this book is in fact just a concatenation of already published papers from which mathematical equations were removed. Two years later, after the success of the first book, Poincaré published in 1904 a second book entitled *The Value of Science.*[86] The construction of this second book is similar to the first one. In it, Poincaré improved his idea that there is a need for a new mechanics (Ref. 86, Par. 197):

> From all these results, if they were confirmed, would arise an entirely new mechanics, which would be, above all, characterized by this fact, that no velocity could surpass that of light, because bodies would oppose an increasing inertia to the causes which would tend to accelerate their motion; and this inertia would become infinite when one approached the velocity of light.

This assertion results from crucial experiments conducted by Walter Kaufmann (1871–1947)[68]:

> I was able to report about an experiment with the result that the ratio $\frac{e}{m}$ of Becquerel rays would decrease with increased velocity, and m would increase if one assumes e as constant, namely it increases the quicker, the more the velocity V would approach the speed of light c. Such a behavior is theoretically given from the equation of energy of a quickly moving electric charge.

Such an experiment was motivated by the theoretical study developed by Abraham[67] and Kaufman.[68]

The year of 1904 was also the year of an important conference in Saint Louis (Missouri) where Poincaré gave a talk entitled *L'état actuel et l'avenir de la physique mathématique*,[73] which was included in his second book *The Value of Science.* As we already mentioned, the concept of "principle" was very important for Poincaré (Ref. 81, p. 491):

> The principles of mechanics are presented to us under two different aspects. On the one hand, there are truths founded on experiment, and verified approximately as far as almost isolated systems are concerned; on the other hand, they are postulated applicable to the whole of the universe and regarded as rigorously true. If these postulates possess a generality and a certainty which are lacking in experimental truths from which they were deduced, it is because they reduce in final analysis to a simple convention that we have a right to make, because we are certain beforehand that

> no experiment can contradict it. [...] We admit it because certain experiments have shown us that it will be convenient, and thus is explained how experiments have built up the principle of mechanics, and why, moreover, it cannot reverse them.

Retrospectively, in 1895, Poincaré only evoked "*the principle of* [*conservation of*] *energy*" in his lectures at La Sorbonne on the mathematical theory of light.[72] In his text on the principles of mechanics,[81] Poincaré discussed five principles:

- The principle of inertia;
- The law of acceleration;
- The principle of reaction;
- The principle of relative motion;
- The principle of energy.

These principles summarized all of Poincaré's questioning about light propagation arising from Fizeau's and Michelson and Morley's experiments. The principle of relative motion was enunciated as follows (Ref. 81, p. 477):

> The movement of any system whatever ought to obey the same laws, whether it is referred to fixed axes or to the movable axes which are implied in uniform motion in a straight line.

This is clearly the invariance of the laws under a change of reference frame, when one is related to the other by a constant velocity. Nevertheless, in 1900, the transformation from one frame to the other was not known. The "classical" composition law for velocities was clearly not working for explaining optical experiments such as Michelson and Morley's. In his 1904 talk at Saint Louis, Poincaré discussed six principles ranked as follows (Ref. 73, p. 306):

(i) The conservation of energy;
(ii) the degradation of energy;
(iii) the equality of action and reaction;
(iv) the principle of relativity;
(v) the conservation of mass;
(vi) the principle of least action.

Poincaré's discussion of these six principles expressed all his doubts. We could say that this was the year of questionings induced by a crisis. The law of acceleration, in fact the fundamental principle of dynamics, was no longer discussed but the principle of least action was added. The degradation of energy was introduced by Sadi Carnot (1796–1832).[87] Energy seemed to be conserved but it is degraded when its form (mechanical, electrical, thermal, etc.) was changed. The principle of action and reaction was not verified by Lorentz's theory. The principle of relativity was enunciated in a slightly different way compared to 1901

(Ref. 73, p. 306):

> The laws of physical phenomena must be the same for a stationary as well as for an observer carried along in a uniform motion of translation; so that we have not and cannot have any means of discerning whether or not we are carried along in such a motion.

This principle remains the key point to solve the problem posed by light propagation. In 1904, it was not yet possible to apply them to optical experiments. Kaufmann's experiments seemed to contradict the conservation of mass since the mass was varying with the velocity. Even the conservation of energy was subject to doubt, due to Pierre Curie and Albert Laborde's experiments on radioactive elements which showed that a significant amount of energy was emitted.[88] Poincaré then concluded his talk as follows (Ref. 73, p. 324):

> Perhaps, too, we shall have to construct an entirely new mechanics that we only succeed in catching a glimpse of, where, inertia increasing with velocity, the velocity of light would become an impassable limit. The ordinary mechanics, more simple, would remain a first approximation, since it would be true for velocities not too great, so that the old dynamics would still be found under the new.

Less than one year after his talk given at Saint-Louis, Poincaré proposed his own version of the dynamics for electrified particles and, more particularly, the dynamics of the electron resulting from his analysis of the theories published by his predecessors. He based his investigations on Lorentz's theory[64] and Langevin's theory.[89] Poincaré quickly stated that, although interesting because it was only based on electromagnetic forces and binding forces, it is not compatible with the principle of relativity, as initially shown by Lorentz and then by himself.[90]

Poincaré's theory was published in two papers, one short note read at the Academy of Sciences (Paris) on June 5, 1905 (see Ref. 91) which is only a summary of Poincaré's findings due to the length limitation to publish in Comptes-Rendus de l'Académie des Sciences, and one full paper, submitted on July 23rd, 1905 to the Rendiconti del Circolo Mathematico di Palermo and which was only published in 1906. No doubt that most of the second paper was already written when Poincaré read his contribution on June 5, since he wrote "*I show*" (but no proof was given in the short note, in contrast to what was included in the full paper), "*this is what I determined*". Poincaré was not used to announcing a result without at least a sketch of the proof. This is why we preferred to directly describe the full paper submitted on 23 July.

Poincaré started by recalling that this entire problem was coming from the stellar aberration and some optical experiments, as the one conducted by Michelson and Morley, related to the driving of the ether. He clearly mentioned the principle of relativity (Ref. 90, p. 129):

> It seems that this impossibility of demonstrating an experimental evidence for absolute motion of the Earth is a general law of nature; we are naturally led to admit this law, which we will call the Principle of Relativity and admit it without restriction.

He then stated that FitzGerald's and Lorentz's contraction was sufficient when the principle of relativity was taken in all its generality.[90]

It was thus of importance to find how the principle of relativity can be applied to electromagnetic phenomena, that is, to what is today called the Maxwell equations. This is presented in Poincaré's paper as follows (Ref. 90, p. 130):

> The idea of Lorentz can be summarized as follows: If we can bring the whole system to a common translation, without modification of any of the apparent phenomena, it is because the equations of the electromagnetic medium are not altered by certain transformations, which we will call Lorentz transformations; two systems, one motionless, the other in translation, thus become exact images of one another.

Indeed, one of the results provided in Lorentz's paper is that the Maxwell equations are left invariant under the "*Lorentz transformation*", which is the modern expression of what the principle of relativity tells us when applied to Maxwell equations. Poincaré already tried to show this in his lecture performed in 1899 and published in 1901,[71] but failed. His best results were obtained with the 1892 Lorentz's theory. Lorentz developed his second theory to get such an invariance (and failed too) but, for him, this was only an intermediary step to explain optical experiments: Consequently, he did not give any comment about this result in his paper.

In contrast to this, the invariance of Maxwell's equations in Poincaré's 1905 contribution is central. Moreover, he showed that Lorentz transformations — a name given by Poincaré in Refs. 90 and 91 — are shown to form a group, a mathematical concept usually used by mathematicians and on which Poincaré had largely worked and published in 1881 (see Refs. 92–95 and also see Gray for a review on Poincaré's contribution to group theory[96]). When properly applied, this transformation provides the right composition law for velocities, a point missed by Lorentz. From the mathematical point of view, it allows one to understand what Poincaré had in mind when he wrote (Ref. 91, p. 576)

> the results which I obtained are in agreement with those of Lorentz on all important points; I was only led to modify and supplement them in some points of detail; one will further see the differences which are of secondary importance.

He "*just*" corrected the composition law for velocities and its consequences in Lorentz's theory. The equations for electromagnetic media are nothing else than the Maxwell equations. Poincaré adopted Lorentz's specific system of units, in particular he used the electric displacement **D** and the magnetic force **H** in order to eliminate the factors 4π in the formulas. Poincaré also choose the units of length and time so that the speed of light was equal to 1 — as he mentioned in the letter sent to Lorentz in May 1905, in which he already gave the corrected transformation[65,66] — a convention very useful to emphasize the underlying symmetry between the electric and magnetic fields. The Maxwell equations used by Poincaré translated in the modern vectorial form are thus (Ref. 90, p. 132)

$$\begin{cases} \mathbf{I} = \dfrac{d\mathbf{D}}{dt} + \rho\mathbf{V} = \boldsymbol{\nabla}\wedge\mathbf{H}, & \dfrac{d\mathbf{H}}{dt} = -\boldsymbol{\nabla}\wedge\mathbf{D} \\ \dfrac{d\rho}{dt} + \boldsymbol{\nabla}\cdot\rho\mathbf{V} = 0, & \boldsymbol{\nabla}\cdot\mathbf{D} = \rho \end{cases}, \tag{59}$$

where **I** designates the current, **V** is the velocity of the electrons, **H** is the magnetic force, **D** is the electric displacement and ρ is the electric density. Poincaré did not add to this set of equations the common

$$\boldsymbol{\nabla}\cdot\mathbf{H} = 0, \tag{60}$$

perhaps because, when he expressed **D** and **H** in terms of the scalar potential Ψ and the vector potential **A**, he obtained (Ref. 29, p. 132)

$$\begin{cases} \mathbf{D} = -\dfrac{d\mathbf{A}}{dt} - \boldsymbol{\nabla}\Psi \\ \mathbf{H} = \boldsymbol{\nabla}\wedge\mathbf{A} \end{cases}. \tag{61}$$

Since the divergence of a rotation is necessarily null, by definition Eq. (60) holds. Nevertheless, Eq. (60) was explicitly justified by Poincaré when he investigated Lorentz's electromagnetic theory (Ref. 29, pp. 426–427). Poincaré checked all these fundamental equations before any attempt to demonstrate their invariance under the Lorentz transformation.

Poincaré also clarified his notations for partial derivatives:

> Our functions can be viewed (i) either being dependent on the five variables x, y, z, t, ϵ in such a way that one always remains at the same place when only t and ϵ are varied: We will thus designate their derivatives by ordinary "d"; (ii) or being dependent on the five variables x, y, z, t, ϵ in such a way that one always follows a single given electron when only t and ϵ are varied: We will designate their derivatives by "∂".

Thus, Poincaré expressed the magnetic and electric fields in terms of the scalar potential Ψ and the vector potential **A**. This means that the electromagnetic field results from a four-dimensional potential. From this, and using the gauge condition

$$\partial_t\Psi + \boldsymbol{\nabla}\cdot\mathbf{A} = 0, \tag{62}$$

the Maxwell equations (59) can be rewritten in the very compact form (Ref. 90, p. 132)

$$\begin{cases} \Box\,\Psi = -\rho \\ \Box\,\mathbf{A} = -\rho\mathbf{V}, \end{cases} \tag{63}$$

where

$$\Box = \nabla^2 - \frac{d^2}{dt^2}$$

is the d'Alembertian as named after Lorentz.[64] Poincaré thus added the Lorentz force (Ref. 90, p. 132)

$$\mathbf{F}_L = \rho\mathbf{D} + \rho(\mathbf{\nabla} \wedge \mathbf{H}). \tag{64}$$

All these equations were thus transformed by applying the "*remarkable transformation discovered by Lorentz*" (Ref. 90, p. 132)

$$T_L = \left| \begin{array}{l} t' = \gamma l(t + \beta x) \\ x' = \gamma l(x + \beta t) \\ y' = ly \\ z' = lz \end{array} \right., \tag{65}$$

where l and β are two arbitrary constants used by Lorentz, and where (Ref. 90, p. 132)

$$\gamma = \frac{1}{\sqrt{1 - \beta^2}} \tag{66}$$

is the contraction coefficient introduced by Lorentz. The Lorentz transformation was the missing key element Poincaré needed to show the invariance of the Maxwell equations. He was thus rewarded with this transformation for having pushed Lorentz to write a second theory.

Poincaré then introduced the inverse transformation (Ref. 90, p. 132)

$$T_L^{-1} = \left| \begin{array}{l} t = \dfrac{\gamma}{l}(t' - \beta x') \\ x = \dfrac{\gamma}{l}(x' - \beta t') \\ y = \dfrac{y'}{l} \\ z = \dfrac{z'}{l} \end{array} \right. \tag{67}$$

and also introduced the right composition law for velocities (Ref. 90, p. 133):

$$\left| \begin{aligned} V'_x &= \frac{dx'}{dt'} = \frac{d(x+\beta t)}{d(t+\beta x)} = \frac{V_x+\beta}{1+\beta V_x} \\ V'_y &= \frac{dy'}{dt'} = \frac{dy}{\gamma d(t+\beta x)} = \frac{V_y}{\gamma(1+\beta V_x)} \\ V'_z &= \frac{dz'}{dt'} = \frac{dz}{\gamma d(t+\beta x)} = \frac{V_z}{\gamma(1+\beta V_x)} \end{aligned} \right. . \tag{68}$$

This is one of the very relevant contributions by Poincaré to the principle of relativity: He was the first to understand that the "*classical composition law for velocities*" no longer applies when the Lorentz transformation is used. Poincaré was thus able to obtain the continuity on the new electric density ρ' when the electric charge of the electron is kept constant (Ref. 90, p. 134):

$$\frac{d\rho'}{dt'} + \rho' \boldsymbol{\nabla} \cdot \mathbf{V}' = 0. \tag{69}$$

A departure from the second theory of Lorentz was remarked by Poincaré: The value for this new density was found by Lorentz to be (Ref. 90, p. 133)

$$\rho' = \frac{1}{kl^3}\rho, \tag{70}$$

that Poincaré corrected into (Ref. 90, p. 133)

$$\rho' = \frac{k}{l^3}\rho(1+\beta V_x). \tag{71}$$

He was also able to express the new scalar and vector potentials (Ref. 90, p. 134)

$$\begin{cases} \Box' \, \Psi' = -\rho' \\ \Box' \, \mathbf{A}' = -\rho' \mathbf{V}', \end{cases} \tag{72}$$

whose components are (Ref. 90, p. 134)

$$\left| \begin{aligned} \Psi' &= \frac{k}{l}(\Psi + \beta A_x) \\ A'_x &= \frac{k}{l}(A_x + \beta \Psi) \\ A'_y &= \frac{1}{l} A_y \\ A'_z &= \frac{1}{l} A_z \end{aligned} \right. . \tag{73}$$

The new magnetic and electric fields *in the moving frame* thus satisfy the equations (Ref. 90, p. 134)

$$\begin{cases} \mathbf{D}' = -\dfrac{d\mathbf{A}'}{dt'} - \boldsymbol{\nabla}'\Psi' \\ \mathbf{H}' = \boldsymbol{\nabla}' \wedge \mathbf{A}' \end{cases}, \tag{74}$$

which led to the Maxwell equations rewritten in the form (Ref. 90, p. 135)

$$\begin{cases} \dfrac{d\mathbf{D}'}{dt'} + \rho'\mathbf{V}' = \boldsymbol{\nabla}' \wedge \mathbf{H}', & \dfrac{d\mathbf{H}'}{dt'} = -\boldsymbol{\nabla}' \wedge \mathbf{D}' \\ \dfrac{d\rho'}{dt'} + \boldsymbol{\nabla}' \cdot \rho'\mathbf{V}' = 0, & \boldsymbol{\nabla}' \cdot \mathbf{D}' = \rho' \end{cases}. \tag{75}$$

The Maxwell equations are thus left invariant under the Lorentz transformation. Poincaré ended his proof by checking that the Lorentzian force is also left invariant, that is (Ref. 90, p. 135),

$$\mathbf{F}'_L = \rho'\mathbf{D}' + \rho'(\boldsymbol{\nabla}' \wedge \mathbf{H}').$$

Poincaré thus obtained the invariance of Maxwell's equations under Lorentz transformation, a result he was looking for at least since 1899 (he already tried to obtain it in his 1899 lecture).

In order to complete his study, Poincaré wanted to show that the Lorentz transformation defined a group, an extremely important condition, according to his statements written in 1895.[97] He thus introduced four infinitesimal generators (Ref. 90, p. 145)

$$\left| \begin{aligned} T_0 &= \mathbf{x} \cdot \boldsymbol{\nabla} + t\frac{d}{dt} \\ T_x &= t \cdot \frac{d}{dx} + x \cdot \frac{d}{dt} \\ T_y &= t \cdot \frac{d}{dy} + y \cdot \frac{d}{dt} \\ T_z &= t \cdot \frac{d}{dz} + z \cdot \frac{d}{dt}, \end{aligned} \right.$$

corresponding to the temporal component and the three spatial coordinates, respectively. Poincaré then established that the form of the Lorentz equations are independent of the choice for axes. From this, he introduced a continuous group, that he named the "*Lorentz group*", and which admits four infinitesimal transformations (Ref. 90, p. 146):

(i) The transformation T_0 which is permutable with all others;
(ii) the three transformations T_x, T_y, T_z;
(iii) the three rotations $[T_x, T_y]$, $[T_y, T_z]$, $[T_z, T_x]$.

Any transformation of this group can always be decomposed into a scaling transformation of the form (Ref. 90, p. 146)

$$\left| \begin{array}{l} t' = lt \\ x' = lx \\ y' = ly \\ z' = lz \end{array} \right. \tag{76}$$

and a linear transformation that does not alter the quadratic form (Ref. 90, p. 146)

$$x^2 + y^2 + z^2 - t^2. \tag{77}$$

Poincaré also showed that this last quantity is invariant under the action of the group because it is annihilated by the infinitesimal generators T_0, T_x, T_y, T_z. He added that any transformation under the form (Ref. 90, p. 146)

$$T = \left| \begin{array}{l} t' = kl(t + \beta x) \\ x' = kl(x + \beta t) \\ y' = ly \\ z' = lz \end{array} \right. , \tag{78}$$

preceded and followed by a suitable rotation is a transformation of the Lorentz group. In his second theory, Lorentz proposed $l \approx 1$ modulo a second order quantity. Poincaré showed that $l = 1$ is required for having a transformation belonging to the Lorentz group after a rotation by π around the y-axis, (Ref. 90, p. 163).

8. Einstein's 1905 Contribution

The first contribution by Albert Einstein (1879–1955) to the theory of special relativity was published in 1905 in the *Annalen der Physik*. In the Introduction, the principle of relativity is stated as follows (Ref. 98, pp. 891–892):

> The unsuccessful attempts to discover any motion of the Earth relatively to the "light-medium" suggest that the phenomena of electrodynamics as well as of mechanics possess no properties corresponding to the ideal of absolute rest. They suggest rather that, as has already been shown for the first order of small quantities, the same laws of electrodynamics and optics will be valid for all frames of reference for which the equations of mechanics hold good. We will raise this conjecture (the purport of which will hereafter be called the *Principle of Relativity*) to the status of a postulate, and also introduce another postulate, which is only apparently irreconcilable with the former, namely that light is always propagated in empty space with a definite velocity c which is independent of the state of motion of the emitting body.

> These two postulates suffice for the attainment of a simple and consistent theory of the electrodynamics of moving bodies based on Maxwell's theory for bodies at rest. The introduction of a "luminiferous ether" will prove to be superfluous inasmuch as the view here to be developed will not require an "absolutely stationary space" provided with special properties, nor assign a velocity-vector to a point of the empty space in which electromagnetic processes take place.

The postulates used by Einstein to construct his theory are thus (i) the principle of relativity and (ii) the postulate according which the light velocity c is constant and enunciated as (Ref. 98, p. 895):

> Any ray of light moves in the "stationary" system of co-ordinates with the determined velocity c, whether the ray be emitted by a stationary or by a moving body.

Einstein used in fact the electrodynamic theory developed by Hertz: He thus based his theory on the "*kinematics of rigid bodies*" — corresponding to the coordinate system — and which has to be compared to the "*rigid system of bodies*" used by Hertz (Ref. 54, p. 246 and also see the quotation in p. 40).

Einstein needed to define what is meant by simultaneity and to explain how to get synchronous clocks as Poincaré did.[80] He also introduced the idea that "*the length of the moving rod measured from the stationary system* [is] *different from* [...] *the length of the rod in the moving system*" (Ref. 98, p. 896): This is equivalent to the length contraction introduced by FitzGerald and Lorentz that we previously discussed.

Einstein then investigated "*the transformation of coordinates and time from a stationary system to another system which is in uniform motion of translation relatively to the former*" (Ref. 98, p. 897). The unavowed aim of this section is to establish the transformation of coordinates which will leave invariant Maxwell's equations when one switches from a frame at rest to a moving frame. Maxwell's equations were established for describing electromagnetic phenomena. The coordinate transformation — the so-called Lorentz transformation — was obtained by Lorentz by investigating these equations. In contrast to this, Einstein only considered one of the coordinate systems "k" which "*has a constant velocity* [V] *in the direction of the x-axis of the other which is a stationary system* "K"" (Ref. 98, p. 897). He defined the relative position of the two systems as follows (Ref. 98, p. 897):

> Any time t of the stationary system "K" corresponds to a definite position of the axes of the moving system, which are always parallel to the axes of the stationary system. By t, we always mean the time in the stationary system.

A point in the stationary system "K" corresponds to a point in the moving system "k" according to (Ref. 98, p. 898)

> any system of values (x, y, z, t) which completely defines the position and time of an event in the stationary system, there corresponds a system of values (x', y', z', t') determining that event relatively to the system "k", and our task is now to find the system of equations connecting these quantities.

In order to obtain these equations, Einstein stated that "*if* $\xi = x - Vt$, *it is clear that a point at rest in the system* "k" *must have a system of values* (ξ, y, z) *which are independent of time.*" (Ref. 98, p. 898). In this expression, V is the velocity of system "k" with respect to the stationary system "K" and ξ is a given position along the x'-axis in the moving system "k". After some unclear considerations leading to (Ref. 98, p. 899)

$$\frac{\partial t'}{\partial \xi} + \frac{V}{c^2 - V^2}\frac{\partial t'}{\partial t} = 0, \tag{79}$$

Einstein expressed time t' in the moving system "k" as (Ref. 98, p. 899)

$$t' = a\left(t - \frac{V\xi}{c^2 - V^2}\right), \tag{80}$$

where "*a is a function* $\varphi(V)$ *at present unknown, and where for brevity it is assumed that at the origin of* "*k*", $t' = 0$ *when* $t = 0$" (Ref. 98, p. 899).

Einstein then proposed a coordinate transformation between the resting system "K" and the moving system "k" under the form (Ref. 98, p. 900):

$$\left|\begin{array}{l} t' = \varphi(V)\gamma\left(t - \dfrac{Vx}{c^2}\right) \\ x' = \varphi(V)\gamma(x - Vt) \\ y' = \varphi(V)y \\ z' = \varphi(V)z \end{array}\right. , \tag{81}$$

where $\varphi(V)$ is not yet known and $\gamma = \frac{1}{\sqrt{1-\frac{V^2}{c^2}}}$ is the contraction coefficient introduced by Lorentz. Once he showed that $\varphi(V) = 1$, he obtained the final coordinate transformation (Ref. 98, p. 902)

$$\left|\begin{array}{l} t' = \gamma\left(t - \dfrac{Vx}{c^2}\right) \\ x' = \gamma(x - Vt) \\ y' = y \\ z' = z \end{array}\right. . \tag{82}$$

He then wanted to prove that (Ref. 98, p. 901)

> any ray of light, measured in the moving system, is propagated with the velocity c, if, as we have assumed, this is the case in the stationary system; for we have not as yet furnished the proof that the principle of the constancy of the velocity of light is compatible with the principle of relativity.

However, Einstein already used it when he wrote expression (80) for time t' in the moving system "k"; indeed, he previously wrote that (Ref. 98, p. 899)

> light (as required by the principle of the constancy of the velocity of light, in combination with the principle of relativity) is also propagated with velocity c when measured in the moving system.

Indeed, he explicitly already used $x = ct$ and $x' = ct'$ for getting his coordinate transformation. We have thus necessarily

$$x^2 + y^2 + z^2 = c^2t^2 \tag{83}$$

and

$$x'^2 + y'^2 + z'^2 = c^2t'^2. \tag{84}$$

The second part of Einstein's paper is devoted to electrodynamics and, more particularly, to the "*transformation of the Maxwell–Hertz equations for empty space*" (Ref. 98, p. 907). The equations actually used by Einstein are those (with a correction in the minus sign which he applied to Hertz's equation (13) as it is done today and not to Eq. (12)) proposed by Hertz (Ref. 50, p. 138) (see Eqs. (12) and (13), p. 35 of this paper) for describing electromagnetic fields in the ether: One can therefore question in which sense the ether is superfluous in Einstein's theory. Indeed Hertz's equations (12) and (13) that Einstein used were written for etherous molecules and not for bodies at rest or in motion.

Applying his coordinate transformation to Hertz's equations, Einstein then obtained — without any detail — the corresponding equations in a coordinate system moving with a velocity V with respect to the resting system, that is (Ref. 98, p. 907),

$$\begin{cases} \dfrac{1}{c}\dfrac{\partial E_x}{\partial t'} = \gamma\dfrac{\partial}{\partial y'}\left(H_z - \dfrac{V}{c}E_y\right) - \gamma\dfrac{\partial}{\partial z'}\left(H_y + \dfrac{V}{c}E_z\right) \\ \dfrac{1}{c}\gamma\dfrac{\partial}{\partial t'}\left(E_y - \dfrac{V}{c}H_z\right) = \dfrac{\partial H_x}{\partial z'} - \gamma\dfrac{\partial}{\partial x'}\left(H_z - \dfrac{V}{c}E_y\right) \\ \dfrac{1}{c}\gamma\dfrac{\partial}{\partial t'}\left(E_z + \dfrac{V}{c}H_y\right) = \gamma\dfrac{\partial}{\partial x'}\left(H_y + \dfrac{V}{c}E_z\right) - \dfrac{\partial H_x}{\partial y'} \end{cases} \tag{85}$$

and (Ref. 98, p. 908)

$$\begin{cases} \dfrac{1}{c}\dfrac{\partial H_x}{\partial t'} = \gamma\dfrac{\partial}{\partial z'}\left(E_y - \dfrac{V}{c}H_z\right) - \gamma\dfrac{\partial}{\partial y'}\left(E_z + \dfrac{V}{c}H_y\right) \\ \dfrac{1}{c}\gamma\dfrac{\partial}{\partial t'}\left(H_y + \dfrac{V}{c}E_z\right) = \gamma\dfrac{\partial}{\partial x'}\left(E_z + \dfrac{V}{c}H_y\right) - \dfrac{\partial E_x}{\partial z'}. \\ \dfrac{1}{c}\gamma\dfrac{\partial}{\partial t'}\left(H_z - \dfrac{V}{c}E_y\right) = \dfrac{\partial E_x}{\partial y'} - \gamma\dfrac{\partial}{\partial x'}\left(E_y - \dfrac{V}{c}H_z\right) \end{cases} \tag{86}$$

Then using the principle of relativity requiring that "*the Maxwell–Hertz equations for empty space (sic) hold in system "K", they also hold in system "k"*" (Ref. 98, p. 908). In system "k", Einstein should get

$$\frac{1}{c}\frac{\partial \mathbf{E}'}{\partial t'} = \boldsymbol{\nabla}' \wedge \mathbf{H}' \tag{87}$$

and

$$\frac{1}{c}\frac{\partial \mathbf{H}'}{\partial t'} = -\boldsymbol{\nabla}' \wedge \mathbf{E}'. \tag{88}$$

Einstein then identified term to term these last two equations with Eqs. (85) and (86). In other words, he just wrote what should be the transformed electric and magnetic fields — similar to those found by Poincaré (Ref. 90, p. 135) — for verifying that his "Maxwell–Hertz equations" obey to the principle of relativity. The key point is to switch from the original Maxwell–Hertz equations to the transformed Eqs. (85) and (86) about which there is no indication how Einstein got them (the application of the coordinate transformation to the electric and magnetic fields is not straightforward at all).

9. Conclusion

Special relativity clearly came from the description of optical phenomena in moving bodies. Such a problem was investigated at least for 200 years before 1905. Investigating the various descriptions of these phenomena led us to show that all theories were dual, combining particles and waves. When a wave theory is required for describing the experimental observations, this is always a mechanical explanation in terms of light particles interacting with etherous molecules which was proposed if provided. It is interesting to note that all optical phenomena are well described without any mention to the ether but were never explained without it. We also showed that, contrary to what Einstein claimed, the ether is not removed from his 1905 paper since he used the equations written by Hertz for the electromagnetic fields in the ether. At many times, the ether blurred the discussion but, if it can be easily omitted for describing the experimental facts, one still has many difficulties to explain how light is propagated (or what a photon is).

As shown by Poincaré, the electromagnetic theory describing Michelson and Morley's experiment is the theory proposed by Maxwell (and not the one by Hertz). Poincaré thus showed, by proposing the correct addition law for velocities the invariance of Maxwell's equations when one switches from a resting frame to a moving frame or, more exactly, between two frames presenting a uniform translation between each other. Various attempts were also published by Voigt, Larmor, Lorentz and Einstein. Only Lorentz and Poincaré clearly worked in a four-dimensional formulation by using the vector- and the scalar-potential for describing the electromagnetic fields.

Acknowledgments

We wish to thank James Nester for his careful reading of the paper and his comments that helped us to greatly improve its contents.

Appendices

A.1. *Fizeau's experiments*

In order to check Fresnel's theory, Hippolyte Fizeau (1819–1896) conducted an experiment to determine how the ether could be entrained. He retained three main hypotheses[58]:

(i) The ether adheres, as it was fixed to the molecules of the body and, consequently, shares motions that may be imposed on this body.
(ii) The ether is free and independent, and is not driven by the body in its motion.
(iii) A portion of the ether would be free, while the other portion would be fixed to the molecules of the body and would solely share in its motion.

The experiment conducted by Fizeau is sketched in Fig. A.1. The fringe shift resulting from the experiment was $\Delta p = 0.4$. Fizeau then compared this value to the fringe shift which should result from the different hypotheses on the state of the ether. Depending on the assumption retained, the expected shift of fringes Δp is more or less large. The first hypothesis ($\Delta p = 0.92$) corresponds to a total driving of the ether by the motion of the Earth: The light velocity is thus deeply affected. The second hypothesis ($\Delta p = 0$) is associated with no driving and the light velocity is constant. Finally, the third hypothesis ($\Delta p = 0.46$) corresponds to a partial driving and the light velocity is partly affected. Fizeau's result thus validated Fresnel's hypothesis for a partial driving of the ether by the Earth. As a consequence, the motion of transparent bodies induces a change in the light velocity according to their refractive property.

A.2. *Michelson and Morley's experiments*

The first experiment to determine the relative motion between the Earth and the ether using light propagation was suggested by Maxwell in 1878 in an article entitled

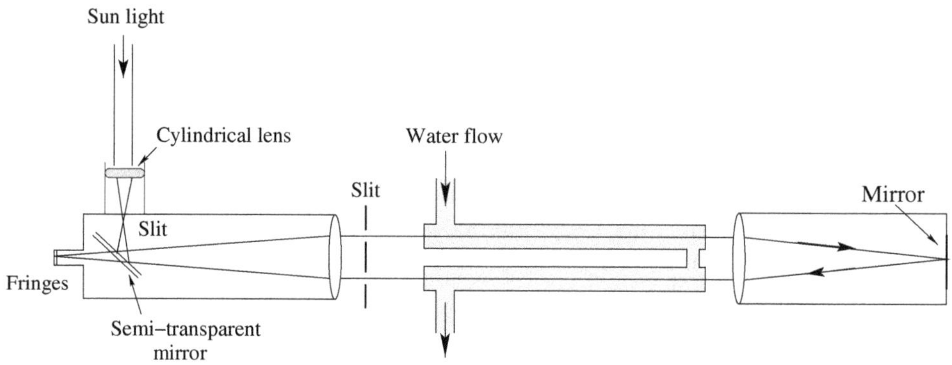

Fig. A.1. Sketch of the experimental setup conducted by Fizeau.

Aether for the *Encyclopedia Britannica* (Ref. 99, p. 270).

> If it was possible to determine the light velocity by observing the time it takes to travel between one station and another on Earth's surface, we might, by comparing the observed velocities in opposite directions, determine the velocity of aether with respect to these terrestrial stations. All methods, however, by which it is practicable to determine the velocity of light from terrestrial experiments depend on the measurement of the time required for the double journey from one station to the other and back again, and the increase of this time on account of a relative velocity of the ether equal to that of the Earth in its orbit would be only about one hundred millionth part of the whole time of transmission, and would therefore be quite insensible.

Although Maxwell was not too confident in the success of his experiment since the difference was depending on the squared ratio of velocities, Albert Michelson (1852–1931) built a similar experiment[100]:

> If, therefore, an apparatus is so constructed as to permit two pencils of light, which have traveled over paths at right angles to each other, to interfere, the pencil which have traveled in the direction of the Earth's motion, will in reality travel $\frac{4}{100}$ of a wavelength farther than it would have done, were the Earth at rest. The other pencil being at right angles to the motion would not be affected. If now, the apparatus be revolved through 90° so that the second pencil is brought into the direction of the Earth's motion, its path will have lengthened $\frac{4}{100}$ wave-lengths. The total change in the position of the interference bands would be $\frac{8}{100}$ of the distance between the bands, a quantity easily measurable.

But Michelson's first results did not show any fringe shift and refuted George Stokes's hypothesis for a stationary ether.[101] Considering that his experimental

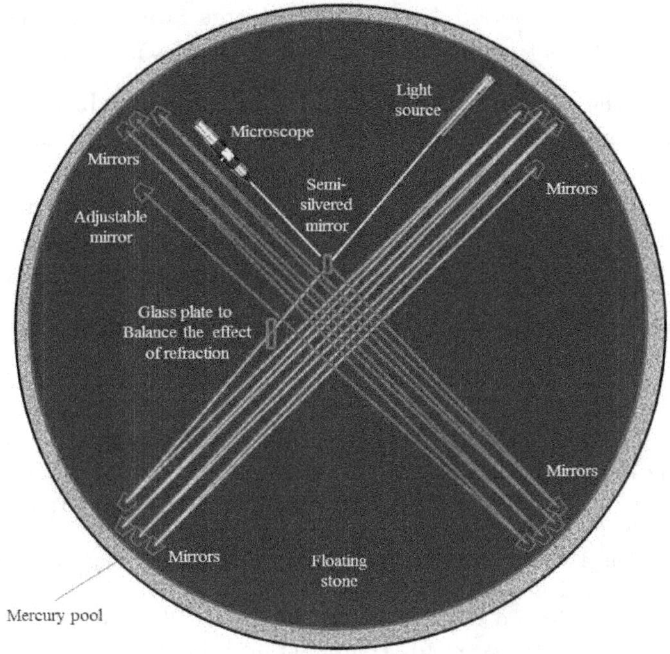

Fig. A.2. Sketch of the experimental device conducted by Michelson and Morley in 1887. (For color version, see page I-CP1.)

errors were sufficient to mask the fringe shift, Michelson built a second experiment with Edward Morley (1838–1923). The same device (Fig. A.2) was placed on a massive stone floating on a mercury bath in order to remove any vibrating perturbation. They also augmented by ten times the length of the two optical paths (11 m in length) by multiple reflections on the mirrors.[59]

In spite of the great care spent to build this second experiment, the result was the same as for the first one: No fringe shift. Michelson and Morley thus concluded (Ref. 59, p. 341):

> It appears, from all that precedes, reasonably certain that, if there be any relative motion between the Earth and the luminiferous ether, it must be small; quite small enough entirely to refute Fresnel's explanation of aberration. Stokes has given a theory of aberration which assumes the ether at the Earth's surface to be at rest with regard to the latter, and only requires in addition that the relative velocity have a potential; but Lorentz shows that these conditions are incompatible. Lorentz then proposes a modification which combines some ideas of Stokes and Fresnel, and assumes, the existence of a potential, together with Fresnel's coefficient. If now it were legitimate to conclude from the present work that the ether is at rest with regard to Earth's surface according to Lorentz there could not be a velocity potential, and his own theory also fails."

In 1889, George FitzGerald (1851–1901), George Stoney's nephew, suggested that the sole hypothesis to explain Michelson and Morley's experiment[102]

> [...] is that the length of material body changes, according as they are moving through the ether or across it, by an amount depending on the square of the ratio of their velocity to that of light. We know that electric forces are affected by the motion of the electrified bodies relative to the ether, and it seems a not improbable supposition that the molecular forces are affected by the motion, and that the size of a body alters consequently.

Such a contraction of the length was enough to explain the experimental facts.

References

1. N. Cusano, *De docta ignorantia* (1440), *On Learned Ignorance*, 2nd edn. (The Arthur J. Banning Press, Minneapolis, 1985).
2. W. Turner and G. Bruno, *The Catholic Encyclopedia* (Robert Appleton Company, New York, 1913).
3. G. Bruno, *The Ash Wednesday Supper* (1584), Translated by S. L. Jaki (University of Toronto Press, Toronto, 1995).
4. V. Messager, R. Gilmore and C. Letellier, *Contemp. Phys.* **53** (2012) 397.
5. Aristotle, *On the Heavens* (-350), Translated by J. L. Stocks (Global Grey, 2014).
6. G. E. R. Lloyd, *Early Greek Science*: *Thales to Aristotle* (W. W. Norton and New York, 1970).
7. P. Duhem, *Le système du monde*: *Histoire des doctrines cosmologiques de Platon à Copernic, Tome I, Les sphères homocentriques d'Eudoxe* (Librairie Scientifique A. Hermann et Fils, Paris, 1913).
8. R. Grosseteste, On Light (De Luce), wrote between 1228–1232, Translated by C. C. Riedl (Marquette University Press, Milwaukee, 1942).
9. R. Descartes, *Les Météores* (1637), *Discourse on Method, Optics, Geometry, and Meteorology*, Translated by P. J. Olscamp (Hackett Publishing, 2001).
10. N. Malebranche, *Oeuvres complètes*, Tome i, *De la recherche de la vérité*, Livre III: *de l'entendement ou de l'esprit* (Librairie de Spaia, Paris, 1837).
11. M. Kopernik, *De revolutionibus orbium coelestium* (1543), *On the Revolutions of the Heavently Spheres*, Translated by C. G. Wallis (St. John's College Bookstore, Annapolis, 1939).
12. G. Galilei, *Dialogo sopra i due massimi sistemi del mondo* (1632), *Dialogue Concerning the Two Chief World Systems*, Translated by S. Drake (Modern Library, 2001).
13. G. Galilei, *Discorsi e dimostrazioni matematiche intorno a due nuove scienze attinenti alla meccanica e ai moti locali* (1638), *Discourses and Mathematical Demonstrations Relating to Two New Sciences*, Translated by H. Crew and A. de Salvio (MacMillian, 1914).
14. O. Roemer, *Journal des Sçavans* (1676) 233.
15. J. Bradley, *Philos. Trans.* **35** (1727–1728) 637.
16. C. Huygens, *Traité de la lumière* (P. van der Aa, Leiden, 1690).
17. T. Young, *Philos. Trans. R. Soc. Lond.* **92** (1802) 12.
18. T. Young, *Philos. Trans. R. Soc. Lond.* **94** (1804) 1.
19. I. Newton, *Opticks*: *Or a Treatise of the Reflections, Refractions, Inflections and Colors of Light* (1704), reprint (Dover, 1952).

20. J.-B. Biot, *Traité de physique expérimentale et mathématique*, Tome I (Déterville Libraire, Paris, 1816).
21. J. Michell, *Philos. Trans. R. Soc. Lond.* **74** (1784) 35.
22. F. Arago, *C. R. Acad. Sci.* **36** (1853) 38.
23. A. Fresnel, Lettre de A. Fresnel à F. Arago (1818), Sur l'influence du mouvement terrestre dans quelques phénomènes d'optique, in *Oeuvres Complètes*, Théorie de la lumière: Cinquième Section, Questions diverses d'optique, N xlix, MM. H. de Senarmont, E. Verdet et L. Fresnel (Imprimerie impériale, Paris, 1866–1870).
24. A. Fresnel, Premier Mémoire sur la diffraction (October 15, 1815), in *Oeuvres complètes d'Augustin Fresnel*, M. M. H. de Senarmont, E. Verdet et L. Fresnel (Imprimerie impériale, Paris, 1866–1870).
25. A. Fresnel, Théorie de la lumière: Troisième section, Exposition systématique de la théorie des ondulations et controverse (1822), N xxxi, *in Oeuvres Complètes d'Augustin Fresnel*, Tome II, MM. H. de Senarmont, E. Verdet et L. Fresnel (Imprimerie impériale, Paris, 1866–1870).
26. A.-M. Ampère, Théorie mathématique des phénomènes électro-dynamiques uniquement déduite de l'expérience (1820–22–23–25), in *Mémoire de l'Académie Royale des Sciences de l'Institut de France*, Tome vi (1823) (F. Didot, Paris, 1827), pp. 175–388.
27. M. Faraday, *Philos. Trans. R. Soc.* **122** (1832) 125.
28. W. Weber, Determinations of electrodynamic measure: Concerning a universal law of electrical action, in *Treatise at the Founding of the Royal Scientific Society of Saxony on the day of the 200th Anniversary Celebration of Leibniz's Birthday* (Prince Jablonowski Society, Leipzig, 1846), pp. 211–378. Translated by S. P. Johnson and edited by L. Hecht and A. K. T. Assis, http://www.ifi.unicamp.br/assis/(2007), pp. 1–143.
29. H. Poincaré, *Electricité et Optique*: *ii Les théories de Helmholtz et les expériences de Hertz* (G. Carré Ed, Paris, 1891).
30. M. Faraday, *Philos. Trans. R. Soc.* **136** (1846) 1.
31. W. Weber and R. Kohlrausch, *Pogg. Ann.* **99** (1856) 10.
32. G. Kirchhoff, *Philos. Mag.* **37** (1850) 463.
33. G. Kirchhoff, *Philos. Mag.* **13** (1857) 393.
34. G. Kirchhoff, *Pogg. Ann.* **102** (1857) 529.
35. W. Weber, *Abhandlungen der Königlich Sächsischen Gesellschaft der Wissenschaften* **6** (1864) 571.
36. J. C. Maxwell, *Philos. Trans. R. Soc. Lond.* **155** (1865) 459.
37. H. Fizeau, *C. R. Acad. Sci.* **29** (1849) 92.
38. G. G. Stoney, *Sci. Proc. R. Dublin Soc.* **3** (1881–1883) 51.
39. H. Helmholtz, *Journal für die reine und angewandte Mathematik* **72** (1870) 57.
40. F. E. Neumann, *Physikalische Abhandlungen der Königlichen Akademie der Wissenschaften zu Berlin* (1847) 1.
41. H. Helmholtz, *Über die Erhaltung der kraft* (On the Conservation of Force) (Bruck und Verlag von G. Reimer, Berlin, 1847).
42. H. Helmholtz, *Journal für die reine und angewandte Mathematik* **72** (1874) 273.
43. J. C. Maxwell, *A Treatise of Electricity and Magnetism* (Clarenton, Oxford, 1875), reprinted by Dover (1954).
44. H. Hertz, *Ann. Phys.* **270** (1888) 155.
45. H. Hertz, *Ann. Phys.* **270** (1888) 551 [On the finite velocity of propagation of electromagnetic actions (1888), Translated by D. E. Jones, pp. 107–123 in Ref. 47].
46. H. Hertz, *Annalen der Physik und Chemie* **34** (1888) 609 [On electromagnetic waves in air and their reflection, Translated by D. E. Jones, pp. 124–136 in Ref. 47].

47. H. Hertz, *Electric Waves Being Researches on the Propagation of Electric Action with Finite Velocity Through Space*, Translated by D. E. Jones (MacMillan, 1893), Reprinted by Dover (1962).
48. H. Poincaré, *C. R. Hebd. Seances Acad. Sci.* **111** (1890) 322.
49. H. Hertz, Lettre de H. Hertz à H. Poincaré (1890), http://www.henripoincarepapers.univ-lorraine-fr.
50. H. Hertz, *Annalen der Physik und Chemie* **36** (1889) 1 [The forces of electric oscillations, treated according to Maxwell's theory, Translated by D. E. Jones, pp. 137–159 in Ref. 47].
51. O. Heaviside, *Philos. Mag.* **22** (1886) 118.
52. H. A. Lorentz, La théorie électromagnétique de Maxwell et son application aux corps mouvants, *Archives Néerlandaises des Sciences Exactes et Naturelles* **25** (1892) 363.
53. H. Hertz, *Annalen der Physik und Chemie* **40** (1890) 577 [On the fundamental equations of electromagnetics for bodies at rest, Translated by D. E. Jones, pp. 195–240 in Ref. 47].
54. H. Hertz, *Annalen der Physik und Chemie* **45** (1890) 28 [On the fundamental equations of electromagnetics for bodies in motion, Translated by D. E. Jones, pp. 241–268 in Ref. 47].
55. O. Heaviside, Electrical Papers, Vol. I (pp. 429–556) and Vol. II (pp. 39–151), London, MacMillan and Co. (1892).
56. W. Voigt, Ueber das Dopplersche Princip, *Gött. Nachr.* **2** (1887) 41, [reprinted in *Physikalische Zeitschrift* **16** (1915) 381, Translated by Wikisource, On the principle of Doppler].
57. W. Voigt, *Nachrichten von d. Königl. Gesellsch. d. Wissen. u. d. Georg-Augusts Universität z. Göttingen* **8** (1887) 177.
58. H. Fizeau, *C. R. Acad. Sci.* **33** (1851) 349.
59. A. A. Michelson and E. W. Morley, *Am. J. Sci.* **34** (1887) 333.
60. H. A. Lorentz, *Verslagen en Mededeelingen der Koninklijke Akademie van Wetenschappen Amsterdam* **1** (1892) 74. Translated as The relative motion of the Earth and the ether, in *Collected Papers*, Vol. 4 (Martinus Nijhoff, The Hague, 1934–1939), pp. 220–223.
61. H. A. Lorentz, *Versuch einer Theorie der electrischen und optischen Erscheinungen in bewegten Körpern*, Translated by Wikisource, *Attempt of Theory of Electrical and Optical Phenomena in Moving Bodies* (E. J. Brill, Leiden, 1895).
62. H. A. Lorentz, *Proc. R. Netherlands Acad. Arts Sci.* **1** (1899) 427.
63. H. A. Lorentz, Lettre de Lorentz à Poincaré (1901).
64. H. A. Lorentz, *Proc. R. Netherlands Acad. Arts Sci.* **6** (1904) 809.
65. H. Poincaré, Lettre de Poincaré à Lorentz (05–1905), ALS 3p. H. A. Lorentz papers, inv. nr. 62, Noord-Hollands Archief., Henri Poincaré papers, Archives Henri Poincaré.
66. H. Poincaré, Lettre de Poincaré à Lorentz (05–1905), ALS 2p. H.A. Lorentz papers, inv. nr. 62, Noord-Hollands Archief., *Henri Poincaré papers*, Archives Henri Poincaré.
67. M. Abraham, *Phys. Z.* **4** (1902) 57.
68. W. Kaufmann, *Phys. Z.* **4** (1902) 54.
69. J. Larmor, *Philos. Trans. R. Soc. A* **190** (1897) 205.
70. J. Larmor, *Aether and Matter: A Development of the Dynamical Relations of the Aether to Material Systems* (Cambridge University Press, 1900).
71. H. Poincaré, *Electricité et Optique: La lumière et les Théories électrodynamiques* (Gauthier-Villars, Paris, 1901).

72. H. Poincaré, *Optique et Electricité: Leçons sur la théorie mathématique de la lumière* (G. Carré Editeur, Paris, 1889).
73. H. Poincaré, *Bulletin des Sciences Mathématiques* **28** (1904) 302.
74. K. Popper, *Realism and the Aim of Science* (Routledge, London, 1985).
75. H. Poincaré, *Rev. Gén. Sci. Pures Appl.* **11** (1900) 1163.
76. H. Poincaré, *Electricité et Optique: i Les théories de Maxwell et la théorie électromagnétique de la lumière* (G. Carré Ed, Paris, 1890).
77. A. Cornu, Détermination de la vitesse de la lumière d'après des expériences exécutées en 1874 entre l'observatoire et Montlhéry, *Annales de l'Observatoire de Paris*, xiii (Gauthier-Villars, Paris, 1876).
78. H. Poincaré, *L' éclairage Electrique* **5** (1895) 5; **5** (1895) 385.
79. H. von Helmholtz, *Ann. Phys.* **48** (1893) 389; **48** (1893) 723.
80. H. Poincaré, *Revue de Mathématique et de Morale* **6** (1898) 1.
81. H. Poincaré, *Bibliothèque du Congrès International de Philosophie* **3** (1900) 457.
82. H. Poincaré, *Archives Néerlandaises Des Sciences Exactes Et Naturelles* **5** (1900) 252.
83. A. Einstein, *Phys. Z.* **18** (1917) 121.
84. A. Einstein, *Ann. Phys.* **18** (1905) 639.
85. H. Poincaré, *La Science et l'Hypothèse* (Flammarion, Paris, 1902).
86. H. Poincaré, *La valeur de la Science* (1904), Translated in English, *The Foundations of Science* (The Value of Science) (Science Press, New York, 1913).
87. S. Carnot, *Réflexions sur la puissance motrice du feu et sur les machines* (Bachelier Libraire, Paris, 1824), reprinted by J. Gabay, Paris (2005).
88. P. Curie and A. Laborde, *C. R. Acad. Sci.* **136** (1903) 673.
89. P. Langevin, *C. R. Acad. Sci.* **140** (1905) 1171.
90. H. Poincaré, *Rendiconti del Circolo Matematico di Palermo* **21** (1906) 129.
91. H. Poincaré, *C. R. Acad. Sci.* **140** (1905) 1504.
92. H. Poincaré, *C. R. Acad. Sci.* **92** (1881) 333.
93. H. Poincaré, *C. R. Acad. Sci.* **92** (1881) 395.
94. H. Poincaré, *C. R. Acad. Sci.* **92** (1881) 698.
95. H. Poincaré, *Acta Math.* **1** (1882) 1.
96. J. Gray, *Nieuw Archief voor Wiskunde* **13** (2012) 178.
97. H. Poincaré, *Revue de Mathématique et de Morale* **3** (1895) 631.
98. A. Einstein, *Ann. Phys.* **17** (1905) 891 — Translated by W. Perrett and G. B. Jeffery, On the electrodynamics of moving bodies, in *The Principle of Relativity: Original Papers* (Dover, 1952).
99. J. C. Maxwell, Aether, in *Encyclopedia Britannica* (1878), in *Scientific Papers*, Tome 2, Cambridge University Press (1890), pp. 568–572.
100. A. A. Michelson, *Am. J. Sci.* **22** (1850) 120.
101. G. G. Stokes, *Philos. Mag.* **29** (1846) 6.
102. G. FitzGerald, *Science* **13** (1889) 390.

Chapter 2

Genesis of general relativity — A concise exposition*

Wei-Tou Ni

School of Optical-Electrical and Computer Engineering,
University of Shanghai for Science and Technology,
516, Jun Gong Rd., Shanghai 200093, P. R. China

weitou@gmail.com

This short exposition starts with a brief discussion of situation before the completion of special relativity (Le Verrier's discovery of the Mercury perihelion advance anomaly, Michelson–Morley experiment, Eötvös experiment, Newcomb's improved observation of Mercury perihelion advance, the proposals of various new gravity theories and the development of tensor analysis and differential geometry) and accounts for the main conceptual developments leading to the completion of the general relativity (CGR): gravity has finite velocity of propagation; energy also gravitates; Einstein proposed his equivalence principle and deduced the gravitational redshift; Minkowski formulated the special relativity in four-dimentional spacetime and derived the four-dimensional electromagnetic stress–energy tensor; Einstein derived the gravitational deflection from his equivalence principle; Laue extended Minkowski's method of constructing electromagnetic stress-energy tensor to stressed bodies, dust and relativistic fluids; Abraham, Einstein, and Nordström proposed their versions of scalar theories of gravity in 1911–13; Einstein and Grossmann first used metric as the basic gravitational entity and proposed a "tensor" theory of gravity (the "Entwurf" theory, 1913); Einstein proposed a theory of gravity with Ricci tensor proportional to stress–energy tensor (1915); Einstein, based on 1913 Besso–Einstein collaboration, correctly derived the relativistic perihelion advance formula of his new theory which agreed with observation (1915); Hilbert discovered the Lagrangian for electromagnetic stress–energy tensor and the Lagrangian for the gravitational field (1915), and stated the Hilbert variational principle; Einstein equation of GR was proposed (1915); Einstein published his foundation paper (1916). Subsequent developments and applications in the next two years included Schwarzschild solution (1916), gravitational waves and the quadrupole formula of gravitational radiation (1916, 1918), cosmology and the proposal of cosmological constant (1917), de Sitter solution (1917) and Lense–Thirring effect (1918).

Keywords: General relativity; Einstein equivalence principle; Minkowski formalism; stress-energy tensor; Hilbert variational principle; cognition and history of science.

PACS Number(s): 01.65.+g, 04.20−q, 04.80.Cc

*This review is dedicated to the memory of my father Fu-Yuan Ni (1915.11.28–2016.03.24).

1. Prelude — Before 1905

General Relativity (GR) was fast in its acceptance in the world community. This was not the case for Newtonian gravitation.[1] We quote from the beginning of Chapter V on Gravitation of Vol. II from Whittaker[2]: "We have seen (cf. Vol. I, pp. 29–31) that for many years after its first publication, the Newtonian doctrine of gravitation was not well received. Even in Newton's own University of Cambridge, the textbook of physics in general use during the first quarter of the 18th century was still Cartesian: while all the great mathematicians of the Continent — Huygens in Holland, Leibnitz in Germany, Johann Bernoulli in Switzerland, Cassini in France — rejected the Newtonian theory altogether".

"This must not be set down entirely to prejudice: many well-informed astronomers believed, apparently with good reason, that the Newtonian law was not reconcilable with the observed motions of the heavenly bodies. They admittedly that it explained satisfactorily the first approximation to the planetary orbit, namely that they are ellipses with the sun in one focus: but by the end of seventeenth century much was known observationally about the departures from elliptic motion, or *inequalities* as they are called, which were presumably due to mutual gravitational interaction: and some of these seemed to resist every attempt to explain them as consequences of the Newtonian law".

The most serious one was the *Great inequality of Jupiter and Saturn.* In the same page, Whittaker continued: "A comparison of the ancient observations cited by Ptolemy in the Almagest with those of the earlier astronomers of Western Europe and their more recent successors, showed that for centuries past the mean motion, or average angular velocity round the sun, of Jupiter, had been continually increasing, while the mean motion of Saturn had been continually decreasing". According to Kepler's[3] third law, the orbit of Jupiter must be shrinking and the orbit of Saturn must be expanding. This stimulates the development of celestial mechanics. Euler and Lagrange made significant advances. In 1784, Laplace found that the *Great inequality* is not a secular inequality but a periodic inequality of 929-year long period due to nearly commeasurable orbital periods of Jupiter and Saturn. Calculation agreed with observations. The issue was completely solved. For a more thorough study of the history of the *Great inequality of Jupiter and Saturn*, see the doctoral thesis of Curtis Wilson.[4]

In 1781, Herschel discovered the planet Uranus. Over years, Uranus persistently wandered away from its expected Newtonian path. In 1834, Hussey suggested that the deviation is due to perturbation of an undiscovered planet. In 1846, Le Verrier predicted the position of this new planet. On 25, September 1846, Galle and d'Arrest found the new planet, Neptune, within one degree of arc of Le Verrier's calculation. This symbolized the great achievement of Newton's theory.[5]

With the discovery of Neptune, Newton's theory of gravitation was at its peak. As the orbit determination of Mercury reached 10^{-8}, relativistic effect of gravity showed up. In 1859, Le Verrier discovered the anomalous perihelion advance of Mercury.[6]

Anomalous perihelion advance of Mercury. In 1840, Arago suggested to Le Verrier to work on the subject of Mercury's motion. Le Verrier published a provisional theory in 1843. It was tested at the 1848 transit of Mercury and there was not close agreement. As to the cause, Le Verrier[7] wrote "Unfortunately, the consequences of the principle of gravitation have not be deduced in many particulars with a sufficient rigor: we will not be able to decide, when faced with a disagreement between observation and theory, whether this results completely from analytical errors or whether it is due in part to the imperfection of our knowledge of celestial physics".[7,8]

In 1859, Le Verrier[6] published a more sophisticated theory of Mercury's motion. This theory was sufficiently rigorous for any disagreement with observation to be taken quite confidently as indicating a new scientific fact. In this paper, he used two sets of observations — a series of 397 meridian observations of Mercury taken at the Paris Observatory between 1801 and 1842, and a set of observations of 14 transits of Mercury. The transit data are more precise and the uncertainty is of the order of $1''$. The calculated planetary perturbations of Mercury are listed in Table 1.[6,8] In addition to these perturbations, there is a $5025''$/century general precession in the observational data due to the precession of equinox. The fit of observational data with theoretical calculations has discrepancies. These discrepancies turned out to be due to relativistic-gravity effects. Le Verrier attributed these discrepancies to an additional $38''$ per century anomalous advance in the perihelion of Mercury.[7]

Newcomb[9] in 1882, with improved calculations and data set, obtained $42''.95$ per century anomalous perihelion advance of Mercury. The value more recently (1990) was $(42''.98 \pm 0.04)$/century.[10] At present, ephemeris fitting reached 10^{-4} precision. See Ref. 11 and references therein.

Michelson–Morley experiment. According to Newton's second law of motion and Galilean transformation, light velocity would change in a moving frame. However, this is not the experimental finding of Michelson and Morley in 1887[12]: "Considering the motion of the earth in its orbit only, this displacement should be $2Dv^2/V^2 = 2D \times 10^{-8}$. The distance D was about eleven meters, or 2×10^7 wavelengths of yellow light; hence the displacement to be expected was 0.4 fringe. The actual displacement was certainly less than the twentieth part of this kind, and probably less than the 40th part. But since the displacement is proportional to the square of the velocity, the relative velocity of the earth and the ether is probably less than one sixth the earth's orbital velocity, and certainly less than one-fourth". D is the optical

Table 1. Planetary perturbations of the perihelion of Mercury.[6,8]

Venus	$280''.6$/century
Earth	$83''.6$/century
Mars	$2''.6$/century
Jupiter	$152''.6$/century
Saturn	$7''.2$/century
Uranus	$0''.1$/century
Total	$526''.7$/century

path length in one arm of the multi-reflection Michelson–Morley interferometer sat on the granite floating in liquid mercury; v is the velocity of earth relative to ether; V is the light velocity. In modern Michelson–Morley experiments, one measures the frequency changes $\Delta\nu/\nu$ of two perpendicular Fabry–Perot cavities. The most precise experiment by Nagel *et al.*[13] measured the changes $\Delta\nu/\nu$ of two cryogenic cavities to be $(9.2 \pm 10.7) \times 10^{-19}$ (95% confidence interval), a nine order improvement to the original Michelson–Morley experiment.

Eötvös experiment. In 1889, Eötvös[14] used a torsion balance with different types of sample materials to significantly improve on the test of the Galileo equivalence principle (the equivalence of gravitational mass and inertial mass; the universality of free fall)[15] to a precision of 1 in 20 million (5×10^{-8}). The most recent terrestrial experiments of Washington group used torsion-balance to compare the differential accelerations of beryllium–aluminum and beryllium–titanium test-body pairs with precisions at the part in 10^{13} level and confirmed the Galileo equivalence principle.[16] The first space experiment Microscope (MICRO-Satellite à trainée Compensée pour l'Observation du Principle d'Équivalence)[17,18] has been in orbit since 26 April, 2016 with the aim of improving the test accuracy to one part in 10^{15} level and is performing functional tests successfully.[18] The Microscope test masses are made of alloys of Platinum–Rhodium (PtRh10 – 90% Pt, 10% Rh) and Titanium–Aluminum–Vanadium (TA6V – 90% Ti, 6% Al, 4% V), while the REF test masses are made of the same PtRh10 alloy. The weak equivalence for photons are confirmed with precisions at the part in 10^{38} level in astrophysical and cosmological observations on electromagnetic wave propagation.[19]

The discovery of Mercury perihelion advance anomaly undermined Newton's gravitation theory while the null results of Michelson and Morley undermined the Galilean invariance and Newton's dynamics. The foundation of Newton's world system and classical physics needed to be replaced. The precise verification of weak equivalence principle and realization that the phenomena are the same in a uniformly moving boat and on ground made it easier to advance one step in cognition to comprehend and formulate Einstein Equivalence Principle (EEP) (the phenomena in a falling elevator are the same as in free space).

In the last half of the 19th century, efforts to account for the anomalous perihelion advance of Mercury explored two general directions: (i) searching for a putative planet 'Vulcan' or other matter inside Mercury's orbit; and (ii) postulating an *ad hoc* modified gravitational force law. Both these directions proved unsuccessful. Proposed modifications of the gravitational law included Clairaut's force law (of the form $A/r^2 + B/r^4$), Hall's hypothesis (that the gravitational attraction is proportional to the inverse of distance to the $(2 + \delta)$ power instead of the square), and velocity-dependent force laws. The reader is referred to Ref. 8 for a thorough study of the history related to the Mercury's perihelion advance.

A compelling solution to this problem had to await the development of GR. When GR is taken as the correct theory for predicting corrections to Newton's theory, we understand why when the observations reached an accuracy of the order of 1″ per century (transit observations), a discrepancy would be seen. Over a century, Mercury orbits around the Sun 400 times, amounting to a total angle of 5×10^8 arcsec. The fractional relativistic correction (perihelion advance anomaly) of Mercury's orbit is of order $\xi G_{\rm N} M_{\rm Sun}/dc^2$ with d being the distance of Mercury to the Sun and ξ a parameter of order one depending on theory; for GR with $\xi = 3$, it is 8×10^{-8}. Therefore, the general relativistic correction for perihelion advance is about 40 arc sec per century. As the orbit determination of Mercury reached an accuracy of order 10^{-8}, the relativistic corrections to Newtonian gravity became manifest.

We thus see how gravitational anomalies can lead either to the discovery of missing matter or to a modification of the fundamental theory for gravity.

It is not totally coincidental that Le Verrier not only predicted the position of a new planet, but also discovered the Mercury perihelion advance anomaly as astronomical observation were refined and accumulated for a century.

Michelson–Morley experiment inspired the consideration of new covariant formulation of electromagnetism under reference frame transformation. Michelson–Morley experiment, with various proposals and developments, led eventually to the approximate transformation theory of Lorentz,[20] and the principle of relativity of Poincaré.[21–23] In 1901, Poincaré[24] performed a rigorous mathematical and physical analysis of various variants of the electrodynamic theory; in the introduction, he wrote (English translation from pp. 48–49 of Ref. 25): "Although none of these theories seems to me fully satisfactory, each one contains without any doubt a part of the truth and comparing them maybe instructive. From all of them, Lorentz theory seems to me the one which describes in the better way the facts". What Poincaré used as a criterion of satisfaction is whether the principle of relative motion is fully satisfied (Ref. 21, p. 477): The movement of any system whatever ought to obey the same laws, whether it is referred to fixed axes or to the movable axes which are implied in uniform motion in a straight line (English translation from p. 63 of Ref. 25). This is clearly the invariance of the laws under a change of reference frame, when one is related to the other by a constant velocity.[25] Nevertheless, in 1900, the transformation from one frame to the other was not known. The "classical" composition law for velocities was clearly not working for explaining optical experiments such as Michelson and Morley's.[25]

In 1902, Poincaré called the principle of relative motion as the principle of relativity.[22,23] In 1904, Poincaré gave a talk entitled *L'état actuel et l'avenir de la physique mathématique* to the scientific congress at the Saint Louis World Fair and stated the Principle of Relativity[23,26] as "*The laws of physical phenomena must be the same for a fixed observer and for an observer in rectilinear and uniform motion so that we have no possibility of perceiving whether or not we are dragged in such a motion*". In the same year, *Lorentz*[20] *formulated an approximate transformation theory which satisfied the principle of relativity and agreed with all the experiments to their precision at that time.*

In 1905, *using the principle of relativity, Poincaré*[27] *arrived at the exact invariant transformation (Poincaré called it the Lorentz transformation) and completed the transformation theory of special relativity.* In a subsequent paper, Einstein[28] also arrived at the exact Lorentz transformation and completed the transformation theory of special relativity. Thus, the special theory of relativity was born. For a more complete study of the history of the development of special theory of relativity, we refer the readers to Messager and Letellier's review.[25]

In addition to the transformation theory of special theory of relativity, Einstein[29] *made a cognition advance in postulating and ascertaining the general mass–energy equivalence relation:* $E = mc^2$. For a brief history of the genesis of the mass–energy equivalence relation, we quote Whittaker (pp. 51–52 of Ref. 2):

"We have now to trace the gradual emergence of one of the greatest discoveries of the twentieth century, namely, the connection of mass and energy".

"As we have seen,[1] Thomson in 1881 arrived at the result that a charged spherical conductor moving in a straight line behaves as if it had an additional mass of amount $(4/3c^2)$ times the energy of its electrostatic field.[2] In 1900 Poincaré,[3] referring to the fact that in free aether the electromagnetic momentum is $(1/c^2)$ times the Poynting flux of energy, suggested that electromagnetic energy might possess mass density equal to $(1/c^2)$ times the energy density: that is to say, $E = mc^2$ where E is energy and m is mass: and he remarked that if this were so, then a Hertz oscillator, which sends out electromagnetic energy preponderantly in one direction, should recoil as a gun does when it is fired. In 1904, Hasenöhrl[4] (1874–1915) considered a hollow box with perfectly reflecting walls filled with radiation, and found that when it is in motion there is an (*continued to next page*) apparent addition to its mass, of amount $(8/3c^2)$ times the energy possessed by the radiation when the box is at rest: in the following year[1] he corrected this to $(4/3c^2)$ times the energy possessed by the radiation when the box is at rest[2]; that is, he agreed with Thomson's $E = (3/4)mc^2$ rather than with Poincaré's $E = mc^2$. In 1905, Einstein[3] asserted that when a body is losing energy in the form of radiation its mass is diminished approximately (i.e. neglecting quantities of the fourth-order) by $(1/c^2)$ times the energy lost. He remarked that it is not essential that the energy loss by the body should consist of radiation, and suggested the general conclusion, in agreement with Poincaré, that the mass of a body is a measure of its energy content: if the energy changes by E ergs, the mass changes in the same sense by E/c^2 grams. In the following year he claimed[4] that this law is the necessary and sufficient condition that the law of conservation of motion of the center of gravity should be valid for systems in which electromagnetic as well as mechanical processes are taking place". (We refer the readers to Ref. 2 for footnotes and references in the quotation except noting that the Einstein's two references are Refs. 29 and 30 and that further studies of Fermi (1922), Wilson (1936), von Mosengeil (1907) and Planck (1907) corrected both cases with $E = (3/4)mc^2$ to agree with $E = mc^2$.)

Table 2. Historical steps toward synthesis of a new theory of gravity (post Newtonian theory) before the genesis of special relativity in 1905.

Year	Reference	Historical step
1859	Le Verrier[6]	Discovery of Mercury perihelion advance anomaly
1882	Newcomb[9]	Improved measurement of the Mercury perihelion advance anomaly
1887	Michelson and Morley[12]	Michelson–Morley experiment
1889	Eötvös[14]	Eötvös experiment to test WEP to 10^{-8} level
1864 on	See, e.g. Roseveare[8]	The proposals of various new gravity theories
1854–1900	Riemann,[33] Klein,[35] Ricci and Levi-Civita[40]	The development of differential geometry and tensor analysis
1887–1904	Lorentz,[20] and various authors	Approximate transformation theory of special relativity
1900–1904	Poincaré[21–23]	Principle of relativity
1905	Poincaré,[27] Einstein[28]	Exact transformation theory of special relativity
1905	Einstein[29]	$E = mc^2$ in special relativity

Further developments in special relativity. The development of special relativity continued after 1905. Planck in 1906[31] obtained the relativistic formulas of kinetic energy and momentum of a material particle. Minkowski in 1907 derived the four-dimensional covariant formulation of the Maxwell's equations together with the four-dimensional stress–energy tensor of electromagnetic field.[32] We will address more of these developments relevant to the genesis of GR.

Differential geometry and tensor calculus. In 1854, Riemann[33] founded Riemannian geometry. Metric was the fundamental entity in Riemannian geometry. Christoffel's[34] introduced covariant differentiation. In the 1872 Erlangen program, Klein[35] first gave a generalized definition of geometry and cleared indicated the essential nature of a vector under the group of rotations of orthogonal axes in three-dimensional space. Various authors[36–39] drew attentions to symmetric tensors of rank 2, scalars and tensors of rank 2. From 1887 onwards, Ricci–Curbastro generalized the theory to tensor calculus for transformations in curved space of any dimensions. It became widely known when Ricci (Ricci-Curbastro) and Levi-Civita[40] published their memoir describing it in 1900. These developments greatly facilitated the development of GR.

Table 2 lists important historical steps toward synthesis of a new theory of gravity (post Newtonian theory) agreeing with experiment/observation before the genesis of special relativity in 1905.

2. The Period of Searching for Directions and New Ingredients: 1905–1910

The genesis of GR can be roughly divided into 3 periods: (i) 1905–1910, *the period of searching for directions and ingredients*; (ii) 1911–1914, *the period of various trial theories*; (iii) 1915–1916, *the synthesis and consolidation.* In the prelude we have seen that Newton's gravitation theory needs to be replaced. In this section, we first discuss some ingredients of it followed by searching for directions and new ingredients towards genesis of a new gravitation theory.

Ingredients of Newton's theory. Newton's theory of gravity is an inverse law with active gravitational mass proportional to passive gravitational mass and active gravitational mass also proportional to inertial mass. With appropriate choice of units, the gravitational force $F_{1\to 2}$ acting on body 2 from body 1 can be written in the form

$$\mathbf{F}_{1\to 2} = G_{\mathrm{N}} m_{\mathrm{a}1} m_{\mathrm{p}2} \frac{\mathbf{n}_{1\leftarrow 2}}{r_{12}} = m_{\mathrm{iner}2} \mathbf{a}_2, \tag{1}$$

where $m_{\mathrm{a}1}$ is the active gravitational mass of body 1, $m_{\mathrm{p}2}$ the passive gravitational mass of body 2, $\mathbf{n}_{1\leftarrow 2}$ the unit vector from body 2 to body 1, r_{12} the distance between body 1 and body 2, $m_{\mathrm{iner}2}$ the inertial mass of body 2, $\mathbf{a}_2$ the acceleration of body 2 and G_{N} the universal Newton constant. The Galileo weak equivalence principle dictates the equality of passive gravitational mass and the inertial mass, i.e. $m_{\mathrm{p}} = m_{\mathrm{iner}} \equiv m$ while Newton's third law of motion dictates the equality of passive gravitational mass and the active gravitational mass, i.e. $m_{\mathrm{p}} = m_{\mathrm{a}} = m$. Hence, (1) becomes

$$F_{1\to 2} = G_{\mathrm{N}} m_1 m_2 \frac{n_{1\leftarrow 2}}{r_{12}} = m_2 \mathbf{a}_2. \tag{2}$$

The action is instant. In Newton's original form the theory is an action-at-a-distance theory. In potential theory form, the gravitational potential $\Phi(\mathbf{x}, t)$ for a mass distribution $\rho(\mathbf{x}, t)$ satisfies the Poison equation:

$$\nabla^2 \Phi(\mathbf{x}, t) = 4\pi G_{\mathrm{N}} \rho(\mathbf{x}, t). \tag{3}$$

The left-hand side of (3) depends on the gravitational field while the right-hand side depends on the gravitating source. In the field approach, to reach a new theory of gravity we may need to replace both the left-hand side and right-hand side.

Finite velocity of propagation. It is natural for Poincaré who reached the exact transformation theory in agreement with the principle of relativity to also think about how to reconcile gravity. Poincaré[27,41] pointed out that for principle of relativity to be true, gravity must be propagated with speed of light, and mentioned gravitational-wave propagating with the speed of light based on Lorentz invariance. He attempted to formulate an action-at-a-distance theory of gravity with finite propagation velocity compatible with principle of relativity, but was unsuccessful.

All energy must gravitate. As we mentioned in the last section, Planck[31] obtained the relativistic formulas of kinetic energy and momentum of a material particle in 1906. Since energy is equivalent to mass and has inertia, it must gravitate according to the equivalence of the inertia mass and the gravitational mass which was verified to great precision by Eötvös experiment. Hence, Planck[42] postulated that all energy must gravitate in 1907 and made another step toward a new theory of gravity.

EEP. Einstein,[43] in the last part (Principle of Relativity and Gravitation) of his Comprehensive 1907 essay on relativity, proposed the complete physical equivalence of a homogeneous gravitational field to a uniformly accelerated reference system: "We consider two systems of motion, Σ_1 and Σ_2. Suppose Σ_1 is accelerated in the direction of its X-axis, and γ is the magnitude (constant in time) of this acceleration. Suppose Σ_2 is at rest, but situated in a homogeneous gravitational field, which imparts to all objects an acceleration $-\gamma$ in the direction of the X-axis. As far as we know, the physical laws with respect to Σ_1 do not differ from those with respect to Σ_2, this derives from the fact that all bodies are accelerated alike in the gravitational field. We have therefore no reason to suppose in the present state of our experience that the systems Σ_1 and Σ_2 differ in any way, and will therefore assume in what follows the complete physical equivalence of the gravitational field and the corresponding acceleration of the reference system".

From this equivalence, Einstein derived clock and energy redshifts in a gravitational field. The reasoning is clear and simple: two observers at different location of the uniform gravitational field can be equivalently considered in an accelerated frame. In the equivalent accelerated frame there are Doppler shift. This gives redshift/blueshift in the gravitational field. When applied to a spacetime region where inhomogeneities of the gravitational field can be neglected, this equivalence dictates the behavior of matter in gravitational field. The postulate of this equivalence is called the EEP. EEP is the cornerstone of the gravitational coupling of matter and nongravitational fields in GR and in metric theories of gravity. EEP fixes local physics to be special relativistic.

Local physics in Newtonian gravity also observed this equivalence principle formally except here the local physics is Newtonian mechanics, not special relativity (Here the transformation to the accelerated frame is through a non-Galilean transformation. See, e.g. Ref. 19 and references therein for details.).

Four-dimensional spacetime formulation and the Minkowski metric. On 21 December 1907, Minkowski read before the Academy "Die Grundgleichungen für die elektromagnetischen Vorgänge in bewegten Körpern" (The fundamental equations for electromagnetic processes in Moving bodies)[32] (See also Ref. 44). In this paper, Minkowski put Maxwell equations into geometric form in four-dimensional spacetime with Lorentz covariance using Cartesian coordinates x, y, z and imaginary time it and numbering them as $x_1 \equiv x, x_2 \equiv y$, $x_3 \equiv z$ and $x_4 \equiv it$. Minkowski defined the four-dimensional excitation in terms of D and H, and the four-dimensional field strength in terms of E and B.

Maxwell equations in Minkowski form was soon written in integral form by Hargreaves[45] and devoted a detailed investigation by Bateman[46] and Kottler.[47]

In 1909, Bateman[46] worked on the electrodynamic equations. He used time coordinate t instead of x_4, and studied integral equations and the invariant transformation groups. He considered specifically transformations that leave the invariance of the differential (form) equation:

$$(dx)^2 + (dy)^2 + (dz)^2 - (dt)^2 = 0 \tag{4}$$

and included conformal transformations in addition to Lorentz transformations, therefore he went one step forward toward general coordinate invariance. He did use more general (indefinite) metric from coordinate transformations in his study of electromagnetic equation.

With the definition $x^1 \equiv x$, $x^2 \equiv y$, $x^3 \equiv z$ and $x^0 \equiv t$, Eq. (4) can be written as

$$(dx^1)^2 + (dx^2)^2 + (dx^3)^2 - (dx^0)^2 = -\eta_{ij} dx^i dx^j = 0, \tag{5}$$

where the Minkowski metric η_{ij} is defined as

$$\eta_{kl} = \begin{pmatrix} 1 & 0 & 0 & 0 \\ 0 & -1 & 0 & 0 \\ 0 & 0 & -1 & 0 \\ 0 & 0 & 0 & -1 \end{pmatrix}, \tag{6a}$$

with its inverse η^{kl}

$$\eta^{kl} = \begin{pmatrix} 1 & 0 & 0 & 0 \\ 0 & -1 & 0 & 0 \\ 0 & 0 & -1 & 0 \\ 0 & 0 & 0 & -1 \end{pmatrix}. \tag{6b}$$

In (5) and this paper, we use Einstein convention of summing over repeated indices. Minkowski metric is used in raising and lowering covariant and contravariant indices in special relativity.

With indefinite metric, one has to distinguish covariant and contravariant tensors and indices. Aware of this, one can readily put Maxwell equations into covariant form without using imaginary time. Following Minkowski[32] but using real time coordinate, in terms of Minkowski four-dimensional field strength F_{kl} (E, B) and four-dimensional excitation (density) H^{ij} (D, H)

$$F_{kl} = \begin{pmatrix} 0 & E_1 & E_2 & E_3 \\ -E_1 & 0 & -B_3 & B_2 \\ -E_2 & B_3 & 0 & -B_1 \\ -E_3 & -B_2 & B_1 & 0 \end{pmatrix}, \tag{7a}$$

$$H^{ij} = \begin{pmatrix} 0 & -D_1 & -D_2 & -D_3 \\ D_1 & 0 & -H_3 & H_2 \\ D_2 & H_3 & 0 & -H_1 \\ D_3 & -H_2 & H_1 & 0 \end{pmatrix}. \tag{7b}$$

Maxwell equations can be expressed in Minkowski form as

$$H^{ij}{}_{,j} = -4\pi J^i, \tag{8a}$$

$$e^{ijkl} F_{jk,l} = 0, \tag{8b}$$

where J^k is the charge 4-current density (ρ, J) and e^{ijkl} the completely anti-symmetric tensor density (Levi-Civita symbol) with $e^{0123} = 1$. "," means partial derivation. In vacuum, the relation of Minkowski four-dimensional field strength F_{kl} (E, B) and four-dimensional excitation (density) H^{ij} (D, H) is

$$H^{ij} = \eta^{ik}\eta^{jl}F_{kl} = \frac{1}{2}(\eta^{ik}\eta^{jl} - \eta^{il}\eta^{jk})F_{kl}, \quad \text{i.e. } H^{ij} = F^{ij}. \tag{9}$$

Four-dimensional electromagnetic stress–momentum–energy tensor. In the same paper, Minkowski[32] derived the four-dimensional electromagnetic stress–momentum–energy (or stress–energy or energy–momentum) tensor $T^{(\mathrm{EM})}{}_i{}^j$ of rank 2:

$$T^{(\mathrm{EM})}{}_i{}^j = \left(\frac{1}{16\pi}\right)\delta_i{}^j F_{kl}F^{kl} - \left(\frac{1}{4\pi}\right)F_{il}F^{jl}, \tag{10}$$

with

$$T^{(\mathrm{EM})}{}_0{}^0 = \left(\frac{1}{8\pi}\right)[(E)^2 + (B)^2], \tag{11}$$

the electromagnetic energy density discovered by W. Thomson (Kelvin) in 1853;

$$T^{(\mathrm{EM})}{}_0{}^\mu = -\left(\frac{1}{4\pi}\right)F_{0\nu}F^{\mu\nu} = \left(\frac{1}{4\pi}\right)(\boldsymbol{E} \times \boldsymbol{B})^\mu, \tag{12}$$

$(1/c)$ times the electromagnetic energy flux discovered by Poynting and Heaviside in 1884;

$$T^{(\mathrm{EM})}{}_\mu{}^0 = -\left(\frac{1}{4\pi}\right)F_{\mu\nu}F^{0\nu} = \left(\frac{1}{4\pi}\right)(\boldsymbol{E} \times \boldsymbol{B})_\mu \left(= -\left(\frac{1}{4\pi}\right)(\boldsymbol{E} \times \boldsymbol{B})^\mu\right), \tag{13}$$

$(-c)$ times the electromagnetic momentum density discovered by J. J. Thomson in 1893;

$$\begin{aligned} T^{(\mathrm{EM})}{}_\mu{}^\nu &= \left(\frac{1}{16\pi}\right)\delta_\mu{}^\nu F_{kl}F^{kl} - \left(\frac{1}{4\pi}\right)F_{\mu\alpha}F^{\nu\alpha} \\ &= \left(\frac{1}{8\pi}\right)\{\delta_\mu{}^\nu[(E)^2 + (B)^2] - \eta_{\mu\alpha}[(E)^\alpha(B)^\nu + (B)^\alpha(E)^\nu]\}, \end{aligned} \tag{14}$$

the electromagnetic stress discovered by Maxwell in 1873. Here we use Greek indices to run from 1 to 3.

The importance of constructing the four-dimensional electromagnetic stress–momentum–energy tensor is that it was the first four-dimensional stress–momentum–energy tensor ever constructed. For electromagnetic energy to gravitate, it should enter the right-hand side of the new covariant (3). However, electromagnetic energy is only the (0, 0) component of four-dimensional stress–momentum–energy tensor, other components should enter the right-hand also to make it covariant.

Directions and new ingredients. During 1905–1910, directions and new ingredients were formed for a new theory of gravity. We had finite propagation velocity, all energy gravitating, EEP, spacetime formulation of special relativity, indefinite metric, and four-dimensional covariant electromagnetic stress–momentum–energy (stress–energy) tensor. Two crucial steps are (i) the generalization of the principle of relativity to include situation in gravity, i.e. the EEP; and (ii) the spacetime formulation of the (special) relativity theory using Minkowski metric and its generalization to the general concept of indefinite spacetime metric. EEP means the local physics is special relativistic. Then one can ask what is gravity. It must be how various local physics are connected. We have special relativity from locality to locality and gravity describes how they are connected. (A mathematical natural description is a four-dimensional base manifold with special relativity as fibre attached to each (world) point in the base manifold, and gravity is the connection bundle or the metric which induces the connection bundle.) Although this logic seems compelling, the full metric as dynamical gravitational entity was not used until 1913. A test of EEP was derived by Einstein: the gravitational redshift. It has been an important test of relativistic gravity which people try to improve the accuracy constantly.

3. The Period of Various Trial Theories: 1911–1914

Basic formulas of (pseudo-)Riemannian geometry. Here we summarize some basic formulas used in developing a new theory of gravity for straightening out the convention and notation. First, a (pseudo)-Riemannian manifold is endowed with a metric g_{ij}. The metric g_{ij} is related to the line element ds as:

$$ds^2 = g_{ij} dx^i dx^j. \tag{15}$$

If the metric g_{ij} is positive definite, the geometry is Riemannian. If the metric g_{ij} is indefinite, the geometry is pseudo-Riemannian. g^{ij} is the matrix inverse of g_{ij} and they are used to raise and lower covariant and contravariant indices. For our case, the geometry is pseudo-Riemannian. We use the MTW[48] conventions with signature -2; this is also the convention used in Ref. 49. Latin indices run from 0 to 3; Greek indices run from 1 to 3. The Christoffel connection $\Gamma^i{}_{jk}$ of the metric is given by

$$\Gamma^i{}_{jk} = \frac{1}{2} g^{il} (g_{lj},_k + g_{lk},_j - g_{jk},_l). \tag{16}$$

With Christoffel connection, one can define covariant derivative. The Riemannian curvature tensor $R^i{}_{jkl}$, the Ricci curvature tensor R_{jl}, the scalar curvature R and the Einstein tensor G_{jl} are defined as:

$$R^i{}_{jkl} = \Gamma^i{}_{jl},_k - \Gamma^i{}_{jk},_l + \Gamma^i{}_{km} \Gamma^m{}_{lj} - \Gamma^i{}_{lm} \Gamma^m{}_{jk}; \quad R_{jl} = R^i{}_{jil};$$
$$R = g^{jl} R_{jl}; \quad G_{jl} = R_{jl} - \left(\frac{1}{2}\right) g_{jl} R. \tag{17}$$

Gravitational deflection of light and EEP. Extending his work on gravitational redshift, Einstein[50] derived light deflection in gravitational field using EEP in 1911. He argued that since light is a form of energy, light must gravitate and the velocity of light must depend on the gravitational potential. He obtained that light passing through the limb of the Sun would be gravitationally deflected by 0.83 arc s. This is very close to the value 0.84 arc sec derived by Soldner[51] in 1801 assuming that light is corpuscular in Newtonian theory of gravitation. This prediction was half the value of GR. Before 1919, there were four expeditions intent to measure the gravitational deflection of starlight (in 1912, 1914, 1916 and 1918); because of bad weather or war, the first three expeditions failed to obtain any results, the results of 1918 expedition was never published.[52] In 1919, the observation of gravitational deflection of light passing near the Sun during a solar eclipse[53] confirmed the relativistic deflection of light and made GR famous and popular.

Stress-energy tensor. In 1911, Laue[54,55] extended Minkowski's method of constructing electromagnetic stress-energy tensor to stressed bodies, dust and relativistic fluids.

Gravity theories with 'variable velocity of light' and scalar theories of gravity. Accepting that the velocity of light depends on gravitational potential, Abraham[56] postulated that the negative gradient indicates the direction of gravitational force and worked out a theory of gravity. Einstein[57] worked out a somewhat different theory. These gravity theories with 'variable velocity of light' led to the proposals of conformally flat scalar theories of Nordström.[58–60]

The equation corresponding to Eq. (3) in Newtonian theory for electromagnetism is

$$\left(\frac{1}{c^2}\right)\left(\frac{\partial^2 A_i}{\partial t^2}\right) - \nabla^2 A_i = A_{i,j}{}^{j} = 4\pi J_i, \tag{18}$$

with gauge condition

$$A_{i,}{}^{i} = 0. \tag{19}$$

Here A_i is the electromagnetic 4-potential guaranteed locally by (8b) such that $F_{ij} = A_{j,i} - A_{i,j}$. To incorporate the finite propagation speed with light velocity into the gravitation field equation, one could just replace (3) with

$$\left(\frac{1}{c^2}\right)\left(\frac{\partial^2 \Phi^*}{\partial t^2}\right) - \nabla^2 \Phi^* = \eta^{ij}\Phi^*{}_{,ij} = -4\pi G_{\mathrm{N}}\rho^*(x,t), \tag{20}$$

where $\Phi^*(x,t)$ is a new gravitational field entity. Φ^* could be a scalar field, a vector field or a tensor field or some combination of them. If Φ^* is a scalar field, ρ^* must be a scalar; in the weak field and slow motion limit, one must be able to approximate Φ^* and ρ^* by Φ and ρ. Let us illustrate with Einstein's theory with 'variable velocity of light'.

In the original formulation of Einstein,[57] the equation of motion for particles was derived from the variational principle

$$\delta \int ds = 0, \tag{21}$$

where

$$ds^2 = c^2 dt^2 - dx^2 - dy^2 - dz^2 \tag{22}$$

and where c is a scalar function which Einstein regarded as the velocity of light in the metric (22). Einstein postulated that c depends on the scalar field φ in the following way:

$$c^2 = c_0^2 - 2\varphi \tag{23}$$

and that φ is generated by ρ^* through the wave equation

$$\left(\frac{1}{c^2}\right)\left(\frac{\partial^2\varphi}{\partial t^2}\right) - \nabla^2\varphi(\mathbf{x},t) = -4\pi G_{\mathrm{N}}\rho^*(\mathbf{x},t). \tag{24}$$

By choosing suitable units, we can set $c_0 = 1$; and by postulating that Einstein's ds^2 is the "physical metric", we can bring the theory into the form:

$$ds^2 = (c_0^2 - 2\varphi)dt^2 - dx^2 - dy^2 - dz^2, \tag{25a}$$

$$\eta^{ij}\varphi_{,ij} = 4\pi G_{\mathrm{N}}\rho^*. \tag{25b}$$

Note that in (25b) as well in (20), we have the *a priori* (nondynamical) geometric element η^{ij} to make the equation fully coordinate covariant. More precisely, Eq. (25a) also contains *a priori* geometric elements — a flat-space metric and a time direction. This makes the theory a stratified theory with conformally flat space slices. For more detailed discussions, see Refs. 61 and 62.

The physical metric can always be transformed locally into the Lorentz form

$$ds^2 = c_0^2 d\underline{t}^2 - d\underline{x}^2 - d\underline{y}^2 - d\underline{z}^2, \tag{26}$$

where $d\underline{t}$ is the proper time interval and $dl = (d\underline{x}^2 + d\underline{y}^2 + d\underline{z}^2)^{1/2}$, the proper-length element. Since light trajectories all lie on null cones of this metric, the velocity of light as measured using the physical metric is always c_0 — as it must be for any theory that satisfy the EEP.

This theory did not agree with the Mercury perihelion advance observation. However, it led to the conformally flat theories of Nordström.[58–60]

The field equations of Nordström's second theory[59,60] can be written as

$$C_{ijkl} = 0, \tag{27}$$

$$R = 24\pi\left(\frac{G_{\mathrm{N}}}{c^4}\right)T, \tag{28}$$

where C_{ijkl} is the Weyl conformal tensor and R is the curvature scalar both constructed from the metric g_{ij}. T is the trace contraction of stress–energy tensor. The field equations (27) and (28) are geometric and make no reference to any gravitational fields except the physical metric g_{ij}. However, they guarantee the existence of a flat spacetime metric η_{ij} (prior geometry in the language of Ref. 48) and a scalar field related to g_{ij} by

$$g_{ij} = \varphi^2\eta^{ij}; \tag{29}$$

and they allow φ to be calculated from the variational principle

$$\delta\int\left[L_{\mathrm{I}} - \left(\frac{1}{3}\right)R(-g)^{1/2}\right]d^4x = 0, \tag{30}$$

where $g = \det(g_{ij})$ and L_{I} is the interaction Lagrangian density of matter with gravity (see Ref. 61 for more details). Expressed in terms of φ, the field equation (28) becomes

$$\eta^{ij}\varphi_{ij} = -\left(\frac{4\pi G_N}{c^4}\right) T\varphi^3 \tag{31a}$$

or

$$\eta^{ij}\varphi_{i,j}\varphi^{-1} = -\left(\frac{4\pi G_{\mathrm{N}}}{c^4}\right) T_{\mathrm{flat}}. \tag{31b}$$

Equation (31) is Nordström's original field equation,[59,60] while Eq. (28) is the Einstein–Fokker version.[63] This second Nordström theory did not agree with the Mercury perihelion advance observation either. From (14), we note that the trace contraction of electromagnetic energy tensor vanishes. Therefore, in this theory the electromagnetic energy does not contribute to the generation of gravitational field. Neither the gravitational field gives light deflection. Somehow nature did not choose this way. Nature chose to make the whole metric dynamic.

Tensor theory of gravity. In 1913, Einstein and Grossmann turned into tensor theory of gravity making full use of the metric. They tried to incorporate all the ingredients discussed in the last section into their "Entwurf (outline)" theory[64] and proposed the following equation using the metric g_{ij} as dynamical entity for the gravitational field:

$$\textit{Part of Ricci tensor } R_{ij} \propto T_{ij}. \tag{32}$$

Since the left-hand side did not contain all the terms of the Ricci tensor, it is not covariant. In 1913, Besso and Einstein[65] worked out a Mercury perihelion advance formula in the "Einstein–Grossmann Entwurf" theory,[64] but the calculation contained an error and the result did not agree with the Mercury perihelion advance observation. Nevertheless, the "Entwurf" theory is an important landmark in the genesis of GR.

Einstein became versed at differential geometry and tensor analysis in 1914. A quote of Einstein's October 1914 writing on "The formal foundation of the general theory of relativity"[66] showed the situation: in the abstract "In recent years I have worked, in part with my friend Grossmann, on a generalization of the theory of relativity. During these investigations, a kaleidoscopic mixture of postulates from physics and mathematics has been introduced and used as heuristical tools; as a consequence it is not easy to see through and characterize the theory from a formal point of view, that is, only based upon these papers. The primary objective of the present paper is to close this gap. In particular, it has been possible to obtain the equations of the gravitational field in a purely covariance-theoretical manner (Section D). I also tried to give simple derivations of the basic laws of absolute differential calculus — in part, they are probably new ones (Section B) — in order to allow the reader to get a complete grasp of the theory without having to read other, purely mathematical tracts. As an illustration of the mathematical methods, I derived the Eulerian equations of hydrodynamics and the field equations of the electrodynamics of moving bodies (section C). Section E shows that Newton's theory of gravitation follows from the general theory as an approximation. The mist elementary features of the present theory are also derived inasfar as they are characteristic of a Newtonian (static) gravitational field (curvature of light rays, shift of spectral lines)".

In the 1911–1914 period, various trial theories, based largely on the ingredients and directions set in the previous period 1905–1910, emerged and led step by step towards the synthesis of GR.

4. The Synthesis and Consolidation: 1915–1916

Einstein's big step. Continued along the direction set in the "Entwurf" theory, Einstein[67,68] reached the following equation for GR in 1915:

$$R_{ij} \propto T_{ij}. \tag{33}$$

Subsequently, Einstein[69] corrected an error made in his collaboration with Besso of 1913[65] and obtained a value of Mercury perihelion advance from his new equation (33) in agreement with the observation.[9] Apparently, this correct calculation played a significant role in the final genesis of GR. The divergence of T_{ij} vanishes. However, the divergence of R_{ij} does not vanish unless T vanishes or is constant. Since the trace $T^{(\mathrm{EM})}$ of electromagnetic stress–energy tensor does vanish, Einstein argued that:[68]

"One now has to remember that by our knowledge "matter" is not to be perceived as something primitively given or physically plain. There even are those, and not just a few, who hope to reduce matter to purely electrodynamic processes, which of course would have to be done in a theory more completed than Maxwell's electrodynamics. Now let us just assume that in such completed electrodynamics scalar of the energy tensor also would vanish! Would the result, shown above, prove that matter cannot be constructed in this theory? I think I can answer this question in the negative, because it might very well be that in "matter", to which the previous expression relates, gravitational fields do form an important constituent. In that case, ΣT_μ^μ can appear positive for the entire structure while in reality only $\Sigma(T_\mu^\mu + t_\mu^\mu)$ is positive and ΣT_μ^μ vanishes everywhere. *In the following we assume the conditions* $\Sigma T_\mu^\mu = 0$ *really to be generally true*".

Hilbert variational principle. Shortly after Einstein obtained (33), Hilbert[70] proposed the variational principle for gravitational field:

$$\delta \int [L_{\mathrm{I}}^{(\mathrm{EM})} + L^{(\mathrm{GRAV})}] d^4x = 0, \tag{34}$$

with the Lagrangian densities $L_{\mathrm{I}}^{(\mathrm{EM})}$ and $L^{(\mathrm{GRAV})}$ given by

$$L_{\mathrm{I}}^{(\mathrm{EM})} \propto F_{kl} F^{kl} (-g)^{1/2}, \tag{35a}$$

$$L^{(\mathrm{GRAV})} \propto R(-g)^{1/2}. \tag{35b}$$

The variation of integral of (35a) plus the Lagrangian density term of electromagnetic 4-current interaction with electromagnetic 4-potential gives the Maxwell equations. The variation of the integral of (35a) with respect to the metric gives the electromagnetic stress–energy tensor. The variation of (35b) together with (35a) with respect to the metric would give:

$$R_{ij} - \frac{1}{2} g_{ij} R \propto T_{ij}^{(\mathrm{EM})}. \tag{36a}$$

With more general Lagrangian L_{I} such as Mie's Lagrangian instead of $L_{\mathrm{I}}^{(\mathrm{EM})}$, the variation of its integral and Hilbert's gravitational integral with respect to the metric would give

$$R_{ij} - \frac{1}{2} g_{ij} R \propto T_{ij} \tag{36b}$$

Here T_{ij} is given using Hilbert's variation-with-respect-to-the-metric definition.

Einstein equation. After Hilbert's work,[70] Einstein[71] soon corrected his field equation (33) on 25, November 1915 to

$$R_{ij} \propto T_{ij}^{(\mathrm{EM})} - \frac{1}{2} g_{ij} T. \tag{37}$$

Equation (36a), or Equation (36b) with $T_{ij}^{(\mathrm{EM})}$ replaced by T_{ij}, i.e.

$$R_{ij} - \frac{1}{2} g_{ij} R \propto T_{ij} \tag{38}$$

and Eq. (37) are equivalent to each other: by taking a trace contraction of either equation, one has

$$R \propto -T \tag{39}$$

and the equivalence becomes clear. Since variation principle became common, the Einstein equation is normally written in the form (38) nowadays. With the proportional constant inserted, the Einstein equation is

$$G_{ij} = R_{ij} - \frac{1}{2} g_{ij} R = 8\pi \left(\frac{G_{\mathrm{N}}}{c^4} \right) T_{ij}. \tag{40}$$

In 1916, Einstein[72] wrote a foundational paper on GR. In the same year, Einstein performed a linear approximation in the weak field and obtained the quadrupole radiation formula[73]; major errors were corrected in his 1918 paper[74] while a factor of 2 was corrected by Eddington.[75] (For later controversial issues on gravitational wave and the quadrupole formula, see e.g. Refs. 76 and 77.) Einstein thought that quantum effects must modify GR in his first paper on linear approximation and gravitational waves[73] although he switched to a different point of view working on the unification of electromagnetism and gravitation in the 1930s. The merging of GR and quantum theory is an important issue. For a brief history of ideas and prospects, see e.g. Ref. 78.

In 1916, Schwarzschild discovered an exact spherical solution (Schwarzschild solution) of Einstein equation.[79,80]

In 1917, Einstein[81] postulated the cosmological principle, applied GR to cosmology and proposed the cosmological constant; de Sitter[82–85] followed in the same

Table 3. Historical steps in the genesis of GR since the genesis of special relativity.

Year	Reference	Historical step
1905–1906	Poincaré[27,41]	Attempt to formulate an action at a distance theory of gravity with finite propagation velocity compatible with principle of relativity
1907	Planck[42]	All energy must gravitate
1907–1908	Einstein[43]	Generalized principle of relativity (EEP) and the prediction of gravitational redshift
1907–1908	Minkowski[32,44]	Covariant spacetime formulation of electromagnetism and the derivation of four-dimensional electromagnetic stress–energy tensor
1909–1910	Bateman[46]	Introducing indefinite spacetime metric
1911	Einstein[50]	Using EEP to derive deflection of light in gravitational field
1911	Laue[54,55]	Stress–energy tensor of matter
1911–1912	Abraham,[56] Einstein[57]	Theories with 'variable velocity of light'
1912	Nordström[58]	Norström's first theory
1913	Nordstrom[59,60]	Norström's second theory
1913	Einstein and Grossman[64]	"Entwurf (Outline)" theory
1914	Einstein and Fokker[63]	Covariant formulation of Norström's second theory
1914	Einstein[66]	Einstein versed at covariant formulation
1915	Einstein[67–69]	Source restricted Einstein equation
1915	Hilbert[70]	Hilbert variational principle
1915	Einstein[71]	Einstein equation
1916	Einstein[72]	Einstein's foundation paper of general relativity
1916	Einstein[73]	Approximate solution and gravitational waves
1916	Schwarzschild[78]	Exact spherical solutions of Einstein equation
1917	Einstein[81]	Cosmological principle, cosmology and cosmological constant
1917	de Sitter[82–85]	de Sitter inflationary solution (cosmology)
1918	Einstein[74,75]	Quadrapole radiation formula
1918	Lense–Thirring[93]	Lense–Thirring gravitomagnetic effect

year with an inflationary solution to cosmology. Ever since the genesis of GR, it went hand-in-hand with the development of cosmology. For a brief history of this connection and mutual development, see e.g. Ref. 86; for recent reviews on various topics in cosmology, see e.g. Refs. 87–92.

In 1918, Lense and Thirring[93] discovered the frame-dragging effect in GR.

After one hundred years of developments, GR becomes indispensable in precision measurement, astrophysics, cosmology and theoretical physics. The first direct detections of gravitational waves[94,95] in the centennial of the genesis of GR truly celebrate this occasion.

The route to GR is indeed guided by covariance. However, when GR is reached and covariance is fully understood, the principle of covariance could accommodate various things, scalars, vectors, *a priori* objects, etc. and various theories of gravity. It is probably a minimax principle that worked in nature: When an entity is needed, it should saturate its maximal capacity.

In Table 3, we list historical steps in the genesis of GR discussed in the last section and this section.

5. Epilogue

The study of histories gives inspiration. The study of the genesis of GR is clearly so. It is fortunate that most of the records are intact and the step-by-step development are transparent. We hope that this short exposition presents the flavor and some insights of the development. The genesis of GR was a community effort with Einstein clearly dominated the scene. Knowledge accrues gradually most of the time. Cognition sometimes comes in somewhat bigger steps. The cognition that energy must gravitates and that the EEP must be valid are such examples. Initial cognition needs consolidation and development. EEP indicates that local physics must be special relativistic. Minkowski's spacetime formulation indicates that local physics must have an indefinite metric. To take metric as basic entity for gravitation took a few years through studying gravitational redshift, gravitational light deflection, theories of gravity with "variable velocity of light" and more scalar theories of gravity. Eventually, metric as a full dynamic entity for gravitation emerged in 1913. It took a couple of years to master this approach for leading to the genesis of GR. During different phases of genesis, the first discovered general relativistic effect — the Mercury perihelion advance anomaly played a key role.

Acknowledgments

I would like to thank Science and Technology Commission of Shanghai Municipality (STCSM-14140502500) and Ministry of Science and Technology of China (MOST-2013YQ150829, MOST-2016YFF0101900) for supporting this work in part.

References

1. I. Newton, *Philosophiae Naturalis Principia Mathematica* (Streater, London, 1687).
2. E. Whittaker, *A History of the Theories of Aether and Electricity II. The Modern Theories* (Philosophical Library, 1954; American Institute of Physics, 1987).
3. J. Kepler, *Astronomia Nova de Motibus Stellae Martis* (Prague, 1609); *Harmonice Mundi* (Linz, 1619).
4. C. Wilson, The great inequality of Jupiter and Saturn from Kepler to Laplace, Ph. D. thesis, St. John's College, Annapolis, Maryland, January 10, 1984, www.docin.com/p-1627154943.html.
5. P. Moore, *The Story of Astronomy*, 5th rev. ed. (Grosset & Dunlap Publishing, New York, 1977).
6. U. J. J. Le Verrier, Theorie du mouvement de Mercure, *Ann. Observ. imp. Paris (Mém.)* **5** (1859) 1.
7. U. J. J. Le Verrier, Nouvelles recherches sur les mouvements des planètes, *C. R. Acad. Sci. Paris* **29** (1849); the English translation of the quote in the text is from Ref. 8.
8. N. T. Roseveare, *Mercury's Perihelion from Le Verrier to Einstein* (Clarendon Press, Oxford, 1982).
9. S. Newcomb, Discussion and results of observations on transits of Mercury from 1677 to 1881, *Astr. Pap. Am. Ephem. Naut. Alm.* **1** (1882) 367.
10. I. I. Shapiro, Solar system tests of GR: Recent results and present plans in *General Relativity and Gravitation: Proc. 12th Int. Conf. General Relativity and Gravitation,*

University of Colorado at Boulder, July 2–8, 1989, eds. N. Ashby, D. F. Bartlett and W. Wyss (Cambridge, Cambridge University Press, 1990), pp. 313–330.
11. W.-T. Ni, Solar-system tests of the relativistic gravity, in *One Hundred Years of General Relativity: From Genesis and Foundations to Gravitational Waves, Cosmology and Quantum Gravity*, Chap. 8, eds. by W.-T. Ni (World Scientific, Singapore, 2016); *Int. J. Mod. Phys. D* **25** (2016) 1630003.
12. A. A. Michelson and E. W. Morley, On the relative motion of the Earth and the luminiferous ether, *Am. J. Sci.* **34** (1887) 333.
13. M. Nagel, S. R. Parker, E. V. Kovalchuk, P. L. Stanwix, J. G. Hartnett, E. N. Ivanov, A. Peters and M. E. Tobar, Direct terrestrial test of Lorentz symmetry in electrodynamics to 10^{-18}, *Nat. Commun.* **6** (2015) 8174.
14. R. V. Eötvös, *Math. Naturwiss. Ber. Ungarn* **8** (1889) 65.
15. G. Galilei, *Discorsi e Dimostrazioni Matematiche Intorno a due Muove Scienze* (Elzevir, Leiden, 1638).
16. T. A. Wagner, S. Schlamminger, J. H. Gundlach and E. G. Adelberger, Torsion-balance tests of the weak equivalence principle, *Class. Quantum Grav.* **29** (2012) 184002.
17. P. Touboul, M. Rodrigues, G. Métris and B. Tatry, MICROSCOPE, testing the equivalence principle in space, *C. R. Acad. Sci. Ser.* IV **2**(9) (2001) 1271.
18. MICROSCOPE Collaboration, CNES ONERA cooperation first ultra-precise measurements from Microscope (September 27, 2016) https://presse.cnes.fr/en/cnes-onera-cooperation-first-ultra-precise-measurements-microscope.
19. W.-T. Ni, Equivalence principles, spacetime structure and the cosmic connection, in *One Hundred Years of General Relativity: From Genesis and Foundations to Gravitational Waves, Cosmology and Quantum Gravity*, Chap. 5, ed. W.-T. Ni (World Scientific, Singapore, 2016); *Int. J. Mod. Phys. D* **25** (2016) 1630002.
20. H. A. Lorentz, *Kon. Neder. Akad. Wet. Amsterdam. Versl. Gewone Vergad. Wisen Natuurkd. Afd.* **6** (1904) 809.
21. H. Poincaré, *Bibl. Congr. Int. Philos.* **3** (1900) 457.
22. H. Poincaré, *La Science et l'Hypothése* (Flammarion, Paris, 1902) (in German); *Science and Method* (Dover Publications Inc., London, 1952).
23. H. Poincaré, L'état et l'avenir de la physique mathematique, *Bull. Sci. Math.* **28** (1904) 302; the English translation in the text is from Ref. 26.
24. H. Poincaré, *Electricité et Optique: La Lumiére et les Théories Électrodynamiques* (Gauthier-Villars, Paris, 1901).
25. V. Messager and C. Letellier, A genesis of special relativity, in *One Hundred Years of General Relativity: From Genesis and Foundations to Gravitational Waves, Cosmology and Quantum Gravity*, Chap. 1, ed. W.-T. Ni (World Scientific, Singapore, 2016); *Int. J. Mod. Phys. D* **25** (2015) 1530024.
26. C. Marchal, *Sciences* **97** (1997) 2.
27. H. Poincaré, Sur la dynamique de l'electron, *C. R. Acad. Sci.* **140** (1905) 1504.
28. A. Einstein, Zur elektrodynamik bewegter Körper [On the Electrodynamics of Moving Bodies], *Ann. Phys.* **17** (1905) 891.
29. A. Einstein, Ist die Trägheit eines Körpers von seinem Energieinhalt abhängig? *Ann. Phys.* **18** (1905) 639.
30. A. Einstein, Das Prinzip von der Erhaltung der Schwerpunktsbewegung und die Trägheit der Energie, *Ann. Phys.* **20** (1906) 627.
31. M. Planck, *Verh. Dtsch. Phys. Ges.* **8** (1906) 136.
32. H. Minkowski, Die Grundgleichungen für die elektromagnetischen Vorgänge in bewegten Körpern, *Königliche Gesellschaft der Wissenschaften zu Gttingen.*

Mathematisch-Physikalische Klasse. Nachrichten, pp. 53–111 (1908); this paper was read before the Academy on 21 December 1907; (English translation) The fundamental equations for electromagnetic processes in Moving bodies, translated from German by Meghnad Saha and Wikisource, Available at: http://en.wikisource.org/wiki/Translation:The_Fundamental_Equations_for_Electro.

33. B. Riemann, Über die Hypothesen welche der Geometrie zu Grunde liegen (On the hypotheses which underlie geometry), lecture at Göttingen (1854).
34. E. B. Christoffel, *J. Math.* **70** (1869) 241.
35. F. Klein, *Programm zum Eintritt in die Philosophische Fakultät der Universität. zu Erlangen*, Erlangen, A. Deichert (1872); Reprinted in 1893 in *Math. Ann.* **63**, and in Klein's *Ges. Math. Abhandl. I*, 460.
36. C. Niven, *Trans. R. Soc. Edinburgh* **27** (1874) 473.
37. W. Thomson, *Philos. Trans.* **146** (1856) 481.
38. W. J. M. Rankine, *Philos. Trans.* **146** (1856) 261.
39. J. W. Gibbs, *Vector Analysis* (New Haven, 1881–1884), p. 57.
40. G. Ricci and T. Levi-Civita, Méthodes de calcul différentiel absolu et leurs applications, *Math. Ann.* **54** (1900) 125.
41. H. Poincaré, Sur la dynamique de l'electron, *Rend. Circ. Mat. Palermo* **21** (1906) 129.
42. M. Planck, Zur dynamik bewegter systeme, *Sitz. K.-Preuss. Akad. Wiss.* (1907) 542 (Specially at p. 544) (in German); On the dynamics of moving systems (1907) (in German); M. Planck, translated from German by Wikisource, Availabel at: https://en.wikisource.org/wiki/Translation:On_the_Dynamics_of_Moving_Systems.
43. A. Einstein, Über das Relativitätprinzip und die aus demselben gezogenen Folgerungen, *Jahrb. Radioakt. Elektron.* **4** (1907) 411 (in German); Corrections by Einstein in *Jahrb. Radioakt. Elektronik* **5** (1908) 98; English translations by H. M. Schwartz in *Am. J. Phys.* **45** (1977) 811.
44. H. Minkowski, *Raum und Zeit, Phys. Z.* **10** (1909) 104 (in German); H. A. Lorentz, A. Einstein, H. Minkowski and H. Weyl, *The Principle of Relativity*, translated by W. Perrett and G. B. Jeffery (Dover, New York, 1952).
45. R. Hargreaves, Integral forms and their connection with physical equations, *Camb. Philos. Trans.* **21** (1908) 107.
46. H. Bateman, The transformation of the electrodynamical equations, *Proc. Camb. Math. Soc., Ser. 2* **8** (1910) 223.
47. F. Kottler, Über die Raumzeitlinien der Minkowski'schen Weit, *Sitz. Akad. Wiss. Wien, Math.-Naturw. Kl. Abt. IIa* **121** (1912) 1688.
48. C. W. Misner, K. S. Thorne and J. A. Wheeler, *Gravitation* (Freeman, San Francisco, 1973).
49. K. Kuroda, W.-T. Ni and W.-P. Pan, Gravitational waves: Classification, methods of detection, sensitivities, and sources, in *One Hundred Years of General Relativity: From Genesis and Empirical Foundations to Gravitational Waves, Cosmology and Quantum Gravity*, Chap. 10, ed. W.-T. Ni (World Scientific, Singapore, 2016); *Int. J. Mod. Phys. D* **24** (2015) 1530031.
50. A. Einstein, Über den einfluß der Schwerkraft auf die Ausbreitung des Lichtes, *Ann. Phys.* **35** (1911) 898; translated "On the Influence of Gravitation on the Propagation of Light" in *The collected papers of Albert Einstein*, Vol. 3: *The Swiss years: writings*, 1909–1911 (Princeton University Press, Princeton, New Jersey, 1994), Anna Beck translator.
51. J. G. von Soldner, On the deviation of a light ray from its motion along a straight line through the attraction of a celestial body which it passes close by, *Astronomishes Jahrbuch für das Jahr 1804* (C. F. E. Späthen, Berlin, 1801), pp. 161–172.

52. J. Earman and C. Glymour, Relativity and eclipses: The British eclipse expeditions of 1919 and their predecessors, in *Historical Studies in the Physical Sciences*, Vol. 11 (1980), pp. 49–88.
53. F. Dyson, A. Eddington and C. Davidson, A determination of the deflection of light by the Sun's gravitational field, from observations made at the total eclipse of May 29, 1919, *Philos. Trans. R. Soc.* **220A** (1920) 291.
54. M. von Laue, Zur Dynamik der Relativitätstheorie, *Annalen der Physik* **35** (1911) 524–542.
55. M. von Laue, *Das Relativitätsprinzip* (Friedrich Vieweg und Sohn, 1911).
56. M. Abraham, *Lincei Atti* **20** (1911) 678.
57. A. Einstein, *Ann. Phys.* **38** (1912) 355, 443.
58. G. Nordström, Relativitätsprinzip und Gravitation, *Phys. Zeit.* **13** (1912) 1126–1129.
59. G. Nordström, Zur Theorie der Gravitation vom Standpunkt der Relativitätsprinzip, *Ann. Phys.* **42** (1913) 533.
60. G. Nordström, Die Fallgesetze und Planeten bewegung in der Relativitätstheorie, *Ann. Phys.* **43** (1914) 1101.
61. W.-T. Ni, Theoretical frameworks for testing relativistic gravity, IV: A compendium of metric theories of gravity and their post-Newtonian limits, *Astrophys. J.* **176** (1972) 769.
62. G. J. Whitrow and G. E. Morduch, Relativistic theories of gravitation, in *Vistas in Astronomy*, ed. A. Beer, Vol. 6 (Pergamon Press, Oxford, 1965), pp. 1–67.
63. A. Einstein and A. D. Fokker, *Ann. Phys.* **44** (1914) 321.
64. A. Einstein and M. Grossmann, Entwurf einer verallgemeinerten Relativitätstheorie und einer Theorie der Gravitation [Outline of a Generalized Theory of Relativity and of a Theory of Gravitation], *Z. Math. Phys.* **62** (1913) 225.
65. A. Einstein and M. Besso, Manuscript on the motion of the perihelion of Mercury (1913) 360, (in German), in *The Collected Papers of Albert Einstein, Vol.* 4: *The Swiss Years*: *Writings, 1912–1914 Albert Einstein*, eds. M. J. Klein, A. J. Kox, J. Renn and R. Schulmann (Princeton University Press, 1995), see also the editorial note on p. 344; available at http://press.princeton.edu/einstein/digital/.
66. A. Einstein, Die formale Grundlage der allgemeinen Relativitätstheorie, *Sitz. ber. Akad. Wiss.* **1914** (1914) 1030 (in German) *The Collected Papers of Albert Einstein, Vol.* 6: *The Berlin Years*: *Writings, 1914–1917 Albert Einstein* (English translation supplement), eds. by M. J. Klein, A. J. Kox, J. Renn and R. Schulmann (Princeton University Press, 1995), pp. 30–84; available at: http://press.princeton.edu/einstein/digital/.
67. A. Einstein, Zür allgemeinen Relativitätstheorie, *Sitzber. Preuss. Akad. Wiss.* **1915** (1915) 778 (in German); in *The Collected Papers of Albert Einstein, Vol.* 6: *The Berlin Years*: *Writings, 1914–1917 Albert Einstein* (English translation supplement), eds. M. J. Klein, A. J. Kox, J. Renn and R. Schulmann (Princeton University Press, 1995), pp. 98–107; available at: http://press.princeton.edu/einstein/digital/.
68. A. Einstein, Zür allgemeinen Relativitätstheorie (Nachtrag), *Sitz. ber. Preuss. Akad. Wiss.* (1915) 799 (in German); in *The Collected Papers of Albert Einstein, Vol.* 6: *The Berlin Years*: *Writings, 1914–1917 Albert Einstein* (English translation supplement) pp. 108-110, eds. M. J. Klein, A. J. Kox, J. Renn and R. Schulmann (Princeton University Press, 1995); available at: http://press.princeton.edu/einstein/digital/.
69. A. Einstein, Erklärung der Perihelbewegung des Merkur aus allgemeinen Relativitätstheorie, *Sitz. ber. Preuss. Akad. Wiss.* (1915) 831 (in German); English translation in *The Collected Papers of Albert Einstein, Vol.* 6: *The Berlin Years*: *Writings, 1914–1917 Albert Einstein* (English translation supplement) eds. by M. J. Klein,

A. J. Kox, J. Renn and R. Schulmann (Princeton University Press, 1995), pp. 112–116, available at: http://press.princeton.edu/einstein/digital/.

70. D. Hilbert, Die Grundlagen der Physik, *K. Ges. Wiss. Gött. Nachr. Math.-Phys. K.* (1915) 395.
71. A. Einstein, Die Feldgleichungen der Gravitation, *Sitz. ber. Preuss. Akad. Wiss.* (1915) 844 (in German); *The Collected Papers of Albert Einstein, Vol.* 6: *The Berlin Years*: *Writings, 1914–1917 Albert Einstein* (English translation supplement) eds. M. J. Klein, A. J. Kox, J. Renn and R. Schulmann (Princeton University Press, 1995) pp. 117–120, available at: http://press.princeton.edu/einstein/digital/.
72. A. Einstein, Die Grundlage der allgemeinen Relätivitatstheorie, *Ann. Phys.* **49** (1916) 769.
73. A. Einstein, Näherungsweise Integration der Feldgleichungen der Gravitation, *Sitz. ber Preuss. Akad. Wiss.* (1916) 688 (in German) (translated by Alfred Engel) in *The Collected Papers of Albert Einstein, Vol.* 6: *The Berlin Years*: *Writings, 1914–1917* (English translation supplement, Doc. 32 (Princeton University Press, 1997), pp. 201–210, Available at: http://einsteinpapers.press.princeton.edu/vol6-trans).
74. A. Einstein, Über Gravitationswellen, *Sitz. ber. K. Preuss. Akad. Wiss.* (1918) (in German) 154; (in German) (translated by Alfred Engel) in *The Collected Papers of Albert Einstein, Vol.* 7: *The Berlin Years*: *Writings, 1918–1921* (English translation supplement, Doc. 1 On gravitational waves (Princeton University Press), pp. 9–27, Available at: http://einsteinpapers.press.princeton.edu/vol7-trans).
75. A. S. Eddington, The propagation of gravitational waves, *Proc. R. Soc. Lond. A* **102** (1922) 268.
76. D. Kennefick, *Traveling at the Speed of Thought*: *Einstein and the Quest for Gravitational Waves* (Princeton University Press, Princeton, 2007).
77. C.-M. Chen, J. M. Nester and W.-T. Ni, A brief history of gravitational wave research, *Chin. J. Phys.* (2016), Available at: http://dx.doi.org/10.1016/j.cjph.2016.10.014.
78. S. Carlip, D.-W. Chiou, W.-T. Ni and R. Woodard, Quantum gravity: A brief history of ideas and some prospects, in *One Hundred Years of General Relativity*: *From Genesis and Foundations to Gravitational Waves, Cosmology and Quantum Gravity*, ed. W.-T. Ni (World Scientific, Singapore, 2016); *Int. J. Mod. Phys. D* **25** (2015) 1530028.
79. K. Schwarzschild, "Über das Gravitationsfeld eines Massenpunktes nach der Einsteinschen Theorie", *Sitzungsberichte der Königlich Preussischen Akademie der Wissenschaften* **7** (1916) 189–196; English Translation, S. Antoci and A. Loinger, "On the gravitational field of a mass point according to Einstein's theory", arXiv:physics/9905030.
80. C. Heinicke and F. W. Hehl, Schwarzschild and Kerr solutions of Einstein's field equation: An introduction, in *One Hundred Years of General Relativity*: *From Genesis and Foundations to Gravitational Waves, Cosmology and Quantum Gravity*, Chap. 3, ed. W.-T. Ni (World Scientific, Singapore, 2016); *Int. J. Mod. Phys. D* **25** (2015) 1530006.
81. A. Einstein, Kosmologische Betrachtungen zur allgemeinen Relativitaetstheorie, *Sitz. ber. K. Preuss. Akad. Wiss. Berlin* **1** (1917) 142.
82. W. de Sitter, On the relativity of inertia: Remarks concerning Einstein's latest hypothesis, *Proc. Kon. Ned. Akad. Wet.* **19** (1917) 1217.
83. W. de Sitter, The curvature of space, *Proc. Kon. Ned. Akad. Wet.* **20** (1917) 229.
84. W. de Sitter, *Proc. Kon. Ned. Akad. Wet.* **20** (1917) 1309.
85. W. de Sitter, *Mon. Not. R. Astron. Soc.* **78** (1917) 3.

86. M. Bucher and W.-T. Ni, General relativity and cosmology, in *One Hundred Years of General Relativity: From Genesis and Foundations to Gravitational Waves, Cosmology and Quantum Gravity*, Chap. 13, ed. W.-T. Ni (World Scientific, Singapore, 2016); *Int. J. Mod. Phys. D* **24** (2015) 1530030.
87. M. Davis, Cosmic structure in *One Hundred Years of General Relativity: From Genesis and Empirical Foundations to Gravitational Waves, Cosmology and Quantum Gravity*, ed. W.-T. Ni (World Scientific, Singapore, 2016); *Int. J. Mod. Phys. D* **23** (2014) 1430011.
88. M. Bucher, Physics of the cosmic microwave background anisotropy, in *One Hundred Years of General Relativity: From Genesis and Empirical Foundations to Gravitational Waves, Cosmology and Quantum Gravity*, ed. W.-T. Ni (World Scientific, Singapore, 2016); *Int. J. Mod. Phys. D* **24** (2015) 1530004.
89. X. Meng, Y. Gao and Z. Han, SNe Ia as a cosmological probe, in *One Hundred Years of General Relativity: From Genesis and Empirical Foundations to Gravitational Waves, Cosmology and Quantum Gravity*, ed. W.-T. Ni (World Scientific, Singapore, 2016); *Int. J. Mod. Phys. D* **24** (2015) 1530029.
90. T. Futamase, Gravitational lensing in cosmology, in *One Hundred Years of General Relativity: From Genesis and Empirical Foundations to Gravitational Waves, Cosmology and Quantum Gravity*, ed. W.-T. Ni (World Scientific, Singapore, 2016); *Int. J. Mod. Phys. D* **24** (2015) 530011.
91. K. Sato and J. Yokoyama, Inflationary cosmology: First 30+ Years, in *One Hundred Years of General Relativity: From Genesis and Empirical Foundations to Gravitational Waves, Cosmology and Quantum Gravity*, ed. W.-T. Ni (World Scientific, Singapore, 2016); *Int. J. Mod. Phys. D* **24** (2015) 1530025.
92. D. Chernoff and H. Tye, Inflation, string theory and cosmic strings, in *One Hundred Years of General Relativity: From Genesis and Empirical Foundations to Gravitational Waves, Cosmology and Quantum Gravity*, ed. W.-T. Ni (World Scientific, Singapore, 2016); *Int. J. Mod. Phys. D* **24** (2015) 1530010.
93. Lense and H. Thirring, *Phys. Z.* **19** (1918) 156 (in German); *Gen. Relativ. Gravit.* **16** (1984) 712.
94. B. P. Abbott, LIGO Scientific and Virgo Collab., Observation of gravitational waves from a binary black hole merger, *Phys. Rev. Lett.* **116** (2016) 061102.
95. B. P. Abbott, LIGO Scientific and Virgo Collab., GW151226: Observation of gravitational waves from a 22-solar-mass binary black hole coalescence, *Phys. Rev. Lett.* **116** (2016) 241103.

Chapter 3

Schwarzschild and Kerr solutions of Einstein's field equation: An Introduction

Christian Heinicke[*,‡] and Friedrich W. Hehl[*,†,§]

[*]*Institute for Theoretical Physics, University of Cologne, 50923 Köln, Germany*

[†]*Department of Physics and Astronomy, University of Missouri, Columbia, MO 65211, USA*

[‡]*christian.heinicke@t-online.de*

[§]*hehl@thp.uni-koeln.de*

Starting from Newton's gravitational theory, we give a general introduction into the spherically symmetric solution of Einstein's vacuum field equation, the Schwarzschild(–Droste) solution, and into one specific stationary axially symmetric solution, the Kerr solution. The Schwarzschild solution is unique and its metric can be interpreted as the exterior gravitational field of a spherically symmetric mass. The Kerr solution is only unique if the multipole moments of its mass and its angular momentum take on prescribed values. Its metric can be interpreted as the exterior gravitational field of a suitably rotating mass distribution. Both solutions describe objects exhibiting an *event horizon*, a frontier of no return. The corresponding notion of a black hole is explained to some extent. Eventually, we present some generalizations of the Kerr solution.

Keywords: General relativity; Kerr and Schwarzschild solutions; black holes; gravito-electromagnetism; torsion.

PACS Number(s): 04.50.−h, 02.70.Wz, 01.65.+g, 01.30.Rr, 04.20.−q, 04.70.Bw

1. Prelude[a]

In Sec. 1.1, *we provide some background material on Newton's theory of gravity and, in Sec.* 1.2, *on the flat and gravity-free Minkowski space of special relativity theory. Both theories were superseded by Einstein's gravitational theory, general relativity. In Sec.* 1.3, *we supply some machinery for formulating Einstein's field equation without and with the cosmological constant.*

1.1. *Newtonian gravity*

Newton's gravitational theory is described — in particular tidal gravitational forces — and applied to a spherically symmetric body (a "star").

[a]Parts of Secs. 1 and 2 are adapted from our presentation[83] in Falcke *et al.*[56]

Gravity exists in all bodies universally and is proportional to the quantity of matter in each [...] If two globes gravitate towards each other, and their matter is homogeneous on all sides in regions that are equally distant from their centers, then the weight of either globe towards the other will be inversely as the square of the distance between the centers.

Isaac Newton[136] (1687)

The gravitational force of a point-like mass m_2 on a similar one of mass m_1 is given by Newton's attraction law

$$\mathbf{F}_{2\to 1} = -G\frac{m_1 m_2}{|\mathbf{r}|^2}\frac{\mathbf{r}}{|\mathbf{r}|}, \tag{1}$$

where G is Newton's gravitational constant (CODATA, 2010),

$$G \overset{\text{SI}}{=} 6.67384(80) \times 10^{-11}\frac{(\text{m/s})^4}{N}.$$

The vector $\mathbf{r} := \mathbf{r}_1 - \mathbf{r}_2$ points from m_2 to m_1, see Fig. 1.

According to *actio* = *reactio* (Newton's third law), we have $\mathbf{F}_{2\to 1} = -\mathbf{F}_{1\to 2}$. Thus, a complete symmetry exists of the gravitational interaction of the two masses onto each other. Let us now distinguish the mass m_2 as field-generating active gravitational mass and m_1 as (point-like) passive test-mass. Accordingly, we introduce a hypothetical *gravitational field* as describing the force per unit mass ($m_2 \hookrightarrow M, m_1 \hookrightarrow m$)

$$\mathbf{f} := \frac{\mathbf{F}}{m} = -\frac{GM}{|\mathbf{r}|^2}\frac{\mathbf{r}}{|\mathbf{r}|}. \tag{2}$$

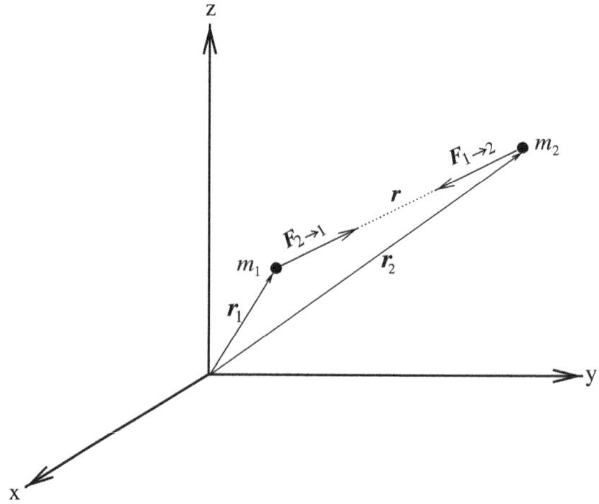

Fig. 1. Two mass points m_1 and m_2 attracting each other in three-dimensional space, Cartesian coordinates x, y, z.

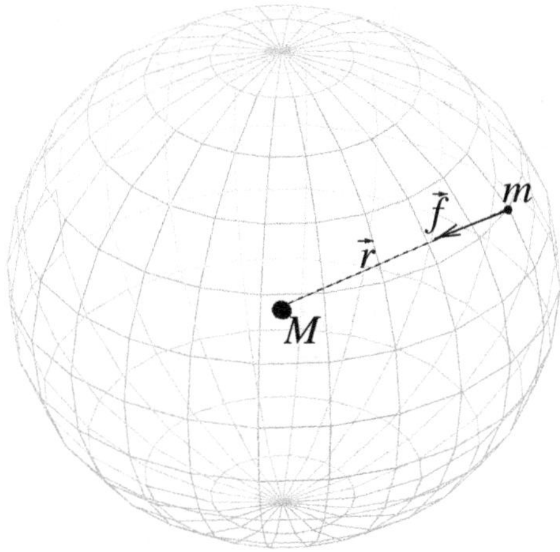

Fig. 2. The "source" M attracts the test mass m.

With this definition, the force acting on the test-mass m is equal to *field strength* $\times$ *gravitational charge* (mass) or $\mathbf{F}_{M\to m} = m\mathbf{f}$, in analogy to electrodynamics. The active gravitational mass M is thought to emanate a gravitational field which is always directed to the center of M and has the same magnitude on every sphere with M as center, see Fig. 2. Let us now investigate the properties of the gravitational field (2). Obviously, there exists a potential

$$\phi = -G\frac{M}{|\mathbf{r}|}, \quad \mathbf{f} = -\boldsymbol{\nabla}\phi. \tag{3}$$

Accordingly, the gravitational field is curl-free: $\boldsymbol{\nabla} \times \mathbf{f} = 0$.

By assumption it is clear that the source of the gravitational field is the mass M. We find, indeed

$$\boldsymbol{\nabla} \cdot \mathbf{f} = -4\pi GM\delta^3(\mathbf{r}), \tag{4}$$

where $\delta^3(\mathbf{r})$ is the three-dimensional (3D) delta function. By means of the *Laplace operator* $\Delta := \boldsymbol{\nabla}\cdot\boldsymbol{\nabla}$, we infer for the gravitational potential

$$\Delta\phi = 4\pi GM\delta^3(\mathbf{r}). \tag{5}$$

The term $M\delta^3(\mathbf{r})$ may be viewed as the mass density of a point mass. Equation (5) is a second order linear partial differential equation for ϕ. Thus, the gravitational potential generated by several point masses is simply the linear superposition of the respective single potentials. Hence, we can generalize the *Poisson equation* (5) straightforwardly to a continuous matter distribution $\rho(\mathbf{r})$

$$\Delta\phi = 4\pi G\rho. \tag{6}$$

This equation interrelates the source ρ of the gravitational field with the gravitational potential ϕ and thus completes the quasi-field theoretical description of Newton's gravitational theory.

We speak here of *quasi*-field theoretical because the field ϕ as such represents a convenient concept. However, it has no *dynamical* properties, no genuine degrees of freedom. The Newtonian gravitational theory is an *action at a distance* theory (also called *mass-interaction theory*). When we remove the source, the field vanishes instantaneously. Newton himself was very unhappy about this consequence. Therefore, he emphasized the preliminary and purely descriptive character of his theory. But before we liberate the gravitational field from this constraint by equipping it with its own degrees of freedom within the framework of general relativity theory, we turn to some properties of the Newtonian theory.

A very peculiar fact characteristic to the gravitational field is that the acceleration of a freely falling test-body does not depend on the mass of this body but only on its position within the gravitational field. This comes about because of the equality (in suitable units) of the gravitational and the inertial mass

$$\overset{\text{inertial}}{m}\ \ddot{\mathbf{r}} = \mathbf{F} = \overset{\text{grav}}{m}\ \mathbf{f}. \tag{7}$$

This equality has been well tested since Galileo's time by means of pendulum and other experiments with an ever increasing accuracy, see Will.[189]

In order to allow for a more detailed description of the structure of a gravitational field, we introduce the concept of *tidal force*. This can be best illustrated by means of Fig. 3. In a spherically symmetric gravitational field, for example, two test-masses will fall radially toward the center and thereby get closer and closer. Similarly, a spherical drop of water is deformed to an ellipsoidal shape because the gravitational force at its bottom is bigger than at its top, which has a greater distance to the source. If the distance between two freely falling test masses is relatively small, we can derive an explicit expression for their relative acceleration by means of a Taylor expansion. Consider two mass points with position vectors **r** and

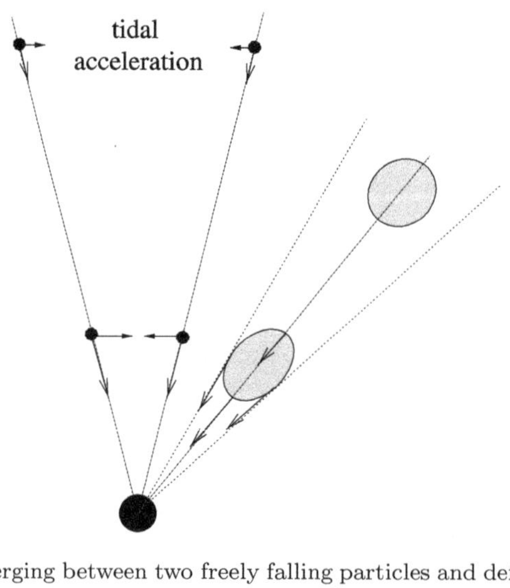

Fig. 3. Tidal forces emerging between two freely falling particles and deforming a spherical body.

$\mathbf{r} + \delta\mathbf{r}$, with $|\delta\mathbf{r}| \ll 1$. Then the relative acceleration reads

$$\delta\mathbf{a} = [\mathbf{f}(\mathbf{r} + \delta\mathbf{r}) - \mathbf{f}(\mathbf{r})] = \delta\mathbf{r} \cdot (\boldsymbol{\nabla} \boldsymbol{f}). \tag{8}$$

We may rewrite this according to (the sign is conventional, $\partial/\partial x^a =: \partial_a$, $x^1 = x$, $x^2 = y$, $x^3 = z$)

$$K_{ab} := -(\boldsymbol{\nabla}\mathbf{f})_{ab} = -\partial_a f_b, \quad a, b = 1, 2, 3. \tag{9}$$

We call K_{ab} the *tidal force* matrix. The vanishing curl of the gravitational field is equivalent to its symmetry, $K_{ab} = K_{ba}$. Furthermore, $K_{ab} = \partial_a \partial_b \phi$. Thus, the Poisson equation becomes

$$\sum_{a=1}^{3} K_{aa} = \operatorname{trace} K = 4\pi G \rho. \tag{10}$$

Accordingly, in vacuum K_{ab} is trace-free.

Let us now investigate the gravitational potential of a homogeneous *star* with constant mass density $\rho_\odot$ and total mass $M_\odot = (4/3)\pi R_\odot^3 \rho_\odot$. For our Sun, the radius is $R_\odot = 6.9598 \times 10^8$ m and the total mass is $M = 1.989 \times 10^{30}$ kg.

Outside the sun (in the idealized picture we are using here), we have vacuum. Accordingly, $\rho(\mathbf{r}) = 0$ for $|\mathbf{r}| > R_\odot$. Then the Poisson equation reduces to the *Laplace equation*

$$\Delta\phi = 0, \quad \text{for } r > R_\odot. \tag{11}$$

In 3D polar coordinates, the r-dependent part of the Laplacian has the form $(1/r^2)\partial_r(r^2\partial_r)$. Thus, (11) has the solution

$$\phi = \frac{\alpha}{r} + \beta, \tag{12}$$

where α and β are integration constants. Requiring that the potential tends to zero as r goes to infinity, we get $\beta = 0$. The integration constant α will be determined from the requirement that the force should change smoothly as we cross the star's surface, that is, the interior and exterior potential and their first derivatives have to be matched continuously at $r = R_\odot$.

Inside the star we have to solve

$$\Delta\phi = 4\pi G \rho_\odot, \quad \text{for } r \leq R_\odot. \tag{13}$$

We find

$$\phi = \frac{2}{3}\pi G \rho_\odot r^2 + \frac{C_1}{r} + C_2, \tag{14}$$

with integration constants C_1 and C_2. We demand that the potential in the center $r = 0$ has a finite value, say ϕ_0. This requires $C_1 = 0$. Thus

$$\phi = \frac{2}{3}\pi G \rho_\odot r^2 + \phi_0 = \frac{GM(r)}{2r} + \phi_0, \tag{15}$$

where we introduced the *mass function* $M(r) = (4/3)\pi r^3 \rho_\odot$ which measures the total mass inside a sphere of radius r.

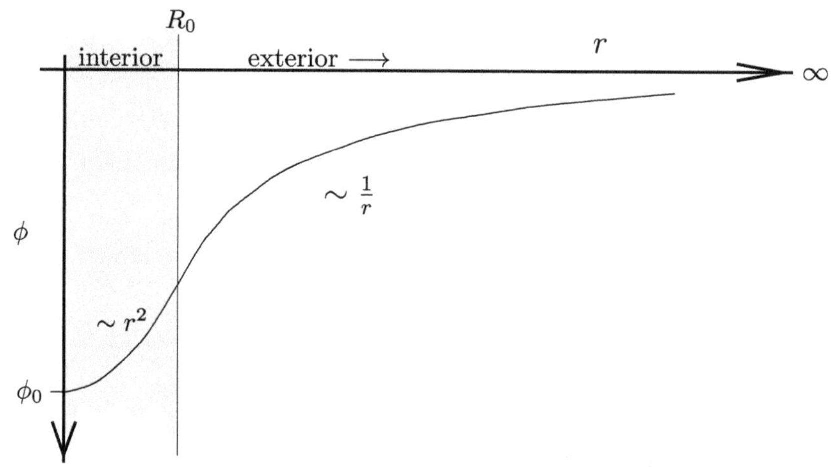

Fig. 4. Newtonian potential of a homogeneous star.

Continuous matching of ϕ and its first derivatives at $r = R_\odot$ finally yields

$$\phi(\mathbf{r}) = \begin{cases} -G\dfrac{M_\odot}{|\mathbf{r}|} & \text{for } |\mathbf{r}| \geq R_\odot, \\ G\dfrac{M_\odot}{2R_\odot^3}|\mathbf{r}|^2 - \dfrac{3GM_\odot}{2R_\odot} & \text{for } |\mathbf{r}| < R_\odot. \end{cases} \tag{16}$$

The slope of this curve indicates the magnitude of the gravitational force, the curvature (second derivative) the magnitude of the tidal force (or acceleration).

1.2. *Minkowski space*

When, in a physical experiment, gravity can be safely neglected, we seem to live in the flat Minkowski space of special relativity theory. We introduce the metric of the Minkowksi space and rewrite it in terms of so-called null coordinates, that is, we use light rays for a parametrization of Minkowski space.

> *Henceforth space by itself, and time by itself, are doomed to fade away into mere shadows, and only a kind of union of the two will preserve an independent reality.*
>
> Hermann Minkowski (1908)

It was Minkowski who welded space and time together into spacetime, thereby abandoning the observer-independent meaning of spatial and temporal distances.

Instead, the spatio-temporal distance, the line element

$$ds^2 = -c^2 dt^2 + dx^2 + dy^2 + dz^2$$

is distinguished as the invariant measure of spacetime. The Poincaré (or inhomogeneous Lorentz) transformations form the invariance group of this spacetime metric. The principle of the constancy of the speed of light is embodied in the equation $ds^2 = 0$. Suppressing one spatial dimension, the solutions of this equation can be regarded as a double cone. This light cone visualizes the paths of all possible light rays arriving at or emitted from the cone's apex. Picturing the light cone structure, and thereby the causal properties of spacetime, will be our method for analyzing the meaning of the Schwarzschild and the Kerr solution.

1.2.1. *Null coordinates*

We first introduce so-called null coordinates. The Minkowski metric (with $c = 1$), in spherical polar coordinates reads

$$ds^2 = -dt^2 + dr^2 + r^2(d\theta^2 + \sin^2\theta d\phi^2) = -dt^2 + dr^2 + r^2 d\Omega^2. \tag{17}$$

We define *advanced* and *retarded null coordinates* according to

$$v := t + r, \quad u := t - r, \tag{18}$$

and find

$$ds^2 = -dvdu + \frac{1}{4}(v-u)^2 d\Omega^2. \tag{19}$$

In Fig. 5 we show the Minkowski spacetime in terms of the new coordinates. Incoming photons, that is, point-like particles with velocity $\dot{r} = -c = -1$, move on paths with $v =$ const. Correspondingly, we have for outgoing photons $u =$ const. The special relativistic wave equation is solved by any function $f(u)$ and $f(v)$. The surfaces $f(u) =$ const. and $f(v) =$ const. represent the wavefronts which evolve with the velocity of light. The trajectory of every material particle with $\dot{r} < c = 1$ has to remain inside the region defined by the surface $r = t$. In an (r,t)-diagram, this surface is represented by a cone, the so-called *light cone*. Any point in the *future light cone* $r = t$ can be reached by a particle or signal with a velocity less than c. A given spacetime point P can be reached by a particle or signal from the spacetime region enclosed by the *past light cone* $r = -t$.

1.2.2. *Penrose diagram*

We can map, following Penrose, the infinitely distant points of spacetime into finite regions by means of a conformal transformation which leaves the light cones intact. Then we can display the whole infinite Minkowski spacetime on a (finite) piece of

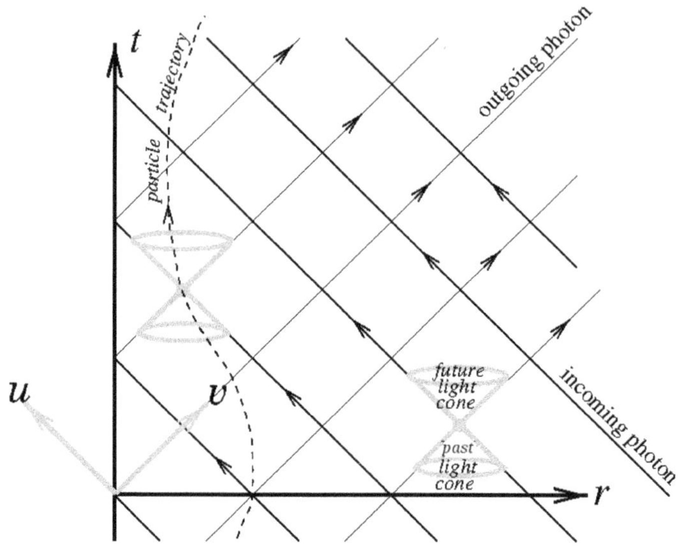

Fig. 5. Minkowski spacetime in null coordinates.

paper. Accordingly, introduce the new coordinates

$$\tilde{v} := \arctan v, \quad \tilde{u} := \arctan u, \quad \text{for } -\frac{\pi}{2} \leq (\tilde{v}, \tilde{u}) \leq +\frac{\pi}{2}. \tag{20}$$

Then the metric reads

$$ds^2 = \frac{1}{\cos^2 \tilde{v}} \frac{1}{\cos^2 \tilde{u}} \left[-d\tilde{v} d\tilde{u} + \frac{1}{4} \sin^2(\tilde{v} - \tilde{u}) d\Omega^2 \right]. \tag{21}$$

We can go back to time- and space-like coordinates by means of the transformation

$$\tilde{t} := \tilde{v} + \tilde{u}, \quad \tilde{r} := \tilde{v} - \tilde{u}, \tag{22}$$

see (18). Then the metric reads

$$ds^2 = \frac{-d\tilde{t}^2 + d\tilde{r}^2 + \sin^2 \tilde{r} d\Omega^2}{4 \cos^2 \dfrac{\tilde{t} + \tilde{r}}{2} \cos^2 \dfrac{\tilde{t} - \tilde{r}}{2}}, \tag{23}$$

that is, up to the function in the denominator, it appears as a flat metric. Such a metric is called conformally flat (it is conformal to a static Einstein cosmos). The back-transformation to our good old Minkowski coordinates reads

$$t = \frac{1}{2} \left(\tan \frac{\tilde{t} + \tilde{r}}{2} + \tan \frac{\tilde{t} - \tilde{r}}{2} \right), \tag{24}$$

$$r = \frac{1}{2} \left(\tan \frac{\tilde{t} + \tilde{r}}{2} - \tan \frac{\tilde{t} - \tilde{r}}{2} \right). \tag{25}$$

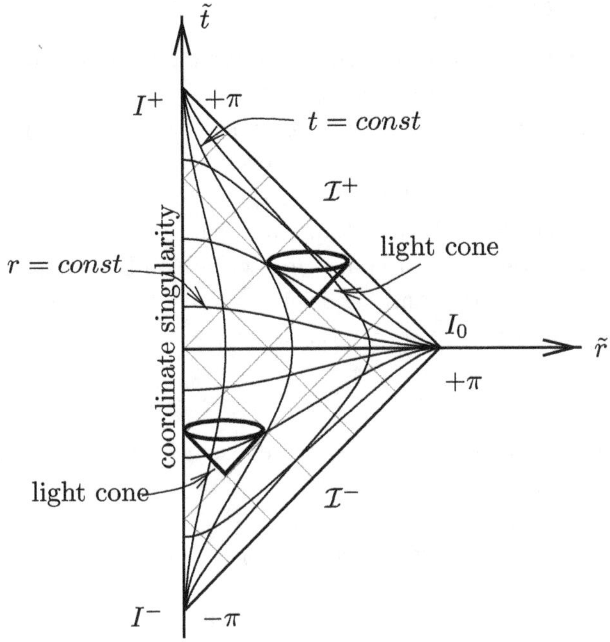

Fig. 6. Penrose diagram of Minkowski spacetime.

Our new coordinates $\tilde{t}, \tilde{r}$ extend only over a finite range of values, as can be seen from (24) and (25). Thus, in the Penrose diagram of a Minkowski spacetime, see Fig. 6, we can depict the whole Minkowski spacetime, with a coordinate singularity along $\tilde{r} = 0$. All trajectories of uniformly moving particles (with velocity smaller than c) emerge from one single point, past infinity I^-, and all will eventually arrive at the one single point I^+, namely at future infinity. All incoming photons have their origin on the segment $\mathcal{I}^-$ (script I^- or "scri minus"), light-like past-infinity, and will run into the coordinate singularity on the $\tilde{t}$-axis. All outgoing photons arise from the coordinate singularity and cease on the line $\mathcal{I}^+$, light-like future infinity ("scri plus"). The entire spacelike infinity is mapped into the single point I^0. For later reference we collect these notions in a table (Table 1).

Now, we have a really compact picture of the Minkowski space. Next, we would like to proceed along similar lines in order to obtain an analogous picture for the Schwarzschild spacetime.

Table 1. The different infinities in Penrose diagrams.

I^-	Timelike past infinity	Origin of all particles
I^+	Timelike future infinity	Destination of all particles
I_0	Spacelike infinity	Inaccessible for all particles
$\mathcal{I}^-$	Lightlike past infinity	Origin of all light rays
$\mathcal{I}^+$	Lightlike future infinity	Destination of all light rays

1.3. *Einstein's field equation*

We display our notations and conventions for the differential geometric tools used to formulate Einstein's field equation.

We assume that our readers know at least the rudiments of general relativity (GR) as represented, for instance, in Einstein's *Meaning of Relativity*,[50] which we still recommend as a gentle introduction into GR. More advanced readers may then want to turn to Rindler[165] and/or to Landau–Lifshitz.[106]

We assume a 4D Riemannian spacetime with (Minkowski-)Lorentz signature $(-+++)$, see Misner, Thorne, and Wheeler.[127] Thus, the metric field, in arbitrary holonomic coordinates x^μ, with $\mu = 0, 1, 2, 3$, reads

$$\mathbf{g} \equiv ds^2 = g_{\mu\nu} dx^\mu \otimes dx^\nu . \tag{26}$$

By partial differentiation of the metric, we can calculate the Christoffel symbols (Levi-Civita connection)

$$\Gamma^\mu{}_{\alpha\beta} := \frac{1}{2} g^{\mu\gamma} \left(\partial_\alpha g_{\beta\gamma} + \partial_\beta g_{\gamma\alpha} - \partial_\gamma g_{\alpha\beta}\right). \tag{27}$$

This empowers us to determine the geodesics (curves of extremal length) of the Riemannian spacetime

$$\frac{D^2 x^\alpha}{D\tau^2} := \frac{d^2 x^\alpha}{d\tau^2} + \Gamma^\alpha{}_{\mu\nu} \frac{dx^\mu}{d\tau} \frac{dx^\nu}{d\tau} = 0. \tag{28}$$

This equation can be read as a vanishing of the 4D covariant acceleration. If we define the four-velocity $u^\alpha := dx^\alpha / d\tau$, then the geodesics can be rewritten as

$$\frac{Du^\alpha}{D\tau} = \frac{du^\alpha}{d\tau} + \Gamma^\alpha{}_{\mu\nu} u^\mu u^\nu = 0. \tag{29}$$

In a neighborhood of any given point in spacetime, we can introduce Riemannian normal coordinates, which are such that the Christoffels vanish at that point. In order to find a tensorial measure of the gravitational field, we have to go one differentiation order higher. By partial differentiation of the Christoffels, we find the Riemann curvature tensor[b]

$$R^\mu{}_{\nu\alpha\beta} := 2(\partial_{[\alpha} \Gamma^\mu{}_{|\nu|\beta]} + \Gamma^\mu{}_{\sigma[\alpha} \Gamma^\sigma{}_{|\nu|\beta]}). \tag{30}$$

The curvature is doubly antisymmetric, its two index pairs commute, and its totally antisymmetric piece vanishes

$$R_{(\mu\nu)\alpha\beta} = 0, \quad R_{\mu\nu(\alpha\beta)} = 0; \quad R_{\mu\nu\alpha\beta} = R_{\alpha\beta\mu\nu}; \quad R_{[\mu\nu\alpha\beta]} = 0. \tag{31}$$

[b]Always symmetrizing of indices is denoted by parentheses, $(\alpha\beta) := \{\alpha\beta + \beta\alpha\}/2!$, antisymmetrization by brackets $[\alpha\beta] := \{\alpha\beta - \alpha\beta\}/2!$, with corresponding generalizations $(\alpha\beta\gamma) := \{\alpha\beta\gamma + \beta\gamma\alpha + \gamma\alpha\beta + \cdots\}/3!$, etc. Indices standing between two vertical strokes $|\ |$ are excluded from the (anti)symmetrization process, see Schouten.[170]

If we define collective indices $A, B, \ldots = 1, \ldots, 6$ for the antisymmetric index pairs according to the rule $\{01, 02, 03; 23, 31, 12\} \to \{1, 2, 3; 4, 5, 6\}$, then the algebraic symmetries of (31) can be rephrased as

$$R_{AB} = R_{BA}, \quad \text{trace}(R_{AB}) = 0. \tag{32}$$

Thus, in 4D the curvature can be represented as a trace-free symmetric 6×6-matrix. Hence, it has 20 independent components.

With the curvature tensor, we found a tensorial measure for the gravitational field. Freely falling particles move along geodesics of Riemannian spacetime. What about the tidal accelerations between two freely falling particles? Let the "infinitesimal" vector n^α describe the distance between two particles moving on adjacent geodesics. A standard calculation,[127] linear to the order of n, yields the *geodesic deviation equation*

$$\frac{D^2 n^\alpha}{D\tau^2} = u^\beta u^\gamma R^\alpha{}_{\beta\gamma\delta} n^\delta. \tag{33}$$

This equation describes the relative acceleration of neighboring particles, similar as (8) and (9) in the Newtonian case. The role of the tidal matrix K_{ab} is taken over by $\mathcal{K}^\alpha{}_\delta := u^\beta u^\gamma R^\alpha{}_{\beta\gamma\delta}$.

By contraction of the curvature, we can define the second rank Ricci tensor $R_{\mu\nu}$ and the curvature scalar R, respectively

$$R_{\mu\nu} := R^\alpha{}_{\mu\alpha\nu}, \quad R := g^{\mu\nu} R_{\mu\nu}. \tag{34}$$

For convenience, we can also introduce the Einstein tensor $G_{\mu\nu} := R_{\mu\nu} - \frac{1}{2} g_{\mu\nu} R$. The curvature with its 20 independent components can be irreducibly decomposed into smaller pieces according to $20 = 10+9+1$. The Weyl curvature tensor $C_{\alpha\beta\gamma\delta}$ is trace-free and has 10 independent components, whereas the trace-free Ricci tensor has nine components and the curvature scalar just 1.

Now we have all the tools for displaying Einstein's field equation. With G as Newton's gravitational constant and c as velocity of light, we define Einstein's gravitational constant $\kappa := 8\pi G/c^4$. Then, the *Einstein field equation* with cosmological constant Λ reads

$$R_{\mu\nu} - \frac{1}{2} g_{\mu\nu} R + \Lambda g_{\mu\nu} = \kappa T_{\mu\nu}. \tag{35}$$

The source on the right-hand side is the energy–momentum tensor of matter. The *vacuum* field equation, without cosmological constant, simply reduces to $R_{\mu\nu} = 0$. Mostly this equation will keep us busy in this paper. A vanishing Ricci tensor implies that only the Weyl curvature $C_{\alpha\beta\gamma\delta} \neq 0$. Accordingly, the vacuum field in GR (without Λ) is represented by the Weyl tensor.

Equation (35) represents a generalization of the Poisson equation (10). There, the contraction of the tidal matrix is proportional to the mass density. In GR, the contraction of the curvature tensor is proportional to the energy–momentum tensor.

The physical mass is denoted by M. Usually, we use the *mass parameter*, $m := \frac{GM}{c^2}$. The Schwarzschild radius reads $r_S := 2m = \frac{2GM}{c^2}$. Usually we put $c = 1$

and $G = 1$. We make explicitly use of G and c as soon as we stress analogies to Newtonian gravity or allude to observational data.

2. The Schwarzschild Metric (1916)

Spatial spherical symmetry is assumed and a corresponding exact solution for Einstein's theory searched for. After a historical outline (Sec. 2.1), we apply the equivalence principle to a freely falling particle and try to implement that on top of the Minkowskian line element. In this way, we heuristically arrive at the Schwarzschild metric (Sec. 2.2). In Sec. 2.3, we display the Schwarzschild metric in six different classical coordinate systems. We outline the concept of a Schwarzschild black hole in Sec. 2.4. In Secs. 2.5 and 2.6, we construct the Penrose diagram for the Schwarzschild(-Kruskal) spacetime. We add electric charge to the Schwarzschild solution in Sec. 2.7. The interior Schwarzschild metric, with matter, is addressed in Sec. 2.8.

> *It is quite a wonderful thing that from such an abstract idea the explanation of the Mercury anomaly emerges so inevitably.*
>
> Karl Schwarzschild[171] (1915)

2.1. *Historical remarks*

The genesis of the Schwarzschild solution (1915/16) *is described. In particular, we show that Droste, a bit later than Schwarzschild, arrived at the Schwarzschild metric independently. He put the Schwarzschild solution into that form in which we use it today.*

The first exact solution of Einstein's field equation was born in hospital. Unfortunately, the circumstances were more tragic than joyful. The astronomer Karl Schwarzschild joined the German army right at the beginning of World War I and served in Belgium, France and Russia. At the end of the year 1915, he was admitted to hospital with an acute skin disease. There, not far from the Russian front, enduring the distant gunfire, he found time to "stroll through the land of ideas" of Einstein's theory, as he puts it in a letter to Einstein[c] dated 22 December 1915. According to this letter, Schwarzschild started out from the approximate solution in Einstein's "perihelion paper", published November 25th. Since presumably letters from Berlin to the Russian front took a few days, Schwarzschild[172] found the solution within about a fortnight. Fortunately, the premature field equation of the "perihelion paper" is correct in the vacuum case treated by Schwarzschild.

[c]The letters from and to Einstein can be found in Einstein's Collected Works,[51] see also Schwarzschild's Collected Works.[171]

In February 1916, Schwarzschild[173] submitted the spherically symmetric solution with matter — the "interior Schwarzschild solution" — now based on Einstein's final field equation. In March 1916, he was sent home where he passed away on 11 May 1916.

The field equation used by Schwarzschild requires $\det g = -1$. To fulfill this condition, he uses modified polar coordinates (Schwarzschild's original notation used),

$$x_1 = \frac{r^3}{3}, \quad x_2 = -\cos\theta, \quad x_3 = \phi, \quad x_4 = t.$$

The spherically symmetric ansatz then reads

$$ds^2 = f_4 dx_4^2 - f_1 dx_1^2 - f_2 \frac{dx_2^2}{1 - x_2^2} - f_3 dx_3^2 (1 - x_2^2),$$

where f_1 to f_4 are functions of x_1 only. The solution turns out to be

$$f_1 = \frac{1}{R^4} \frac{1}{1 - \frac{\alpha}{R}}, \quad f_2 = f_3 = R^2, \quad f_4 = 1 - \frac{\alpha}{R}, \quad R = (r^3 + \alpha^3)^{1/3}.$$

In this paper, as well as in his letter to Einstein, he eventually returns to the usual spherical polar coordinates,

$$ds^2 = \left(1 - \frac{\alpha}{R}\right) dt^2 - \frac{dR^2}{1 - \frac{\alpha}{R}} - R^2 (d\theta^2 + \sin^2\theta d\phi^2), \quad R = (r^3 + \alpha^3)^{1/3}.$$

This looks like the Schwarzschild metric we are familiar with. One should note, however, that the singularity at $R = \alpha$ is (as we know today) a coordinate singularity, it corresponds to $r = 0$. In the early discussion, the meaning of such a singularity was rather obscure. Flamm[60] in his 1916 article on embedding constant time slices of the Schwarzschild metric into Euclidean space mentions "the oddity that a point mass has an finite circumference of $2\pi\alpha$".

In 1917, Weyl[188] talks of the "inside" and "outside" of the point mass and states that "in nature, evidently, only that piece of the solution is realized which does not touch the singular sphere." In Hilbert's[86] opinion, the singularity $R = \alpha$ indicates the illusiveness of the concept of a pointlike mass. A point mass is just the limiting case of a spherically symmetric mass distribution. Illuminating the interior of "Schwarzschild's sphere" took quite a while and it was the discovery of new coordinates which brought first elucidations. Lanczos,[104] in 1922, clearly speaks out that singularities of the metric components do not necessarily have physical significance since they may vanish in appropriate coordinates. However, it took another 38 years to find a maximally extended fully regular coordinate system for the Schwarzschild metric. We will become acquainted with these Kruskal/Szekeres coordinates in Sec. 2.5.

Schwarzschild's solution, published in the widely read minutes of the Prussian Academy, communicated by Einstein himself, nearly instantly triggered further investigations of the gravitational field of a point mass. Already in March 1916, Reissner,[163] a civil engineer by education, published a generalization of the Schwarzschild metric, including an electrical charge; this was later completed by Weyl[188] and by Nordström.[140] Today it is called Reissner–Nordström solution.

Nevertheless, one should not ignore the Dutch twin of Schwarzschild's solution. On 27 May 1916, Droste[47] communicated his results on "the field of a single centre in Einstein's theory of gravitation, and the motion of a particle in that field" to the Dutch Academy of Sciences. He presents a very clear and easy to read derivation of the metric and gives a quite comprehensive analysis of the motion of a point particle. Since 1913, he had been working on general relativity under the supervision of Lorentz at Leiden University. *Published in Dutch*, Droste's results are fairly unknown today. Einstein, probably informed by his close friend Ehrenfest, rather appreciated Droste's work, praising the graceful mathematical style. Weyl[188] also cites Droste, but in Hilbert's[86] second communication the reference is not found. Einstein, Hilbert, and Weyl always allude to "Schwarzschild's solution".

After Droste took his Ph.D. in 1916, he worked as school teacher and eventually became professor for mathematics in Leiden. He never resumed his work on Einstein's theory and his name faded from the relativistic memoirs. In Leiden, people like Lorentz, de Sitter, Nordström, or Fokker learned about the gravitational field of a point mass primarily from Droste's work. Thus, the name "Schwarzschild–Droste solution" would be quite justified from a historical point of view.

The importance of the Schwarzschild metric is made evident by the Birkhoff[16] theorem[d]: For vanishing cosmological constant, the unique spherically symmetric vacuum spacetime is the Schwarzschild solution, which can be expressed most conveniently in Schwarzschild coordinates, see Table 3, entry 1. Thus, a spherically symmetric body is static (outside the horizon). In particular, it cannot emit gravitational radiation. Moreover, the asymptotic Minkowskian behavior of the Schwarzschild solution is dictated by the solution itself, it is *not* imposed from the outside.

2.2. *Approaching the Schwarzschild metric*

We start from an ansatz for the metric of an accelerated motion in the radial direction and combine it, in the sense of the equivalence principle, with the free-fall velocity of a particle in a Newtonian gravitational field. In this way, we find a curved metric that, after a coordinate transformation, turns out to be the Schwarzschild metric.

[d]The "Birkhoff" theorem was discovered by Jebsen,[95] Birkhoff,[16] and Alexandrow.[4] For more details on Jebsen, see Johansen and Ravndal.[96] The objections of Ehlers and Krasiński[48] appear to us as nitpicking.

Einstein, in his 1907 *Jahrbuch* article,[49] suggests the generalization of the relativity principle to arbitrarily accelerated reference frames.

A plausible notion of a (local) rest frame in general relativity is a frame where the coordinate time is equal to the proper time (for an observer spatially at rest, of course). For a purely radial motion, the following metric would be an obvious ansatz, see also Ref. 185:

$$ds^2 = -dt^2 + [dr + f(r)dt]^2 + r^2 d\Omega^2, \quad \text{with } d\Omega^2 := d\theta^2 + \sin^2\theta d\phi^2.$$

For $d\phi = 0$, $d\theta = 0$, and $dr/dt = -f(r)$, we have $ds^2 = -dt^2$. Thereby, $-f(r)$ is identified as a kind of "radial infall velocity". Note also that constant time-slices, $dt = 0$, are Euclidean.

In Newtonian gravity, a particle falling from infinity toward the origin picks up a velocity

$$\frac{dr}{dt} = v = -\sqrt{2\Phi(r)} = -\sqrt{\frac{2GM}{r}} \Leftrightarrow \frac{1}{2}mv^2(r) = m\Phi(r) = m\frac{GM}{r}. \tag{36}$$

Here, Φ is the absolute value of the Newtonian potential of a spherical body with mass M.

Hence, in some Newtonian limit, we demand $f(r) \to \sqrt{2\Phi}$. This leads to the metric

$$ds^2 = -dt^2 + (dr + \sqrt{2\psi}dt)^2 + r^2 d\Omega^2, \tag{37}$$

where we allow for an arbitrary potential $\psi = \psi(r)$. This metric generates curvature. The calculations can be conveniently done even by hand. The Ricci tensor reads

$$R_0{}^0 = R_1{}^1 = \frac{1}{r}\partial_r\partial_r(r\psi) = 0, \quad R_2{}^2 = R_3{}^3 = \frac{2\partial_r(r\psi)}{r^2} = 0.$$

The equations $R_0{}^0 = 0 = R_1{}^1$ are mere integrability conditions of the $R_2{}^2 = 0 = R_3{}^3$ relations. Hence, $r\psi$ is determined by its first order approximation alone and reads

$$\psi = \frac{\alpha}{r},$$

with α as an unknown constant so far. By construction, we have

$$\frac{dr}{dt} = -\sqrt{2\psi} = -\sqrt{\frac{2\alpha}{r}} \stackrel{!}{=} -\sqrt{\frac{2GM}{r}} \Rightarrow \alpha = GM =: m.$$

The metric (37), expanding the parenthesis and collecting the terms in front of dt^2, reads

$$ds^2 = -\left(1 - \frac{2GM}{r}\right) dt^2 + 2\sqrt{\frac{2GM}{r}} dt dr + dr^2 + r^2 d\Omega^2. \tag{38}$$

Using different methods, this metric was derived by Gullstrand[70] in May 1921. Gullstrand claimed to have found a new spherically symmetric solution of Einstein's field equation. In his opinion,[e] this showed the ambiguity of Einstein's field equation. However, the metric is of the form

$$ds^2 = -Adt^2 + 2Bdtdr + dr^2 + r^2 d\Omega^2, \quad A := 1 - \frac{2GM}{r}, \quad B := \sqrt{\frac{2GM}{r}},$$

and can be diagonalized by completing the square via

$$ds^2 = -A\left(dt - \frac{B}{A}dr\right)^2 + \left(1 + \frac{B^2}{A}\right)dr^2 + r^2 d\Omega^2.$$

Introducing a new time coordinate,

$$dt_S := dt - \frac{B}{A}dr$$

or, explicitly,

$$t_S = t - \left(2r\sqrt{\frac{2GM}{r}} - 4GM\,\mathrm{Artanh}\sqrt{\frac{2GM}{r}}\right),$$

we arrive at (A and B re-substituted)

$$ds^2 = -\left(1 - \frac{2GM}{r}\right)dt_S^2 + \left(1 - \frac{2GM}{r}\right)^{-1}dr^2 + r^2 d\Omega^2.$$

In contrast to what Gullstrand was aiming at, he "just" rederived the Schwarzschild metric.

Later, applying a coordinate transformation to the Schwarzschild metric, Painlevé[147] obtained the metric (38) independently and presented his result in October 1921. His aim was to demonstrate the vacuity of ds^2 by showing that an exact solution does not determine the physical geometry and is therefore meaningless. In a letter (7th December, 1921) to Painlevé, Einstein stresses on the contrary *the meaninglessness of the coordinates*! In the words of Einstein himself (our translation): "... merely results obtained by eliminating the coordinate dependence can claim an objective meaning."

In the subsequent section, we will meet the Schwarzschild metric in many different coordinate systems. All of them have their merits and their shortcomings.

Using Gullstrand–Painlevé coordinates for the Schwarzschild metric does not change the physics, of course. However, as a coordinate system it is what Gustav Mie[126] calls a *sensible* coordinate system. In contrast to many other coordinate systems, the physics looks quite like we are used to. As an example, we analyze the motion of a radial infalling particle in Schwarzschild and *Gullstrand–Painlevé coordinates.*

[e]Gullstrand, who was a member of the Nobel committee, was responsible that Einstein did not get his Nobel prize for relativity theory. He thought that GR is untenable.

The equations of motion for point particles in general relativity are obtained via the geodesic equation (28). It can be shown that this equation is equivalent to the solution of the variational principle $\delta \int_{x^\alpha} ds^2 = \delta \int \dot{x}^\alpha \dot{x}^\beta g_{\alpha\beta} d\tau^2$. We choose the proper time τ for the parametrization of the curve, the dot denotes the derivative with respect to τ. In the present context, we are only interested in the velocity of particles along ingoing geodesics ("freely falling particles"). For time-like geodesics we have $-1 = \frac{ds^2}{d\tau^2}$. This allows the algebraic determination of $\dot{r}$ provided we know $\dot{t}$. Since we consider static metrics here, t is a cyclic variable and $\left(\frac{\partial}{\partial \dot{t}} \frac{ds^2}{d\tau^2}\right) = K =$ const. The constant is determined by the boundary condition $\dot{r} = 0$ for $r \to \infty$.

The difference between the coordinate systems appears in the first line of Table 2: In Gullstrand–Painlevé coordinates, the coordinate velocity of a freely infalling particle increases smoothly toward the center. Nothing special happens at $r = 2GM$. From a given position, the particle will plunge into the center in a finite time. Even numerically this looks quite Newtonian. In contrast, the velocity with respect to Schwarzschild coordinates approaches zero as the particle approaches $r = 2GM$. Hence, the particle apparently will not be able to go further than $r = 2GM$.

For the Gullstrand–Painlevé metric for incoming light the radial coordinate velocity is always larger in magnitude than -1, at $r = 2GM$ it is -2, for outgoing rays it vanishes at $r = 2GM$ and is negative for $r < 2GM$.

Taking the mere numerical values is misleading. Contemplate for incoming light

$$\frac{\left(\frac{dr}{dt}\right)_{\text{particle}}}{\left(\frac{dr}{dt}\right)_{\text{light}}} = \frac{1}{1+\sqrt{\frac{r}{2GM}}} \leq 1.$$

So the particle is always slower than light, however it approaches the velocity of light when approaching $r = 0$.

Table 2. Velocities in different coordinate systems.[f]

	Schwarzschild	Gullstrand–Painlevé
Particles		
Coordinate velocity $\frac{dr}{dt}$	$\pm(1-\frac{2GM}{r})\sqrt{\frac{2GM}{r}}$	$-\sqrt{\frac{2GM}{r}}$
Proper velocity $\frac{dr}{d\tau}$	$\pm\sqrt{\frac{2GM}{r}}$	$\pm\sqrt{\frac{2GM}{r}}$
Light rays		
Coordinate velocity $\frac{dr}{dt}$	$\pm(1-\frac{2GM}{r})$	$\pm 1-\sqrt{\frac{2GM}{r}}$

[f] The velocities of outgoing particles are valid only for the boundary condition specified. The *coordinate* velocity for outgoing particles in GP coordinates does not fit in our table and is thus suppressed.

The Gullstrand–Painlevé form of the metric is regular at the surface $r = 2GM$. This shows that it is not any kind of barrier, but this observation was not made until much later, see Eisenstadt.[52]

2.3. *Six classical representations of the Schwarzschild metric*

As we mentioned, a coordinate system should be chosen according to its convenience for describing a certain situation. In the following table (Table 3), we collect six widely used forms of the Schwarzschild metric.

Table 3. The Schwarzschild metric in various coordinates.

Schwarzschild metric in various coordinates		Coordinate transformation	Characteristic properties
Schwarzschild $ds^2 = -(1-\frac{2m}{r})dt^2 + \frac{1}{1-\frac{2m}{r}}dr^2 + r^2 d\Omega^2$	(t, r, θ, ϕ)	—	Area of spheres $r =$ const. is the "Euclidean" $4\pi r^2$
Isotropic $ds^2 = -\left(\frac{1-m/2\bar{r}}{1+m/2\bar{r}}\right)^2 dt^2 + \left(1+\frac{m}{2\bar{r}}\right)^4 \times \left(d\bar{r}^2 + \bar{r}^2 d\Omega^2\right)$	$(t, \bar{r}, \theta, \phi)$	$r = \bar{r}\left(1+\frac{m}{2\bar{r}}\right)^2$	Constant-curvature time slices
Eddington–Finkelstein $ds^2 = -(1-\frac{2m}{r})dv^2 + 2dvdr + r^2 d\Omega^2$	(u, r, θ, ϕ)	$v = t + r + 2m \times \ln\left\|\frac{r}{2m} - 1\right\|$	Ingoing light rays: $dv = 0$
Kerr–Schild $ds^2 = (\eta_{\alpha\beta} + 2m\ell_\alpha\ell_\beta)dx^\alpha dx^\beta$; $\ell_\alpha = \frac{1}{\sqrt{r}}\left(1, \frac{x}{\sqrt{r}}, \frac{y}{\sqrt{r}}, \frac{z}{\sqrt{r}}\right)$	$(\bar{t}, x, y, z)$	$\bar{t} = v - r$; $r^2 = x^2 + y^2 + z^2$	"Cartesian" coordinates
Lemaître $ds^2 = -dT^2 + \frac{2m}{r}dR^2 + r^2 d\Omega^2$, $r = \left[\frac{2\sqrt{2m}}{3}(R-T)\right]^{\frac{2}{3}}$	(T, R, θ, ϕ)	$dT = dt + \sqrt{\frac{2m}{r}}\frac{1}{1-\frac{2m}{r}}dr$; $dR = dt + \sqrt{\frac{r}{2m}}\frac{1}{1-\frac{2m}{r}}dr$	Infalling particles: $dR = 0$
Gullstrand–Painlevé $ds^2 = -(1-\frac{2m}{r})d\tilde{t}^2 + 2\sqrt{\frac{2m}{r}}d\tilde{t}dr + dr^2 + r^2 d\Omega^2$	$(\tilde{t}, r, \theta, \phi)$	$dt = d\tilde{t} - \frac{dr}{\sqrt{\frac{r}{2m}} - \sqrt{\frac{2m}{r}}}$	Infalling particles: $dr = -\sqrt{\frac{2m}{r}}d\tilde{t}$

2.4. *The concept of a Schwarzschild black hole*

We first draw a simple picture of a black hole. The event horizon and the stationary limit emerge as characteristic features. These are subsequently defined in a more mathematical way.

In 1783, John Michell communicated his thoughts *on the means of discovering the Distance, magnitude, etc. of the fixed stars, in consequence of the diminuation of the velocity of their light* ...[125] to the Royal Society in London. In the context of Newton's particle theory of light, he calculated that sufficiently massive stars exhibit a gravitational attraction to such vast an amount that even light could not escape. A few years later (1796) Pierre–Simon Laplace published similar ideas.

In modern notation, we may reconstruct the arguments as follows. We throw a mass m from the surface of the Earth, assuming that there were no air, in upward direction with an initial velocity v. It will always fall back, unless its initial velocity reaches a sufficiently high value v_{escape} providing the mass with such a kinetic energy that it can overpower the gravitational attraction of the Earth. Energy conservation yields then immediately the formula

$$v_{\text{escape}} = \sqrt{\frac{2GM_{\oplus}}{R_{\oplus}}},$$

where G is Newton's gravitational constant and M and $R_{\oplus}$ the mass and the radius of the spherically conceived Earth, respectively.

For the Earth we find $v_{\text{escape}} \approx 11.2\,\text{km/s}$. If we now compress the Earth appreciably (thought experiment!) until the escape velocity coincides with the speed of light $v_{\text{escape}} = c$, its compressed "Schwarzschild" radius becomes $r_{\oplus} = 2GM_{\oplus}/c^2 \approx 1\,\text{cm}$. For the Sun, with its mass $M_{\odot}$, we have[g]

$$r_{\odot} = \frac{2GM_{\odot}}{c^2} \approx 3\text{km}.$$

At any smaller radius, the light will be confined to the corresponding body. This is an intuitive picture of a spherically symmetric invisible "black hole".[h]

It is very intriguing to see how far-sighted Michell anticipated the status of today's observational black hole physics:

> *If there should really exist in nature bodies, whose density is not less than that of the sun, and whose diameters are more than* 500 *times the diameter of the sun, since their light could not arrive at us*; [...] *we could have no information from sight*; *yet, if any other luminous bodies should happen to revolve about them we might still perhaps from the motions of these revolving bodies infer the existence of the central ones with some degree of probability* ...

[g]For the sake of clarity, we display here the speed of light c explicitly.

[h]The phrase "black hole" is commonly associated with Wheeler (1968). It appears definitely earlier in the literature: In the January 1964 edition of the *Science News Letter*, the journalist Ann Ewing entitled her report at the meeting of the American Association for the Advancement of Science in Cleveland "*Black Holes*" *in space*. And if you have a look into an arbitrary English language dictionary published before ca. 1970, you will learn that "black hole" refers to a notorious dungeon in Calcutta (now Kolkata) in the 18th century, apparently a place of no return.

This could be a verdict on the current observations of the black hole Sgr A* ("Sagittarius A-star") in the center of our Milky Way — and this is *not* a thought experiment — for a popular account, see Sanders.[168] Sgr A* has a mass of about $4 \times 10^6 M_\odot$. Thus, its Schwarzschild radius is far from being minute, it is about $3 \times 4 \times 10^6$ km or about 17 solar radii.

A cautionary remark has to be made, though, see Penrose.[150] Newtonian gravity c has no absolute meaning like in special relativity. It is conceivable that the speed of light in strong Newtonian gravitational fields could be larger than c. Consequently, the Michell type argument becomes only pertinent if c is the maximal speed for all phenomena like in the Minkowski space of special relativity, or, if gravity is involved, in the Riemannian space of GR.

Let us follow the way of visualizing the black hole concept by means of everyday physics a bit further: We explore the Schwarzschild and, later in Sec. 3.4, the Kerr spacetime by boat. Schwarzschild spacetime is mimicked by a hole in a lake in which the surrounding water plunges simply radially without whirling around (Monticello Dam, California). The water flowing toward the hole will drag our boat to the center. Our boat may move around quite freely as long as the current is weak.

However, at some distance from the hole, the current becomes so strong that our boat, engines working at their maximum power, merely can keep its position. This is the *stationary limit*. In the case of our circularly symmetric water hole the stationary limit forms a ring. Bad for the boat: The stationary limit is also the ring of no return. At best, the boat remains at its position, it never will escape. Any millimeter across the stationary limit will doom the boat, it will be inevitably sucked into the throat. Accordingly, the stationary limit coincides in this spherically symmetric case with the so-called *event horizon*.

2.4.1. *Event horizon*

In 1958, Finkelstein[59] characterized the surface $r = 2m$ as a "semi-permeable membrane" in spacetime, that is, a surface which can be crossed only in one direction.

Fig. 7. Not quite seriously: "Schwarzschild" (left) versus "Kerr" (right). (For color version, see page I-CP2.)

As soon as our boat has passed the event horizon, it can never come back. This property can be formulated in an invariant way: The light cones at each point of the surface have to nestle tangentially to the membrane. In 1964, Penrose[149] termed the null cone which divides observable from unobservable regions an *event horizon.* Mathematically speaking, the event horizon is characterized by having *tangent vectors* which are *light-like* or null at all points. Therefore, the event horizon is a *null hypersurface.* This is what is meant by a *trapped surface,*[29] see Figs. 8 and 9, left image: a compact, spacelike, two-dimensional submanifold with the property that outgoing future-directed light rays converge in both directions everywhere on the submanifold. All these characterizations quite intuitively show up in the Penrose–Kruskal diagram to be discussed later.

In view of the preceding paragraph, we define a black hole as a region of spacetime separated from infinity by an event horizon, see Carroll[29] and Brill.[22]

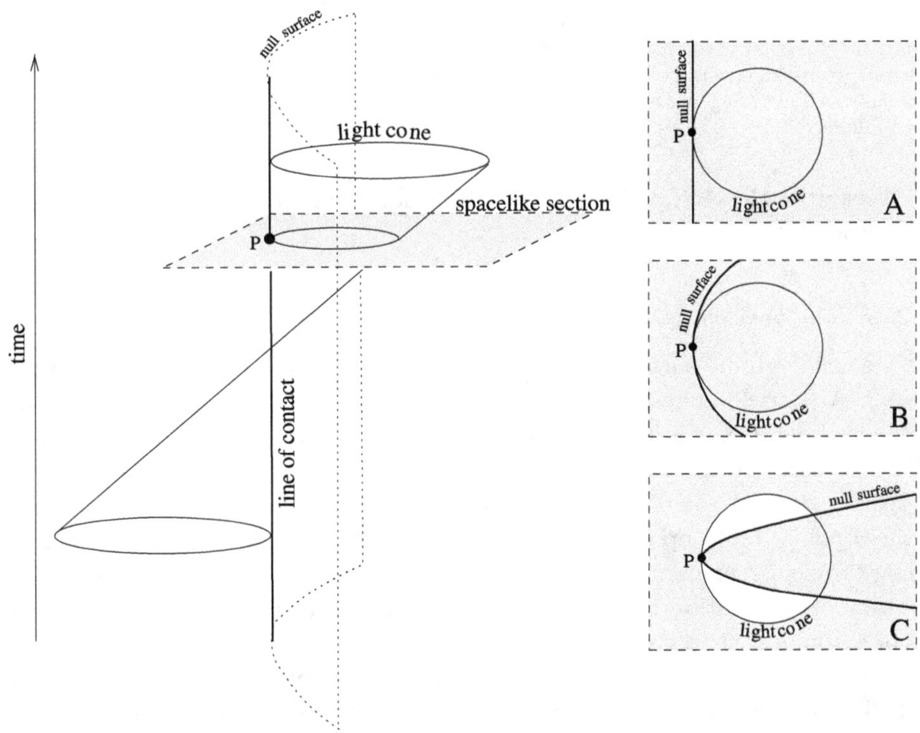

Fig. 8. A null hypersurface is not necessarily an event horizon: Imagine a light cone that touches a hypersurface along the line of contact. Thus, the light cone is tangent as well as normal (in a spacetime sense) to the surface. Consequently, all such surfaces are null hypersurfaces. In the cases A and B, the light cone is entirely trapped inside the surface. Case A suggests that the surface does not close in a finite region, therefore the enclosed volume is not compact. Case B represents a (part of a) circle, which encloses all tangential light cones, and this forms an (black hole) event horizon. In case C, the light cone intersects the hypersurface. The white domain is outside the null surface but inside the light cone and, thus, reachable from within the enclosed domain.

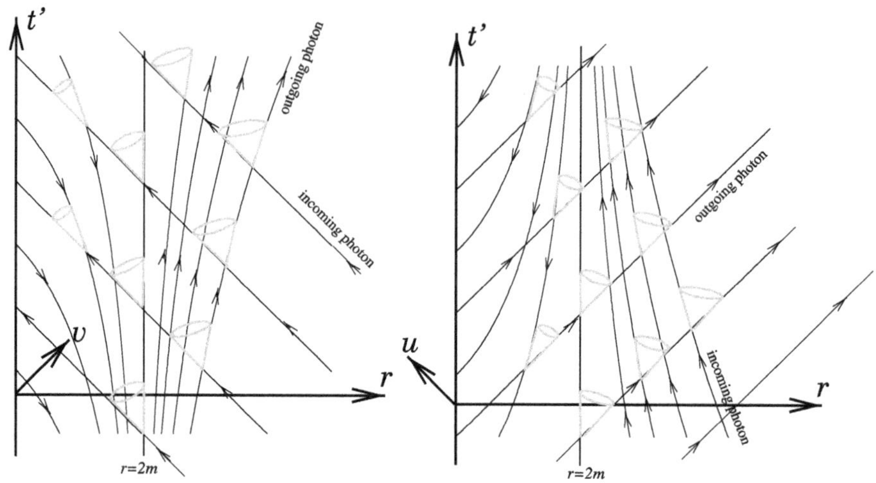

Fig. 9. In- and outgoing Eddington–Finkelstein coordinates (where we introduce t' with $v = t' + r$, $u = t' - r$). The arrows indicate the direction of the original Schwarzschild coordinate time (and thereby the direction of the Killing vector ∂_t). The *left* figure illustrates a *black hole*: All incoming photons traverse the event horizon and terminate in the singularity. The *right* figure illustrates a *white hole:* All outgoing photons emerge from the singularity, cross the horizon, and propagate out to infinity.

Observational evidence in favor of black holes was reviewed by Narayan and McClintock.[129]

2.4.2. *Killing horizon*

The stationary limit surface is rendered more precise in the notion of a *Killing horizon.* A particle at rest (with respect to the infinity of an asymptotically flat, stationary spacetime) is to be required to follow the trajectories of the timelike Killing vector.[i] However, if we have a Killing vector K describing a stationary spacetime, then at some points K may become lightlike, that is $K^\mu K_\mu = 0$. If all these points build up a hypersurface Σ, then this null hypersurface is called a Killing horizon. Apparently, this notion is of a local character, in contrast to the definition of an event horizon, the definition of which refers to events in the future, it is of a nonlocal character, see Fig. 8.

As we will see for the Schwarzschild black hole, see Fig. 9, outside the black hole the Killing vector is timelike, that is, $K^\mu K_\mu < 0$, on the Killing horizon it becomes

[i] Using the definitions of the covariant derivative and of the Christoffel symbols, we can derive the following equation for an arbitrary vector K,

$$K^\alpha \partial_\alpha g_{\mu\nu} = 2\nabla_{(\mu} K_{\nu)} - 2g_{\alpha(\mu}\partial_{\nu)} K^\alpha. \tag{39}$$

Assuming K^α and $g_{\mu\nu}$ to be constant in time, demands $\nabla_{(\mu} K_{\nu)} = 0$. Hence K has to be a Killing vector. In this coordinate system, we have $K^\alpha K_\alpha = g_{00}$. Although K acts as time translation, it is not necessarily timelike!

null $K^\mu K_\mu = 0$ (by definition of the horizon), and inside it becomes spacelike $K^\mu K_\mu > 0$.

In the Schwarzschild case it will turn out that the event horizon and Killing horizon coincide, in the Kerr case they separate.

2.4.3. *Surface gravity*

From the definition of the Killing horizon it can be shown[29] that the quantity

$$\kappa^2 := -\frac{1}{2}(\nabla_\mu K_\nu)(\nabla^\mu K^\nu)|_\Sigma \tag{40}$$

is constant on the Killing horizon and positive. The quantity κ is called *surface gravity*. In simple cases, it has the interpretation of an acceleration or gravitational force per unit mass on the horizon. In the Schwarzschild spacetime it takes the value $\kappa = 1/4m$, which is the acceleration of a particle with unit mass as seen from infinity, compare with the Newtonian "field strength" (2) for $r = 2m$:

$$f = \frac{GM}{r^2} = \frac{m}{(2m)^2} = \frac{1}{4m}. \tag{41}$$

In general, there is no such simple interpretation.

2.4.4. *Infinite redshift*

Another property associated with the surface $K^\mu K_\nu = 0$ is the infinite redshift. In view of the relation for the general relativistic time delay,

$$\tau_0(\mathbf{x}_B) = \frac{\sqrt{g_{tt}(\mathbf{x}_B)}}{\sqrt{g_{tt}(\mathbf{x}_A)}}\tau_0(\mathbf{x}_A).$$

$g_{tt} \to 0$ can be interpreted as follows. Consider $\tau_0(\mathbf{x}_B)$ the time measured by a clock B resting well away from the Killing horizon, whereas clock A with $\tau_0(\mathbf{x}_A)$ is nearly at the Killing horizon. If $g_{tt}(\mathbf{x}_A) \to 0$ we get $\tau_0(\mathbf{x}_B) \to \infty$. From the point of view of clock B, clock A's last signal, right before A hits the Killing horizon, will not reach B in a finite time, that is, never. To put it a little bit different: Signals sent with respect to A with constant frequency arrive increasingly delayed at B. For B the frequency approaches zero. This is called infinite redshift.

Let us work out these ideas for the Schwarzschild solution and let us take "photons" in spacetime instead of boats on a lake.

2.5. *Using light rays as coordinate lines*

Schwarzschild coordinates exhibit a coordinate singularity at $r = 2m$. *This obstructs the discussion of the event horizon considerably. As we have seen, light rays penetrate the horizon without difficulty. This suggests to use light rays as coordinate lines. Therefore, we introduce in- and outgoing Eddington–Finkelstein coordinates. By combining both, we arrive at Kruskal–Szekeres coordinates, which provide a regular coordinate system for the whole Schwarzschild spacetime.*

2.5.1. *Eddington–Finkelstein coordinates*

In relativity, light rays, the quasi-classical trajectories of photons, are null geodesics. In special relativity, this is quite obvious, since in Minkowski space the geodesics are straight lines and "null" just means $v = c$. A more rigorous argument involves the solution of the Maxwell equations for the vacuum and the subsequent determination of the normals to the wave surface (rays) which turn out to be null geodesics. This remains valid in general relativity. Null geodesics can be easily obtained by integrating the equation $0 = ds$. We find for the Schwarzschild metric, specializing to radial light rays with $d\phi = 0 = d\theta$

$$t = \pm\left(r + 2m\ln\left|\frac{r}{2m} - 1\right|\right) + \text{const.} \tag{42}$$

If we denote with r_0 the solution of the equation $r + 2m\ln\left|\frac{r}{2m} - 1\right| = 0$, we have for the t-coordinate of the light ray $t(r_0) =: v$. Hence, if $r = r_0$, we can use v to label light rays. In view of this, we introduce v and u

$$v := t + r + 2m\ln\left|\frac{r}{2m} - 1\right|, \tag{43}$$

$$u := t - r - 2m\ln\left|\frac{r}{2m} - 1\right|. \tag{44}$$

Then ingoing null geodesics are described by $v =$ const., outgoing ones by $u =$ const., see Fig. 9. We define *ingoing Eddington–Finkelstein coordinates* by replacing the "Schwarzschild time" t by v. In these coordinates (v, r, θ, ϕ), the metric becomes

$$ds^2 = -\left(1 - \frac{2m}{r}\right)dv^2 + 2dvdr + r^2 d\Omega^2. \tag{45}$$

For radial null geodesics $ds^2 = d\theta = d\phi = 0$, we find two solutions of (45), namely $v =$ const. and $v = 4m\ln|r/2m - 1| + 2r +$ const. The first one describes infalling photons, i.e. t increases if r approaches 0. At $r = 2m$, there is no singular behavior any longer for incoming photons. Ingoing Eddington–Finkelstein coordinates are particular useful in order to describe the gravitational collapse. Analogously, for outgoing null geodesics take (u, r, θ, ϕ) as new coordinates. In these *outgoing Eddington–Finkelstein coordinates* the metric reads

$$ds^2 = -\left(1 - \frac{2m}{r}\right)du^2 - 2dudr + r^2 d\Omega^2. \tag{46}$$

Outgoing light rays are now described by u = const., ingoing light rays by $u = -(4m \ln|r/2m - 1| + 2r)$ + const. In these coordinates, the hypersurface $r = 2m$ (the "horizon") can be recognized as a null hypersurface (its normal is null or lightlike) and as a semi-permeable membrane.

2.5.2. *Kruskal–Szekeres coordinates*

Next we try to combine the advantages of in- and outgoing Eddington–Finkelstein coordinates in the hope to obtain a fully regular coordinate system of the Schwarzschild spacetime. Therefore, we assume coordinates (u, v, θ, ϕ). Some (computer) algebra yields the corresponding representation of the metric:

$$ds^2 = -\left(1 - \frac{2m}{r(u,v)}\right) du dv + r^2(u,v) d\Omega^2. \tag{47}$$

Unfortunately, we still have a coordinate singularity at $r = 2m$. We can get rid of it by reparametrizing the surfaces u = const. and v = const. via

$$\tilde{v} = \exp\left(\frac{v}{4m}\right), \quad \tilde{u} = -\exp\left(-\frac{u}{4m}\right). \tag{48}$$

In these coordinates, the metric reads [$r = r(\tilde{u}, \tilde{v})$ is implicitly given by (48) and (44), (43), $r_S = 2m$]

$$ds^2 = -\frac{4r_S^3}{r(\tilde{u},\tilde{v})} \exp\left(-\frac{r(\tilde{u},\tilde{v})}{2m}\right) d\tilde{v} d\tilde{u} + r^2(\tilde{u},\tilde{v}) d\Omega^2. \tag{49}$$

Again, we go back from $\tilde{u}$ and $\tilde{v}$ to time and spacelike coordinates

$$\tilde{t} := \frac{1}{2}(\tilde{v} + \tilde{u}), \quad \tilde{r} := \frac{1}{2}(\tilde{v} - \tilde{u}). \tag{50}$$

In terms of the original Schwarzschild coordinates we have

$$\tilde{r} = \sqrt{\left|\frac{r}{2m} - 1\right|} \exp\left(\frac{r}{4m}\right) \cosh \frac{t}{4m}, \tag{51}$$

$$\tilde{t} = \sqrt{\left|\frac{r}{2m} - 1\right|} \exp\left(\frac{r}{4m}\right) \sinh \frac{t}{4m}. \tag{52}$$

The Schwarzschild metric

$$ds^2 = \frac{4r_S^3}{r} \exp\left(-\frac{r}{2m}\right) \left(-d\tilde{t}^2 + d\tilde{r}^2\right) + r^2 d\Omega^2, \tag{53}$$

in these *Kruskal–Szekeres* coordinates $(\tilde{t}, \tilde{r}, \theta, \phi)$, behaves regularly at the gravitational radius $r = 2m$. If we substitute (53) into the Einstein equation (via computer algebra), then we see that it is a solution for all $r > 0$. Equations (51) and (52) yield

$$\tilde{r}^2 - \tilde{t}^2 = \left|\frac{r}{2m} - 1\right| \exp\left(\frac{r}{2m}\right). \tag{54}$$

Thus, the transformation is valid only for regions with $|\tilde{r}| > \tilde{t}$. However, we can find a set of transformations which cover the entire $(\tilde{t}, \tilde{r})$-space. They are valid in different domains, indicated here by I, II, III and IV, to be explained below:

$$\text{(I)}\quad \begin{cases} \tilde{t} = \sqrt{\dfrac{r}{2m} - 1}\, \exp\!\left(\dfrac{r}{4m}\right) \sinh \dfrac{t}{4m} \\ \tilde{r} = \sqrt{\dfrac{r}{2m} - 1}\, \exp\!\left(\dfrac{r}{4m}\right) \cosh \dfrac{t}{4m} \end{cases}, \tag{55}$$

$$\text{(II)}\quad \begin{cases} \tilde{t} = \sqrt{1 - \dfrac{r}{2m}}\, \exp\!\left(\dfrac{r}{4m}\right) \cosh \dfrac{t}{4m} \\ \tilde{r} = \sqrt{1 - \dfrac{r}{2m}}\, \exp\!\left(\dfrac{r}{4m}\right) \sinh \dfrac{t}{4m} \end{cases}, \tag{56}$$

$$\text{(III)}\quad \begin{cases} \tilde{t} = -\sqrt{\dfrac{r}{2m} - 1}\, \exp\!\left(\dfrac{r}{4m}\right) \sinh \dfrac{t}{4m} \\ \tilde{r} = -\sqrt{\dfrac{r}{2m} - 1}\, \exp\!\left(\dfrac{r}{4m}\right) \cosh \dfrac{t}{4m} \end{cases}, \tag{57}$$

$$\text{(IV)}\quad \begin{cases} \tilde{t} = -\sqrt{1 - \dfrac{r}{2m}}\, \exp\!\left(\dfrac{r}{4m}\right) \cosh \dfrac{t}{4m} \\ \tilde{r} = -\sqrt{1 - \dfrac{r}{2m}}\, \exp\!\left(\dfrac{r}{4m}\right) \sinh \dfrac{t}{4m} \end{cases}. \tag{58}$$

The inverse transformation is given by

$$\left(\frac{r}{2m} - 1\right) \exp\!\left(\frac{r}{2m}\right) = \tilde{r}^2 - \tilde{t}^2, \tag{59}$$

$$\frac{t}{4m} = \begin{cases} \operatorname{Artanh}\dfrac{\tilde{t}}{\tilde{r}}, & \text{for (I) and (III)}, \\ \operatorname{Artanh}\dfrac{\tilde{r}}{\tilde{t}}, & \text{for (II) and (IV)}. \end{cases} \tag{60}$$

The Kruskal–Szekeres coordinates $(\tilde{t}, \tilde{r}, \theta, \phi)$ cover the entire spacetime, see Fig. 10. By means of the transformation equations we recognize that we need two Schwarzschild coordinate systems in order to cover the same domain. Regions (I) and (III) both correspond each to an asymptotically flat universe with $r > 2m$. Regions (II) and (IV) represent two regions with $r < 2m$. Since $\tilde{t}$ is a time coordinate, we see that the regions are time reversed with respect to each other. Within these regions, real physical singularities (corresponding to $r = 0$) occur along the curves $\tilde{t}^2 - \tilde{r}^2 = 1$. From the form of the metric we can infer that radial light-like geodesics (and therewith the light cones $ds = 0$) are lines with slope 1. This makes the discussion of the causal structure particularly simple.

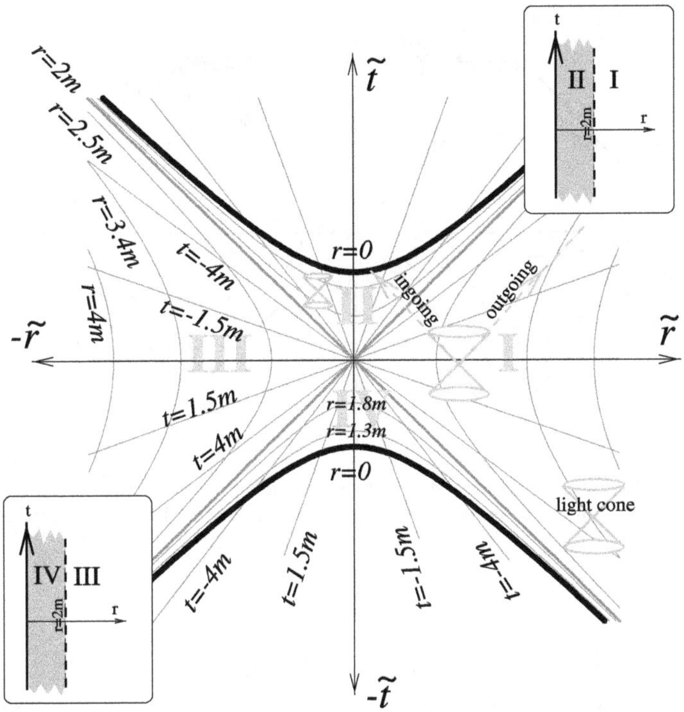

Fig. 10. Kruskal–Szekeres diagram of the Schwarzschild spacetime.

2.6. *Penrose–Kruskal diagram*

We represent the Schwarzschild spacetime in a manner analogous to the Penrose diagram of the Minkowski spacetime. To this end, we proceed along the same line as in the Minkowskian case.

First, we switch again to null-coordinates $v' = \tilde{t} + \tilde{r}$ and $u' = \tilde{t} - \tilde{r}$ and perform a conformal transformation which maps infinity into the finite (again, by means of the tangent function). Finally, we return to a time-like coordinate $\hat{t}$ and a space-like coordinate $\hat{r}$. We perform these transformations all in one according to

$$\tilde{t} + \tilde{r} = \tan \frac{\hat{t} + \hat{r}}{2}, \tag{61}$$

$$\tilde{t} - \tilde{r} = \tan \frac{\hat{t} - \hat{r}}{2}. \tag{62}$$

The Schwarzschild metric then reads

$$ds^2 = \frac{r_S^3}{r(\hat{r}, \hat{t})} \frac{\exp\left(-\frac{r(\hat{r}, \hat{t})}{2m}\right)\left(-d\hat{t}^2 + d\hat{r}^2\right)}{\cos^2 \frac{\hat{t} + \hat{r}}{2} \cos^2 \frac{\hat{t} - \hat{r}}{2}} + r^2(\hat{t}, \hat{r}) d\Omega^2, \tag{63}$$

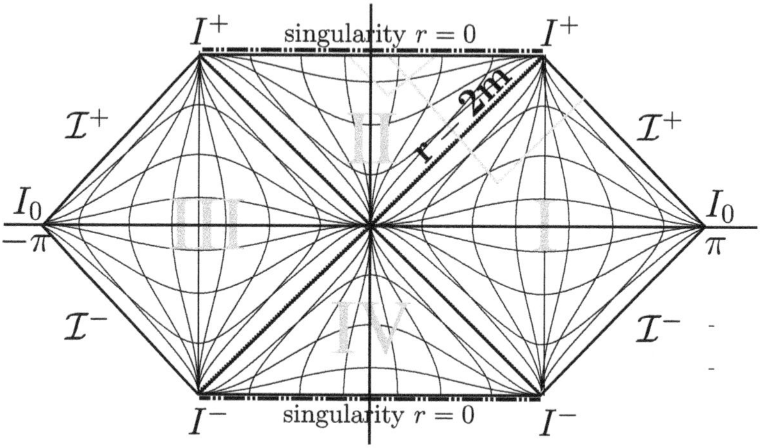

Fig. 11. Penrose–Kruskal diagram of the Schwarzschild spacetime. Region II corresponds to a black hole, region IV to a white hole. Regions I and III correspond to two universes.

where the function $r(\hat{t}, \hat{r})$ is implicitly given by

$$\left(\frac{r}{2m} - 1\right) \exp\left(\frac{r}{2m}\right) = \tan \frac{\hat{t} + \hat{r}}{2} \tan \frac{\hat{t} - \hat{r}}{2}. \tag{64}$$

The corresponding Penrose–Kruskal diagram is displayed in Fig. 11. The notations for the different infinities can be extracted from Table 1. In contrast to Minkowski space, light rays and particles may not escape to infinity, but enter the black hole (II). Likewise, light rays and particles may not emerge from infinity, but from the white hole (IV).

2.7. *Adding electric charge and the cosmological constant: Reissner–Nordström*

As mentioned in the historical remarks, soon after Schwarzschild's solution, the first generalizations, including electric charge and the cosmological constant were published. We can be even quicker. We already calculated the Ricci tensor for the Gullstrand–Painlevé ansatz. If we use the well-known energy–momentum tensor for a point charge,[81] the field equation may be written as[j]

$$R_\mu{}^\nu - \frac{1}{2} R \delta_\mu^\nu + \Lambda \delta_\mu^\nu = \kappa \lambda_0 \frac{q^2}{2r^4} \mathrm{diag}(-1, -1, 1, 1). \tag{65}$$

Taking the trace, we find $R = 4\Lambda$ and arrive at

$$R_2{}^2 = R_3{}^3 = \frac{2\partial_r(r\psi)}{r^2} = \Lambda - \frac{q^2}{r^4}. \tag{66}$$

[j]Einstein's gravitational constant is denoted by κ, $\lambda_0 = \sqrt{\frac{\varepsilon_0}{\mu_0}}$ is the admittance of the vacuum. With $c = 1$ and $G = 1$ we have $\kappa\lambda_0 = 2$.

This equation can be integrated elementarily

$$2\psi = \frac{1}{3}\Lambda r^2 - \frac{q^2}{r^2} + \frac{2\alpha}{r}. \tag{67}$$

This function also solves the remaining two field equations. The integration constant α is again the mass m. Substituted into (37) and transformed to Schwarzschild coordinates ($f = 1 - 2\psi$) the solution reads

$$ds^2 = -f(r)\,dt^2 + \frac{dr^2}{f(r)} + r^2 d\Omega^2, \tag{68}$$

with

$$f(r) := 1 - \frac{2m}{r} + \frac{q^2}{r^2} - \frac{\Lambda}{3}r^2. \tag{69}$$

A detailed derivation using Schwarzschild coordinates and computer algebra can be found in Puntigam *et al.*[158]

A discussion of the Reissner–Nordström(-de Sitter) solution can be found in Griffiths and Podolsky,[69] for example. We only remark that we recover the Schwarzschild solution for $q = 0$ and $\Lambda = 0$. The algebraic structure of the solution is identical to the Schwarzschild case. Thus, we find, in general, a singularity at $r = 0$. However, a pure cosmological solution, $m = 0, q = 0$ and $\Lambda \neq 0$, possesses no singularity and no horizon! On the other hand, an electrically charged black hole, $\Lambda = 0$, exhibits *two* horizons,

$$f(r) = 0 \Leftrightarrow r_\pm = m \pm \sqrt{m^2 - q^2}. \tag{70}$$

In this respect, the charged black hole shows some similarities to a rotating (Kerr) black hole. We will pick up this discussion in Sec. 3.4.

2.8. *The interior Schwarzschild solution and the TOV equation*

In the last section, we investigated the gravitational field outside a spherically symmetric mass-distribution. Now it is time to have a look inside matter, see Adler *et al.*[1] Of course, in a first attempt, we have to make decisive simplifications on the internal structure of a star. We will consider cold catalyzed stellar material during the later phase of its evolution which can be reasonably approximated by a perfect fluid. The typical mass densities are in the range of $\approx 10^7\,\mathrm{g/cm^3}$ (white dwarfs) or $\approx 10^{14}\,\mathrm{g/cm^3}$ (neutron stars, e.g. pulsars). In this context, we assume vanishing angular momentum.

We start again from a static and spherically symmetric metric

$$ds^2 = -\mathrm{e}^{A(r)}c^2dt^2 + \mathrm{e}^{B(r)}dr^2 + r^2 d\Omega^2 \tag{71}$$

and the energy–momentum tensor

$$T_{\mu\nu} = \left(\rho + \frac{p}{c^2}\right) u_\mu u_\nu + p g_{\mu\nu}, \tag{72}$$

where $\rho = \rho(r)$ is the spherically symmetric mass density and $p = p(r)$ the pressure (isotropic stress). This has to be supplemented by the equation of state which, for a simple fluid, has the form $p = p(\rho)$.

We compute the nonvanishing components of the field equation by means of computer algebra as (here $\kappa = 8\pi G/c^4$ is Einstein's gravitational constant and $()' = d/dr$)

$$-e^B \kappa r^2 c^2 \rho + e^B + B'r - 1 = 0, \tag{73}$$

$$-e^B \kappa p r^2 - e^B + A'r + 1 = 0, \tag{74}$$

$$-4e^B \kappa p r + 2A''r + (A')^2 r - A'B'r + 2A' - 2B' = 0. \tag{75}$$

The (ϕ, ϕ)-component turns out to be equivalent to the (θ, θ)-component. For convenience, we define a *mass function* $m(r)$ according to

$$e^{-B} =: 1 - \frac{2m(r)}{r}. \tag{76}$$

We can differentiate (76) with respect to r and find, after substituting (73), a differential equation for $m(r)$ which can be integrated, provided $\rho(r)$ is assumed to be known

$$m(r) = \int_0^r \frac{\kappa}{2} \rho c^2 \xi^2 d\xi. \tag{77}$$

Differentiating (74) and using all three components of the field equation, we obtain a differential equation for A

$$A' = -\frac{2p'}{p + \rho c^2}. \tag{78}$$

We can derive an alternative representation of A' by substituting (76) into (74). Then, together with (78), we arrive at the *Tolman–Oppenheimer–Volkoff* (TOV) equation

$$p' = -\frac{(\boldsymbol{\rho} c^2 + p)\left(\mathbf{m} + \frac{\kappa p r^3}{2}\right)}{\mathbf{r}(\mathbf{r} - 2m)}. \tag{79}$$

Terms that survive in the Newtonian limit are emphasized by boldface letters. The system of equations consisting of (77), (78), the TOV equation (79), and the equation of state $p = p(\rho)$ forms a complete set of equations for the unknown functions $A(r), \rho(r), p(r)$, and $m(r)$, with

$$ds^2 = -e^{A(r)} c^2 dt^2 + \frac{dr^2}{1 - \frac{2m(r)}{r}} + r^2 d\Omega^2. \tag{80}$$

These differential equations have to be supplemented by initial conditions.

In the center of the star, there is, of course, no enclosed mass. Hence, we demand $m(0) = 0$. The density has to be finite at the origin, i.e. $\rho(0) = \rho_c$, where ρ_c is the density of the central region. At the surface of the star, at $r = R_\odot$, we have to match matter with vacuum. In vacuum, there is no pressure which requires $p(R_\odot) = 0$. Moreover, the mass function should then yield the total mass of the star, $m(R_\odot) := GM_\odot/c^2$. Finally, we have to match the components of the metric. Therefore, we have to demand $\exp[A(r_0)] = 1 - 2m(R_\odot)/R_\odot$.

Equations (73), (74), (75) and certain regularity conditions which generalize our boundary conditions, that is,

- regularity of the geometry at the origin,
- finiteness of central pressure and density,
- positivity of central pressure and density,
- positivity of pressure and density,
- monotonic decrease of pressure and density,

impose conditions on the functions ρ and p. Then, even without the explicit knowledge of the equation of state, the general form of the metric can be determined. For recent work, see Rahman and Visser[162] and the literature given there.

We can obtain a simple solution, if we assume a constant mass density

$$\rho = \rho(r) = \text{const.} \tag{81}$$

One should mention here that ρ is not the physically observable fluid density, which results from an appropriate projection of the energy–momentum tensor into the reference frame of an observer. Thus, this model is not as unphysical as it may look at the first. However, there are serious but more subtle objections which we will not discuss further in this context.

When $\rho =$ const., we can explicitly write down the mass function (77)

$$m(r) = \frac{r^3}{2\widehat{R}^2}, \quad \text{with } \widehat{R} = \sqrt{\frac{3}{\kappa\rho c^2}}, \quad m_\odot := \frac{R_\odot^3}{2\widehat{R}^2}. \tag{82}$$

This allows immediately to determine one metric function

$$e^B = \frac{1}{1 - \dfrac{r^2}{\widehat{R}^2}}. \tag{83}$$

The TOV equation (79) factorizes according to

$$\frac{dp}{dr} = -\frac{1}{2}(\rho c^2 + p)(1 + \kappa\widehat{R}^2 p)\frac{r}{\widehat{R}^2 - r^2}. \tag{84}$$

It can be elementarily solved by separation of variables

$$p(r) = \rho c^2 \frac{\sqrt{\widehat{R}^2 - R_\odot^2} - \sqrt{\widehat{R}^2 - r^2}}{\sqrt{\widehat{R}^2 - r^2} - 3\sqrt{\widehat{R}^2 - R_\odot^2}}. \tag{85}$$

Using (78) as $A' = -2\left[\ln(p + \rho c^2)\right]'$ and continuous matching to the exterior, eventually yields the *interior and exterior Schwarzschild solution* for a spherically symmetric body[173]

$$ds^2 = \begin{cases} -\left(\frac{3}{2}\sqrt{1-\frac{R_\odot^2}{\widehat{R}^2}} - \frac{1}{2}\sqrt{1-\frac{r^2}{\widehat{R}^2}}\right)^2 c^2 dt^2 + \frac{1}{1-\frac{r^2}{\widehat{R}^2}} dr^2 + r^2 d\Omega^2, & r \le R_\odot, \\ -\left(1-\frac{2m_\odot}{r}\right) c^2 dt^2 + \frac{1}{1-\frac{2m_\odot}{r}} dr^2 + r^2 d\Omega^2, & r > R_\odot. \end{cases} \tag{86}$$

The solution is only defined for $R_\odot < \widehat{R}$. For the Sun[k] we have $M_\odot \approx 2 \times 10^{30}\,\mathrm{kg}$, $R_\odot \approx 7 \times 10^8\,\mathrm{m}$ and accordingly $\rho_\odot \approx 1.4 \times 10^3\,\mathrm{kg/m^3}$. This leads to $\widehat{R} \approx 3 \times 10^{11}\,\mathrm{m}$, that is, the radius of the star $R_\odot$ is much smaller than $\widehat{R}$: $R_\odot < \widehat{R}$. Hence, the square roots in (86) remain real.

The condition $R_\odot < \widehat{R}$ suggests that a sufficiently massive object cannot be stable since no static gravitational field seems possible. This conjecture can be further supported. Even before reaching $\widehat{R}$, the central pressure becomes infinite

$$p(0) \to \infty \quad \text{for } R_\odot \to \sqrt{\frac{8}{9}}\widehat{R}, \quad \text{or} \quad m_\odot \to \frac{4}{9} R_\odot. \tag{87}$$

If there is no static solution and the situation remains spherically symmetric, we are forced to the conclusion that such a mass distribution must radially collapse. Either in an infinite time or to a single point in space. With reasonable simplifications, it was first shown by Oppenheimer and Snyder[145] that the second alternative is true: A very massive object collapses to a black hole. As various singularity theorems show today, this behavior is indeed generic, see Chruściel *et al.*[38] and Sec. 3.10.

[k] To ascertain the consistency of dimensions and units, we recollect the basic definitions:

$$[G] = \frac{(\mathrm{m/s})^4}{\mathrm{N}} = \frac{\mathrm{m}^3}{\mathrm{kg\,s^2}}, \qquad \kappa = \frac{8\pi G}{c^4}.$$

The mass M carries the unit kg, the mass *parameter* has the dimension of a length:

$$m := \frac{GM}{c^2}, \qquad [m] = \frac{\mathrm{m^3 kg\,s^2}}{\mathrm{kg\,s^2\,m^2}} = \mathrm{m}.$$

The definition of $m(r)$ in Eq. (77) is consistent. For $\rho = \mathrm{const.}$, we have

$$m(r) = \frac{\kappa}{2}\rho c^2 \frac{1}{3} r^3 = \frac{G}{c^2}\frac{4}{3}\pi r^3 \rho = \frac{GM(r)}{c^2}.$$

Here ρ denotes the physical mass density, $[\rho] = \mathrm{kg/m^3}$. Thus

$$M(r) := \frac{4}{3}\pi r^3 \rho$$

is the physical mass with the unit kg.

3. The Kerr Metric (1963)

After some historical reminiscences (Sec. 3.1), we point out how one can arrive at the Kerr metric (Sec. 3.2). For that purpose, we derive, in cylindrical coordinates, the four corresponding partial differential equations and explain how this procedure leads to the Kerr metric. In Sec. 3.3, we display the Kerr metric in three classical coordinate systems. Thereafter, we develop the concept of the Kerr black hole (Sec. 3.4). In Secs. 3.5–3.7, we depict and discuss the geometrical/kinematical properties of the Kerr metric. Subsequently, in Sec. 3.8, we turn to the multipole moments of the mass and the angular momentum of the Kerr metric, stressing analogies to electromagnetism. In Sec. 3.9, we present the Kerr–Newman solution with electric charge. Eventually, in Sec. 3.10, we wonder in which sense the Kerr black hole is distinguished from the other stationary axially symmetric vacuum spacetimes, and, in Sec. 3.11, we mention the rotating disk metric of Neugebauer–Meinel as a relevant interior solution with matter.

> *... When I turned to Alfred Schild, who was still sitting in the armchair smoking away, and said "Its rotating!" he was even more excited than I was. I do not remember how we celebrated, but celebrate we did!*
>
> Roy P. Kerr (2009)

3.1. *Historical remarks*

The search for axially symmetric solutions of the Einstein equation started in 1917 *with static and was extended in* 1924 *to stationary metrics. It culminated in* 1963 *with the discovery of the Kerr metric.*

The Schwarzschild solution, as we have seen, describes the gravitational field of a spherically symmetric body. Obviously, most planets, moons, and stars rotate so that spherical symmetry is lost and one spatial direction is distinguished by the three-dimensional angular momentum vector $\boldsymbol{J}$ of the body. Hence, the next problem to attack was to search for the gravitational field of a massive rotating body.

When one considers a *static* and axially symmetric situation — this is the case if the body does not carry angular momentum — then one can choose the rotation axis as the z-axis of a cylindrical polar coordinate system: $x^1 = z$, $x^2 = \rho$ and $x^3 = \phi$. Then static axial symmetry means that the components of the metric $g_{\mu\nu} = g_{\mu\nu}(z, \rho)$ do not depend on the time t and the azimuthal angle ϕ (we have here one timelike and one spacelike Killing vector[1]).

[1]*Remark on Killing vectors*: Consider a point P of spacetime with coordinates x^α. We specify a direction ξ^μ at P. If we have a flat Minkowski space, the components $g_{\mu\nu}$ of the metric, given in Cartesian coordinates, would not change under a motion in the ξ-direction. However, in a curved spacetime, the $g_{\mu\nu}$ will change in general. If ξ^μ fulfills the Killing equations (see Stephani[179])

$$\nabla_\mu \xi_\nu + \nabla_\nu \xi_\mu = 0, \tag{88}$$

Already in 1917, Weyl[188] started to investigate static axially symmetric vacuum solutions of Einstein's field equation. He took cylindrical coordinates and proposed the following "canonical" form of the static axisymmetric vacuum line element[m]:

$$ds^2 = f dt^2 - \left\{ h(dz^2 + d\rho^2) + \frac{\rho^2 d\phi^2}{f} \right\}. \tag{89}$$

Here $f = f(z, \rho)$ and $h = h(z, \rho)$ and $(x^0 = t, x^1 = z, x^2 = \rho, x^3 = \phi)$. Weyl was led, in analogy to Newton's theory, to a Poisson equation and found thereby a family of static cylindrically symmetric solutions that could be understood as the exterior field of a line distribution of mass along the rotation axis. Similar investigations were undertaken by Levi-Civita[108] (1917/19).

In the year 1918, Lense and Thirring[107] investigated a rotating body. They specified the energy–momentum tensor of a slowly rotating ball of matter of homogeneous density and integrated the Einstein equation in lowest approximation. They found, for a ball rotating around the z-axis of a spatial Cartesian coordinate system, the linearized Schwarzschild solution in isotropic coordinates, see Table 2, together with two new "gravitomagnetic" correction terms in off-diagonal components of the metric (κ is Einstein's gravitational constant)

$$ds^2 = \underbrace{\left(1 - \frac{2\kappa M}{r}\right) dt^2 - \left(1 + \frac{2\kappa M}{r}\right)(dx^2 + dy^2 + dz^2)}_{\text{linearized Schwarzschild}} \underbrace{- \frac{4\kappa J_z}{r^3}(xdy - ydx)dt}_{\text{gravitomagnetic term}}. \tag{90}$$

This is valid for $\kappa M \ll r$ and $\kappa J_z \ll r^2$. This gravitomagnetic effect ("the Lense–Thirring effect") is typical for GR: in Newton's theory a rotating rigid ball has the same gravitational field as a nonrotating one. Gravitomagnetism is alien to Newton's gravitational theory.

In the meantime, the Lense–Thirring effect has been experimentally confirmed by the Gravity Probe B experiment, see Everitt *et al.*[55] They took a gyroscope in a satellite falling freely around the (rotating) Earth. The spin axis of the gyroscope pointed to a fixed guide star. Because of the gravitomagnetic term in (90), the gyroscope executed a (very small) *Lense–Thirring precession.*[n] This can be understood as an interaction of the spin of the gyroscope with the spin of the Earth (spin–spin interaction). Since the gyroscope moves along a 4D geodesic of a spacetime curved by the mass of the Earth, an additional *geodetic precession* occurs that has to be

with ∇ as covariant derivative operator, then ξ^μ is called a *Killing vector*, and this vector specifies a direction under which the metric does not change. The Schwarzschild metric is static, that is, it has one timelike Killing vector along the time coordinate. Furthermore, it is spherically symmetric and thus has three additional spacelike Killing vectors. In the Weyl case, because of the axial symmetry around the z-axis, two of those spacelike Killing vectors get lost. Left over in the Weyl case are the two Killing vectors, one timelike ${}^{(1)}\xi^t = \partial_t$ and one spacelike ${}^{(2)}\xi^\phi = \partial_\phi$.

[m] Weyl used $\rho \to r$, $\phi \to \vartheta$.

[n] For related experiments, see Ciufolini *et al.*[34,35] and Iorio *et al.*[89,92] A recent comprehensive review was given by Will.[189] A textbook presentation may be found in Ohanian and Ruffini.[143]

experimentally separated from the Lense–Thirring term. The geodetic precession had already been derived earlier by *de Sitter*[43] in 1916.[o]

In spherical polar coordinates we have $ydx - xdy = r^2 \sin^2\theta d\phi$. Thus, the gravitomagnetic cross-term in (90) may be rewritten as $(4\kappa J_z \sin^2\theta/r)dtd\phi$. A comparison with (89) shows that the canonical Weyl form of the static metric is too narrow for describing rotating bodies.

From 1919 on, there appeared further articles on axisymmetric solutions. Levi-Civita[p] (1919) reacted to Weyl's article, and Bach[8] (1922) pushed the Lense–Thirring line element to the second order in the approximation.

Then, in 1924, Lanczos[105] extending the Weyl ansatz, started to investigate *stationary*[q] solutions. He found an exact solution for uniformly rotating dust. However, his work was apparently partially overlooked. Later, Akeley[2,3] (1931), Andress[6] (1930) and, in a more definite form, Lewis[109] (1932) generalized the static Weyl metric to a stationary one by taking into account the gravitomagnetic term of Lense–Thirring. Lewis (1932) wrote, in cylindrical polar coordinates ($x_1 \rightsquigarrow \rho, x_2 \rightsquigarrow z$),

$$ds^2 = fdt^2 - (e^\mu dx_1{}^2 + e^\nu dx_2{}^2 + ld\phi^2) - 2mdtd\phi. \tag{91}$$

He found some exact solutions, typically for rotating cylinders, but not for rotating balls. It became definitely clear that, in the axially symmetric case, we may have many different exact vacuum solutions, in contrast to the case of spherical symmetry with, according to the Birkhoff theorem, the Schwarzschild solution as being unique.

Not much later, van Stockum[184] (1937) determined the gravitational field of an infinite rotating cylinder of dust particles, thereby recovering the Lanczos solution, inter alia. He fitted one of the interior matter solutions of Lewis to an exterior vacuum solution. Continuing on this line of research, Papapetrou[148] (1953) started from the Andress-Lewis line element, putting it in a slightly different form, suitable for all stationary axisymmetric vacuum solutions:

$$ds^2 = -e^\mu(d\rho^2 + dz^2) - ld\phi^2 - 2md\phi dt + fdt^2. \tag{92}$$

The functions μ, l, m and f depend only on ρ and z. Papapetrou integrated the field equations and found exact stationary rotating vacuum solutions. However, his solution carried either *mass and no angular momentum* or *angular momentum and no mass.* Thus,[148] "this solution is very special and physically of little interest."

A year later, a new result was published, which gave the problem of finding solutions for a rotating ball a new direction. Petrov[152] (1954), from Kazan, classified algebraically the Einstein vacuum field, that is, the Weyl curvature tensor, according to its eigenvalues and eigenvectors. This information reached the West, in the time

[o]De Sitter had applied it to the Earth–Moon system conceived as a gyroscope precessing around the Sun (the rotation of which can be neglected). This effect can nowadays be measured by Lunar Laser Ranging, see Will.[189]

[p]See Ref. 108, note 8 with the subtitle "*Soluzioni binarie di Weyl*".

[q]Stationary spacetimes are those that admit a time-like Killing vector. Static spacetimes are stationary spacetimes for which this Killing vector is hypersurface orthogonal. Physically this implies time reversal invariance and thus the absence of rotation.

of the Cold War, with some delay. A bit later, Pirani[154] (1957) developed a related formalism. It was the Petrov classification and the picking of a suitable class for the gravitational field of an isolated body (Petrov class D, with two double principal null directions) that finally led to the discovery of the Kerr solution during 1963, ten years after the unphysical solutions of Papapetrou.

Accordingly, it turned out to be a formidable task to find an exact solution for a rotating ball and it was only found nearly half a century after the publication of Einstein's field equation, namely in 1963 by Roy Kerr,[98] a New Zealander, who worked at the time in Texas within the research group of Alfred Schild. It is a two-parameter solution of Einstein's vacuum field equation with mass M and rotation (or angular momentum) parameter $a := J/M$.

The story of the discovery of the Kerr solution was told by Kerr himself at a conference on the occasion of his 70th birthday.[99] A decisive starting point of Kerr's investigations was, as mentioned, the Petrov classification. Melia, in his popular book[124] "Cracking the Einstein Code", which does not contain any mathematical formula — apart from those appearing in two copies of Kerr's notes and on a blackboard in another figure — has told this fascinating battle for solving Einstein's equation, see also the Kerr story in Ferreira.[58]

Dautcourt[40] discussed the work of people who were involved in this search for axially symmetric solutions but who were not so fortunate as Kerr. In particular, Dautcourt himself got this problem handed over from Papapetrou in 1959 as a subject for investigation. He used the results of Papapetrou (1953). Dautcourt's scholarly article is an interesting complement to Melia's book. In particular, it becomes clear that the (Lanczos–Akeley–Andress–Lewis-)Papapetrou line element (92) was the correct ansatz for the stationary axially symmetric metric and the Kerr metric is a special case therefrom. The Papapetrou approach with the line element (92) was later, after Kerr's discovery, brought to fruition by Ernst[53] and by Kramer and Neugebauer.[100]

3.2. *Approaching the Kerr metric*

We derive a second order partial differential equation, the Ernst equation, that governs the stationary axially symmetric metrics in Einstein's theory. Subsequently, we sketch how the Kerr solution emerges as a simple case therefrom.

3.2.1. *Papapetrou line element and vacuum field equation*

In more modern literature, the Papapetrou line element (92), which describes some rotation around the axis with $\rho = 0$, is usually parametrized as follows[r]:

$$\begin{aligned} ds^2 &= f(dt - \omega d\phi)^2 - f^{-1}[e^{2\gamma}(d\rho^2 + dz^2) + \rho^2 d\phi^2], \\ &\quad t \in (-\infty, \infty), \quad \rho \in [0, \infty), \quad z \in (-\infty, \infty), \quad \phi \in [0, 2\pi); \end{aligned} \tag{93}$$

[r]See Ernst,[53] Buchdahl,[23] de Felice and Clarke,[57] Quevedo,[160] O'Neill,[144] Stephani *et al.*,[180] Eq. (19.21), Griffiths and Podolský,[69] and Sternberg.[181]

we assume $f > 0$. We compute the vacuum field equation of this metric. Nowadays we can do this straightforwardly with the assistance of a computer algebra system. During the 1960s, when this work was mainly done, there were no computer algebra systems around. Hearn[79] released the computer algebra system REDUCE in 1968. Back then, one had to be in command of huge computer resources in order to bring the underlying computer language LISP to work. Today, Reduce can run on every laptop, for other computer algebra systems, see Grabmeier *et al.*[68] and Wolfram.[191]

Because of its efficiency, we will use Schrüfer's Reduce-package EXCALC, which was built for manipulating expressions within the calculus of exterior forms. For that purpose, we reformulate the metric (93) in terms of an orthonormal coframe of four one-forms $\vartheta^\alpha = e_i{}^\alpha dx^i$, with the unknown functions $f = f(\rho, z)$, $\omega = \omega(\rho, z)$, and $\gamma = \gamma(\rho, z)$, namely

$$\vartheta^{\hat{0}} = f^{\frac{1}{2}}(dt - \omega d\phi) = e_i{}^{\hat{0}} dx^i = f^{\frac{1}{2}}\,(dx^0 - \omega dx^3), \tag{94}$$

$$\vartheta^{\hat{1}} = f^{-\frac{1}{2}} e^\gamma d\rho = e_i{}^{\hat{1}} dx^i = f^{-\frac{1}{2}} e^\gamma dx^1, \tag{95}$$

$$\vartheta^{\hat{2}} = f^{-\frac{1}{2}} e^\gamma dz = e_i{}^{\hat{2}} dx^i = f^{-\frac{1}{2}} e^\gamma dx^2, \tag{96}$$

$$\vartheta^{\hat{3}} = f^{-\frac{1}{2}} \rho d\phi = e_i{}^{\hat{3}} dx^i = f^{-\frac{1}{2}} \rho dx^3. \tag{97}$$

Because of the orthonormality of the coframe ϑ^α, we have

$$ds^2 \equiv g = +\vartheta^{\hat{0}} \otimes \vartheta^{\hat{0}} - \vartheta^{\hat{1}} \otimes \vartheta^{\hat{1}} - \vartheta^{\hat{2}} \otimes \vartheta^{\hat{2}} - \vartheta^{\hat{3}} \otimes \vartheta^{\hat{3}}. \tag{98}$$

Equations (94)–(98) are equivalent to (93).

The corresponding computer code, as input for Reduce-Excalc, reads as follows:

```
pform f=0, omega=0, gamma=0 $
fdomain f=f(rho,z), omega=omega(rho,z), gamma=gamma(rho,z);

coframe  o(0) = sqrt(f) * (d t - omega * d phi),
         o(1) = sqrt(f)**(-1) * exp(gamma) * d rho,
         o(2) = sqrt(f)**(-1) * exp(gamma) * d z,
         o(3) = sqrt(f)**(-1) * rho * d phi
    with signature (1,-1,-1,-1);
```

Isn't that simple enough? From this data, the Einstein equation is calculated, with the Einstein tensor $G^\mu{}_\nu$. The complete, fairly trivial program is documented in Appendix. Note in particular that we used a LaTeX interface allowing us to output the expressions directly in LaTeX. This computer output — without changing anything of the formulas — after some post-editing for display purposes, reads as follows:

$$\begin{aligned} G^0{}_0 := {} & (4 \cdot \partial_{\rho,\rho} f \cdot f \cdot \rho^2 - 5 \cdot \partial_\rho f^2 \cdot \rho^2 + 4 \cdot \partial_\rho f \cdot f \cdot \rho + 4 \cdot \partial_{z,z} f \cdot f \cdot \rho^2 \\ & - 5 \cdot \partial_z f^2 \cdot \rho^2 - 4 \cdot \partial_{\rho,\rho}\gamma \cdot f^2 \cdot \rho^2 - 4 \cdot \partial_{z,z}\gamma \cdot f^2 \cdot \rho^2 + 3 \cdot \partial_\rho \omega^2 \cdot f^4 \\ & + 3 \cdot \partial_z \omega^2 \cdot f^4)/(4 \cdot e^{2\cdot\gamma} \cdot f \cdot \rho^2), \end{aligned} \tag{99}$$

$$\begin{aligned}G^3{}_0 :=\ & f\cdot(2\cdot\partial_\rho f\cdot\partial_\rho\omega\cdot\rho+2\cdot\partial_z f\cdot\partial_z\omega\cdot\rho+\partial_{\rho,\rho}\omega\cdot f\cdot\rho-\partial_\rho\omega\cdot f\\ &+\partial_{z,z}\omega\cdot f\cdot\rho)/(2\cdot e^{2\cdot\gamma}\cdot\rho^2), \qquad (100)\\ G^1{}_1 :=\ & (\partial_\rho f^2\cdot\rho^2-\partial_z f^2\cdot\rho^2-4\cdot\partial_\rho\gamma\cdot f^2\cdot\rho-\partial_\rho\omega^2\cdot f^4\\ &+\partial_z\omega^2\cdot f^4)/(4\cdot e^{2\cdot\gamma}\cdot f\cdot\rho^2), \qquad (101)\\ G^1{}_2 :=\ & (\partial_\rho f\cdot\partial_z f\cdot\rho^2-2\cdot\partial_z\gamma\cdot f^2\cdot\rho-\partial_\rho\omega\cdot\partial_z\omega\cdot f^4)/(2\cdot e^{2\cdot\gamma}\cdot f\cdot\rho^2), \qquad (102)\\ G^3{}_3 :=\ & (-\partial_\rho f^2\cdot\rho^2-\partial_z f^2\cdot\rho^2-4\cdot\partial_{\rho,\rho}\gamma\cdot f^2\cdot\rho^2-4\cdot\partial_{z,z}\gamma\cdot f^2\cdot\rho^2\\ &-\partial_\rho\omega^2\cdot f^4-\partial_z\omega^2\cdot f^4)/(4\cdot e^{2\cdot\gamma}\cdot f\cdot\rho^2). \qquad (103)\end{aligned}$$

This calculation of the Einstein tensor by machine did not require more than about 15 min, including the programming and the typing in. For sample programs, see Socorro *et al.*[176] and Stauffer *et al.*[178]

Inspecting these equations, it becomes immediately clear that the numerator of (100) does not depend on γ. In order to get a better overview, we abbreviate the partial derivatives of Reduce $\partial_\rho f$ by subscripts, f_ρ, and drop the superfluous multiplication dots of Reduce. We find

$$G^3{}_0 = 0 \to 0 = f\left(\omega_{\rho\rho}+\omega_{zz}-\frac{1}{\rho}\omega_\rho\right)+2(f_\rho\omega_\rho+f_z\omega_z). \qquad (104)$$

Moreover, by subtracting (103) from (99), we find another equation free of γ

$$G^0{}_0 - G^3{}_3 = 0 \to 0 = f\left(f_{\rho\rho}+\frac{1}{\rho}f_\rho+f_{zz}\right)-f_\rho^2-f_z^2+\frac{f^4}{\rho^2}(\omega_\rho^2+\omega_z^2). \qquad (105)$$

Left over are Eqs. (101) and (102), which can be resolved with respect to the first derivatives of γ

$$G^1{}_1 = 0 \to \gamma_\rho = \frac{\rho}{4f^2}(f_\rho^2-f_z^2)+\frac{f^2}{4\rho}(\omega_z^2-\omega_\rho^2), \qquad (106)$$

$$G^1{}_2 = 0 \to \gamma_z = \frac{\rho}{2f^2}f_\rho f_z-\frac{f^2}{2\rho}\omega_\rho\omega_z. \qquad (107)$$

Collected, we have these four equations determining the stationary axisymmetric vacuum metric

$$0 = f\left(f_{\rho\rho}+\frac{1}{\rho}f_\rho+f_{zz}\right)-f_\rho^2-f_z^2+\frac{f^4}{\rho^2}(\omega_\rho^2+\omega_z^2), \qquad (108)$$

$$0 = f\left(\omega_{\rho\rho}+\omega_{zz}-\frac{1}{\rho}\omega_\rho\right)+2(f_\rho\omega_\rho+f_z\omega_z), \qquad (109)$$

$$\gamma_\rho = \frac{\rho}{4f^2}(f_\rho^2-f_z^2)+\frac{f^2}{4\rho}(\omega_z^2-\omega_\rho^2), \qquad (110)$$

$$\gamma_z = \frac{\rho}{2f^2}f_\rho f_z-\frac{f^2}{2\rho}\omega_\rho\omega_z. \qquad (111)$$

Let us underline how effortless — under computer assistance — we arrived at these four partial differential equations (PDEs) for determining stationary axially symmetric solutions of Einstein's field equation.

3.2.2. *Ernst equation* (1968)

It is one step ahead, before we arrive at a still more convincing form of these PDEs. After some attempts, one recognizes that (109) can be written as

$$\left(\frac{f^2}{\rho}\omega_\rho\right)_\rho + \left(\frac{f^2}{\rho}\omega_z\right)_z = 0. \tag{112}$$

With the ansatz ($\Omega = \Omega(\rho, z)$),

$$\Omega_z = \frac{f^2}{\rho}\omega_\rho, \quad \Omega_\rho = -\frac{f^2}{\rho}\omega_z, \tag{113}$$

Equation (112) is identically fulfilled. We substitute (113) into (108)

$$f\left(f_{\rho\rho} + \frac{1}{\rho}f_\rho + f_{zz}\right) - f_\rho^2 - f_z^2 + \Omega_\rho^2 + \Omega_z^2 = 0. \tag{114}$$

Since (109) is already exploited, we can find Ω by differentiating the Ω's in (113) with respect to z and ρ, respectively, and by adding the emergent expressions ($\omega_{\rho z} = \omega_{z\rho}$)

$$f\left(\Omega_{\rho\rho} + \frac{1}{\rho}\Omega_\rho + \Omega_{zz}\right) - 2f_\rho\Omega_\rho - 2f_z\Omega_z = 0. \tag{115}$$

Equations (108) and (115) can be put straightforwardly into a vector analytical form, if we recall that our functions do not depend on the angle ϕ[s]

$$f\Delta f - (\boldsymbol{\nabla} f)\cdot\boldsymbol{\nabla} f + (\boldsymbol{\nabla}\Omega)\cdot\boldsymbol{\nabla}\Omega = 0, \tag{116}$$

$$f\Delta\Omega - 2(\boldsymbol{\nabla} f)\cdot\boldsymbol{\nabla}\Omega = 0. \tag{117}$$

The last equation can also be written as $\boldsymbol{\nabla}\cdot(f^{-2}\boldsymbol{\nabla}\Omega) = 0$. Equations (116) and (117) liberate ourselves from the cylindrical coordinates, that is, this expression is

[s]In cylindrical coordinates, we have for a vector $\mathbf{V}$ and a scalar s the following formulas, see Jackson[94]

$$\boldsymbol{\nabla}\cdot\mathbf{V} = \frac{1}{\rho}\partial_\rho(\rho V_1) + \partial_z V_2 + \frac{1}{\rho}\partial_\phi V_3, \quad \nabla^2 s \equiv \Delta s = \frac{1}{\rho}\partial_\rho(\rho\partial_\rho s) + \partial_z^2 s + \frac{1}{\rho^2}\partial_\phi^2 s,$$

$$\boldsymbol{\nabla} s = \mathbf{e_1}\partial_\rho s + \mathbf{e_2}\partial_z s + \mathbf{e_3}\frac{1}{\rho}\partial_\phi s, \quad \boldsymbol{\nabla} s\cdot\boldsymbol{\nabla} s = (\partial_\rho s)^2 + (\partial_z s)^2 + \frac{1}{\rho^2}(\partial_\phi s)^2.$$

now put in form independent of the specific 3D coordinates. With the potential ($i^2 = -1$)

$$\mathcal{E} := f + i\Omega, \tag{118}$$

which was found by Ernst[53] and Kramer and Neugebauer,[100] we find the Ernst equation[53]

$$(\mathrm{Re}\,\mathcal{E})\Delta\mathcal{E} = \boldsymbol{\nabla}\mathcal{E} \cdot \boldsymbol{\nabla}\mathcal{E}, \tag{119}$$

or, in components

$$(\mathrm{Re}\,\mathcal{E})\left[\frac{\partial^2\mathcal{E}}{\partial z^2} + \frac{1}{\rho}\frac{\partial}{\partial\rho}\left(\rho\frac{\partial\mathcal{E}}{\partial\rho}\right)\right] = \left(\frac{\partial\mathcal{E}}{\partial z}\right)^2 + \left(\frac{\partial\mathcal{E}}{\partial\rho}\right)^2. \tag{120}$$

The "Re" denotes the real part of a complex quantity. Under stationary axial symmetry — the corresponding metric is displayed in (93) — the Ernst equation (119), together with Eqs. (118), (113), (110) and (111), are equivalent to the vacuum Einstein field equation.

3.2.3. *From Ernst back to Kerr*

This reduces the problem of axial symmetry to the solution of the second order PDE (119). This method, which came along only five years after Kerr's publication, led to many new exact solutions, amongst them the Kerr solution (1963) as one of the simplest cases. We are only going to sketch how one arrives at the Kerr solution eventually. We follow here closely Buchdahl.[23]

One introduces a new complex potential ξ by

$$\mathcal{E} =: \frac{\xi - 1}{\xi + 1}. \tag{121}$$

Then the Ernst equation becomes

$$(\xi\bar{\xi} - 1)\Delta\xi = 2\bar{\xi}\boldsymbol{\nabla}\xi \cdot \boldsymbol{\nabla}\xi, \tag{122}$$

where the overline denotes complex conjugation. If one has a solution of this equation, we can determine the functions f, ω and γ by

$$f = \mathrm{Re}\frac{\xi - 1}{\xi + 1}, \tag{123}$$

$$\omega_\rho = -2\rho\frac{\mathrm{Im}[(\bar{\xi}+1)^2\xi_z]}{(\xi\bar{\xi}-1)^2}, \quad \omega_z = 2\rho\frac{\mathrm{Im}[(\bar{\xi}+1)^2\xi_\rho]}{(\xi\bar{\xi}-1)^2}, \tag{124}$$

$$\gamma_\rho = \rho\frac{\xi_\rho\bar{\xi}_\rho - \xi_z\bar{\xi}_z}{(\xi\bar{\xi}-1)^2}, \quad \gamma_z = 2\rho\frac{\mathrm{Re}(\xi_\rho\bar{\xi}_z)}{(\xi\bar{\xi}-1)^2}. \tag{125}$$

For rotating bodies, *spherical prolate coordinates* x, y, with a constant k, are much more adapted

$$\rho = k(x^2-1)^{\frac{1}{2}}(1-y^2)^{\frac{1}{2}}, \quad z = kxy. \tag{126}$$

It turns out that one simple potential solving the Ernst equation, with the constants p and q, is

$$\xi = px - iqy \quad \text{with } p^2 + q^2 = 1. \tag{127}$$

It leads to the Kerr metric. For this purpose, one has to introduce the redefined constants $m := k/p$ (mass) and $a := kq/p$ (angular momentum per mass) and to execute subsequently the transformations $px = (\tilde{\rho}/m) - 1$ and $qy = (a/m)\cos\theta$ to the new coordinates $\tilde{\rho}$ and θ. Then one arrives at the Kerr metric in Boyer–Lindquist coordinates, which is displayed in the table on the next page. For more detail, compare, for instance, the books of Buchdahl,[23] Islam,[93] Heusler,[85] Meinel *et al.*[123] or Griffiths *et al.*[69] By similar techniques, a Kerr solution with a topological defect was found by Bergamini *et al.*[14]

Incidentally, in the context of the Ernst equation, Geroch made the following interesting conjecture: A subset of all stationary axially symmetric vacuum spacetimes, including all of its asymptotically flat members, that is, in particular the Kerr solution, can be obtained from Minkowski space by transformations generated by an infinite-dimensional Lie group. This conjecture was "proved" by Hauser and Ernst,[76] see also Ref. 75. However, the proof contained a mistake that was subsequently corrected in Ref. 77.

Starting from 4D ellipsoidal coordinates, Dadhich[39] gave a heuristic derivation of the Kerr metric by requiring, amongst other things, that light propagation should be influenced by gravity.

3.3. *Three classical representations of the Kerr metric*

We collected these three classical versions of the Kerr metric in Table 4, see also Visser.[186] Three more coordinate systems should at least be mentioned:

- *Pretorius and Israel*[157] double null coordinates:
  Very convenient to tackle the initial value problem
- *Doran*[46] coordinates:
  Gullstrand–Painlevé like; useful in analog gravity
- *Debever/Plebański-Demiański*[155] coordinates:
  Components of the metric are rational polynomials; convenient for (computer assisted) calculations.

As input for checking the *Kerr* solution, we use the *orthonormal coframe*[181]

$$\vartheta^{\hat{0}} := \frac{\sqrt{\epsilon\Delta}}{\rho}(dt - a\sin^2\theta d\phi), \tag{128}$$

$$\vartheta^{\hat{1}} := \frac{\rho}{\sqrt{\epsilon\Delta}}dr, \tag{129}$$

$$\vartheta^{\hat{2}} := \rho d\theta, \tag{130}$$

$$\vartheta^{\hat{3}} := \frac{\sin\theta}{\rho}[(r^2 + a^2)d\phi - adt]. \tag{131}$$

Table 4. Kerr metric: The three classical representations.

Kerr–Schild (t, x, y, z) Cartesian background

$$ds^2 = -dt^2 + dx^2 + dy^2 + dz^2 + \frac{2mr^3}{r^4 + a^2z^2}\left(dt + \frac{r(x\,dx + y\,dy)}{a^2 + r^2} + \frac{a(y\,dx - x\,dy)}{a^2 + r^2} + \frac{z}{r}\,dz\right)^2$$

$$x^2 + y^2 + z^2 = r^2 + a^2\left(1 - \frac{z^2}{r^2}\right), \quad r = r(x, y, z)$$

$$\begin{aligned} x &= (r\cos\phi + a\sin\phi)\sin\theta \\ y &= (r\sin\phi - a\cos\phi)\sin\theta \\ z &= r\cos\theta \end{aligned}$$

Boyer–Lindquist (t, r, θ, ϕ) Schwarzschild like

$$ds^2 = -\left(1 - \frac{2mr}{\rho^2}\right)dt^2 - \frac{4mra\sin^2\theta}{\rho^2}\,dt\,d\phi + \frac{\rho^2}{\Delta}\,dr^2 + \rho^2 d\theta^2 + \left(r^2 + a^2 + \frac{2mra^2\sin^2\theta}{\rho^2}\right)\sin^2\theta\,d\phi^2$$

$$\rho^2 := r^2 + a^2\cos^2\theta \quad \Delta := r^2 - 2mr + a^2 = (r - r_+)(r - r_-)$$

$$dv = dt + \frac{r^2 + a^2}{\Delta}\,dr$$

$$d\varphi = d\phi + \frac{a}{\Delta}\,dr$$

Kerr original (v, r, θ, φ) Eddington–Finkelstein like

$$ds^2 = -\left(1 - \frac{2mr}{\rho^2}\right)(dv - a\,\sin^2\theta\,d\varphi)^2 + 2(dv - a\sin^2\theta\,d\varphi)(dr - a\sin^2\theta\,d\varphi) + \rho^2(d\theta^2 + \sin^2\theta\,d\varphi^2)$$

$$r_{\mathrm{E}\pm} := m \pm \sqrt{m^2 - a^2\cos^2\theta} \quad r_\pm := m \pm \sqrt{m^2 - a^2}$$

We introduced the sign function, which is convenient for discussing the different regions in the Penrose–Carter diagram

$$\epsilon = \begin{cases} +1 & \text{for } r > r_+ \text{ or } r < r_-, \\ -1 & \text{for } r_- < r < r_+. \end{cases} \tag{132}$$

The metric can then be written in terms of the coframe as

$$ds^2 \equiv g = \epsilon(-\vartheta^{\hat{0}} \otimes \vartheta^{\hat{0}} + \vartheta^{\hat{1}} \otimes \vartheta^{\hat{1}}) + \vartheta^{\hat{2}} \otimes \vartheta^{\hat{2}} + \vartheta^{\hat{3}} \otimes \vartheta^{\hat{3}}. \tag{133}$$

From Table 4 it is not complicated to read off the Schwarzschild and the Lense–Thirring metric as special cases. In comparison to the Schwarzschild metric, the Kerr solution includes a new parameter a which will be related to the angular momentum. However, it should be noted that, by setting $a = 0$, the Kerr metric reduces to the Schwarzschild metric, as it should be ($\rho^2 \to r^2$ and $\Delta \to r^2 - 2mr$):

$$ds^2 = -\left(1 - \frac{2mr}{r^2}\right) dt^2 + \frac{r^2}{r^2 - 2mr} dr^2 + r^2 d\theta^2 + r^2 \sin^2\theta d\phi^2. \tag{134}$$

By canceling r^2 in the dr^2-term, we immediately recognize the Schwarzschild metric.

Considering the parameter a we should note the following fact. For small values of the parameter a, where we may neglect terms of the order of a^2, we arrive at ($\rho^2 \to r^2$ and $\Delta \to r^2 - 2mr$) and

$$ds^2 = -\left(1 - \frac{2m}{r}\right) dt^2 + \frac{1}{1 - \frac{2m}{r}} dr^2 + r^2 \left(d\theta^2 + \sin^2\theta d\phi^2\right) - \frac{4ma \sin^2\theta}{r} dt d\phi. \tag{135}$$

Since, in spherical coordinates we have $ydx - xdy = r^2 \sin^2\theta d\phi$, the cross-term may be rewritten as $\frac{4ma\sin^2\theta}{r} dtd\phi = \frac{4ma}{r^3}(xdz - ydx)$. Thus, in the limiting case $a^2 \ll 1$, the Kerr metric yields the Lense–Thirring metric, provided we identify $ma = J_z$.

3.4. *The concept of a Kerr black hole*

We come back to our Fig. 7 with "Schwarzschild" versus "Kerr". The Kerr spacetime may be visualized by a vortex, where the water of the lake spirals toward the center. Much of the above said for the Schwarzschild case is still valid. However, one important difference occurs. The stationary limit and the event horizon separate, which will be illustrated by corresponding graphical representations.

In case of a vortex, the flow velocity of the in-spiraling water has two components. The radial component which drags the boat toward the center whereas the additional angular component forces the boat to circle around the center. Again, the *stationary limit* is defined by the distance at which the boat ultimately can withstand the radial and circular drag of the water flow. Beyond the stationary limit the situation is not as hopeless as in the Schwarzschild case. Using all its power, the boat may brave the inward flow. But then it has not enough power to overcome the angular drag and is forced to orbit the center. By means of a clever spiral course the boat may even escape beyond the stationary limit. The stationary limit is not necessary an event horizon. At some distance, nearer to the center than the stationary limit, also the pure radial flow of water will exceed the power of the boat. There, inside the stationary limit, is the event horizon.

In order to investigate the structure of the Kerr spacetime, we first look at "strange behavior" of the metric components in Boyer–Lindquist coordinates. The following cases can be distinguished:

$$\begin{array}{ll} \Delta = 0 & g_{rr} \text{ becomes singular,} \\ \rho^2 = 2mr & g_{tt} \text{ vanishes,} \\ \rho^2 = 0 & g_{rr} \text{ and } g_{\theta\theta} \text{ vanish, the other components are singular.} \end{array}$$

As we have extensively discussed in the previous section, singularities of components of the metric *may* signify physical effects but, on the other hand, may only be due to "defective" coordinates. Thus, we will proceed along similar lines to investigate the nature of these singularities.

We will not address the geodesics of the Kerr metric in detail. For an elementary discussion the reader is referred to Frolov and Novikov[64] and to the more advanced discussion in Hackmann *et al.*[72]

3.4.1. *Depicting Kerr geometry*

We draw a picture of the spatial appearances and relations of the various horizons and the singularity of the Kerr metric. From outside to inside these are, explicitly,

$$\begin{array}{rl} \text{outer ergosurface} & r_{E+} := m + \sqrt{m^2 - a^2\cos^2\theta} \\ & \qquad\qquad \updownarrow \text{ joined at polar axis} \\ \text{event horizon} & r_{+} := m + \sqrt{m^2 - a^2} \\ & \qquad\qquad \updownarrow \text{ merge for } a \longrightarrow m \\ \text{Cauchy horizon} & r_{-} := m - \sqrt{m^2 - a^2} \\ & \qquad\qquad \updownarrow \text{ joined at polar axis} \\ \text{inner ergosurface} & r_{E-} := m - \sqrt{m^2 - a^2\cos^2\theta} \\ & \qquad \uparrow \text{ lies on the rim for } \theta = \pi/2 \\ \text{singularity} & r = 0. \end{array} \tag{136}$$

For $a = 0$, inner ergosurface and Cauchy horizon vanish, whereas outer ergosurface and event horizon merge to the Schwarzschild horizon. To visualize the various surfaces we use Kerr–Schild quasi-Cartesian coordinates. The radial coordinate r of the Boyer–Lindquist coordinates is related to the coordinates x, y, z of the Kerr–Schild coordinates via, see Table 4

$$x^2 + y^2 + \frac{r^2 + a^2}{r^2} z^2 = r^2 + a^2, \quad z = r\cos\theta. \tag{137}$$

Substituting $r = 0$, $r = r_\pm$, $r = r_{E\pm}$, and a little bit of algebra yields:

- **Singularity** $r = 0$

Since $r = 0$ leads to $z = 0$, we get the equation of a circle of radius a in the equatorial plane

$$x^2 + y^2 = a^2. \tag{138}$$

For $a = 0$, the ring collapses to the Schwarzschild singularity.

A closer inspection shows that the structure of the singularity is more complex.[66,110]

- **Horizons** $r = r_\pm$

  In this case we arrive at the equation for an oblate (for $a < m$) ellipsoid

$$\frac{x^2}{a_1^2} + \frac{y^2}{a_2^2} + \frac{z^2}{a_3^2} = 1, \tag{139}$$

  where

$$a_1^2 = a_2^2 = r_\pm^2 + a^2 > a_3^2 = \frac{1}{r_\pm^2}. \tag{140}$$

- **Ergosurfaces** $r = r_{E\pm}(\theta)$

  Things are a little bit more involved in this case because r is not constant. We can also derive a "ellipsoid-like" equation (for $a \leq m$)

$$\frac{x^2}{a_1^2(\theta)} + \frac{y^2}{a_2^2(\theta)} + \frac{z^2}{a_3^2(\theta)} = 1, \tag{141}$$

  now with

$$a_1^2(\theta) = a_2^2(\theta) = r_{E\pm}^2(\theta) + a^2, \quad a_3^2(\theta) = \frac{1}{r_{E\pm}^2(\theta)}. \tag{142}$$

  The θ-dependence will deform the ellipsoid. On the equatorial plane with $\theta = \pi$ we have $r_{E\pm}^2 = 0$. Hence, $a_1 = a_2 = a$ and a_3 diverges. This results in a nonregular rim on which the ring singularity is located.

  For $a > m$, the $r_{E\pm}$ is partly not defined, since the term under the square-root changes sign, and becomes negative if

$$\cos(\theta) = \frac{m}{a}. \tag{143}$$

  This defines two rings with $\theta_1 = \arccos(m/a)$, $\theta_2 = \pi - \theta_1$ and $r = m$. As a consequence, the outer ergosurface only extends to these rings from the outside, and the inner ergosurface up to the rings from the inside. The outcome is a kind of torus. The center-facing side is constituted by a part of the inner ergosurface (along with the ring singularity), whereas the outside facing parts are given by a part of the outer ergosurface.

  An extensive discussion of the embedding of the ergosurfaces into Euclidean space, together with corresponding Mathematica-programs, can be found in Ref. 114.

The surfaces are visualized in Fig. 12. We did not use a faithful embedding but rescaled axes in order to achieve better visibility.

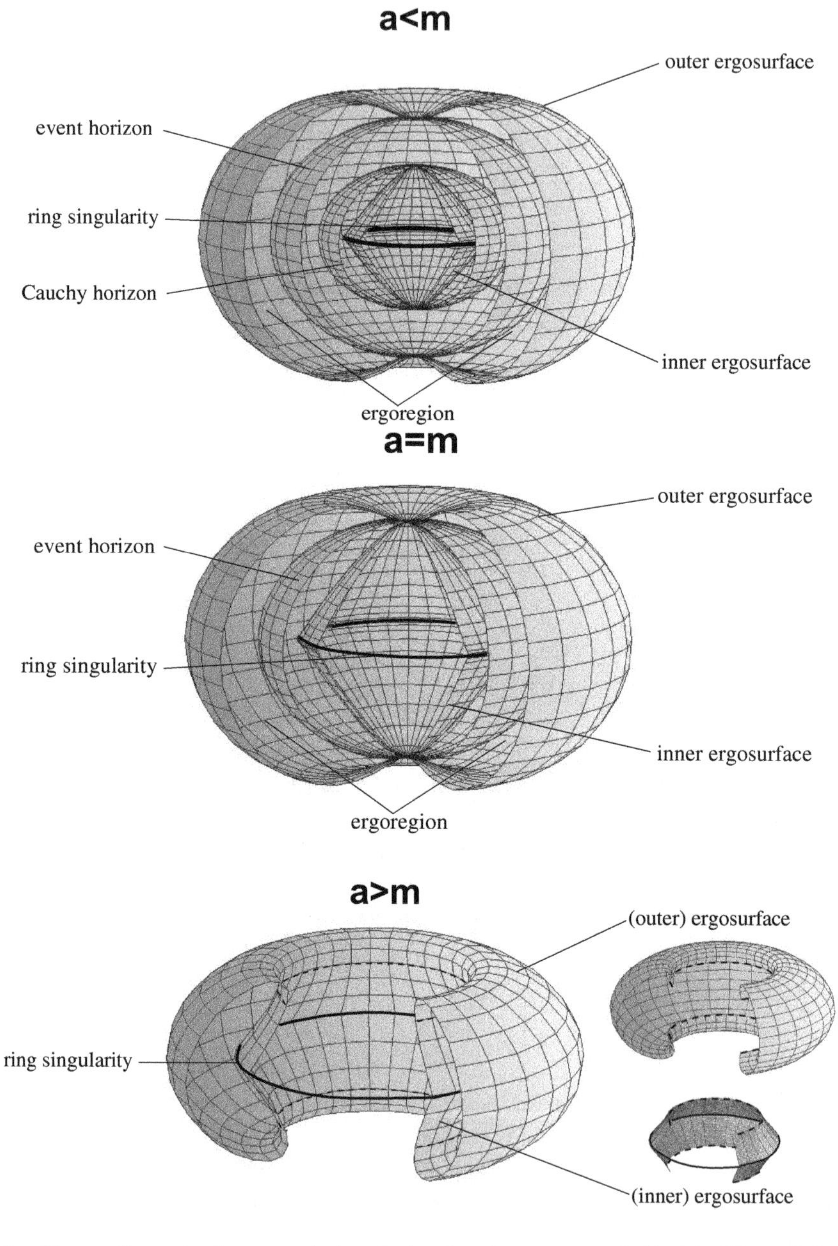

Fig. 12. Ergosurfaces, horizons, and singularity for slow, extremal ("critical"), and fast Kerr black holes. (For color version, see page I-CP2.)

The presence of the term $\sqrt{m^2 - a^2}$ requires the distinction of three different cases dependent on the values of the mass parameter m and the angular momentum parameter a:

$$m > a \rightsquigarrow \text{slow rotation} \quad m = a \rightsquigarrow \text{critical rotation} \quad m < a \rightsquigarrow \text{fast rotation.}$$

The slow rotating case shows the richest structure. Both ergosurfaces and both horizons are present and distinct from each other. As a approaches m, the event and the Cauchy horizon draw nearer and nearer. At critical rotation, $a = m$, both horizons merge into one single event horizon with $r = m$. Eventually, for fast rotation $a > m$, the event horizon disappears and reveals the naked ring singularity which now is located at the inner edge of the now toroidal shaped (outer) ergosurface.

3.5. *The ergoregion*

We explore the region between the outer ergosurface and the event horizon. There it is not possible to stand still, anything has to rotate, even the event horizon. The compulsory rotation in the ergoregion allows one to extract energy from the black hole. This so-called Penrose process leads to black hole thermodynamics.

3.5.1. *Constrained rotation*

The outer ergosurface, $r = R_{E_+}$, is defined by the equation $g_{00}(r_{E_+}) = 0$. Thus, it is a surface of infinite redshift and a Killing horizon. For a third characterization of the ergosurfaces we have to deal not only with radial but also with rotational motion. Consider the Kerr metric in Boyer–Lindquist coordinates with $dr = d\theta = 0$

$$ds^2 = g_{tt}dt^2 + 2g_{t\phi}dtd\phi + g_{\phi\phi}d\phi^2,$$

or, after dividing by dt^2,

$$\left(\frac{ds}{dt}\right)^2 = g_{tt} + 2g_{t\phi}\frac{d\phi}{dt} + g_{\phi\phi}\left(\frac{d\phi}{dt}\right)^2.$$

The explicit form of the metric components is not needed here. Note that $\Omega = \frac{d\phi}{dt}$ is the angular velocity with respect to a distant observer

$$\left(\frac{ds}{dt}\right)^2 = g_{tt} + 2g_{t\phi}\Omega + g_{\phi\phi}\Omega^2. \tag{144}$$

The worldline of a particle has to be timelike, $ds^2 < 0$. Since the last equation is quadratic in Ω, this is only possible between the roots $ds^2 = 0$,

$$\Omega_{\text{min/max}} := -\frac{g_{t\phi}}{g_{\phi\phi}} \mp \sqrt{\left(\frac{g_{t\phi}}{g_{\phi\phi}}\right)^2 - \frac{g_{tt}}{g_{\phi\phi}}}.$$

What does

$$\Omega_{\text{min}} < \Omega < \Omega_{\text{max}}$$

mean? In flat Minkowski spacetime (with Cartesian coordinates), $\Omega_{\text{min/max}} = \pm 1$ implies that a particle, e.g. may freely circle around a point, restricted only by the condition $|v| = |r \cdot \Omega| < c$. In the Kerr spacetime, at $r = r_{E_\pm}$, the smallest possible value of Ω becomes 0. The particle may just stay at rest, but can rotate only in one direction, namely in direction of the angular momentum of the black hole. Beyond r_{E_+}, Ω is forced to be larger than zero: The particle must co-rotate with the black hole.

The preceding statement is only correct for radially infalling particles. In general, the influence of the rotating black hole on the motion of particles is more complex.[165]

3.5.2. *Rotation of the event horizon*

The behavior of $\Omega_\pm$ on the event horizon is quite remarkable. By using the identities

$$2mr_\pm = r_\pm^2 + a^2, \quad \rho^2 - r^2 = a^2\cos^2\theta = a^2 - a^2\sin^2\theta, \tag{145}$$

one finds for the event horizon ($r = r_+$),

$$\Omega_+ = \Omega_- = \Omega_H := \frac{a}{2mr_+}. \tag{146}$$

To interpret this result we use (144) and write

$$g_{\mu\nu}l^\mu l^\nu|_{r=r_+} = 0, \quad \text{with } l^\mu = (1, 0, 0, \Omega_H). \tag{147}$$

The integral lines of $l^\mu = \dot{x}^\mu$,

$$x^\mu = (t, r_+, \theta_0, \Omega_H t), \tag{148}$$

define a lightlike hypersurface rotating with a uniform angular velocity: The event horizon of a Kerr black hole rotates "rigidly" with Ω_H, see in this context Frolov and Frolov.[63] A consequence of this finding is discussed in the next paragraph.

3.5.3. *Penrose process and black hole thermodynamics*

The (outer) ergosurface is a Killing horizon, not an event horizon. It is possible for particles to pass from the inside to the outside. This allows for a peculiar scenario: Since inside the Killing horizon the particle is forced to spin around, it picks up an additional rotational energy. This energy can be partly extracted by means of the Penrose process. An infalling particle traverses the Killing horizon, picks up rotational energy and subsequently decays into two parts. If one part plunges into the event horizon, the other part, carrying away some of the rotational energy, can return to the outside of the Killing horizon. Thus, the region between Killing and event horizon is justly labeled as "ergoregion" (from Greek ergon = work).

The observation that energy can also be extracted from the black hole gave rise to black hole thermodynamics. The next question is then how the parameters change if the black hole is infinitesimally disturbed. It was Bekenstein[13] who established a relation between the variations of the mass, the angular momentum, and the area of the event horizon. Using the coframe (186)–(190), for $\lambda = 0$, we find for the area of the event horizon

$$A = \int_{\substack{r=r_+\\ t=\text{const.}}} \vartheta^{\hat{2}} \wedge \vartheta^{\hat{3}} = \int_{\substack{r=r_+\\ t=\text{const.}}} \sin\theta(r^2+a^2)d\theta \wedge d\phi = 4\pi(r_+^2 + a^2). \quad (149)$$

We can rewrite (149), using (145) and $J = ma$,

$$A = 8\pi m r_+ = 8\pi(m^2 - \sqrt{m^4 - J^2}). \quad (150)$$

The differential of this equation is

$$dA = \frac{\partial A}{\partial m}dm + \frac{\partial A}{\partial J}dJ = \frac{8\pi}{\kappa}dm - \frac{8\pi}{\kappa}\Omega_H dJ, \quad (151)$$

with

$$\kappa = \frac{1}{2m}\frac{\sqrt{m^4-J^2}}{m^2+\sqrt{m^4-J^2}}, \qquad \Omega_H = \kappa\frac{J}{\sqrt{m^4-J^2}}. \quad (152)$$

The parameter Ω_H is the angular velocity of the horizon (146). The parameter κ is the surface gravity. Equation (151) can be rewritten as

$$dm = \frac{\kappa}{8\pi}dA + \Omega_H dJ. \quad (153)$$

The infinitesimal change of the mass, dm, is proportional to the infinitesimal change of the energy, dE. The term $\Omega_H dJ$ describes the infinitesimal change of the rotational energy. This suggests the identification of (153) with the first law of thermodynamics. The analogy is still more compelling by observing that, for a given black hole of initial (or irreducible) mass m, the area of the horizon is always increasing. Even by exercising a Penrose process, which extracts rotational energy from the black hole, a fragment of the incoming particle will fall into the black hole thereby increasing its mass and, in turn, the area of the horizon. Accordingly, the area A of the horizon behaves formally as if it is proportional to an entropy S and the surface gravity κ as if it is proportional to a temperature T. In fact, the *Hawking temperature* and the *Bekenstein–Hawking entropy* turn out to be

$$T = \frac{\hbar}{2\pi k_{\text{B}}}\kappa, \qquad S = \frac{1}{4G\hbar}A, \quad (154)$$

with k_{B} as the Boltzmann constant. Equation (153) together with its thermodynamical interpretation (154) can be considerably generalized thereby establishing the new discipline of "black hole thermodynamics", see Heusler[85] and Carlip.[28]

3.6. *Beyond the horizons*

In the Schwarzschild spacetime, event horizon and Killing horizon coincide. In the Kerr spacetime, for $m > a$, there is an outer Killing horizon, an event horizon, an inner Killing horizon and an inner horizon. So far, all the coordinate systems we used for the Kerr metric show singularities at the outer and inner horizons $r = r_\pm$. The construction of a regular coordinate system is possible along the same lines as for the Schwarzschild metric. Of course, the corresponding calculations are much more involved for the Kerr case. Therefore, we will give more a kind of heuristic approach to motivate Kruskal-like coordinates for the Kerr metric.

3.6.1. *Using light rays as coordinate lines*

Our first task is to construct Eddington–Finkelstein like coordinates for the Kerr metric by considering radial light rays. We restrict ourselves to the case $\theta = 0 = \phi$. The Kerr metric in Boyer–Lindquist coordinates reduces to ($\theta = 0 \to \rho^2 = r^2 + a^2 = \Delta + 2mr$):

$$ds^2 = -\frac{\Delta}{\rho^2}dt^2 + \frac{\rho^2}{\Delta}dr^2.$$

Hence, for in-/out-going light rays, $ds^2 = 0$, we find

$$dt = \pm\frac{\rho^2}{\Delta}dr = \pm\frac{r^2 + a^2}{(r - r_+)(r - r_-)}dr$$

or, explicitly,

$$\pm t = \int dr\frac{r^2 + a^2}{(r - r_-)(r - r_+)} = r + \frac{r_+^2 + a^2}{r_+ - r_-}\ln|r - r_+| - \frac{r_-^2 + a^2}{r_+ - r_-}\ln|r - r_-| + \text{const.} \tag{155}$$

Unlike in the Schwarzschild spacetime, there form **two** event horizons, at $r = r_-$ and $r = r_+$, respectively. However, as $a \to 0$, r_- goes to 0, whereas r_+ approaches $2m$ and the Schwarzschild situation is reproduced.

We next focus on the (Boyer–Lindquist) coordinates (t, r) and how the horizons etc. will appear in terms of the new coordinates. The other coordinates and the regularity of the metric is not addressed. However, all the details can be found in the literature, see Refs. 20, 30 and 78. Using (155) analogously to (42), we introduce Eddington–Finkelstein like coordinates for Kerr,

$$v := t + r + \sigma_+ \ln|r - r_+| - \sigma_- \ln|r - r_-|, \tag{156}$$

$$u := t - r - \sigma_+ \ln|r - r_+| + \sigma_- \ln|r - r_-|, \tag{157}$$

where (according to the notation in Ref. 20)

$$\sigma_\pm := \frac{r_\pm^2 + a^2}{r_+ - r_-} = \frac{mr_\pm}{\sqrt{m^2 - a^2}}. \tag{158}$$

Again, we can get rid of the coordinate singularity by rescaling u and v analogously to (48). Since we have two horizons, $r = r_+$ and $r = r_-$, we have to decide with respect to which singularity we rescale. We first choose r_+ and define, see (48),

$$\tilde{v} := \exp\left(\frac{v}{2\sigma_+}\right) = \frac{|r - r_+|^{\frac{1}{2}}}{|r - r_-|^{\frac{\nu}{2}}}e^{\frac{r+t}{2\sigma_+}}, \tag{159}$$

$$\tilde{u} := -\exp\left(-\frac{u}{2\sigma_+}\right) = -\frac{|r - r_+|^{\frac{1}{2}}}{|r - r_-|^{\frac{\nu}{2}}}e^{\frac{r-t}{2\sigma_+}}, \tag{160}$$

with

$$\nu := \frac{\sigma_-}{\sigma_+} = \frac{r_-}{r_+} > 1. \tag{161}$$

Again, we go back to time- and space-like coordinates, exactly like in (50)

$$\tilde{t} := \frac{1}{2}(\tilde{v} + \tilde{u}), \quad \tilde{r} := \frac{1}{2}(\tilde{v} - \tilde{u}). \tag{162}$$

Then we work out the four coordinate patches exactly like (55)–(60). We arrive at a Kruskal like coordinate system. However, there arises an important difference: The coordinate system still is singular for $r = r_-$. This can be most easily seen from the analog to (59), the inverse transformation to r, which now reads

$$\tilde{r}^2 - \tilde{t}^2 = -\tilde{v}\tilde{u} = \frac{r - r_+}{(r - r_-)^\nu} e^{\frac{r}{\sigma_+}}. \tag{163}$$

The horizon $r = r_+$ is regular in this coordinate system and is described by $\tilde{r} = \pm\tilde{t}$. The transformation(s) are valid in the domain $r_- < r < +\infty$

$r = r_+$	$: \tilde{r} = \pm\tilde{t}$	as for Schwarzschild
$r \to +\infty$	$: \tilde{r}^2 - \tilde{t}^2 \to +\infty$	particularly $\tilde{r} \to \pm\infty$ for $\tilde{t} = 0$
$r \to r_-$	$: \tilde{r}^2 - \tilde{t}^2 \to -\infty$	particularly $\tilde{t} \to \pm\infty$ for $\tilde{r} = 0$
$r = r_{E_+}$	$: \tilde{r}^2 - \tilde{t}^2 = \text{const.} > 0$	hyperbolas in I, II patches.

In contrast to the Schwarzschild case, the full upper and lower halfplanes of the $(\tilde{r}, \tilde{t})$ plane is covered. It is not limited by the hyperbolas of the Schwarzschild singularity $r = 0$!

We can regularize with respect to r_- by introducing

$$\tilde{v} := -\exp\left(-\frac{v}{2\sigma_-}\right), \quad \tilde{u} := \exp\left(\frac{u}{2\sigma_-}\right). \tag{164}$$

Now we find

$$\tilde{r}^2 - \tilde{t}^2 = \frac{r - r_-}{(r - r_+)^{\frac{1}{\nu}}} e^{-\frac{r}{\sigma_-}}. \tag{165}$$

This coordinate system covers the domain $-\infty < r < r_+$. Like the first coordinate system, it contains also the region between the horizons, $r_- < r < r_+$. This time, $r \geq r_+$ is excluded.

$r = r_-$	$: \tilde{r} = \pm\tilde{t}$	as above
$r \to -\infty$	$: \tilde{r}^2 - \tilde{t}^2 \to \infty$	particularly $\tilde{t} \to \pm\infty$ for $\tilde{r} = 0$
$r \to r_+$	$: \tilde{r}^2 - \tilde{t}^2 \to -\infty$	particularly $\tilde{r} \to \pm\infty$ for $\tilde{t} = 0$
$r = r_{E_-}$	$: \tilde{r}^2 - \tilde{t}^2 = \text{const.} > 0$	hyperbola in I*, II* patches
$r = 0$	$: \tilde{r}^2 - \tilde{t}^2 = -\frac{r_-}{r_+} < 0$	hyperbola in I*, II* patches.

Again, the whole $(\tilde{r}, \tilde{t})$ plane is covered. Note that the spacetime extends beyond the ring(!) singularity.

3.7. Penrose–Carter diagram and Cauchy horizon

We compactify the Kruskal-like coordinate system for Kerr, yielding conformal Penrose–Carter diagrams. We discuss the analytical extension and the role of the inner horizon as Cauchy horizon.

In order to draw a *Penrose–Carter diagram* for the Kerr spacetime, we compactify the coordinates via the tangent function like in Sec. 2.6. The result looks at first quite similar to Schwarzschild in Fig. 11. However, the cut-off at $r = 0$ vanishes. Figures 13 and 14 both show the entire compactified $(\tilde{r}, \tilde{t})$-space.

The two coordinate sets overlap in the region between the horizons. Thus, the corresponding coordinate patches have to be identified. And we can even draw beyond that Patch II is identified with patch IV*, II* with another patch IV**. And so on: We find an infinite sequence of coordinate systems. Formally, this constitutes a *maximal analytic extension* of the Kerr spacetime. Alas, there are good reasons for not believing in such vast an extension.

The Kerr metric is a vacuum solution of Einstein's field equation — it describes a *totally empty* spacetime. To render it physically meaningful, we should regard it as the spacetime structure generated by a sensible physical source. One may ask then, why a single source should produce an infinite number of spacetimes. And it is even worse. The regions beyond the Cauchy horizon are exceptionally badly behaved.

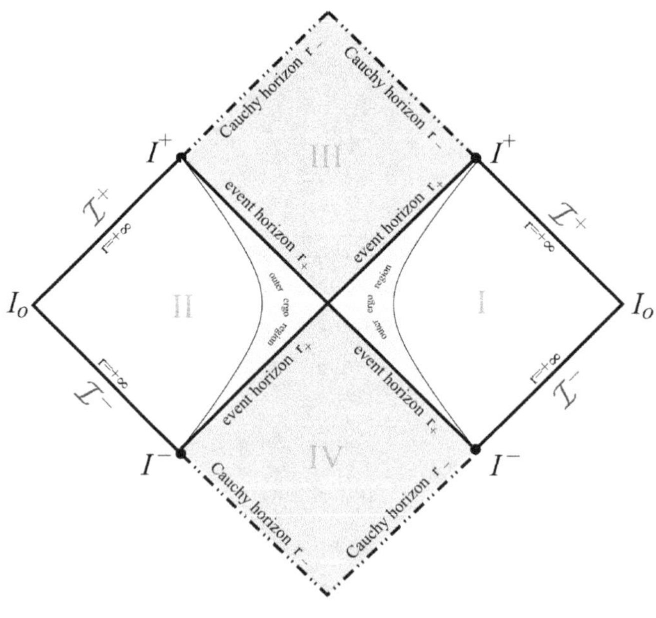

Fig. 13. The Penrose diagram for the Kerr spacetime for $r > r_-$.

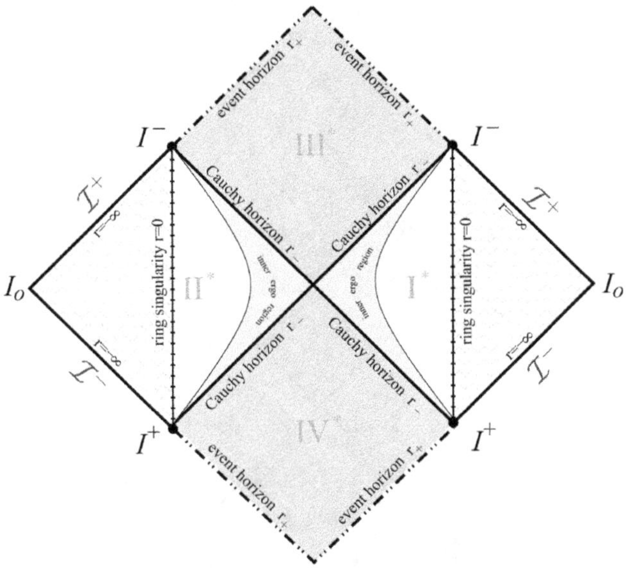

Fig. 14. The Penrose diagram for the Kerr spacetime for $r < r_-$.

Consider the Cauchy surface in regions I+II of Fig. 15. All light rays and particle trajectories from the past intersect this surface only once. Then the field equations will tell us their future development, see Franzen,[62] for example. In Fig. 15, this is roughly indicated by the little light cone. However, even *total* knowledge of the world in I+II does not determine what might be going on in regions I*+II*. That is why $r = r_-$ is called a Cauchy horizon, see Fig. 16. Thus, I* and II* are not only beyond the Cauchy horizon but also beyond predictable, sound physics. Moreover, the zigzagged region beyond the singularity is physically doubtful. In this region, the asymptotics is reversed, see the permutation of I^+ and I^-. As a consequence, the asymptotic mass in I* picks up a minus sign as compared to I. So the *same source* possesses a positive mass $+m$ in I and a negative mass $-m$ in I*, which seems strange. Moreover, it turns out that these regions are crowded with closed timelike curves. The whole extension is not globally hyperbolic. Thus one should restrict to the "diamond of sound physics", I+II+III+IV. To do this consistently, one has to devise a physical mechanism preventing traveling beyond the Cauchy horizon, that is, the Cauchy horizon should become singular in some sense (cosmic censorship, see Penrose[150]).

3.8. *Gravitoelectromagnetism, multipole moments*

The curvature tensor of the Kerr metric is calculated. By squaring it suitably, we find the two quadratic curvature invariants. Subsequently, we determine the gravitoelectric and the gravitomagnetic multipole moments of the Kerr metric, and we mention the Simon–Mars tensor the vanishing of which leads to the Kerr metric.

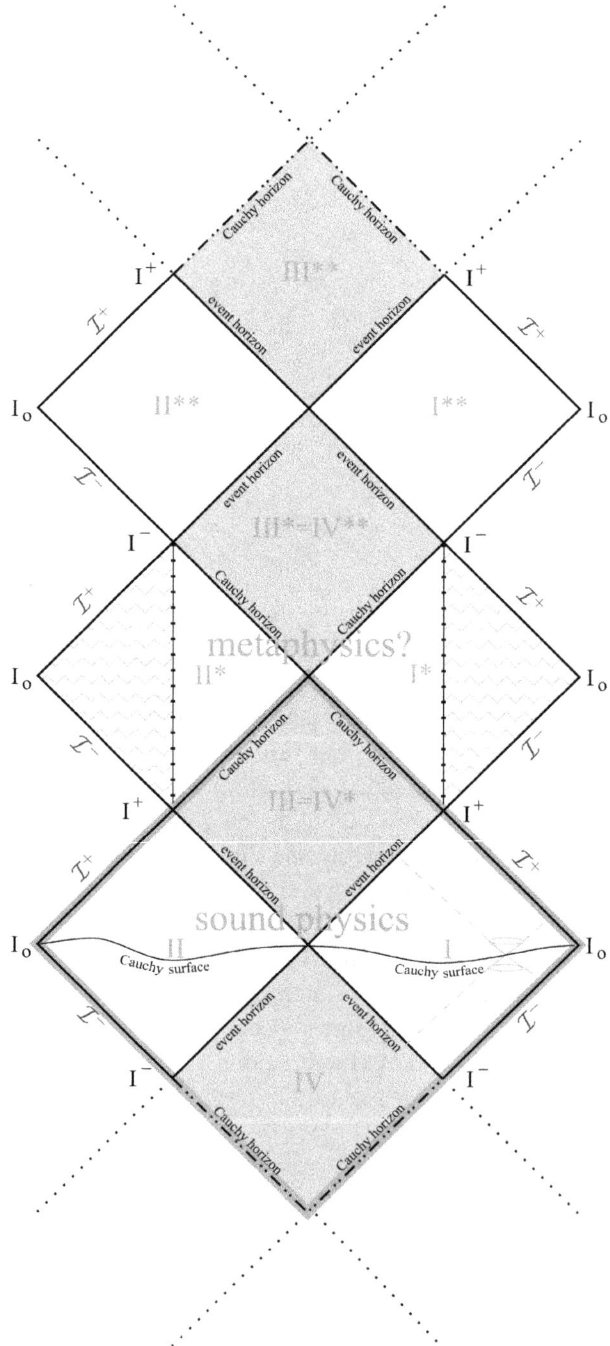

Fig. 15. Maximal analytic extension of the Kerr spacetime. (For color version, see page I-CP3.)

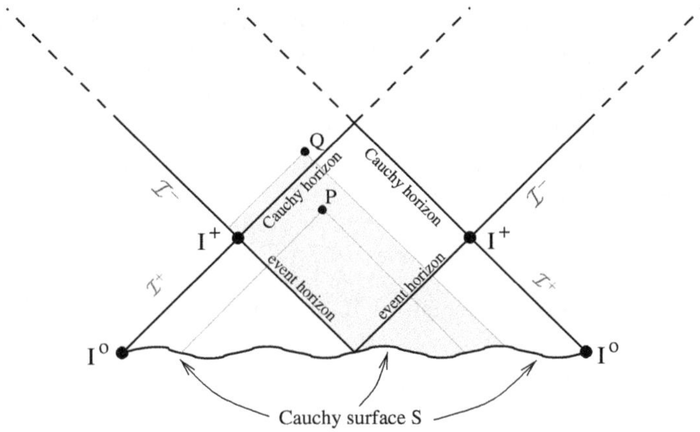

Fig. 16. Cauchy horizon: The causal past of a point P *outside* the *Cauchy horizon* of S is entirely determined by the information given on the Cauchy surface S. A point Q *inside* the Cauchy horizon receives also information from $\mathcal{I}^-$. Evidently, initial data on S are not sufficient to uniquely determine events at point Q. The surface that separates the two regions "causally determined by S" and "not causally determined by S" is called Cauchy horizon.

The analogy between gravity and electrostatics became apparent when the Coulomb law was discovered in 1785. The gravitational and the electrostatic forces both obeyed an inverse-square law, with the difference that the mass can only be positive whereas the electric charge exists with both signs. Equal electric charges repel, opposite ones attract; in contrast, gravity is always attractive.

In 1820, electromagnetism was discovered by Oersted, and the emerging unified theory, called "electrodynamics" by Ampère, eventually found its expression in the Maxwell equations of 1864. Besides the electric field $\boldsymbol{E}$ related to charge, we have the magnetic field $\boldsymbol{B}$ related to *moving* charge. These fields, together with the electric and magnetic excitations $\boldsymbol{D}$ and $\boldsymbol{H}$, respectively, obey the Maxwell equations.

Newton's gravitational theory was only superseded in 1915/16 by Einstein's gravitational theory, general relativity. However, already in the 1870s physicists began to speculate whether, besides Newton's "gravitoelectric" field, related to mass at rest, there may also exist a new "gravitomagnetic" field, accompanying moving mass; for more details and references see Mashhoon.[116] As we saw above, these speculations became a solid basis in general relativity. In (90), the gravitomagnetic Lense–Thirring term surfaced, which found solid experimental verification in the meantime. Thus, we can speak with justification of gravitoelectromagnetism[116] (GEM), a notion which can guide our intuition, see in this context also Ni and Zimmermann.[137]

3.8.1. *Gravitoelectromagnetic field strength*

Electrodynamics is a linear theory, GR a nonlinear one. Still, if we take a linearized version of GR, there are those strong analogies between electrodynamics and gravitodynamics, as worked out, for instance, nicely in Rindler's[165] book. However, the analogies go even further, as pointed out particularly by Mashhoon.[116] Even in an arbitrary gravitational field, if referred to a Fermi propagated reference frame with coordinates $(T, \mathbf{X})$, GEM is a useful concept. If we apply the geodesic deviation equation (33) to such a frame, the gravitoelectromagnetic field strength, representing the tidal forces, turns out to be[116,t]

$$^{\mathrm{GEM}}F_{\alpha\beta} = -R_{\alpha\beta 0i}(T)X^i. \tag{166}$$

If we develop (33) up to the order linear in the velocity $\mathbf{V} := d\mathbf{X}/dT$, we find

$$\frac{d^2X^i}{dT^2} = -R_{0i0j}X^j + 2R_{ik0j}X^jV^k = -{}^{\mathrm{GEM}}F_{i0} - 2\,{}^{\mathrm{GEM}}F_{ki}V^k. \tag{167}$$

Now we recall that in electrodynamics the electric and the magnetic fields $\mathbf{E}$ and $\mathbf{B}$, respectively, are accommodated in the 4D electromagnetic field strength tensor according to

$$(F_{\alpha\beta}) = \begin{pmatrix} 0 & -E_1 & -E_2 & -E_3 \\ \diamond & 0 & B_3 & -B_2 \\ \diamond & \diamond & 0 & B_1 \\ \diamond & \diamond & \diamond & 0 \end{pmatrix} = -(F_{\beta\alpha}). \tag{168}$$

The diamond symbol $\diamond$ denotes matrix elements already known because of the antisymmetry of the matrix involved. The corresponding two-form reads $F = \frac{1}{2}F_{\alpha\beta}dx^\alpha \wedge dx^\beta$. Keeping (168) in mind, Eq. (167) can be rewritten as a vector equation

$$\frac{d^2\mathbf{X}}{dT^2} = -{}^{\mathrm{gr}}\mathbf{E} - 2\mathbf{V}\times{}^{\mathrm{gr}}\mathbf{B}. \tag{169}$$

In accordance with the equivalence principle, this equation of motion is *independent* of the mass. The analogy with electromagnetism requires that the gravito*electric* charge, in terms of the mass m, is -1 and the gravito*magnetic* charge -2. In electrodynamics, both quantities are $+1$. The difference comes from the vector nature of the electromagnetic potential A_α as compared to the tensor nature of the gravitational potential $g_{\alpha\beta}$, that is, helicity 1 as compared to helicity 2. The relation between the gravitomagnetic to the gravitoelectic charge, that is, the *gyrogravitomagnetic ratio*, is two: $^{\mathrm{gr}}\gamma = 2$. Note that in Gravity Probe-B the authors specify the gyrogravitomagnetic ratio as 1. However, their gyros carried only orbital angular momentum rather than spin angular momentum. Hence, this is to be expected. For more detailed discussions on this difference, see Refs. 80 and 138.

[t] Alternatively, we could generalize the Newtonian tidal force matrix of (9) to the gravitoelectric and gravitomagnetic tidal force matrices, $\mathcal{E}_{ij} = R_{i0j0}$ and $\mathcal{B}_{ij} = \epsilon_{ikl}R_{klj0}$, respectively, see Scheel and Thorne.[169] Both matrices are symmetric and trace-free. Note that $^{\mathrm{GEM}}F_{\alpha\beta}$ is an antisymmetric 4×4 matrix and $\mathcal{E}$ and $\mathcal{B}$ are both symmetric trace-free 3×3 matrices.

It has been pointed out by Ni[139] that the "measurement of the gyrogravitational ratio of [a] particle would be a further step[138] toward probing the microscopic origin of gravity. GP-B serves as a starting point for the measurement of the gyrogravitational factor of particles."

3.8.2. *Quadratic invariants*

In electrodynamics, we have two quadratic invariants[81]:

$$\frac{1}{2}F_{\alpha\beta}F^{\alpha\beta} = \star(\star F \wedge F) = B^2 - E^2, \quad \frac{1}{2}F_{\alpha\beta}F^{*\alpha\beta} = \star(F \wedge F) = 2\mathbf{E}\cdot\mathbf{B}, \tag{170}$$

where we used for the tensor dual the notation $F^{*\alpha\beta} := \frac{1}{2}\varepsilon^{\alpha\beta\gamma\delta}F_{\gamma\delta}$. We also employed the very concise notation of exterior calculus with the Hodge star operator.[u]

The first invariant is proportional to the Maxwell vacuum Lagrangian and is an ordinary scalar, whereas the second one corresponds to a surface term and is a pseudoscalar (negative parity).

Turn now directly to the Kerr metric and list for this example the tidal gravitational forces, which are represented by the curvature tensor. With its 20 independent components, it can be represented by a trace-free symmetric 6×6 matrix, see (32). The collective indices $A, B, \ldots = 1, \ldots, 6$ are defined as follows: $\{\hat{t}\hat{r}, \hat{t}\hat{\theta}, \hat{t}\hat{\phi}; \hat{\theta}\hat{\phi}, \hat{\phi}\hat{r}, \hat{r}\hat{\theta}\} \rightarrow \{1, 2, 3, 4, 5, 6\}$. We throw the orthonormal Kerr coframe (128) to (133) into our computer and out pops the 6×6 curvature matrix

$$(R_{AB}) = \begin{pmatrix} -2\mathbb{E} & 0 & 0 & 2\mathbb{B} & 0 & 0 \\ \circ & \mathbb{E} & 0 & 0 & -\mathbb{B} & 0 \\ \circ & \circ & \mathbb{E} & 0 & 0 & -\mathbb{B} \\ \circ & \circ & \circ & 2\mathbb{E} & 0 & 0 \\ \circ & \circ & \circ & \circ & -\mathbb{E} & 0 \\ \circ & \circ & \circ & \circ & \circ & -\mathbb{E} \end{pmatrix} = (R_{BA}), \tag{171}$$

with

$$\mathbb{E} := mr\frac{r^2 - 3a^2\cos^2\theta}{(r^2 + a^2\cos^2\theta)^3}, \quad \mathbb{B} := ma\cos\theta\frac{3r^2 - a^2\cos^2\theta}{(r^2 + a^2\cos^2\theta)^3}. \tag{172}$$

It is straightforward to identify $\mathbb{E}$ as the gravitoelectric and $\mathbb{B}$ as the gravitomagnetic component of the curvature. This is in accordance with (166).

[u]The Hodge star $\star\omega$ of a p-form $\omega = (1/p!)\omega_{\mu_1\cdots\mu_p}dx^{\mu_1}\wedge\cdots\wedge dx^{\mu_p}$ is an $(n-p)$-form $\star\omega$, with the components $(\star\omega)_{\mu_1\cdots\mu_{n-p}} = (1/p!)\varepsilon^{\nu_1\cdots\nu_p}{}_{\mu_1\cdots\mu_{n-p}}\omega_{\nu_1\cdots\nu_p}$, where ε is the totally antisymmetric unit tensor and n the dimension of the space, see Eq. (C.2.90) in Ref. 81.

It is obvious how we should continue. Our gravitoelectromagnetic invariants will be[v]

$$K := \frac{1}{2} R_{\alpha\beta\gamma\delta} R^{\alpha\beta\gamma\delta} = - \star (\star R_{\alpha\beta} \wedge R^{\alpha\beta}), \tag{173}$$

$$\mathcal{P} := \frac{1}{4} \varepsilon^{\gamma\delta\mu\nu} R_{\alpha\beta\gamma\delta} R^{\alpha\beta}{}_{\mu\nu} = \star (R_{\alpha\beta} \wedge R^{\alpha\beta}). \tag{174}$$

Again, our program determines the *Kretschmann*[w] scalar K and the *Chern-Pontryagin* pseudoscalar $\mathcal{P}$ to be[19]

$$K = -24(\mathbb{B}^2 - \mathbb{E}^2), \quad \mathcal{P} = -48\,\mathbb{E}\mathbb{B}. \tag{175}$$

The similarity to (170) is impressive. The GEM analogy quite apparently applies to the full nonlinear theory. The results in (175), partly in more involved representations, can be found in the literature, see, for instance, the books of de Felice and Clarke[42] and of Ciufolini and Wheeler,[36] but compare also de Felice and Bradley,[41] Henry,[84] and Cherubini *et al.*[32]

Thus, the quadratic invariants K and $\mathcal{P}$ confirm that the Kerr metric is the exterior field of a rotating mass distribution. In order to get more information about this distribution, we proceed, like in electrodynamics, and look into the gravitoelectromagnetic multipole moments of this rotating mass.

3.8.3. *Gravitomagnetic clock effect of Mashhoon, Cohen et al.*

According to the results of Lense–Thirring, the rotation of the Sun changes the spacetime around it by inducing gravitomagnetic effects. As we saw above, in a similar way the temporal structure around a Kerr metric is affected by the angular momentum of the Kerr source. Thus, a gravitomagnetic clock effect should emerge,[x] the measurability of which requires very accurate clocks. The effect can be demonstrated by two clocks that move on equatorial orbits, one in prograde and the other in retrograde orbit around the Kerr metric. It turns out[117] that the *prograde* equatorial clock is *slower* than the retrograde one. This is not necessarily what our intuition would tell us. It is connected with the fact that the dragging of frames in a Kerr metric can sometimes turn out to be an "antidragging", thus making this notion less intuitive,[165] as we already recognized in Sec. 3.5.

[v] In exterior calculus, we have the Euler four-form $E := R_{\alpha\beta} \wedge \star R^{\alpha\beta}$, with $K = \star E$. Analogously, we have the Chern–Pontryagin four-form $P := -R^{\alpha}{}_{\beta} \wedge R^{\beta}{}_{\alpha}$, which is an exact form, with $\mathcal{P} := \star P$, see e.g. Obukhov *et al.*[141]

[w] Usually in the literature,[36,42] the Kretschmann scalar is defined as $R_{\alpha\beta\gamma\delta} R^{\alpha\beta\gamma\delta}$, even though the electrodynamics analogy would suggest to include the factor 1/2.

[x] This was first predicted by Cohen and Mashhoon[37] and worked out in greater detail by Mashhoon *et al.*,[117,118] see also Bonnor and Steadman[18] and the review papers in the workshop of Lämmerzahl *et al.*[103] In a similar way, there emerges also a gravitomagnetic *time delay*, see Ciufolini *et al.*[33]

Generalizations of this clock effect were studied, for example, by Hackmann and Lämmerzahl.[71] The recent discussion of the *Clocks around Sgr A**, by Angélil and Saha[7] is, in effect, just one more manifestation of the gravitomagnetic clock effect.

3.8.4. *Multipole moments: Gravitoelectric and gravitomagnetic ones*

In Newton's theory, one gets a good idea about a mass distribution and its gravitational field by determining the multipole moments of the mass distribution M. In GR, because of the existence gravitomagnetism, we have to expect a new type of multipole moments, namely the moments J of the angular momentum distribution.

If a stationary axially symmetric line element of the form (93) is asymptotically flat, then it is possible[180] to define two sets of multipole moments, the gravitoelectric moments M_s ("mass multipole moments") and the gravitomagnetic moments J_s ("angular momentum multipole moments"), for $s = 0, 1, 2, \ldots$. These moments were found by Geroch[67] for the static and by Hansen[73] for the stationary case. They were reviewed by Quevedo[160] and used for constructing new exact solutions by Quevedo and Mashhoon.[159,161] Hansen computed the multipole moments for the Kerr solution and found

$$s = 0 \qquad M_0 = -m \qquad J_1 = ma \tag{176}$$

$$s = 1 \qquad M_2 = ma^2 \qquad J_3 = -ma^3 \tag{177}$$

$$s = 2 \qquad M_4 = -ma^4 \qquad J_5 = ma^5 \tag{178}$$

$$s = 3\ldots \qquad M_6 = ma^6\ldots \qquad J_7 = -ma^7. \tag{179}$$

More compactly, we have

$$M_{2s} = (-1)^{s+1} m a^{2s}, \quad M_{2s+1} = 0; \tag{180}$$

$$J_{2s} = 0, \quad J_{2s+1} = (-1)^s m a^{2s+1}. \tag{181}$$

It is possible to introduce normalized multipole moments, see Meinel *et al.*,[123] such that for Kerr we have $\widetilde{M}_s + i\widetilde{J}_s = m(ia)^s$. Then the mass monopole $\widetilde{M}_0 = m$ is positive. Apparently, the Kerr metric has a simple multipolar structure or, formulated differently, only very specific matter distributions can represent the interior of the Kerr metric.

Quevedo[160] compiled a number of theorems which illustrate the use of the multipole moments:

(i) A stationary spacetime is *static* if and only if all its gravito*magnetic* multipole moments vanish (Xanthopoulos, 1979).
(ii) A static metric is flat if and only if all its gravito*electric* multipole moments vanish (Xanthopoulos, 1979).
(iii) A stationary metric is axisymmetric if and only if all its multipole moments are axisymmetric (Gürsel, 1983).

(iv) Two metrics with the same multipole moments have the same geometry at large distances from the source (Beig and Simon, 1981; Kundu, 1981; Van den Bergh and Wils, 1985).
(v) Any stationary, axisymmetric, asymptotically flat solution of Einstein's vacuum equation approaches the Kerr solution asymptotically (Beig and Simon, 1980).
(vi) Any static, axisymmetric, asymptotically flat vacuum solution approaches the Schwarzschild solution asymptotically (Beig, 1980).
In the formulation of Stephani *et al.*[180]
(vii) A given asymptotically flat stationary vacuum spacetime is uniquely characterized by its multiple moments.

We recognize that the knowledge of the multipole moments provides a lot of insight into the physical properties of an exact solution.

From the point of view of the Kerr solution Theorem (v), see Beig and Simon,[12] is perhaps the most interesting one. It underlines the central importance of the Kerr solution. The considerations in the context of Theorem (v) were further developed by Simon.[174,175] On the three-dimensional spatial slices of a stationary axially symmetric metric, he defined the 3D "*Simon tensor*",[15] a kind of complexified generalized Cotton-Bach tensor.[65] The vanishing of the Simon tensor then leads to the multipole moments of the Kerr solution. Later, Mars,[112] see also Mars[111] and Mars and Senovilla,[113] generalized this approach and was led to the 4D "*Simon-Mars tensor*". In Ionescu and Klainerman,[88] one can find a more extended discussion of the Simon-Mars tensor, see also Wong.[192] More recently, Bäckdahl and Valiente Kroon[9] have proposed replacing the Simon–Mars tensor by another measure of "non-Kerrness", namely a scalar parameter.

3.9. *Adding electric charge and the cosmological constant: Kerr–Newman metric*

Enriching the Kerr metric by an electric charge is straightforwardly possible. We start from the metric (133) with coframe (128) to (132). This coframe can accommodate the Kerr, the Schwarzschild, and the Reissner–Nordström solutions. The different forms of the function Δ suggest how a charged Kerr solution should look like

Schwarzschild	(m)	$\rho = r^2$	$\Delta = r^2 - 2mr$		
Reissner–Nord.	$(m,\, q)$	$\rho = r^2$	$\Delta = r^2 - 2mr$	$+q^2$	
Kerr	$(m,\, a)$	$\rho = r^2 + a^2\cos^2\theta$	$\Delta = r^2 - 2mr$		$+a^2$
Kerr–Newman	$(m,\, a,\, q)$	$\rho = r^2 + a^2\cos^2\theta$	$\Delta = r^2 - 2mr$	$+q^2$	$+a^2$

Charging the Schwarzschild solution is achieved by adding q^2 to the function Δ. Since the charged Kerr solution should encompass the Reissner–Nordström solution, we tentatively keep the term q^2 for the case $a \neq 0$. Now, we can indeed find a potential

$$A = -\frac{qr}{\rho^2}(dt - a\sin^2\theta d\phi), \tag{182}$$

such that the Einstein–Maxwell equations are fulfilled. The potential describes a line-like charge distribution at $\rho = 0$, that is, on the ring singularity of the Kerr spacetime, which is quite satisfying.[135] This charged Kerr solution was first worked out by Newman *et al.*[134] (1965), using "methods which transcend logic", as Ernst[54] puts it. He, in turn, proceeded from (120). Replacing[y] ξ by $\sqrt{1 - qq^*}\xi$ generates a solution of the Einstein–Maxwell equations with potential $A_t + iA_\phi = q/(\xi + 1)$.

The Kerr and the Kerr–Newman solution behave quite similarly. We can adopt most of the discussion of the Kerr metric by substituting $a^2 + q^2$ for a^2.

We can further generalize the Kerr–Newman metric to include also a cosmological constant, see Sec. 4.1, and even more parameters, see Fig. 17.

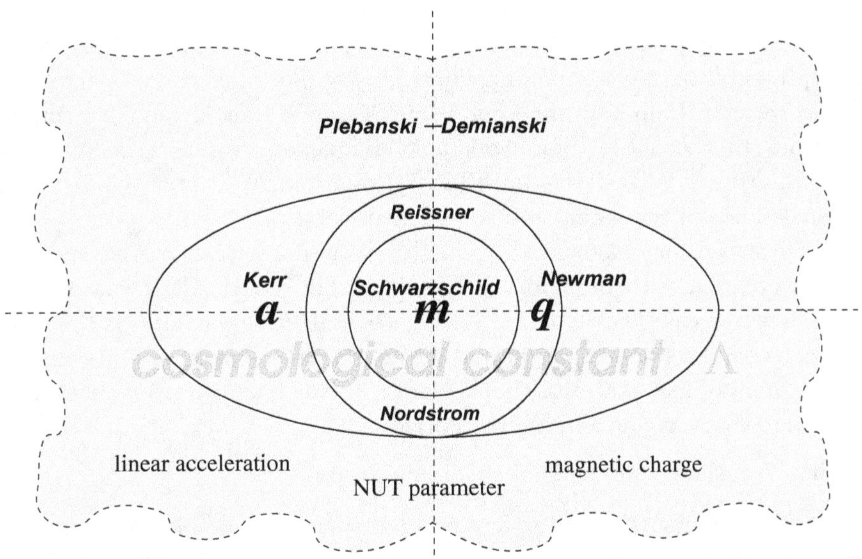

Fig. 17. Schematics of Petrov D solutions: The spherically symmetric Schwarzschild solution with mass parameter m is located in the center. Adding an electric charge q brings us to the Reissner–Nordström solution. It is still spherically symmetric, but adds a second horizon. The distance between the horizons increases with the charge q. Setting the black hole into rotation, the angular momentum parameter emerges, $a \neq 0$, and reduces the spherical symmetry to an axial one. An oblate ergosurface (two, actually) forms. Event horizon and ergosurface meet at the polar axis, the equatorial distance increases with a. All these solutions can be *deSittered*, that is, a cosmological constant Λ is added. All presented solutions are subcases of the Plebanski–Demianski solution, which adds three more parameters.[69]

[y]Here, q is *not* the charge but a complex parameter in the solution of the Ernst equation.

3.10. *On the uniqueness of the Kerr black hole*

The Kerr black hole, up to some technical assumptions, is the unique solution for the stationary, axially symmetric case. We point to some of the literature where these results can be found.

Because of the Birkhoff theorem, the Schwarzschild solution (mass parameter m) represents the general spherically symmetric solution of the Einstein vacuum field equation. The analogous is true in the Einstein–Maxwell case for the two parameter Reissner–Nordström solution (mass and charge parameters m and q, respectively). Thus, for spherical symmetry, we have a fairly simple situation.

In contrast, in the axially symmetric case, there does *not* exist a generalized Birkhoff theorem. The two-parameter Kerr solution (mass and rotation parameters m and a, respectively), is just a particular solution for the axially symmetric case. As we saw in Sec. 3.8, the Kerr solution has very simple gravitoelectric and gravitomagnetic multipole moments (180, 181). Numerous solutions are known that represent the exterior of matter distributions with different multipole moments. The analogous is valid for the three parameter Kerr–Newman solution (parameters m, a, q), see Stephani *et al.*[180] and Griffiths and Podolský.[69]

However, one can show under quite general conditions that the Kerr–Newman metric represents the most general asymptotically flat, stationary electro-vacuum *black hole solution* ("no-hair theorem"), see Meinel's short review.[122] Important contributions to the subject of black hole uniqueness were originally made by Israel,[90,91] Carter,[30,31] Hawking and Ellis,[78] Robinson,[166,167] and Mazur[120] (1967–1982), for details see the recent review of Chruściel *et al.*[38]

More recently Neugebauer and Meinel[130,133] found a constructive method for proving the uniqueness theorem for the Kerr black hole metric. This was extended to the Kerr–Newman case by Meinel.[121] By inverse scattering techniques, they showed how one can construct the Ernst potential of the Kerr(–Newman) solution amongst the asymptotically flat, stationary, and axially symmetric (electro–)vacuum spacetimes surrounding a connected Killing horizon.

Let us then eventually pose the following questions[27]:

(i) Are axially symmetric, stationary vacuum solutions outside some matter distribution "Kerr"? The answer is "certainly not", and it makes sense to figure out ways to characterize the Kerr metric, see Sec. 3.8.
(ii) Is the Kerr solution the unique axially symmetric, stationary vacuum black hole? The answer is essentially "yes" (modulo some technical issues) — see, for example Mazur.[120]

The general tendency in the recent development of the subject is to use additional scalar[z] or other matter fields. They weaken the uniqueness theorems, which is probably not too surprising.

[z]Recently, Herdeiro and Radu took as source for the Einstein field equation a massive complex scalar field (without self-interaction). They found numerically a generalization of the Kerr solution, which may be of some relevance to astrophysics.[193,194]

Let us conclude with a quotation that may make you curious to learn still more about the beauty of the Kerr metric: We have many different axially symmetric solutions. The Kerr solution is characterized by "*stationary, axially symmetric, asymptotically flat, Petrov type D vacuum solution of the vanishing of the Simon tensor, admitting a rank-2 Killing–Stäckel (KS) tensor of Segre type* [(11)(11)] *constructed from a (nondegenerate) rank-2 Killing–Yano (KY) tensor*", see Hinoui *et al.*[87]

3.11. *On interior solutions with material sources*

To match the Kerr (vacuum) metric to a material source consistently is one of the big unsolved problems. Only the rotating disc solution of Neugebauer and Meinel provides some hope.

This section is added in order to draw your attention to an unsolved problem, to the solution of which you might want to contribute. Find a realistic material source for the Kerr metric in the sense of an exact solution. Many unsuccessful attempts have been made, see the early review of Krasiński[101] of 1978. More recently, in 2006, Krasiński[156] concludes "*that a bright new idea is needed, as opposed to routine standard tricks tested so far.*" This statement was not made lightheartedly, Krasiński knows what he is talking about.

Many axially symmetric vacuum solutions were constructed. Quevedo and Mashhoon,[161] for example, deformed the multipole moments of the Kerr(–Newman) metric and constructed appropriate solutions of the Einstein(–Maxwell) equation that describe the exterior gravitational field of a (charged) rotating mass. It is always the hope that somebody may find a suitable matter distribution with the multipole moments of the Kerr solution — but this did not happen so far. For another approach see Marsh.[115]

We are only aware of one exact solution that fits into this general context: It is the *infinitesimally thin and rigidly rotating dust* solution of Neugebauer and Meinel[131,132] (1993). It is an exact analytical solution of the Einstein equation *with* matter. It depends on two independent parameters, the radius ρ_0 of the disk and its angular velocity Ω. Petroff and Meinel[151] developed, by means of an iterative procedure, a post-Newtonian approximation of the solution that helps to understand the Newtonian limit.

We recall that in electrostatics in flat space, for example, we prescribe an electric charge distribution and we are used to solve the corresponding boundary value problem within Maxwell's theory. Similarly, Neugebauer and Meinel specified a very thin rotating disk of dust and solved the boundary value problem within GR. This is a well-defined procedure. The problem is, however, that within a nonlinear theory, such as GR, it is extremely hard to implement. Remarkably, for certain parameter values, the gravitational field of the disk approach the extremal Kerr case. Accordingly, there exists a certain relation to the Kerr problem. The desideratum would be to find a rotating matter distribution the external field of which coincides with the *complete* Kerr field.

Driven by the fact that the electrically charged Kerr solution, the Kerr–Newman solution, has a g-factor of 2, exactly like the electron (see also Pfister and King[153]), Burinskii[24,25] speculated that a soliton like solution of the Dirac equation may be the source of the Kerr metric, see also Burinskii and Kerr.[26] Is that the "bright new idea" Krasiński was talking about? We do not know but a hard check of the Burinskii ansatz seems worthwhile.

4. Kerr Beyond Einstein

In generalizations of Einstein's theory of gravity, the Riemannian geometry of spacetime is often extended to a more general geometrical framework. We describe two such examples in which the Kerr metric still plays a vital role.

4.1. *Kerr metric accompanied by a propagating linear connection*

We display the Kerr metric with cosmological constant that, together with an explicitly specified torsion, represents an exact vacuum solution of the two field equations of the Poincaré gauge theory of gravity with quadratic Lagrangian.

In gauge theories of gravitation, see Blagojević *et al.*,[17] the linear connection becomes a field that is at least partially independent from the metric. It can be either metric-compatible, then it is a connection with values in the Lie-algebra of the Lorentz group $SO(1,3)$ and the geometry is called a Riemann–Cartan geometry, or it can be totally independent, then it resides in a so-called metric-affine space and the connection is $GL(4,R)$-valued. For simplicity, we concentrate here on the former case, the Poincaré gauge theory of gravity, but the latter case is also treated in the literature.[11,187]

Let us shortly sketch the theory. Gauging the Poincaré group leads to a spacetime with torsion T^α and curvature $R^{\alpha\beta}$ (Riemann–Cartan geometry[aa]):

$$T^\alpha := D\vartheta^\alpha = d\vartheta^\alpha + \Gamma_\beta{}^\alpha \wedge \vartheta^\beta = \frac{1}{2} T_{ij}{}^\alpha dx^i \wedge dx^j , \tag{183}$$

$$R^{\alpha\beta} := d\Gamma^{\alpha\beta} - \Gamma^{\alpha\gamma} \wedge \Gamma_\gamma{}^\beta = -R^{\beta\alpha} = \frac{1}{2} R_{ij}{}^{\alpha\beta} dx^i \wedge dx^j . \tag{184}$$

Besides the coframe one-form ϑ^α, the Lorentz connection one-form $\Gamma^{\alpha\beta} = \Gamma_i{}^{\alpha\beta} dx^i = -\Gamma^{\beta\alpha}$ is a second field variable of the gauge theory. For a Riemannian space, torsion $T^\alpha = 0$ and $\Gamma^{\alpha\beta}$ becomes the Levi-Civita connection.

We choose a model Lagrangian quadratic in torsion and curvature, in actual fact (for $\hbar = 1, c = 1$),

$$V = -\frac{1}{2\kappa}(T^\alpha \wedge \vartheta^\beta) \wedge \star(T_\beta \wedge \vartheta_\alpha) - \frac{1}{2\varrho} R^{\alpha\beta} \wedge \star R_{\alpha\beta} , \tag{185}$$

with Einstein's gravitational constant κ (dimension length-squared) and a dimensionless *strong* gravity coupling constant ϱ. One can calculate the two vacuum field

[aa]Experimental limits of a possible torsion of spacetime were recently specified in a remarkable paper by Obukhov *et al.*,[142] see also the literature given there.

equations by varying with respect to ϑ^α and $\Gamma^{\alpha\beta}$. In 1988, for these two field equations, a Kerr metric with torsion[10] was found as an exact solution.

We display here the orthonormal coframe and the torsion: The coframe ϑ^α, in terms of Boyer–Lindquist coordinates (t, r, θ, ϕ), reads (in the conventions used in Ref. 10)

$$\vartheta^{\hat{0}} := \frac{\sqrt{\Delta}}{\rho}(dt + a\sin^2\theta d\phi), \tag{186}$$

$$\vartheta^{\hat{1}} := \frac{\rho}{\sqrt{\Delta}}dr, \tag{187}$$

$$\vartheta^{\hat{2}} := \frac{\rho}{\sqrt{F}}d\theta, \tag{188}$$

$$\vartheta^{\hat{3}} := \frac{\sqrt{F}\sin\theta}{\rho}[adt + (r^2 + a^2)d\phi]. \tag{189}$$

As before, we have $\rho^2 := r^2 + a^2\cos^2\theta$. However, the other structure functions pick up a cosmological constant λ:

$$F := 1 + \frac{1}{3}\lambda a^2\cos^2\theta, \quad \Delta := r^2 + a^2 - 2Mr - \frac{1}{3}\lambda r^2(r^2 + a^2). \tag{190}$$

The corresponding metric is called a Kerr–de Sitter metric. The coframe is orthonormal. Then the metric reads

$$g = -\vartheta^{\hat{0}} \otimes \vartheta^{\hat{0}} + \vartheta^{\hat{1}} \otimes \vartheta^{\hat{1}} + \vartheta^{\hat{2}} \otimes \vartheta^{\hat{2}} + \vartheta^{\hat{3}} \otimes \vartheta^{\hat{3}}. \tag{191}$$

It is a characteristic feature of these exact solutions that even though the Lagrangian (185) does *not* carry a cosmological constant, in the coframe and the metric there emerges such a constant, namely $\lambda := -3\varrho/(4\kappa)$. This could be of potential importance for cosmology.

The torsion T^α of this stationary axially symmetric solution of the Poincaré gauge theory reads ($\vartheta^{\alpha\beta} := \vartheta^\alpha \wedge \vartheta^\beta$)

$$\begin{aligned}
T^{\hat{0}} &= \frac{\rho}{\sqrt{\Delta}}\left[-v_1\vartheta^{\hat{0}\hat{1}} + \frac{\rho}{\sqrt{\Delta}}[v_2(\vartheta^{\hat{0}\hat{2}} - \vartheta^{\hat{1}\hat{2}}) + v_3(\vartheta^{\hat{0}\hat{3}} - \vartheta^{\hat{1}\hat{3}})] - 2v_4\vartheta^{\hat{2}\hat{3}}\right], \\
T^{\hat{1}} &= T^{\hat{0}}, \\
T^{\hat{2}} &= \frac{\rho}{\sqrt{\Delta}}[v_5(\vartheta^{\hat{0}\hat{2}} - \vartheta^{\hat{1}\hat{2}}) + v_4(\vartheta^{\hat{0}\hat{3}} - \vartheta^{\hat{1}\hat{3}})], \\
T^{\hat{3}} &= \frac{\rho}{\sqrt{\Delta}}[-v_4(\vartheta^{\hat{0}\hat{2}} - \vartheta^{\hat{1}\hat{2}}) + v_5(\vartheta^{\hat{0}\hat{3}} - \vartheta^{\hat{1}\hat{3}})],
\end{aligned} \tag{192}$$

with the following gravitoelectric and gravitomagnetic functions:

$$v_1 = \frac{M}{\rho^4}(r^2 - a^2\cos^2\theta), \quad v_5 = \frac{Mr^2}{\rho^4}, \tag{193}$$

$$v_2 = -\frac{Ma^2 r\sin\theta\cos\theta}{\rho^5}\sqrt{F}, \quad v_3 = \frac{Mar^2\sin\theta}{\rho^5}\sqrt{F}, \quad v_4 = \frac{Mar\cos\theta}{\rho^4}. \tag{194}$$

Metric and torsion of this exact solution are closely interwoven. Note, in particular, that the leading gravitoelectric part in the torsion, for small a, is $\sim M/r^2$, a definitive Coulombic behavior proportional to the mass. For $a = 0$, we find a Schwarzschild–de Sitter solution with torsion.

One may legitimately ask, why is it that the Lagrangian (185) yields an exact solution with a Kerr–de Sitter metric? The answer is simple: The Lagrangian was devised such that the torsion square-piece, in lowest order in κ, encompasses a Newtonian approximation. This is already sufficient in order to enable the existence of a Kerr–de Sitter metric. One could even add another torsion-square piece to V for getting an Einsteinian approximation, but this is not even necessary. Thus, only a Newtonian limit of some kind seems necessary for the emergence of the Kerr structure.

4.2. *Kerr metric in higher dimensions and in string theory*

There also exist Schwarzschild and Kerr metrics in higher dimensional spacetimes. These investigations are mainly motivated by supergravity and string theory.

Tangherlini[183] (1963) started to investigate higher-dimensional Schwarzschild solutions, with $n-1$ spatial dimensions. He studied the ("planetary") orbits in an n-dimensional Schwarzschild field ("Sun") and found that only for $n = 4$ we have stable orbits, see also Ortin.[146] According to Tangherlini, this is then the only case that is interesting for physics. Nowadays, however, many physicists hypothesize that higher dimensions do exist because string theory suggests it.

Somewhat later, Myers and Perry[128] (1986) generalized these considerations to higher-dimensional Kerr metrics. In the meantime a plethora of such higher-dimensional objects have been found, see Allahverdizadeh *et al.*[5] and Frolov and Zelnikov.[64] Recently Keeler *et al.*[97] investigated, in the context of string theory, the separability of Klein–Gordon or Dirac fields on top of a higher-dimensional Kerr type solutions. Lately Brihaye *et al.*,[21] for example, discussed the exact solution of a 5D Myers–Perry black holes as coupled to a to a massive scalar field. The physical interpretations of these results remain to be seen.

Acknowledgments

We are grateful to Wei-Tou Ni (Hsin-chu) for inviting us to contribute to his Einstein volume. Georg Dautcourt (Berlin), Anne Franzen (Utrecht), Bahram Mashhoon (Columbia, Missouri), Reinhard Meinel (Jena), Gernot Neugebauer (Jena), Hernando Quevedo (Mexico City), and Walter Simon (Vienna) helped us to understand gravitoelectromagnetism, the Kerr solution, and more. We would like to thank Jens Boos (Cologne), Alberto Favaro (Oldenburg/Treviso), Eva Hackmann (Bremen), Carlos Herderiro (Aveiro), Gerald Marsh (Chicago), Yuri Obukhov (Moscow) and Herbert Pfister (Tübingen) for valuable remarks. Our three referees, Steve Carlip (Davis), James Nester (Chung-li), and Dah-Wei Chiou (Taipei) helped us to avoid a number of slips. Our sincere thanks for their numerous suggestions. We thank Mikael Rågstedt (Djursholm), librarian of the Mittag–Leffler Institute, for sending us a copy of Gullstand's original paper[70] and for providing information about Gullstrand.

Appendix

A.1. *Exterior calculus and computer algebra*

We want to use as input the Papapetrou metric (93). We take the equivalent representation in the form of the orthonormal coframe of Eqs. (94)–(98). How such a Reduce–Excalc program can be set up, is demonstrated in Stauffer *et al.*[178] and in Socorro *et al.*,[176] for the Einstein three-form, see Heinicke[82]:

```
%***************************************************
% Coframe of Andress-Lewis-Papapetrou-Buchdahl metric
%***************************************************
% file Buchdahl03.exi, 29 July 2014, fwh & chh
% in "Buchdahl03.exi";

load_package excalc;
off exp$
pform f=0, omega=0, gamma=0 $
fdomain f=f(rho,z), omega=omega(rho,z), gamma=gamma(rho,z) ;

coframe  o(0) = sqrt(f) * (d t - omega * d phi),
         o(1) = sqrt(f)**(-1) * exp(gamma) * d rho,
         o(2) = sqrt(f)**(-1) * exp(gamma) * d z,
         o(3) = sqrt(f)**(-1) * rho * d phi
    with signature (1,-1,-1,-1);

displayframe;
frame e$

%***************************************************
%    Connection, curvature, and Einstein forms
%***************************************************
pform conn1(a,b)=1, curv2(a,b)=2$
antisymmetric conn1, curv2$
factor o(0), o(1), o(2), o(3)$

conn1(-a,-b) := (1/2)*( e(-a)_|d o(-b) - e(-b)_|d o(-a)
                      - (e(-a)_|(e(-b)_|d o(-c))) * o(c))$
curv2(-a,b) := d conn1(-a,b) - conn1(-a,c) ^ conn1(-c,b)$

% Einstein tensor = Einstein 0-form
```

```
pform einstein3(a)=3, einstein0(a,b)=0$
symmetric einstein0$
einstein3(-a) := -(1/2) * curv2(b,-c) ^ # (o(-a) ^ o(-b) ^ o(c))$
einstein0(a,-b):= #( o(a) ^ einstein3(-b))$

on exp, gcd$
factor ^$
on nero;

einstein0(a,-b):= #( o(a) ^ einstein3(-b));

off nero;

% by inspection, we find
einstein0(1,-1) + einstein0(2,-2); % equals 0
einstein0(0,-0) - einstein0(3,-3); % eliminates gamma

out "Buchdahl03.exo";


load_package tri;
on tex;
on TeXBreak;
einstein0(a,-b):=einstein0(a,-b);
off tex;
einstein0(a,-b):=einstein0(a,-b);
omega:=0;
einstein0(a,-b):=einstein0(a,-b);

shut "Buchdahl03.exo";
;end;
```

References

1. R. Adler, M. Bazin and M. Schiffer, *Introduction to General Relativity*, 2nd edn. (McGraw-Hill, New York, 1975), p. 31.
2. E. S. Akeley, The axially symmetric stationary gravitational field, *Philos. Mag.* **11** (1931) 322–330.
3. E. S. Akeley, The rotating fluid in the relativity theory, *Philos. Mag.* **11** (1931) 330–344.
4. W. Alexandrow, Über den kugelsymmetrischen Vakuumvorgang in der Einsteinschen Gravitationstheorie, *Ann. Phys. (Berlin)* **377** (1923) 141–152.
5. M. Allahverdizadeh, J. Kunz and F. Navarro-Lerida, Charged rotating black holes in higher dimensions, *J. Phys. Conf. Ser.* **314** (2011) 012109, arXiv:1012.5052.

6. W. R. Andress, Some solutions of Einstein's gravitational equations for systems with axial symmetry, *Proc. R. Soc.* (*London*) *A* **126** (1930) 592–602.
7. R. Angélil and P. Saha, Clocks around Sgr A*, *Mon. Not. R. Astron. Soc.* **444** (2014) 3780–3791, arXiv:1408.0283.
8. R. Bach, Neue Lösungen der Einsteinschen Gravitationsgleichungen. A. Das Feld in der Umgebung eines langsam rotierenden kugelähnlichen Körpers von beliebiger Masse in 1. und 2. Annäherung, *Mathematische Zeitschrift* **13** (1922) 119–133.
9. T. Bäckdahl and J. A. V. Kroon, Constructing "non-Kerrness" on compact domains, *J. Math. Phys.* **53** (2012) 042503, arXiv:1111.6019.
10. P. Baekler, M. Gürses, F. W. Hehl and J. D. McCrea, The exterior gravitational field of a charged spinning source in the Poincaré gauge theory: A Kerr–Newman metric with dynamic torsion, *Phys. Lett. A* **128** (1988) 245–250.
11. P. Baekler and F. W. Hehl, Rotating black holes in metric-affine gravity, *Int. J. Mod. Phys. D* **15** (2006) 635–668, arXiv:gr-qc/0601063.
12. R. Beig and W. Simon, The stationary gravitational field near spatial infinity, *Gen. Relativ. Gravit.* **12** (1980) 1003–1013.
13. J. D. Bekenstein, Extraction of energy and charge from a black hole, *Phys. Rev. D* **7** (1973) 949–953.
14. R. Bergamini and S. Viaggiu, A novel derivation for Kerr metric in Papapetrou gauge, *Class. Quantum Grav.* **21** (2004) 4567–4573, arXiv:gr-qc/0305035.
15. D. Bini and R. T. Jantzen, Stationary spacetimes and the Simon tensor, *Nuovo Cim. B* **119** (2004) 863–873, arXiv:gr-qc/0411051.
16. G. D. Birkhoff, *Relativity and Modern Physics* (Harvard University Press, Cambridge, MA, 1923), p. 253.
17. M. Blagojević and F. W. Hehl (eds.), *Gauge Theories of Gravitation*: *A Reader with Commentaries* (Imperial College Press, London, 2013).
18. W. B. Bonnor and B. R. Steadman, The gravitomagnetic clock effect, *Class. Quantum Grav.* **16** (1999) 1853–1861.
19. J. Boos, Plebański-Demiański solution of general relativity and its expressions quadratic and cubic in curvature: Analogies to electromagnetism, *Int. J. Mod. Phys. D* **24** (2015) 1550079, arXiv: 1412.1958.
20. R. H. Boyer and R. W. Lindquist, Maximal analytic extension of the Kerr metric, *J. Math. Phys.* **8** (1967) 265–281.
21. Y. Brihaye, C. Herdeiro and E. Radu, Myers–Perry black holes with scalar hair and a mass gap, *Phys. Lett. B* **739** (2014) 1–7, arXiv:1408.5581.
22. D. Brill, History of a black hole horizon, *Grav. Cosmol.* **20** (2014) 165–170.
23. H. A. Buchdahl, *Seven Simple Lectures on General Relativity Theory* (Wiley, New York, 1981).
24. A. Burinskii, Complex structure of the four-dimensional Kerr geometry: Stringy system, Kerr theorem, and Calabi–Yau twofold, *Adv. High Energy Phys.* **2013** (2013) 509749, arXiv:1211.6021.
25. A. Burinskii, Emergence of the Dirac equation in the solitonic source of the Kerr spinning particle, *Grav. Cosmol.* **21** (2015) 28–34, arXiv:1404.5947.
26. A. Burinskii and R. P. Kerr, Nonstationary Kerr congruences, arXiv:gr-qc/9501012.
27. S. Carlip, Private communication (September 2014).
28. S. Carlip, Black hole thermodynamics, *Int. J. Mod. Phys. D* **23** (2014) 1430023, arXiv:1410.1486.
29. S. Carroll, *Spacetime and Geometry*: *An Introduction to General Relativity* (Addison Wesley, San Francisco, 2004).

30. B. Carter, *Black Hole Equilibrium States*, In Ref. 45, pp. 57–214 (1973).
31. B. Carter, *Global and Local Problems Solved by the Kerr Metric*, In Ref. 190, pp. 95–114 (2009).
32. C. Cherubini, D. Dini, S. Capozziello and R. Ruffini, Second order scalar invariants of the Riemann tensor: Applications to black hole spacetimes, *Int. J. Mod. Phys. D* **11** (2002) 827–841, arXiv:gr-qc/0302095.
33. I. Ciufolini, S. Kopeikin, B. Mashhoon and F. Ricci, On the gravitomagnetic time delay, *Phys. Lett. A* **308** (2003) 101–109, arXiv:gr-qc/0210015.
34. I. Ciufolini and E. C. Pavlis, A confirmation of the general relativistic prediction of the Lense–Thirring effect, *Nature* **431** (2004) 958–960.
35. I. Ciufolini, E. Pavlis, F. Chieppa, E. Fernandes-Vieria and J. Pérez-Mercader, Test of general relativity and the measurement of the Lense–Thirring effect with two earth satellites, *Science* **279** (1998) 2100–2104.
36. I. Ciufolini and J. A. Wheeler, *Gravitation and Inertia* (Princeton University Press, Princeton, NJ, 1995).
37. J. M. Cohen and B. Mashhoon, Standard clocks, interferometry, and gravitomagnetism, *Phys. Lett. A* **181** (1993) 353–358.
38. P. T. Chruściel, J. L. Lopes Costa and M. Heusler, Stationary black holes: Uniqueness and beyond, *Living Rev. Rel.* **15**(7) (2012) 1–69, arXiv:1205.6112.
39. N. Dadhich, A novel derivation of the rotating black hole metric, *Gen. Relativ. Gravit.* **45** (2013) 2383–2388, arXiv:1301.5314.
40. G. Dautcourt, Race for the Kerr field, *Gen. Relativ. Gravit.* **41** (2009) 1437–1454, arXiv:0807.3473.
41. F. de Felice and M. Bradley, Rotational anisotropy and repulsive effects in the Kerr metric, *Class. Quantum Grav.* **5** (1988) 1577–1585.
42. F. de Felice and C. J. S. Clarke, *Relativity on Curved Manifolds* (Cambridge University Press, Cambridge, UK, 1990).
43. W. de Sitter, On Einstein's theory of gravitation, and its astronomical consequences, first paper and second paper, *Mon. Not. R. Astron. Soc.* **76** (1916) 699–728; W. de Sitter, *Mon. Not. R. Astron. Soc.* **77** (1916) 155–184.
44. C. DeWitt and B. DeWitt (eds.), *Relativity, Groups and Topology*, Les Houches Summer School 1963 (Gordon and Breach, New York, 1964).
45. C. DeWitt and B. S. DeWitt (eds.), *Black Holes*, Les Houches Summer School 1972 (Gordon and Breach, New York, 1973).
46. C. Doran, New form of the Kerr solution, *Phys. Rev. D* **61** (2000) 067503, arXiv:gr-qc/9910099.
47. J. Droste, The field of a single centre in Einstein's theory of gravitation, and the motion of a particle in that field, *Gen. Relativ. Gravit.* **34** (2002) 1545–1563; [translation of the Dutch original of 1916].
48. J. Ehlers and A. Krasiński, Comment on the paper by J. T. Jebsen reprinted in Gen. Relativ. Gravit. 37, 2253–2259 (2005), *Gen. Relativ. Gravit.* **38** (2006) 1329–1330.
49. A. Einstein, *Über das Relativitätsprinzip und die aus demselben gezogenen Folgerungen*, In Ref. 177, pp. 411–462 (1907).
50. A. Einstein, *The Meaning of Relativity*, 5th edn. (Princeton University Press, Princeton, NJ, 1992) [first published in 1922].
51. A. Einstein, *The Collected Papers of Albert Einstein, Vol. 8: The Berlin Years: Correspondence* (1914–1918, Part A: 1914–1917, Part B: 1918), eds. R. Schulmann *et al.* (Princeton University Press, Princeton, NJ, 1998).
52. J. Eisenstaedt, The early interpretation of the Schwarzschild solution, in *Einstein and the History of General Relativity*, D. Howard and J. Stachel (eds.) (Birkhäuser, Boston, 1989), pp. 213–233.

53. F. J. Ernst, New formulation of the axially symmetric gravitational field problem, *Phys. Rev.* **167** (1968) 1175–1179.
54. F. J. Ernst, New formulation of the axially symmetric gravitational field problem. II, *Phys. Rev.* **168** (1968) 1415–1417; Erratum **172** (1968) 1850.
55. C. W. F. Everitt *et al.*, Gravity Probe B: Final results of a space experiment to test general relativity, *Phys. Rev. Lett.* **106** (2011) 221101, arXiv:1105.3456.
56. H. Falcke and F. W. Hehl (eds.), *The Galactic Black Hole, Lectures on General Relativity and Astrophysics* (IOP Publishing, Bristol, UK, 2003).
57. F. de Felice and C. J. S. Clarke, *Relativity on Curved Manifolds* (Cambridge University Press, Cambridge, UK, 1990).
58. P. G. Ferreira, *Perfect Theory: A Century of Geniuses and the Battle over General Relativity* (Houghton Mifflin Harcourt, Boston, MA, 2014).
59. D. Finkelstein, Past-future asymmetry of the gravitational field of a point particle, *Phys. Rev.* **110** (1958) 965–967.
60. L. Flamm, Beiträge zur Einsteinschen Gravitationstheorie, *Phys. Zeits.* **17** (1916) 448–454.
61. J. Fleischer, J. Grabmeier, F. W. Hehl and W. Küchlin (eds.), *Computer Algebra in Science and Engineering* (World Scientific, Singapore, 1995).
62. A. Franzen, Boundedness of massless scalar waves on Reissner–Nordström interior backgrounds, arXiv:1407.7093.
63. A. V. Frolov and V. P. Frolov, Rigidly rotating zero-angular-momentum observer surfaces in the Kerr spacetime, *Phys. Rev. D* **90** (2014) 124010, arXiv:1408.6316.
64. V. P. Frolov and A. Zelnikov, *Introduction to Black Hole Physics* (Oxford University Press, Oxford, UK, 2011).
65. A. García, F. W. Hehl, C. Heinicke and A. Macías, The Cotton tensor in Riemannian space-times, *Class. Quantum Grav.* **21** (2004) 1099–1118, arXiv:gr-qc/0309008.
66. H. Garcia-Compean and V. S. Manko, Are known maximal extensions of the Kerr and Kerr–Newman spacetimes physically meaningful, and analytic? *Progr. Theor. Exper. Phys.* 043E02 (2015), arXiv:1205.5848.
67. R. P. Geroch, Multipole moments. II. Curved space, *J. Math. Phys.* **11** (1970) 2580–2588.
68. J. Grabmeier, E. Kaltofen and V. Weispfenning (eds.), *Computer Algebra Handbook. Foundations, Applications, Systems* (Springer, Berlin, 2003).
69. J. B. Griffiths and J. Podolský, *Exact Space-Times in Einstein's General Relativity* (Cambridge University Press, Cambridge, UK, 2009).
70. A. Gullstrand, *Allgemeine Lösung des statischen Einkörperproblems in der Einsteinschen Gravitationstheorie* (General solution of the static one-body problem in Einstein's gravitational theory), *Arkiv Mat. Astron. Fys.* **16** (1922) 1–15 [presented on 25 May 1921].
71. E. Hackmann and C. Lämmerzahl, A generalized gravitomagnetic clock effect, *Phys. Rev. D* **90** (2014) 044059, arXiv:1406.6232.
72. E. Hackmann, C. Lämmerzahl, Y. N. Obukhov, D. Puetzfeld and I. Schaffer, Motion of spinning test bodies in Kerr spacetime, *Phys. Rev. D* **90** (2014) 064035, arXiv:1408.1773.
73. R. O. Hansen, Multipole moments of stationary spacetimes, *J. Math. Phys.* **15** (1974) 46–52.
74. A. Harvey, *On Einstein's Path: Essays in Honor of Engelbert Schucking* (Springer, New York, 1999).
75. I. Hauser and F. J. Ernst, A homogeneous Hilbert problem for the Kinnersley–Chitre transformations, *J. Math. Phys.* **21** (1980) 1126–1149.

76. I. Hauser and F. J. Ernst, Proof of a Geroch conjecture, *J. Math. Phys.* **22** (1981) 1051–1063.
77. I. Hauser and F. J. Ernst, *A New Proof of an Old Conjecture*, In Ref. 164, pp. 165–214 (1987).
78. S. W. Hawking and G. F. R. Ellis, *The Large Scale Structure of Spacetime* (Cambridge University Press, Cambridge, UK, 1973).
79. A. C. Hearn, *Reduce User's Manual, Version 3.5,* RAND Publication CP78 (Rev. 10/93). The RAND Corporation, Santa Monica, CA 90407-2138, USA (1993).
80. F. W. Hehl, A. Macías, E. W. Mielke and Y. N. Obukhov, *On the Structure of the Energy–Momentum and the Spin Currents in Dirac's Electron Theory*, In Ref. 74, pp. 257–274 (1999), arXiv:gr-qc/9706009.
81. F. W. Hehl and Yu. N. Obukhov, *Foundations of Classical Electrodynamics: Charge, Flux, and Metric* (Birkhäuser, Boston, 2003).
82. C. Heinicke, The Einstein 3-form G_α and its equivalent 1-Form L_α in Riemann–Cartan space, *Gen. Relativ. Gravit.* **33** (2001) 1115–1130, arXiv:gr-qc/0012037.
83. C. Heinicke and F. W. Hehl, *The Schwarzschild Black Hole: A General Relativistic Introduction*, In Ref. 56, pp. 3–34 (2003).
84. R. C. Henry, Kretschmann scalar for a Kerr–Newman black hole, *Astrophys. J.* **535** (2000) 350–353, arXiv:astro-ph/9912320.
85. M. Heusler, *Black Hole Uniqueness Theorems* (Cambridge University Press, Cambridge, UK, 1996).
86. D. Hilbert, Die Grundlagen der Physik (Zweite Mitteilung), *Nachr. Gesell. Wissensch. zu Göttingen, Math.-Phys. Klasse* (1917), pp. 53–76.
87. K. Hinoue, T. Houri, C. Rugina and Y. Yasui, General Wahlquist metrics in all dimensions, *Phys. Rev. D* **90** (2014) 024037, arXiv:1402.6904.
88. A. D. Ionescu and S. Klainerman, On the uniqueness of smooth, stationary black holes in vacuum, *Invent. Math.* **175** (2009) 35–102, arXiv:0711.0040.
89. L. Iorio, An assessment of the systematic uncertainty in present and future tests of the Lense–Thirring effect with satellite laser ranging, *Space Sci. Rev.* **148** (2009) 363–381, arXiv:0809.1373.
90. W. Israel, Event horizons in static vacuum space-times, *Phys. Rev.* **164** (1967) 1776–1779.
91. W. Israel, Event horizons in static electrovac space-times, *Commun. Math. Phys.* **8** (1968) 245–260.
92. L. Iorio, H. I. M. Lichtenegger, M. L. Ruggiero and C. Corda, Phenomenology of the Lense–Thirring effect in the solar system, *Astrophys. Space Sci.* **331** (2011) 351–395, arXiv:1009.3225.
93. J. N. Islam, *Rotating Fields in General Relativity* (Cambridge University Press, Cambridge, UK, 1985).
94. J. D. Jackson, *Classical Electrodynamics*, 3rd edn. (Wiley, New York, 1999).
95. J. T. Jebsen, On the general spherically symmtric solutions of Einstein's gravitational equations in vacuo, *Gen. Relativ. Gravit.* **37** (2005) 2253–2259; [reprinted, the original was published in German in 1921].
96. N. V. Johansen and F. Ravndal, On the discovery of Birkhoff's theorem, *Gen. Relativ. Gravit.* **38** (2006) 537–540, arXiv:physics/0508163.
97. C. Keeler and F. Larsen, Separability of black holes in string theory, *J. High Energy Phys.* **1210** (2012) 152, arXiv:1207.5928.
98. R. P. Kerr, Gravitational field of a spinning mass as an example of algebraically special metrics, *Phys. Rev. Lett.* **11** (1963) 237–238.
99. R. P. Kerr, *Discovering the Kerr and Kerr–Schild Metrics*, In Ref. 190, pp. 38–72 (2009), arXiv:0706.1109.

100. D. Kramer and G. Neugebauer, Zu axialsymmetrischen stationären Lösungen der Einsteinschen Feldgleichungen für das Vakuum, *Commun. Math. Phys.* **10** (1968) 132–139.
101. A. Krasiński, Ellipsoidal space-times, sources for the Kerr metric, *Ann. Phys. (NY)* **112** (1978) 22–40.
102. A. Krasiński, G. F. R. Ellis and M. A. H. MacCallum (eds.), *Golden Oldies in General Relativity. Hidden Gems* (Springer, Heidelberg, 2013).
103. C. Lämmerzahl, C. W. F. Everitt and F. W. Hehl (eds.), *Gyros, Clocks, Interferometers: Testing Relativistic Gravity in Space*, Lecture Notes in Physics, Vol. 562 (Springer, Berlin, 2001).
104. K. Lanczos, Ein vereinfachendes Koordinatensystem für die Einsteinschen Gravitationsgleichungen, *Phys. Zeits.* **23** (1922) 537–539.
105. K. Lanczos, Über eine stationäre Kosmologie im Sinne der Einsteinschen Gravitationstheorie, *Z. Physik* **21** (1924) 73–110.
106. L. D. Landau and E. M. Lifshitz, *The Classical Theory of Fields, Course of Theoretical Physics*, Vol. 2 [transl. from the Russian], 4th edn. (Elsevier, Amsterdam, 1975), p. 281.
107. J. Lense and H. Thirring, Über den Einfluss der Eigenrotation der Zentralkörper auf die Bewegung der Planeten und Monde nach der Einsteinschen Gravitationstheorie, *Physikalische Zeitschrift* **19** (1918) 156–163, an English translation is provided in Ref. 119, see also Ref. 102, Chapter 6.
108. T. Levi-Civita, ds^2 einsteiniani in campi newtoniani, *Rendiconti della R. Academia dei Lincei* **27–29** (1917–1919).
109. T. Lewis, Some special solutions of the equations of axially symmetric gravitational fields, *Proc. R. Soc. London A* **136** (1932) 176–192.
110. V. S. Manko and H. Garcia-Compean, Nondisk geometry of $r = 0$ in Kerr–de Sitter and Kerr–Newman–de Sitter spacetimes, *Phys. Rev. D* **90** (2014) 047501, arXiv:1406.3434.
111. M. Mars, A Space-time characterization of the Kerr metric, *Class. Quantum Grav.* **16** (1999) 2507–2523, arXiv:gr-qc/9904070.
112. M. Mars, Uniqueness properties of the Kerr metric, *Class. Quantum Grav.* **17** (2000) 3353–3373, arXiv:gr-qc/0004018.
113. M. Mars and J. M. M. Senovilla, A spacetime characterization of the Kerr-NUT-(A)de Sitter and related metrics, arXiv:1307.5018.
114. G. E. Marsh, The infinite red-shift surfaces of the Kerr and the Kerr–Newman solutions of the Einstein field equations, arXiv:gr-qc/0702114.
115. G. E. Marsh, Rigid rotation and the Kerr metric, arXiv:1404.5297.
116. B. Mashhoon, Gravitoelectromagnetism: A brief review, in: *The Measurement of Gravitomagnetism: A Challenging Enterprise*, L. Iorio (ed.) (Nova Science, New York, 2007), pp. 29–39, arXiv:gr-qc/0311030.
117. B. Mashhoon, F. Gronwald and H. I. M. Lichtenegger, Gravitomagnetism and the clock effect, *Lect. Notes Phys.* **562** (2001) 83–108, see Ref. 103, arXiv:gr-qc/9912027.
118. B. Mashhoon, F. Gronwald and D. S. Theiss, On measuring gravitomagnetism via spaceborne clocks: A gravitomagnetic clock effect, *Ann. Phys. (Berlin)* **8** (1999) 135–152, arXiv:gr-qc/9804008
119. B. Mashhoon, F. W. Hehl and D. S. Theiss, On the gravitational effects of rotating masses: The Thirring–Lense papers, *Gen. Relativ. Gravit.* **16** (1984) 711–749.
120. P. O. Mazur, Proof of uniqueness of the Kerr–Newman black hole solution, *J. Phys. A* **15** (1982) 3173–3180.
121. R. Meinel, Constructive proof of the Kerr–Newman black hole uniqueness including the extreme case, *Class. Quantum Grav.* **29** (2012) 035004, arXiv:1108.4854.

122. R. Meinel, A physical derivation of the Kerr–Newman black hole solution, in: *1st Karl Schwarzschild Meeting on Gravitational Physics*, P. Nicolini *et al.* (eds.), *Springer Proc. Phys.* **170** (2016) 53–61, arXiv:1310.0640.
123. R. Meinel, M. Ansorg, A. Kleinwächter, G. Neugebauer and D. Petroff, *Relativistic Figures of Equilibrium* (Cambridge University Press, Cambridge, UK, 2008).
124. F. Melia, *Cracking the Einstein Code. Relativity and the Birth of Black Hole Physics* (University of Chicago Press, Chicago, 2009).
125. J. Michell, On the means of discovering the distance, magnitude, etc. of the fixed stars, *Philos. Trans. R. Soc. London* **74** (1784) 35–57.
126. G. Mie, Die Einführung eines vernunftgemäßen Koordinatensystems in die Einsteinsche Gravitationstheorie und das Gravitationsfeld einer schweren Kugel, *Ann. Phys. (Berlin)* **367** (1920) 47–74.
127. C. W. Misner, K. S. Thorne and J. A. Wheeler, *Gravitation* (Freeman, San Francisco, 1973).
128. R. C. Myers and M. J. Perry, Black holes in higher dimensional space-times, *Ann. Phys. (NY)* **172** (1986) 304–347.
129. R. Narayan and J. E. McClintock, Observational evidence for black holes, *General Relativity and Gravitation: A Centennial Perspective*, eds. A. Ashtekar *et al.* (Cambridge University Press, Cambridge, 2015), pp. 133–147, arXiv:1312.6698.
130. G. Neugebauer, Rotating bodies as boundary value problems, *Ann. Phys. (Berlin)* **9** (2000) 342–354.
131. G. Neugebauer and R. Meinel, The Einsteinian gravitational field of a rigidly rotating disk of dust, *Astrophys. J.* **414** (1993) L97–L99, see also Sec. 2.3 in Ref. 123.
132. G. Neugebauer and R. Meinel, General relativistic gravitational field of a rigidly rotating disk of dust: Solution in terms of ultraelliptic functions, *Phys. Rev. Lett.* **75** (1995) 3046–3047, arXiv:gr-qc/0302060.
133. G. Neugebauer and R. Meinel, Progress in relativistic gravitational theory using the inverse scattering method, *J. Math. Phys.* **44** (2003) 3407–3429, arXiv:gr-qc/0304086.
134. E. T. Newman, E. Couch, K. Chinnapared, A. Exton, A. Prakash and R. Torrence, Metric of a rotating, charged mass, *J. Math. Phys.* **6** (1965) 918–919.
135. E. T. Newman and A. I. Janis, Note on the Kerr spinning-particle metric, *J. Math. Phys.* **6** (1965) 915–917.
136. I. Newton, *The Principia: Mathematical Principles of Natural Philosophy*, translation by B. I. Cohen, A. Whitman and J. Budenz (Univeristy of California Press, Berkeley, 1999).
137. W. T. Ni and M. Zimmermann, Inertial and gravitational effects in the proper reference frame of an accelerated, rotating observer, *Phys. Rev. D* **17** (1978) 1473–1476.
138. W. T. Ni, Searches for the role of spin and polarization in gravity, *Rep. Prog. Phys.* **73** (2010) 056901, arXiv:0912.5057.
139. W. T. Ni, Rotation, equivalence principle, and GP-B experiment, *Phys. Rev. Lett.* **107** (2011) 051103, arXiv:1105.4305.
140. G. Nordström, On the energy of the gravitational field in Einstein's theory, *Proc. Kon. Ned. Akad. Wet.* **20** (1918) 1238–1245.
141. Y. N. Obukhov and F. W. Hehl, On the relation between quadratic and linear curvature Lagrangians in Poincaré gauge gravity, *Acta Phys. Polon. B* **27** (1996) 2685–2694, arXiv:gr-qc/9602014.
142. Y. N. Obukhov, A. J. Silenko and O. V. Teryaev, Spin-torsion coupling and gravitational moments of Dirac fermions: Theory and experimental bounds, *Phys. Rev. D* **90** (2014) 124068, arXiv:1410.6197.

143. H. C. Ohanian and R. Ruffini, *Gravitation and Spacetime*, 3rd edn. (Cambridge University Press, New York, NY, 2013).
144. B. O'Neill, *The Geometry of Kerr Black Holes* (Peters, Wellesley, MA, 1995).
145. J. R. Oppenheimer and H. Snyder, On continued gravitational contraction, *Phys. Rev.* **56** (1939) 455–459.
146. T. Ortín, *Gravity and Strings* (Cambridge University Press, Cambridge, UK, 2004).
147. P. Painlevé, La mécanique classique et la théorie de la relativité (Classical mechanics and relativity theory), *C. R. Acad. Sci. (Paris)* **173** (1921) 677–680 [presented on 24 October 1921].
148. A. Papapetrou, Eine rotationssymmetrische Lösung in der allgemeinen Relativitätstheorie, *Ann. Phys. (Berlin)* **447** (1953) 309–315.
149. R. Penrose, *Conformal Treatment of Infinity*, In Ref. 44, pp. 563–584 (1964); reprinted in *Gen. Rel. Grav.* **43** (2011) 901–922.
150. R. Penrose, *The Road to Reality. A Complete Guide to the Laws of the Universe* (Knopf, New York, 2005).
151. D. Petroff and R. Meinel, The post-Newtonian approximation of the rigidly rotating disc of dust to arbitrary order, *Phys. Rev. D* **63** (2001) 064012, arXiv:gr-qc/0101081.
152. A. Z. Petrov, The classification of spaces defining gravitational fields, [Translation from the Russian original of 1954] in Ref. 102, Chapter 4, see also *Gen. Rel. Grav.* **32** (2000) 1665–1685.
153. H. Pfister and M. King, The gyromagnetic factor in electrodynamics, quantum theory and general relativity, *Class. Quantum Grav.* **20** (2003) 205–213.
154. F. A. E. Pirani, Invariant formulation of gravitational radiation theory, *Phys. Rev.* **105** (1957) 1089–1099.
155. J. F. Plebański and M. Demiański, Rotating, charged uniformly accelerated mass in general relativity, *Ann. Phys. (NY)* **98** (1976) 98–127.
156. J. Plebański and A. Krasiński, *An Introduction to General Relativity and Cosmology* (Cambridge University Press, Cambridge, UK, 2006).
157. F. Pretorius and W. Israel, Quasi-spherical light cones of the Kerr geometry, *Class. Quantum Grav.* **15** (1998) 2289–2301.
158. R. A. Puntigam, E. Schrüfer and F. W. Hehl, The use of computer algebra in Maxwell's theory, in Ref. 61, pp. 195–211 (1995), arXiv:gr-qc/9503023.
159. H. Quevedo, General static axisymmetric solution of Einstein's vacuum field equations in prolate spheroidal coordinates, *Phys. Rev. D* **39** (1989) 2904–2911.
160. H. Quevedo, Multipole moments in general relativity — Static and stationary vacuum solutions, *Fortschr. Phys. Progr.-Phys.* **38** (1990) 733–840.
161. H. Quevedo and B. Mashhoon, Generalization of Kerr spacetime, *Phys. Rev. D* **43** (1991) 3902–3906.
162. S. Rahman and M. Visser, Spacetime geometry of static fluid spheres, *Class. Quantum Grav.* **19** (2002) 935–952, arXiv:gr-qc/0103065.
163. H. Reissner, Über die Eigengravitation des elektrischen Feldes nach der Einsteinschen Theorie, *Ann. Phys. (Berlin)* **355** (1916) 106–120.
164. W. Rindler and A. Trautman (eds.), *Gravitation and Geometry, Festschrift for I. Robinson*, Bibliopolis, Naples, Italy (1987).
165. W. Rindler, *Relativity, Special, General, and Cosmological* (Oxford University Press, Oxford, UK, 2001).
166. D. C. Robinson, Uniqueness of the Kerr black hole, *Phys. Rev. Lett.* **34** (1975) 905–906.
167. D. C. Robinson, *Four Decades of Black Hole Uniqueness Theorems*, In Ref. 190, pp. 115–142 (2009).

168. R. H. Sanders, *Revealing the Heart of the Galaxy, the Milky Way and Its Black Hole* (Cambridge University Press, Cambridge, UK, 2014).
169. M. A. Scheel and K. S. Thorne, Geometrodynamics: The nonlinear dynamics of curved spacetime, *Phys. Usp.* **57** (2014) 342–351, [*Usp. Fiz. Nauk* **184**(4) (2014) 367–378].
170. J. A. Schouten, *Ricci-Calculus*, 2nd edn. (Springer, Berlin, 1954).
171. K. Schwarzschild, *Gesammelte Werke/Collected Works*, Vols. 1–3, ed. H. H. Voigt (Springer, Berlin, 1992).
172. K. Schwarzschild, *Über das Gravitationsfeld eines Massenpunktes nach der Einsteinschen Theorie* (On the gravitational field of a mass point according to Einstein's theory), Sitzungsber. Preuss. Akad. Wiss. Physik-Math. Kl. (1916), pp. 189–196.
173. K. Schwarzschild, *Über das Gravitationsfeld einer Kugel aus inkompressibler Flüssigkeit nach der Einsteinschen Theorie* (On the gravitational field of a ball of an incompressible fluid according to Einstein's theory), *Sitzungsber. Preuss. Akad. Wiss. Physik-Math. Kl.* (1916), pp. 424–434.
174. W. Simon, Characterizations of the Kerr metric, *Gen. Relativ. Gravit.* **16** (1984) 465–476.
175. W. Simon, The multiple expansion of stationary Einstein–Maxwell fields, *J. Math. Phys.* **25** (1984) 1035–1038.
176. J. Socorro, A. Macias and F. W. Hehl, Computer algebra in gravity: Reduce-Excalc programs for (non-)Riemannian space-times. I, *Comput. Phys. Commun.* **115** (1998) 264–283, arXiv:gr-qc/9804068.
177. J. Stark (ed.), *Jahrbuch der Radioaktivität und Elektronik*, Vol. 4 (Hirzel, Leipzig, 1907).
178. D. Stauffer, F. W. Hehl, N. Ito, V. Winkelmann and J. G. Zabolitzky, *Computer Simulation and Computer Algebra. Lectures for Beginners*, 3rd edn. (Springer, Berlin, 1993).
179. H. Stephani, *General Relativity. An Introduction to the Theory of the Gravitational Field*, 2nd edn. (Cambridge University Press, Cambridge, UK, 1990).
180. H. Stephani, D. Kramer, M. A. H. MacCallum, C. Hoenselaers and E. Herlt, *Exact Solutions to Einstein's Field Equations*, 2nd edn. (Cambridge University Press, Cambridge, UK, 2003).
181. S. Sternberg, *Curvature in Mathematics and Physics* (Dover, Mineola, NY, 2012).
182. N. Straumann, *General Relativity*, 2nd edn. (Springer, Dordrecht, 2013).
183. F. R. Tangherlini, Schwarzschild field in n dimensions and the dimensionality of space problem, *Nuovo Cim.* **27** (1963) 636–651.
184. W. J. van Stockum, The gravitational field of a distribution of particles rotating about an axis of symmetry, *Proc. R. Soc. Edinburgh A* **57** (1937) 135–154.
185. M. Visser, Heuristic approach to the Schwarzschild geometry, *Int. J. Mod. Phys. D* **14** (2005) 2051–2068, arXiv:gr-qc/0309072.
186. M. Visser, *The Kerr Spacetime — A Brief Introduction*, in Ref. 190, pp. 3–37 (2009), arXiv:0706.0622.
187. E. J. Vlachynsky, R. Tresguerres, Y. N. Obukhov and F. W. Hehl, An axially symmetric solution of metric-affine gravity, *Class. Quantum Grav.* **13** (1996) 3253–3259, arXiv:gr-qc/9604035.
188. H. Weyl, Zur Gravitationstheorie, *Ann. Phys. (Berlin)* **359** (1917) 117–145; an English translation is provided in *Gen. Relativ. Gravit.* **44** (2012) 779–810.
189. C. M. Will, The confrontation between general relativity and experiment, *Living Rev. Rel.* **17**(4) (2014), arXiv:1403.7377.

190. D. L. Wiltshire, M. Visser and S. M. Scott (eds.), *The Kerr Spacetime, Rotating Black Holes in General Relativity* (Cambridge University Press, Cambridge, UK, 2009).
191. S. Wolfram, *Mathematica, Mathematica Edition: Version* (Wolfram Research, Inc., Champaign, IL, 2013).
192. W. W. Wong, A space-time characterization of the Kerr–Newman metric, *Annales Henri Poincaré* (*Birkhäuser, Basel*) **10** (2009) 453–484, arXiv:0807.1904.
193. C. A. R. Herdeiro and E. Radu, Kerr black holes with scalar hair, *Phys. Rev. Lett.* **112** (2014) 221101, arXiv:1403.2757.
194. C. A. R. Herdeiro and E. Radu, Asymptotically flat black holes with scalar hair: a review, *Int. J. Mod. Phys. D* **24** (2015) 1542014, arXiv:1504.08209.

Chapter 4

Gravitational energy for GR and Poincaré gauge theories: A covariant Hamiltonian approach

Chiang-Mei Chen*,†,‖, James M. Nester*,†,‡,§,** and Roh-Suan Tung¶,††

*Department of Physics, National Central University, Chungli 32054, Taiwan

†Center for Mathematics and Theoretical Physics, National Central University, Chungli 32054, Taiwan

‡Graduate Institute of Astronomy, National Central University, Chungli 32054, Taiwan

§Leung Center for Cosmology and Particle Astrophysics, National Taiwan University, Taipei 10617, Taiwan

¶Institute of Advanced Studies, Nanyang Technological University, Singapore

‖cmchen@phy.ncu.edu.tw

**nester@phy.ncu.edu.tw

††roh.suan.tung@gmail.com

Our topic concerns a long standing puzzle: The energy of gravitating systems. More precisely we want to consider, for gravitating systems, how to best describe energy–momentum and angular momentum/center-of-mass momentum (CoMM). It is known that these quantities cannot be given by a local density. The modern understanding is that (i) they are quasi-local (associated with a closed 2-surface), (ii) they have no unique formula, (iii) they have no reference frame independent description. In the first part of this work, we review some early history, much of it not so well known, on the subject of gravitational energy in Einstein's general relativity (GR), noting especially Noether's contribution. In the second part, we review (including some new results) much of our covariant Hamiltonian formalism and apply it to Poincaré gauge theories of gravity (PG), with GR as a special case. The key point is that the Hamiltonian boundary term has two roles, it determines the quasi-local quantities, and furthermore, it determines the boundary conditions for the dynamical variables. Energy–momentum and angular momentum/CoMM are associated with the geometric symmetries under Poincaré transformations. They are best described in a local Poincaré gauge theory. The type of spacetime that naturally has this symmetry is Riemann–Cartan spacetime, with a metric compatible connection having, in general, both curvature and torsion. Thus our expression for the energy–momentum of physical systems is obtained via our covariant Hamiltonian formulation applied to the PG.

Keywords: Quasi-local energy; Hamiltonian boundary term.

PACS Number(s): 04.20.Cv, 04.20.Fy

1. Introduction

How to give a meaningful description of energy–momentum for gravitating systems (hence for all physical systems) has been an outstanding fundamental issue since Einstein began his search for his gravity theory, general relativity (GR). It is deeply connected to the essential nature of not only geometric gravity but of all the fundamental interactions — their inherent gauge nature. Noether's paper that includes her two famous theorems relating global symmetries to conserved quantities and local gauge symmetries to differential identities was originally motivated by this very issue. She showed that gravitational energy has no proper local description. So investigators only found various expressions which were inherently nontensorial (reference frame dependent *pseudotensors*). They have two inherent ambiguities: (i) There are many possible expressions, (ii) they are noncovariant — reference frame dependent. The modern view is that energy–momentum is *quasi-local* (associated with a closed 2-surface). Quasi-local proposals have analogous ambiguities. These ambiguities can be clarified by the Hamiltonian approach. From a first-order Lagrangian for quite general differential form fields, we have constructed a spacetime covariant Hamiltonian formalism, which incorporates the Noether conserved currents and differential identities. The Hamiltonian that dynamically evolves a spatial region includes a boundary term. The explicit form of the boundary term depends on the boundary conditions and also on an appropriate reference choice. With a suitable vector field, it gives expressions for the quasi-local quantities (energy–momentum, angular momentum/center-of-mass momentum, CoMM) and also quasi-local energy flux. A geometric gauge theory perspective provides the most appropriate dynamical variables. The geometry is Riemann–Cartan, with, in general, both curvature and torsion. For the PG (GR is a special case) with general source and gauge fields, we identified a preferred Hamiltonian boundary expression along with a procedure for finding a "best matched" reference. With this one can obtain values for the quasi-local energy–momentum and angular momentum/CoMM.[a]

Our topic here concerns the localization of energy–momentum. The main aim of our research program has actually been to better understand the Hamiltonian for dynamic spacetime geometry, especially the role of the Hamiltonian boundary term. It turns out that this sheds much light on the issue of the localization of energy.[88] A number of different ideas will be fit together to give a good picture of this long standing puzzle. In addition to being mindful of Noether's results, we will use a Hamiltonian approach combined with a local gauge theory view of dynamic spacetime geometry.

This present work is largely just an application of Noether's result. We will begin with some early history (much of it not so well known) regarding energy in

[a]For an alternative to our Hamiltonian approach to energy–momentum and angular momentum for the PG see Ref. 66.

the context of GR, especially Noether's contribution. Next, we will show how in GR pseudotensors are connected with the Hamiltonian and introduce the quasi-local idea. Following this, we make some brief remarks about gravity, geometry, connection and gauge. We then introduce and give a short review of our main tool: differential forms. We develop in some detail variational principles with differential form fields. With this, we can give simple examples of applications of the two Noether theorems. We then introduce the first-order formalism followed by our Hamiltonian formulation and the 3+1 split. The Hamiltonian boundary term and its important roles are discussed next. Asymptotic fall offs for the fields are noted. We explain why Riemann–Cartan geometry is appropriate for our purposes. Variational principles for dynamic spacetime geometry with quite general sources are developed, including the Noether conserved currents and differential identities. We present a first-order and Hamiltonian formulation for the PG along with the Hamiltonian boundary term and identify our preferred Hamiltonian boundary term for these dynamic spacetime geometry theories. We also include a prescription for choosing the necessary reference values that are needed for the quasi-local energy–momentum and angular momentum expressions.

2. Background

As this present work approached its final form, we received some very good news: All of the volumes of the Einstein papers published so far[b] — both the originals and the English translations[38] — are now freely available online.[37] An examination of Einstein's dozens of papers on gravity during the period 1913–1918, as well as his extensive correspondence with his contemporaries on the topic of gravity, shows that most of them include a significant consideration of the topic of gravitational energy.

2.1. *Some brief early history*

We have only begun to look into the historical development of the modern ideas regarding gravitational energy; the topic merits much deeper study. Here, we can only give a brief report, relying on the Einstein papers as well as some of the many good historical investigations available, in particular regarding energy–momentum conservation.[13,18,138]

We will rely on the Hamiltonian formalism applied to dynamical variables that are related to a local gauge theory of spacetime symmetry approach. (Earman[32] has given a very interesting discussion on how the Hamiltonian approach connects with the gauge theory perspective.)

It seems not so well known that gravitational energy, or more precisely the proper description of the energy of gravitating systems (i.e. all real physical systems), has played a large role in the development of 20th century physics.

[b]At present up to 1923.

In the years 1912–1915, Einstein, when he was searching for satisfactory field equations, used a form of the equations that explicitly included an energy–momentum density for the gravitational field and were designed to satisfy the principle of energy–momentum conservation.[c] Thus, an expression for the Einstein energy–momentum pseudotensor already existed even before he found the correct field equations. It should be appreciated that general covariance brought with it features that had never before been encountered in any theory. (Indeed there is still controversy up to the present day.[16,94,95,124]) For a couple of years, Einstein very much doubted that a generally covariant theory could be found[d]; he proposed that energy conservation would select the preferred physical coordinate frame. Initially Hilbert followed Einstein in this belief (see the proofs of his first note in Ref. 107).

Although Einstein began using variational principles in 1914[33] this was not his path to the field equations. Hilbert was the first to identify a generally covariant Lagrangian[e] (proportional to the Riemannian scalar curvature). He also constructed (in a complicated way that was not easy to understand) his "conserved energy vector," a vector with vanishing divergence associated with the general coordinate invariance (i.e. diffeomorphism invariance) of his Lagrangian.

Einstein's energy–momentum pseudotensor was criticized[36] for giving "unphysical" values (Schrödinger[116] noted that one could choose the coordinates to give a vanishing value outside a fluid sphere, and Bauer[2] noted that one could choose the coordinates to give a vanishing energy value for empty Minkowski space).

Lorentz, Levi-Civita and Klein argued that the Einstein curvature tensor $G_{\mu\nu}$ was the only proper gravitational energy–momentum density; hence one should regard the Einstein equation in the form

$$-\frac{1}{\kappa}G_{\mu\nu} + T_{\mu\nu} = 0, \tag{1}$$

as describing the vanishing sum of gravitational and material energy–momentum. (This idea has been advanced more recently by Cooperstock.[26]) In our modern perspective for GR their idea is quite correct — but a *density* is not the whole story. There is more to energy–momentum than just a density.

2.2. *From Einstein's correspondence*

Here are some excerpts from Einstein's correspondence concerning the Einstein pseudotensor, Hilbert's energy vector and Noether's contribution.[13,14,16,74,91,110] They reveal the difficulties and the extent of understanding these people had at that time. All these are quoted from the Einstein papers[37,38] Vol. 8.

[c]See Refs. 60–62, 93, 97 and 125 for discussions of how Einstein found his field equations.[34]

[d]One reason was his famous "hole" argument.[60,97]

[e]Einstein and Hilbert had quite different agendas[111,114,136,141]; Hilbert in his Foundation of Physics papers, based on the work of Einstein and Mie, was using his axiomatic method with the objective of finding a unified field theory of gravity and electromagnetism.[15,27,57,107,108,113]

"Highly esteemed Colleague, ... I am sitting over your relativity paper, ..., and am honestly toiling over it. I do admire your method, as far as I have understood it. But at certain points I cannot progress and therefore ask that you assist me with brief instructions. ... I still do not grasp the energy principle at all, not even as a statement" (Doc. 221 to Hilbert 25 May 1916).

"Your explanation of Eq. (6) of your paper was charming. Why do you make it so hard for poor mortals by withholding the technique behind your ideas? ... In your paper everything is understandable to me now except for the energy theorem. Please do not be angry with me that I ask you about this again. ... How is this cleared up? It would suffice, of course, if you would charge Miss Noether with explaining this to me" (Doc. 223 to Hilbert 30 May 1916).

"My $\mathfrak{t}_\sigma^\mu$'s are being rejected by everyone as unkosher" (Doc. 503 to Hilbert 12 April 1918).

"... Only (24) is an identity ... The relations here are exactly analogous to those for nonrelativistic theories" (Doc. 480 to Klein 13 March 1918).

"I have succeeded in discovering the organic formation law for Hilbert's energy vector" (Doc. 588 from Klein 15 July 1918).

"The only thing I was unable to grasp in your paper is the conclusion at the top of page 8 that ε^σ was a vector" (Doc. 638 to Klein 22 Oct 1918).

"Thank you very much for the transparent proof, which I understood completely" (Doc. 646 to Klein 8 Nov 1918).

"... Meanwhile, with Miss Noether's help, I understand that the proof for the vector character of ε^σ from "higher principles" as I had sought was already given by Hilbert on pp. 6, 7 of his first note, although in a version that does not draw attention to the essential point" (Doc. 650 from Klein 10 Nov 1918).

Briefly, after a couple of years Klein clarified Hilbert's energy–momentum "vector"; he related it to Einstein's pseudotensor, but (as we will discuss in more detail shortly) disagreed with Einstein's physical interpretation of divergenceless expressions.[f] Enlisted by Hilbert and Klein, it was Emmy Noether who solved the primary puzzle regarding gravitational energy.

[f]For these investigators, these things were not as easy as they are for us today; in particular the Bianchi identity and its contracted version were not known to these people,[97,112] so they had, in effect, to rediscover that identity from effectively the diffeomorphism invariance of the Lagrangian.

2.3. *Noether's contribution*

If one had to describe 20th century physics in one word, a good choice would be *symmetry*. Most of the new theoretical physics ideas involved symmetry. Essentially they can be seen as applications of Noether's theorems. Briefly, Noether's first theorem associates conserved quantities with global symmetries, and Noether's second theorem concerns local symmetries: It is the mathematical foundation of the modern gauge theories. Unfortunately her work[g] was largely overlooked for about 50 years.[74] Why did Noether make her investigation? To clarify the issue of gravitational energy.

Klein was looking into Einstein's theory and the relationship between Einstein's pseudotensor and Hilbert's energy vector. Some of the correspondence between Hilbert and Klein was published in a paper.[71] We quote some excerpts[h]:

Klein wrote

> You know that Miss Noether advises me continually regarding my work, and that in fact it is only thanks to her that I have understood these questions. When I was speaking recently to Miss Noether about my result concerning your energy vector, she was able to inform me that she had derived the same result on the basis of developments of your note (and thus not from the simplified calculations of my section 4) more than a year ago, and that she had then put all of that in a manuscript (which I was subsequently able to read). She simply did not set it out as forcefully as I recently did at the Mathematical Society (22 January [1918]).

Hilbert responded

> I fully agree in fact with your statements on the energy theorems: Emmy Noether, on whom I have called for assistance more than a year ago to clarify this type of analytical questions concerning my energy theorem, found at that time that the energy components that I had proposed — as well as those of Einstein — could be formally transformed, using the Lagrange differential equations (4) and (5) of my first note, into expressions whose divergence vanishes identically, that is to say, without using the Lagrange equations (4) and (5).

Also

> Indeed I believe that in the case of general relativity, i.e. in the case of the general invariance of the Hamiltonian function, the energy equations which in your opinion correspond to the energy equations of the theory

[g]For discussions see Refs. 13, 14, 16, 74, 91 and 110.

[h]We do not know of any English translation of Klein's papers; the translations of the following excerpts are quoted from Ref. 74, p. 66.

> of orthogonal invariance do not exist at all; I can even call this fact a characteristic of the general theory of relativity.

What Hilbert here calls the Hamiltonian function we now refer to as the Lagrangian. Noether wrote her 1918 paper to clarify the situation.

2.4. *Noether's result*

Many have heard of Noether's theorems, but the full scope of what she actually did is not so generally well known. So, we take this opportunity to quote (from the highly recommended book of Kosmann-Schwartzbach[74]) in full her key results. (It should be mentioned that her Lagrangians were quite general, they could depend on any finite number of derivatives.)

Theorem I. *If the integral I is invariant under a (finite continuous group with ρ parameters) G_ρ, then there are ρ linearly independent combinations among the Lagrangian expressions which become divergences — and conversely, that implies the invariance of I under a (group) G_ρ. The theorem remains valid in the limiting case of an infinite number of parameters.*

Theorem II. *If the integral I is invariant under a (an infinite continuous group) $G_{\infty\rho}$ depending on arbitrary functions and their derivatives up to order σ, then there are ρ identities among the Lagrangian expressions and their derivatives up to order σ. Here as well the converse is valid.*

Furthermore she has another important result, although it follows easily from Theorem II, in our opinion, both because of its importance and the fact that it was the key issue motivating her investigation, it could have been set off and called Theorem III:

Given I invariant under the group of translations, then the energy relations are improper if and only if I is invariant under an infinite group which contains the group of translations as a subgroup.

Regarding this latter result, she ends her paper with the remarks

> As Hilbert expresses his assertion, the lack of a proper law of energy constitutes a characteristic of the "general theory of relativity." For that assertion to be literally valid, it is necessary to understand the term "general relativity" in a wider sense than is usual, and to extend it to the aforementioned groups that depend on n arbitrary functions.[27]

The footnote that ends her paper is also of interest:

> [27] This confirms once more the accuracy of Klein's remark that the term "relativity" as it is used in physics should be replaced by "invariance with respect to a group."

Her result regarding the lack of a proper law of energy applies not just to Einstein's GR theory, but in fact to all geometric theories of gravity: for all such theories, there is no proper conserved energy–momentum density.

As a well-known textbook expresses it:

> Anyone who looks for a magic formula for "local gravitational energy–momentum" is looking for the right answer to the wrong question. Unhappily, enormous time and effort were devoted in the past to trying to "answer this question" before investigators realized the futility of the enterprise (Misner, Thorne and Wheeler, *Gravitation*,[82] p. 467).

3. The Noether Energy–Momentum Current Ambiguity

Let us begin our technical discussion by first reviewing some background.

As we will soon show, a well-known result is that from a classical field Lagrangian density, $\mathcal{L}(\varphi_A, \partial_\mu \varphi_A)$, via Noether's first theorem, the translational symmetry of Minkowski spacetime leads to a simple formula for the "conserved" *canonical energy–momentum density*

$$T^\mu{}_\nu := \delta^\mu_\nu \mathcal{L} - \frac{\partial \mathcal{L}}{\partial \partial_\mu \varphi_A} \partial_\nu \varphi_A. \tag{2}$$

The divergence of this expression satisfies the identity[i]

$$\partial_\mu T^\mu{}_\nu \equiv \frac{\delta \mathcal{L}}{\delta \varphi_A} \partial_\nu \varphi_A, \tag{3}$$

which can easily be directly verified using the definition of the *Euler–Lagrange variational derivative*

$$\frac{\delta \mathcal{L}}{\delta \varphi_A} := \frac{\partial \mathcal{L}}{\partial \varphi_A} - \partial_\mu \left(\frac{\partial \mathcal{L}}{\partial \partial_\mu \varphi_A} \right). \tag{4}$$

The canonical energy–momentum density is a conserved current in the sense that "on shell" (i.e. when the Euler–Lagrange field equations are satisfied: $\delta \mathcal{L}/\delta \varphi_A = 0$) its divergence vanishes.

The above energy–momentum density has the usual conserved current ambiguity:

$$T'^\mu{}_\nu := T^\mu{}_\nu + \partial_\lambda U^{[\mu\lambda]}{}_\nu, \tag{5}$$

is likewise conserved but defines different energy–momentum values. Essentially, one can always adjust by a "curl" a divergence free current.

[i] A consequence of assuming that the Lagrangian depends on position only through the field φ_A.

At first thought, one might be inclined to follow the rule of sticking with the results obtained directly from the Lagrangian and the above formula. But sometimes the results so obtained are not so suitable physically.

A simple example in Minkowski space is the Lagrangian density

$$\mathcal{L} = -\frac{1}{4}F^{\mu\nu}F_{\mu\nu}, \quad F_{\mu\nu} = \partial_\mu A_\nu - \partial_\nu A_\mu. \tag{6}$$

If one regards this Lagrangian density according to the above paradigm as a function of A_μ and $\partial_\mu A_\nu$, then the above formula leads directly to the conserved expression

$$T^\mu{}_\nu = F^{\mu\alpha}\partial_\nu A_\alpha - \frac{1}{4}\delta^\mu_\nu F^{\alpha\beta}F_{\alpha\beta}. \tag{7}$$

Now the above Lagrangian density (up to a suitable overall scaling coefficient that can be set aside for the purposes of this section) can be used to describe Maxwell electrodynamics. As is physically appropriate, this Lagrangian density and the field equations obtained from it are gauge invariant under the local "gauge" transformation $A_\mu \to A_\mu + \partial_\mu \chi$. However, the above canonical energy–momentum density *is not gauge invariant* (nevertheless, as we will see later, it can still be useful physically). Naturally, one would generally prefer to have a gauge invariant energy–momentum density for electrodynamics. In this particular case, there are several ways that one can find an alternative to (7): (i) One can exploit the abovementioned freedom (5) and thereby find "by hand" an "improved" gauge invariant expression, (ii) one could regard the Lagrangian as being a function of a one-form A and its differential (this is really the proper way to treat electromagnetism — but then one needs an extension of the above classical field theory formalism that can accommodate form fields; we will discuss such a formalism below) or (iii) one can consider that physically any time one has material energy–momentum one must also have gravity: The gravitational equations will include an unambiguous gauge invariant energy–momentum density.[j] From a specific gravity theory, one gets a specific formula for the energy–momentum density. In this way, the ambiguity in the canonical energy–momentum density for any classical field can be entirely removed when we consider their gravitational effects. Specifically, in the case of GR, knowing the curvature gives, via the Einstein tensor, the *symmetric Hilbert energy–momentum density*. In particular for the electromagnetism example this is

$$T^\mu{}_\nu = F^{\mu\alpha}F_{\nu\alpha} - \frac{1}{4}\delta^\mu_\nu F^{\alpha\beta}F_{\alpha\beta}, \tag{8}$$

which is no doubt a good choice for the energy–momentum density for Maxwell electrodynamics; in fact it is, as we shall see, the same as the energy–momentum density that one obtains by regarding the vector potential as a one-form.

[j]The existence of a gravitational field will reveal the location of a source with energy–momentum even if it has not otherwise been detected. An important example of this is — assuming gravity is well described by GR — from astronomical observations it seems that there is a large amount "dark matter" in the universe. Clearly, the issue of the proper description of energy for gravitating systems can have major consequences for our conception of the physical world.

While for electromagnetism one has another criteria (gauge invariance) that can be used to arrive at a physically suitable energy–momentum density, for most other sources one can only turn to gravity to identify a unique energy–momentum density. Hence, gravity *uniquely* detects the energy–momentum density of its sources. It may thus seem somewhat ironic that *for gravity itself* there is no proper energy–momentum density.

4. Pseudotensors

The Einstein Lagrangian for GR differs from the Hilbert scalar curvature Lagrangian by a certain total divergence which removes the 2nd derivatives of the metric[k]:

$$\begin{aligned} 2\kappa\mathcal{L}_{\rm E}(g_{\alpha\beta},\partial_\mu g_{\alpha\beta}) &:= -\sqrt{-g}g^{\beta\sigma}\Gamma^\alpha{}_{\gamma\mu}\Gamma^\gamma{}_{\beta\nu}\delta^{\mu\nu}_{\alpha\sigma} \\ &\equiv \sqrt{-g}R - \partial_\mu(\sqrt{-g}g^{\beta\sigma}\Gamma^\alpha{}_{\beta\gamma}\delta^{\mu\gamma}_{\alpha\sigma}) \\ &=: 2\kappa\mathcal{L}_{\rm H}(g_{\alpha\beta},\partial_\mu g_{\alpha\beta},\partial_{\mu\nu}g_{\alpha\beta}) - \partial_\mu\mathcal{K}^\mu. \end{aligned} \tag{9}$$

They give the same field equations. The Einstein pseudotensor can be obtained from $\mathcal{L}_{\rm E}$ using the aforementioned formula for the canonical energy–momentum tensor (2):

$$\mathfrak{t}^\mu_{{\rm E}\nu} := \delta^\mu_\nu\mathcal{L}_{\rm E} - \frac{\partial\mathcal{L}_{\rm E}}{\partial\partial_\mu g_{\alpha\beta}}\partial_\nu g_{\alpha\beta}. \tag{10}$$

(Here, following tradition, gothic letters indicate densities.) We have from (3) using the Einstein equation

$$\partial_\mu(\mathfrak{t}^\mu_{{\rm E}\nu}) \equiv \frac{\delta\mathcal{L}_{\rm E}}{\delta g_{\alpha\beta}}\partial_\nu g_{\alpha\beta} \equiv -(2\kappa)^{-1}\sqrt{-g}G^{\alpha\beta}\partial_\nu g_{\alpha\beta} = -\frac{1}{2}\mathfrak{T}^{\alpha\beta}\partial_\nu g_{\alpha\beta}. \tag{11}$$

Hence, using the vanishing covariant divergence of the material energy–momentum,

$$0 = \nabla_\mu(\mathfrak{T}^\mu{}_\nu) = \partial_\mu(\mathfrak{T}^\mu{}_\nu) - \Gamma^\gamma{}_{\nu\mu}\mathfrak{T}^\mu{}_\gamma = \partial_\mu(\mathfrak{T}^\mu{}_\nu) - \frac{1}{2}\mathfrak{T}^{\alpha\beta}\partial_\nu g_{\alpha\beta}, \tag{12}$$

we obtain

$$\partial_\mu(\mathfrak{T}^\mu{}_\nu + \mathfrak{t}^\mu_{{\rm E}\nu}) = 0, \tag{13}$$

a vanishing ordinary divergence, i.e. a conserved total energy–momentum "current." Here, we assumed the vanishing of the covariant divergence of the material energy–momentum tensor and used Einstein's equations to obtain an ordinary divergence conserved current. But one can argue the other way around, as Einstein did in 1916.

[k]Here, $\Gamma^\alpha{}_{\beta\gamma} := \frac{1}{2}g^{\alpha\lambda}(\partial_\beta g_{\lambda\gamma} + \partial_\gamma g_{\lambda\beta} - \partial_\lambda g_{\beta\gamma})$ is the well-known Christoffel/Levi-Civita connection, $\delta^{\mu\nu}_{\beta\sigma} := 2\delta^\mu_{[\beta}\delta^\nu_{\sigma]}$ and $\kappa := 8\pi G/c^4$.

4.1. *Einstein, Klein and superpotentials*

Einstein[35] obtained results of the form

$$\partial_\mu(\mathfrak{T}^\mu{}_\nu + \mathfrak{t}^\mu_{\mathrm{E}\nu}) = 0, \tag{14}$$

$$\mathfrak{T}^\mu{}_\nu + \mathfrak{t}^\mu_{\mathrm{E}\nu} = \partial_\lambda \mathfrak{s}^{\mu\lambda}{}_\nu, \tag{15}$$

$$\partial_{\mu\lambda}\mathfrak{s}^{\mu\lambda}{}_\nu \equiv 0, \tag{16}$$

$$\mathfrak{s}^{\mu\lambda}{}_\nu := g^{\mu\alpha}\frac{\partial \mathcal{L}_\mathrm{E}}{\partial\partial_\lambda g^{\nu\alpha}}. \tag{17}$$

Klein regarded the first three relations as mathematical identities and argued that energy–momentum conservation in GR was fundamentally different from that in classical mechanics.[13,71] Einstein did not agree with either of these statements; he regarded only (16) as an identity, which he obtained using a general coordinate invariance argument. Now taking the divergence of (15) using (16) gives (14), which — reversing the computation in the previous subsection — leads to (12). In this way Einstein showed that the local coordinate invariance identity plus his field equations — which are equivalent to (15) — gives the conservation of material energy momentum, without ever having to use any matter field equations. This type of argument is referred to as *automatic conservation of the source* (see MTW,[82] Sec. 17.1); effectively it uses a Noether second theorem type of argument to obtain current conservation. Weyl used the same type of argument for the conservation of the electromagnetic current in his seminal gauge theory papers,[147,148] whereas modern field theory books generally use Noether's first theorem in connection with current conservation.[12]

The identity (16) is equivalent to the contracted Bianchi identity, $\nabla_\mu G^\mu{}_\nu \equiv 0$, where $G^\mu{}_\nu$ is the Einstein curvature tensor. In those days the *Bianchi identity*, $\nabla_{[\mu}R^{\alpha\beta}{}_{\nu\gamma]} \equiv 0$, was not generally well known (it was first used in GR by Levi-Civita in 1917[112]). For any Lagrangian constructed out of the metric and its derivatives, it is now well known that local diffeomorphism invariance (with $\delta g_{\mu\nu} = \pounds_\xi g_{\mu\nu} = \nabla_\mu\xi_\nu + \nabla_\nu\xi_\mu$) of the associated action leads to a divergence identity:

$$\int \frac{\delta\mathcal{L}}{\delta g_{\mu\nu}}\delta g_{\mu\nu} d^4x = 0 \Rightarrow \nabla_\mu \frac{\delta\mathcal{L}}{\delta g_{\mu\nu}} \equiv 0. \tag{18}$$

In general such identities can involve higher derivatives of the curvature, however for the Hilbert scalar curvature Lagrangian of GR this Noether second theorem type argument yields a divergence which coincides with the contraction of the Bianchi identity.

By the way, Einstein had been using essentially the set of energy–momentum conservation relations (14)–(17) for some years in connection with the (noncovariant) "Entwurf" equations that he worked out with Marcel Grossmann.[39] However, his Lagrangian for that scheme (see CPAE,[37,38] Vol. 6, Doc. 2) was not — up to an exact differential — diffeomorphically invariant, so that (16) was in that case a relation that selected a preferred set of coordinates.

Since for the Einstein Lagrangian (16) is an identity, one might wonder why did Einstein (and others) favor an invariance argument rather than just directly calculating it? It turns out that the invariance argument is considerably easier. Let us look into this a little further. The detailed form of (17) can be written out with the help of some formulas given by Tolman.[137] From his Eqs. (87.5) and (89.9), we have

$$\mathfrak{s}^{\mu\lambda}{}_{\nu} = -\mathfrak{g}^{\alpha\mu} W^{\lambda}{}_{\nu\alpha} + \frac{1}{2}\delta^{\mu}_{\nu}\mathfrak{g}^{\alpha\beta} W^{\lambda}{}_{\alpha\beta}, \tag{19}$$

$$W^{\lambda}{}_{\alpha\beta} := \frac{\partial \mathcal{L}_{\mathrm{E}}}{\partial \partial_{\lambda}\mathfrak{g}^{\alpha\beta}} = -\Gamma^{\lambda}{}_{\alpha\beta} + \delta^{\lambda}_{(\alpha}\Gamma^{\sigma}{}_{\beta)\sigma}. \tag{20}$$

To show directly that this expression satisfies (16) is not short or simple. The only published calculation that we know of is in Møller,[83] some 40 years later.

If $\mathfrak{s}^{\mu\lambda}{}_{\nu}$ were antisymmetric in its upper indices the identity would be trivially satisfied. Stated another way, if (14) is to be satisfied then there should exist a *superpotential* $\mathfrak{U}^{\mu\lambda}{}_{\nu} \equiv \mathfrak{U}^{[\mu\lambda]}{}_{\nu}$ such that

$$\mathfrak{T}^{\mu}{}_{\nu} + \mathfrak{t}^{\mu}{}_{\nu} = \partial_{\lambda}\mathfrak{U}^{\mu\lambda}{}_{\nu}. \tag{21}$$

Einstein's $\mathfrak{s}^{\mu\lambda}{}_{\nu}$ is not antisymmetric; it is not the right kind of potential. A suitable superpotential for the Einstein pseudotensor was found over 20 years later by Freud[43]:

$$\mathfrak{U}_{\mathrm{F}}^{\mu\lambda}{}_{\nu} := -\mathfrak{g}^{\beta\sigma}\Gamma^{\alpha}{}_{\beta\gamma}\delta^{\mu\lambda\gamma}_{\alpha\sigma\nu}. \tag{22}$$

To obtain this, he did not follow Einstein's path. He started with the basics, the Einstein equations, and rearranged them using some formulas from Weyl's book[146] and some complicated identities he found in Pauli's 1921 encyclopedia article.[99] Later, we will give a simple derivation of Freud's superpotential using a better technique.

4.2. *Other GR pseudotensors*

The presence of a nonvanishing energy–momentum density necessarily produces gravity (i.e. the curvature of spacetime). In curved spacetime, the total source energy–momentum tensor satisfies (12). Without the second term we would have an expression suitable for integrating to obtain a conservation law. The second term represents a local interaction exchanging energy–momentum between the source and the gravitational field. To have a good conservation law, we would like to rewrite (12) in the form of (14) for some suitable *gravitational energy–momentum density* $\mathfrak{t}^{\mu}{}_{\nu}$. In fact this can be done in an infinite number of ways, *and*, in all cases, the quantity $\mathfrak{t}^{\mu}{}_{\nu}$ is not a tensor. (For some good overviews of such pseudotensors and their properties see Refs. 44, 64, 130 and 138.)

Here is a construction. Select an object (referred to as a *superpotential*) with suitable symmetries: $\mathfrak{U}^{\nu\lambda}{}_{\mu} \equiv \mathfrak{U}^{[\nu\lambda]}{}_{\mu}$. Now use it to split the Einstein tensor, defining a *gravitational energy–momentum pseudotensor* according to

$$2\kappa \mathfrak{t}^{\mu}{}_{\nu} := -2\sqrt{-g}G^{\mu}{}_{\nu} + \partial_{\lambda}\mathfrak{U}^{\mu\lambda}{}_{\nu}. \tag{23}$$

Then Einstein's equation, $G^{\mu}{}_{\nu} = \kappa T^{\mu}{}_{\nu}$, takes a form (analogous to Maxwell's equation) with a *total* effective energy–momentum pseudotensor as its source:

$$\partial_{\lambda}\mathfrak{U}^{\mu\lambda}{}_{\nu} = 2\kappa(\mathfrak{t}^{\mu}{}_{\nu} + \mathfrak{T}^{\mu}{}_{\nu}). \tag{24}$$

The essential feature of such source expressions is that they are equated to a derivative of a *superpotential* in such a way that their divergence *automatically* vanishes. Thus, as a consequence of the symmetry of $\mathfrak{U}^{\mu\lambda}{}_{\nu}$ we have (similar to Maxwell's theory) "automatic conservation of the source": $\partial_{\mu}(\mathfrak{T}^{\mu}{}_{\nu} + \mathfrak{t}^{\mu}{}_{\nu}) \equiv 0$. This expression can be integrated to define the total conserved energy–momentum within any volume V:

$$P_{\nu}(V) := \int_V (\mathfrak{T}^{\mu}{}_{\nu} + \mathfrak{t}^{\mu}{}_{\nu})d^3\Sigma_{\mu} = \frac{1}{2\kappa}\int_V \partial_{\lambda}\mathfrak{U}^{\mu\lambda}{}_{\nu}d^3\Sigma_{\mu} \equiv \frac{1}{2\kappa}\oint_{\partial V} \mathfrak{U}^{\mu\lambda}{}_{\nu}\frac{1}{2}d^2S_{\mu\lambda}. \tag{25}$$

From the volume integral on the left, one would expect that the results would be highly ambiguous — depending on the choice of reference frame throughout the volume of interest. However, from the last surface form, one can see that the situation is not nearly so bad. The result does not depend on the choice of reference frame within the volume, it is *quasi-local*, i.e. it depends on the fields and choice of reference frame only on the boundary. It should be noted that (for any given reference frame on the boundary) the value of $P_{\nu}(V)$ is well defined by the above integral. Its value, however, comes from a mixture of physics and a quasi-local reference frame; still it can be useful if one is mindful of its nature.

There are some variations on the above formulation. The classical pseudotensorial total energy–momentum density complexes, $\mathcal{T}^{\mu}{}_{\nu} := \mathfrak{T}^{\mu}{}_{\nu} + \mathfrak{t}^{\mu}{}_{\nu}$, all follow from suitable superpotentials according to one of the patterns

$$2\kappa\mathcal{T}^{\mu}{}_{\nu} = \partial_{\lambda}\mathfrak{U}^{\mu\lambda}{}_{\nu}, \quad 2\kappa\mathcal{T}^{\mu\nu} = \partial_{\lambda}\mathfrak{U}^{\mu\lambda\nu}, \quad 2\kappa\mathcal{T}^{\mu\nu} = \partial_{\alpha\beta}\mathfrak{H}^{\alpha\mu\beta\nu}, \tag{26}$$

where the superpotentials have certain symmetries which automatically guarantee conservation: Specifically $\mathfrak{U}^{\mu\lambda}{}_{\nu} \equiv \mathfrak{U}^{[\mu\lambda]}{}_{\nu}$, $\mathfrak{U}^{\mu\lambda\nu} \equiv \mathfrak{U}^{[\mu\lambda]\nu}$, while $\mathfrak{H}^{\alpha\mu\beta\nu}$ has the algebraic symmetries of the Riemann tensor (this latter form yields a *symmetric* pseudotensor and, hence, a simpler conservation of angular momentum description, see MTW Ref. 82, §20.3). We have already considered the Einstein total energy–momentum density which follows from the Freud superpotential (22). For completeness, we list the other well-known ones. The Bergmann–Thompson,[6] Landau–Lifshitz,[77] Papapetrou,[98] Weinberg[145] (also used in MTW[82]) and Møller[83] total energy–momentum complex expressions can be obtained respectively from

$$\mathfrak{U}^{\mu\lambda\nu}_{\rm BT} := g^{\nu\delta}U^{\mu\lambda}_{\rm F}{}_{\delta}, \tag{27}$$

$$\mathfrak{U}_{\rm LL}^{\mu\lambda\nu} := |g|^{\frac{1}{2}} U_{\rm BT}^{\mu\lambda\nu}, \quad \text{equivalently } \mathfrak{H}_{\rm LL}^{\alpha\mu\beta\nu} := |g| \delta^{\mu\alpha}_{ma} g^{a\beta} g^{m\nu}, \tag{28}$$

$$\mathfrak{H}_{\rm P}^{\alpha\mu\beta\nu} := \delta^{\mu\alpha}_{ma} \delta^{\nu\beta}_{nb} \bar{g}^{ab} (|g|^{\frac{1}{2}} g^{mn}), \tag{29}$$

$$\mathfrak{H}_{\rm W}^{\alpha\mu\beta\nu} := \delta^{\mu\alpha}_{ma} \delta^{\nu\beta}_{nb} |\bar{g}|^{\frac{1}{2}} \bar{g}^{ab} \left(-\bar{g}^{mc} \bar{g}^{nd} + \frac{1}{2} \bar{g}^{mn} \bar{g}^{cd} \right) g_{cd}, \tag{30}$$

$$\mathfrak{U}_{\rm M}^{\mu\lambda}{}_{\nu} := -|g|^{\frac{1}{2}} g^{\beta\sigma} \Gamma^{\alpha}{}_{\beta\nu} \delta^{\mu\lambda}_{\alpha\sigma} \equiv |g|^{\frac{1}{2}} g^{\beta\mu} g^{\lambda\delta} (\partial_\beta g_{\delta\nu} - \partial_\delta g_{\beta\nu}). \tag{31}$$

Here, $\bar{g}^{ab}$ is the Minkowski metric, all indices in these expressions refer to spacetime and range from 0 to 3, otherwise our conventions follow MTW.[82]

People have often looked askance at such pseudotensors, e.g. the above quote from MTW and Schrödinger refers to them as "sham."[117] As we noted, there are no doubt two unsatisfactory aspects: (i) Which of the many possible expressions should one use? (ii) and which *quasi-local* (in view of (25)) reference frame should be used. On the other hand, one should also be mindful that (a) they do provide a description of energy–momentum conservation, (b) they (like connection coefficients) really are geometric objects, with well defined values in each reference frame (this issue has been rigourously addressed using fiber bundle formulations[42,104,130]).

All of these pseudotensors (except for Møller's) give the expected total energy–momentum values at spatial infinity. On the other hand, *none* of them give the desired positivity of energy for small vacuum regions to lowest nonvanishing order,[121] however a set of new pseudotensors depending on several parameters with this desirable property has been constructed.[120] How can one understand the physical significance of these various pseudotensors? We have found a way using the Hamiltonian approach.

4.3. *Pseudotensors and the Hamiltonian*

To see how one can be led to the Hamiltonian, one merely needs to redo the calculation of (25) as an identity ("off shell"). For some fixed reference frame, with a (constant in the present reference frame) vector field Z^μ inserted we find[l]

$$\begin{aligned} -Z^\mu P_\mu(V) &:= -\int_V Z^\mu \mathfrak{T}^\nu{}_\mu \sqrt{-g} d^3\Sigma_\nu \\ &\equiv \int_V \left[Z^\mu \sqrt{-g} \left(\frac{1}{\kappa} G^\nu{}_\mu - T^\nu{}_\mu \right) - \frac{1}{2\kappa} \partial_\lambda \left(Z^\mu \mathfrak{U}^{\nu\lambda}{}_\mu \right) \right] d^3\Sigma_\nu \\ &\equiv \int_V Z^\mu \mathcal{H}^{\rm GR}_\mu + \oint_{S=\partial V} \mathcal{B}^{\rm GR}(Z) \equiv H(Z, V). \end{aligned} \tag{32}$$

Here, $\mathcal{H}^{\rm GR}_\mu$ can be recognized as the covariant expression which, when expressed in terms of the appropriate canonical variables, is just the ADM Hamiltonian density

[l]The sign in this expression is dictated by the condition for positive energy determined by the Hamiltonian using our local Minkowski signature convention: $P_\mu = (-E/c, \vec{p})$.

(i.e. the superhamiltonian and supermomentum), see, e.g. Refs. [1, 59] and MTW[82] Chap. 21. The expression includes a *Hamiltonian boundary term*, a 2-surface integral of $\mathcal{B}^{\rm GR}(Z) = -Z^\mu(1/2\kappa)\mathfrak{U}^{\nu\lambda}{}_\mu(1/2)d^2S_{\nu\lambda}$, i.e. it is entirely determined by the superpotential. The value of the Hamiltonian on a solution is entirely determined by this boundary term; the initial value constraints ensure that the Hamiltonian density in the spatial volume integral vanishes "on shell" (i.e. when the field equations are satisfied). In a similar way the value given by any pseudotensor can be regarded as the value of the Hamiltonian with a certain boundary term.[19] From the Hamiltonian variation, as we will discuss below, one gets important information that tames the ambiguity in the boundary term — namely boundary conditions — and thereby determines the physical significance of the various quasi-local values. The energy–momentum values obtained for the various pseudotensors can all be regarded as values of the Hamiltonian with different boundary conditions.

5. The Quasi-Local View

The modern idea, due to Penrose[100] in 1982, is that energy–momentum is *quasi-local*: i.e. it is associated with a closed 2-surface (while the pseudotensor energy–momentum complexes always had this property, its essential importance became much more appreciated after this work of Penrose which introduced this convenient term). There is a comprehensive review of this topic: Szabados (2009).[131] The many recent works cited in this review show that this is still a topic of considerable interest. In a brief summary one can find the statement:

> "... contrary to the high expectations of the 1980s, finding an appropriate quasi-local notion of energy–momentum has proven to be surprisingly difficult. Nowadays, the state of the art is typically postmodern: Although there are several promising and useful suggestions, we not only have no ultimate, generally accepted expression for the energy–momentum and especially for the angular momentum, but there is not even a consensus in the relativity community on general questions... or on the list of the criteria of reasonableness of such expressions."

However if one takes a more specific approach, one can come to a more satisfactory conclusion. In particular, the Hamiltonian view quite changes the prospects, especially when used along with a gauge perspective.

6. Currents as Generators

Noether's work was entirely Lagrangian based. Her results can be taken a further step when they are combined with the Hamiltonian formulation. As we will see, the Hamiltonian formulation offers a handle on the Noether current ambiguity.

One key feature can be seen already in Hamiltonian mechanics. A quantity Q conserved under the time evolution generated by a Hamiltonian $H = H(q,p)$ is more than just a conserved quantity, it is also *the canonical generator* of a one parameter transformation on phase space $(q(\lambda), p(\lambda))$ which is a symmetry of the Hamiltonian.

$$0 = \frac{dQ}{dt} = [Q, H] \Rightarrow \frac{dH}{d\lambda} = [H, Q] = 0. \tag{33}$$

In Hamiltonian field theory, the conserved currents are the generators of the associated symmetry. In particular, the generator of a local spacetime "translation" (an infinitesimal diffeomorphism) is the Hamiltonian; energy–momentum is the associated conserved quantity. Conversely, for spacetime translations, the associated Noether conserved current expression (i.e. the energy–momentum density) is the Hamiltonian density — the canonical generator of spacetime displacements. As we will see, because it can be varied this translation generator gives a handle on the associate conserved current ambiguity. The Lagrangian formulation affords no such handle, because in terms of Lagrangian variables the translation current is not a generator that can be varied.

7. Gauge and Geometry

For the early history of gauge theory see O'Raifeartaigh.[96] Briefly, the milestone works are Hermann Weyl's treatments of electromagnetism: Weyl (1918),[147] Weyl (1929),[148] then the generalization to non-Abelian groups by Yang and Mills (1954)[154] and Utiyama (1956, 1959).[139,140] Explicitly treating gravity as a gauge theory was pioneered by Utiyama,[139,140] using the Lorentz group and Riemannian geometry. Sciama[115] also used the Lorentz group but with Riemann–Cartan geometry (i.e. nonvanishing torsion). Kibble[67] put things in their proper place, he gauged the Poincaré group (i.e. the inhomogeneous Lorentz group, including translations).

For accounts of gravity as a spacetime symmetry gauge theory, see Hehl and coworkers,[45,52–54,56] Mielke[80] and Blagojević.[7] A comprehensive reader with summaries, discussions, and many reprints has recently appeared: Blagojević and Hehl.[8] For the observational constraints on torsion see Ni (2010).[90]

To us it is rather surprising that the idea of regarding gravity as a gauge theory is not better known. Examined more closely, one finds that gravity played an important role in the argument used in both of the above mentioned seminal works of Weyl, and thus in all of the above — except for the Yang–Mills paper. Furthermore, later in 1974 Yang himself published a paper[152] where he proposed a certain treatment of gravity as a gauge theory.[m]

[m]The aforementioned reader includes a chapter with a critical discussion of Yang's gauge theory of gravity. Recently Yang was asked about his 1974 paper; he said: "I do not believe that paper is correct."[153]

According to our understanding, properly speaking, GR can be understood as the original gauge theory. After all, it was the first physical theory where local gauge freedom (in the guise of general coordinate invariance) played a key role.[n]

The conserved quantities, energy–momentum and angular momentum/CoMM are associated with the geometric symmetry of Minkowski spacetime, the spacetime translations and Lorentz rotations, i.e. the Poincaré group. Furthermore this group is used to classify physical particles according to mass and spin. So a local Poincaré gauge theory is quite appropriate both geometrically and physically.

To give a good account, one should also be mindful of the parallel development of the closely related concept of a connection in differential geometry. Here, we just briefly mention that the main ideas were due to Hessenberg, Levi-Civita, Schouten, Weyl, Cartan, Ehresmann, and Koszul; for discussions of connections see Nomizu,[92] Kobayashi and Nomizu[72] and Spivak.[123]

As we will see in more detail, Riemann–Cartan (with a metric and a metric compatible connection, having both curvature and torsion) is the most appropriate geometry for a dynamic spacetime geometry theory: its local symmetries are just those of the local Poincaré group. So in this presentation, we will be considering the Poincaré gauge theories of gravity (PG); GR is included in this class as a special case.

8. Dynamical Spacetime Geometry and the Hamiltonian

We will consider geometric gravity theories with both a metric and an *a priori* metric compatible connection. Both curvature and torsion are allowed. The variational principles are developed. The Noether symmetries and the associated conserved quantities and differential identities are discussed. From a first-order Lagrangian formalism using differential forms, we construct a spacetime covariant Hamiltonian formalism. The Hamiltonian boundary term gives appropriate expressions for the quasi-local quantities, energy–momentum, angular momentum and CoMM, as well as quasi-local energy flux. The formalism easily specializes to teleparallel theory and Einstein's GR.

The Hamiltonian approach reveals certain aspects of a theory, including the constraints, gauges, and degrees-of-freedom, as well as expressions for energy–momentum and angular momentum. However, the usual ADM approach achieves this at a heavy cost: The loss of manifest 4D-covariance. Our alternative approach is complementary: A major benefit is manifestly 4D-covariant expressions for the quasi-local quantities: Energy–momentum and angular momentum/CoMM.

[n]It is true that the electrodynamics potentials along with their gauge freedom were known long before GR (in fact a Lagrangian which is locally gauge invariant had already been presented[118]), but this gauge invariance was not seen as having any important role in connection with the nature of the interaction, the conservation of current, or a differential identity — until the seminal work of Weyl, which post-dated (and was inspired by) GR.

8.1. *The main ideas*

The Hamiltonian for physical systems and dynamic spacetime geometry generates the evolution of a spatial region along a vector field. It includes a boundary term which determines the boundary conditions and supplies the value of the Hamiltonian. The Hamiltonian value gives the quasi-local quantities: Energy–momentum and angular momentum/CoMM. A spacetime gauge theory perspective identifies suitable geometric variables. We found a certain preferred Hamiltonian boundary term. The Hamiltonian boundary term depends not only on the dynamical variables but also on their reference values; they determine the ground state — the state with vanishing quasi-local quantities. To determine the "best matched" reference metric and connection values for our preferred boundary term, we propose on the boundary 2-surface: (i) 4D isometric matching and (ii) extremizing the energy.

8.2. *Some comments*

Before we begin our technical discussion of our work, let us make a few general comments. We work in 4D spacetime, but most of this can be extended to other dimensions in a straightforward fashion (except for the reference construction). The class of dynamical Lagrangians that we will consider does not allow for any derivatives of curvature or torsion. Our concerns are entirely classical.

We focus here on Riemann–Cartan spacetime geometry (i.e. spacetimes with a metric and a metric compatible connection, having both curvature and torsion) and the PG; our general analysis can be specialized both to Riemannian geometry (vanishing torsion) and teleparallel geometry (vanishing curvature); it includes GR and the teleparallel equivalent of GR as two special cases. Here, we assume a metric compatible connection; elsewhere we will present the generalization which includes nonmetricity. The extension to nonmetricity and the special case of teleparallel geometry each offer further insight into gravitational energy; we believe those insights are best appreciated when compared to the results presented here for the Riemann–Cartan geometry with the PG.

9. Differential Forms

In this work, we mainly use differential forms.[41,55,142] The reader may wonder why we use this less widely familiar idiom. The simple brief explanation is that they have some qualities that are technically very convenient for our needs. Differential forms are multiplied using the (graded Grassmann) wedge product. They can be differentiated using d, the *exterior differential*, a graded derivation which enjoys the property $d^2 \equiv 0$, so a differential equation $d\alpha = \beta$ has the integrability condition $0 = d\beta$, furthermore $d\beta = 0 \Rightarrow \beta = d\alpha$, at least locally. The integrals of forms satisfy the general boundary theorem[o]

$$\int_U d\beta \equiv \oint_{\partial U} \beta. \tag{34}$$

Also, they are well suited to representing interacting physical fields, especially gauge fields, and, as we shall see, they give a succinct representation of the main geometric objects: Connection, curvature, coframe, torsion, etc. Moreover, as will be explained below, they are quite convenient for the essential $3+1$ spacetime decomposition of derivatives that is needed for a dynamical Hamiltonian formulation.[84]

Regarding notation, here the contraction (or *interior* product) with a vector field is denoted by $i_X\alpha(\cdot,\cdot,\ldots) := \alpha(X,\cdot,\ldots)$ (some authors use the notation of left contraction: $X\rfloor\alpha$). The *Lie derivative* on forms is given by $\pounds_X \equiv i_X d + d i_X$, it has the nice property $d\pounds_X \equiv \pounds_X d$. We are concerned here with the case of 4D spacetime, which has a local Minkowski structure, having a metric with Lorentz signature. The metric determines the unit volume 4-form η with components $\eta_{\mu\nu\alpha\beta} = \eta_{[\mu\nu\alpha\beta]}$, $\eta_{0123} = \sqrt{|g|}$ which is used to construct the Hodge dual that maps k-forms to $(4-k)$-forms.

From the coframe ϑ^α, one can construct a basis for k-forms $\vartheta^{\alpha\beta\ldots} := \vartheta^\alpha\wedge\vartheta^\beta\wedge\cdots$ and a useful dual basis $\eta^{\alpha\beta\ldots} := *\vartheta^{\alpha\beta\ldots}$. They are related by various identities, especially

$$\vartheta^\rho \wedge \eta_{\mu\nu\lambda} \equiv \delta^\rho_\lambda \eta_{\mu\nu} + \delta^\rho_\mu \eta_{\nu\lambda} + \delta^\rho_\nu \eta_{\lambda\mu}, \tag{35}$$

$$\vartheta^{\alpha\beta} \wedge \eta_{\mu\nu\lambda} \equiv \delta^{\alpha\beta}_{\mu\nu} \eta_\lambda + \delta^{\alpha\beta}_{\nu\lambda} \eta_\mu + \delta^{\alpha\beta}_{\lambda\mu} \eta_\nu, \tag{36}$$

$$\vartheta^\rho \wedge \eta_{\mu\nu} \equiv \delta^\rho_\nu \eta_\mu - \delta^\rho_\mu \eta_\nu. \tag{37}$$

Maxwell's electrodynamics is a good example of the utility of differential forms. Charge identifies the charge-current 3-form (density):

$$Q(V) = \int_V J. \tag{38}$$

Charge is conserved:

$$dJ = 0 \Rightarrow J = dH. \tag{39}$$

The electromagnetic field is represented by a 2-form F. An elementary way to see why this is appropriate is to examine the motion of a point test charge. One should begin with kinematics in Minkowski space. Consider the motion of a point particle as a function of proper time: $x^\mu = x^\mu(\tau)$. The 4-velocity $v^\mu := dx^\mu/d\tau$ has constant magnitude: $v^\mu v_\mu = -c^2$ so the 4-acceleration is Lorentz orthogonal to the 4-velocity. Hence, the 4-force must be orthogonal to the 4-velocity. Consequently

[o]This generalization of the fundamental theorem of calculus is often referred to as the generalized Stokes theorem. Special cases include the Ostrogradsky–Gauss and Stokes theorem of vector analysis.

the 4-force must depend on the velocity. The simplest case is for a 4-force linear in the 4-velocity. Thus the simplest dynamical law has the form

$$\frac{dp_\mu}{d\tau} = qF_{\mu\nu}v^\nu, \tag{40}$$

where $p_\mu := mv_\mu$ is the 4-momentum, q is a coupling constant and $F_{\mu\nu}$ is some tensor field which is *antisymmetric*, i.e. it is a 2-form. The Lorentz force law of electrodynamics has this form. The force law identifies a certain field strength 2-form F which includes the electric and magnetic fields. Conservation of magnetic flux through a closed 2 surface $S = \partial V$ gives

$$0 = \oint_S F = \int_V dF, \quad \Rightarrow dF = 0 \Leftrightarrow F = dA, \tag{41}$$

and there are local gauge transformations: $A \to A + d\chi$. The vacuum constitutive relation is $H = *F/Z_0$ (Z_0 is the vacuum impedance). This covariant formulation for Maxwell's electrodynamics is valid for all dynamic geometry gravity theories and does not depend upon using a particular set of units, for a detailed, comprehensive and instructive presentation see Hehl and Obukhov.[55]

10. Variational Principle for Form Fields

Why do we use variational principles?[76] The answer is pragmatics: *Because they work.* With appropriate symmetries, they give consistent interacting field equations along with conserved Noether currents for all the desired quantities. As far as we know, all the known good dynamical evolution equations for the fundamental interacting classical field theories have a variational formulation.

In the usual formulations, most dynamical fields satisfy second-order equations. We refer to such formulations as *second-order*.

Let φ^A be some kind of vector field. The label "A" stands for some collection of indices, e.g. spinor, spacetime, isospin. Allow φ^A to also be a differential form of rank f where $f = 0, 1, 2,$ or 3, e.g. $\varphi^A = \frac{1}{2}\varphi^A_{\mu\nu}\vartheta^\mu \wedge \vartheta^\nu = \frac{1}{2}\varphi^A_{ij}dx^i \wedge dx^j$ for $f = 2$.

The *Lagrangian density* is a 4-form:

$$\mathcal{L} = \mathcal{L}(\varphi^A, d\varphi^A). \tag{42}$$

Note that there is no explicit appearance of the coordinates x^i or the coordinate partials ∂_i; $d\varphi^A$ is an $(f+1)$-form which geometrically includes partial derivatives of the components of φ^A, but only in an antisymmetric combination. (Here, we explicitly consider just one f-form field. The generalization to include several fields of different grades, is straightforward.)

Our convention is to vary fields off to the left (other conventions would differ only by some signs). The variation of $\mathcal{L}$ is thus

$$\delta\mathcal{L} = \delta d\varphi^A \wedge \frac{\partial\mathcal{L}}{\partial d\varphi^A} + \delta\varphi^A \wedge \frac{\partial\mathcal{L}}{\partial\varphi^A}. \tag{43}$$

This implicitly defines $\partial\mathcal{L}/\partial d\varphi^A$ as a $(3-f)$-form and $\partial\mathcal{L}/\partial\varphi^A$ as a $(4-f)$-form. Next, interchange the order (i.e. $\delta\, d = d\,\delta$) to get

$$\begin{aligned}\delta\mathcal{L} &= d\delta\varphi^A \wedge \frac{\partial\mathcal{L}}{\partial d\varphi^A} + \delta\varphi^A \wedge \frac{\partial\mathcal{L}}{\partial\varphi^A}\\ &\equiv d\left(\delta\varphi^A \wedge \frac{\partial\mathcal{L}}{\partial d\varphi^A}\right) - (-1)^f\delta\varphi^A \wedge d\left(\frac{\partial\mathcal{L}}{\partial d\varphi^A}\right) + \delta\varphi^A \wedge \frac{\partial\mathcal{L}}{\partial\varphi^A}.\end{aligned} \tag{44}$$

(Upon integration over some spacetime region this last step is "integration by parts," with the total differential term becoming a boundary term). From the above, it follows that the basic variational relation

$$\delta\mathcal{L} = d(\delta\varphi^A \wedge p_A) + \delta\varphi^A \wedge \frac{\delta\mathcal{L}}{\delta\varphi^A} \tag{45}$$

can be regarded as implicitly defining the *conjugate field momentum* and the Euler–Lagrange *variational derivative*, which have the respective explicit definitions

$$p_A := \frac{\partial\mathcal{L}}{\partial d\varphi^A}, \tag{46}$$

$$\frac{\delta\mathcal{L}}{\delta\varphi^A} := \frac{\partial\mathcal{L}}{\partial\varphi^A} - \varsigma d\left(\frac{\partial\mathcal{L}}{\partial d\varphi^A}\right). \tag{47}$$

A small price for using form fields is the appearance of occasional sign factors like $\varsigma := (-1)^f$.

10.1. *Hamilton's principle*

Our first application of (45) is Hamilton's principle (*the principle of least action*). Let the action within a region U be given by $S := \int_U \mathcal{L}$. Then

$$\begin{aligned}\delta S &\equiv \int_U \delta\mathcal{L}\\ &\equiv \int_U d(\delta\varphi^A \wedge p_A) + \delta\varphi^A \wedge \frac{\delta\mathcal{L}}{\delta\varphi^A}\\ &\equiv \int_U \delta\varphi^A \wedge \frac{\delta\mathcal{L}}{\delta\varphi^A} + \oint_{\partial U} \delta\varphi^A \wedge p_A.\end{aligned} \tag{48}$$

Now require the action S to be extreme (i.e. $\delta S = 0$) for $\delta\varphi^A$ vanishing on the boundary of U. This yields the field equation $\delta\mathcal{L}/\delta\varphi^A = 0$.

10.2. *Compact representation*

For a compact general discussion, it is convenient to suppress the field component index.[87] (This could be represented in matrix notation; our basic fields φ and their differential could be regarded as row vectors.) The Lagrangian then has the form $\mathcal{L} = \mathcal{L}(\varphi, d\varphi)$ and the variational scheme proceeds as

$$\begin{aligned}\delta\mathcal{L} &= d\delta\varphi \wedge \frac{\partial\mathcal{L}}{\partial d\varphi} + \delta\varphi \wedge \frac{\partial\mathcal{L}}{\partial d\varphi}\\ &= d\left(\delta\varphi \wedge \frac{\partial\mathcal{L}}{\partial d\varphi}\right) + \delta\varphi \wedge \left[-\varsigma\, d\left(\frac{\partial\mathcal{L}}{\partial d\varphi}\right) + \frac{\partial\mathcal{L}}{\partial\varphi}\right],\end{aligned} \tag{49}$$

hence the key Lagrangian variational identity takes the form

$$\delta\mathcal{L} \equiv d(\delta\varphi \wedge p) + \delta\varphi \wedge \frac{\delta\mathcal{L}}{\delta\varphi}. \tag{50}$$

In this succinct alternative to (45), the conjugate momentum and the variational derivative can be regarded as form-valued column vector fields.

11. Some Simple Examples of the Noether Theorems

Here, we present simple examples of Noether's two theorems.[91] Later, we shall use the same types of arguments in more complicated situations.

11.1. *Noether's first theorem: Energy–momentum*

Further applications of the basic variational identity (45) or (50) yield the Noether theorems. Their applications to physical systems are discussed in many works, e.g. Konopleva and Popov.[73] Here, we introduce our particular use of them using two specific important cases.

Noether's first theorem states that for a constant parameter symmetry, there is a conserved current.

For our concerns the most important example is the conservation of energy–momentum. As our specific relevant simple case exemplifying the argument, we specialize in this subsection to Minkowski spacetime, which is homogeneous and thus naturally has a geometric symmetry under translations. Dynamically let us assume symmetry also of the action under constant translations. The symmetry depends on a continuous parameter; it is sufficient to consider the infinitesimal case. Geometrically an infinitesimal translation corresponds to a constant vector field Z. Under such a transformation, the change in the components of form fields is given by the Lie derivative. We have:

$$\Delta\varphi = -\pounds_Z\varphi = -(di_Z + i_Z\, d)\varphi, \tag{51}$$

$$\Delta\mathcal{L} = -\pounds_Z\mathcal{L} = -di_Z\mathcal{L}. \tag{52}$$

Equation (50) under these specific variations (i.e. replacing the general δ by these specific changes) should be an identity (since the Lagrangian $\mathcal{L}$ depends on the position only through the fields φ). Rearranging leads to

$$\Delta\mathcal{L} - d(\Delta\varphi \wedge p) \equiv \Delta\varphi \wedge \frac{\delta\mathcal{L}}{\delta\varphi}. \tag{53}$$

Because of (52) the l.h.s. of Eq. (53) is a total differential — a total differential which moreover vanishes if the Euler–Lagrange field equations are imposed:

$$d(-i_Z\mathcal{L} + \pounds_Z\varphi \wedge p) \equiv -\pounds_Z\varphi \wedge \frac{\delta\mathcal{L}}{\delta\varphi}. \tag{54}$$

This identifies a *conserved current density* (a 3-form with vanishing differential *on shell*, i.e. when the field equations are satisfied) called the generalized *canonical stress energy–momentum density* (3-form):

$$\mathcal{T}(Z) := i_Z\mathcal{L} - \pounds_Z\varphi \wedge p. \tag{55}$$

For a zero-form field, it takes the special shape $T^\alpha{}_\mu Z^\mu\,\eta_\alpha$ where (with $L = *\mathcal{L}$)

$$T^\alpha{}_\mu = \delta^\alpha_\mu L - \partial_\mu\varphi\frac{\partial L}{\partial\partial_\alpha\varphi}, \tag{56}$$

which is the well-known expression mentioned earlier (2).

11.2. *Noether's second theorem: Gauge fields*

Note: For the rest of the present section, there is no need to restrict the spacetime geometry in any way. Our considerations apply quite generally.

Now, we want to consider invariance under *local* gauge transformations:

$$\Delta\varphi = \alpha^p\,\varphi\,T_p, \tag{57}$$

where the α^p are position dependent parameters and the T_p are the matrix generators (in our representation the fields are on the left and the matrix on the right) of the gauge group. Replace $d\varphi$ by the *gauge covariant differential*:

$$D\varphi := d\varphi + A^p \wedge \varphi\,T_p, \tag{58}$$

containing a certain compensating field, the *gauge vector potential* (a.k.a. the *gauge connection one-form*): $A^p = A^p{}_j dx^j$, which has a special nonhomogeneous gauge transformation:

$$\Delta A^p = -D\alpha^p := -(d\alpha^p + A^q C^p{}_{qr}\,\alpha^r), \tag{59}$$

where $C^p{}_{qr}$ are the gauge group structure constants: $[T_q, T_r] = C^p{}_{qr}T_p$. Then

$$\begin{aligned}\Delta(D\varphi) &:= (\Delta D)\varphi + D\Delta\varphi \\ &= -(D\alpha^p)\wedge\varphi T_p + D(\alpha^p\varphi T_p) \\ &= \alpha^p(D\varphi)T_p.\end{aligned} \tag{60}$$

Thus $D\varphi$ transforms just like φ.

Rather than starting with a Lagrangian 4-form of the type $\mathcal{L} = \mathcal{L}(\varphi, d\varphi, A^p, dA^p)$ and then discovering that the variables $A^p, d\varphi, dA^p$ can only appear in nice covariant combinations, let us proceed more covariantly, beginning with the *Lagrangian 4-form*

$$\mathcal{L} = \mathcal{L}(\varphi, D\varphi, A^p, F^p), \tag{61}$$

where F^p is the *field strength* or *gauge curvature* 2-form:

$$F^p := dA^p + \frac{1}{2} C^p{}_{qr}\, A^q \wedge A^r. \tag{62}$$

Since A^p is still allowed to appear independently in (61), there is no loss of generality.

The variation of $\mathcal{L}$ (61) is

$$\delta\mathcal{L} = d\left(\delta\varphi \wedge \frac{\partial\mathcal{L}}{\partial D\varphi} + \delta A^p \wedge \frac{\partial\mathcal{L}}{\partial F^p}\right) + \delta\varphi \wedge \frac{\delta\mathcal{L}}{\delta\varphi} + \delta A^p \wedge \frac{\delta\mathcal{L}}{\delta A^p}. \tag{63}$$

Now we are set for an example of Noether's 2nd theorem — that for each local invariance, there is a differential identity.

Assume that $\mathcal{L}$ (61) is invariant under the special changes $\Delta\varphi, \Delta A$ of Eqs. (57) and (59). From (63) we then have the identity

$$0 \equiv d\left(\alpha^p \varphi T_p \wedge \frac{\partial\mathcal{L}}{\partial D\varphi} - D\alpha^p \wedge \frac{\partial\mathcal{L}}{\partial F^p}\right) + \alpha^p \varphi \wedge T_p \frac{\delta\mathcal{L}}{\delta\varphi} - D\alpha^p \wedge \frac{\delta\mathcal{L}}{\delta A^p}. \tag{64}$$

The second term in the parenthesis may be rewritten as $-d(\alpha^p \frac{\partial\mathcal{L}}{\partial F^p}) + \alpha^p D\frac{\partial\mathcal{L}}{\partial F^p}$, then using $d^2 \equiv 0$ gives

$$0 \equiv d\left[\alpha^p\left(\varphi T_p \wedge \frac{\partial\mathcal{L}}{\partial D\varphi} + D\frac{\partial\mathcal{L}}{\partial F^p}\right)\right] + \alpha^p \varphi T_p \wedge \frac{\delta\mathcal{L}}{\delta\varphi} - D\alpha^p \wedge \frac{\delta\mathcal{L}}{\delta A^p}, \tag{65}$$

$$\begin{aligned} &\equiv D\alpha^p\left(\varphi T_p \wedge \frac{\partial\mathcal{L}}{\partial D\varphi} + D\frac{\partial\mathcal{L}}{\partial F^p} - \frac{\delta\mathcal{L}}{\delta A^p}\right) \\ &\quad + \alpha^p\left[D\left(\varphi T_p \wedge \frac{\partial\mathcal{L}}{\partial D\varphi} + D\frac{\partial\mathcal{L}}{\partial F^p}\right) + \varphi T_p \wedge \frac{\delta\mathcal{L}}{\delta\varphi}\right]. \end{aligned} \tag{66}$$

For a *local symmetry* the quantities α^p and $D\alpha^p$ are pointwise independent; their coefficients must vanish separately. The coefficient of α^p identifies a Noether I type conserved current:

$$J_p := \varphi T_p \wedge \frac{\partial\mathcal{L}}{\partial D\varphi} + D\frac{\partial\mathcal{L}}{\partial F^p}, \tag{67}$$

which satisfies the "conservation" law

$$DJ_p \equiv -\varphi T_p \wedge \frac{\delta\mathcal{L}}{\delta\varphi}. \tag{68}$$

The r.h.s. vanishes "on shell" (i.e. when the field equations are satisfied).

From the coefficient of $D\alpha^p$, we obtain an *algebraic identity* relating the Noether I current to a variational derivative:

$$J_p \equiv \frac{\delta \mathcal{L}}{\delta A^p}, \tag{69}$$

thereby the Noether I current conservation becomes a *differential identity*

$$D\frac{\delta \mathcal{L}}{\delta A^p} \equiv -\varphi T_p \wedge \frac{\delta \mathcal{L}}{\delta \varphi}, \tag{70}$$

between the variational derivatives. Note that to obtain these results, there is no need for the explicit form of the field equations.

Another way to argue is to replace the last term in (65) with a total differential minus a compensating term, bringing that relation into the form

$$0 \equiv d\left[\alpha^p \left(\varphi T_p \wedge \frac{\partial \mathcal{L}}{\partial D\varphi} + D\frac{\partial \mathcal{L}}{\partial F^p} - \frac{\delta \mathcal{L}}{\delta A^p}\right)\right] + \alpha^p \left(\varphi T_p \wedge \frac{\delta \mathcal{L}}{\delta \varphi} + D\frac{\delta \mathcal{L}}{\delta A^p}\right). \tag{71}$$

If one integrates this over any region, the total differential term gives rise to an integral over the boundary. To have a vanishing value for all possible gauge parameters with small support, the coefficient of the gauge parameter everywhere within the region and the coefficient of the gauge parameter everywhere on the boundary must both vanish identically. This again yields (70) and

$$\varphi T_p \wedge \frac{\partial \mathcal{L}}{\partial D\varphi} + D\frac{\partial \mathcal{L}}{\partial F^p} - \frac{\delta \mathcal{L}}{\delta A^p} \equiv 0, \tag{72}$$

which is equivalent to (68) with (69).

11.3. *Field equations with local gauge theory*

It should be noted that the Noether invariance argument yields the differential identities just found involving the Euler–Lagrange expressions without any need to have the explicit form of the Euler–Lagrange expressions. Of course if one explicitly computes the Euler–Lagrange expressions, one could go on to verify these identities directly. Furthermore, if one has the Euler–Lagrange expressions, one could (probably not so easily) directly discover such identities, even if one was not aware of the local symmetry.

To compute the field equations, the explicit variations

$$\delta D\varphi = D\delta\varphi + \delta A^p \wedge \varphi T_p, \tag{73}$$

$$\delta F^p = d\delta A^p + C^p{}_{qr} A^q \wedge \delta A^r = D\delta A^p, \tag{74}$$

are needed. The variation of the Lagrangian 4-form $\mathcal{L}$ (63) is

$$\begin{aligned}\delta \mathcal{L} &= \delta D\varphi \wedge \frac{\partial \mathcal{L}}{\partial D\varphi} + \delta\varphi \wedge \frac{\partial \mathcal{L}}{\partial \varphi} + \delta F^p \wedge \frac{\partial \mathcal{L}}{\partial F^p} + \delta A^p \wedge \frac{\partial \mathcal{L}}{\partial A^p} \\ &= (D\delta\varphi + \delta A^p \wedge \varphi T_p) \wedge \frac{\partial \mathcal{L}}{\partial D\varphi} + \delta\varphi \wedge \frac{\partial \mathcal{L}}{\partial \varphi} + D\delta A^p \wedge \frac{\partial \mathcal{L}}{\partial F^p} + \delta A^p \wedge \frac{\partial \mathcal{L}}{\partial A^p}\end{aligned}$$

$$= D\left(\delta\varphi \wedge \frac{\partial \mathcal{L}}{\partial D\varphi} + \delta A^p \wedge \frac{\partial \mathcal{L}}{\partial F^p}\right) + \delta\varphi \wedge \left(-\varsigma D\frac{\partial \mathcal{L}}{\partial D\varphi} + \frac{\partial \mathcal{L}}{\partial \varphi}\right)$$

$$+ \delta A^p \wedge \left(D\frac{\partial \mathcal{L}}{\partial F^p} + \frac{\partial \mathcal{L}}{\partial A^p} + \varphi T_p \wedge \frac{\partial \mathcal{L}}{\partial D\varphi}\right). \tag{75}$$

Comparing the explicit form of $\delta\mathcal{L}/\delta A^p$ found here with (69) and (67) shows that (69) means

$$\frac{\partial \mathcal{L}}{\partial A^p} \equiv 0. \tag{76}$$

Thus, local gauge invariance means: no explicit dependence on the gauge potential; all dependence on A^p comes through $D\varphi$ and F^p. Furthermore, if one makes the usual *minimal coupling* assumption,

$$\mathcal{L} = \mathcal{L}_A(A^p, F^p) + \mathcal{L}_\varphi(\varphi, D\varphi, A^p), \tag{77}$$

the identities (76) and (70) apply separately to each term. Hence, the Lagrangian 4-form must have the simpler form

$$\mathcal{L} = \mathcal{L}_A(F^p) + \mathcal{L}_\varphi(\varphi, D\varphi), \tag{78}$$

and the differential identity (70) becomes the two identities

$$D\frac{\delta \mathcal{L}_A}{\delta A^p} \equiv 0, \quad D\frac{\delta \mathcal{L}_\varphi}{\delta A^p} \equiv -\varphi T_p \wedge \frac{\delta \mathcal{L}_\varphi}{\delta \varphi}. \tag{79}$$

The first relation is explicitly

$$0 \equiv D^2\frac{\partial \mathcal{L}_A}{\partial F^p} \equiv -F^q C^r{}_{qp} \wedge \frac{\partial \mathcal{L}_A}{\partial F^r}. \tag{80}$$

The latter is a kind of gauge current "conservation," as the r.h.s. vanishes since

$$\frac{\delta \mathcal{L}_\varphi}{\delta \varphi} \equiv \frac{\delta \mathcal{L}}{\delta \varphi} = 0 \tag{81}$$

on shell. In more detail, this gauge current "conservation" relation has the form

$$0 = DJ_p = dJ_p - A^q C^r{}_{qp} \wedge J_r. \tag{82}$$

Thus, it has some similarities to the vanishing covariant differential of the material energy–momentum. Just as in that case, one can rearrange the field equation to obtain a conserved gauge *pseudocurrent.*

We have gone into considerable detail in this relatively simple example. We have done this to prepare the reader, because we are going to use a very similar argumentation in connection with the rather more complicated case involving gravity and the dynamic spacetime geometric symmetries associated with energy–momentum and angular momentum. It will be seen that almost every step used in our later argument and every expression has an analogue with what we have done in this subsection.

12. First-Order Formulation

In this section, we discuss the general formulation of the first-order formalism; the spacetime geometry has no restrictions.

We proceed from the action principle. Any action principle can be rewritten in an equivalent form, which (following, e.g. ADM[1] and Kuchař[75]) we refer to as *first-order*; this is the most convenient form for our purposes. Here, we present a simple argument (essentially the same Legendre transform idea as is used in classical mechanics to construct the Hamiltonian) which is applicable to a large class of second-order Lagrangians.

Given a second-order Lagrangian 4-form $\mathcal{L}(\varphi, d\varphi)$, we define its associated canonical momentum in the usual way:

$$p := \frac{\partial \mathcal{L}}{\partial d\varphi}(\varphi, d\varphi). \tag{83}$$

Next, we define a 4-form by

$$\Lambda(\varphi, d\varphi, p) := d\varphi \wedge p - \mathcal{L}. \tag{84}$$

Now consider the variation of Λ:

$$\begin{aligned}\delta\Lambda &= \delta(d\varphi)\wedge p + d\varphi\wedge\delta p - \delta\mathcal{L}\\ &= \delta(d\varphi)\wedge\left(p - \frac{\partial\mathcal{L}}{\partial d\varphi}\right) + d\varphi\wedge\delta p - \delta\varphi\wedge\frac{\partial\mathcal{L}}{\partial\varphi}\\ &= d\varphi\wedge\delta p - \delta\varphi\wedge\frac{\partial\mathcal{L}}{\partial\varphi}.\end{aligned} \tag{85}$$

Which shows that Λ can be regarded as a function only of φ, p.[p]

This construction takes one from the usual second-order Lagrangian to a *first-order* type of variational principle:

$$\mathcal{L}^{1\text{st}}(\varphi, d\varphi, p) = d\varphi\wedge p - \Lambda(\varphi, p), \tag{86}$$

where φ and p are now regarded as being independent variables and are varied independently. Varying (86) gives

$$\begin{aligned}\delta\mathcal{L}^{1\text{st}} &= \delta d\varphi\wedge p + d\varphi\wedge\delta p - \delta\Lambda\\ &= d(\delta\varphi\wedge p) - \varsigma\delta\varphi\wedge dp + d\varphi\wedge\delta p - \delta\varphi\wedge\frac{\partial\Lambda}{\partial\varphi} - \frac{\partial\Lambda}{\partial p}\wedge\delta p, \qquad (87)\\ \therefore\quad \delta\mathcal{L}^{1\text{st}} &= d(\delta\varphi\wedge p) + \delta\varphi\wedge\frac{\delta\mathcal{L}^{1\text{st}}}{\delta\varphi} + \frac{\delta\mathcal{L}^{1\text{st}}}{\delta p}\wedge\delta p.\end{aligned}$$

[p]The procedure becomes technically somewhat more complicated if (83) cannot be inverted for $d\varphi$ in terms of φ, p. In that case, one must introduce some additional variables that appear in Λ only algebraically and thus function as Lagrange multipliers introducing some algebraic constraints. We will not go into such complications in our general development here. Later in our treatment of Einstein's GR, we will see a concrete example. Examples of how this has been dealt with in field theory in practice can be found in Refs. 46, 128 and 129.

(*Note*: We find it more convenient to vary our momentum fields p off to the right. This reduces a little the number of appearances of the sign factor ς, and merely amounts to a sign convention on the definition of $\partial\Lambda/\partial p$.)

Using independent p and φ variations gives a pair of first-order field equations for the differentials of the fields:

$$0 = \frac{\delta\mathcal{L}^{1\text{st}}}{\delta\varphi} = -\varsigma dp - \frac{\partial\Lambda}{\partial\varphi}, \quad 0 = \frac{\delta\mathcal{L}^{1\text{st}}}{\delta p} = d\varphi - \frac{\partial\Lambda}{\partial p}. \tag{88}$$

13. The Hamiltonian and the 3 + 1 Spacetime Split

Here, we introduce the Hamiltonian and the spacetime split. In this introductory subsection, we use for motivation some well-known elementary expressions in Minkowski spacetime. In the subsequent subsections, the spacetime geometry is quite general.

A key feature of the canonical Hamiltonian formulation is that the field equations are decomposed into two sets: The initial value constraint equations and the dynamic equations. A familiar example which illustrates many of the ideas is Maxwell's vacuum electrodynamics.[q] The 4-covariant equations were given earlier (39) and (41): $d*F = Z_0 J, dF = 0$, or in tensor index form

$$\partial_\mu(\sqrt{-g}F^{\nu\mu}) = Z_0\sqrt{-g}J^\nu, \quad \partial_{[\alpha}F_{\mu\nu]} = 0. \tag{89}$$

They split (in Minkowski spacetime) into the familiar initial value constraints (spatial projections, with no time derivatives):

$$\nabla\cdot\mathbf{E} = \frac{\rho}{\epsilon_0}, \quad \nabla\cdot\mathbf{B} = 0, \tag{90}$$

and the time projections, a pair of dynamic equations:

$$\dot{\mathbf{B}} + \nabla\times\mathbf{E} = 0, \quad \nabla\times\mathbf{B} = \mu_0\mathbf{J} + \dot{\mathbf{E}}, \tag{91}$$

which contain the first time derivatives of the dynamical fields linearly.

The canonical Hamiltonian form of these equations is in terms of the 4-vector potential (which satisfies $F = dA$ and splits into the scalar and vector potential). The familiar vector form is

$$\text{constraint} \quad \nabla\cdot\mathbf{E} = \frac{\rho}{\epsilon_0}, \tag{92}$$

$$\text{dynamic} \quad \dot{\mathbf{A}} = -\mathbf{E} - \nabla\Phi, \quad -\dot{\mathbf{E}} = -\nabla\times(\nabla\times\mathbf{A}) + \mu_0\mathbf{J}. \tag{93}$$

The scalar potential field appears here, but it has no evolution equation; it can be chosen freely. This *gauge* freedom affects the evolution of an "unphysical" part of

[q] Z_0 is the vacuum impedance, $\varepsilon_0 = (Z_0 c)^{-1}$ is the vacuum permittivity, and $\mu_0 = Z_0 c^{-1}$ is the vacuum permeability. Here, we are taking for simplicity $\mu_0\varepsilon_0 = c^{-2} = 1$ in relativistic spacetime units.

the vector potential. Considering this along with the constraint on **E** (92), one finds that the electromagnetic field has two physical degrees of freedom.

13.1. *Canonical Hamiltonian formalism*

The canonical Hamiltonian formalism[85] is of interest because it clearly reveals the constraints, gauges, and degrees of freedom, as well as the total energy–momentum — and it offers a practical way to numerically calculate solutions.

The dynamical theories of interest all have constraints. The canonical formalism for constrained Hamiltonian systems was developed mainly by Dirac[29–31] and Bergmann.[4,5] For a general discussion see, e.g. Hanson *et al.*,[47] Sundermeyer.[128,129] Rosenfeld[109] seems to have been the first to consider a Hamiltonian approach to GR, but this early work was not followed up. As far as we know Pirani *et al.*[103] were the next to address the issue. Dirac gave a rather complete treatment in 1958.[29,30] The treatment by Arnowitt Deser & Misner (ADM)[1] has come to be regarded as the standard. For a basic discussion see MTW Ref 82, Chap. 21 or Isenberg and Nester.[59] For some critical comparison, see Kiriushcheva and Kuzmin.[70] Going beyond Einstein's theory, a remarkable "if constraint" formalism was developed for the PG by Blagojević and Nikolić[9] to deal with a conditionally degenerate kinetic Hessian. They use the Dirac type of approach; so to construct the Hamiltonian one must first find the primary constraints — which depend on the *conditional* degeneracies of the Legendre transformation. The "if constraint" technique is a marvelous way to manage the technicalities involved in constructing the Hamiltonian. In our first-order approach, in contrast, one can readily formally construct the Hamiltonian and the Hamiltonian equations, however (in line with the principle of the "conservation of difficulties") a suitably adapted version of the "if" constraint technique will still be needed when one actually tries to solve the dynamical and constraint equations. The first-order approach as used in the covariant Hamiltonian formalism allows one to investigate the general formalism and, in particular, to find covariant expressions for the "conserved" quantities while postponing dealing with such technical details.

13.2. *The differential form of the spacetime decomposition*

Note: For the rest of this section, the spacetime geometry is quite general.

A feature of this standard approach is the loss of manifest 4-covariance. Now a Hamiltonian formulation essentially requires that the time derivatives to be singled out from the spatial derivatives, so in this sense it cannot be truly 4-covariant. The usual approach, however, departs far more from 4-covariance than is necessary, all the indices are $3 + 1$ projected, leading to much extra bookkeeping. In the ADM approach, the spacetime metric is replaced by the spatial metric and the *lapse* and *shift*. However only the derivatives ∂_μ really need to be projected.

Since interaction fields are one-form fields, this means decomposing the exterior differential d, decomposing d will inevitably involve decomposing the differential form. One of the reasons for using differential forms is in how nicely they decompose in this fashion.

Begin from the basic first-order form Lagrangian (86). Its variation (88) identifies the first-order Euler–Lagrange expressions (88). According to Hamilton's principle, the first-order Euler–Lagrange expressions should vanish. This gives us our first-order field equations.

We want to extract the "time derivative" of p and φ, i.e. the change with respect to an evolution parameter (which we refer to as time) as seen by observers who move along some fixed congruence of worldlines. This change is given by the Lie derivative in the direction of the (fixed) vector field Z tangent to the congruence: i.e. $\partial_t := \pounds_Z$. The Lie derivative on the components of differential forms is given by a simple neat expression: $\pounds_Z = di_Z + i_Z d$. Using this, from the differential we can extract the "time" derivative:

$$i_Z d\beta = \pounds_Z \beta - di_Z \beta = \dot{\beta} - di_Z \beta. \tag{94}$$

(The congruence need not actually be timelike. Indeed, what we are doing here does not require a metric tensor. Even when one has a metric whether the vector field is "timelike" is not an important issue, our whole Hamiltonian formalism is linear in the spacetime displacement vector field Z, so by considering the difference between two timelike displacements, one could get a spacelike displacement. A metrically timelike displacement is important when one actually tries to find a physical solution to the equations; for evolution one wants hyperbolic equations).

The description of "time" also includes the idea of "instants of time." Geometrically this is a set (foliation) of (nonintersecting) 3D hypersurfaces (we usually think of them as being spacelike). Locally, we can always choose adapted coordinates: $x^\mu = \{t, x^k\}$, where $k = 1, 2, 3$ so that the spacelike hypersurfaces are Σ_t with $t =$ constant. With respect to these adapted coordinates, Z is the directional derivative in the time direction: $Z = \partial_t$. Note that $i_Z dt \equiv 1$.

From these considerations, we are led to define the "*time*" and "*space*" projections of differential forms. We use the notations

$$\hat{\alpha} := i_Z \alpha, \quad \underline{\alpha} := \alpha - dt \wedge \hat{\alpha}, \tag{95}$$

to indicate the "time" component and the "spatial" part of a form. These projections have simple expressions in terms of adapted coordinates.[r]

Thus a general form decomposes according to

$$\alpha = dt \wedge \hat{\alpha} + \underline{\alpha}. \tag{96}$$

[r]We have long used this type of decomposition beginning with[59,84]; see Ref. 81 for a similar technique.

In our formalism i_Z and t are thought of as freely (except that $i_Z dt = 1$) chosen covariant fields, then the decomposition of α is essentially covariant. With this notation, the differential decomposes according to $d\alpha = dt \wedge \widehat{d\alpha} + \underline{d\alpha}$. From (94)

$$\widehat{d\alpha} = \dot{\alpha} - d\hat{\alpha}; \tag{97}$$

thus we can extract the part with the time derivative.

It is convenient to decompose the differential operator d itself. In view of the adapted coordinate expression $d = dx^\mu \wedge \partial_\mu = dt \wedge \partial_t + dx^k \wedge \partial_k$, we define the decomposition as

$$d = dt \wedge \hat{d} + \underline{d}, \quad \text{with } \hat{d} := \pounds_Z. \tag{98}$$

Now we can examine the spacetime decomposition of our first-order field equations (88). We first consider the time projections, which include all the time derivatives:

$$\widehat{\left(\frac{\delta \mathcal{L}^{1\text{st}}}{\delta \varphi}\right)} = -\varsigma(\underline{\dot{p}} - \underline{d}\hat{p}) - \widehat{\left(\frac{\partial \Lambda}{\partial \varphi}\right)} = 0, \tag{99}$$

$$\widehat{\left(\frac{\delta \mathcal{L}^{1\text{st}}}{\delta p}\right)} = (\underline{\dot{\varphi}} - \underline{d}\hat{\varphi}) - \widehat{\left(\frac{\partial \Lambda}{\partial p}\right)} = 0. \tag{100}$$

they are the dynamic equations for $\underline{p}$ and $\underline{\varphi}$. In order to use these equations to evolve $\underline{p}$, $\underline{\varphi}$, we generally need to know $\hat{p}$ and $\hat{\varphi}$, which normally are provided by the initial value constraints: The spatial restriction of (88):

$$\underline{\left(\frac{\delta \mathcal{L}^{1\text{st}}}{\delta \varphi}\right)} = -\varsigma \underline{dp} - \underline{\left(\frac{\partial \Lambda}{\partial \varphi}\right)} = 0, \tag{101}$$

$$\underline{\left(\frac{\delta \mathcal{L}^{1\text{st}}}{\delta p}\right)} = \underline{d\varphi} - \underline{\left(\frac{\partial \Lambda}{\partial p}\right)} = 0. \tag{102}$$

If these two equations can be solved for $\hat{p}$ and $\hat{\varphi}$ all is well and good (in that case the two equations are, in Bergmann's terminology, *second class constraints*). They then define $\hat{p}$ and $\hat{\varphi}$ for all time as functions which depend on $\underline{\varphi}$, $\underline{p}$, $\underline{dp}$ and $\underline{d\varphi}$. How to proceed for the case where these quantities cannot be found from the constraints is best understood from concrete examples. For our purposes of this work, the already-discussed Maxwell electrodynamic example is sufficient. In that case, there is some undetermined gauge freedom.

13.3. *Spacetime decomposition of the variational formalism*

We decomposed the equations. One could decompose the Lagrangian or its variation. Our approach easily relates these alternatives, as can be seen from the following:

$$
\begin{array}{rcl}
\mathcal{L}^{\rm 1st} = d\varphi \wedge p - \Lambda & \xrightarrow{3+1} & \underline{\widehat{\mathcal{L}^{\rm 1st}}} = (\underline{\dot{\varphi}} - \underline{d\hat{\varphi}}) \wedge \underline{p} - (\underline{\hat{\Lambda}} + \varsigma \underline{d\varphi} \wedge \underline{\hat{p}}), \qquad (103)\\
\delta \Big\downarrow & & \Big\downarrow \delta \\
\delta\mathcal{L}^{\rm 1st} = d(\delta\varphi \wedge p) & \xrightarrow{3+1} & \dfrac{d}{dt}(\delta\underline{\varphi} \wedge \underline{p}) - \underline{d}(\delta\underline{\hat{\varphi}} \wedge \underline{p} + \varsigma\delta\underline{\varphi} \wedge \underline{\hat{p}}) \\
+\,\delta\varphi \wedge \dfrac{\delta\mathcal{L}^{\rm 1st}}{\delta\varphi} & & +\delta\underline{\hat{\varphi}} \wedge \underline{\left(\dfrac{\delta\mathcal{L}^{\rm 1st}}{\delta\varphi}\right)} + \delta\underline{\varphi} \wedge \varsigma \underline{\widehat{\left(\dfrac{\delta\mathcal{L}^{\rm 1st}}{\delta\varphi}\right)}} \\
+\,\dfrac{\delta\mathcal{L}^{\rm 1st}}{\delta p} \wedge \delta p & & -\varsigma\underline{\left(\dfrac{\delta\mathcal{L}^{\rm 1st}}{\delta p}\right)} \wedge \underline{\delta\hat{p}} + \underline{\widehat{\left(\dfrac{\delta\mathcal{L}^{\rm 1st}}{\delta p}\right)}} \wedge \underline{\delta p}, \qquad (104)\\
\text{extract} \Big\downarrow & & \Big\downarrow \text{extract} \\
\dfrac{\delta\mathcal{L}^{\rm 1st}}{\delta\varphi} & \xrightarrow{3+1} & \underline{\left(\dfrac{\delta\mathcal{L}^{\rm 1st}}{\delta\varphi}\right)},\ \underline{\widehat{\left(\dfrac{\delta\mathcal{L}^{\rm 1st}}{\delta\varphi}\right)}}, \qquad (105)\\
\dfrac{\delta\mathcal{L}^{\rm 1st}}{\delta p} & \xrightarrow{3+1} & \underline{\left(\dfrac{\delta\mathcal{L}^{\rm 1st}}{\delta p}\right)},\ \underline{\widehat{\left(\dfrac{\delta\mathcal{L}^{\rm 1st}}{\delta p}\right)}}. \qquad (106)
\end{array}
$$

For our objectives here we will not need to use this projection into the space and time parts of form expressions very much. Our intention here was to include enough of the details so that the reader can have some confidence that this formalism can yield a proper Hamiltonian description. From what we have discussed, it can be seen that dynamical equations in this first-order covariant form already contain both the constraint and dynamical evolution equations. It should be noted that the first line of the above set of relations (103) shows how the Hamiltonian can be simply extracted from the first-order Lagrangian.

14. The Hamiltonian and Its Boundary Term

In this section, we establish some of our main formal results concerning the covariant Hamiltonian and its boundary term. The geometry is quite general. The energy, as well as the other conserved quantities, of a physical system can be identified with the value of the Hamiltonian. In particular for a gravitating system, the associated Hamiltonian is proportional to the field equations which vanish on-shell. Therefore the corresponding conserved quantities are determined by the Hamiltonian boundary term. The choice of Hamiltonian boundary term is associated with the specific boundary condition.[19,21–24]

14.1. *The translational Noether current*

The action should not depend on the particular way points are labeled. Thus it should be invariant under diffeomorphisms, in particular, infinitesimal diffeomorphisms — a displacement along some vector field Z. From a gauge theory perspective, such displacements are a "local translation." Under a local translation, quantities change according to the Lie derivative. Hence, for a diffeomorphism invariant action, the key variational relation (88) should be identically satisfied when the variation operator δ is replaced by the Lie derivative $\pounds_Z$ ($\equiv di_Z + i_Z d$):

$$di_Z\mathcal{L}^{1\text{st}} \equiv \pounds_Z\mathcal{L}^{1\text{st}} \equiv d(\pounds_Z\varphi \wedge p) + \pounds_Z\varphi \wedge \frac{\delta\mathcal{L}^{1\text{st}}}{\delta\varphi} + \frac{\delta\mathcal{L}^{1\text{st}}}{\delta p} \wedge \pounds_Z p. \tag{107}$$

This simply means that $\mathcal{L}^{1\text{st}}$ is a 4-form which depends on position only through the fields φ, p. (According to our understanding this is only possible if the set of fields in $\mathcal{L}^{1\text{st}}$ includes some dynamic spacetime geometric variables: gravity.)

From (107) it directly follows that the 3-form

$$\mathcal{H}(Z) := \pounds_Z\varphi \wedge p - i_Z\mathcal{L}^{1\text{st}}, \tag{108}$$

satisfies the identity

$$-d\mathcal{H}(Z) \equiv \pounds_Z\varphi \wedge \frac{\delta\mathcal{L}^{1\text{st}}}{\delta\varphi} + \frac{\delta\mathcal{L}^{1\text{st}}}{\delta p} \wedge \pounds_Z p; \tag{109}$$

thus it is a conserved "current" *on shell* (i.e. when the field equations are satisfied). Substituting (86) into (108) gives the explicit expression

$$\mathcal{H}(Z) \equiv d(i_Z\varphi \wedge p) + \varsigma i_Z\varphi \wedge dp + \varsigma d\varphi \wedge i_Z p + i_Z\Lambda, \tag{110}$$

thus this conserved *Noether translation current* can be written as a 3-form linear in the displacement vector plus a total differential:

$$\mathcal{H}(Z) =: Z^\mu\mathcal{H}_\mu + d\mathcal{B}(Z). \tag{111}$$

Compare the differential of this expression, $d\mathcal{H}(Z) \equiv dZ^\mu \wedge \mathcal{H}_\mu + Z^\mu d\mathcal{H}_\mu$, with (109); equating the dZ^μ coefficient on both sides reveals that

$$Z^\mu\mathcal{H}_\mu \equiv -i_Z\varphi \wedge \frac{\delta\mathcal{L}^{1\text{st}}}{\delta\varphi} + \varsigma\frac{\delta\mathcal{L}^{1\text{st}}}{\delta p} \wedge i_Z p. \tag{112}$$

Thus, as a consequence of *local* diffeomorphism invariance, $\mathcal{H}_\mu$ vanishes *on shell*; hence conservation of the translational Noether current (109) reduces to a differential identity between Euler–Lagrange expressions. This an instance of Noether's second theorem, and, moreover, it is exactly the sort of case to which Hilbert's remark regarding the lack of a proper energy law applies.

From the above it follows quite generally, as we remarked earlier in the special case of GR (32), the *value* of the conserved quantity, $-P(Z, V)$, associated with a 3D region V is determined by a 2-surface integral over the boundary, i.e. it is *quasi-local*:

$$-P(Z,V) := \int_V \mathcal{H}(Z) = \oint_{\partial V} \mathcal{B}(Z). \tag{113}$$

For *any* choice of Z this expression defines a conserved quasi-local quantity. What do these values mean? As we shall see in detail later, for a suitable timelike (spacelike) quasi-translation displacement on the boundary the expression defines a quasi-local energy (momentum), and for a suitable quasi-rotation (boost) it defines a quasi-local angular momentum (CoMM). However it must be noted that, like all other conserved currents, the translational current is likewise subject to the usual ambiguity: one can add *by hand* the differential of any 2-form and still have a conserved current. But that amounts to being able to adjust $\mathcal{B}$ freely, consequently one could obtain almost any quasi-local value. The Hamiltonian perspective brings this freedom under physical control. As we shall show, the *first-order* translational current 3-form is something more: it is *the generator of local diffeomorphisms*, i.e. *the Hamiltonian.*

14.2. *The Hamiltonian formulation*

From the first-order field equations (88), by contraction with a "time evolution vector field" Z, we get a pair of Hamiltonian-like evolution equations for the "time derivatives": $\pounds_Z\varphi$, $\pounds_Z p$. A key identity involving these time derivatives is revealed by comparing two relations. Consider the projection of the Lagrangian 4-form $i_Z\mathcal{L}^{1\text{st}}$, which from (108) is just $\pounds_Z\varphi \wedge p - \mathcal{H}(Z)$; its variation is

$$\begin{aligned} \delta i_Z\mathcal{L}^{1\text{st}} &\equiv \delta(\pounds_Z\varphi \wedge p) - \delta\mathcal{H}(Z) \\ &\equiv \delta(\pounds_Z\varphi) \wedge p + \pounds_Z\varphi \wedge \delta p - \delta\mathcal{H}(Z) \\ &\equiv \pounds_Z\delta\varphi \wedge p + \pounds_Z\varphi \wedge \delta p - \delta\mathcal{H}(Z) \\ &\equiv \pounds_Z(\delta\varphi \wedge p) - \delta\varphi \wedge \pounds_Z p + \pounds_Z\varphi \wedge \delta p - \delta\mathcal{H}(Z). \end{aligned} \tag{114}$$

Compare this with the projection of $\delta\mathcal{L}^{1\text{st}}$ (88) along Z:

$$\begin{aligned} i_Z\delta\mathcal{L}^{1\text{st}} &\equiv i_Z d(\delta\varphi \wedge p) + i_Z\left(\delta\varphi \wedge \frac{\delta\mathcal{L}^{1\text{st}}}{\delta\varphi} + \frac{\delta\mathcal{L}^{1\text{st}}}{\delta p} \wedge \delta p\right) \\ &\equiv \pounds_Z(\delta\varphi \wedge p) - d i_Z(\delta\varphi \wedge p) + i_Z\left(\delta\varphi \wedge \frac{\delta\mathcal{L}^{1\text{st}}}{\delta\varphi} + \frac{\delta\mathcal{L}^{1\text{st}}}{\delta p} \wedge \delta p\right). \end{aligned} \tag{115}$$

Since Z is not varied, the two relations are identical: $\delta i_Z\mathcal{L}^{1\text{st}} \equiv i_Z\delta\mathcal{L}^{1\text{st}}$; consequently,

$$\delta\mathcal{H}(Z) \equiv -\delta\varphi \wedge \pounds_Z p + \pounds_Z\varphi \wedge \delta p + d i_Z(\delta\varphi \wedge p) - i_Z\left(\delta\varphi \wedge \frac{\delta\mathcal{L}^{1\text{st}}}{\delta\varphi} + \frac{\delta\mathcal{L}^{1\text{st}}}{\delta p} \wedge \delta p\right). \tag{116}$$

The last term vanishes "on shell." This relation identifies the Noether translational current $\mathcal{H}(Z)$ as the *Hamiltonian* 3-*form* (i.e. density), as the following considerations show. The integral of $\mathcal{H}(Z)$ over a 3D region,

$$H(Z,\Sigma) := \int_\Sigma \mathcal{H}(Z), \tag{117}$$

is the Hamiltonian which displaces this region along Z, since the integral of its variation:

$$\delta H(Z,\Sigma) = \int_\Sigma \delta\mathcal{H}(Z), \tag{118}$$

yields, from (116), "on shell" (then the last bracketed term vanishes) the Hamilton equations:

$$\pounds_Z\varphi = \frac{\delta H(Z,\Sigma)}{\delta p}, \qquad \pounds_Z p = -\frac{\delta H(Z,\Sigma)}{\delta\varphi}, \tag{119}$$

if the boundary term in the variation of the Hamiltonian *vanishes*. In this case that means when $\delta\varphi$ vanishes on $\partial\Sigma$. Technically, the variational derivatives of the Hamiltonian $H(Z,\Sigma)$ displayed in (119) are only defined for variations satisfying this boundary condition. In other words, this Hamiltonian is "well-defined," i.e. *functionally differentiable*, only on the phase space of fields satisfying the particular boundary condition $\delta\varphi|_{\partial\Sigma} = 0$.

14.3. *Boundary terms: The boundary condition and reference*

In some important cases, the fields of physical interest do not satisfy the boundary condition naturally inherited from the Lagrangian,[21] this happens in particular for the spacetime metric of an asymptotically flat region. A modified formulation is needed to deal with this.

One alternative is to modify the Lagrangian 4-form itself by a total differential. This strategy has often been adopted, beginning with Einstein (9) and including many of the Hamiltonian formulations.[1,30,103] But such a modification is necessarily noncovariant. For our formalism, we want to keep our Lagrangian covariant. Furthermore a Lagrangian boundary term would modify the boundary condition on the whole 3D boundary of the spacetime region, thus inducing the same type of modification on the spatial boundary at large spatial distances as on the initial time hypersurface. However we want the freedom to adjust the boundary condition on the 2D boundary of the spacelike region $\partial\Sigma$ independently of the type of initial conditions imposed within the initial time hypersurface Σ_t. Thus for our objectives we turn to the Hamiltonian boundary term.[s]

Note that the Hamiltonian (111) has two distinct parts; each plays a distinct role. The proper density $Z^\mu\mathcal{H}_\mu$, although it has vanishing value on shell, generates

[s]In the end, it turns out that our favored Hamiltonian boundary term for GR is related to one induced by a Lagrangian boundary term.[21]

the equations of motion, whereas the boundary term $\mathcal{B}(Z)$ determines not only the *quasi-local value* (113) but also the *boundary condition*. Now we should make note of a very important fact: The boundary term can be adjusted — without changing the Hamilton equations or the conservation property (109). Thus one can replace the 2-form $\mathcal{B}(Z) = i_Z\varphi \wedge p$ inherited from the Lagrangian by another.

Such an adjustment is in one respect just a special case of the conserved Noether current ambiguity (i.e. for any 2-form χ, J and $J' := J + d\chi$ are both conserved currents (3-forms) if $dJ = 0$, even though they define different conserved values).

However here, in this Hamiltonian case, any such adjustment modifies — *in parallel* — not only the value of the quasi-local quantities but also the spatial boundary conditions. Thus the boundary term ambiguity is under physical control: Each distinct choice of the quasi-local expression given by the Hamiltonian boundary term is associated with a physically distinct boundary condition.

In order to accommodate suitable boundary conditions we found that, in general, one needs to introduce on the boundary for each of the dynamical fields certain reference values $\bar{p}$, $\bar{\varphi}$, which represent the ground state of the field — the "vacuum" (or background field) values. This is necessary in particular for fields whose natural ground state is nonvanishing; the spacetime metric is such a field.

We take our boundary terms to be linear in $\Delta\varphi := \varphi - \bar{\varphi}$, $\Delta p := p - \bar{p}$, so that they (and thus all the quasi-local quantities) vanish if the fields take on the ground state (reference) values.[t] We presume that the reference values (like Z) are not varied: $\delta\bar{\varphi} = 0$ and $\delta\bar{p} = 0$, consequently $\delta\Delta\varphi = \delta\varphi$, $\delta\Delta p = \delta p$.

14.4. *Covariant-symplectic Hamiltonian boundary terms*

To find an improved Hamiltonian boundary term for (110) first drop the one inherited from the Lagrangian, examine the boundary term generated in the variation of the 3-form part of the Hamiltonian (110); it is $-i_Z\varphi \wedge \delta p + \varsigma\delta\varphi \wedge i_Z p$. This invites us to add a suitable complimentary boundary term. In this way, we were led to the boundary terms[21,23,24]

$$\mathcal{B}(Z) := i_Z \begin{Bmatrix} \varphi \\ \bar{\varphi} \end{Bmatrix} \wedge \Delta p - \varsigma\Delta\varphi \wedge i_Z \begin{Bmatrix} p \\ \bar{p} \end{Bmatrix}. \tag{120}$$

Then the associated variational Hamiltonian boundary term becomes

$$\delta\mathcal{H}(Z) \sim d\left[\begin{Bmatrix} i_Z\delta\varphi \wedge \Delta p \\ -i_Z\Delta\varphi \wedge \delta p \end{Bmatrix} + \varsigma \begin{Bmatrix} -\Delta\varphi \wedge i_Z\delta p \\ \delta\varphi \wedge i_Z\Delta p \end{Bmatrix}\right]. \tag{121}$$

Here, *for each bracket independently* one may choose either the upper or lower term, which represent essentially a choice of Dirichlet (fixed field) or Neumann

[t] Some authors use the terminology *regularize*.

(fixed momentum) boundary conditions for the space and time parts of the fields separately.[u]

In each of these cases, the boundary term in the Hamiltonian variation has a certain *symplectic* structure which pairs certain *control* quantities — i.e. the independent variables — with certain associated *response* quantities — the dependent variables. (For discussions of this paradigm of the symplectic structure associated with variational principles which we have found to be illuminating see Refs. 68 and 69). The symmetry of the above expressions under an interchange of "control" and "response," formally $\delta \to \Delta, \Delta \to -\delta$, is noteworthy.

Thus, although it is not so well known, when the issue is examined one can readily see that there are many choices of boundary conditions and consequently really many different expressions for energy in classical field theory. This is true especially for gravitating fields. Actually this sort of thing is not unusual in physics; in particular one can compare the situation with that in thermodynamics (which has several physically meaningful energies: internal, enthalpy, Gibbs and Helmholtz).

Nevertheless, it should be noted that one of our boundary term expressions stands out: For any field which allows trivial reference values, $\bar{\varphi} = 0 = \bar{p}$, one boundary term choice vanishes (the lower choice in each bracket). Such fields, with this choice of boundary condition, make no explicit contribution to the quasi-local boundary term. This particular boundary term has another virtue: For any field with gauge freedom, it is the only gauge invariant choice. Thus there is a certain preferred boundary expression — and thus *a preferred boundary condition* — for this large class of fields, a class which includes all the physical fields of the standard model. There is, however, a quite important exception: *Gravity*, more specifically, any gravity theory formulated in terms of dynamic spacetime geometry which includes the spacetime metric as a dynamical field. The natural reference choice for the metric *is not* a vanishing metric tensor but rather the nonvanishing Minkowski metric. Consequently one must have, in general, a nonvanishing Hamiltonian boundary term.

15. Standard Asymptotics

This section is concerned with suitable asymptotic conditions for our classical fields at spatial and null infinity. It also includes a discussion of energy flux.

For spatial infinity, the issue of asymptotic conditions was first investigated in GR by Regge and Teitelboim,[106] with later refinements by Beig and Ó Murchadha[3] and then Szabados.[132,134] We have developed a similar idea for general fields.

[u]There are more complicated possibilities, "mixed" choices involving some linear combination of the upper and lower expressions.[119] We do not have any specific physical examples, but mixed boundary conditions may be of interest in certain cases.

15.1. *Spatial infinity*

For finite regions, these boundary terms in the variation of the Hamiltonian tell us exactly what needs to be held fixed (i.e. "controlled"). For asymptotically flat regions, however, one should take into account the asymptotic fall off rates. The various boundary terms we have constructed enable the Hamiltonian to be well defined on the phase space of fields with suitable asymptotic behavior for all typical physical fields.

For the fields, it is sufficient[v] to take the respective asymptotic fall offs for even and odd parity terms to be

$$\Delta\varphi \approx \mathcal{O}^{+}\left(\frac{1}{r}\right) + \mathcal{O}^{-}\left(\frac{1}{r^2}\right), \quad \Delta p \approx \mathcal{O}^{-}\left(\frac{1}{r^2}\right) + \mathcal{O}^{+}\left(\frac{1}{r^3}\right). \tag{122}$$

Parity here means the parity of the components in an asymptotically Cartesian reference frame. The 2-surface area element has odd parity, so even parity 2-forms automatically have vanishing 2-surface integral.

For asymptotically flat spaces, the displacement should asymptotically be a Minkowski Killing vector, i.e. an infinitesimal Poincaré displacement. It is sufficient to take

$$Z^{\mu} \approx Z^{\mu}_{0} + \lambda^{\mu}_{0\,\nu} x^{\nu} + \mathcal{O}^{+}\left(\frac{1}{r}\right) + \mathcal{O}^{-}(1), \tag{123}$$

where (in terms of asymptotically Minkowski coordinates) Z^{μ}_{0} is a constant translation parameter and $\lambda^{\mu\nu}_{0} = \lambda^{[\mu\nu]}_{0}$ is a constant asymptotic infinitesimal Lorentz boost/rotation parameter.

With the asymptotics (122) and (123), it is straightforward to check that for any of the boundary term choices (120) all of the quasi-local quantities have finite values; furthermore, for any of the choices all of our Hamiltonians are differentiable on the specified phase space — since the respective boundary terms in the variations of the Hamiltonians (121) vanish asymptotically. Thus our Hamiltonians are generally well defined on a large phase space which includes physically interesting solutions. At asymptotically-flat spatial infinity, the aforementioned asymptotics are physically reasonable. Our considerations naturally straightforwardly extend to asymptotically anti-de Sitter spaces. Here the details are omitted.

15.2. *Null infinity*

Let us next consider what can be expected if the boundary of our 2-surface $\partial\Sigma$ approaches future null infinity. Long range radiation fields (e.g. electromagnetism)

[v]Sufficient, but not necessary. When examined in detail it can be seen that one really only needs conditions on certain combinations of the components, but it is not in the spirit of our treatment to break fields up into, e.g. components parallel to and perpendicular to some specific boundary surface, etc. Here we are satisfied with a formalism that includes a large class of fields. We leave to more specific investigations finding the largest acceptable phase space with the weakest conditions.

have slower fall offs, like $\Delta p \approx d\varphi \approx O(1/r)$. Then the boundary terms in the variation of our various Hamiltonians will not vanish, so the Hamiltonian is no longer functionally differentiable. This seeming calamity is actually providential — it is directing us to additional physics contained within the formalism, namely energy flux expressions.[24,84,151]

15.3. *Energy flux*

For the flux of "energy" there is a special way of calculating — the analog of the classical mechanics calculation (for conservative Hamiltonian systems) of

$$\delta H = \dot{q}^k \delta p_k - \dot{p}_k \delta q^k \Rightarrow \dot{E} := \dot{H} \equiv 0, \tag{124}$$

under the replacement $\delta \to d/dt$, where the remarkable cancelation is a consequence of the particular (symplectic) form of the Hamiltonian variation. In the present case from (116) under the replacement $\delta \to \pounds_Z$ the same type of symplectic calculation occurs, and we are left with the respective contributions from our various boundary term choices (121)

$$\pounds_Z \mathcal{H}(Z) = d\left[\begin{Bmatrix} i_Z \pounds_Z \varphi \wedge \Delta p \\ -i_Z \Delta\varphi \wedge \pounds_Z p \end{Bmatrix} + \varsigma \begin{Bmatrix} -\Delta\varphi \wedge i_Z \pounds_Z p \\ \pounds_Z \varphi \wedge i_Z \Delta p \end{Bmatrix}\right]. \tag{125}$$

(We are presuming that $\pounds_Z \bar{\varphi}$, $\pounds_Z \bar{p}$ vanish).

In particular, as we mentioned earlier, for all fields with vanishing reference, there is a standard choice of Hamiltonian boundary term, namely the one that vanishes. The corresponding energy flux expression is

$$\pounds_Z \mathcal{H}(Z) = d(-i_Z \varphi \wedge \pounds_Z p + \varsigma \pounds_Z \varphi \wedge i_Z p). \tag{126}$$

16. Application to Electromagnetism

To illustrate these ideas in a familiar setting, we briefly consider vacuum electromagnetism in Minkowski space (for the complete details see Ref. 24).

For electromagnetism in Minkowski space, the formalism developed above, with the important exception of the "on shell" vanishing of $\mathcal{H}_\mu$, is still applicable. A first-order Lagrangian 4-form for the (source free) Maxwell one-form (the U(1) gauge potential) A is

$$\mathcal{L}^{\text{1st}}_{\text{EM}} = dA \wedge \mathcal{P} - \frac{1}{2} Z_0 * \mathcal{P} \wedge \mathcal{P}, \tag{127}$$

which yields the pair of first-order equations

$$d\mathcal{P} = 0, \quad dA - Z_0 * \mathcal{P} = 0. \tag{128}$$

These are just the vacuum Maxwell equations with $*\mathcal{P} = Z_0^{-1} F := Z_0^{-1} dA$; hence $\mathcal{P} = -Z_0^{-1} * F$, and $d * F = 0$. (Here Z_0 is the vacuum impedance, which has the value $\mu_0 = \varepsilon_0^{-1}$ in our relativistic units in which $c = 1$. With our conventions, our conjugate momentum field $\mathcal{P}$ turns out to be the negative of H which was introduced earlier in (39)). With $Z = \partial_t$ and the decomposition $A = (-\phi, A_k)$, we find that $i_Z F = i_Z dA = \pounds_Z A - d i_Z A$ corresponds to $F_{0k} = \dot{A}_k + \partial_k \phi = -E_k$. The magnetic field strength is $F_{ij} := \partial_i A_j - \partial_j A_i =: \epsilon_{ijk} B^k$. Hence, $\mathcal{P}_{0i} = -Z_0^{-1} * F_{0i} = -\mu_0^{-1} B_i$, and $\mathcal{P}_{ij} = -\lambda_0 * F_{ij} = -\varepsilon_0 \epsilon_{ijk} E^k$. The natural reference choice is $\bar{A} = 0 = \bar{\mathcal{P}}$.

The Hamiltonian 3-form is

$$\mathcal{H}^{\mathrm{EM}}(Z) = -i_Z A\, d\mathcal{P} - dA \wedge i_Z \mathcal{P} + i_Z \left(\frac{1}{2} Z_0 * \mathcal{P} \wedge \mathcal{P} \right) + d\mathcal{B}^{\mathrm{EM}}. \tag{129}$$

In the usual tensor index notation, the volume density part has the form

$$\mathcal{H}^{\mathrm{EM}} = \phi \partial_k \pi^k + \frac{1}{2}(\partial_i A_j - \partial_j A_i)\epsilon^{ijk} H_k + \frac{1}{2\varepsilon_0}\pi^k \pi_k - \mu_0 \frac{1}{2} H^k H_k, \tag{130}$$

where the momentum conjugate to the 3-vector potential is given the name π^k (it works out to have in the usual terminology the value $-\varepsilon_0 E^k$, i.e. $-D^k$). By varying H^k one obtains $\mu_0 H^k = \frac{1}{2}\epsilon^{kij}(\partial_i A_j - \partial_j A_i) = B^k$, a 2nd class constraint that could be used to eliminate the magnetic field, then the Hamiltonian volume density would correspond to the familiar energy density $\frac{1}{2}(\varepsilon_0 E^2 + \mu_0 B^2)$ plus a gauge generating term, $\phi \partial_k \pi^k$, which vanishes "on shell"; the scalar potential in this term acts as a Lagrange multiplier to enforce the (first class) Gauss constraint $\partial_k D^k = 0$.

Let us just consider two boundary term choices, namely our preferred choice with vanishing boundary term, and the above Hamiltonian 3-form with the boundary term

$$\mathcal{B}^{\mathrm{EM}} = i_Z A \mathcal{P} = -\phi \pi^k dS_k. \tag{131}$$

These two are actually both well known physically, the former corresponds to the energy density from the gauge invariant energy–momentum tensor (8), and the latter is the energy density of the electromagnetic canonical energy–momentum tensor (7). Here, our interest is not in the field equations but in the total differential term which, upon integration, becomes a boundary term indicating the boundary condition. Briefly, for the choice with vanishing Hamiltonian boundary term, the total derivative term in the variation of the Hamiltonian is

$$-d(i_Z A \delta \mathcal{P} + \delta A \wedge i_Z \mathcal{P}) \simeq \partial_k(\phi \delta \pi^k - \epsilon^{kij} \delta A_i H_j), \tag{132}$$

which tells us that one should hold fixed on the boundary of the dynamical region the normal component of the electric field and the surface parallel components of the vector potential (the gauge independent part of which determines the normal component of the magnetic field). On the other hand, for the Hamiltonian including the boundary term (131), one finds that the total differential in the variation of the Hamiltonian is now

$$d(i_Z \delta A \mathcal{P} - \delta A \wedge i_Z \mathcal{P}) \simeq -\partial_k(\delta \phi \pi^k + \epsilon^{kij} \delta A_i H_j). \tag{133}$$

This is the same boundary condition in the vector potential/magnetic sector but now for the electric sector, one should instead hold fixed on the boundary of the region the scalar potential.

The physical meaning of such boundary conditions are well known. Fixing the normal component of the electric field on the boundary corresponds to fixing the surface charge density. An instructive physical example is a parallel plate capacitor. One can use a battery to charge up a capacitor with a moveable dielectric. Disconnect the battery and measure the work needed to remove/insert the dielectric (the potential varies but the charge is fixed, no current or power flow). Alternatively leave the battery connected and measure the work needed to displace the dielectric — now the potential is fixed but the charge varies, so current and hence power flows. The respective boundary terms in the variation of the Hamiltonian are $\phi\delta\pi^k dS_k$ and $-\delta\phi\pi^k dS_k$. Both boundary condition choices are physically meaningful corresponding to real situations.

Nevertheless for electrodynamics one expression stands out, the one with vanishing boundary term. This choice is the only one in which the value of the Hamiltonian is *gauge invariant.* Moreover, this is the only *non-negative* Hamiltonian density. Consequently the associated energy has a lower bound and the system has a natural vacuum or ground state: zero energy for vanishing fields. The value of the Hamiltonian with this boundary term can be interpreted as the *internal energy*, whereas the other expressions can be regarded as including some additional energy on the boundary of the system associated with maintaining the boundary condition. The associated electromagnetic energy flux expression from our formula (126) reduces to just the usual Poynting energy flux:

$$\begin{aligned} \pounds_Z \mathcal{H}^{\mathrm{EM}} &= d[-i_Z A \wedge (di_Z + i_Z d)\mathcal{P} - (di_Z + i_Z d)A \wedge i_Z \mathcal{P}] \\ &\equiv -d(i_Z F \wedge i_Z \mathcal{P}) \\ &= d(-E_i H_j \, dx^i \wedge dx^j). \end{aligned} \tag{134}$$

Clearly this choice, associated with fixing the normal components of the electric and magnetic fields on the boundary, is preferred; it is the one suitable for most physical applications. It gives the usual energy density and Poynting energy flux. Similarly, for all other fields — *except for dynamic spacetime geometry* — there is available a standard Hamiltonian (the one with *vanishing boundary term contribution*) associated with a certain preferred boundary condition.

17. Geometry: Covariant Differential Formulation

In the discussion of our covariant Hamiltonian approach, up to this point (except as was specified for a couple of specific examples), there has been no need to make any restriction on the type of geometry for our manifold. Here in this section we discuss the specific sort of dynamic spacetime geometry that we will consider and relate it to the gauge theory paradigm.

The covariant Hamiltonian formulation can apply to general theories of dynamical geometry. Standard references for differential geometry are Kobayashi and Nomizu[72] and Spivak.[123]

17.1. Metric and connection

For the dynamical spacetimes that we consider, there are two basic geometric ideas: a *metric tensor* $g = g_{\mu\nu}\vartheta^\mu \otimes \vartheta^\nu$ (which determines *length* and *angle*), and *parallel*, associated with *parallel transport*, *covariant derivative* and *connection*, i.e. we consider *metric-affine* geometry.

The metric gives the *causal* structure, arc length, area and volume. Furthermore it is used to raise and lower indexes (i.e. it determines a specific isomorphism between tangent and cotangent vectors). It also provides the paths of extremal length (geodesics). For the 4D spacetimes of interest here, the metric has the Lorentz signature. Given such a metric, there is a naturally defined associated symmetry group, the group of local Lorentz transformations: $L \in SO(1,3) \Rightarrow g(LX, LY) = g(X,Y)$.

For the other structure, let e_μ for $\mu = 0,1,2,3$ be a basis for spacetime vector fields. The *covariant differential* ∇ of each basis vector is a vector valued one-form, hence some linear combination of the e_β's with one-form coefficients:

$$\nabla e_\beta = e_\alpha \Gamma^\alpha{}_\beta, \quad \Gamma^\alpha{}_\beta = \Gamma^\alpha{}_{\beta i}\, dx^i, \tag{135}$$

called the *connection one-forms.*

The covariant differential of a vector field $V = e_\mu V^\mu$ is

$$\nabla V = \nabla(e_\mu V^\mu) = (\nabla e_\mu)V^\mu + e_\mu \nabla V^\mu = e_\nu \Gamma^\nu{}_\mu V^\mu + e_\mu dV^\mu =: e_\mu DV^\mu. \tag{136}$$

Its components are determined by the operator D:

$$DV^\mu := dV^\mu + \Gamma^\mu{}_\nu \wedge V^\nu, \tag{137}$$

which extends, as indicated, to vector valued forms.

The notation automatically antisymmetrizes:

$$\begin{aligned}\nabla^2 V &= \nabla^2(e_\mu V^\mu) = \nabla(e_\mu DV^\mu) = e_\mu D^2 V^\mu \\ &= e_\mu \left[d(dV^\mu + \Gamma^\mu{}_\nu \wedge V^\nu) + \Gamma^\mu{}_\lambda \wedge (dV^\lambda + \Gamma^\lambda{}_\sigma \wedge V^\sigma)\right] \\ &= e_\mu (d\Gamma^\mu{}_\nu + \Gamma^\mu{}_\lambda \wedge \Gamma^\lambda{}_\nu) \wedge V^\nu = e_\mu R^\mu{}_\nu \wedge V^\nu,\end{aligned} \tag{138}$$

where

$$R^\mu{}_\nu := d\Gamma^\mu{}_\nu + \Gamma^\mu{}_\lambda \wedge \Gamma^\lambda{}_\nu = \frac{1}{2} R^\mu{}_{\nu ij} dx^i \wedge dx^j. \tag{139}$$

is the *curvature 2-form*

Exterior covariant differential form notation treats some, but not all, indices as differential forms. Rather than work with the operator ∇ on geometric objects we often find it more convenient to work with D on their coefficients. The covariant differential D can be extended to operate on a *tensor valued form* of any type, e.g.

$$DP^\alpha{}_\beta = dP^\alpha{}_\beta + \Gamma^\alpha{}_\gamma \wedge P^\gamma{}_\beta - \Gamma^\gamma{}_\beta \wedge P^\alpha{}_\gamma, \tag{140}$$

$$D^2 P^\alpha{}_\beta = R^\alpha{}_\gamma \wedge P^\gamma{}_\beta - R^\gamma{}_\beta \wedge P^\alpha{}_\gamma. \tag{141}$$

Generically, arranging the components of a tensor valued form φ as a row vector we get

$$D\varphi = d\varphi + \Gamma^{\alpha}{}_{\beta} \wedge \varphi\sigma_{\alpha}{}^{\beta}, \tag{142}$$

$$D^2\varphi = R^{\alpha}{}_{\beta} \wedge \varphi\sigma_{\alpha}{}^{\beta}, \tag{143}$$

for some appropriate representation matrix $\sigma_{\alpha}{}^{\beta}$.

The special case of the vector whose components are the coframe one-forms $\vartheta^{\alpha} = e^{\alpha}{}_{i}dx^{i}$ yields the *torsion* 2-form:

$$T^{\alpha} := D\vartheta^{\alpha} := d\vartheta^{\alpha} + \Gamma^{\alpha}{}_{\beta} \wedge \vartheta^{\beta} = \frac{1}{2}T^{\alpha}{}_{ij}dx^{i} \wedge dx^{j}. \tag{144}$$

On the coframe ϑ^{α}, D^2 gives an important special case of (143):

$$DT^{\alpha} = D^2\vartheta^{\alpha} = R^{\alpha}{}_{\beta} \wedge \vartheta^{\beta}, \tag{145}$$

which is known as the *first Bianchi identity*.

Applying D to the curvature 2-form $R^{\alpha}{}_{\beta}$ gives

$$\begin{aligned} DR^{\alpha}{}_{\beta} &:= dR^{\alpha}{}_{\beta} + \Gamma^{\alpha}{}_{\gamma} \wedge R^{\gamma}{}_{\beta} - \Gamma^{\gamma}{}_{\beta} \wedge R^{\alpha}{}_{\gamma} \\ &\equiv d(d\Gamma^{\alpha}{}_{\beta} + \Gamma^{\alpha}{}_{\gamma} \wedge \Gamma^{\gamma}{}_{\beta}) + \Gamma^{\alpha}{}_{\gamma} \wedge (d\Gamma^{\gamma}{}_{\beta} + \Gamma^{\gamma}{}_{\sigma} \wedge \Gamma^{\sigma}{}_{\beta}) \\ &\quad - \Gamma^{\gamma}{}_{\beta} \wedge (d\Gamma^{\alpha}{}_{\gamma} + \Gamma^{\alpha}_{\sigma} \wedge \Gamma^{\sigma}{}_{\gamma}) \equiv 0, \end{aligned} \tag{146}$$

by explicit calculation. This is the *second Bianchi identity*.

With $g_{\mu\nu}$ the metric tensor, $Dg_{\mu\nu}$ defines the *nonmetricity* 1-form.

$$Dg_{\mu\nu} := dg_{\mu\nu} - \Gamma^{\lambda}{}_{\mu}g_{\lambda\nu} - \Gamma^{\lambda}{}_{\nu}g_{\mu\lambda}. \tag{147}$$

Correspondingly, we have

$$D^2g_{\mu\nu} = -R^{\lambda}{}_{\mu}g_{\lambda\nu} - R^{\lambda}{}_{\nu}g_{\mu\lambda}. \tag{148}$$

17.2. *Riemann–Cartan geometry*

Here we are interested in particular in the special case where the geometry can be regarded as a local gauge theory of an appropriate spacetime symmetry group. With due consideration given to the understanding of both the geometry of and physics in Minkowski spacetime, the appropriate choice for the symmetry group is the inhomogeneous Lorentz group, generally referred to as the Poincaré group. This is both the symmetry group for the spacetime of special relativity and the group used to classify elementary particles in terms of mass and spin. The group is a semidirect product of the translation group and the group of rotations/Lorentz boosts. The Noether conserved quantities associated with these global symmetries are energy–momentum and angular momentum/CoMM. The type of spacetime geometry with local Poincaré symmetry is known as Riemann–Cartan geometry.

In Riemann–Cartan geometry, the connection is assumed to be *a priori metric compatible*, $Dg_{\mu\nu} = 0$, via (148) this gives $R_{\alpha\beta} = R_{[\alpha\beta]}$ (i.e. a Lorentz Lie algebra valued two-form). For our purposes, it is convenient to use the orthonormal frame gauge condition, then the metric components are constant and $dg_{\mu\nu} = 0$, so, via (148), this gives $\Gamma_{\alpha\beta} = \Gamma_{[\alpha\beta]}$ (i.e. a *Lorentz Lie algebra valued one-form*). The geometry has in general nonvanishing torsion and curvature. The metric information is encoded in the orthonormal coframe ϑ^μ, which has local Lorentz gauge freedom.

Riemannian geometry is a special case with vanishing torsion, $T^\alpha = 0$; such a connection is called symmetric. Then the geometry is given by the curvature, which is generally nonvanishing, so the parallel transport is path dependent. On the other hand another special case is *teleparallel* geometry, which has a vanishing curvature 2-form, $R^\alpha{}_\beta = 0$. This is referred to as *flat*. Parallel transport is then path independent, nevertheless it is generally nontrivial — being dependent on the torsion, which is generally nonvanishing.

17.3. *Regarding geometry and gauge*

Note the respective similarities in the form of the commutator of the gauge covariant derivatives for the flat space U(1) phase case ($\nabla_\mu\phi = \partial_\mu\phi + ieA_\mu\phi$) and the Yang–Mills case ($\nabla_\mu\psi = \partial_\mu\psi + iqA^p_\mu T_p\psi$) compared with the spacetime geometric case:

$$[\nabla_\mu, \nabla_\nu]\phi = ieF_{\mu\nu}\phi, \tag{149}$$

$$[\nabla_\mu, \nabla_\nu]\psi = iqF^p{}_{\mu\nu}T_p\psi, \tag{150}$$

$$[\nabla_\mu, \nabla_\nu]V^\alpha = R^\alpha{}_{\beta\mu\nu}V^\beta - T^\gamma{}_{\mu\nu}\nabla_\gamma V^\alpha. \tag{151}$$

Here, ∇_μ is the *covariant derivative*; the latter relation is called the *Ricci identity*. On the right hand side the respective gauge field strengths appear. One can see that for spacetime the curvature is the Lorentz field strength, and the torsion is the spacetime "translational" field strength, associated with the generator of infinitesimal translations, the directional derivative.

Riemann–Cartan geometry is ideally suited to admit an interpretation as a local gauge theory of the symmetry group of Minkowski space, the Poincaré group. (In the standard Riemannian GR formulations, the torsion is *a priori* assumed to vanish, then gravity does not look much like a local spacetime symmetry gauge theory. Teleparallel geometry can be regarded as a gauge theory for translations).

We see that, when suitably formulated, gravity has both Lorentz/rotational and translational "vector potentials" which are similar to those of the Maxwell/Yang–Mills theories.

17.4. *On the affine connection and gauge theory*

The "connection" one-forms for "translations" and "Lorentz" transformations can be packaged together in a way that offers some further insight into their essential similarities and differences.

The Poincaré transformations on Minkowski spacetime

$$V^{\alpha'} = \Lambda^{\alpha'}{}_{\beta} V^{\beta} + A^{\alpha'}, \tag{152}$$

can be conveniently represented in matrix form as

$$\begin{pmatrix} V' \\ 1 \end{pmatrix} = \begin{pmatrix} \Lambda & A \\ 0 & 1 \end{pmatrix} \begin{pmatrix} V \\ 1 \end{pmatrix} = \begin{pmatrix} \Lambda V + A \\ 1 \end{pmatrix}. \tag{153}$$

Then the matrix product

$$\begin{pmatrix} \Lambda_1 & A_1 \\ 0 & 1 \end{pmatrix} \begin{pmatrix} \Lambda_2 & A_2 \\ 0 & 1 \end{pmatrix} = \begin{pmatrix} \Lambda_1\Lambda_2 & \Lambda_1 A_2 + A_1 \\ 0 & 1 \end{pmatrix}, \tag{154}$$

reflects the semi-direct product structure. This matrix representation for infinitesimal Poincaré transformations has the Lie algebra

$$\left[\begin{pmatrix} l_1 & a_1 \\ 0 & 0 \end{pmatrix}, \begin{pmatrix} l_2 & a_2 \\ 0 & 0 \end{pmatrix} \right] = \begin{pmatrix} [l_1, l_2] & l_1 a_2 - l_2 a_1 \\ 0 & 0 \end{pmatrix}. \tag{155}$$

Now a connection can be viewed as a Lie algebra valued one-form. The spacetime "translation" and "Lorentz" connections can thus be neatly packaged in terms of the above Poincaré Lie algebra matrix representation:

$$\omega := \begin{pmatrix} \Gamma & \vartheta \\ 0 & 0 \end{pmatrix}. \tag{156}$$

The associated "curvature" Lie algebra valued 2-form

$$\Omega := d\omega + \omega \wedge \omega = \begin{pmatrix} d\Gamma + \Gamma \wedge \Gamma & d\vartheta + \Gamma \wedge \vartheta \\ 0 & 0 \end{pmatrix} = \begin{pmatrix} R & T \\ 0 & 0 \end{pmatrix}, \tag{157}$$

includes the spacetime curvature and torsion 2-forms in one package. Furthermore, the Bianchi identity for this Poincaré Lie algebra curvature matrix,

$$\begin{aligned} 0 = D\Omega &:= d\Omega + \omega \wedge \Omega - \Omega \wedge \omega \\ &= \begin{pmatrix} dR + \Gamma \wedge R - R \wedge \Gamma & dT + \Gamma \wedge T - R \wedge \vartheta \\ 0 & 0 \end{pmatrix} \\ &= \begin{pmatrix} DR & DT - R \wedge \vartheta \\ 0 & 0 \end{pmatrix}, \end{aligned} \tag{158}$$

unifies the first and second spacetime Bianchi identities.

This packaging shows similarities between the connection and coframe one-forms and the curvature and torsion 2-forms, but also some clear differences inherited from the semi-direct product structure of the Poincaré group. The gauge theories of Yang–Mills[154] and Utiyama[139,140] also have the $\Omega = d\omega + \omega \wedge \omega$ and $D\Omega \equiv 0$ form, but the groups do not have a semi-direct product structure.

Although we find this formulation quite helpful for seeing how the coframe plays the role of the "vector potential for translations," we will not use it below in our treatment of the PG. For our purposes, we consider local Poincaré transformations to be Lorentz transformations of the coframe plus local spacetime diffeomorphisms.

18. Variational Principles for Dynamic Spacetime Geometry

In this section, we develop the second-order variational principle for gravitating material and internal gauge fields along with their associated Noether currents and differential identities. The spacetime is assumed to have Riemann–Cartan geometry, i.e. we are considering the Poincaré gauge theory of gravity (PG).

We are considering geometric gravity that can be regarded as a gauge theory for the Poincaré group. Several authors have considered such theories, see, e.g. Refs. 7, 8, 47, 52, 56 and 80.

We wish to consider the conserved Noether currents and differential identities as well as the field equations for dynamic spacetime geometry and gauge interactions. Here, in this section, we first work with the usual second-order type Lagrangian, since for that case certain expressions take a simpler form and the arguments are more transparent. Having established these results, we can then present more briefly the analogous first-order version which is the basis for our covariant Hamiltonian expressions.

18.1. *The Lagrangian and its variation*

Rather than beginning with the Lagrangian 4-form of the type[w]

$$\mathcal{L} = \mathcal{L}(\varphi, \vartheta^\mu, \Gamma^\alpha{}_\beta, A^p; d\varphi, d\vartheta^\mu, d\Gamma^\alpha{}_\beta, dA^p), \tag{159}$$

we take the more covariant form

$$\mathcal{L} = \mathcal{L}(\varphi, \vartheta^\mu, \Gamma^\alpha{}_\beta, A^p; D\varphi, T^\mu, R^\alpha{}_\beta, F^p), \tag{160}$$

which is no less general. Here, φ is a generic f-form source field with total covariant differential (the factor ordering is suitable for a matrix representation with φ as a row "vector")

$$D\varphi = d\varphi + \Gamma^\alpha{}_\beta \wedge \varphi\sigma_\alpha{}^\beta + A^p \wedge \varphi T_p. \tag{161}$$

The torsion 2-form, curvature 2-form and gauge field strength were given earlier (144), (139) and (62). The respective variations are

$$\delta D\varphi = D\delta\varphi + \delta\Gamma^\alpha{}_\beta \wedge \varphi\,\sigma_\alpha{}^\beta + \delta A^p \wedge \varphi\,T_p, \tag{162}$$

$$\delta T^\mu = D\delta\vartheta^\mu + \delta\Gamma^\mu{}_\nu \wedge \vartheta^\nu, \tag{163}$$

$$\delta R^\alpha{}_\beta = d\delta\Gamma^\alpha{}_\beta + \delta\Gamma^\alpha{}_\lambda \wedge \Gamma^\lambda{}_\beta + \Gamma^\alpha{}_\lambda \wedge \delta\Gamma^\lambda{}_\beta = D\delta\Gamma^\alpha{}_\beta, \tag{164}$$

$$\delta F^p = d\delta A^p + C^p{}_{qr} A^q \wedge \delta A^r = D\delta A^p. \tag{165}$$

[w]This is sufficiently general to include all the fundamental fields of the standard model, with φ including the Higgs and the fermions and A^p the $U(1) \times SU(2) \times SU(3)$ gauge vector potential.

The variation of the total Lagrangian 4-form (160) is:

$$\delta\mathcal{L} = \delta D\varphi \wedge \frac{\partial\mathcal{L}}{\partial D\varphi} + \delta T^\mu \wedge \frac{\partial\mathcal{L}}{\partial T^\mu} + \delta R^\alpha{}_\beta \wedge \frac{\partial\mathcal{L}}{\partial R^\alpha{}_\beta} + \delta F^p \wedge \frac{\partial\mathcal{L}}{\partial F^p}$$

$$+ \delta\varphi \wedge \frac{\partial\mathcal{L}}{\partial\varphi} + \delta\vartheta^\mu \wedge \frac{\partial\mathcal{L}}{\partial\vartheta^\mu} + \delta\Gamma^\alpha{}_\beta \wedge \frac{\partial\mathcal{L}}{\partial\Gamma^\alpha{}_\beta} + \delta A^p \wedge \frac{\partial\mathcal{L}}{\partial A^p}, \tag{166}$$

$$= (D\delta\varphi + \delta\Gamma^\alpha{}_\beta \wedge \varphi\sigma_\alpha{}^\beta + \delta A^p \wedge \varphi T_p) \wedge \frac{\partial\mathcal{L}}{\partial D\varphi} + \delta\varphi \wedge \frac{\partial\mathcal{L}}{\partial\varphi}$$

$$+ (D\delta\vartheta^\mu + \delta\Gamma^\mu{}_\nu \wedge \vartheta^\nu) \wedge \frac{\partial\mathcal{L}}{\partial T^\mu} + \delta\vartheta^\mu \wedge \frac{\partial\mathcal{L}}{\partial\vartheta^\mu}$$

$$+ D\delta\Gamma^\alpha{}_\beta \wedge \frac{\partial\mathcal{L}}{\partial R^\alpha{}_\beta} + \delta\Gamma^\alpha{}_\beta \wedge \frac{\partial\mathcal{L}}{\partial\Gamma^\alpha{}_\beta}$$

$$+ D\delta A^p \wedge \frac{\partial\mathcal{L}}{\partial F^p} + \delta A^p \wedge \frac{\partial\mathcal{L}}{\partial A^p}. \tag{167}$$

"Integrating by parts" and rearranging gives

$$\delta\mathcal{L} = D\left(\delta\varphi \wedge \frac{\partial\mathcal{L}}{\partial D\varphi} + \delta\vartheta^\mu \wedge \frac{\partial\mathcal{L}}{\partial T^\mu} + \delta\Gamma^\alpha{}_\beta \wedge \frac{\partial\mathcal{L}}{\partial R^\alpha{}_\beta} + \delta A^p \wedge \frac{\partial\mathcal{L}}{\partial F^p}\right)$$

$$+ \delta\varphi \wedge \left(-\varsigma D\frac{\partial\mathcal{L}}{\partial D\varphi} + \frac{\partial\mathcal{L}}{\partial\varphi}\right) + \delta\vartheta^\mu \wedge \left(D\frac{\partial\mathcal{L}}{\partial T^\mu} + \frac{\partial\mathcal{L}}{\partial\vartheta^\mu}\right)$$

$$+ \delta\Gamma^\alpha{}_\beta \wedge \left(D\frac{\partial\mathcal{L}}{\partial R^\alpha{}_\beta} + \frac{\partial\mathcal{L}}{\partial\Gamma^\alpha{}_\beta} + \varphi\sigma_\alpha{}^\beta \wedge \frac{\partial\mathcal{L}}{\partial D\varphi} + \vartheta^\beta \wedge \frac{\partial\mathcal{L}}{\partial T^\alpha}\right)$$

$$+ \delta A^p \wedge \left(D\frac{\partial\mathcal{L}}{\partial F^p} + \frac{\partial\mathcal{L}}{\partial A^p} + \varphi T_p \wedge \frac{\partial\mathcal{L}}{\partial D\varphi}\right), \tag{168}$$

which yields the conjugate momenta and Euler–Lagrange variational derivatives according to the pattern:

$$\delta\mathcal{L} = d(\delta\varphi \wedge p + \delta\vartheta^\alpha \wedge \tau_\alpha + \delta\Gamma^\alpha{}_\beta \wedge \rho_\alpha{}^\beta + \delta A^p \wedge \mathcal{P}_p)$$

$$+ \delta\varphi \wedge \frac{\delta\mathcal{L}}{\delta\varphi} + \delta\vartheta^\alpha \wedge \frac{\delta\mathcal{L}}{\delta\vartheta^\alpha} + \delta\Gamma^\alpha{}_\beta \wedge \frac{\delta\mathcal{L}}{\delta\Gamma^\alpha{}_\beta} + \delta A^p \wedge \frac{\delta\mathcal{L}}{\delta A^p}. \tag{169}$$

This is the key variational relation. According to Hamilton's principle, the second-order field equations are the vanishing of the Euler–Lagrange expressions named in (169) and explicitly displayed in (168).

18.2. *Local gauge symmetries, Noether currents and differential identities*

The Lagrangian 4-form $\mathcal{L}$ (160) should be "gauge" invariant under the local spacetime gauge transformations:

$$\Delta\varphi = l^\alpha{}_\beta\,\varphi\sigma_\alpha{}^\beta + \alpha^p\,\varphi T_p - \pounds_Z\varphi, \tag{170}$$

$$\Delta\vartheta^\mu = l^\mu{}_\nu\,\vartheta^\nu - \pounds_Z\vartheta^\mu, \tag{171}$$

$$\Delta\Gamma^\alpha{}_\beta = -Dl^\alpha{}_\beta - \pounds_Z\Gamma^\alpha{}_\beta, \tag{172}$$

$$\Delta A^p = -D\alpha^p - \pounds_Z A^p, \tag{173}$$

where the 6 $l^\alpha{}_\beta$ control an infinitesimal Lorentz rotation of the spacetime frame ϑ^α, the α^p control an "internal" gauge and Z is a spacetime vector field which determines a "local translation." The Lie derivative $\pounds_Z$ is given by $di_Z + i_Z d$ on the components of our fields. This action on the components and on the basis one-forms correctly represents the Lie derivative on geometric objects. Under (170)–(173) the Lagrangian 4-form $\mathcal{L}$ (160), if it depends on position only through the indicated fields, should change according to

$$\Delta\mathcal{L} = -\pounds_Z\mathcal{L} = -di_Z\mathcal{L}, \tag{174}$$

which happens to be a total differential because $\mathcal{L}$ is a 4-form. With the special variations given by (170)–(174) Eq. (169) should be satisfied identically.

One may collect all the total differential terms on the l.h.s. of the identity giving:

$$d\mathcal{J}(l^\alpha{}_\beta, \alpha^p, Z) \equiv \Delta\varphi \wedge \frac{\delta\mathcal{L}}{\delta\varphi} + \Delta\vartheta^\mu \wedge \frac{\delta\mathcal{L}}{\delta\vartheta^\mu} + \Delta\Gamma^\alpha{}_\beta \wedge \frac{\delta\mathcal{L}}{\delta\Gamma^\alpha{}_\beta} + \Delta A^p \wedge \frac{\delta\mathcal{L}}{\delta A^p}, \tag{175}$$

where

$$\begin{aligned}\mathcal{J}(l^\alpha{}_\beta, \alpha^p, Z) := -i_Z\mathcal{L} - \Delta\varphi \wedge \frac{\partial\mathcal{L}}{\partial D\varphi} - \Delta\vartheta^\mu \wedge \frac{\partial\mathcal{L}}{\partial T^\mu}\\ - \Delta\Gamma^\alpha{}_\beta \wedge \frac{\partial\mathcal{L}}{\partial R^\alpha{}_\beta} - \Delta A^p \wedge \frac{\partial\mathcal{L}}{\partial F^p},\end{aligned} \tag{176}$$

is the *generalized total current* 3-form. Using (170)–(173) we have in more detail (recall the canonical stress tensor (55) and $E := \frac{\partial L}{\partial \dot q}\dot q - L$)

$$\begin{aligned}\mathcal{J}(l^\alpha{}_\beta, \alpha^p, Z) := -i_Z\mathcal{L} + (\pounds_Z\varphi - l^\alpha{}_\beta\varphi\sigma_\alpha{}^\beta - \alpha^p\varphi T_p) \wedge \frac{\partial\mathcal{L}}{\partial D\varphi}\\ + (\pounds_Z\vartheta^\alpha - l^\alpha{}_\beta\vartheta^\beta) \wedge \frac{\partial\mathcal{L}}{\partial T^\alpha} + (\pounds_Z\Gamma^\alpha{}_\beta + Dl^\alpha{}_\beta) \wedge \frac{\partial\mathcal{L}}{\partial R^\alpha{}_\beta}\\ + (\pounds_Z A^p + D\alpha^p) \wedge \frac{\partial\mathcal{L}}{\partial F^p}.\end{aligned} \tag{177}$$

The second Noether theorem differential identities may be obtained from (175) by comparing the coefficients of $\alpha^p, l^\alpha{}_\beta, Z^\mu, d\alpha^p, dl^\alpha{}_\beta, dZ^\mu$ on both sides. The results will be covariant, *but* the computation will not be manifestly so — because of the Lie derivative terms. However, it so happens that the Lie derivative differs from a covariant operation only by a gauge transformation, as the following short computations reveal:

$$\pounds_Z\varphi := di_Z\varphi + i_Z d\varphi \equiv Di_Z\varphi + i_Z D\varphi - \Gamma^\alpha{}_\beta(Z)\varphi\sigma_\alpha{}^\beta - A^p(Z)\varphi T_p, \tag{178}$$

$$\pounds_Z\vartheta^\mu := di_Z\vartheta^\mu + i_Z d\vartheta^\mu \equiv Di_Z\vartheta^\mu + i_Z D\vartheta^\mu - \Gamma^\mu{}_\nu(Z)\vartheta^\nu, \tag{179}$$

$$\pounds_Z\Gamma^\alpha{}_\beta := di_Z\Gamma^\alpha{}_\beta + i_Z d\Gamma^\alpha{}_\beta \equiv i_Z R^\alpha{}_\beta + D(\Gamma^\alpha{}_\beta(Z)), \tag{180}$$

$$\pounds_Z A^p := di_Z A^p + i_Z dA^p \equiv i_Z F^p + D(A^p(Z)), \tag{181}$$

(in the last two relations on the r.h.s., the D is formally defined by treating $A^p(Z)$ and $\Gamma^\alpha{}_\beta(Z)$ as tensors). Thus the translation vector field Z induces a modification to the gauge parameters

$$l'^\alpha{}_\beta := l^\alpha{}_\beta + \Gamma^\alpha{}_\beta(Z), \quad \alpha'^p := \alpha^p + A^p(Z), \tag{182}$$

which effectively replaces the noncovariant $\pounds_Z$ by the "covariant Lie derivative" defined by

$$L_Z := Di_Z + i_Z D, \tag{183}$$

on the "normal" fields $\varphi, \vartheta^\alpha$, and by

$$L_Z\Gamma^\alpha{}_\beta := i_Z R^\alpha{}_\beta, \quad L_Z A^p := i_Z F^p, \tag{184}$$

on the connection one-forms.

The gauge transformations (170)–(173) then take the manifestly covariant form:

$$\Delta\varphi = l'^\alpha{}_\beta\varphi\sigma_\alpha{}^\beta + \alpha'^p\varphi T_p - L_Z\varphi = l'^\alpha{}_\beta\varphi\sigma_\alpha{}^\beta + \alpha'^p\varphi T_p - Di_Z\varphi - i_Z D\varphi, \tag{185}$$

$$\Delta\vartheta^\mu = l'^\mu{}_\nu\vartheta^\nu + 0 - L_Z\vartheta^\mu = l'^\mu{}_\nu\vartheta^\nu + 0 - DZ^\mu - i_Z D\vartheta^\mu, \tag{186}$$

$$\Delta\Gamma^\alpha{}_\beta = -Dl'^\alpha{}_\beta + 0 - L_Z\Gamma^\alpha{}_\beta = -Dl'^\alpha{}_\beta + 0 - i_Z R^\alpha{}_\beta, \tag{187}$$

$$\Delta A^p = 0 - D\alpha'^p - L_Z A^p = 0 - D\alpha'^p - i_Z F^p. \tag{188}$$

If one specializes to matter fields which *are not* form fields — which is the only kind of matter that we know of physically — then one can see here a striking pattern which supports our identification of the geometric gauge fields. From the r.h.s. expressions, one finds that most of the fields transform *algebraically* under gauge infinitesimal internal gauge transformations α^p; the only field that transforms with $D\alpha^p$ is the internal connection one-form (a.k.a. gauge vector potential) A^p. Similarly, most of the fields transform algebraically in $l'^\alpha{}_\beta$; the only field which transforms according to the differential $Dl'^\alpha{}_\beta$ is the spacetime connection one-form $\Gamma^\alpha{}_\beta$. Moreover (in the "physical" 0-form matter case) the only field having a differential — DZ^μ — rather than an algebraic transformation formula under the infinitesimal spacetime displacement Z^μ is the coframe one-form ϑ^μ. This is one more way of seeing that the coframe one-form can be identified as the "translational gauge vector potential."

With the above reparametrization, the generalized total current 3-form takes the form:

$$\begin{aligned}\mathcal{J}(l'^{\alpha}{}_{\beta},\alpha'^{p},Z) := -i_Z\mathcal{L} + (L_Z\varphi - l'^{\alpha}{}_{\beta}\varphi\sigma_{\alpha}{}^{\beta} - \alpha'^{p}\varphi T_p)\wedge\frac{\partial\mathcal{L}}{\partial D\varphi}\\ + (L_Z\vartheta^{\alpha} - l'^{\alpha}{}_{\beta}\vartheta^{\beta})\wedge\frac{\partial\mathcal{L}}{\partial T^{\alpha}} + (L_Z\Gamma^{\alpha}{}_{\beta} + Dl'^{\alpha}{}_{\beta})\wedge\frac{\partial\mathcal{L}}{\partial R^{\alpha}{}_{\beta}}\\ + (L_ZA^p + D\alpha'^p)\wedge\frac{\partial\mathcal{L}}{\partial F^p}.\end{aligned}\tag{189}$$

To extract the differential identities from (175) it is best to write $\mathcal{J}(l'^{\alpha}{}_{\beta},\alpha'^{p},Z)$ as terms algebraically linear in $l'^{\alpha}{}_{\beta},\alpha'^{p},Z$ plus a total differential:

$$\begin{aligned}\mathcal{J}(l'^{\alpha}{}_{\beta},\alpha'^{p},Z) := -i_Z\mathcal{L} + d\left(i_Z\varphi\wedge\frac{\partial\mathcal{L}}{\partial D\varphi} + i_Z\vartheta^{\mu}\wedge\frac{\partial\mathcal{L}}{\partial T^{\mu}} + l'^{\alpha}{}_{\beta}\frac{\partial\mathcal{L}}{\partial R^{\alpha}{}_{\beta}} + \alpha'^{p}\frac{\partial\mathcal{L}}{\partial F^p}\right)\\ + i_ZD\varphi\wedge\frac{\partial\mathcal{L}}{\partial D\varphi} + i_ZD\vartheta^{\mu}\wedge\frac{\partial\mathcal{L}}{\partial T^{\mu}} + i_ZR^{\alpha}{}_{\beta}\wedge\frac{\partial\mathcal{L}}{\partial R^{\alpha}{}_{\beta}}\\ + i_ZF^p\wedge\frac{\partial\mathcal{L}}{\partial F^p} + \varsigma i_Z\varphi\wedge D\frac{\partial\mathcal{L}}{\partial D\varphi} - i_Z\vartheta^{\mu}\wedge D\frac{\partial\mathcal{L}}{\partial T^{\mu}}\\ - l'^{\alpha}{}_{\beta}\left(D\frac{\partial\mathcal{L}}{\partial R^{\alpha}{}_{\beta}} + \varphi\sigma_{\alpha}{}^{\beta}\wedge\frac{\partial\mathcal{L}}{\partial D\varphi} + \vartheta^{\beta}\wedge\frac{\partial\mathcal{L}}{\partial T^{\alpha}}\right)\\ - \alpha'^{p}\left(D\frac{\partial\mathcal{L}}{\partial F^p} + \varphi T_p\wedge\frac{\partial\mathcal{L}}{\partial D\varphi}\right).\end{aligned}\tag{190}$$

The total differential will later be related to the total energy–momentum. For now merely note that it does not contribute to the l.h.s. of (175) as $d^2 \equiv 0$.

Internal gauge symmetry. With $Z = 0 = l'^{\alpha}{}_{\beta}$ the general result (190) reduces to the expression we considered earlier for internal gauge symmetry of the Yang–Mills type (64), and we again obtain just as before (recall the argument leading to (76) and (70)):

$$\frac{\partial\mathcal{L}}{\partial A^p} \equiv 0, \quad D\frac{\delta\mathcal{L}}{\delta A^p} + \varphi T_p\wedge\frac{\delta\mathcal{L}}{\delta\varphi} \equiv 0. \tag{191}$$

Local Lorentz gauge symmetry. In the same fashion, with $Z = 0 = \alpha'^{p}$ we obtain from (175) and (185)–(187)

$$\begin{aligned}d\mathcal{J}(l'^{\alpha}{}_{\beta},0,0) := -d\left\{l'^{\alpha}{}_{\beta}\left(\frac{\delta\mathcal{L}}{\delta\Gamma^{\alpha}{}_{\beta}} - \frac{\partial\mathcal{L}}{\partial\Gamma^{\alpha}{}_{\beta}}\right)\right\}\\ \equiv l'^{\alpha}{}_{\beta}\varphi\sigma_{\alpha}{}^{\beta}\wedge\frac{\delta\mathcal{L}}{\delta\varphi} + l'^{\alpha}{}_{\beta}\vartheta^{\beta}\wedge\frac{\delta\mathcal{L}}{\delta\vartheta^{\alpha}} + (-Dl'^{\alpha}{}_{\beta})\wedge\frac{\delta\mathcal{L}}{\delta\Gamma^{\alpha}{}_{\beta}}.\end{aligned}\tag{192}$$

Equating the coefficients of $l'^{\alpha}{}_{\beta}$ and $Dl'^{\alpha}{}_{\beta}$ (keeping in mind that $l'^{\alpha}{}_{\beta}$ is antisymmetric) leads to the algebraic and differential identities:

$$\frac{\partial\mathcal{L}}{\partial\Gamma^{\alpha}{}_{\beta}} \equiv 0, \tag{193}$$

$$D\frac{\delta\mathcal{L}}{\delta\Gamma^{[\alpha\beta]}} + \varphi\sigma_{\alpha\beta}\wedge\frac{\delta\mathcal{L}}{\delta\varphi} + \vartheta_{[\beta}\wedge\frac{\delta\mathcal{L}}{\delta\vartheta^{\alpha]}} \equiv 0. \tag{194}$$

Formally these two relations are quite similar to those found for the internal symmetries (191). Consequently local Lorentz symmetry is in certain respects rather like an internal gauge symmetry.

The conditions $\partial\mathcal{L}/\partial A^p \equiv 0 \equiv \partial\mathcal{L}/\partial\Gamma^{\alpha}{}_{\beta}$ mean (as we expected) that how the internal and Lorentz gauge potentials can appear is quite restricted — i.e. only via the covariant derivative or the associated field strength. With these conditions, we note that the generalized current has the neat form

$$\mathcal{J}(l'^{\alpha}{}_{\beta}, \alpha'^{p}, Z^{\mu}) = Z^{\mu}\mathcal{J}_{\mu} - l'^{\alpha}{}_{\beta}\frac{\delta\mathcal{L}}{\delta\Gamma^{\alpha}{}_{\beta}} - \alpha'^{p}\frac{\delta\mathcal{L}}{\delta A^{p}} + d\mathcal{B}. \tag{195}$$

Local diffeomorphism invariance. This was already considered for general theories in first-order form in Sec. 14.1. Here, we consider local translations in second-order form while distinguishing between the source, gauge and geometric variables. This case again has some similarities to the internal and Lorentz symmetries but also some striking differences. With $l'^{\alpha}{}_{\beta} = 0 = \alpha'^{p}$ in (175) we have

$$\begin{aligned} d\mathcal{J}(0,0,Z^{\mu}) &= d(Z^{\mu}\mathcal{J}_{\mu}) = DZ^{\mu}\wedge\mathcal{J}_{\mu} + Z^{\mu}D\mathcal{J}_{\mu} \\ &\equiv -(Di_Z\varphi + i_Z D\varphi)\wedge\frac{\delta\mathcal{L}}{\delta\varphi} - (Di_Z\vartheta^{\mu} + i_Z D\vartheta^{\mu})\wedge\frac{\delta\mathcal{L}}{\delta\vartheta^{\mu}} \\ &\quad - i_Z R^{\alpha}{}_{\beta}\wedge\frac{\delta\mathcal{L}}{\delta\Gamma^{\alpha}{}_{\beta}} - i_Z F^{p}\wedge\frac{\delta\mathcal{L}}{\delta A^{p}}. \end{aligned} \tag{196}$$

The coefficient of Z^{μ} on both sides of (196) is the differential identity associated with translation invariance (related to energy–momentum) while the coefficient of DZ^{μ} provides a new algebraic expression for $\mathcal{J}_{\mu}$ in terms of the Euler–Lagrange variations:

$$\mathcal{J}_{\mu} \equiv -\varphi_{\mu}\wedge\frac{\delta\mathcal{L}}{\delta\varphi} - \frac{\delta\mathcal{L}}{\delta\vartheta^{\mu}}. \tag{197}$$

For ordinary (i.e. 0-form valued) matter fields the first term vanishes, leading to the especially simple and intuitively reasonable relation:

$$\mathcal{J}_{\mu} \equiv -\frac{\delta\mathcal{L}}{\delta\vartheta^{\mu}}. \tag{198}$$

We emphasize that such a neat formula for the translational current 3-form is only possible if one regards the coframe as a dynamical variable. It is also noteworthy that this remarkable relation — which does not restrict how the translation gauge potential appears in the action — was obtained from the coefficient of DZ^{μ}, whereas the corresponding relation in both the internal and local Lorentz cases put quite severe limits, viz. (191a), (193), on how those gauge potentials could appear.

The identity (197) permits $\mathcal{J}(l'^{\alpha}{}_{\beta}, \alpha'^{p}, Z)$ (195) to be written in the nice symmetrical form

$$\mathcal{J}(l'^{\alpha}{}_{\beta}, \alpha'^{p}, Z) \equiv -i_Z\varphi \wedge \frac{\delta\mathcal{L}}{\delta\varphi} - i_Z\vartheta^{\mu}\frac{\delta\mathcal{L}}{\delta\vartheta^{\mu}} - l'^{\alpha}{}_{\beta}\frac{\delta\mathcal{L}}{\delta\Gamma^{\alpha}{}_{\beta}} - \alpha'^{p}\frac{\delta\mathcal{L}}{\delta A^{p}} + d\left(i_Z\varphi \wedge \frac{\partial\mathcal{L}}{\partial D\varphi} + i_Z\vartheta^{\mu}\frac{\partial\mathcal{L}}{\partial T^{\mu}} + l'^{\alpha}{}_{\beta}\frac{\partial\mathcal{L}}{\partial R^{\alpha}{}_{\beta}} + \alpha'^{p}\frac{\partial\mathcal{L}}{\partial F^{p}}\right). \tag{199}$$

Restricting to the case where the source fields are 0-forms, we have

$$\mathcal{J}(l'^{\alpha}{}_{\beta}, \alpha'^{p}, Z) \equiv -Z^{\mu}\frac{\delta\mathcal{L}}{\delta\vartheta^{\mu}} - l'^{\alpha}{}_{\beta}\frac{\delta\mathcal{L}}{\delta\Gamma^{\alpha}{}_{\beta}} - \alpha'^{p}\frac{\delta\mathcal{L}}{\delta A^{p}} + d\left(Z^{\mu}\frac{\partial\mathcal{L}}{\partial T^{\mu}} + l'^{\alpha}{}_{\beta}\frac{\partial\mathcal{L}}{\partial R^{\alpha}{}_{\beta}} + \alpha'^{p}\frac{\partial\mathcal{L}}{\partial F^{p}}\right). \tag{200}$$

This result displays the purely gauge nature of the current — it is especially noteworthy that there is no explicit appearance of the source field or its field equation.

Note that if we take the variational derivatives as field equations, the numerical value of $\mathcal{J}$ will be entirely from the total differential term. When the 3-form $\mathcal{J}$ is integrated over a 3D region this total differential becomes, via the fundamental boundary theorem, (34), an integral over the 2D boundary of the region. In other words, the value is *quasi-local.*

Our results here are an application of Noether's ideas. As expected from her 2nd theorem, with a local symmetry the conserved current becomes a differential identity. Furthermore, we also displayed here detailed results that exactly reflect her remarks about verifying and generalizing Hilbert's assertion regarding the lack of a proper energy–momentum. Our generalized current expressions (199) and (200) are linear combinations of Euler–Lagrange expressions plus a total differential. They have the usual conserved current ambiguity: The total differential does not contribute to the conservation law, it can be adjusted without affecting the conservation property, nevertheless it affects the value of the associated conserved quantity. The second-order Lagrangian formalism has no way to control this ambiguity.

18.3. *Interpretation of the differential identities*

Let us now specialize and consider the customary "minimal coupled" decomposition:

$$\mathcal{L}_{\text{total}} = \mathcal{L}_{\theta\Gamma}(\cdot, \vartheta^{\mu}, \cdot, \cdot; \cdot, T^{\mu}, R^{\alpha}{}_{\beta}, \cdot) + \mathcal{L}_{A}(\cdot, \vartheta^{\mu}, \cdot, \cdot; \cdot, \cdot, \cdot, F^{p}) + \mathcal{L}_{\varphi}(\varphi, \vartheta^{\mu}, \cdot, \cdot; D\varphi, \cdot, \cdot, \cdot). \tag{201}$$

Each of these separate Lagrangian pieces is a scalar valued 4-form. So, the Noether identities we have obtained can be applied to each piece. However, it must be kept in mind that for the separate pieces our variational derivatives are, in general, no longer field equations. There is, however, one exception:

$$\frac{\delta \mathcal{L}_\varphi}{\delta \varphi} = \frac{\delta \mathcal{L}_{\text{total}}}{\delta \varphi}, \tag{202}$$

which vanishes "on shell." It should also be noted that for each separate piece of $\mathcal{L}_{\text{total}}$ many of the terms in the various identities (191), (194), (196) and (197) will vanish trivially. Consider (191): for $\mathcal{L}_{\vartheta\Gamma}$ it is trivial; for $\mathcal{L}_A$ and $\mathcal{L}_\varphi$ it reduces to the results obtained earlier (79).

Let us introduce some suitable names for certain expressions. Specifically for the material source, the *energy–momentum* and *spin density* 3-forms are defined respectively by

$$\mathfrak{T}^\varphi_\mu := \frac{\partial \mathcal{L}_\varphi}{\partial \vartheta^\mu}, \quad \mathfrak{S}^{\beta\alpha}_\varphi := 2\frac{\partial \mathcal{L}_\varphi}{\partial \Gamma_{\alpha\beta}}, \tag{203}$$

with an analogous expression defining $\mathfrak{T}^A_\mu$.

Then (194) for $\mathcal{L}_\varphi$ and $\mathcal{L}_A$ becomes, respectively,

$$D\mathfrak{S}^{\alpha\beta}_\varphi + \vartheta^\alpha \wedge \mathfrak{T}^\beta_\varphi - \vartheta^\beta \wedge \mathfrak{T}^\alpha_\varphi \equiv 2\varphi\sigma^{\alpha\beta} \wedge \frac{\delta \mathcal{L}_\varphi}{\delta \varphi} = 0 \quad \text{(on shell)}, \tag{204}$$

$$\vartheta^\alpha \wedge \mathfrak{T}^\beta_A - \vartheta^\beta \wedge \mathfrak{T}^\alpha_A \equiv 0. \tag{205}$$

The physical interpretation of these concerns angular momentum conservation (or more precisely the exchange of angular momentum with the gravitational field). The first term in (204) is the (covariant) divergence of the source spin density, the next two terms (the anti-symmetric part of the energy–momentum density) describe the change in "orbital" angular momentum. The second relation is an identity which shows that gauge fields have symmetric energy–momentum densities and trivial angular momentum conservation relations. For $\mathcal{L}_{\vartheta\Gamma}$ (194) reduces to the identity

$$-R^\gamma{}_\alpha \wedge \frac{\partial \mathcal{L}_{\vartheta\Gamma}}{\partial R^{\gamma\beta}} - R^\gamma{}_\beta \wedge \frac{\partial \mathcal{L}_{\vartheta\Gamma}}{\partial R^{\alpha\gamma}} + \vartheta_{[\beta} \wedge \left(D\frac{\partial \mathcal{L}_{\vartheta\Gamma}}{\partial T^{\alpha]}} + \frac{\partial \mathcal{L}_{\vartheta\Gamma}}{\partial \vartheta^{\alpha]}} \right) \equiv 0, \tag{206}$$

which is satisfied iff $\mathcal{L}_{\vartheta\Gamma}$ is a local Lorentz scalar.

Now let us consider the differential identity part of (196). For a 0-form material source field, it takes the form

$$D\mathfrak{T}^\varphi_\mu - i_{e_\mu} T^\nu \wedge \mathfrak{T}^\varphi_\nu - \frac{1}{2} i_{e_\mu} R^{\alpha\beta} \wedge \mathfrak{S}_{\beta\alpha} \equiv D_\mu \varphi \frac{\delta \mathcal{L}_\varphi}{\delta \varphi} = 0 \quad \text{(on shell)}. \tag{207}$$

When the source field equation is satisfied, this relation describes the exchange of material energy–momentum with the gravitational field. The analogous expression for the gauge field Lagrangian is a little simpler:

$$D\mathfrak{T}^A_\mu - i_{e_\mu} T^\nu \wedge \mathfrak{T}^A_\nu + i_{e_\mu} F^p \wedge D\frac{\partial \mathcal{L}_A}{\partial F^p} \equiv 0, \tag{208}$$

the interpretation is similar.

Finally, for the differential identity of (196) applied to $\mathcal{L}_{\vartheta\Gamma}$, after some straightforward calculation, we have the identity

$$D\frac{\partial\mathcal{L}_{\vartheta\Gamma}}{\partial\vartheta^\mu} - i_{e_\mu}T^\nu \wedge \left(D\frac{\partial\mathcal{L}_{\vartheta\Gamma}}{\partial T^\nu} + \frac{\partial\mathcal{L}_{\vartheta\Gamma}}{\partial\vartheta^\nu}\right)$$
$$- i_{e_\mu}R^{\alpha\beta} \wedge D\frac{\partial\mathcal{L}_{\vartheta\Gamma}}{\partial R^{\alpha\beta}} - DT^\alpha \wedge i_{e_\mu}\frac{\partial\mathcal{L}_{\vartheta\Gamma}}{\partial T^\alpha} \equiv 0, \tag{209}$$

which is satisfied if $\mathcal{L}_{\vartheta\Gamma}$ is a scalar valued 4-form constructed out of the coframe, torsion and curvature. This is the PG identity that plays a role analogous to that of the contracted Bianchi identity in GR.

To obtain these detailed results, we used the definition of the various Euler–Lagrange expressions given in (168) and (169).

19. First-Order Form and the Hamiltonian

Here, for our general PG with generic matter and gauge sources, we briefly present the first-order form along with the associated Noether currents and differential identities and then the associated covariant Hamiltonian including our preferred boundary term which yields our quasi-local quantities.

19.1. *First-order Lagrangian and local gauge symmetries*

For certain purposes we find a first-order formulation convenient. The *first-order form* of our variational principle for geometry and gauge is

$$\mathcal{L}^{1\text{st}} = D\varphi \wedge p + D\vartheta^\alpha \wedge \tau_\alpha + R^{\alpha\beta} \wedge \rho_{\alpha\beta} + F^p \wedge \mathcal{P}_p$$
$$- \Lambda(\varphi, \vartheta^\alpha, \Gamma^{\alpha\beta}, A^p; p, \tau_\alpha, \rho_{\alpha\beta}, \mathcal{P}_p), \tag{210}$$

with $\Gamma^{\alpha\beta}$, $R^{\alpha\beta}$ and $\rho_{\alpha\beta}$ being *a priori* antisymmetric. The variation takes the pattern

$$\delta\mathcal{L}^{1\text{st}} =: d(\delta\varphi \wedge p + \delta\vartheta^\alpha \wedge \tau_\alpha + \delta\Gamma^{\alpha\beta} \wedge \rho_{\alpha\beta} + \delta A^p \wedge \mathcal{P}_p)$$
$$+ \delta\varphi \wedge \frac{\delta\mathcal{L}^{1\text{st}}}{\delta\varphi} + \delta\vartheta^\alpha \wedge \frac{\delta\mathcal{L}^{1\text{st}}}{\delta\vartheta^\alpha} + \delta\Gamma^{\alpha\beta} \wedge \frac{\delta\mathcal{L}^{1\text{st}}}{\delta\Gamma^{\alpha\beta}} + \delta A^p \wedge \frac{\delta\mathcal{L}^{1\text{st}}}{\delta A^p}$$
$$+ \frac{\delta\mathcal{L}^{1\text{st}}}{\delta p} \wedge \delta p + \frac{\delta\mathcal{L}^{1\text{st}}}{\delta\tau_\alpha} \wedge \delta\tau^\alpha + \frac{\delta\mathcal{L}^{1\text{st}}}{\delta\rho_{\alpha\beta}} \wedge \delta\rho_{\alpha\beta} + \frac{\delta\mathcal{L}^{1\text{st}}}{\delta\mathcal{P}_p} \wedge \delta\mathcal{P}_p, \tag{211}$$

where

$$\frac{\delta\mathcal{L}^{1\text{st}}}{\delta\varphi} = -\varsigma Dp - \frac{\partial\Lambda}{\partial\varphi}, \quad \frac{\delta\mathcal{L}^{1\text{st}}}{\delta p} = D\varphi - \frac{\partial\Lambda}{\partial p}, \tag{212}$$

$$\frac{\delta\mathcal{L}^{1\text{st}}}{\delta\vartheta^\alpha} = D\tau_\alpha - \frac{\partial\Lambda}{\partial\vartheta^\alpha}, \quad \frac{\delta\mathcal{L}^{1\text{st}}}{\delta\tau_\alpha} = D\vartheta^\alpha - \frac{\partial\Lambda}{\partial\tau_\alpha}, \tag{213}$$

$$\frac{\delta\mathcal{L}^{1\mathrm{st}}}{\delta\Gamma^{\alpha\beta}} = D\rho_{\alpha\beta} - \frac{\partial\Lambda}{\partial\Gamma^{\alpha\beta}} + \varphi\sigma_{\alpha\beta}p + \vartheta_{[\beta}\wedge\tau_{\alpha]}, \quad \frac{\delta\mathcal{L}^{1\mathrm{st}}}{\delta\rho_{\alpha\beta}} = R^{\alpha\beta} - \frac{\partial\Lambda}{\partial\rho_{\alpha\beta}}, \tag{214}$$

$$\frac{\delta\mathcal{L}^{1\mathrm{st}}}{\delta A^p} = D\mathcal{P}_p - \frac{\partial\Lambda}{\partial A^p} + \varphi T_p p, \quad \frac{\delta\mathcal{L}^{1\mathrm{st}}}{\delta\mathcal{P}_p} = F^p - \frac{\partial\Lambda}{\partial\mathcal{P}_p}. \tag{215}$$

It is instructive to compare these (and subsequent) relations with the corresponding ones in the previous section. Here, we have twice as many fields and thus twice as many Euler–Lagrange expressions, but, on the other hand, the Euler–Lagrange expressions are all linear and much simpler.

The first-order formulation is convenient for imposing certain *constraints.* In particular in order to impose one of the conditions

$$R^\alpha{}_\beta \equiv 0 \quad \text{teleparallel connection,} \tag{216}$$

$$T^\alpha \equiv 0 \quad \text{symmetric connection,} \tag{217}$$

in the first-order formalism we need merely take the potential Λ to be independent of the corresponding conjugate momenta. The momentum then functions as a Lagrange multiplier which imposes the constraint. The related "coordinate" equation then loses its dynamical significance and instead becomes a relation for determining the multiplier.

19.2. *Generalized Hamiltonian and differential identities*

To obtain the Noether differential identities in this mode we need, in addition to (185)–(188), the gauge transformations of the conjugate momenta:

$$\Delta p = -l'^{\alpha\beta}\sigma_{\alpha\beta}p - \alpha'^p T_p p - L_Z p, \tag{218}$$

$$\Delta\tau_\beta = -l'^\alpha{}_\beta\tau_\alpha - L_Z\tau_\beta, \tag{219}$$

$$\Delta\rho_\alpha{}^\beta = l'^\beta{}_\gamma\rho_\alpha{}^\gamma - l'^\gamma{}_\alpha\rho_\gamma{}^\beta - L_Z\rho_\alpha{}^\beta, \tag{220}$$

$$\Delta\mathcal{P}_q = -\mathcal{P}_p C^p{}_{qr}\,\alpha'^r - L_Z\mathcal{P}_q, \tag{221}$$

where $L_Z = Di_Z + i_Z D$. These results were deduced from relations like

$$\Delta(p\wedge D\varphi) = \Delta p\wedge D\varphi + p\wedge\Delta(D\varphi), \tag{222}$$

using the fact that $p\wedge D\varphi$ is a scalar valued 4-form.

As before $\Delta\mathcal{L}$ is a total differential: $-di_Z\mathcal{L}$. Taking the total differential terms to the l.h.s. in (211) gives

$$\begin{aligned} &d\mathcal{H}(l'^\alpha{}_\beta, \alpha'^p, Z) \\ &\quad\equiv \Delta\varphi\wedge\frac{\delta\mathcal{L}}{\delta\varphi}^{1\mathrm{st}} + \Delta\vartheta^\alpha\wedge\frac{\delta\mathcal{L}^{1\mathrm{st}}}{\delta\vartheta^\alpha} + \Delta\Gamma^{\alpha\beta}\wedge\frac{\delta\mathcal{L}^{1\mathrm{st}}}{\delta\Gamma_{\alpha\beta}} + \Delta A^p\wedge\frac{\delta\mathcal{L}^{1\mathrm{st}}}{\delta A^p} \\ &\qquad + \frac{\delta\mathcal{L}}{\delta p}^{1\mathrm{st}}\wedge\Delta p + \frac{\delta\mathcal{L}^{1\mathrm{st}}}{\delta\tau_\alpha}\wedge\Delta\tau_\alpha + \frac{\delta\mathcal{L}^{1\mathrm{st}}}{\delta\rho_{\alpha\beta}}\wedge\Delta\rho_{\alpha\beta} + \frac{\delta\mathcal{L}^{1\mathrm{st}}}{\delta\mathcal{P}_p}\wedge\Delta\mathcal{P}_p, \end{aligned} \tag{223}$$

where the first-order "generalized current" is

$$\begin{aligned}
&\mathcal{H}(l'^{\alpha}{}_{\beta}, \alpha'^{p}, Z) \\
&\quad := -i_Z\mathcal{L}^{1\text{st}} - \Delta\varphi \wedge p - \Delta\vartheta^{\alpha} \wedge \tau_{\alpha} - \Delta\Gamma^{\alpha\beta} \wedge \rho_{\alpha\beta} - \Delta A^{p} \wedge \mathcal{P}_{p}, && (224)\\
&\quad = i_Z\Lambda - i_Z(D\varphi \wedge p + D\vartheta^{\alpha} \wedge \tau_{\alpha} + R^{\alpha\beta} \wedge \rho_{\alpha\beta} + F^{p} \wedge \mathcal{P}_{p}) \\
&\qquad + (L_Z\varphi - l'^{\alpha}{}_{\beta}\varphi\sigma_{\alpha}{}^{\beta} - \alpha'^{p}\varphi T_{p}) \wedge p + (L_Z\vartheta^{\alpha} - l'^{\alpha}{}_{\beta}\vartheta^{\beta}) \wedge \tau_{\alpha} \\
&\qquad + (L_Z\Gamma^{\alpha\beta} + Dl'^{\alpha\beta}) \wedge \rho_{\alpha\beta} + (L_Z A^{p} + D\alpha'^{p}) \wedge \mathcal{P}_{p}, && (225)\\
&\quad = i_Z\Lambda + \varsigma D\varphi \wedge i_Z p - D\vartheta^{\alpha} \wedge i_Z\tau_{\alpha} - R^{\alpha\beta} \wedge i_Z\rho_{\alpha\beta} - F^{p} \wedge i_Z\mathcal{P}_{p} \\
&\qquad + (Di_Z\varphi - l'^{\alpha}{}_{\beta}\varphi\sigma_{\alpha}{}^{\beta} - \alpha'^{p}\varphi T_{p}) \wedge p + (Di_Z\vartheta^{\alpha} - l'^{\alpha}{}_{\beta}\vartheta^{\beta}) \wedge \tau_{\alpha} \\
&\qquad + Dl'^{\alpha\beta} \wedge \rho_{\alpha\beta} + D\alpha'^{p} \wedge \mathcal{P}_{p}, && (226)\\
&\quad = i_Z\Lambda + \varsigma D\varphi \wedge i_Z p - D\vartheta^{\alpha} \wedge i_Z\tau_{\alpha} - R^{\alpha\beta} \wedge i_Z\rho_{\alpha\beta} - F^{p} \wedge i_Z\mathcal{P}_{p} \\
&\qquad + \varsigma i_Z\varphi \wedge Dp - i_Z\vartheta^{\alpha} D\tau_{\alpha} \\
&\qquad - l'^{\alpha\beta}(D\rho_{\alpha\beta} + \varphi\sigma_{\alpha\beta} \wedge p + \vartheta_{[\beta} \wedge \tau_{\alpha]}) - \alpha'^{p}(D\mathcal{P}_{p} + \varphi T_{p} \wedge p) \\
&\qquad + D(i_Z\varphi \wedge p + i_Z\vartheta^{\alpha}\tau_{\alpha} + l'^{\alpha\beta}\rho_{\alpha\beta} + \alpha'^{p}\mathcal{P}_{p}). && (227)
\end{aligned}$$

This expression is not just a Noether current, it is, as we already justified quite generally in Sec. 14.2, *the (generalized) Hamiltonian* 3-*form*, i.e. the canonical generator for internal, local Lorentz *and* local spacetime displacements (which includes the time evolution for any choice of time). Nevertheless, like the second-order current, one still has the various differential identities.

Local internal gauge symmetry. Equating coefficients of α'^{p} and $D\alpha'^{p}$ in (223) gives the algebraic and differential identities:

$$\frac{\partial\Lambda}{\partial A^{p}} \equiv 0, \quad -D\frac{\delta\mathcal{L}^{1\text{st}}}{\delta A^{q}} \equiv \varphi T_{q} \wedge \frac{\delta\mathcal{L}^{1\text{st}}}{\delta\varphi} - \frac{\delta\mathcal{L}^{1\text{st}}}{\delta p} \wedge T_{q}p - \frac{\delta\mathcal{L}^{1\text{st}}}{\delta\mathcal{P}^{p}} \wedge \mathcal{P}_{r}C^{r}{}_{pq}, \tag{228}$$

which can be compared with (191). The significance is the same, but now the r.h.s. includes also Euler–Lagrange variations w.r.t. momentum variables. If these equations are imposed, the expression has the same form as (191).

Local Lorentz symmetry. Equating coefficients of $l'^{\alpha\beta}$ and $Dl'^{\alpha\beta}$ gives the algebraic and differential identities:

$$\frac{\partial\Lambda}{\partial\Gamma^{\alpha\beta}} \equiv 0, \tag{229}$$

$$\begin{aligned}
-D\frac{\delta\mathcal{L}^{1\text{st}}}{\delta\Gamma^{\alpha\beta}} \equiv \varphi\sigma_{[\alpha\beta]} \wedge \frac{\delta\mathcal{L}^{1\text{st}}}{\delta\varphi} + \vartheta_{[\beta} \wedge \frac{\delta\mathcal{L}^{1\text{st}}}{\delta\vartheta^{\alpha]}} - \frac{\delta\mathcal{L}^{1\text{st}}}{\delta p} \wedge \sigma_{[\alpha\beta]}p \\
+ \frac{\delta\mathcal{L}^{1\text{st}}}{\delta\tau^{[\alpha}} \wedge \tau_{\beta]} + \frac{\delta\mathcal{L}^{1\text{st}}}{\delta\rho_{\gamma}{}^{\beta}} \wedge \rho_{\alpha\gamma} - \frac{\delta\mathcal{L}^{1\text{st}}}{\delta\rho_{\gamma}{}^{\alpha}} \wedge \rho_{\beta\gamma},
\end{aligned} \tag{230}$$

which should be compared with (193) and (194). The significance is the same, but now the r.h.s. of the latter includes also several Euler–Lagrange variations w.r.t.

momentum variables. If these relations are imposed, the expression has the same form as (194).

Local translation symmetry. With the decomposition $\mathcal{H}(0,0,Z^\mu) = Z^\mu \mathcal{H}_\mu + dB$,

$$\begin{aligned} &D\mathcal{H}(0,0,Z^\mu) \\ &\quad = DZ^\mu \wedge \mathcal{H}_\mu + Z^\mu D\mathcal{H}_\mu \\ &\quad \equiv -L_Z\varphi \wedge \frac{\delta\mathcal{L}^{1\text{st}}}{\delta\varphi} - L_Z\vartheta^\alpha \wedge \frac{\delta\mathcal{L}^{1\text{st}}}{\delta\vartheta^\alpha} - L_Z\Gamma^{\alpha\beta} \wedge \frac{\delta\mathcal{L}^{1\text{st}}}{\delta\Gamma^{\alpha\beta}} - L_Z A^p \wedge \frac{\delta\mathcal{L}^{1\text{st}}}{\delta A^p} \\ &\qquad - \frac{\delta\mathcal{L}^{1\text{st}}}{\delta p} \wedge L_Z p - \frac{\delta\mathcal{L}^{1\text{st}}}{\delta\tau_\alpha} \wedge L_Z\tau_\alpha - \frac{\delta\mathcal{L}^{1\text{st}}}{\delta\rho_{\alpha\beta}} \wedge L_Z\rho_{\alpha\beta} - \frac{\delta\mathcal{L}^{1\text{st}}}{\delta\mathcal{P}_p} \wedge L_Z\mathcal{P}_p. \end{aligned} \tag{231}$$

From the coefficient of Z^μ, we obtain a differential identity involving $\mathcal{H}_\mu$; this relation includes the conservation of energy–momentum.

We also get a new algebraic identity giving $\mathcal{H}_\mu$ in terms of variational derivatives: (compare (197)):

$$\begin{aligned} \mathcal{H}_\mu \equiv -\varphi_\mu \wedge \frac{\delta\mathcal{L}^{1\text{st}}}{\delta\varphi} - \frac{\delta\mathcal{L}^{1\text{st}}}{\delta\vartheta^\mu} + \varsigma\frac{\delta\mathcal{L}^{1\text{st}}}{\delta p} \wedge p_\mu - \frac{\delta\mathcal{L}^{1\text{st}}}{\delta\tau_\alpha} \wedge \tau_{\alpha\mu} \\ - \frac{\delta\mathcal{L}^{1\text{st}}}{\delta\rho_{\alpha\beta}} \wedge \rho_{\alpha\beta\mu} - \frac{\delta\mathcal{L}^{1\text{st}}}{\delta\mathcal{P}_p} \wedge \mathcal{P}_{p\mu}. \end{aligned} \tag{232}$$

When the momentum relations are imposed, this reduces to the corresponding expression for the second-order Noether translational current (197). Inserting this new result into (227) gives an expression for the Hamiltonian 3-form in terms of variational coefficients (which has the same form as (199) if the momentum relations are imposed)

$$\begin{aligned} &\mathcal{H}(l'^{\alpha\beta}, \alpha'^p, Z^\mu) \\ &\quad \equiv -i_Z\varphi \wedge \frac{\delta\mathcal{L}^{1\text{st}}}{\delta\varphi} - i_Z\vartheta^\mu \frac{\delta\mathcal{L}^{1\text{st}}}{\delta\vartheta^\mu} - l'^{\alpha\beta}\frac{\delta\mathcal{L}^{1\text{st}}}{\delta\Gamma^{\alpha\beta}} - \alpha'^p\frac{\delta\mathcal{L}^{1\text{st}}}{\delta A^p} \\ &\qquad + \varsigma\frac{\delta\mathcal{L}^{1\text{st}}}{\delta p} \wedge i_Z p - \frac{\delta\mathcal{L}^{1\text{st}}}{\delta\tau_\alpha} \wedge i_Z\tau_\alpha - \frac{\delta\mathcal{L}^{1\text{st}}}{\delta\rho_{\alpha\beta}} \wedge i_Z\rho_{\alpha\beta} - \frac{\delta\mathcal{L}^{1\text{st}}}{\delta\mathcal{P}_p} \wedge i_Z\mathcal{P}_p \\ &\qquad + D(i_Z\varphi \wedge p + i_Z\vartheta^\alpha\tau_\alpha + l'^{\alpha\beta}\rho_{\alpha\beta} + \alpha'^p\mathcal{P}_p). \end{aligned} \tag{233}$$

This generalized current expression is again an example of applying Noether's analysis. In accord with the second theorem, for local symmetries, the current conservation expression becomes a differential identity. Again, we have a detailed expression that reflects Noether's remarks regarding a more general version of Hilbert's assertion concerning the absence of a proper energy–momentum conservation law. Moreover there is again the Noether current ambiguity regarding the total differential term. However, as explained in Sec. 14, the first-order current is also the Hamiltonian, the canonical generator of local transformations including spacetime displacements. The total differential (boundary) term in the Hamiltonian can and should be adjusted, as we have discussed in general terms earlier in Sec. 14.4. Then, when varied, the chosen boundary term in the variation of the Hamiltonian gives the associated boundary conditions. Thereby the usual Noether current ambiguity is fixed by the chosen boundary condition.

19.3. *General geometric Hamiltonian boundary terms*

Specializing our general Hamiltonian boundary term expression (120) to our present variables, with the preferred choice for the material and internal gauge fields leads to boundary term expressions which explicitly contain only the geometric variables:

$$\mathcal{B}(Z) = i_Z \begin{Bmatrix} \vartheta^\alpha \\ \bar{\vartheta}^\alpha \end{Bmatrix} \Delta\tau_\alpha + \Delta\vartheta^\alpha \wedge i_Z \begin{Bmatrix} \tau_\alpha \\ \bar{\tau}_\alpha \end{Bmatrix} + i_Z \begin{Bmatrix} \Gamma^\alpha{}_\beta \\ \bar{\Gamma}^\alpha{}_\beta \end{Bmatrix} \Delta\rho_\alpha{}^\beta + \Delta\Gamma^\alpha{}_\beta \wedge i_Z \begin{Bmatrix} \rho_\alpha{}^\beta \\ \bar{\rho}_\alpha{}^\beta \end{Bmatrix}, \tag{234}$$

where the upper or lower line in each bracket is to be selected. A special case of this expression, (upper, lower, upper, upper) with $\bar{\tau}_\alpha = 0$, was first proposed in 1991.[85] With the above boundary term, the total differential term in $\delta\mathcal{H}(Z)$ has the symplectic form

$$\mathcal{C}(Z) = \begin{Bmatrix} i_Z\delta\vartheta^\alpha \wedge \Delta\tau_\alpha \\ -i_Z\Delta\vartheta^\alpha \wedge \delta\tau_\alpha \end{Bmatrix} + \begin{Bmatrix} \Delta\vartheta^\alpha \wedge i_Z\delta\tau_\alpha \\ -\delta\vartheta^\alpha \wedge i_Z\Delta\tau_\alpha \end{Bmatrix} + \begin{Bmatrix} i_Z\delta\Gamma^\alpha{}_\beta \wedge \Delta\rho_\alpha{}^\beta \\ -i_Z\Delta\Gamma^\alpha{}_\beta \wedge \delta\rho_\alpha{}^\beta \end{Bmatrix} + \begin{Bmatrix} \Delta\Gamma^\alpha{}_\beta \wedge i_Z\delta\rho_\alpha{}^\beta \\ -\delta\Gamma^\alpha{}_\beta \wedge i_Z\Delta\rho_\alpha{}^\beta \end{Bmatrix}. \tag{235}$$

The general geometric Hamiltonians evolve gauge dependent quantities, including the frame and connection coefficients. Consequently they naturally include terms that are gauge dependent. These terms are in particular those with factors of $i_Z\Gamma^\alpha{}_\beta$. However, the value of the Hamiltonian boundary term will then include a contribution to the energy, etc. due to the choice of frame gauge. From such a boundary term, one could still get the correct physical value, but only if one takes on the boundary the particular frame gauge $\pounds_Z\vartheta^\alpha = 0$, which means one may need a different frame for the energy–momentum and angular momentum components. For the purposes of obtaining directly a physical value for the observable quantities, one must separate the frame gauge dependent term into a physical energy plus a gauge dependent unphysical energy. This issue was first considered by Hecht,[48,49] and he discovered a suitable remedy. The frame gauge dependent part can be separated using the identity

$$\pounds_Z\vartheta^\alpha \equiv di_Z\vartheta^\alpha + i_Zd\vartheta^\alpha \equiv Di_Z\vartheta^\alpha + i_ZD\vartheta^\alpha - i_Z\Gamma^\alpha{}_\beta\vartheta^\beta. \tag{236}$$

With the aid of this relation one can get frame gauge independent boundary terms for the quasi-local quantities. Thus, to drop the contribution from the frame gauge, one should make the replacement

$$i_Z\Gamma^\alpha{}_\beta \equiv D_\beta Z^\alpha + (i_Z T^\alpha)_\beta - (\pounds_Z \vartheta^\alpha)_\beta \to \tilde{D}_\beta Z^\alpha, \tag{237}$$

where we have introduced the convenient *transposed* connection:

$$\tilde{D}Z^\alpha := DZ^\alpha + i_Z T^\alpha, \tag{238}$$

which naturally shows up whenever one expresses Lie derivative expressions in terms of a covariant derivative. In addition to our argument above, this replacement has been justified using (i) theoretical analysis,[48,49] (ii) calculations for exact solutions,[50] (iii) holonomic variables (for GR)[22] and (iv) via a manifestly covariant Hamiltonian formulation.[87] In our presentation here, we used the coframe both for convenience and for its local translational gauge role; the reference just cited provides a completely frame independent — manifestly covariant — alternative approach to the covariant Hamiltonian formalism.

19.4. *Quasi-local boundary terms*

With the aforementioned replacement, we get our general set of symplectic quasi-local quantity boundary terms for the PG:

$$\begin{aligned}\mathcal{B}(Z) = {} & i_Z \begin{Bmatrix} \vartheta^\alpha \\ \bar{\vartheta}^\alpha \end{Bmatrix} \Delta\tau_\alpha + \Delta\vartheta^\alpha \wedge i_Z \begin{Bmatrix} \tau_\alpha \\ \bar{\tau}_\alpha \end{Bmatrix} \\ & + \begin{Bmatrix} \tilde{D}_\beta Z^\alpha \\ \bar{\tilde{D}}_\beta Z^\alpha \end{Bmatrix} \Delta\rho_\alpha{}^\beta + \Delta\Gamma^\alpha{}_\beta \wedge i_Z \begin{Bmatrix} \rho_\alpha{}^\beta \\ \bar{\rho}_\alpha{}^\beta \end{Bmatrix},\end{aligned} \tag{239}$$

where the upper or lower line in each bracket is to be selected. As in thermodynamics, there are several kinds of "energy," each corresponds to the work done in a different (ideal) physical process.[63,69]

19.5. *A preferred choice*

The cases of the PG that have been studied are mostly those where the first-order potential is quadratic in the momentum fields, this leads to a linear relation between the momenta and field strengths, and corresponds to second-order quasi-linear equations for the geometric potentials. The natural reference choice is Minkowsi spacetime, for which $\bar{\tau}_\mu$ vanishes and $\bar{\tilde{D}} = \bar{D}$. For this class of theories, other things being equal, we would favor from among the set (239) the Hamiltonian boundary term quasi-local choice (upper, lower, lower, upper), i.e.

$$\mathcal{B}(Z) = i_Z\vartheta^\alpha \tau_\alpha + \Delta\Gamma^\alpha{}_\beta \wedge i_Z\rho_\alpha{}^\beta + \bar{D}_\beta Z^\alpha \Delta\rho_\alpha{}^\beta, \tag{240}$$

which leads to the following boundary term in the variation of the Hamiltonian:

$$\mathcal{C}(Z) = i_Z(\delta\vartheta^\alpha \wedge \tau_\alpha - \Delta\Gamma^\alpha{}_\beta \wedge \delta\rho_\alpha{}^\beta). \tag{241}$$

This corresponds to imposing boundary conditions on the coframe and the momentum conjugate to the connection. The associated energy flux expression is

$$\pounds_Z \mathcal{H}(Z) = d i_Z(\pounds_Z \vartheta^\alpha \wedge \tau_\alpha - \Delta\Gamma^\alpha{}_\beta \wedge \pounds_Z \rho_\alpha{}^\beta). \tag{242}$$

Regarding the *total* energy–momentum and angular momentum/CoMM, the expression (240) matches the expression (with Minkowski reference) for the PG Hamiltonian boundary term at spatial infinity that was first proposed by Hecht in 1993.[48]

19.6. *Einstein's GR*

Within the PG context, the special case of Riemannian geometry can be reached by imposing vanishing torsion with a Lagrange multiplier. In the general first-order formulation it is sufficient to take the potential to be independent of the coframe conjugate momentum τ_μ.

A first-order Lagrangian for vacuum[x] GR is

$$\mathcal{L}_{\mathrm{GR}} = R^{\alpha\beta} \wedge \rho_{\alpha\beta} + D\vartheta^\mu \wedge \tau_\mu - V^{\alpha\beta} \wedge \left(\rho_{\alpha\beta} - \frac{1}{2\kappa}\eta_{\alpha\beta}\right), \tag{243}$$

which uses a Lagrange multiplier field $V^{\alpha\beta} \equiv V^{[\alpha\beta]}$ to give the connection's conjugate momentum field a value which depends on the orthonormal frame:

$$\rho_{\alpha\beta} - \frac{1}{2\kappa}\eta_{\alpha\beta} = 0. \tag{244}$$

Variation of (243) w.r.t. the coframe, connection and their respective momenta fields gives the (vacuum) equations:

$$\delta\vartheta^\mu : 0 = D\tau_\mu + V^{\alpha\beta} \wedge \frac{1}{2\kappa}\eta_{\alpha\beta\mu}, \tag{245}$$

$$\delta\Gamma^{\alpha\beta} : 0 = D\rho_{\alpha\beta} + \vartheta_{[\beta} \wedge \tau_{\alpha]}, \tag{246}$$

$$\delta\tau_\mu : 0 = D\vartheta^\mu, \tag{247}$$

$$\delta\rho_{\alpha\beta} : 0 = R^{\alpha\beta} - V^{\alpha\beta}. \tag{248}$$

As expected (247) gives vanishing torsion. From the differential of (244) one gets

$$D\rho_{\alpha\beta} = \frac{1}{2\kappa} D\vartheta^\mu \wedge \eta_{\alpha\beta\mu}, \tag{249}$$

[x]For our purposes concerning the Hamiltonian boundary term, we consider here for simplicity just the vacuum case. The results will apply to all situations where the boundary of the region of interest is in the vacuum region, outside of the domain of the matter fields. That should include most of the cases of physical interest.

which vanishes, hence (246) reduces to $\vartheta_{[\beta} \wedge \tau_{\alpha]} = 0$, from which it follows that τ_μ vanishes. Then (245) with (248) reduces to the vanishing of the Einstein 3-form:

$$0 = \frac{1}{2\kappa} R^{\alpha\beta} \wedge \eta_{\alpha\beta\mu} = -\frac{1}{\kappa} G^{\nu}{}_{\mu} \eta_{\nu}. \tag{250}$$

By the way, had we included a suitable source, we would have obtained

$$0 = \frac{1}{2\kappa} R^{\alpha}{}_{\beta} \wedge \eta_{\alpha}{}^{\beta}{}_{\mu} + \mathfrak{T}_{\mu}, \tag{251}$$

where $\mathfrak{T}_\mu$ is the Hilbert energy–momentum 3-form. Using $D\eta_{\alpha}{}^{\beta}{}_{\mu} = 0$, this relation can be rearranged as follows[122]:

$$\begin{aligned} 0 &= \frac{1}{2\kappa} \left[d(\Gamma^{\alpha}{}_{\beta} \wedge \eta_{\alpha}{}^{\beta}{}_{\mu}) + \Gamma^{\alpha}{}_{\beta} \wedge d\eta_{\alpha}{}^{\beta}{}_{\mu} + \Gamma^{\alpha}{}_{\gamma} \wedge \Gamma^{\gamma}{}_{\beta} \wedge \eta_{\alpha}{}^{\beta}{}_{\mu} \right] + \mathfrak{T}_{\mu} \\ &\equiv \frac{1}{2\kappa} \left[d(\Gamma^{\alpha}{}_{\beta} \wedge \eta_{\alpha}{}^{\beta}{}_{\mu}) - \Gamma^{\alpha}{}_{\gamma} \wedge \Gamma^{\gamma}{}_{\beta} \wedge \eta_{\alpha}{}^{\beta}{}_{\mu} + \Gamma^{\alpha}{}_{\beta} \wedge \Gamma^{\gamma}{}_{\mu} \wedge \eta_{\alpha}{}^{\beta}{}_{\gamma} \right] + \mathfrak{T}_{\mu}. \end{aligned} \tag{252}$$

The rearrangement identifies a certain superpotential 2-form,

$$\mathfrak{U}_{\mu} = -\Gamma^{\alpha}{}_{\beta} \wedge \eta_{\alpha}{}^{\beta}{}_{\mu}, \tag{253}$$

and gravitational energy–momentum pseudotensor 3-form,

$$2\kappa \mathfrak{t}_{\mu} = -\Gamma^{\alpha}{}_{\gamma} \wedge \Gamma^{\gamma}{}_{\beta} \wedge \eta_{\alpha}{}^{\beta}{}_{\mu} + \Gamma^{\alpha}{}_{\beta} \wedge \Gamma^{\gamma}{}_{\mu} \wedge \eta_{\alpha}{}^{\beta}{}_{\gamma}. \tag{254}$$

These manipulations and the resultant expressions are meaningful in both orthonormal and holonomic frames. In orthonormal frames, they give the expressions for the so-called *tetrad-teleparallel gauge current*,[28] while in holonomic frames we have obtained neat form versions of the Freud superpotential (22) and the Einstein pseudotensor (10). There is a remarkable contrast between the simple short calculation given here for these quantities and the long complicated ones discussed in Sec. 4 that were done in the past using tensor calculus. Via rearrangements of the field equations analogous to (252), generalized pseudotensors can be found for the PG.[86]

19.7. *Preferred boundary term for GR*

Over 20 years ago using the covariant Hamiltonian symplectic boundary term approach, we proposed a quasi-local boundary term for GR[23] (an equivalent quasi-local expression obtained from a Noether argument using a global background reference which was proposed at about the same time by Katz *et al.*[65,79]):

$$\mathcal{B}(Z) = \frac{1}{2\kappa} (\Delta\Gamma^{\alpha}{}_{\beta} \wedge i_Z \eta_{\alpha}{}^{\beta} + \bar{D}_{\beta} Z^{\alpha} \Delta\eta_{\alpha}{}^{\beta}); \quad \eta^{\alpha\beta\cdots} := *(\vartheta^{\alpha} \wedge \vartheta^{\beta} \wedge \cdots). \tag{255}$$

(This has the form of Hecht's PG expression restricted to GR and natural extended to a boundary that need not be at infinity). The boundary term in the variation of the Hamiltonian has the form

$$\delta\mathcal{H}(Z) \sim di_Z(\Delta\Gamma^\alpha{}_\beta \wedge \delta\eta_\alpha{}^\beta), \tag{256}$$

Since $\eta^{\alpha\beta} = *(\vartheta^\alpha \wedge \vartheta^\beta)$, this corresponds to fixing the orthonormal coframe ϑ^μ (and thus the metric) on the boundary. The energy flux expression is

$$\pounds_Z\mathcal{H}(Z) = di_Z(\Delta\Gamma^\alpha{}_\beta \wedge \pounds_Z\eta_\alpha{}^\beta). \tag{257}$$

Like many other boundary term choices, at spatial infinity it gives the ADM,[1] MTW,[82] Regge–Teitelboim,[106] Beig–Ó Murchadha,[3] Szabados[132] energy, momentum, angular momentum, CoMM.

It has some special virtues:

(i) At null infinity: The Bondi–Trautman energy and the Bondi energy flux,[24]
(ii) It is "covariant,"
(iii) It is positive: at least for spherical solutions[58] and large spheres,
(iv) For small spheres it is a positive multiple of the Bel–Robinson tensor,[87]
(v) First law of thermodynamics for black holes,[21]
(vi) For spherical solutions it has the hoop property,[58]
(vii) For a suitable choice of reference it vanishes for Minkowski space.

20. A "Best Matched" Reference

In this section, we turn to the second ambiguity that is inherent in quasi-local energy–momentum expressions: The choice of reference. Minkowski space is the natural choice, but one still needs to choose a specific Minkowski space. Recently we proposed (i) *4D isometric matching on the boundary and* (ii) *energy optimization as criteria for selecting the "best matched" reference on the boundary of the quasi-local region.*

Note: For all other fields, it is appropriate to choose vanishing reference values as the reference ground state — the vacuum. But for geometric gravity, the standard ground state is the nonvanishing Minkowski metric, so a nontrivial reference is essential. One still needs to specify exactly which Minkowski space.

Reference values can be determined by choosing, in a neighborhood of the desired spacelike boundary 2-surface S, 4 smooth functions $y^i = y^i(x^\mu)$, $i = 0, 1, 2, 3$ with $dy^0 \wedge dy^1 \wedge dy^2 \wedge dy^3 \neq 0$ and then defining a Minkowski reference by

$$\bar{g} = -(dy^0)^2 + (dy^1)^2 + (dy^2)^2 + (dy^3)^2. \tag{258}$$

Geometrically this is equivalent to finding a diffeomorphism for a neighborhood of the 2-surface into Minkowski space. The associated reference connection is the pullback of the flat Minkowski connection:

$$\bar{\Gamma}^\alpha{}_\beta = x^\alpha{}_i(\bar{\Gamma}^i{}_j y^j{}_\beta + dy^i{}_\beta) = x^\alpha{}_i dy^i{}_\beta. \tag{259}$$

Here $x^\alpha{}_i$ is the inverse of $y^i{}_\beta$, where $dy^i = y^i{}_\beta\vartheta^\beta$.

With these standard Minkowski coordinates y^i, a Killing field of the reference has the form $Z^k = Z_0^k + \lambda_0^k{}_l y^l$, where $\lambda_0^{kl} = \lambda_0^{[kl]}$, with Z_0^k and λ_0^{kl} being constants. The 2-surface integral of any one of our Hamiltonian boundary terms then has a value linear in these constant values:

$$\oint_S \mathcal{B}(Z) = -Z_0^k p_k(S) + \frac{1}{2}\lambda_0^{kl} J_{kl}(S). \tag{260}$$

This implicity determines not only a quasi-local energy–momentum but also a quasi-local angular momentum/CoMM. When specialized to GR the integrals $p_k(S)$, $J_{kl}(S)$ in the spatial asymptotic limit agree with accepted expressions for these quantities: MTW[82] §20.2 and Regge–Teitelboim[106] with the refinements of Beig–Ó Murchadha[3] and Szabados.[132] For the PG at spatial infinity, Hecht[49] compared in detail his expression with the other proposed expressions, e.g. Ref. 10. At spatial infinity, with the asymptotics (122), all of our PG symplectic boundary terms (239) will give the same values as Hecht's expression (240).

For energy–momentum, one takes Z to be a translational Killing field of the Minkowski reference. Then the second term in our preferred quasi-local boundary expressions (240) and (255) vanishes.[y] With $Z^k = Z_0^k =$ constant our preferred quasi-local expressions now take the form

$$\mathcal{B}^{\rm PG}(Z) = Z_0^k x^\mu{}_k [\tau_\mu + (\Gamma^\alpha{}_\beta - x^\alpha{}_j\, dy^j{}_\beta) \wedge i_{e_\mu} \rho_\alpha{}^\beta], \tag{261}$$

$$\mathcal{B}^{\rm GR}(Z) = Z_0^k x^\mu{}_k (\Gamma^\alpha{}_\beta - x^\alpha{}_j dy^j{}_\beta) \wedge \eta_{\mu\alpha}{}^\beta. \tag{262}$$

20.1. *The choice of reference*

To be completed, our Hamiltonian boundary term and the quasi-local energy–momentum/angular momentum proposal needs a prescription for choosing a reference on the boundary. There are several options; one needs a choice suited to the application.

For an extended region, one may want a global background (see Refs. 101 and 102 for this approach in GR). Consider for example solar system applications, more specifically say we want to calculate using our quasi-local energy flux formula the tidal energy flux between Jupiter and its moon Io, that is believed to be responsible for Io's volcanos. (This has already been done by several methods.[11,40,105]) On the other hand, if one wishes to study a given metric expressed in a specific coordinate system, the analytic approach[78] may be a good choice.

To explicitly determine the specific values of the quasi-local quantities, one needs some good way to choose the reference. Minkowski spacetime is the natural choice, especially for asymptotically flat spacetimes. However, as noted above, almost any four functions will determine some Minkowski reference. With such freedom, one

[y] For GR the second term in (255) also vanishes for 4D isometric matching on S, a condition we shall use below.

can still get almost any value for the quasi-local quantities. This freedom is the quasi-local version of the second type of ambiguity mentioned in the introduction.

Recently we proposed a program[89] to fix the "best" choice for a *quasi-local reference*, i.e. one that is determined by the dynamical fields on the boundary. It has two parts: 4D isometric matching and optimization of a certain quantity. Here we present it along with a promising alternative optimization.[127] For GR we have already found that our new procedure works well for an important special case: A certain class of axisymmetric spacetimes[126] which includes the Kerr metric.

For the PG, not so much has been done yet. While PG energy–momentum and angular momentum calculations at both spatial and future null infinity were done long ago,[50,51] and gave, in particular, the expected results for the Kerr metric, as far as we know there have not yet been any genuine *quasi-local* (i.e. for a finite region) PG calculations. This is not so surprising. In general the big obstacle is the 2D isometric embedding, which we are about to discuss. For spherical symmetry, all the calculations at least for GR can easily be done analytically. For the aforementioned class of axisymmetric metrics, the 2D embedding problem has an algebraic solution. But the boundary expressions are already sufficiently complicated that the quasi-local energy integral for GR could only be evaluated numerically. Now that it is known how to do the case of axisymmetric GR the way is open for truly quasi-local PG energy and angular momentum calculations. For the PG, it seems that numerical calculations will be unavoidable.

20.2. *Isometric matching of the 2-surface*

We first recall an important procedure that has been used: Isometric matching of the 2-surface S. This can be expressed in terms of quasi-spherical foliation adapted coordinates t, r, θ, φ as

$$g_{AB} \doteq \bar{g}_{AB} = \bar{g}_{ij} y^i_A y^j_B = -y^0_A y^0_B + \delta_{ab} y^a_A y^b_B, \tag{263}$$

where S is given by constant values of t, r, and $a, b = 1, 2, 3$ while A, B range over θ, φ. We use $\doteq$ to indicate a relation which holds only on the 2-surface S. Equation (263) is three conditions on the four functions y^i. One can regard y^0 as the free choice. From a classic closed 2-surface into $\mathbb{R}^3$ embedding theorem — as long as S and $y^0(x^\mu)$ are such that on S

$$g'_{AB} := g_{AB} + y^0_A y^0_B, \tag{264}$$

is convex — one has a unique embedding. Wang and Yau have discussed in detail this type of embedding of a 2-surface into Minkowski controlled by one function in their recent quasi-local work.[25,143,144]

Unfortunately, although there is a unique embedding, there is generally no explicit analytic formula except in special simple cases, such as spherical symmetry or axisymmetry. The lack of an explicit formula for the solution of this 2D isometric embedding prevents exact quasi-local calculations in general.

20.3. *Complete 4D isometric matching*

Our "new" proposal[z] is: Complete 4D isometric matching on S: $g \doteq \bar{g}$, a part of which is still the just discussed 2D isometric embedding.

In view of isometric matching, there should be a Lorentz transformation which on the boundary brings the dynamical coframe into line with the reference frame:

$$\vartheta^i := L^i{}_\alpha \vartheta^\alpha \doteq dy^i. \tag{265}$$

The integrability condition for this equation is

$$d\vartheta^i|_S = 0. \tag{266}$$

This 2-form equation restricted to the 2-surface gives 4 conditions on the 6 parameter set of Lorentz transformations $L^i{}_\alpha$. Thus 4D isometric matching has $6 - 4 = 2$ degrees of freedom. They can be identified as the choice of time embedding function y^0 in (263) plus a boost parameter α in the plane normal to S.

20.4. *Optimal energy*

To fix the remaining two isometric matching parameters, one can regard the quasi-local value as a measure of the difference between the dynamical and the reference boundary values. This value will be a functional of y^0, α. The critical points of this functional determine the distinguished choices for these two functions.

Previously we proposed[89] taking the optimal "best matched" embedding as the one which gives an extreme value to the associated invariant mass $m^2 = -p_i p_j \bar{g}^{ij}$. This should determine the reference up to a Poincaré transformation.

This is a reasonable condition, but, unfortunately, not so practical. The invariant mass is a sum of four terms, each quadratic in an integral over S. Note, however, that using the Poincaré freedom, one can get the same m value in the center-of-momentum frame from p_0. This leads us to our new proposal: Take the preferred reference as one that gives a critical value to the quasi-local *energy* given by (260) and (262) or (261) with $Z^k = Z_0^k = \delta_0^k$. We expect this much simpler optimization to give the same reference geometry as that obtained from using m^2.

Based on some physical and practical computational arguments, it seems reasonable to expect a unique solution in general. In a numerical calculation in principle one could just calculate the energy values given by (260) and (261) or (262) with $Z^k = \delta_0^k$ for a great many choices of y^0, α subject to the 4D isometric matching constraint (265) and the integrability condition (266), and then note the energy critical points.

For GR, this "best matching" procedure already gave reasonable quasi-local energy results for spherically symmetric systems.[20,78,149,150] For the Schwarzschild

[z]For GR, this was proposed by Szabados at a workshop in Hsinchu, Taiwan in 2000. He has since investigated it in detail.[133]

metric, the "best matched" quasi-local energy has the value first found by Brown and York,[17]

$$E(r) = \frac{2m}{1 + \sqrt{1 - \dfrac{2m}{r}}}, \tag{267}$$

using a closely related boundary term and a 2-surface embedding into $\mathbb{R}^3$. Now we also have sensible results for certain axisymmetric systems including the Kerr metric.[126] For the surface of constant Boyer–Lindquist r, the angular momentum is simply a constant independent of r, equal to its usual asymptotic value, J. The quasi-local energy integral, however, is not so simple and can only be evaluated numerically.

21. Concluding Discussion

In addition to her two key theorems regarding symmetry, Emmy Noether in her 1918 paper also proved that for diffeomorphically invariant gravity, there is no proper total energy–momentum density. In other words there is no covariant total energy momentum density tensor for gravitating systems. In Ch. 20 of Ref. 82, Misner, Thorne and Wheeler discuss this feature as a consequence of the equivalence principle and argue that only the total energy–momentum of gravitating systems is meaningful. But clearly the gravitational interaction is local and does allow for the *local* exchange of energy–momentum. To account for this various noncovariant expressions called pseudotensors (each generated by a certain superpotential) have been proposed. There thus arose two ambiguities: which expression? in which reference frame? The modern idea is *quasi-local*: energy–momentum is associated with a closed 2-surface.

One approach, which is the one we have used, is via the Hamiltonian. With the aid of differential forms and a first-order variational formulation, we have developed a covariant Hamiltonian formalism. The Hamiltonian boundary term plays key roles: It determines the boundary conditions and the quasi-local values. We have shown that this approach includes all the traditional pseudotensors while clarifying the ambiguities: Each superpotential is a possible Hamiltonian boundary term which is associated with a specific boundary condition, and the reference frame becomes a choice of the reference values (ground state) on the boundary.

One can regard gravity as a local gauge theory for the global symmetry group for Minkowski spacetime: The Poincaré group. The appropriate geometry is *Riemann–Cartan*: The *curvature* is the field strength for Lorentz transformations and the *torsion* is the field strength for local translations (infinitesimal diffeomporhisms). For comparison, we included in our discussion a general internal gauge field. In this way, we can identify the analog of the internal gauge vector potential. For the local Lorentz symmetry, it is the (metric compatible) *connection one-form*; for the local translation symmetry, it is the orthonormal *coframe*. We developed in considerable detail the associated Lagrangian and Hamiltonian Noether currents and differential identities, so one can compare the similarities and differences of the spacetime gauge symmetry expressions with those of a generic internal symmetry. Briefly, the Lorentz rotation sector is quite similar to that of an internal gauge symmetry, but the translation symmetry has both striking similarities and differences. This would be much less clear if we had opted for a formulation which does not include the coframe as a dynamical variable. In the approach of Blagojević and Hehl,[8] the use of the orthonormal frame is motivated by the need to describe spin; here our basic motivation is in terms of the coframe's fundamental gauge role regarding local spacetime translations.

The geometric/gauge symmetry approach is helpful in identifying a good expression for our Hamiltonian boundary term expression for quasi-local quantities. Our preferred expression for GR turns out to correspond to fixing the metric on the boundary — which is obviously a reasonable boundary condition choice.

A notable feature of the Hamiltonian boundary term for dynamic geometry is that it necessarily depends on the choice of some nondynamical reference values (this is a manifestation of Noether's result regarding nonexistence of a proper energy–momentum density). One can get almost any quasi-local value if one allows free rein in the choice of reference. One needs to fix a nondynamical reference frame, but only on the boundary of the region. This can be compared to choosing some flat plane to map a part of the curved surface of the Earth. One could slice the plane through the surface of the Earth; for a spherical Earth the planar geometry would exactly match on a circle. Similarly for spacetime. On the 2D boundary of a spatial region, one can exactly match the curved 4D spacetime metric with a flat Minkowski spacetime metric. Detailed analysis shows that there is still two degrees of freedom. We proposed that a good way to fix these was by considering the critical points of our quasi-local expression. There might be other sensible options, but the main point is that a reasonable choice is available.

Our principal results, the Hamiltonian boundary terms that are our preferred quasi-local energy–momentum expressions (240) and (255) for the PG and GR, were obtained by considering the Hamiltonian, geometry, Noether symmetry, and spacetime gauge theory. The harmonious combination of all of these perspectives makes a strong case for the results. Nevertheless, it should be noted that one can also be led to this result from other perspectives. Regarding GR, essentially the same expression has been obtained (i) via a Noether approach with a global reference[65,79] and (ii) via a symplectic covariant Hamiltonian approach using the metric in a holonomic frame.[22] For the PG (including GR as a special case), the same preferred expression was found via an entirely frame independent manifestly covariant Hamiltonian formalism.[87] Although in principle there are unlimited number of possible Hamiltonian boundary term quasi-local energy–momentum expressions — which are in a formal sense all of equal status — in practice one can — with very good reasons — discover that one expression stands out.

Acknowledgments

We would first like to thank Prof. F. W. Hehl for his many recommendations that have been very helpful in our efforts to improve the quality of this presentation. J. M. N. would like to express his appreciation for the hospitality at the Institute of Physics, Academia Sinica, Taipei 115, Taiwan, the Mathematical Sciences Center, Tsinghua University, Beijing, China 100084, and the Morningside Center of Mathematics, Academy of Mathematics and System Science, Chinese Academy of Sciences, Beijing 100190, China. Part of this work was developed at those institutions during visits in 2013 and 2014.

C.M.C. was supported by the Ministry of Science and Technology of the R.O.C. under the grant MOST 102-2112-M-008-015-MY3.

References

1. R. Arnowitt, S. Deser and C. W. Misner, The dynamics of general relativity, in *Gravitation: An Introduction to Current Research*, ed. L. Witten (Wiley, New York, 1962), pp. 227–265; *Gen. Relativ. Gravit.* **40** (2008) 1997, arXiv:gr-qc/0405109.
2. H. Bauer, "Uber die Energiekomponenten des Gravitationsfeldes," *Phys. Zeitschrift* **19** (1918) 163.
3. R. Beig and N. Ó Murchadha, "The Poincare Group as the Symmetry Group of Canonical General Relativity," *Ann. Phys. (N.Y.)* **174** (1987) 463.
4. P. G. Bergmann, "Non-Linear Field Theories," *Phys. Rev.* **75** (1949) 680.
5. P. G. Bergmann, "Conservation Laws in General Relativity as the Generators of Coordinate Transformations," *Phys. Rev.* **112** (1958) 287.
6. P. G. Bergmann and R. Thomson, "Spin and Angular Momentum in General Relativity," *Phys. Rev.* **89** (1953) 400.
7. M. Blagojević, *Gravitation and Gauge Symmetries* (Institute of Physics, Bristol, 2002).
8. M. Blagojević and F. W. Hehl, *Gauge Theories of Gravitation* (Imperial College Press, London, 2013).
9. M. Blagojević and I. A. Nikolić, "Hamiltonian Dynamics of Poincare Gauge Theory: General Structure in the Time Gauge," *Phys. Rev. D* **28** (1983) 2455; I. A. Nikolić, "Dirac's Hamiltonian Structure of $R + R^2 + T^2$ Poincare Gauge Theory of Gravity Without Gauge Fixing," *Phys. Rev. D* **30** (1984) 2508.
10. M. Blagojević and M. Vasilić, "Asymptotic Symmetry and Conserved Quantities in the Poincare Gauge Theory of Gravity," *Class. Quantum Grav.* **5** (1988) 1241.
11. I. S. Booth and J. D. E. Creighton, "A Quasilocal Calculation of Tidal Heating," *Phys. Rev. D* **62** (2000) 067503, arXiv:gr-qc/0003038.
12. K. Brading, "Which Symmetry? Noether, Weyl, and Conservation of Electric Charge", *Studies History Philos. Modern Phys.* **33** (2002) 3.
13. K. Brading, A note on general relativity, energy conservation, and Noether's theorems, in *The Universe of General Relativity*, eds. A. J. Kox and J. Eisenstadt (Einstein Studies, Vol. 11) (Birkhäuser, Boston, 2005), pp. 125–135.
14. K. Brading and H. R. Brown, Symmetries and Noether's theorems, in *Symmetries in Physics: Philosophical Reflections*, eds. K. Brading and E. Castellani (Cambridge University Press, Cambridge, 2003), pp. 89–109.
15. K. Brading and T. A. Ryckman, "Hilbert's 'Foundations of Physics': Gravitation and Electromagnetism within the Axiomatic Method", *Studies History Philos. Sci B: Modern Phys.* **39** (2008) 102.

16. H. R. Brown and K. Brading, "General Covariance from the Perspective of Noether's Theorems," *Diálogos* **79** (2002) 59.
17. J. D. Brown and J. W. York, Jr., "Quasilocal Energy and Conserved Charges Derived from the Gravitational Action," *Phys. Rev. D* **47** (1993) 1407, arXiv:gr-qc/9209012.
18. C. Cattani and M. De Maria, Conservation laws and gravitational waves in general relativity (1915–1918), in *The Attraction of Gravitation: New Studies in the History of General Relativity*, eds. J. Earman, M. Janssen and J. D. Norton (Birkhäuser, Boston, 1993), pp. 63–87.
19. C.-C. Chang, J. M. Nester and C.-M. Chen, "Pseudotensors and Quasilocal Gravitational Energy Momentum," *Phys. Rev. Lett.* **83** (1999) 1897, arXiv:gr-qc/9809040.
20. C.-M. Chen, J.-L. Liu, J. M. Nester and M.-F. Wu, "Optimal Choices of Reference for Quasi-local Energy," *Phys. Lett. A* **374** (2010) 3599, arXiv:0909.2754 [gr-qc].
21. C.-M. Chen and J. M. Nester, "Quasilocal Quantities for GR and other Gravity Theories," *Class. Quantum Grav.* **16** (1999) 1279, arXiv:gr-qc/9809020.
22. C.-M. Chen and J. M. Nester, "A Symplectic Hamiltonian Derivation of Quasilocal Energy Momentum for GR," *Grav. Cosmol.* **6** (2000) 257, arXiv:gr-qc/0001088.
23. C.-M. Chen, J. M. Nester and R.-S. Tung, "Quasilocal Energy Momentum for Gravity Theories," *Phys. Lett. A* **203** (1995) 5, arXiv:gr-qc/9411048.
24. C.-M. Chen, J. M. Nester and R.-S. Tung, "The Hamiltonian Boundary Term and Quasi-local Energy Flux," *Phys. Rev. D* **72** (2005) 104020, arXiv:gr-qc/0508026.
25. P.-N. Chen, M.-T. Wang and S.-T. Yau, "Conserved Quantities in General Relativity: From the Quasi-Local Level to Spatial Infinity", *Commun. Math. Phys.* **338** (2015) 31.
26. F. I. Cooperstock, "Energy Localization in General Reletivity: A New Hypothesis," *Foundations Phys.* **22** (1992) 1011.
27. L. Corry, *David Hilbert and the Axiomatization of Physics (1898–1918): From Grundlagen der Geometrie to Grundlagen der Physik* (Kluwer Academic Publishers, Dordrecht/Boston/London, 2004).
28. V. C. de Andrade, L. C. T. Guillen and J. G. Pereira, "Gravitational Energy Momentum Density in Teleparallel Gravity," *Phys. Rev. Lett.* **84** (2000) 4533, arXiv:gr-qc/0003100.
29. P. A. M. Dirac, *Proc. Roy. Soc. (London) A* **246** (1958) 326.
30. P. A. M. Dirac, "The Theory of Gravitation in Hamiltonian Form," *Poc. Roy. Soc. (London) A* **246** (1958) 333.
31. P. A. M. Dirac, *Lectures on Quantum Mechanics* (Belfer, Yeshiva Univ., 1964).
32. J. Earman, Tracking down gauge: An Ode to the constrained Hamiltonian formalism, in *Symmetries in Physics: Philosophy and Reflections*, eds. K. Brading and E. Castellani (Cambridge University Press, Cambridge, 2003), pp. 140–162.
33. A. Einstein, *Zeitschrift für Mathematik und Physik* **63** (1914) 215, English translation "Covariance properties of the field equations of the theory of gravitation based on the generalized theory of relativity," in Refs. 37, 38, Vol. 6, Doc. 2, pp. 6–15.
34. A. Einstein, *Königlich Preussische Akademie der Wissenschaften (Berlin). Sitzungsberichte* (1915) 4, 11, 18, 25 November, English translation in Refs. 37, 38, Vol. 6, Doc. 21, pp. 98–107, Doc. 22, pp. 108–110, Doc. 25, pp. 112–116, Doc. 25, pp. 117–120.
35. A. Einstein, *Königlich Preussische Akademie der Wissenschaften (Berlin). Sitzungsberichte* (1916) 26 October, English translation "Hamilton's principle and the general theory of relativity," in Refs. 37, 38, Vol. 6, Doc. 41, pp. 240–246 and *The Principle of Relativity* (Dover, New York, 1952), pp. 165–173.
36. A. Einstein, *Phys. Zeitschrift* **19** (1918) 115, English translation "Note on E. Schrödingers paper 'The energy components of the gravitational field"', in Refs. 37, 38, Vol. 7, Doc. 2, pp. 28–30.

37. The Collected Papers of Albert Einstein: http://einsteinpapers.press.princeton.edu.
38. A. Einstein, *The Collected Papers of Albert Einstein* (Princeton University Press, Princeton, 1987ff).
39. A. Einstein and M. Grossmann, *Zeitschrift für Mathematik und Physik* **62** (1914) 225, English translation: "Outline of a generalized theory of relativity and of a theory of gravitation," in Refs. 37, 38, Vol. 4, Doc. 13, pp. 151–188.
40. M. Favata, "Energy Localization Invariance of Tidal Work in General Relativity," *Phys. Rev. D* **63** (2001) 064013, arXiv:gr-qc/0008061.
41. T. Frankel, *The Geometry of Physics: An Introduction*, 2nd edn. (Cambridge Univ. Press, 2004).
42. J. Frauendiener, "Geometric Description of Energy-momentum Pseudotensors," *Class. Quantum Grav.* **6** (1989) L237.
43. Ph. Freud, "Uber die Ausdrücke der Gesamtenergie und des Gesamtimpulses eines materiellen Systems in der allgemeinen Relativitätstheorie," *Ann. Math.* **40** (1939) 417.
44. J. N. Goldberg, "Conservation Laws in General Relativity," *Phys. Rev.* **111** (1958) 315.
45. F. Gronwald and F. W. Hehl, Erice 1995, Quantum gravity, pp. 148–198, arXiv:gr-qc/9602013.
46. A. Hanson, T. Regge and C. Teitelboim, *Constrained Hamiltonian Systems* (Accademia Nazionale Dei Lincei, Rome, 1976).
47. K. Hayashi and T. Shirafuji, "Gravity from Poincare Gauge Theory of the Fundamental Particles. I. Linear and Quadratic Lagrangians," *Prog. Theor. Phys.* **64** (1980) 866; II. Equations of motion for test bodies and various limits, p. 883; III. Weak field approximation, p. 1435; IV. Mass and energy of particle spectrum, p. 2222.
48. R. D. Hecht, "Erhaltungsgrößen in der Poincaré-Eichtheorie der gravitation", Ph.D. thesis, University of Cologne (1993).
49. R. D. Hecht, "Mass and Spin of Poincare Gauge Theory," *Gen. Relativ. Gravit.* **27** (1995) 537, arXiv:gr-qc/9501035.
50. R. D. Hecht and J. M. Nester, "A New Evaluation of PGT Mass and Spin," *Phys. Lett. A* **180** (1993) 324.
51. R. D. Hecht and J. M. Nester, "An Evaluation of the Mass and Spin at Null Infinity for the PGT and GR Gravity Theories," *Phys. Lett. A* **217** (1996) 81.
52. F. W. Hehl, "Four lectures on Poincaré gauge theory," in *Proc. 6th Course of the Int. School of Cosmology and Gravitation on Spin Torsion and Supergravity*, eds. P. G. Bergmann and V. de Sabbatta (Plenum, New York, 1980).
53. F. W. Hehl, "Gauge theory of gravity and spacetime," arXiv:1204.3672 [gr-qc].
54. F. W. Hehl, J. D. McCrea, E. W. Mielke and Y. Ne'eman, "Metric Affine Gauge Theory of Gravity: Field Equations, Noether Identities, World Spinors, and Breaking of Dilation Invariance," *Phys. Rep.* **258** (1995) 1, arXiv:gr-qc/9402012.
55. F. W. Hehl and Yu. N. Obukhov, *Foundations of Classical Electrodynamics* (Birkhäuser, Boston, 2003).
56. F. W. Hehl, P. von der Heyde, D. Kerlick and J. M. Nester, "General Relativity with Spin and Torsion: Foundations and Prospects," *Rev. Mod. Phys.* **48** (1976) 393.
57. D. Hilbert, "The Foundations of Physics," *Math.-Phy. Klasse. Issue* (8) (1916), 395–407; (1917) 53–76. For English translation see Ref. 107, pp. 1003–1015, 1017–1038.
58. F.-H. Ho and N. Xie, "Positivity and Hoop Properties for Chen-Nester-Tung Quasi-local Energy with Analyic Reference in Spherical Symmetry," *Chinese J. Phys.* **52** (2014) 1432, arXiv:1309.1024 [gr-qc].

59. J. Isenberg and J. M. Nester, "Canonical Gravity," in *General Relativity and Gravitation: One Hundred Years after the Birth of Albert Einstein*, ed. A. Held (Plenum, New York, 1980).
60. M. Janssen, "Of Pots and Holes: Einstein's Bumpy Road to General Relativity," *Ann. Phys.* (*Leipzig*) **14** (Supp.) (2005) 58.
61. M. Janssen, J. Norton, J. Renn, T. Sauer and J. Stachel, *The Genesis of General Relativity*, Vol. 1. *Einstein's Zurich Notebook*: *Introduction and Source*; Vol. 2. *Einstein's Zurich Notebook*: *Commentary and Essays* (Springer, Berlin, 2007).
62. M. Janssen and J. Renn (2007). "Untying the Knot: How Einstein found his way back to field equations discarded in the Zurich notebook", in *The Genesis of General Relativity, Vol. 2, Einstein's Zurich Notebook*: *Commentary and Essays*, eds. J. Renn *et al.* (Springer, Dordrecht, 2007), pp. 839–922.
63. J. Jezierski and J. Kijowski, "The Localization of Energy in Gauge Field Theories and in Linear Gravitation," *Gen. Relativ. Gravit.* **22** (1990) 1283.
64. B. Julia and S. Silva, "Current and Superpotentials in Classical Gauge Covariant Theories I. Local Results with Applications to Perfect Fluids and General Relativity," *Class. Quantum Grav.* **15** (1998) 2173, arXiv:gr-qc/9804029.
65. J. Katz, J. Bičák and D. Lynden-Bell, "Relativistic Conservation Laws and Integral Constraints for Large Cosmological Perturbations," *Phys. Rev. D* **55** (1997) 5957, arXiv:gr-qc/0504041.
66. T. Kawai, "Energy Momentum and Angular Momentum in Poincare Gauge Theory of Gravity," *Prog. Theor. Phys.* **79** (1988) 920.
67. T. W. B. Kibble, "Lorentz Invariance and the Gravitational Field," *J. Math. Phys.* **2** (1961) 212.
68. J. Kijowski, "A Simple Derivation of Canonical Structure and Quasi-local Hamiltonians in General Relativity," *Gen. Relativ. Gravit.* **29** (1997) 307.
69. J. Kijowski and W. M. Tulczyjew, *A Symplectic Framework for Field Theories*, Lecture Notes in Physics, Vol. 107 (Springer-Verlag, Berlin, New York, 1979).
70. N. Kiriushcheva and S. V. Kuzmin, "The Hamiltonian Formulation of General Relativity: Myths and Reality," *Central Eur. J. Phys.* **9** (2011) 576, arXiv:0809.0097 [gr-qc].
71. F. Klein, "Zu Hilberts erster note über die grundlagen der physik", *Königliche Gesselschaft der Wissenschaften zu Göttingen. Mathematisch-physikalische Klasse. Nachrichten* (1917) 469; (1918) 171. Reprinted in F. Klein, *Gesammelte mathematische Abhandlungen*, Vol. 1 (Springer, Berlin, 1921), pp. 553–585.
72. S. Kobayashi and K. Nomizu, *Foundations of Differential Geometry*, 2 volumes (Interscience, New York, 1963, 1969).
73. N. P. Konopleva and V. N. Popov, *Gauge Fields* (Harwood, New York, 1980).
74. Y. Kosmann-Schwarzbach, *The Noether Theorems: Invariance and Conservation Laws in the Twentieth Century* (Springer, New York, 2011).
75. K. Kuchař, "Dynamics of Tensor Fields in Hyperspace. III," *J. Math. Phys.* **17** (1976) 801.
76. C. Lanczos, *The Variational Principles of Mechanics* (University of Toronto Press, Toronto, 1949, 1962).
77. L. D. Landau and E. M. Lifshitz, *The Classical Theory of Fields*, 2nd edn. (Addison-Wesley, Reading, MA, 1962).
78. J.-L. Liu, C.-M. Chen and J. M. Nester, "Quasi-local Energy and the Choice of Reference," *Class. Quantum Grav.* **28** (2011) 195019, arXiv:1105.0502 [gr-qc].
79. D. Lynden-Bell, J. Katz and J. Bičák, "Mach's Principle from the Relativistic Constraint Equations," *Mon. Not. R. Astron. Soc.* **272** (1995) 150.

80. E. W. Mielke, *Geometrodynamics of Gauge Fields* (Akademie-Verlag, Berlin, 1987).
81. E. W. Mielke and R. P. Wallner, "Mass and Spin of Double Dual Solutions in Poincaré Gauge Theory", *Nuovo Cimento B* **101** (1988) 607, erratum *B* **102** (1988) 555.
82. C. W. Misner, K. S. Thorne and J. A. Wheeler, *Gravitation* (W. H. Freeman, San Francisco, 1973).
83. C. Møller, "On the Localization of the Energy of a Physical System in General Theory of Relativity," *Ann. Phys. (N.Y.)* **4** (1958) 347.
84. J. M. Nester, The Gravitational Hamiltonian, in *Asymptotic Behavior of Mass and Space-time Geometry*, ed. F. Flaherty, Lecture Notes in Physics, Vol. 202 (Springer, Berlin, 1984), pp. 155–163.
85. J. M. Nester, "A Covariant Hamiltonian for Gravity Theories," *Mod. Phys. Lett. A* **6** (1991) 2655.
86. J. M. Nester, "General Pseudotensors and Quasilocal Quantities," *Class. Quantum Grav.* **21** (2004) S261.
87. J. M. Nester, "A Manifestly Covariant Hamiltonian Formalism for Dynamical Geometry," *Prog. Theor. Phys. Suppl.* **172** (2008) 30.
88. J. M. Nester, "Gravitational energy," in *Gravitation and Astrophysics, Proceedings of the 9th Asia-Pacific International Conference*, eds. J. Luo, Z.-B. Zhou, H. C. Yeh and J. P. Hsu (World Scientific, Singapore, 2010), pp. 193–212.
89. J. M. Nester, C.-M. Chen, J.-L. Liu and G. Sun, "A reference for the covariant Hamiltonian term," in *Relativity and Gravitation: 100 Years after Einstien in Prague*, eds. J. Bicak and T. Ledvinka (Springer, Berlin, 2014), pp. 177–184, arXiv:1210.6148 [gr-qc].
90. W.-T. Ni, "Searches for the Role of Spin and Polarization in Gravity," *Rep. Prog. Phys.* **73** (2010) 056901, arXiv:0912.5057 [gr-qc].
91. E. Noether, *Königliche Gessellschaft der Wissenschaften zu Göttingen. Mathematisch-physikalische Klasse. Nachrichten* (1918) 235–257. English translation: "Invariant Variational Problems" in Tavel (1971) Ref. 135 and Kosmann-Schwarzbach (2011) Ref. 74.
92. K. Nomizu, *Lie Groups and Differential Geometry* (The Mathematical Society of Japan, Tokyo, 1956).
93. J. D. Norton, "How Einstein found His field equations, 1912–1915," in *Einstein and the History of General Relativity*, eds. D. Howard and J. Stachel (Birkhäuser, Boston, 1989), pp. 101–159.
94. J. D. Norton, "General Covariance and the Foundations of General Relativity: Eight Decades of Dispute," *Rep. Prog. Phys.* **56** (1993) 791.
95. J. D. Norton, "General covariance, gauge theories and the Kretschmann objection," in *Symmetries in Physics: Philosophical Reflections*, eds. K. Brading and E. Castellani (Cambridge University Press, Cambridge, 2003), pp. 110–123.
96. L. O'Raifeartaigh, *The Dawning of Gauge Theory* (Princeton University Press, Princeton, 1997).
97. A. Pais, *Subtle is the Lord: The Science and Life of Albert Einstein* (Clarendon, Oxford, 1982).
98. A. Papapetrou, "Einstein's Theory of Gravitation and Flat Space," *Proc. Roy. Irish Acad.* (Sect. A) **52A** (1948) 11.
99. W. Pauli, *Theory of Relativity* (Pergamon, London, New York, Paris, Los Angeles, 1958).
100. R. Penrose, "Quasi-local Mass and Angular Momentum in General Relativity," *Proc. R. Soc. London A* **381** (1982) 53.

101. A. N. Petrov, "Nonlinear perturbations and conservation laws on curved backgrounds in GR and other metric theories," in *Classical and Quantum Gravity Research*, Chp. 2, eds. M. N. Christiansen and T. K. Rasmussen (Nova Science Publishers, Hauppauge, 2008), arXiv:0705.0019 [gr-qc].
102. A. N. Petrov and J. Katz, "Conserved Currents, Superpotentials and Cosmological Perturbations," *Proc. R. Soc. London A* **458** (2002) 319; Relativistic conservation laws on curved backgrounds and the theory of cosmological perturbations, arXiv:gr-qc/9911025.
103. F. Pirani, A. Schild and R. Skinner, "Quantization of Einstein's Gravitational Field Equations. II" *Phys. Rev.* **87** (1952) 452.
104. J. B. Pitts, "Gauge-Invariant Localization of Infinitely Many Gravitational Energies from All Possible Auxiliary Structures," *Gen. Relativ. Gravit.* **42** (2010) 601, arXiv:0902.1288 [gr-qc].
105. P. Purdue, "The Gauge Invariance of General Relativistic Tidal Heating," *Phys. Rev. D* **60** (1999) 104054, arXiv:gr-qc/9901086.
106. T. Regge and C. Teitelboim, "Role of Surface Integrals in the Hamiltonian Formulation of General Relativity," *Ann. Phys. (N.Y.)* **88** (1974) 286.
107. J. Renn and M. Schemmel, *The Genesis of General Relativity*, Vol. 4. *Gravitation in the Twilight of Classical Physics: The Promise of Mathematics* (Springer, Berlin, 2007).
108. J. Renn and J. Stachel, "Hilbert's foundation of physics: From a theory of everything to a constituent of general relativity", in Renn and Schemmel (2007), Ref. 107, pp. 858–973.
109. L. Rosenfeld, "Zur Quantelung der Wellenfelder," *Annalen der Physik* **397** (1930) 113.
110. D. E. Rowe, "The Göttingen response to general relativity and Emmy Noether's theorems," in *The Symbolic Universe: Geometry and Physics 1890–1930*, ed. J. Gray (Oxford University Press, Oxford, 1999), pp. 189–233.
111. D. E. Rowe, "Einstein meets Hilbert: At the Crossroads of Physics and Mathematics," *Physics in Perspective* **3** (2001) 379.
112. D. E. Rowe, "Einstein's Gravitational Field Equations and the Bianchi Identities," *Math. Intelligencer* **24**(4) (2002) 57.
113. T. Sauer, "The Relativity of Discovery: Hilbert's First Note on the Foundations of Physics," *Arch. History of Exact Sci.* **53** (1999) 529, [physics/9811050].
114. T. Sauer, "Einstein Equations and Hilbert Action: What is missing on page 8 of the proofs for Hilberts First Communication on the Foundations of Physics?", *Arch. History of Exact Sci.* **59** (2005) 577; reprinted in Renn and Schemmel (2007), Ref. 107, pp. 975–989.
115. D. W. Sciama, "On the analogy between charge and spin in general relativity," in *Recent Developments in General Relativity* (Pergamon, Oxford; PWN, Warsaw, 1962), pp. 415–439.
116. E. Schrödinger, "Die Energiekomponenten des Gravitationsfeldes," *Phys. Zeitschrift* **19** (1918) 4.
117. E. Schrödinger, *Spacetime Structure* (Cambridge University Press, Cambridge, 1950).
118. K. Schwarzschild, "Zur Elektrodynamik. I. Zwei Formen des Prinzips der kleinsten Wirkung in der Elektronentheorie," *Göttingen Nachrichten Math.-Phys. Klasse* (1903) 126.
119. L. L. So, "A Modification of the Chen-Nester Quasilocal Expressions," *Int. J. Mod. Phys. D* **16** (2007) 875, arXiv:gr-qc/0605149.
120. L. L. So and J. M. Nester, "New Positive Small Vacuum Region Gravitational Energy Expressions," *Phys. Rev. D* **79** (2009) 084028, arXiv:0901.2400 [gr-qc].

121. L. L. So, J. M. Nester and H. Chen, "Energy-momentum Density in Small Regions: The Classical Pseudotensors," *Class. Quantum Grav.* **26** (2009) 085004, arXiv:0901.3884 [gr-qc].
122. L. L. So and J. M. Nester, "Gravitational Energy-momentum in Small Regions According to the Tetrad-teleparallel Expressions," *Chinese J. Phys.* **47** (2009) 10, arXiv:0811.4231 [gr-qc].
123. M. Spivak, *A Comprehensive Introduction to Differential Geometry*, Vol. 2 (Publish or Perish, Houston, 1999).
124. J. Stachel, "Einstein's search for general covariance, 1912–1915," in *Einstein and the History of General Relativity*, eds. D. Howard and J. Stachel (Birkhäuser, Boston, 1989).
125. N. Straumann, "Einstein's 'Zurich Notebook' and his Journey to General Relativity," *Annalen Phys.* **523** (2011) 488, arXiv:1106.0900v2 [physics.hist-ph].
126. G. Sun, C.-M. Chen, J.-L. Liu and J. M. Nester, "An Optimal Choice of Reference for the Quasi-Local Gravitational Energy and Angular Momentum," *Chinese J. Phys.* **52** (2014) 111, arXiv:1307.1039 [gr-qc].
127. G. Sun, C.-M. Chen, J.-L. Liu and J. M. Nester, A reference for the gravitational Hamiltonian boundary term, to appear in *Chinese J. Phys.* **53**(6) (2015), arXiv:1307.1510 [gr-qc].
128. K. Sundermeyer, *Constrained Dynamics: With Applications to Yang–Mills Theory, General Relativity, Classical Spin, Dual String Model*, Lecture Notes in Physics, Vol. 169 (Springer, Berlin, 1982).
129. K. Sundermeyer, *Symmetries in Fundamental Physics*, 2nd edn. (Springer, Heidelberg, 2014).
130. L. B. Szabados, "Canonical Pseudotensors, Sparling's Form and Noether Currents," *Class. Quantum Grav.* **9** (1992) 2521.
131. L. B. Szabados, "Quasi-Local Energy-Momentum and Angular Momentum in General Relativity," *Living Rev. Relativ.* **12** (2009) 4.
132. L. B. Szabados, "On the Roots of the Poincare Structure of Asymptotically Flat Space-times," *Class. Quantum Grav.* **20** (2003) 2627, arXiv:gr-qc/0302033.
133. L. B. Szabados, "Quasi-local energy–momentum and angular momentum in GR: The covariant lagrangian approach," unpublished draft (2005).
134. L. B. Szabados, "The Poincaré structure and the centre-of-mass of asymptotically flat space-times," in *Mathematical Relativity: New Ideas and Developments*, eds. J. Frauendiener, D. Giulini and V. Perlick, Lecture Notes in Physics, Vol. 692 (Springer, Berlin, 2006), pp. 157–184.
135. M. A. Tavel, *Transport Theory Statist. Phys.* **1** (1971) 183. English translation of Noether (1918): Ref. [91].
136. I. T. Todorov, "Einstein and Hilbert: The creation of general relativity," arXiv:0504179 [physics].
137. R. C. Tolman, *Relativity Thermodynamics and Cosmology* (Oxford University Press, London, 1934), Eq. 89.3.
138. A. Trautman, "Conservation laws in general relativity," in *Gravitation: An Introduction to Current Research*, ed. L. Witten (Wiley, New York, 1962), pp. 169–198.
139. R. Utiyama, "Invariant Theoretical Interpretation of Interaction," *Phys. Rev.* **101** (1956) 1597.
140. R. Utiyama, "Theory of Invariant Variation and the Generalized Canonical Dynamics," *Prog. Theor. Phys. Suppl.* **9** (1959) 19.
141. V. P. Vizgin, "On the discovery of the gravitational field equations by Einstein and Hilbert: New materials," *Physics-Uspekhi* **44** (2001) 1283.

142. C. von Westenholz, *Differential Forms in Mathematical Physics* (North Holland, Amsterdam, 1978).
143. M.-T. Wang and S.-T. Yau, "Quasilocal mass in general relativity," *Phys. Rev. Lett.* **102** (2009) 021101, arXiv:0804.1174 [gr-qc].
144. M.-T. Wang and S.-T. Yau, "Isometric Embeddings into the Minkowski Space and New Quasi-local Mass," *Commun. Math. Phys.* **288** (2009) 919, arXiv:0805.1370.
145. S. Weinberg, *Gravitation and Cosmology* (Wiley, New York, 1972).
146. H. Weyl, *Space, Time, Matter* (Dover, New York, 1952).
147. H. Weyl, *Sitzungberichte der Königlich-preussischen Akademie der Wissenschaften zu Berlin* **26** (1918) 465; English translation in O'Raifeartaigh [1997], Ref. 96, pp. 24–37.
148. H. Weyl, "Elektron und Gravitation. I," *Zeit. fur Physik* **56** (1929) 330; English translation in O'Raifeartaigh [1997], Ref. 96, pp. 121–144.
149. M.-F. Wu, C.-M. Chen, J.-L. Liu and J. M. Nester, "Optimal Choices of Reference for a Quasi-local Energy: Spherically Symmetric Spacetimes," *Phys. Rev. D* **84** (2011) 084047, arXiv:1109.4738 [gr-qc].
150. M.-F. Wu, C.-M. Chen, J.-L. Liu and J. M. Nester, "Quasi-local Energy for Spherically Symmetric Spacetimes," *Gen. Relativ. Gravit.* **44** (2012) 2401, arXiv:1206.0506 [gr-qc].
151. X.-N. Wu, C.-M. Chen and J. M. Nester, "Quasi-local Energy-momentum and Energy Flux at Null Infinity," *Phys. Rev. D* **71** (2005) 124010, arXiv:gr-qc/0505018.
152. C. N. Yang, "Integral Formalism for Gauge Fields," *Phys. Rev. Lett.* **33** (1974) 445.
153. C.-N. Yang, "Conceptual origin of Maxwell equations and of gauge theory", Colloquium, National Taiwan University, 17 March 2015, (https://www.youtube.com/user/NCTSNorthPhysics/videos).
154. C.-N. Yang and R. Mills, "Conservation of Isotopic Spin and Isotopic Gauge Invariance," *Phys. Rev.* **96** (1954) 191.

Part II. Empirical Foundations

Chapter 5

Equivalence principles, spacetime structure and the cosmic connection

Wei-Tou Ni

Center for Gravitation and Cosmology,
Department of Physics, National Tsing Hua University,
No. 101, Kuang Fu II Rd., Hsinchu, ROC 30013

weitou@gmail.com

After reviewing the meaning of various equivalence principles and the structure of electrodynamics, we give a fairly detailed account of the construction of the light cone and a core metric from the equivalence principle for photons (no birefringence, no polarization rotation and no amplification/attenuation in propagation) in the framework of linear electrodynamics using cosmic connections/observations as empirical support. The cosmic nonbirefringent propagation of photons independent of energy and polarization verifies the Galileo Equivalence Principle (Universality of Propagation) for photons/ electromagnetic wave packets in spacetime. This nonbirefringence constrains the spacetime constitutive tensor to high precision to a core metric form with an Abelian axion degree and a dilaton degree of freedom. Thus comes the metric with axion and dilation. Constraints on axion and dilaton from astrophysical/cosmic propagation are reviewed. Eötvös-type experiments, Hughes–Drever-type experiments, redshift experiments then constrain and tie this core metric to agree with the matter metric, and hence a unique physical metric and universality of metrology. We summarize these experiments and review how the Galileo equivalence principle constrains the Einstein Equivalence Principle (EEP) theoretically. In local physics this physical metric gives the Lorentz/Poincaré covariance. Understanding that the metric and EEP come from the vacuum as a medium of electrodynamics in the linear regime, efforts to actively look for potential effects beyond this linear scheme are warranted. We emphasize the importance of doing Eötvös-type experiments or other type experiments using polarized bodies/polarized particles. We review the theoretical progress on the issue of gyrogravitational ratio for fundamental particles and update the experimental progress on the measurements of possible long range/intermediate range spin–spin, spin–monopole and spin–cosmos interactions.

Keywords: Equivalence principles; spacetime structure; general relativity; classical electrodynamics; polarization; spin.

1. Introduction

In the genesis of general relativity, there are two important cornerstones: the Einstein Equivalence Principle (EEP) and the metric as the dynamic quantity of gravitation (see, e.g. Ref. 1). With research activities on cosmology thriving, people have been looking actively for alternative theories of gravity again for more than

30 years. Recent theoretical studies include scalars, pseudoscalars, vectors, metrics, bimetrics, strings, loops, etc. as dynamic quantities of gravity. It is the aim of this review to look for the foundations of gravity and general relativity, especially from an empirical point of view.

Relativity sprang out from Maxwell–Lorentz theory of electromagnetism. Maxwell equations in Gaussian units are

$$\boldsymbol{\nabla} \cdot \boldsymbol{D} = 4\pi\rho, \tag{1a}$$

$$\boldsymbol{\nabla} \times \boldsymbol{H} - \frac{\partial \boldsymbol{D}}{\partial t} = 4\pi \boldsymbol{J}, \tag{1b}$$

$$\boldsymbol{\nabla} \cdot \boldsymbol{B} = 0, \tag{1c}$$

$$\boldsymbol{\nabla} \times \boldsymbol{E} + \frac{\partial \boldsymbol{B}}{\partial t} = 0, \tag{1d}$$

where $\boldsymbol{D}$ is the displacement, $\boldsymbol{H}$ the magnetic field, $\boldsymbol{B}$ the magnetic induction, $\boldsymbol{E}$ the electric field, ρ the electric charge density and $\boldsymbol{J}$ the electric current density. We use units with the nominal light velocity c equal to 1 (see, e.g. Ref. 2, p. 218 (6.28)). With the sources known, from these equations with eight components we are supposed to be able to solve for the unknown fields $\boldsymbol{D}, \boldsymbol{H}, \boldsymbol{B}$ and $\boldsymbol{E}$ with 12 degrees of freedom. These equations form an under determined system unless we supplement them with relations. These relations are the constitutive relation between $(\boldsymbol{D}, \boldsymbol{H})$ and $(\boldsymbol{E}, \boldsymbol{B})$ (or $(\boldsymbol{D}, \boldsymbol{B})$ and $(\boldsymbol{E}, \boldsymbol{H})$):

$$(\boldsymbol{D}, \boldsymbol{H}) = \chi(\boldsymbol{E}, \boldsymbol{B}), \tag{2}$$

where $\chi(\boldsymbol{E}, \boldsymbol{B})$ is a 6-component functional of $\boldsymbol{E}$ and $\boldsymbol{B}$. With the constitutive relation, the unknown degrees of freedom become 6, the Maxwell equations seem to be over determined. Note that if we take the divergence of (1d), by (1c) it is automatically satisfied. Hence (1c) and (1d) (the Faraday tetrad) have only three independent equations. If we take the divergence of (1b), by (1a) it becomes the continuity equation

$$\boldsymbol{\nabla} \cdot \boldsymbol{J} + \frac{\partial \rho}{\partial t} = 0, \tag{3}$$

a constraint equation on sources. Hence, (1a) and (1b) (the Ampère–Maxwell tetrad) have only three independent equations also. To form a complete system of equations, we need equations governing the action of the electric field and magnetic induction on the charge and current. Lorentz force law provides this link and completes the system:

$$\boldsymbol{F} = m\frac{d\boldsymbol{v}}{dt} = q(\boldsymbol{E} + \boldsymbol{v} \times \boldsymbol{B}), \tag{4}$$

where $\boldsymbol{v}$ is the velocity of the charge and $\boldsymbol{F}$ is the force on it due to electric field and magnetic induction.

In 1908, Minkowski[3,4] put the Maxwell equations into geometric form in four-dimensional spacetime with Lorentz covariance using Cartesian coordinates x, y, z and imaginary time it and numbering them as $x_1 \equiv x, x_2 \equiv y, x_3 \equiv z$ and $x_4 \equiv it$. Minkowski defined the 4-dim excitation ${}^{(\mathrm{Mink})}f$ and the 4-dim field strength ${}^{(\mathrm{Mink})}F$ as

$$ {}^{(\mathrm{Mink})}f \equiv ({}^{(\mathrm{Mink})}f_{hk}) \equiv \begin{bmatrix} 0 & H_z & -H_y & -iD_x \\ -H_z & 0 & H_x & -iD_y \\ H_y & -H_x & 0 & -iD_z \\ iD_y & iD_y & iD_z & 0 \end{bmatrix}, \tag{5a} $$

$$ {}^{(\mathrm{Mink})}F \equiv ({}^{(\mathrm{Mink})}F_{hk}) \equiv \begin{bmatrix} 0 & B_z & -B_y & -iE_x \\ -B_z & 0 & B_x & -iE_y \\ B_y & -B_x & 0 & -iE_z \\ iE_y & iE_y & iE_z & 0 \end{bmatrix}. \tag{5b} $$

In terms of these quantities, Minkowski put the Maxwell equation into the 4-dim covariant form:

$$ {}^{(\mathrm{Mink})}f_{hk,h} = -s_k, \tag{6a} $$

$$ {}^{(\mathrm{Mink})}F^*{}_{hk,h} = 0, \tag{6b} $$

with

$$ F^*{}_{hk} \equiv \left(\frac{1}{2}\right) \underset{\sim}{e}_{hklm} F_{lm}, \text{ and } s_k \text{ the 4-current.} \tag{6c} $$

Here $\underset{\sim}{e}_{hklm} = \pm 1, 0$ is the totally antisymmetric Levi-Civita symbol with $\underset{\sim}{e}_{1234} = +1$. Equations (6a), (6b) are covariant in the sense that for the linear transformations with constant coefficients that leave the form

$$ x_h x_h \equiv (x_1)^2 + (x_2)^2 + (x_3)^2 + (x_4)^2 \tag{7} $$

invariant, the Maxwell equations in the form (6a), (6b) are covariant with the 4-dim excitation ${}^{(\mathrm{Mink})}f$ and the 4-dim field strength ${}^{(\mathrm{Mink})}F$ transforming as 4-dim covariant tensors (covariant V-six-vectors).

Bateman[5] used time coordinate t instead of x_4, and considered transformations that leave the invariance of the differential (form) equation:

$$ (dx)^2 + (dy)^2 + (dz)^2 - (dt)^2 = 0. \tag{8} $$

Hence, he also included conformal transformations in addition to Lorentz transformations and made one step toward general coordinate invariance. With indefinite metric, one has to distinguish covariant and contravariant tensors and indices. Aware of this, one can readily put Maxwell equations into covariant form without using imaginary time.

In terms of field strength $F_{kl}(\boldsymbol{E},\boldsymbol{B})$ and excitation (density) $H^{ij}(\boldsymbol{D},\boldsymbol{H})$:

$$F_{kl} = \begin{bmatrix} 0 & E_1 & E_2 & E_3 \\ -E_1 & 0 & -B_3 & B_2 \\ -E_2 & B_3 & 0 & -B_1 \\ -E_3 & -B_2 & B_1 & 0 \end{bmatrix}, \tag{9a}$$

$$H^{ij} = \begin{bmatrix} 0 & -D_1 & -D_2 & -D_3 \\ D_1 & 0 & -H_3 & H_2 \\ D_2 & H_3 & 0 & -H_1 \\ D_3 & -H_2 & H_1 & 0 \end{bmatrix}. \tag{9b}$$

Maxwell equations can be expressed as

$$H^{ij}{}_{,j} = -4\pi J^i, \tag{10a}$$

$$e^{ijkl} F_{jk,l} = 0, \tag{10b}$$

with the constitutive relation (2) between the excitation and the field in the form:

$$H^{ij} = \chi^{ij}(F_{kl}), \tag{11}$$

where J^k is the charge 4-current density $(\rho, \boldsymbol{J})$ and e^{ijkl} the completely anti-symmetric tensor density (Levi-Civita symbol) with $e^{0123} = 1$ (see, e.g. Ref. 6). Here $\chi^{ij}(F_{kl})$ is a functional with six independent degrees of freedom. For medium with a local linear response or in the linear local approximation, (11) reduced to

$$H^{ij} = \chi^{ijkl} F_{kl}, \tag{12}$$

with χ^{ijkl} the (linear) constitutive tensor density.[6–10] For isotropic dielectric and isotropic permeable medium, the constitutive tensor density has two degrees of freedom; for anisotropic dielectric and anisotropic permeable medium, the constitutive tensor density has 12 degrees of freedom; for general linear local medium (with magnetoelectric response), the constitutive tensor has 21 degrees of freedom.

Introducing the metric g_{ij} as gravitational potential in 1913 (Ref. 11) and versed in general (coordinate-)covariant formalism in 1914,[12] Einstein put the Maxwell equations in general covariant form ($\mathfrak{F}^{ij} = H^{ij}$ in our notation)[12]:

$$\mathfrak{F}^{ij}{}_{,j} = -4\pi J^i, \tag{13a}$$

$$F_{ij,k} + F_{jk,i} + F_{ki,j} = 0. \tag{13b}$$

Shortly after Einstein constructed general relativity, Einstein noticed that the Maxwell equations can be formulated in a form independent of the metric gravitational potential in 1916.[13] Einstein introduced the covariant V-six-vector equations

(13a) and (13b) which are independent of metric gravitational potential. Only the constitutive tensor density χ^{ijkl} is dependent on the metric gravitational potential:

$$\mathfrak{F}^{ij} = (-g)^{1/2} g^{ik} g^{jl} F_{kl}. \tag{14}$$

Noticing Einstein's $\mathfrak{F}^{ij}$ is our $\boldsymbol{H}^{ij}$ and putting (14) in the form of (12), we have

$$\chi^{ijkl} = (-g)^{1/2} \left[\left(\frac{1}{2}\right) g^{ik} g^{jl} - \left(\frac{1}{2}\right) g^{il} g^{kj} \right]. \tag{15}$$

In local inertial frame the metric-induced constitutive tensor (15) is reduced to special-relativistic Minkowski form:

$$\chi^{ijkl} = (-g)^{1/2} \left[\left(\frac{1}{2}\right) \eta^{ik} \eta^{jl} - \left(\frac{1}{2}\right) \eta^{il} \eta^{kj} \right] + O(x^i x^j), \tag{16}$$

which is dictated by the EEP.

In macroscopic medium, the constitutive tensor gives the medium-coupling to electromagnetism; it depends on the (thermodynamic) state of the medium and in turn depends on temperature, pressure etc. In gravity, the constitutive tensor gives the gravity-coupling to electromagnetism; it depends on the gravitational field(s) and in turn depends on the matter distribution and its state.

In gravity, a fundamental issue is how to arrive at the metric from the constitutive tensor through experiments and observations. That is, how to build the metric empirically and test the EEP thoroughly. Are there other degrees of freedom to be explored?

Since ordinary energy compared to Planck energy is very small, in this situation we can assume that the gravitational (or spacetime) constitutive tensor is a linear and local function of gravitational field(s), i.e. (12) holds. Since the second half of 1970s, we have started to use the following Lagrangian density L $(= L_{\mathrm{I}}^{(\mathrm{EM})} + L_{\mathrm{I}}^{(\mathrm{EM-P})})$ with the electromagnetic field Lagrangian $L_{\mathrm{I}}^{(\mathrm{EM})}$ and the field–current interaction Lagrangian $L_{\mathrm{I}}^{(\mathrm{EM-P})}$ given by

$$L_{\mathrm{I}}^{(\mathrm{EM})} = -\left(\frac{1}{16\pi}\right) H^{ij} F_{ij} = -\left(\frac{1}{16\pi}\right) \chi^{ijkl} F_{ij} F_{kl}, \tag{17a}$$

$$L_{\mathrm{I}}^{(\mathrm{EM-P})} = -A_k J^k, \tag{17b}$$

for studying this issue.[14–16] Here $\chi^{ijkl} = -\chi^{jikl} = \chi^{klij}$ is a tensor density of the gravitational fields or matter fields to be investigated, $F_{ij} \equiv A_{j,i} - A_{i,j}$ the electromagnetic field strength tensor with A_i the electromagnetic 4-potential and comma denoting partial derivation, and J^k the charge 4-current density. The Maxwell equations (10a), (10b) or (1a)–(1d) can be derived from this Lagrangian with the relation (12) and (9a), (9b). Using this χ-framework, we have demonstrated the construction of the light cone core metric from the experiments and observations as in Table 1.[17] After presenting the meaning of various equivalence principles in Sec. 2 and the structure of premetric electrodynamics in Sec. 3.1, we give a fairly detailed account of the construction of the metric together with constraints on

Table 1. Constraints on the spacetime constitutive tensor χ^{ijkl} and construction of the spacetime structure (metric + Abelian axion field φ+ dilaton field ψ) from experiments/observations in skewonless case (U: Newtonian gravitational potential). g_{ij} is the particle metric.[17]

Experiment	Constraints	Accuracy
Pulsar signal propagation	$\chi^{ijkl} \to (1/2)(-h)^{1/2}[h^{ik}h^{jl} - h^{il}h^{kj}]\psi + \varphi e^{ijkl}$	10^{-16}
Radio galaxy observation		10^{-32}
Gamma ray burst (GRB)		10^{-38}
Cosmic Microwave Background (CMB) spectrum measurement	$\psi \to 1$	8×10^{-4}
Cosmic polarization rotation (CPR) experiment	$\varphi - \varphi_0(\equiv \alpha) \to 0$	$\lvert\langle\alpha\rangle\rvert < 0.02$, $\langle(\alpha - \langle\alpha\rangle)^2\rangle^{1/2} < 0.03$
Eötvös–Dicke–Braginsky experiments	$\psi \to 1$	$10^{-10}\,U$
	$h_{00} \to g_{00}$	$10^{-6}\,U$
Vessot–Levine redshift experiment	$h_{00} \to g_{00}$	$1.4 \times 10^{-4}\Delta U$
Hughes–Drever-type experiments	$h_{\mu\nu} \to g_{\mu\nu}$	10^{-24}
	$h_{0\mu} \to g_{0\nu}$	10^{-19}–10^{-20}
	$h_{00} \to g_{00}$	10^{-16}

Abelian axions, dilatons and skewons from the equivalence principle for photons in the framework of premetric electrodynamics using cosmic observations as empirical support in Secs. 3.2–3.6. Section 3.7 discusses the special case of spacetime/medium with constitutive tensor induced by asymmetric metric and its special role. Section 3.8 addresses the issue of empirical foundation of the closure relation.

In Sec. 4, we review theorems and relations among various equivalence principles using the χ-framework including particles and the corresponding χ-framework for the nonabelian field. In Sec. 5, we discuss the relation of universal metrology and equivalence principles. In Sec. 6, we review theoretical works on the gyrogravitational effects. In Sec. 7, experimental progress on the measurement of long range/intermediate range spin–spin, spin–monopole and spin–cosmos interactions is updated. In Sec. 8, prospects are discussed.

2. Meaning of Various Equivalence Principles

Our common understanding and formulation of gravity can be simply described in Table 2. Matter produces gravitational field and gravitational field influences matter. In Newtonian theory of gravity,[18] the Galileo Weak Equivalence Principle (WEP I)[19] determines how matter behaves in a gravitational field, and Newton's inverse square law determines how matter produces gravitational field. In a relativistic theory of gravity such as a metric theory, the EEP determines how matter behaves in a gravitational field, and the field equations determine how matter produces gravitational field(s). In Einstein's general relativity, with a suitable choice of the stress–energy tensor, the Einstein equation can imply the EEP. In nonmetric

Table 2. Gravity and electromagnetism.

Matter	$\xrightarrow{\text{produces}}$ Gravitational Field(s)	$\xrightarrow{\text{influence(s)}}$ Matter
Newtonian Gravity	Inverse Square Law	WEP[I]
Relativistic Gravity	Field Equation(s) e.g., Einstein equation	EEP or substitute
Charges	$\xrightarrow{\text{produce}}$ Electromagnetic Field	$\xrightarrow{\text{influences}}$ Charges
Electromagnetism	Maxwell Equations	Lorentz Force Law

theories of gravity, other versions of equivalence principles may be used. The above situations can be summarized in Table 2 together with those for electromagnetism.

From Table 2, we see the crucial role played by equivalence principles in the formulation of gravity. In the following, we start with the ancient concepts of inequivalence and discuss meaning of various equivalence principles. This section is an update of Sec. 2 of Ref. 16.

2.1. *Ancient concepts of inequivalence*

From the observations that heavy bodies fall faster than light ones in the air, ancient people, both in the orient and in the west, believe that objects with different constituents behave differently in a gravitational field. We now know that this is due to the inequivalent responses to different buoyancy forces and air resistances.

2.2. *Macroscopic equivalence principles*

(i) *Galileo equivalence principle* (WEP I)[19]

Using an inclined plane, Galileo (1564–1642) showed that the distance a falling body travels from rest varies as the square of the time. Therefore, the motion is one of constant acceleration. Moreover, Galileo wrote that "the variation of speed in air between balls of gold, lead, copper, porphyry, and other heavy materials is so slight that in a fall of 100 cubits (about 46 m) a ball of gold would surely not outstrip one of copper by as much as four fingers. Having observed this, I came to the conclusion that in a medium totally void of resistance all bodies would fall with the same speed (together)"[19]; thus Galileo had grasped an equivalence in gravity. (Galileo had demonstrated the equivalence in his experiment on the inclined plane around 1592.) The last conjecture is the famous Galileo equivalence principle; it serves as the beginning of our understanding of gravity. More precisely, Galileo equivalence principle states that in a gravitational field, the trajectory of a test body with a given initial velocity is independent of its internal structure and composition (universality of free fall trajectories).

From Galileo's observations, one can arrive at the following two well-known conclusions:

(a) The gravitational force (weight) at the top of the inclined plane and that at a middle point of the inclined plane can be regarded the same to the experimental limits in those days. Hence a falling body experiences a constant force (its weight). The motion of a falling body is one of constant acceleration. Therefore a constant force f induces a motion of constant acceleration a. Hence force and acceleration (not velocity) are closely related. If one changes the inclinations of the plane to get different "dilutions" of gravity, one finds

$$f \propto a \tag{18}$$

for a falling body. From Galileo's observation of the universality of free fall trajectories, we know that the acceleration a is the same for different bodies. But f (weight) is proportional to mass m. Hence for different bodies,

$$\frac{f}{m} \propto a. \tag{19}$$

If one chooses appropriate units, one arrives at

$$f = ma \tag{20}$$

for falling bodies. If one further assumes that all kinds of forces are equivalent in their ability to accelerate and notices the vector nature of forces and accelerations, one would arrive at Newton's second law,

$$\boldsymbol{f} = m\boldsymbol{a}. \tag{21}$$

(b) From Galileo equivalence principle, the gravitational field can be described by the acceleration of gravity $\boldsymbol{g}$. Newton's second law for N particles in external gravitational field $\boldsymbol{g}$ is

$$m_I \frac{d^2\boldsymbol{x}_I}{dt^2} = m_I \boldsymbol{g}(\boldsymbol{x}_I) + \sum_{J=1}^{N} \boldsymbol{F}_{IJ}(\boldsymbol{x}_I - \boldsymbol{x}_J) \quad (I = 1, \ldots, N; J \neq I), \tag{22}$$

where $\boldsymbol{F}_{IJ}$ is the force acting on particle I by particle J. At a point $\boldsymbol{x}_0$, expand $\boldsymbol{g}(\boldsymbol{x}_I)$ as follows

$$\boldsymbol{g}(\boldsymbol{x}_I) = \boldsymbol{g}_0 + \underset{\sim}{\lambda} \cdot (\boldsymbol{x}_I - \boldsymbol{x}_0) + O(|\boldsymbol{x}_I - \boldsymbol{x}_0|^2). \tag{23}$$

Choosing $\boldsymbol{x}_0$ as origin and applying the following non-Galilean spacetime coordinate transformation

$$\boldsymbol{x}' = \boldsymbol{x} - \left(\frac{1}{2}\right) \boldsymbol{g}_0 t^2, \quad t' = t, \tag{24}$$

(22) is transformed to

$$\frac{m_I d^2\boldsymbol{x}'_I}{dt'^2} = \sum_{J=1}^{N} \boldsymbol{F}_{IJ}(\boldsymbol{x}'_I - \boldsymbol{x}'_J) + O(\boldsymbol{x}'_K) \quad (I = 1, \ldots, N; J \neq I). \tag{25}$$

Thus we see that locally the effect of external gravitational field can be transformed away. Thus we arrive at a Strong Equivalence Principle (SEP). Therefore in Newtonian mechanics,

Galileo Weak Equivalence Principle $\Leftrightarrow$ Strong Equivalence Principle.

In the days of Galileo and Newton, the nature of light and radiation was controversial and had to wait for further development to clarify it.

(ii) *The second weak equivalence principle* (WEP II)

Since the motion of a macroscopic test body is determined not only by its trajectory but also by its rotation state, we have proposed from our previous studies[20,21] the following stronger weak equivalence principle to be tested by experiments, which states that in a gravitational field, the motion of a test body with a given initial motion state is independent of its internal structure and composition (universality of free fall motions). By a test body, we mean a macroscopic body whose size is small compared to the length scale of the inhomogeneities of the gravitational field. The macroscopic body can have an intrinsic angular momentum (spin) including net quantum spin.

2.3. *Equivalence principles for photons (wave packets of light)*

(i) *WEP I for photons* (wave packets of light):

In analogue to the Galileo equivalence principle for test bodies, the WEP I for photons states that the spacetime trajectory of light in a gravitational field depends only on its initial position and direction of propagation, does not depend on its frequency (energy) and polarization.

(ii) *WEP II for photons* (wave packets of light):

The trajectory of light in a gravitational field depends only on its initial position and direction of propagation, not dependent of its frequency (energy) and polarization; the polarization state of the light does not change, e.g. no polarization rotation for linear polarized light; and there is no amplification/attenuation of light.

N.B. We consider the propagation (or trajectory) in eikonal approximation, i.e. in geometrical optics approximation. The wavelength must be small (just like a test body) than the inhomogeneity scale of the gravitational field.

2.4. *Microscopic equivalence principles*

The development of physics in the 19th century brought to improved understanding of light and radiations and to the development of special relativity. In 1905, Einstein[22] postulated the equivalence of mass and energy and proposed the famous Einstein formula $E = mc^2$. A natural question came in at this point: How light and radiations behave in a gravitational field? In 1889, Eötvös[23] experiment showed that inertial mass and gravitational mass are equal to a high precision of 10^{-8}. In June, 1907, Planck[24] reasoned that since all energies have inertial properties, all energies must gravitate. This paved the way to include the energy in the formulation of equivalence principle.

N.B. Since the power of EEP only reaches the gradient of gravity potential, it applies only to a region where the second-order gradients or curvature can be neglected. In applying the equivalence principle to wave packets or a microscopic wave function, we have to assume that the extension is limited to such a region. For example, it should not be applied to a long-distance entangled state.

(i) *Einstein equivalence principle*

Two years after the proposal of special relativity and the formula $E = mc^2$, six months after Planck reasoned that all energy must gravitate, Einstein,[25] in the last part (Principle of Relativity and Gravitation) of his comprehensive 1907 essay on relativity, proposed the complete physical equivalence of a homogeneous gravitational field to a uniformly accelerated reference system: "We consider two systems of motion, Σ_1 and Σ_2. Suppose Σ_1 is accelerated in the direction of its X-axis, and γ is the magnitude (constant in time) of this acceleration. Suppose Σ_2 is at rest, but situated in a homogeneous gravitational field, which imparts to all objects an acceleration $-\gamma$ in the direction of the X-axis. As far as we know, the physical laws with respect to Σ_1 do not differ from those with respect to Σ_2, this derives from the fact that all bodies are accelerated alike in the gravitational field. We have therefore no reason to suppose in the present state of our experience that the systems Σ_1 and Σ_2 differ in any way, and will therefore assume in what follows the complete physical equivalence of the gravitational field and the corresponding acceleration of the reference system."[a] From this equivalence, Einstein derived clock and energy redshifts in a gravitational field. When applied to a spacetime region where inhomogeneities of the gravitational field can be neglected, this equivalence dictates the behavior of matter in gravitational field. The postulate of this equivalence is called the EEP. EEP is the cornerstone of the gravitational coupling of matter and nongravitational fields in general relativity and in metric theories of gravity.

EEP is a microscopic principle and may mean slightly different things for different people. To most people, EEP is equivalent to the coma-goes-to-semicolon rule for matter (not including gravitational energy) in gravitational field. Therefore, EEP means that in any and every local Lorentz (inertial) frame, anywhere and

[a]Einstein further clarified the application of this equivalence to inhomogeneous field, e.g. in his book "The Meaning of Relativity" (p. 58, Fifth edition, Princeton University Press, 1955): "... We may look upon the principle of inertia as established, to a high degree of approximation, for the space of our planetary system, provided that we neglect the perturbations due to the sun and planets. Stated more exactly, there are finite regions, where, with respect to a suitably chosen space of reference, material particles move freely without acceleration, and in which the laws of special relativity, which have been developed above, hold with remarkable accuracy. Such regions we shall call "Galilean regions". We shall proceed from the consideration of such regions as a special case of known properties".

anytime in the universe, all the (nongravitational) laws of physics must take on their familiar special-relativistic forms.[26] *That is, local (nongravitational) physics should be universally special relativistic.* In other words, EEP says that the outcome of any local, nongravitational test experiment is independent of the velocity of the apparatus. For example, the fine structure constant $\alpha = e^2/\hbar c$ must be independent of location, time, and velocity.

(ii) *Modified Einstein equivalence principle*

In 1921, Eddington[27] mentioned the notion of an asymmetric affine connection in discussing possible extensions of general relativity. In 1922, Cartan[28] introduced torsion as the anti-symmetric part of an asymmetric affine connection and laid the foundation of this generalized geometry. Cartan[29] proposed that the torsion of spacetime might be connected with the intrinsic angular momentum of matter. In 1921–1922, Stern and Gerlach[30] discovered the space quantization of atomic magnetic moments. In 1925-1926, Goudsmit and Uhlenbeck[31] introduced our present concept of electron spin as the culmination of a series of studies of doublet and triplet structures in spectra. Following the idea of Cartan, Sciama[32,33] and Kibble[34] developed a theory of gravitation which is commonly called the Einstein–Cartan–Sciama–Kibble (ECSK) theory of gravity.

After the works of Utiyama,[35] Sciama[32,33] and Kibble,[34] interest and activities in gauge-type and torsion-type theories of gravity have continuously increased. Various different theories postulate somewhat different interaction of matter with gravitational field(s). In ECSK theory, in Poincaré gauge theories[36,37] and in some other torsion theories, there is a torsion gravitational field besides the usual metric field.[38] In special relativity, if we use a nonholonomic tetrad frame, there is an antisymmetric part of the affine connection. Therefore many people working on torsion theory take the equivalence principle to mean something different from EEP so that torsion can be included. This is most clearly stated in von der Heyde's paper "The Equivalence Principle in the U_4 Theory of Gravitation" [39]: *Locally the properties of special relativistic matter in a noninertial frame of reference cannot be distinguished from the properties of the same matter in a corresponding gravitational field.* This Modified Einstein Equivalence Principle (MEEP) allows for formal inertial effects in a nonholonomic tetrad frame and hence allows torsion. There are two ways to treat the level of coupling of torsion; one can consider torsion on the same level as symmetric affine connection (MEEP I) or one can consider torsion on the same level as curvature tensors (MEEP II). Hehl and von der Heyde[39] hold the second point of view. MEEP I allows torsion. Since torsion is a tensor, it cannot be transformed away in any frame if it is not zero. EEP is equivalent to MEEP I plus no torsion; therefore we have EEP implies MEEP I but MEEP I does not implies EEP. For a test body, curvature effects are neglected; so MEEP II is essentially equivalent to EEP for test bodies. Test bodies with nonvanishing total intrinsic spin feel torques from the torsion field. Hence MEEP I does not imply

WEP II. Moreover MEEP I does not imply WEP I either.[40] Therefore we have the following:

EEP ⇨ MEEP I
*⇑⇓ ⬀ ⬁
WEP II ⇨ WEP I

*WEP II implies EEP is proved for an electromagnetic system in $\chi - g$ framework.[20,21,41] However, for other frameworks, the issue is still open.

2.5. *Equivalence principles including gravity (Strong equivalence principles)*

How does gravitational energy behave in a gravitational field? Is local gravity experiment depending on where and when in the universe it is performed? These involve nonlinear gravity effects.

(i) *WEP I for massive bodies*

This weak equivalence principle says that in a gravitational field, the trajectory of a massive test body with a given initial velocity is also independent of the amount of gravitational self-energy inside the massive body. In Brans–Dicke theory and many other theories, there are violations of this equivalence principle. The violations are called Nordtvedt effects.[42,43] General relativity obeys WEP I for massive bodies in the post-Newtonian limit and for black hole solutions. The nonexistence of Nordtvedt effects is an efficient way to single out purely metric theory among metric theories of gravity (those comply with EEP). From lunar laser ranging experiment and binary pulsar timing observations, the Nordtvedt effect is limited.

(ii) *Dicke's*[44] *strong equivalence principle*

This is a microscopic equivalence principle. It says that the outcome of any local test experiment — gravitational or nongravitational — is independent of where and when in the universe it is performed, and independent of the velocity of the apparatus. If this equivalence principle is valid, the Newtonian gravitational constant G_N should be a true constant. Brans–Dicke theory with its variable "gravitational constant" as measured by Cavendish experiments satisfies EEP but violates SEP. Also, if this equivalence principle is valid, a self-gravitating system in background with length scale much larger than the self-gravitating system should have locally Lorentz invariance in the background, e.g. no preferred-frame effects.[45,46]

The violations of SEP seem to be linked with the violations of WEP I for massive bodies in many cases. It is interesting to know how SEP and WEP I for massive bodies are connected. The violations of SEP may also be connected to the violations of WEP I at some level in some cases.

We note in passing that there are other versions of equivalence principles which we are not able to list them here one-by-one. For recent discussions on equivalence principles, see also Refs. 47 and 48.

2.6. *Inequivalence and interrelations of various equivalence principles*

In the preceding subsections, we have listed and explained various equivalence principles. Logically all these equivalence principles are different. An important issue is that to what extent they are equivalent, and in what situations they are inequivalent. This issue became conspicuous for more than 50 years since Dicke–Schiff redshift controversy. In 1960, Schiff[49] argued as follows: "The Eötvös experiments show with considerable accuracy that the gravitational and inertial masses of normal matter are equal. This means that the ground state eigenvalue of the Hamiltonian for this matter appears equally in the inertial mass and in the interaction of this mass with a gravitational field. It would be quite remarkable if this could occur without the entire Hamiltonian being involved in the same way, in which case a clock composed of atoms whose motions are determined by this Hamiltonian would have its rate affected in the expected manner by a gravitational field." He suggested that EEP and, hence, the metric gravitational redshift are consequences of WEP I. In short, Schiff believes that

$$\text{WEP I} \Rightarrow \text{EEP}.$$

This conjecture is known as Schiff's conjecture. The scope of validity of Schiff's conjecture has great importance in the analysis of the empirical foundations of EEP.

However Dicke[50] held a different point of view and believed that the redshift experiment has independent theoretical significance. In November 1970, the interests in the issue of the validity of Schiff's conjecture were rekindled during a vigorous argument between Schiff and Thorne at the Caltech-JPL Conference on Experimental Tests of Gravitation Theories. In 1973, Thorne, Lee and Lightman[51] analyzed the fundamental concepts and terms involved in detail and gave a plausibility argument supporting Schiff's conjecture. Lightman and Lee[52] proved Schiff's conjecture for electromagnetically interacting systems in a static, spherically symmetric gravitational field using the $TH\varepsilon\mu$ formalism. I found a nonmetric theory which includes pseudoscalar–photon interaction and showed that it is a counterexample to Schiff's conjecture.[53] In 1974, I showed that this counterexample is the only case in a general premetric constitutive tensor formulation of electromagnetism (χ-framework) with standard particle Lagrangian (The whole framework is called the $\chi - g$ framework.).[20,21] This supports that the approach of Schiff is right in the large, although not completely right. In the eikonal approximations of the $\chi - g$ framework, I showed that the first-order gravitational redshifts are metric[21] (so Schiff was right for redshift in this case to first-order). In the latter part of 1970s, I use the $\chi - g$ framework to look into the issue of gravitational coupling to electromagnetism empirically.[14–16,40] In the next section, we will review the progress for this issue. Recently, the significance of redshift experiments is brought up again in the comparison of redshift and atom interferometry experiment.[54,55]

For the SEP, one could ask similar questions. Would WEP I for massive body imply Dicke's SEP? This is a direct extension of Schiff's conjecture. One can call it Schiff's conjecture for massive bodies. There are significant progresses recently. Gérard[56] has worked out a link between the vanishing of Nordtvedt effects and a condition of SEP. Di Casola, Leberati and Sonego[57] have employed WEP I for massive bodies as a sieve for purely metric theories of gravity using variational approach. They also propose the conjecture that SEP is equivalent to the union of WEP I for massive bodies (GWEP in their term) and EEP. Since WEP I does not imply EEP (Schiff's conjecture is incorrect),[20,21,53] we would like to propose to investigate the validity of the following two statements in various frameworks: (i) WEP II for massive bodies is equivalent to Dicke's SEP; (ii) SEP is equivalent to the union of WEP II for massive bodies (GWEP in their term) and EEP.

3. Gravitational Coupling to Electromagnetism and the Structure of Spacetime

3.1. *Premetric electrodynamics as a framework to study gravitational coupling to electromagnetism*

For the ordinary gravitational field, it is a low energy situation compared to Planck energy, as we mentioned in Sec. 1. If we represent the gravitational coupling to electromagnetism by constitutive tensor density, the constitutive tensor density must be linear and local as given by (12), independent of the field strength F_{kl}, dependent only on the gravitational field(s). The constitutive tensor density (12) has three irreducible pieces. Both H^{ij} and F_{kl} are antisymmetric, hence χ^{ijkl} must be antisymmetric in i and j, and k and l. Therefore the constitutive tensor density χ^{ijkl} has 36 (6×6) independent components. A general linear constitutive tensor density χ^{ijkl} in electrodynamics can first be decomposed into two parts, the symmetric part in the exchange of index pairs ij and kl $[(1/2)(\chi^{ijkl} + \chi^{klij})]$ and the antisymmetric part in the exchange of index pairs ij and kl $[(1/2)(\chi^{ijkl} - \chi^{klij})]$. The first part has 21 degrees of freedom and contains the totally antisymmetric part — the axion part (Ax). Subtracting the axion part, the remaining part is the principal part which has 20 degrees of freedom. The second part is the skewon part and has 15 degrees of freedom. The principal part (P), the Abelian axion part (Ax) and the Hehl–Obukhov–Rubilar skewon part (Sk) constitute the three irreducible parts under the group of general coordinate transformations[6]:

$$\chi^{ijkl} = {}^{(\mathrm{P})}\chi^{ijkl} + {}^{(\mathrm{Sk})}\chi^{ijkl} + {}^{(\mathrm{Ax})}\chi^{ijkl} \quad (\chi^{ijkl} = -\chi^{jikl} = -\chi^{ijlk}) \tag{26}$$

with

$${}^{(\mathrm{P})}\chi^{ijkl} = \left(\frac{1}{6}\right)[2(\chi^{ijkl} + \chi^{klij}) - (\chi^{iklj} + \chi^{ljik}) - (\chi^{iljk} + \chi^{jkil})], \tag{27a}$$

$${}^{(\mathrm{Ax})}\chi^{ijkl} = \chi^{[ijkl]} = \varphi e^{ijkl}, \tag{27b}$$

$${}^{(\mathrm{Sk})}\chi^{ijkl} = \left(\frac{1}{2}\right)(\chi^{ijkl} - \chi^{klij}). \tag{27c}$$

Decomposition (26) is unique. If we substitute (26) into (17a), the skewon part does not contribute to the Lagrangian; hence, for Lagrangian based theory, it is skewonless. The systematic study of skewonful cases started in 2002 (see, e.g. Ref. 6).

The complete agreement with EEP for photon sector requires (as locally in special relativity) (i) no birefringence; (ii) no polarization rotation; (iii) no amplification/no attenuation in spacetime propagation. In Secs. 3.2–3.5, we review how cosmic connection/observation of these three conditions on electromagnetic propagation verifies EEP and determination of the spacetime structure in the skewonless case (Lagrangian–based case). In Sec. 3.2, we derive wave propagation and dispersion relations in the lowest eikonal approximation in weak field in the premetric electrodynamics. In Sec. 3.3, we apply it to the determination of the spacetime structure in the skewonless case using no birefringence condition. With no birefringence, any skewonless spacetime constitutive tensor must be of the form

$$\chi^{ijkl} = (-h)^{1/2}\left[\left(\frac{1}{2}\right) h^{ik}h^{jl} - \left(\frac{1}{2}\right) h^{il}h^{kj}\right]\psi + \varphi e^{ijkl}, \tag{28}$$

where h^{ij} is a metric constructed from $\chi^{ijkl}(h = \det(h_{ij})$ and h_{ij} the inverse of $h^{ij})$ which generates the light cone for electromagnetic wave propagation, ψ a dilaton field constructed from χ^{ijkl} and φ an Abelian axion field constructed from χ^{ijkl}. Observations on no birefringence of cosmic propagation of electromagnetic waves constrain the spacetime constitutive tensor to the form (28) to very high precision. In Sec. 3.4, we review the derivation of the dispersion relation of wave propagation in dilaton field and axion field with constitutive relation (28); we show further that with the condition of no polarization rotation and the condition of no amplification/no attenuation satisfied, the axion φ and the dilaton ψ should be constant, i.e. no varying axion field and no varying dilaton field respectively. The EEP for photon sector would then be observed; the spacetime constitutive tensor density would be of metric-induced form. Thus we tie the three observational conditions to EEP and to metric-induced spacetime constitutive tensor density in the photon sector. In Sec. 3.5, we review the empirical constraint on cosmic dilaton field and cosmic axion field. The results are summarized in Table 1. In Sec. 3.6, we apply the dispersion relations derived in Sec. 3.2 to the case of metric induced constitutive tensor with skewons with further discussions. In Sec. 3.7, we discuss the case of spacetime with asymmetric-metric induced constitutive tensor using Fresnel equation. In Sec. 3.8, we review the application of these results to the accuracy of empirical verification of the closure relations in electrodynamics.

3.2. *Wave propagation and the dispersion relation*

The sourceless Maxwell equation (10b) is equivalent to the local existence of a 4-potential A_i such that

$$F_{ij} = A_{j,i} - A_{i,j}, \tag{29}$$

with a gauge transformation freedom of adding an arbitrary gradient of a scalar function to A_i. The Maxwell equation (10a) in vacuum is

$$(\chi^{ijkl}A_{k,l})_{,j} = 0. \tag{30}$$

Using the derivation rule, we have

$$\chi^{ijkl}A_{k,l,j} + \chi^{ijkl}_{,j}A_{k,l} = 0. \tag{31}$$

(i) For slowly varying, nearly homogeneous field/medium, and/or (ii) in the eikonal approximation with typical wavelength much smaller than the gradient scale and time-variation scale of the field/medium, the second term in (31) can be neglected compared to the first term, and we have

$$\chi^{ijkl}A_{k,lj} = 0. \tag{32}$$

This approximation is the lowest eikonal approximation, usually also called the eikonal approximation. In this approximation, the dispersion relation is given by the generalized covariant quartic Fresnel equation (see, e.g. Ref. 6; also Sec. 3.7). It is well-known that axion does not contribute to this dispersion relation[6,14–16,58–61] as we will see in the following. In this subsection, we use this lowest eikonal approximation and follow Ref. 62 to derive dispersion relation in the general linear local constitutive framework. In Sec. 3.4, we keep the second term of (31) and follow Ref. 63 to find out dispersion relations for the case that the dilaton gradient and the axion gradient cannot be neglected.

In the weak field or dilute medium, we assume

$$\chi^{ijkl} = \chi^{(0)ijkl} + \chi^{(1)ijkl} + O(2), \tag{33}$$

where $O(2)$ means second-order in $\chi^{(1)}$. Since the violation from the EEP would be small and/or if the medium is dilute, in the following we assume that

$$\chi^{(0)ijkl} = \left(\frac{1}{2}\right)g^{ik}g^{jl} - \left(\frac{1}{2}\right)g^{il}g^{kj}, \tag{34}$$

and $\chi^{(1)ijkl}$ is small compared with $\chi^{(0)ijkl}$. We can then find a local inertial frame such that g^{ij} becomes the Minkowski metric η^{ij} good to the derivative of the metric. To look for wave solutions, we use eikonal approximation and choose z-axis in the wave propagation direction so that the solution takes the following form:

$$A = (A_0, A_1, A_2, A_3)e^{ikz-i\omega t}. \tag{35}$$

We expand the solution as

$$A_i = [A_i^{(0)} + A_i^{(1)} + O(2)]e^{ikz-i\omega t}. \tag{36}$$

Imposing radiation gauge condition in the zeroth-order in the weak field/dilute medium/weak EEP violation approximation, we find the zeroth-order solution of (36) and the zeroth order dispersion relation satisfying the zeroth-order equation $\chi^{(0)ijkl}A^{(0)}_{k,lj} = 0$ as follows:

$$A^{(0)} = (0, A_1^{(0)}, A_2^{(0)}, 0), \quad \omega = k + O(1). \tag{37}$$

Substituting (36) and (37) into Eq. (32), we have

$$\chi^{(1)ijkl}A^{(0)}_{k,lj} + \chi^{(0)ijkl}A^{(1)}_{k,lj} = 0 + O(2). \tag{38}$$

The $i = 0$ and $i = 3$ components of (38) both give

$$A^{(1)}_0 + A^{(1)}_3 = 2(\chi^{(1)3013} - \chi^{(1)3010})A^{(0)}_1 + 2(\chi^{(1)3023} - \chi^{(1)3020})A^{(0)}_2 + O(2). \tag{39}$$

Since this equation does not contain ω and k, it does not contribute to the determination of the dispersion relation. A gauge condition in the $O(1)$ order fixes the values of $A^{(1)}_0$ and $A^{(1)}_3$.

The $i = 1$ and $i = 2$ components of (38) are

$$\left(\frac{1}{2}\right)(\omega^2 - k^2)A^{(0)}_1 + \chi^{(0)1jkl}A^{(1)}_{k,lj} + \chi^{(0)1jkl}A^{(0)}_{k,lj} = 0 + O(2), \tag{40a}$$

$$-\left(\frac{1}{2}\right)(\omega^2 - k^2)A^{(0)}_2 + \chi^{(0)2jkl}A^{(1)}_{k,lj} + \chi^{(1)2jkl}A^{(0)}_{k,lj} = 0 + O(2). \tag{40b}$$

These two equations determine the dispersion relation and can be rewritten as

$$\left[\left(\frac{1}{2}\right)(\omega^2 - k^2) - k^2 A_{(1)}\right]A^{(0)1} - k^2 B_{(1)}A^{(0)2} = O(2), \tag{41a}$$

$$-k^2 B_{(2)}A^{(0)}_1 + \left[\left(\frac{1}{2}\right)(\omega^2 - k^2) - k^2 A_{(2)}\right]A^{(0)}_2 = O(2), \tag{41b}$$

where

$$A_{(1)} \equiv \chi^{(1)1010} - (\chi^{(1)1013} + \chi^{(1)1310}) + \chi^{(1)1313}, \tag{42a}$$

$$A_{(2)} \equiv \chi^{(1)2020} - (\chi^{(1)2023} + \chi^{(1)2320}) + \chi^{(1)2323}, \tag{42b}$$

$$B_{(1)} \equiv \chi^{(1)1020} - (\chi^{(1)1023} + \chi^{(1)1320}) + \chi^{(1)1323}, \tag{42c}$$

$$B_{(2)} \equiv \chi^{(1)2010} - (\chi^{(1)2013} + \chi^{(1)2310}) + \chi^{(1)2313}. \tag{42d}$$

We note that $A_{(1)}$ *and* $A_{(2)}$ *contain only the principal part of* χ; $B_{(1)}$ *and* $B_{(2)}$ *contain only the principal and skewon part of* χ. *The axion part drops out and does not contribute to the dispersion relation in the eikonal approximation.* The principal part ${}^{(\mathrm{P})}B$ and skewon part ${}^{(\mathrm{Sk})}B$ of $B_{(1)}$ are as follows:

$${}^{(\mathrm{P})}B = \left(\frac{1}{2}\right)(B_{(1)} + B_{(2)}); \quad {}^{(\mathrm{Sk})}B = \left(\frac{1}{2}\right)(B_{(1)} - B_{(2)}). \tag{43}$$

From (43), $B_{(1)}$ and $B_{(2)}$ can be expressed as

$$B_{(1)} = {}^{(\mathrm{P})}B + {}^{(\mathrm{Sk})}B; \quad B_{(2)} = {}^{(\mathrm{P})}B - {}^{(\mathrm{Sk})}B. \tag{44}$$

For Eqs. (41a), (41b) to have nontrivial solutions of $(A_1^{(0)}, A_2^{(0)})$, we must have the following determinant vanish to first-order:

$$\det\begin{bmatrix} \left(\frac{1}{2}\right)(\omega^2 - k^2) - k^2 A_{(1)} & -k^2 B_{(1)} \\ -k^2 B_{(2)} & \left(\frac{1}{2}\right)(\omega^2 - k^2) - k^2 A_{(2)} \end{bmatrix}$$

$$= \left(\frac{1}{4}\right)(\omega^2 - k^2)^2 - \left(\frac{1}{2}\right)(\omega^2 - k^2)k^2(A_{(1)} + A_{(2)})$$

$$+ k^4(A_{(1)}A_{(2)} - B_{(1)}B_{(2)}) = 0 + O(2). \tag{45}$$

The solution of this quadratic equation in ω^2, i.e. the dispersion relation is

$$\omega^2 = k^2[1 + (A_{(1)} + A_{(2)}) \pm ((A_{(1)} - A_{(2)})^2 + 4B_{(1)}B_{(2)})^{1/2}] + O(2), \tag{46}$$

or

$$\omega = k\left[1 + \frac{1}{2}(A_{(1)} + A_{(2)}) \pm \frac{1}{2}((A_{(1)} - A_{(2)})^2 + 4B_{(1)}B_{(2)})^{1/2}\right] + O(2). \tag{47}$$

From (46) the group velocity is

$$v_{\rm g} = \frac{\partial\omega}{\partial k} = 1 + \frac{1}{2}(A_{(1)} + A_{(2)}) \pm \frac{1}{2}((A_{(1)} - A_{(2)})^2 + 4B_{(1)}B_{(2)})^{1/2} + O(2). \tag{48}$$

The quantity under the square root sign is

$$\xi \equiv (A_{(1)} - A_{(2)})^2 + 4B_{(1)}B_{(2)} = (A_{(1)} - A_{(2)})^2 + 4({}^{(\rm P)}B)^2 - 4({}^{(\rm Sk)}B)^2. \tag{49}$$

Depending on the sign or vanishing of ξ, we have the following three cases of electromagnetic wave propagation:

(i) $\xi > 0, (A_{(1)} - A_{(2)})^2 + 4({}^{(\rm P)}B)^2 > 4({}^{(\rm Sk)}B)^2$: There is birefringence of wave propagation;
(ii) $\xi = 0, (A_{(1)} - A_{(2)})^2 + 4({}^{(\rm P)}B)^2 = 4({}^{(\rm Sk)}B)^2$: There are no birefringence and no dissipation/amplification in wave propagation;
(iii) $\xi < 0, (A_{(1)} - A_{(2)})^2 + 4({}^{(\rm P)}B)^2 < 4({}^{(\rm Sk)}B)^2$: There is no birefringence, but there are both dissipative and amplifying modes in wave propagation.

3.2.1. *The condition of vanishing of $B_{(1)}$ and $B_{(2)}$ for all directions of wave propagation*

From the definition (42c), the condition of vanishing of $B_{(1)}$ for wave propagation in the z-axis direction is

$$B_{(1)} = \chi^{(1)1020} + \chi^{(1)1323} - \chi^{(1)1023} - \chi^{(1)1320} = 0. \tag{50}$$

To look for conditions derivable in combination with those from other directions, we do active Lorentz transformations (rotations/boosts). Active rotation R_θ in the

$y-z$ plane with angle θ is

$$\underline{t}=R_\theta t=t,\quad \underline{x}=R_\theta x,\quad \underline{y}=R_\theta y=y\cos\theta+z\sin\theta,\quad \underline{z}=R_\theta z=-y\sin\theta+z\cos\theta. \tag{51}$$

Applying active rotation R_θ (51) to (50), we have

$$\begin{aligned}0&=\underline{\chi}^{(1)1020}+\underline{\chi}^{(1)1323}-\underline{\chi}^{(1)1023}-\underline{\chi}^{(1)1320}\\&=\chi^{(1)1020}+\chi^{(1)1323}-\chi^{(1)1023}-\chi^{(1)1320}\\&\quad+\theta(\chi^{(1)1030}+\chi^{(1)1220}-\chi^{(1)1223}-\chi^{(1)1330})+O(\theta^2),\end{aligned} \tag{52}$$

for small value of θ. From (52) and (50), we have

$$\chi^{(1)1030}+\chi^{(1)1220}-\chi^{(1)1223}-\chi^{(1)1330}=0. \tag{53}$$

Following the same procedure, we repeatedly apply active rotation R_θ to (53) and the resulting equations together with their linear combinations. After performing cyclic permutation $1\to 2\to 3\to 1$ on the upper indices once and twice on some of the resulting equations, we have the following equations (for detailed derivation, see arXiv:1312.3056v1)

$$\chi^{(1)1220}=\chi^{(1)1330}; \tag{54a}$$
$$\chi^{(1)2330}=\chi^{(1)2110}; \tag{54b}$$
$$\chi^{(1)3110}=\chi^{(1)3220}; \tag{54c}$$
$$\chi^{(1)1020}=-\chi^{(1)1323}; \tag{54d}$$
$$\chi^{(1)2030}=-\chi^{(1)2131}; \tag{54e}$$
$$\chi^{(1)3010}=-\chi^{(1)3212}; \tag{54f}$$
$$\chi^{(1)1320}=-\chi^{(1)1230}; \tag{54g}$$
$$\chi^{(1)3210}=-\chi^{(1)3120}; \tag{54h}$$
$$\chi^{(1)2130}=-\chi^{(1)2310}; \tag{54i}$$
$$\chi^{(1)1023}=-\chi^{(1)1320}; \tag{54j}$$
$$\chi^{(1)2031}=-\chi^{(1)2130}; \tag{54k}$$
$$\chi^{(1)3012}=-\chi^{(1)3210}. \tag{54l}$$

From (54g)–(54l), $\chi^{(1)0123}$ is completely anti-symmetric under any permutation of (0123). Among (54g)–(54i) only two are independent; among (54j)–(54l) also only two are independent. For ${}^{(\mathrm{PA})}\chi^{ijkl}$, (54g)–(54l) give two independent conditions. For ${}^{(\mathrm{Sk})}\chi^{ijkl}$, (54g)–(54l) give three independent conditions and $\chi^{(1)0123}$ must vanish.

The derivation of formulas in this subsection from (50) to (54l) is independent of whether χ^{ijkl} is principal, axionic or skewonic. Hence, ${}^{(\mathrm{P})}$(54a)–${}^{(\mathrm{P})}$(54l) hold for ${}^{(\mathrm{P})}\chi^{ijkl}$ with ${}^{(\mathrm{P})}B_{(1)}=0$, ${}^{(\mathrm{A})}$(54a)–${}^{(\mathrm{A})}$(54l) hold for ${}^{(\mathrm{A})}\chi^{ijkl}$ with ${}^{(\mathrm{A})}B_{(1)}=0$, and ${}^{(\mathrm{Sk})}$(54a)–${}^{(\mathrm{Sk})}$(54l) hold for ${}^{(\mathrm{Sk})}\chi^{ijkl}$ with ${}^{(\mathrm{Sk})}B_{(1)}=0$. Here ${}^{(\mathrm{P})}$(54a)–${}^{(\mathrm{P})}$(54l) means (54a)–(54l) with χ substituted by ${}^{(\mathrm{P})}\chi$, ${}^{(\mathrm{A})}$(54a)–${}^{(\mathrm{A})}$(54l) means (54a)–(54l) with χ

substituted by ${}^{(\mathrm{A})}\chi$, and ${}^{(\mathrm{Sk})}$(54a)–${}^{(\mathrm{Sk})}$(54l) means (54a)–(54l) with χ substituted by ${}^{(\mathrm{Sk})}\chi$; similarly for ${}^{(\mathrm{P})}B_{(1)}$, ${}^{(\mathrm{A})}B_{(1)}$ and ${}^{(\mathrm{Sk})}B_{(1)}$. For $B_{(1)} = B_{(2)} = 0$ in all directions, we have ${}^{(\mathrm{P})}B_{(1)} = {}^{(\mathrm{Sk})}B_{(1)} = 0$ in all directions, and hence, both ${}^{(\mathrm{P})}$(54a)–${}^{(\mathrm{P})}$(54l) and ${}^{(\mathrm{Sk})}$(54a)–${}^{(\mathrm{Sk})}$(54l) are valid.

3.2.2. *The condition of ${}^{(\mathrm{Sk})}B_{(1)} = {}^{(\mathrm{P})}B_{(1)} = 0$ and $A_{(1)} = A_{(2)}$ for all directions of wave propagation*

With the condition ${}^{(\mathrm{Sk})}B_{(1)} = {}^{(\mathrm{P})}B_{(1)} = 0$ and $A_{(1)} = A_{(2)}$ for all directions of wave propagation, there is no birefringence for all directions of wave propagation. From section 2.2.1, we have Eqs. (54a)–(54l) holds from the validity of ${}^{(\mathrm{Sk})}B_{(1)} = {}^{(\mathrm{P})}B_{(1)} = 0$ (i.e. $B_{(1)} = 0$) for all directions of wave propagation. From $A_{(1)} = A_{(2)}$ and the definition (42a), (42b), we have

$$\chi^{(1)1010} - (\chi^{(1)1013} + \chi^{(1)1310}) + \chi^{(1)1313} = \chi^{(1)2020} - (\chi^{(1)2023} + \chi^{(1)2320}) + \chi^{(1)2323}. \tag{55}$$

From (54c) for the principal part, the terms in the parentheses on the two sides of the above equation cancel out and we have

$$\chi^{(1)1010} + \chi^{(1)1313} = \chi^{(1)2020} + \chi^{(1)2323}. \tag{56a}$$

Applying active rotation $R_{\pi/2}$ around in the $y - z$ plane to (56a), we obtain

$$\chi^{(1)1010} + \chi^{(1)1212} = \chi^{(1)3030} + \chi^{(1)3232}. \tag{56b}$$

3.3. *Nonbirefringence condition for the skewonless case*

If EEP is observed, photons with different polarizations as test particles shall follow identical trajectories in a gravitational field. Then the photons obey WEP I and there is no birefringence. In this section, we will first derive the core metric formula for the constitutive tensor density from the nonbirefringence condition in the skewonless case (Lagrangian-based case) and then use the cosmological observations to constrain the spacetime constitutive tensor density to this form to ultra-high precision.

From Eq. (49), the condition of nonbirefringence in the skewonless case is

$$A_{(1)} = A_{(2)}, \quad B_{(1)} = B_{(2)} = {}^{(\mathrm{P})}B = 0. \tag{57}$$

With these conditions, (54a)–(54h) and (56a), (56b) are valid and give 10 conditions on 21 independent components of skewonless constitutive tensor density χ^{ijkl}:

$$\chi^{(1)1220} = \chi^{(1)1330}; \tag{58a}$$

$$\chi^{(1)2330} = \chi^{(1)2110}; \tag{58b}$$

$$\chi^{(1)3110} = \chi^{(1)3220}; \tag{58c}$$

$$\chi^{(1)1020} = -\chi^{(1)1323}; \tag{58d}$$
$$\chi^{(1)2030} = -\chi^{(1)2131}; \tag{58e}$$
$$\chi^{(1)3010} = -\chi^{(1)3212}; \tag{58f}$$
$$\chi^{(1)1320} = -\chi^{(1)1230}; \tag{58g}$$
$$\chi^{(1)3210} = -\chi^{(1)3120}; \tag{58h}$$
$$\chi^{(1)1010} + \chi^{(1)1313} = \chi^{(1)2020} + \chi^{(1)2323}; \tag{58i}$$
$$\chi^{(1)1010} + \chi^{(1)1212} = \chi^{(1)3030} + \chi^{(1)3232}. \tag{58j}$$

Define

$$\begin{aligned}
&h^{(1)10} \equiv h^{(1)01} \equiv -2^{(\mathrm{P})}\chi^{(1)1220}; \quad h^{(1)20} \equiv h^{(1)02} \equiv -2^{(\mathrm{P})}\chi^{(1)2330};\\
&h^{(1)30} \equiv h^{(1)03} \equiv -2^{(\mathrm{P})}\chi^{(1)3110};\\
&h^{(1)12} \equiv h^{(1)21} \equiv -2^{(\mathrm{P})}\chi^{(1)1020}; \quad h^{(1)23} \equiv h^{(1)32} \equiv -2^{(\mathrm{P})}\chi^{(1)2030};\\
&h^{(1)31} \equiv h^{(1)13} \equiv -2^{(\mathrm{P})}\chi^{(1)3010};\\
&h^{(1)11} \equiv 2^{(\mathrm{P})}\chi^{(1)2020} + 2^{(\mathrm{P})}\chi^{(1)2121} - h^{(1)00};\\
&h^{(1)22} \equiv 2^{(\mathrm{P})}\chi^{(1)3030} + 2^{(\mathrm{P})}\chi^{(1)3232} - h^{(1)00};\\
&h^{(1)33} \equiv 2^{(\mathrm{P})}\chi^{(1)1010} + 2^{(\mathrm{P})}\chi^{(1)1313} - h^{(1)00},
\end{aligned} \tag{59a}$$

$$\begin{aligned}
\psi \equiv 1 + 2^{(\mathrm{P})}\chi^{(1)1212} + \left(\frac{1}{2}\right)\eta_{00}(h^{(1)00} - h^{(1)11} - h^{(1)22} - h^{(1)33})&\\
- h^{(1)11} - h^{(1)22},&
\end{aligned} \tag{59b}$$

$$\varphi \equiv \chi^{(1)0123} \equiv \chi^{(1)[0123]}. \tag{59c}$$

Note that in these definitions, $h^{(1)00}$ is not defined and is free. Now it is straightforward to show that when (58a)–(58j) are satisfied, then χ can be written to first-order in terms of the fields $h^{(1)ij}, \psi$, and φ with $h^{ij} \equiv \eta^{ij} + h^{(1)ij}$ and $h \equiv \det(h_{ij})$ in the following form:

$$\begin{aligned}
\chi^{ijkl} &= {}^{(\mathrm{P})}\chi^{(1)ijkl} + {}^{(\mathrm{A})}\chi^{(1)ijkl} + {}^{(\mathrm{SkII})}\chi^{(1)ijkl}\\
&= \frac{1}{2}(-h)^{1/2}[h^{ik}h^{jl} - h^{il}h^{kj}]\psi + \varphi e^{ijkl},
\end{aligned} \tag{60}$$

with

$${}^{(\mathrm{P})}\chi^{(1)ijkl} = \frac{1}{2}(-h)^{1/2}[h^{ik}h^{jl} - h^{il}h^{kj}]\psi, \tag{61a}$$

$${}^{(\mathrm{A})}\chi^{(1)ijkl} = \varphi e^{ijkl}. \tag{61b}$$

It is ready to derive the following theorem.

Theorem. *For linear electrodynamics with Lagrangian* (17a), *i.e. with skewonless constitutive relation* (12), *the following three statements are equivalent to first-order in the field:*

(i) $A_{(1)} = A_{(2)}$ *and* ${}^{(\mathrm{P})}B = 0$ *for all directions, i.e. nonbirefringence in electromagnetic wave propagation,*

(ii) (58a)–(58j) *hold,*
(iii) χ^{ijkl} *can be expressed as* (60) *with* (59a)–(59c).

Proof. (i) ⇒ (ii) has been demonstrated in the derivation of (58a)–(58j).

(ii) ⇒ (iii) has also been demonstrated in the derivation of (60) above.

(iii) ⇒ (i) Equation (60) is a Lorentz tensor density equation. If it holds in one Lorentz frame, it holds in any other frame. From this we readily check that $A_{(1)} = A_{(2)}$ and ${}^{(\mathrm{P})}B = 0$ in any new frame with the wave propagation in the z-direction. □

This theorem is a re-statement of the results of our work.[14–16] We note that previously we used the symbol H^{ik} instead of h^{ik}. Because H^{ik} is already used for excitation in this paper, we changed the notation.

We constructed the relation (60) in the weak-violation approximation of EEP in 1981 (Refs. 14–16); Haugan and Kauffmann[58] reconstructed the relation (60) in 1995. After the cornerstone work of Lämmerzahl and Hehl,[59] Favaro and Bergamin[64] finally proved the relation (60) without assuming weak-field approximation (see also Ref. 65).

Polarization measurements of electromagnetic waves from pulsars, from cosmologically distant radio sources and from GRB sources have yielded stringent constraints agreeing with (60) down to 10^{-16}, 10^{-32} and 10^{-38} respectively as shown in Table 1.

Observational constraints from pulsars[15,16]: In 1970s and 1980s pulsar observations gave the best constraints on the birefringence in the propagation. The pulses and micropulses from pulsars with different polarizations are correlated in general structure and timing.[66] No retardation with respect to different polarizations is observed. This means that conditions similar to (57) are satisfied to observational accuracy. For Crab pulsar, the micropulses with different polarizations are correlated in timing to within 10^{-4} s, the distance of the Crab pulsar is 2200 pc, therefore to within $10^{-4}\,\mathrm{s}/(2200 \times 3.26$ light yr.$) = 5 \times 10^{-16}$ accuracy, two conditions similar to (57) are satisfied. In 1981, over 300 pulsars in different directions had been observed. Many of them had polarization data. Combining all of them, (58a)–(58j) were satisfied to an accuracy of $10^{-14} - 10^{-16}$. Since for galactic gravitational field $U \sim 10^{-6}$, according to the procedure of proving the theorem, $\chi^{(1)}/U$ (or χ/U) agrees with that given by (60) to an accuracy of $10^{-8} - 10^{-10}$. At that time, we anticipated that detailed analysis would reveal better results. In 2002, a detailed analysis using X-ray pulsars[67] demonstrated the full procedure. At that time McCulloch, Hamilton, Ables and Hunt[68] had just observed a radio pulsar in the large Magellanic Cloud; Backer, Kulkarni, Helles, Davis and Goss[69] had discovered a millisecond pulsar which rotates 20 times faster than the Crab pulsar. The progress of these observations would potentially give better constraints on some of the conditions (58a)–(58j) due to larger distance or fast period involved.

We also anticipated that analysis of optical and X-ray polarization data from various astrophysical sources would give better accuracy to some of the 10 constraints in (58a)–(58j).

Thus, to high accuracy, photons are propagating in the metric field h^{ik} and two additional (pseudo)scalar fields ψ and φ. A change of h^{ik} to λh^{ik} does not affect χ^{ijkl} in (60) — this corresponds to the freedom of $h^{(1)00}$ in the definition (59a) of $h^{(1)ij}$. Thus we have constrained the general linear constitutive tensor of 21 degrees of freedom from the 10 constraints (58a)–(58j) to 11 degrees of freedom in (60).

Constraints from extragalactic radio-galaxy observations[61]: Analyzing the data from polarization measurements of extragalactic radio sources, Haugan and Kauffmann[58] in 1995 inferred that the resolution for null-birefringence is 0.02 cycle at 5 GHz. This corresponds to a time resolution of 4×10^{-12} s and gives much better constraints. With a detailed analysis and more extragalactic radio observations, (60) would be tested down to 10^{-28} – 10^{-29} at cosmological distances. In 2002, Kostelecky and Mews[70] used polarization measurements of light from cosmologically distant astrophysical sources to yield stringent constraints down to 2×10^{-32}. The electromagnetic propagation in Moffat's nonsymmetric gravitational theory[71,72] fits the $\chi - g$ framework. Krisher,[73] and Haugan and Kauffmann[58] have used the pulsar data and extragalactic radio observations respectively to constrain it.

Constraints from gamma ray burst observations[17]: Recent polarization observations on GRBs give even better constraints on the dispersion relation and non-birefringence in cosmic propagation.[74,75] The observation on the polarized GRB 061122 ($z = 1.33$) gives a lower limit on its polarization fraction of 60% at 68% confidence level (c.l.) and 33% at 90% c.l. in the 250–800 keV energy range.[74] The observation on the polarized GRB 140206A constrains the linear polarization level of the second peak of this GRB above 28% at 90% c.l. in the 200–400 keV energy range[75]; the redshift of the source is measured from the GRB afterglow optical spectroscopy to be $z = 2.739$. GRBs polarization observations have been used to set constraints on various dispersion relations (see, e.g. Refs. 76 and 77 and references therein). These two new GRB observations have larger and better redshift determinations than previous observations. We use them to give better constraints in our case. Since birefringence is proportional to the wave vector k in our case, as gamma ray of a particular frequency (energy) travels in the cosmic spacetime, the two linear polarization eigen-modes would pick up small phase differences. A linear polarization mode from distant source resolved into these two modes will become elliptical polarized during travel and lose part of the linear coherence. The way of gamma ray losing linear coherence depends on the frequency span. For a band of frequency, the extent of losing coherence depends on the distance of travel. The depolarization distance is of the order of frequency band span $\pi\Delta f$ times the integral $I = \int(1 + z(t))dt$ of the redshift factor $(1 + z(t))$ with respect to the time of travel. For GRB 140206A, this is about

$$\pi\Delta f I = \pi\Delta f \int (1 + z(t))dt \approx 1.5 \times 10^{20}\,\text{Hz} \times 0.6 \times 10^{18}\,\text{s} \approx 10^{38}. \qquad (62)$$

Since we do observe linear polarization in the 200–400 kHz frequency band of GRB 140206A with lower bound of 28%, this gives a fractional constraint of about 10^{-38} on a combination of χ's. A similar constraint can be obtained for GRB 061122 (the band width times the redshift is about the same). A more detailed modeling may give better limits. The distribution of GRBs is basically isotropic. When this procedure is applied to an ensemble of polarized GRBs from various directions, the relation (20) would be verified to about 10^{-38}.

Thus, we see that from the pulsar signal propagation, the polarization observations on radio galaxies and the GRB observations the nonbirefringence condition is verified empirically in spacetime propagation with accuracies to $10^{-16}, 10^{-32}$ and 10^{-38}. The accuracies of three observational constraints are summarized in Table 1. The constitutive tensor can be constructed by the procedure in the proof of the theorem in this subsection to be in the core form (60) with accuracy to 10^{-38}. Nonbirefringence (no splitting, no retardation) for electromagnetic wave propagation independent of polarization and frequency (energy) is the statement of Galileo Equivalence Principle for photons or WEP I for photons. Hence WEP I for photons is verified to this accuracy in the spacetime propagation.

In the following subsection, we assume (60) (i.e. (28)) is valid and look into the influence of the axion field and the dilaton field of the constitutive tensor on the dispersion relation.

3.4. *Wave propagation and the dispersion relation in dilaton field and axion field*

We first notice that in the lowest eikonal approximation, the dispersion relation (46) or (47) does not contain the axion piece and does not contain the gradient of fields. Dilaton in (60) goes in this dispersion relation only as an overall scale factor and drops out too.

To derive the influence of the dilaton field and the Abelian axion field on the dispersion relation, one needs to keep the second term in Eq. (31). This has been done for the axion field in Refs. 53, 60, 61, 78–80. Here we follow the treatment in Ref. 63 to develop it for the joint dilaton field and axion field with the constitutive relation (60). Near the origin in a local inertial frame, the constitutive tensor density in dilaton field ψ and axion field φ (Eq. (60)) becomes

$$\chi^{ijkl}(x^m) = \left[\left(\frac{1}{2}\right)\eta^{ik}\eta^{jl} - \left(\frac{1}{2}\right)\eta^{il}\eta^{kj}\right]\psi(x^m) + \varphi(x^m)e^{ijkl} + O(\delta_{ij}x^i x^j), \quad (63)$$

where η^{ij} is the Minkowski metric with signature -2 and δ_{ij} the Kronecker delta. In the local inertial frame, we use the Minkowski metric and its inverse to raise and lower indices. Substituting (63) into the Eq. (31) and multiplying by 2, we have

$$\psi A^{i,j}{}_{,j} - \psi A^{j,i}{}_{,j} + \psi_{,j}\, A^{i,j} - \psi_{,j}\, A^{j,i} + 2\varphi_{,j}e^{ijkl}A_{k,l} = 0. \quad (64)$$

We notice that (64) is both Lorentz covariant and gauge invariant.

We expand the dilaton field $\psi(x^m)$ and the axion field $\varphi(x^m)$ at the 4-point (event) P with respect to the event (time and position) P_0 at the origin as follows:

$$\psi(x^m) = \psi(P_0) + \psi_{,i}(P_0)x^j + O(\delta_{ij}x^i x^j), \tag{65a}$$

$$\varphi(x^m) = \varphi(P_0) + \varphi_{,i}(P_0)x^j + O(\delta_{ij}x^i x^j). \tag{65b}$$

To look for wave solutions, we use eikonal approximation which does not neglect field gradient/medium inhomogeneity. Choose z-axis in the wave propagation direction so that the solution takes the following form:

$$A \equiv (A_0, A_1, A_2, A_3) = (\underline{A}_0, \underline{A}_1, \underline{A}_2, \underline{A}_3)\mathrm{e}^{ikz-i\omega t} = \underline{A}_i \mathrm{e}^{ikz-i\omega t}. \tag{66}$$

Expand the solution as

$$A_i = A_i^{(0)} + A_i^{(1)} + O(2) = [\underline{A}_i^{(0)} + \underline{A}_i^{(1)} + O(2)]\mathrm{e}^{ikz-i\omega t} = \underline{A}_i \mathrm{e}^{ikz-i\omega t}. \tag{67}$$

Now use eikonal approximation to obtain a local dispersion relation. In the eikonal approximation, we only keep terms linear in the derivative of the dilaton field and the axion field; we neglect terms containing the second-order derivatives of the dilaton field or the axion field, terms of $O(\delta_{ij}x^i x^j)$ and terms of mixed second-order, e.g. terms of $O(A_i^{(1)} x^j)$ or $O(A_i^{(1)} \psi_{,j})$; we call all these terms $O(2)$.

Imposing radiation gauge condition in the zeroth-order in the weak field/dilute medium approximation, we find to zeroth-order, (65) is

$$\psi A^{(0)ij}{}_{,j} = 0 \quad \text{or} \quad A^{(0)ij}{}_{,j} = 0, \tag{68}$$

and the corresponding zeroth-order solution and the dispersion relation are

$$A^{(0)}_i = (0, A^{(0)}_1, A^{(0)}_2, 0) = \underline{A}^{(0)}_i e^{ikz-i\omega t} = (0, \underline{A}^{(0)}_1, \underline{A}^{(0)}_2, 0)e^{ikz-i\omega t}, \tag{69a}$$

$$\omega = k + O(1). \tag{69b}$$

Substituting (68) and (69a), (69b) into Eq. (64), we have

$$\psi A^{(0)ij}{}_{,j} + \psi A^{(1)ij}{}_{,j} + \psi A^{(1)ji}{}_{,j} + \psi_{,j} A^{(0)ij}{}_{,} - \psi_{,j} A^{(0)ji}{}_{,} + 2\varphi_{,j} e^{ijkl} A^{(0)}{}_{k,l} = 0 + O(2). \tag{70}$$

The $i = 0$ and $i = 3$ components of (70) both lead to the same modified Lorentz gauge condition in the dilaton field and the axion field in the $O(1)$ order[63]:

$$A^{(1)j}{}_{,j} = -\psi^{-1}(\psi_{,1} - 2\varphi_{,2})A^{(0)1} - \psi^{-1}(\psi_{,2} + 2\varphi_{,1})A^{(0)}_2 + O(2). \tag{71}$$

Since Eq. (71) does not contain ω and k, it does not contribute to the determination of the dispersion relation.

Using the gauge condition (71), we obtain the $i = 1$ and $i = 2$ components of Eq. (70) as

$$(\omega^2 - k^2)\underline{A}_1^{(0)} - ik\underline{A}_1^{(0)}\psi^{-1}(\psi_{,0} + \psi_{,3}) - 2ik\underline{A}_2^{(0)}\psi^{-1}(\varphi_{,0} + \varphi_{,3}) = 0 + O(2), \tag{72a}$$

$$(\omega^2 - k^2)\underline{A}_2^{(0)} - ik\underline{A}_2^{(0)}\psi^{-1}(\psi_{,0} + \psi_{,3}) + 2ik\underline{A}_1^{(0)}\psi^{-1}(\varphi_{,0} + \varphi_{,3}) = 0 + O(2). \tag{72b}$$

These two equations determine the dispersion relation in the dilaton field and the axion field:

$$\det\begin{bmatrix}(\omega^2-k^2)-ik\psi^{-1}(\psi_{,0}+\psi_{,3}) & -2ik\psi^{-1}(\varphi_{,0}+\varphi_{,3})\\ 2ik\psi^{-1}(\varphi_{,0}+\varphi_{,3}) & (\omega^2-k^2)-ik\psi^{-1}(\psi_{,0}+\psi_{,3})\end{bmatrix}$$
$$=[(\omega^2-k^2)-ik\psi^{-1}(\psi_{,0}+\psi_{,3})]^2-4k^2\psi^{-2}(\varphi_{,0}+\varphi_{,3})^2=0+O(2). \tag{73}$$

Its solutions are

$$\omega=k-\left(\frac{i}{2}\right)\psi^{-1}(\psi_{,0}+\psi_{,3})\pm\psi^{-1}(\varphi_{,0}+\varphi_{,3})+O(2)\quad\text{or} \tag{74a}$$

$$k=\omega+\left(\frac{i}{2}\right)\psi^{-1}(\psi_{,0}+\psi_{,3})\pm\psi^{-1}(\varphi_{,0}+\varphi_{,3})+O(2), \tag{74b}$$

with the group velocity $v_g=\partial\omega/\partial k=1$ independent of polarization. When the dispersion relation is satisfied, (72a) and (72b) have two independent solutions for the polarization eigenvectors $\underline{A}^{(0)}_i=(\underline{A}^{(0)}_0,\underline{A}^{(0)}_1,\underline{A}^{(0)}_2,\underline{A}^{(0)}_3)$ with

$$\frac{\underline{A}^{(0)}_1}{\underline{A}^{(0)}_2}=\frac{[2ik\psi^{-1}(\varphi_{,0}+\varphi_{,3})]}{[(\omega^2-k^2)-ik\psi^{-1}(\psi_{,0}+\psi_{,3})]}=\frac{[2ik\psi^{-1}(\varphi_{,0}+\varphi_{,3})]}{[\pm2k\psi^{-1}(\varphi_{,0}+\varphi_{,3})]}=\pm i; \tag{75a}$$

$$\underline{A}^{(0)}_0=\underline{A}^{(0)}_3=0, \tag{75b}$$

for $\omega=k-(i/2)\psi^{-1}(\psi_{,0}+\psi_{,3})\pm\psi^{-1}(\varphi_{,0}+\varphi_{,3})+\mathrm{O}(2)$ respectively. From (75a), the two polarization eigenstates are left circularly polarized state and right circularly polarized state in varying axion. This agrees with and generalizes the electromagnetic wave propagation in axion field as derived earlier.[53,60,61,78–80]

With the dispersion (74), the plane-wave solution (66) propagating in the z-direction is

$$A\equiv(A_0,A_1,A_2,A_3)=(0,\underline{A}^{(0)}_1,\underline{A}^{(0)}_2,0)e^{ikz-i\omega t}=(0,\underline{A}^{(0)}_1,\underline{A}^{(0)}_2,0)$$
$$\times\exp\left[ikz-ikt\pm(-i)\psi^{-1}(\varphi_{,0}t+\varphi_{,3}z)-\left(\frac{1}{2}\right)\psi^{-1}(\psi_{,0}t+\psi_{,3}z)\right], \tag{76}$$

with $\underline{A}^{(0)}_1=\pm\, i\underline{A}^{(0)}_2$. The additional factor acquired in the propagation is $\exp[\pm(-i)\psi^{-1}(\varphi_{,0}t+\varphi_{,3}z)]\times\exp[-(1/2)\psi^{-1}(\psi_{,0}t+\psi_{,3}z)]$. The first part of this factor, i.e. the axion factor $\exp[\pm(-i)\psi^{-1}(\varphi_{,0}t+\varphi_{,3}z)]$ adds a phase in the propagation. The second part of this factor, i.e. the dilaton factor $\exp[-(1/2)\psi^{-1}(\psi_{,0}t+\psi_{,3}z)]$ amplifies or attenuates the wave according to whether $(\{\psi_{,0}t+\psi_{,3}z)$ is less than zero or greater than zero. For the right circularly polarized electromagnetic wave, the effect of the axion field in the propagation from a point $P_1=\{x_{(1)}^{\ i}\}=\{x_{(1)}^{\ 0};x_{(1)}^{\ \mu}\}=\{x_{(1)}^{\ 0},x_{(1)}^{\ 1},x_{(1)}^{\ 2},x_{(1)}^{\ 3}\}$ to another point $P_2=\{x_{(2)}^{\ i}\}=\{x_{(2)}^{\ 0};x_{(2)}^{\ \mu}\}=\{x_{(2)}^{\ 0},x_{(2)}^{\ 1},x_{(2)}^{\ 2},x_{(2)}^{\ 3}\}$ is to add a phase of $\alpha=\psi^{-1}[\varphi(P_2)-\varphi(P_1)](\approx\varphi(P_2)-\varphi(P_1)$ for $\psi\approx1)$ to the wave; for left circularly polarized light, the effect is to add an opposite phase.[53,60,61,78–80] Linearly polarized electromagnetic wave is a superposition of circularly polarized waves. Its polarization vector will then rotate by an angle α. The effect of the dilaton field is to amplify with a factor $\exp[-(1/2)\psi^{-1}(\psi_{,0}t + \psi_{,3}z)] = \exp[-(1/2)((\ln\psi)_{,0}t + (\ln\psi)_{,3}z)] = (\psi(P_1)/\psi(P_2))^{1/2}$. Whether the dilaton field amplifies or attenuates the propagating wave depends on $\psi(P_1)/\psi(P_2) > 1$ or $\psi(P_1)/\psi(P_2) < 1$ respectively.

For plane wave propagating in direction $n^\mu = (n^1, n^2, n^3)$ with $(n^1)^2 + (n^2)^2 + (n^3)^2 = 1$, the solution is

$$\begin{aligned} A(n^\mu) &\equiv (A_0, A_1, A_2, A_3) = (0, \underline{A}_1, \underline{A}_2, \underline{A}_3)\exp(-ikn^\mu x_\mu - i\omega t) \\ &= (0, A_1, A_2, A_3)\exp\Big[-ikn^\mu x_\mu - ikt \pm (-i)\psi^{-1}(\varphi_{,0}\,t - n^\mu \varphi_{,\mu}\, n_\nu x^\nu) \\ &\quad - \left(\frac{1}{2}\right)\psi^{-1}(\psi_{,0}\,t + n^\mu \psi_{,\mu}\, n_\nu x^\nu)\Big], \end{aligned} \tag{77}$$

where $\underline{A}_\mu = \underline{A}^{(0)}_{\ \mu} + n_\mu n^\nu \underline{A}^{(0)}_{\ \nu}$ with $\underline{A}^{(0)}_{\ 1} = \pm i\underline{A}^{(0)}_{\ 2}$ and $\underline{A}^{(0)}_{\ 3} = 0\,[n_\mu \equiv (-n^1, -n^2, -n^3)]$. There are polarization rotation for linearly polarized light due to axion field gradient, and amplification/attenuation due to dilaton field gradient.

The above analysis is local. In the global situation, choose local inertial frames along the wave trajectory and integrate along the trajectory. Since ψ is a scalar, the integration gives $(\psi(P_1)/\psi(P_2))^{1/2}$ as the amplification factor for the propagation in the dilaton field. For small dilaton field variations, the amplification/attenuation factor is equal to $[1-(1/2)(\Delta\psi/\psi)]$ to a very good approximation with $\Delta\psi \equiv \psi(P_2) - \psi(P_1)$. Since this factor does not depend on the wave number/frequency and polarization, it will not distort the source spectrum in propagation, but gives an overall amplification/attenuation factor to the spectrum. The axion field contributes to the phase factor and induces polarization rotation as in previous investigations.[53,60,61,78–80] For $\psi \approx 1$ (constant), the induced polarization rotation agrees with previous results which were obtained without considering dilaton effect. If the dilaton field varies significantly, a ψ-weight needs to be included in the integration.

The complete agreement with EEP for photon sector requires in addition to Galileo equivalence principle (WEP I; nonbirefringence) for photons: (i) no polarization rotation (WEP II); (ii) no amplification/no attenuation in spacetime propagation; (iii) no spectral distortion. With nonbirefringence, any skewonless spacetime constitutive tensor must be of the form (60), hence no spectral distortion. From (60), (i) and (ii) imply that the dilaton ψ and axion φ must be constant, i.e. no varying dilaton field and no varying axion field; the EEP for photon sector is observed; the spacetime constitutive tensor is of metric-induced form. Thus the three observational conditions are tied to EEP and to metric-induced spacetime constitutive tensor in the photon sector.

In the next subsection, we look into the empirical support of no amplification/no attenuation and no polarization rotation conditions.

3.5. *No amplification/no attenuation and no polarization rotation constraints on cosmic dilaton field and cosmic axion field*

In this section, we look into the observations/experiments to constrain the dilaton field contribution and the axion field contribution to spacetime constitutive tensor density.

No amplification/no attenuation constraint on the cosmic field: From Eqs. (76) and (77) in the last section, we have derived that the amplitude and phase factor of propagation in the cosmic dilaton and cosmic Abelian axion field is changed by $(\psi(P_1)/\psi(P_2))^{1/2} \times \exp[ikz - ikt \pm (-i)(\varphi(P_1) - \varphi(P_2))t]$. The effect of dilaton field is to give amplification $(\psi(P_1) - \psi(P_2) > 0)$ or attenuation $(\psi(P_1) - \psi(P_2) < 0)$ to the amplitude of the wave independent of frequency and polarization.

The spectrum of the CMB is well understood to be Planck blackbody spectrum. In the cosmic propagation, this spectrum would be amplified or attenuated by the factor $(\psi(P_1)/\psi(P_2))^{1/2}$. However, the CMB spectrum is measured to agree with the ideal Planck spectrum at temperature $2.7255 \pm 0.0006\,\mathrm{K}$ (Ref. 81) with a fractional accuracy of 2×10^{-4}. The spectrum is also redshifted due to cosmological curvature (or expansion), but this does not change the blackbody character. The measured shape of the CMB spectra does not deviate from Planck spectrum within its experimental accuracy. In the dilaton field the relative increase in power is proportional to the amplitude increase squared, i.e. $\psi(P_1)/\psi(P_2)$. Since the total power of the blackbody radiation is proportional to the temperature to the fourth power T^4, the fractional change of the dilaton field since the last scattering surface of the CMB must be less than about 8×10^{-4} and we have

$$\frac{|\Delta\psi|}{\psi} \leq 4(0.0006/2.7255) \approx 8 \times 10^{-4}. \tag{78}$$

Direct fitting to the CMB data with the addition of the scale factor $\psi(P_1)/\psi(P_2)$ would give a more accurate value.

Constraints on the cosmic polarization rotation and the cosmic axion field: From (77), for the right circularly polarized electromagnetic wave, the propagation from a point P_1 (4-point) to another point P_2 adds a phase of $\alpha = \varphi(P_2) - \varphi(P_1)$ to the wave; for left circularly polarized light, the added phase will be opposite in sign.[53] Linearly polarized electromagnetic wave is a superposition of circularly polarized waves. Its polarization vector will then rotate by an angle α. In the global situation, it is the property of (pseudo)scalar field that when we integrate along light (wave) trajectory the total polarization rotation (relative to no φ-interaction) is again $\alpha = \Delta\varphi = \varphi(P_2) - \varphi(P_1)$ where $\varphi(P_1)$ and $\varphi(P_2)$ are the values of the scalar field at the beginning and end of the wave. The constraints listed on the axion field in Table 1 are from the UV polarization observations of radio galaxies and the CMB polarization observations — 0.02 for Cosmic Polarization Rotation (CPR) mean value $|\langle\alpha\rangle|$ and 0.03 for the CPR fluctuations $\langle(\alpha - \langle\alpha\rangle)^2\rangle^{1/2}$.[82–84]

Additional constraints to have the unique physical metric: From (78) the fractional change of dilaton $|\Delta\psi|/\psi$ is less than about 8×10^{-4} since the time of the last scattering surface of the CMB. Eötvös-type experiments constrain the fractional variation of dilaton to $\sim 10^{-10}U$ where U is the dimensionless Newtonian potential in the experimental environment. Vessot–Levine redshift experiment and Hughes–Drever-type experiments give further constraints.[61] All these constraints are summarized in Table 1. This leads to unique physical metric to high precision for all degrees of freedom except the axion degree of freedom and cosmic dilaton degree of freedom which are only mildly constrained.

3.6. *Spacetime constitutive relation including skewons*[62,17]

In this subsection, we review the present status of empirical tests of full local linear spacetime constitutive tensor density (26) of premetric electrodynamics. Since EEP is verified to a good precision, we are mainly concerned with weak EEP violations and weak additional field, i.e. we are assuming $\chi^{(0)ijkl}$ is metric and the components of $\chi^{(1)ijkl}$ are small in most parts of our treatment. We note that *all the formulas in Sec. 3.2 are valid with or without skewonless assumption.*

In particular, *the condition of* ${}^{(\mathrm{Sk})}B_{(1)} = {}^{(\mathrm{P})}B_{(1)} = 0$ *and* $A_{(1)} = A_{(2)}$ *for all directions of wave propagation* still gives (54a)–(54l) without skewonless assumption.

We do not assume skewonless condition in this subsection. The Hehl–Obukhov–Rubilar skewon field (27c) can be represented as

$$ {}^{(\mathrm{Sk})}\chi^{ijkl} = e^{ijmk}S_m{}^l - e^{ijml}S_m{}^k, \tag{79}$$

where $S_m{}^n$ is a traceless tensor with $S_m{}^m = 0$.[6] From (79), we have

$$ \begin{aligned} &{}^{(\mathrm{Sk})}\chi^{(1)1320} = -S^{(1)}{}_0{}^0 - S^{(1)}{}_2{}^2; \quad {}^{(\mathrm{Sk})}\chi^{(1)1230} = S^{(1)}{}_0{}^0 + S^{(1)}{}_3{}^3; \\ &{}^{(\mathrm{Sk})}\chi^{(1)2310} = S^{(1)}{}_0{}^0 + S^{(1)}{}_1{}^1. \end{aligned} \tag{80}$$

From ${}^{(\mathrm{Sk})}$(54g)–${}^{(\mathrm{Sk})}$(54l), we must have ${}^{(\mathrm{Sk})}\chi^{(1)1320} = {}^{(\mathrm{Sk})}\chi^{(1)1230} = {}^{(\mathrm{Sk})}\chi^{(1)2310} = 0$. From (80) and $\mathrm{Tr}\, S_n{}^m = 0$, then all $S^{(1)}{}_0{}^0$, $S^{(1)}{}_1{}^1$, $S^{(1)}{}_2{}^2$ and $S^{(1)}{}_3{}^3$ must vanish.

From (79) together with ${}^{(\mathrm{Sk})}$(54a)–${}^{(\mathrm{Sk})}$(54f), we have

$$ \begin{aligned} &S^{(1)}{}_3{}^2 = -S^{(1)}{}_2{}^3; \quad S^{(1)}{}_1{}^3 = -S^{(1)}{}_3{}^1; \quad S^{(1)}{}_2{}^1 = -S^{(1)}{}_1{}^2; \\ &S^{(1)}{}_3{}^0 = S^{(1)}{}_0{}^3; \quad S^{(1)}{}_1{}^0 = S^{(1)}{}_0{}^1; \quad S^{(1)}{}_2{}^0 = S^{(1)}{}_0{}^2. \end{aligned} \tag{81}$$

Using the Lorentz metric (h-metric in the locally inertia frame) to raise/lower the indices, we have

$$ S^{(1)mn} = -S^{(1)nm}, \quad S^{(1)}{}_{mn} = -S^{(1)}{}_{nm}. \tag{82}$$

Thus, when ${}^{(\mathrm{Sk})}$(54a)–${}^{(\mathrm{Sk})}$(54l) (nine independent conditions) are satisfied, the skewon degrees of freedom are reduced to 6 (15 – 9) and only Type II skewon field remains.

Under Lorentz (coordinate) transformation, the symmetric part and the antisymmetric part of S^{mn} transform separately. Hence, with the conditions ${}^{(\mathrm{Sk})}B = 0$ for all directions of wave propagation, the skewon field is constrained to Type II.

The reverse is also true: Since ${}^{(\mathrm{SkII})}S_{nm}$ is a tensor, when it satisfy ${}^{(\mathrm{Sk})}B = 0$ for the z-axis of wave propagation, they satisfy ${}^{(\mathrm{Sk})}B = 0$ for *all directions* of wave propagation. Hence we have the lemma:

Lemma. *The following three statements are equivalent*:

(i) ${}^{(\mathrm{Sk})}B = 0$ *for all directions*,
(ii) ${}^{(\mathrm{Sk})}(54\mathrm{a})$–${}^{(\mathrm{Sk})}(54\mathrm{l})$ *hold*,
(iii) ${}^{(\mathrm{Sk})}S_{mn}$ *as defined by* (79) *can be written as* ${}^{(\mathrm{Sk})}S_{mn} = {}^{(\mathrm{SkII})}S_{mn}$ *with* ${}^{(\mathrm{SkII})}S_{nm} = -{}^{(\mathrm{SkII})}S_{mn}$.

Proof. (i) $\Rightarrow$ (ii) has been demonstrated in the derivation of ${}^{(\mathrm{Sk})}(54\mathrm{a})$–${}^{(\mathrm{Sk})}(54\mathrm{l})$.

(ii) $\Leftrightarrow$ (iii) has also been demonstrated in the derivation of (80)–(82) and its reversibility.

(iii) $\Rightarrow$ (i) ${}^{(\mathrm{SkII})}S_{ij}$ is a tensor. If its anti-symmetric property holds in one frame, it holds in any frame. Hence, in any new frame with the propagation in the $\underline{z}$-direction, ${}^{(\mathrm{Sk})}(54\mathrm{a})$–${}^{(\mathrm{Sk})}(54\mathrm{l})$ hold and we have ${}^{(\mathrm{Sk})}\underline{B} = 0$ for propagation in the $\underline{z}$-direction. Since $\underline{z}$-direction can be arbitrary, we have ${}^{(\mathrm{Sk})}\underline{B} = 0$ for all directions. □

The condition of ${}^{(\mathrm{Sk})}B_{(1)} = {}^{(\mathrm{P})}B_{(1)} = 0$ *and* $A_{(1)} = A_{(2)}$ *for all directions of wave propagation* gives (56a), (56b). Define the anti-symmetric metric p^{ij} as follows:

$$
\begin{aligned}
&p^{10} \equiv -p^{01} \equiv 2^{(\mathrm{SkII})}\chi^{(1)1220}; \quad p^{20} \equiv -p^{02} \equiv 2^{(\mathrm{SkII})}\chi^{(1)2330};\\
&p^{30} \equiv -p^{03} \equiv 2^{(\mathrm{SkII})}\chi^{(1)3110}; \quad p^{12} \equiv -p^{21} \equiv 2^{(\mathrm{SkII})}\chi^{(1)1020};\\
&p^{23} \equiv -p^{32} \equiv 2^{(\mathrm{SkII})}\chi^{(1)2030}; \quad p^{31} \equiv -p^{13} \equiv 2^{(\mathrm{SkII})}\chi^{(1)3010};\\
&p^{00} \equiv p^{11} \equiv p^{22} \equiv p^{33} \equiv 0.
\end{aligned}
\tag{83}
$$

It is straightforward to show now that when (54a)–(54l) and (56a)–(56b) are satisfied, then χ can be written to first-order in terms of the fields $h^{(1)ij}, \psi, \varphi$, and p^{ij} with $h^{ij} \equiv \eta^{ij} + h^{(1)ij}$ and $h \equiv \det(h_{ij})$ in the following form:

$$
\begin{aligned}
\chi^{ijkl} &= {}^{(\mathrm{P})}\chi^{(1)ijkl} + {}^{(\mathrm{A})}\chi^{(1)ijkl} + {}^{(\mathrm{SkII})}\chi^{(1)ijkl}\\
&= \frac{1}{2}(-h)^{1/2}[h^{ik}h^{jl} - h^{il}h^{kj}]\psi + \varphi e^{ijkl}\\
&\quad + \frac{1}{2}(-\eta)^{1/2}(p^{ik}\eta^{jl} - p^{il}\eta^{jk} + \eta^{ik}p^{jl} - \eta^{il}p^{jk}),
\end{aligned}
\tag{84}
$$

with

$$
{}^{(\mathrm{P})}\chi^{(1)ijkl} = \frac{1}{2}(-h)^{1/2}[h^{ik}h^{jl} - h^{il}h^{kj}]\psi, \tag{85a}
$$

$$
{}^{(\mathrm{A})}\chi^{(1)ijkl} = \varphi e^{ijkl}, \tag{85b}
$$

$$
{}^{(\mathrm{SkII})}\chi^{(1)ijkl} = \frac{1}{2}(-\eta)^{1/2}(p^{ik}\eta^{jl} - p^{il}\eta^{jk} + \eta^{ik}p^{jl} - \eta^{il}p^{jk}). \tag{85c}
$$

It is ready to derive the following theorem.

Theorem. *For linear electrodynamics with skewonful constitutive relation* (26) *with* ${}^{(\mathrm{Sk})}B = 0$ *satisfied for all directions, the following three statements are equivalent to first-order in the field:*

(i) $A_{(1)} = A_{(2)}$ *and* ${}^{(\mathrm{P})}B = 0$ *for all directions, i.e. nonbirefringence in electromagnetic wave propagation,*
(ii) (58a)–(58j) *hold,*
(iii) χ^{ijkl} *can be expressed as* (84) *with* (85a)–(85c).

The proof is similar to that for theorem in Sec. 3.3[62]; readers could readily figure it out.

When the principal part ${}^{(\mathrm{P})}\chi^{ijkl}$ of the constitutive tensor is induced by metric h^{ij} and dilaton, i.e.

$$ {}^{(\mathrm{P})}\chi^{ijkl} = (-h)^{1/2}\left[\left(\frac{1}{2}\right)h^{ik}h^{jl} - \left(\frac{1}{2}\right)h^{il}h^{kj}\right]\psi, \tag{86} $$

it is easy to check by substitution that

$$ A_{(1)} = A_{(2)} \quad \text{and} \quad {}^{(\mathrm{P})}B_{(1)} = {}^{(\mathrm{P})}B = 0, \tag{87} $$

We have $\xi = -4({}^{(\mathrm{Sk})}B)^2$. The three cases discussed after Eq. (49) reduce to two cases:

(a) $\xi = 0$, ${}^{(\mathrm{Sk})}B = 0$: There are no birefringence and no dissipation/amplification in wave propagation;
(b) $\xi < 0$, ${}^{(\mathrm{Sk})}B \neq 0$: There is no birefringence, but there are both dissipative and amplifying modes in wave propagation.

Now the issue is: When the skewon part of the constitutive tensor is nonzero, what can we say about the spacetime structure empirically?

If ξ is less than zero, i.e. $(A_{(1)} - A_{(2)})^2 + 4({}^{(\mathrm{P})}B)^2 < 4({}^{(\mathrm{Sk})}B)^2$, the dispersion relation (47) is

$$ \omega = k\left[1 + \frac{1}{2}(A_{(1)} + A_{(2)}) \pm \frac{1}{2}(-\xi)^{1/2}i\right] + O(2). \tag{88} $$

The exponential factor in the wave solution (36) is of the form

$$ \exp(ikz - i\omega t) \sim \exp\left[ikz - ik\left(1 + \frac{1}{2}(A_{(1)} + A_{(2)})\right)t\right]\exp\left(\pm\frac{1}{2}(-\xi)^{1/2}kt\right). \tag{89} $$

There are both dissipative and amplifying wave propagation modes. In the small ξ limit, the amplification/attenuation factor $\exp(\pm 1/2(-\xi)^{1/2}kt)$ equals $[1 \pm 1/2(-\xi)^{1/2}kt]$ to a very good approximation. Since this factor depends on the wave number/frequency, it will distort the source spectrum in propagation.

The spectrum of the CMB is well understood to be Planck blackbody spectrum. It is measured to agree with the ideal Planck spectrum at temperature $2.7255 \pm 0.0006\,\mathrm{K}$.[81] The measured shape of the CMB spectra does not deviate from Planck spectrum within its experimental accuracy. The agreement for the overall shape with a fit to Planck plus a linear factor $[1 \pm 1/2(-\xi)^{1/2}kt]$ is to agree with Planck to better than 10^{-4}. Planck Surveyor has nine bands of detection from 30 to 857 GHz.[85] For weak propagation deviation, the amplitude of the wave is increased or decreased linearly as $1/2(-\xi)^{1/2}kt$ depending on frequency. For cosmic propagation, the CMB amplitude change due to redshift (or blueshift) is universal. The frequency (wave number) change is proportional to $(1+z(t))$ with $z(t)$ the redshift factor at time t of propagation. We need to replace kt in the $[1 \pm 1/2(-\xi)^{1/2}kt]$ factor by the integral

$$\int k(t)dt = \int k(t_0)(1+z(t))dt \equiv (1+\langle z(t)\rangle)k(t_0)(t_0 - t_1), \tag{90}$$

with $\langle z(t)\rangle$ the average of $z(t)$ during propagation defined by the last equality of (90), t_0 the present time (the age of our universe) and t_1 the time at the photon decoupling epoch. According to *Planck* 2013 results,[85] the age of our universe t_0 is 13.8 Gyr, the decoupling time t_1 is 0.00038 Gyr, hence $(t_0 - t_1)$ is $\sim$ 13.8 Gyr, and $z(t_1)$ is 1090. Using *Planck* ΛCDM concordance model, the factor $(1 + \langle z(t)\rangle)$ is estimated to be about 3 and the value $(1 + \langle z(t)\rangle)$ $(t_0 - t_1)$ is more than 40 Gyr. The factor $(1 + \langle z(t)\rangle)$ multiply by $(t_0 - t_1)$ is the angular diameter distance D_A at which we are observing the CMB and is equal to the comoving size of the sound horizon at the time of last-scattering, $r_s(z(t_1))$, divided by the observed angular size $\theta_* = r_s/D_A$ from seven acoustic peaks in the CMB anisotropy spectrum. From Planck results, $r_s = 144.75 \pm 0.66$ Mpc and $\theta_* = (1.04148 \pm 0.00066) \times 10^{-2}$. Hence, we have $D_A = r_s/\theta_* = 13898 \pm 64$ Mpc $= 45.328 \pm 0.21$ Gyr. This is consistent with our integral estimation.

For the highest frequency band ω is $2\pi \times 857$ GHz. The amplification/dissipation in fraction is

$$\frac{1}{2}(-\xi)^{1/2}k \times 45.328\,\text{Gyr} = 3.8 \times 10^{30}(-\xi)^{1/2}. \tag{91}$$

For the lowest frequency band ω is $2\pi \times 30$ GHz; the effect is about ±3.5% of (91). From CMB observations that the spectrum is less than 10^{-4} deviation, we have

$$(-\xi)^{1/2} < 2.6 \times 10^{-35}. \tag{92}$$

When the spacetime constitutive tensor is constructed from metric, dilaton and axion plus skewon, the principal part $^{(\mathrm{P})}\chi^{ijkl}$ of the constitutive tensor is given by (86). There are two cases, (a) $^{(\mathrm{Sk})}B = 0$ and (b) $^{(\mathrm{Sk})}B \neq 0$ as mentioned after Eq. (87). For case (a) $\xi = 0$, there are no birefringence and no dissipation/amplification in wave propagation; by the theorem in this subsection, the skewon part must be of Type II. For case (b) $\xi < 0$, $^{(\mathrm{Sk})}B \neq 0$, there are both dissipative and amplifying modes in wave propagation and we can apply (92) from the CMB observations to constrain the skewon part of the constitutive tensor as follows:

$$\frac{1}{2}(-\xi)^{1/2} = |^{(\mathrm{Sk})}B| = \frac{1}{2}|(B_{(1)} - B_{(2)})|$$
$$= |^{(\mathrm{Sk})}\chi^{(1)1020} + {}^{(\mathrm{Sk})}\chi^{(1)1323} - {}^{(\mathrm{Sk})}\chi^{(1)1023} - {}^{(\mathrm{Sk})}\chi^{(1)1320}| < 1.3 \times 10^{-35}, \tag{93}$$

for propagation in the z-direction. Since the CMB observation is omnidirectional, we have the above constraint for many directions. From a few superpositions, we obtain the lemma in this subsection, hence the constraints (54a)–(54l) hold to $\sim$ a few $\times\, 20^{-35}$ and the spacetime skewon field is Type II with Type I skewon field constrained to $\sim$ a few $\times\, 20^{-35}$ cosmologically in the first-order. Thus, *the significant skewon field must be of Type II with six degrees of freedom in the first-order.*

Constraints on the skewon field in the second-order[17]

For metric principal part plus skewon part, we have shown that the Type I skewon part is constrained to $<$ a few $\times 10^{-35}$ in the weak field/weak EEP violation limit. Type II skewon part is not constrained in the first-order. In the second-order Obukhov and Hehl have shown in Sec. IV.A.1 of Ref. 86 that it induces birefringence; since the nonbirefringence observations are precise to 10^{-38} as listed in Table 1, they constrain the Type II skewon part to $\sim 10^{-19}$.[17,86] However, an additional nonmetric induced second-order contribution to the principal part constitutive tensor compensates the Type II skewon birefringence and makes it nonbirefringent.[17] This second-order contribution is just the extra piece to the (symmetric) core metric principal constitutive tensor induced by the antisymmetric part of the asymmetric metric tensor q^{ij}.[17] Table 3 lists various first-order and second-order effects in wave propagation on media with the core metric-based constitutive tensors.[17] In the following subsection, we review the spacetime/medium with constitutive tensor induced from asymmetric metric.

3.7. *Constitutive tensor from asymmetric metric and Fresnel equation*

Eddington,[87] Einstein and Straus,[88] and Schrödinger[89,90] considered asymmetric metric in their exploration of gravity theories. Just like we can build spacetime constitutive tensor from the (symmetric) metric as in metric theories of gravity, we can also build it from the asymmetric metric. Let q^{ij} be the asymmetric metric as follows:

$$\chi^{ijkl} = \frac{1}{2}(-q)^{1/2}(q^{ik}q^{jl} - q^{il}q^{jk}), \tag{94}$$

with $q = \det^{-1}({}^{(\mathrm{S})}q^{ij})$. When q^{ij} is symmetric, this definition reduces to that of the metric theories of gravity. The constitutive law (94) was also put forward by Lindell

Table 3. Various first-order and second-order effects in wave propagation on media with the core metric-based constitutive tensors. ${}^{(\mathrm{P})}\chi^{(\mathrm{c})}$ is the extra contribution due to antisymmetric part of asymmetric metric to the core metric principal part for canceling the skewon contribution to birefringence/amplification-dissipation.[17]

Constitutive tensor density χ^{ijkl}	Birefringence (in the geometric optics approximation)	Dissipation/ amplification	Spectroscopic distortion	CPR
Metric: $(1/2)(-h)^{1/2}[h^{ik}h^{jl} - h^{il}h^{kj}]$	No	No	No	No
Metric + dilaton: $(1/2)(-h)^{1/2}[h^{ik}h^{jl} - h^{il}h^{kj}]\psi$	No (to all orders in the field)	Yes (due to dilaton gradient)	No	No
Metric + Abelian axion: $(1/2)(-h)^{1/2}[h^{ik}h^{jl} - h^{il}h^{kj}] + \varphi e^{ijkl}$	No (to all orders in the field)	No	No	Yes (due to axion gradient)
Metric + dilaton + Abelian axion: $(1/2)(-h)^{1/2}[h^{ik}h^{jl} - h^{il}h^{kj}]\psi + \varphi e^{ijkl}$	No (to all orders in the field)	Yes (due to dilaton gradient)	No	Yes (due to axion gradient)
Metric + Type I skewon	No to first-order	Yes	Yes	No
Metric + Type II skewon	No to first-order; yes to second-order	No to first-order and to second-order	No	No
Metric + ${}^{(\mathrm{P})}\chi^{(\mathrm{c})}$ + Type II skewon	No to first-order; no to second-order	No to first-order and to second-order	No	No
Asymmetric metric induced: $(1/2)(-q)^{1/2}(q^{ik}q^{jl} - q^{il}q^{jk})$	No (to all orders in the field)	No	No	Yes (due to axion gradient)

and Wallen[91] as Q-medium. Resolving the asymmetric metric into symmetric part ${}^{(\mathrm{S})}q^{ij}$ and antisymmetric part ${}^{(\mathrm{A})}q^{ij}$:

$$q^{ij} = {}^{(\mathrm{S})}q^{ij} + {}^{(\mathrm{A})}q^{ij}, \quad \text{with } {}^{(\mathrm{S})}q^{ij} \equiv \frac{1}{2}(q^{ij} + q^{ji}) \quad \text{and} \quad {}^{(\mathrm{A})}q^{ij} \equiv \frac{1}{2}(q^{ij} - q^{ji}), \tag{95}$$

we can decompose the constitutive tensor into the principal part ${}^{(\mathrm{P})}\chi^{ijkl}$, the axion part ${}^{(\mathrm{Ax})}\chi^{ijkl}$ and skewon part ${}^{(\mathrm{Sk})}\chi^{ijkl}$ as follows[62,92]:

$$\chi^{ijkl} = \frac{1}{2}(-q)^{1/2}(q^{ik}q^{jl} - q^{il}q^{jk}) = {}^{(\mathrm{P})}\chi^{ijkl} + {}^{(\mathrm{Ax})}\chi^{ijkl} + {}^{(\mathrm{Sk})}\chi^{ijkl}, \tag{96a}$$

with

$$\begin{aligned} {}^{(\mathrm{P})}\chi^{ijkl} &\equiv \frac{1}{2}(-q)^{1/2}({}^{(\mathrm{S})}q^{ik\,(\mathrm{S})}q^{jl} - {}^{(\mathrm{S})}q^{il\,(\mathrm{S})}q^{jk} \\ &\quad + {}^{(\mathrm{A})}q^{ik\,(\mathrm{A})}q^{jl} - {}^{(\mathrm{A})}q^{il\,(\mathrm{A})}q^{jk} - 2{}^{(\mathrm{A})}q^{[ik\,(\mathrm{A})}q^{jl]}), &(96\text{b}) \\ {}^{(\mathrm{Ax})}\chi^{ijkl} &\equiv (-q)^{1/2\,(\mathrm{A})}q^{[ik\,(\mathrm{A})}q^{jl]}, &(96\text{c}) \\ {}^{(\mathrm{Sk})}\chi^{ijkl} &\equiv \frac{1}{2}(-q)^{1/2}({}^{(\mathrm{A})}q^{ik\,(\mathrm{S})}q^{jl} - {}^{(\mathrm{A})}q^{il\,(\mathrm{S})}q^{jk} + {}^{(\mathrm{S})}q^{ik\,(\mathrm{A})}q^{jl} - {}^{(\mathrm{S})}q^{il\,(\mathrm{A})}q^{jk}). &(96\text{d}) \end{aligned}$$

The axion part ${}^{(\mathrm{Ax})}\chi^{ijkl}$ only comes from the second-order terms of ${}^{(\mathrm{A})}q^{il}$.

Using ${}^{(\mathrm{S})}q^{ij}$ to raise and its inverse to lower the indices, we have as Eq. (16) in Ref. 62

$$S_{ij} = \frac{1}{2}\varepsilon_{ijmk}{}^{(\mathrm{A})}q^{mk}; \quad {}^{(\mathrm{A})}q^{mk} = -\varepsilon^{mkij}S_{ij}, \tag{97}$$

where ε_{ijmk} and ε^{mkij} are respectively the completely antisymmetric covariant and contravariant tensors with $\varepsilon^{0123} = 1$ and $\varepsilon_{0123} = -1$ in local inertial frame. Thus the skewon field S_{ij} from asymmetric metric q^{ik} is antisymmetric and is of Type II.

Dispersion relation in the geometrical optics limit. The dispersion relation for the wave covector q_i of electromagnetic propagation with general constitutive tensor (26) in the geometric-optics limit is given by *the generalized covariant Fresnel equation*[6]:

$$G^{ijkl}(\chi)q_iq_jq_kq_l = 0, \tag{98}$$

where $G^{ijkl}(\chi)(=G^{(ijkl)}(\chi))$ is a completely symmetric fourth-order Tamm–Rubilar (TR) tensor density of weight +1 defined by

$$G^{ijkl}(\chi) \equiv \left(\frac{1}{4!}\right)\underline{e}_{mnpq}\underline{e}_{rstu}\chi^{mnr(i}\chi^{j|ps|k}\chi^{l)qtu}. \tag{99}$$

There are two ways to obtain the TR tensor density (99) for the dispersion relation (98). One way is by straightforward calculation; the other is by covariant method.[92] In the appendix of arXiv:1411.0460v1, we outline the straightforward calculation to obtain the TR tensor density $G^{ijkl}(\chi)$ for the asymmetric metric induced constitutive tensor:

$$G^{ijkl}(\chi) = \left(\frac{1}{8}\right)(-q)^{3/2}\det(q^{ij})q^{(ij}q^{kl)} = \left(\frac{1}{8}\right)(-q)^{3/2}\det(q^{ij})^{(\mathrm{S})}q^{(ij(S)}q^{kl)}. \tag{100}$$

Except for a scalar factor, (100) is the same as for metric-induced constitutive tensor with ${}^{(\mathrm{S})}q_{ij}$ replacing the metric g_{ij} or h_{ij}. Therefore in the geometric optical approximation, there is no birefringence and the unique light cone is given by the metric ${}^{(\mathrm{S})}q_{ij}$.

Constraints on asymmetric-metric induced constitutive tensor.[17] Although the asymmetric-metric induced constitutive tensor leads to a Fresnel equation which is nonbirefringent, it contains an axionic part:

$$ {}^{(\mathrm{Ax})}\chi^{ijkl} \equiv (-q)^{1/2(\mathrm{A})}q^{[ik(\mathrm{A})}q^{jl]} = \varphi e^{ijkl}; \quad \varphi \equiv \left(\frac{1}{4!}\right) e_{ijkl}(-q)^{1/2(\mathrm{A})}q^{[ik(\mathrm{A})}q^{jl]}, \tag{101}$$

which induces polarization rotation in wave propagation. Constraints on CPR and its fluctuation limit the axionic part and therefore also constrain the asymmetric metric. The variation of $\varphi(\equiv (1/4!)e_{ijkl}(-q)^{1/2(\mathrm{A})}q^{[ik(\mathrm{A})}q^{jl]})$ is limited by observations[82–84,60,61] on the CPR to < 0.02 and its fluctuation to < 0.03 since the last scattering surface, and in turn constrains the antisymmetric metric of the spacetime for this degree of freedom. The antisymmetric metric has six degrees of freedom. Further study of the remaining five degrees of freedom experimentally to find either evidence or more constraints would be desired.

Theoretically, there are two issues: one is whether the asymmetric-metric induced constitutive tensors with additional axion piece are the most general nonbirefringent media in the lowest geometric optics limit; the other is what they play in the spacetime structure and in the cosmos.

3.8. *Empirical foundation of the closure relation for skewonless case*[17,62]

There are two equivalent definitions of constitutive tensor which are useful in various discussions (see, e. g. Ref. 6). The first one is to take a dual on the first two indices of χ^{ijkl}:

$$\kappa_{ij}{}^{kl} \equiv \left(\frac{1}{2}\right) \underline{e}_{ijmn}\chi^{mnkl}, \tag{102}$$

where $\underline{e}_{ijmn}$ is the completely antisymmetric tensor density of weight -1 with $\underline{e}_{0123} = 1$. Since $\underline{e}_{ijmn}$ is a tensor density of weight -1 and χ^{mnkl} a tensor density of weight $+1$, $\kappa_{ij}{}^{kl}$ is a (twisted) tensor. From (102), we have

$$\chi^{mnkl} = \left(\frac{1}{2}\right) e^{ijmn}\kappa_{ij}{}^{kl}. \tag{103}$$

With this definition of constitutive tensor ${\kappa_{ij}}^{kl}$, the constitutive relation (12) becomes

$$^*H_{ij} = {\kappa_{ij}}^{kl} F_{kl}, \tag{104}$$

where $^*H_{ij}$ is the dual of H^{ij}, i.e.

$$^*H_{ij} \equiv \left(\frac{1}{2}\right) \underset{\sim}{e}_{ijmn} H^{mn}. \tag{105}$$

The second equivalent definition of the constitutive tensor is to use a 6×6 matrix representation ${\kappa_I}^J$. Since ${\kappa_{ij}}^{kl}$ is nonzero only when the antisymmetric pairs of indices (ij) and (kl) have values (01), (02), (03), (23), (31), (12), these index pairs can be enumerated by capital letters $I, J, \ldots$ from 1 to 6 to obtain ${\kappa_I}^J (\equiv {\kappa_{ij}}^{kl})$. With the relabeling, $F_{ij} \to F_I, H^{ij} \to H^I, \underset{\sim}{e}_{ijmn} \to \underset{\sim}{e}_{IJ}, e^{ijmn} \to e^{IJ}$. We have $F_I = (\boldsymbol{E}, -\boldsymbol{B})$ and $(^*H)_I = (-\boldsymbol{H}, -\boldsymbol{D})$. $\underset{\sim}{e}_{IJ}$ and e^{IJ} can be expressed in matrix form as

$$\underset{\sim}{e}_{IJ} = e^{IJ} = \begin{bmatrix} 0 & \mathbf{I}_3 \\ \mathbf{I}_3 & 0 \end{bmatrix}, \tag{106}$$

where $\mathbf{I}_3$ is the 3×3 unit matrix. In terms of this definition, the constitutive relation (104) becomes

$$^*H_I = 2{\kappa_I}^J F_J, \tag{107}$$

where $^*H_I \equiv {}^*H_{ij} = e_{IJ} H^J$. The axion part ${}^{(\mathrm{Ax})}\chi^{ijkl}$ $(= \varphi e^{ijkl})$ now corresponds to

$${}^{(\mathrm{Ax})}{\kappa_I}^J = \varphi \begin{bmatrix} \mathbf{I}_3 & 0 \\ 0 & \mathbf{I}_3 \end{bmatrix} = \varphi \mathbf{I}_6, \tag{108}$$

where $\mathbf{I}_6$ is the 6×6 unit matrix. The principal part and the Abelian axion part of the constitutive tensor all satisfy the following equation (the skewonless condition):

$$e^{KJ} {\kappa_J}^I = e^{IJ} {\kappa_J}^K. \tag{109}$$

In terms of ${\kappa_{ij}}^{kl}$ and re-indexed ${\kappa_I}^J$, the constitutive tensor (60) is represented in the following forms:

$${\kappa_{ij}}^{kl} = \left(\frac{1}{2}\right) \underset{\sim}{e}_{ijmn} \chi^{mnkl} = \left(\frac{1}{2}\right) \underset{\sim}{e}_{ijmn} (-h)^{1/2} h^{mk} h^{nl} \psi + \varphi {\delta_{ij}}^{kl}, \tag{110}$$

$${\kappa_I}^J = \left(\frac{1}{2}\right) \underset{\sim}{e}_{ijmn} (-h)^{1/2} h^{mk} h^{nl} \psi + \varphi {\delta_I}^J, \tag{111}$$

where ${\delta_{ij}}^{kl}$ is a generalized Kronecker delta defined as

$${\delta_{ij}}^{kl} = {\delta_i}^k {\delta_j}^l - {\delta_i}^l {\delta_j}^k. \tag{112}$$

In the derivation, we have used the formula

$$\underset{\sim}{e}_{ijmn} e^{mnkl} = 2{\delta_{ij}}^{kl}. \tag{113}$$

Let us calculate $\kappa_{ij}{}^{kl}\kappa_{kl}{}^{pq}$ for the constitutive tensor (110):

$$\begin{aligned}\kappa_{ij}{}^{kl}\kappa_{kl}{}^{pq} &= \left[\left(\frac{1}{2}\right)\underline{e}_{ijmn}(-h)^{1/2}h^{mk}h^{nl}\psi + \varphi\delta_{ij}{}^{kl}\right]\\ &\quad\times\left[\left(\frac{1}{2}\right)\underline{e}_{klrs}(-h)^{1/2}h^{rp}h^{sq}\psi + \varphi\delta_{kl}{}^{pq}\right]\\ &= -\left(\frac{1}{2}\right)\delta_{ij}{}^{pq}\psi^2 + 2\delta_{ij}{}^{pq}\varphi^2 + 2\underline{e}_{ijrs}(-h)^{1/2}h^{rp}h^{sq}\varphi\psi\\ &= -\left(\frac{1}{2}\right)\delta_{ij}{}^{pq}\psi^2 + 4\varphi^{(\mathrm{P})}\kappa_{ij}{}^{pq} - 2\delta_{ij}{}^{pq}\varphi^2,\end{aligned} \tag{114}$$

where we have used (113) and the following relations

$$e_{klrs}h^{mk}h^{nl}h^{rp}h^{sq} = e^{mnpq}\det(h^{uv}), \tag{115}$$

$$\det(h^{uv}) = [\det(h_{uv})]^{-1} = h^{-1}, \tag{116}$$

$$\delta_{ij}{}^{kl}\delta_{kl}{}^{pq} = 2\delta_{ij}{}^{pq}. \tag{117}$$

In terms of the six-dimensional index I, Eq. (114) becomes

$$\begin{aligned}\kappa_I{}^J\kappa_J{}^K &= \left(\frac{1}{2}\right)\kappa_{ij}{}^{kl}\kappa_{kl}{}^{pq} = -\left(\frac{1}{4}\right)\psi^2\delta_I{}^K + 2^{(\mathrm{P})}\kappa_I{}^K\varphi - \delta_{ij}{}^{pq}\varphi^2\\ &= -\left(\frac{1}{4}\right)\psi^2\delta_I{}^K + 2^{(\mathrm{P})}\kappa_I{}^K\varphi - \delta_I{}^K\varphi^2.\end{aligned} \tag{118}$$

Thus the matrix multiplication of κ_I^J with itself is a linear combination of itself and the identity matrix, and generates a closed algebra of linear dimension 2. The algebraic relation (118) is a closure relation that generalizes the following closure relation in electrodynamics:

$$\kappa\kappa = (\kappa_I{}^J\kappa_J{}^K) = \left(\frac{1}{6}\right)\mathrm{tr}(\kappa\kappa)\mathbf{I}_6. \tag{119}$$

The matrix multiplication of κ_I^J satisfies the closure relation (119). In case $\varphi = 0$, the Abelian axion part ${}^{(\mathrm{Ax})}\kappa_I^J$ of the constitutive tensor vanishes and (118) reduces to the closure relation (119).

From the nonbirefringence condition (60), we derive the closure relation (118) in a number of algebraic steps which consist of order 100 individual operations of addition/subtraction or multiplication. Equation (60) is empirically verified to 10^{-38}. Therefore Eq. (118) is empirically verified to 10^{-37} (precision 10^{-38} times $100^{1/2}$). Hence, when there are no axion and no dilaton, the closure relation (119) is empirically verified to 10^{-37}. For dilaton is constrained to 8×10^{-4}, if one allow for dilaton, relation (119) is verified to 8×10^{-4} since the last scattering surface of CMB; for axion is constrained to about 10^{-2}, if one allow for axion in addition, relation (119) is verified to about 10^{-2} since the last scattering surface of CMB. As pointed out by Favaro (private communication), the above method could also readily be applied to the other three variants of closure relations (Eqs. (3.2), (3.3), (3.4) in Ref. 92).

The closure relation (119) can also be called idempotent condition for it states that the multiplication of κ by itself goes back essentially to itself. Toupin,[93] Schonberg[94] and Jadczyk[95] in their theoretical approach started from this condition to obtain metric induced constitutive tensor with a dilaton degree of freedom. In this section, we have started with Galileo equivalence principle for photons, i.e. the nonbirefringence condition, to obtain the metric induced core metric form with a dilaton degree of freedom and an axion degree of freedom for the constitutive tensor and then the generalized closure relation (118). We have also shown that (118) is verified empirically to very high precision. Thus in the axionless (and skewonless) case, the birefringence condition and idempotent condition are equivalent and both are verified empirically to high precision.

4. From Galileo Equivalence Principle to Einstein Equivalence Principle

In Sec. 3, we have used equivalence principles in the photon sector to constrain the gravitational coupling to electromagnetism and the structure of spacetime from premetric electrodynamics. In this section, we review and discuss theoretically to what extent Galileo equivalence principle leads to EEP, i.e. Schiff's conjecture.

In 1970s, we used Galileo Equivalence Principle and derived its consequences for an electromagnetic system with Lagrangian density L $(= L_{\mathrm{I}}^{(\mathrm{EM})} + L_{\mathrm{I}}^{(\mathrm{EM-P})} + L_I^{(\mathrm{P})})$ where the electromagnetic field Lagrangian $L_{\mathrm{I}}^{(\mathrm{EM})}$ and the field–current interaction Lagrangian $L_{\mathrm{I}}^{(\mathrm{EM-P})}$ are given by (17a), (17b), and the particle Lagrangian $L_I^{(\mathrm{P})}$ is given by $-\Sigma_I m_I (ds_I)/(dt)\delta(\boldsymbol{x}-\boldsymbol{x}_I)$ with m_I the mass of the Ith particle, s_I its 4-line element from the metric $g_{ij}, \boldsymbol{x}_I$ its position 3-vector, $\boldsymbol{x}$ the coordinate 3-vector, and t the time coordinate[20,21]:

$$\begin{aligned} L &= L_{\mathrm{I}}^{(\mathrm{EM})} + L_{\mathrm{I}}^{(\mathrm{EM-P})} + L_{\mathrm{I}}^{(\mathrm{P})} \\ &= -\left(\frac{1}{16\pi}\right)\chi^{ijkl}F_{ij}F_{kl} - A_k J^k - \Sigma_I m_I \frac{ds_I}{dt}\delta(\boldsymbol{x}-\boldsymbol{x}_I), \end{aligned} \tag{120}$$

$$J^k = \Sigma_I e_I \frac{dx_I^k}{dt}\delta(\boldsymbol{x}-\boldsymbol{x}_I). \tag{120a}$$

Here e_I is the charge of the Ith particle. In (120), only the part of χ^{ijkl} which is symmetric under the interchange of index pairs ij and kl contributes to the Lagrangian, i.e. the constitutive tensor is effectively skewonless. This framework is termed $\chi-g$ framework.

The result of imposing Galileo Equivalence Principle is that the constitutive tensor density χ^{ijkl} can be constrained and expressed in metric form with additional pseudoscalar (axion) field φ:

$$\chi^{ijkl} = (-g)^{1/2}\left[\left(\frac{1}{2}\right)g^{ik}g^{jl} - \left(\frac{1}{2}\right)g^{il}g^{kj}\right] + \varphi e^{ijkl}, \tag{121}$$

where g^{ij} is the metric of the geodesic motions of particles, g_{ij} is the inverse of $g^{ij}, g = \det(g_{ij})$ and e^{ijkl} is the completely anti-symmetric tensor density with $e^{0123} = 1$ as defined in Sec. 3. Hence the metric g^{ij} generates the light cone for electromagnetic wave propagation also. The constraint (121) dictates the gravity coupling to electromagnetic field to be metric plus one additional axionic freedom. With this one axionic freedom the EEP is violated, and therefore the Schiff's conjecture is invalid. However, the spirit of Schiff's conjecture is useful and constrains the gravity coupling effectively. Since the theory with constitutive tensor density (121) does not obey EEP, it is a nonmetric theory.

The theory with $\varphi \neq 0$ is a pseudoscalar theory with important astrophysical and cosmological consequences. Its effect on electromagnetic wave propagation is that the polarization rotation of linearly polarized light is proportional to the difference of the (pseudo)scalar field at the two end points. We have discussed this in detail in Sec. 3.4 and use CPR observations to constrain it. This is an example that investigations in fundamental physical laws lead to implications in cosmology. Investigations of CP problems in high energy physics lead to a theory with a similar piece of Lagrangian with φ the axion field for QCD.[96–103]

In the nonmetric theory with $\chi^{ijkl}(\varphi \neq 0)$ given by Eq. (121),[20,21,40,53] there are anomalous torques on electromagnetic-energy-polarized bodies so that different test bodies will change their rotation state differently, like magnets in magnetic fields. Since the motion of a macroscopic test body is determined not only by its trajectory but also by its rotation state, the motion of polarized test bodies will not be the same. We, therefore, have proposed the following stronger weak equivalence principle (WEP II) to be tested by experiments, which states that in a gravitational field, both the translational and rotational motion of a test body with a given initial motion state is independent of its internal structure and composition (universality of free-fall motion) (Sec. 2.2).[20,21] To put in another way, the behavior of motion including rotation is that in a local inertial frame for test-bodies. If WEP II is violated, then EEP is violated. Therefore from above, in the $\chi - g$ framework, the imposition of WEP II guarantees that EEP is valid. These are the reasons for us to propose WEP II. The $\chi - g$ framework has been extended to nonabelian gauge fields for studying the interrelations of equivalence principles with similar conclusions.[104]

From the empirical side, WEP I for unpolarized bodies is verified to very high precision. However, these experiments only constrain two degrees of freedom of χ's for connecting with gravity coupling of matter. To constrain and connect more degrees of freedom of χ's to gravity coupling of matter, we propose to perform WEP experiments on various polarized test-bodies in 1970s — both electromagnetic polarized and spin polarized test bodies. These polarized experiments are also crucial to probe the role of spin and polarization in gravity. Now with the spacetime constitutive tensor density constrained to the core metric form (60) to ultra-precision 10^{-38}, the polarized WEP experiments will test the gravity–matter interaction more than gravity–radiation interaction. In Sec. 7, we will update our review[61] on the search for the long range/intermediate range spin–spin, spin–monopole and spin–cosmic interactions.

5. EEP and Universal Metrology

EEP states that all local physics are same everywhere at any time in our cosmos. Therefore if we base our metrology everywhere at anytime on local physics with a universal procedure, we have a universal metrology (see, e.g. Refs. 105 and 106). For metrology, we need unit standards. At present all basic standards except for the prototype mass standard are based on physical laws, their fundamental constants and the microscopic properties of matter. The EEP says, in essence, local physics is the same everywhere. Therefore, to the precision of its empirical tests, EEP warrants the universality of these standards and their implementations.

The name Système International d'Unités (International System of Units), with the abbreviation SI, was adopted by the 11th Conférence Générale des Poids et Mesures in 1960. After 1983 redefinition of meter as the length of path traveled by light in a vacuum during a time interval of 1/299792458 of a second, all definition of SI units can be traced to the definition of second and kilogram. The second is defined as the duration of 9 192 631 770 periods of the radiation corresponding to the transition between the two hyperfine levels of the ground state of the cesium-133 atom. The kilogram is the unit of mass; it is equal to the mass of the international prototype of the kilogram (a cylinder of platinum–iridium) (IPK). IPK is the only physical artifact in the definition of SI 7 base units (second, meter, kilogram, ampere, kelvin, mole and candela for 7 base quantities time, length, mass, electric current, thermodynamic temperature, amount of substance and luminous intensity respectively). Although the uncertainty of the mass of IPK is zero by convention, there are evidence that the mass of IPK varies with a fraction of the order of 10^{-8} after storage or cleaning with the estimated relative instability $\delta m/m \approx 5 \times 10^{-8}$ over the past 100 years.[107] When the mass unit is redefined by natural invariants, the SI system will be free of artifacts. In order to ensure continuity of mass metrology, it has been agreed that the relative uncertainty of any new realization must be less than 2×10^{-8} (see, e.g. Ref. 108). Sanchez *et al.*[109] in National Research Council of Canada determined the Planck's constant h using the watt balance to be $6.62607034(12) \times 10^{-34}$ J s within 2×10^{-8} relative uncertainty. NIST has reached 5×10^{-8} relative uncertainty and is building a new watt balance to reach 2×10^{-8} relative uncertainty.[110] The silicon sphere experiment of counting atoms to determine the Avogadro constant reached 3×10^{-8} relative uncertainty (see, e.g. Ref. 108). In 2014, the Avogadro constant N_{A} and derived Planck constant h based on the absolute silicon molar mass measurements with their standard uncertainties are $6.02214076(19) \times 10^{23}\ \mathrm{mol}^{-1}$ and $6.62607017(21) \times 10^{-34}$ J s.[111] The three measurements of NIST,[111] PTB,[112] and NMIJ[113] agree within their stated uncertainties and also agree with the NRC watt balance measurement with 1σ. These experimental progresses set the stage for a new definition of kilogram using Planck constant/Avogadro number. Time is becoming mature to replace all the definitions of units using natural invariants.

In 2018, the 5 SI base quantities — time, length, mass, electric current and thermodynamic temperature — will be replaced by frequency, velocity, action, electric charge and heat capacity, pending upon the expected final resolution of the 26th Conférence Générale des Poids et Mesures (CGPM) (see, e.g. Ref. 110). The two defining constants for frequency and velocity will be the same as the present SI defining constants of time and length. The defining constants for action, electric charge, heat capacity and amount of substance will be the Planck constant h, the elementary charge e, the Boltzmann constant k and the Avogadro constant N_A respectively. The mass unit can be traced to action unit defined by the Planck constant using watt balance or to amount of substance defined by the Avogadro constant based on counting the atoms in a ^{28}Si crystal. In 2018, both methods should reach an uncertainty smaller than 2×10^{-8} to guarantee consistency and continuity. The relative uncertainty of $N_A h$ at present is 7×10^{-10} (CODATA 2010 adjustment[114]) to guarantee consistency at the 2×10^{-8} level.

With the new definition of units based on physical invariants of nature, the applicability becomes wider; as long as the physical laws which the units are based are valid, the standards and metrology are universal. In Sec. 3, we have seen that the unique light cone is experimentally verified to 10^{-38} via gamma ray observations at cosmological distance; it verifies the Galileo equivalence principle for photons/electromagnetic wave packets to this accuracy. This constrains the spacetime (vacuum) constitutive tensor to core metric form with additional dilaton and axion degrees of freedom. In the solar system the variation of the dilaton field is constrained to $10^{-10}U$; in the cosmos, the dilaton field is constrained to 8×10^{-4} (Table 1). The universal metrology system is truly universal with the present accuracies. In case the accuracies are pushed further, we either verify equivalence principles further or discover new physics. Thus we see that universal metrology and equivalence principles go hand-in-hand.

Equivalence principles play very important roles both in the Newtonian theory of gravity and relativistic theories of gravity. The ranges of validity of these equivalence principles or their possible violations give clues and/or constraints to the microscopic origins of gravity. They will be even more important when the precisions of the tests become higher. To pursue further tests of EEP, we have to look into precise experiments and observations in our laboratory, in the solar system, and in diverse astrophysical and cosmological situations. All of these depend on the progress in the field of precision measurement, and demands more precise standards. The constancy of constants is implied by equivalence principles. Their variations give new physics.

The frequency measurement has the best relative uncertainty at present. The optical clocks are reaching relative uncertainties at the 10^{-18} level.[115] When the comparison of optical clocks becomes common, it is anticipated that the frequency standards will go optical. Further improvement in the frequency measurements will have profound impact on precision measurement and gravity experiment. In the realm of gravitational wave detection, the influence will be to enhance the Doppler tracking method and the PTA method.[116] An array of clocks may even become an alternate method for detecting low frequency gravitational waves.

6. Gyrogravitational Ratio

Gyrogravitational effect is defined to be the response of an angular momentum in a gravitomagnetic field produced by a gravitating source having a nonzero angular momentum. Ciufolini and Pavlis[117] have measured and verified this effect with $10 - 30\%$ accuracy for the dragging of the orbit plane (orbit angular momentum) of a satellite (LAGEOS) around a rotating planet (Earth) predicted for general relativity by Lense and Thirring.[118] Gravity Probe B[119] (GP-B) has measured and verified the dragging of spin angular momentum of a rotating quartz ball predicted by Schiff[120] for general relativity with 19% accuracy. GP-B experiment has also verified the Second Weak Equivalence Principle (WEP II) for macroscopic rotating bodies to ultra-precision.[121] On 13 February 2012 the Italian Space Agency (ASI) launched the LARES (LAser RElativity Satellite) satellite with a Vega rocket for improving the measurement of Lense–Thirring effect together with other geodesy satellites.[122] On Earth, GINGER (Gyroscopes IN General Relativity) is a multi-ring-laser array project aimed to measure the Lense–Thirring effect to 1%.[123] When this is achieved, the same technology could be applied to improve the tie between the astronomic reference frame and the solar-system dynamical frame.

Just as in electromagnetism, we can define gyrogravitational factor as the gravitomagnetic moment (response) divided by angular momentum for gravitational interaction. We use macroscopic (spin) angular momentum in GR as standard, its gyrogravitational ratio is 1 by definition. In Ref. 124, we use coordinate transformations among reference frames to study and to understand the Lense–Thirring effect of a Dirac particle. For a Dirac particle, the wave function transformation operator from an inertial frame to a moving accelerated frame is obtained. According to equivalence principle, this gives the gravitational coupling to a Dirac particle. From this, the Dirac wave function is solved and its change of polarization gives the gyrogravitational ratio 1 from the first-order gravitational effects. In a series of papers on spin–gravity interactions and equivalence principle, Obukhov, Silenko and Teryaev[125] have calculated directly the response of the spin of a Dirac particle in gravitomagnetic field and showed that it is the same as the response of a macroscopic spin angular momentum in general relativity (see also Ref. 126 for a derivation in the weak-field limit). Randono has showed that the active frame-dragging of a polarized Dirac particle is the same as that of a macroscopic body with equal angular momentum.[127] All these results are consistent with EEP and the principle of action-equal-to-reaction. However, these findings do not preclude that the gyrogravitational ratio is to be different from 1 in various different theories of gravity, notably torsion theories and Poincaré gauge theories.

What would be the gyrogravitational ratios of actual elementary particles? If they differ from one, they will definitely reveal some inner gravitational structures of elementary particles, just as different gyromagnetic ratios reveal inner electromagnetic structures of elementary particles. These findings would then give clues to the microscopic origin of gravity.

Promising methods to measure particle gyrogravitational ratio include[61]: (i) using spin-polarized bodies (e.g. polarized solid He^3, Dy–Fe, Ho–Fe, or other compounds) instead of rotating gyros in a GP-B type experiment to measure the gyrogravitational ratio of various substances; (ii) atom interferometry; (iii) nuclear spin gyroscopy; (iv) superfluid He^3 gyrometry. Notably, there have been great developments in atom interferometry[128,129] and nuclear gyroscopy.[130] However, to measure particle gyrogravitational ratios the precision is still short by several orders and more developments are required.

7. An Update of Search for Long Range/Intermediate Range Spin–Spin, Spin–Monopole and Spin–Cosmos Interactions

In this section, we update our review[61,131] on the search for the long range/intermediate range spin–spin, spin–monopole and spin–cosmic interactions.

Spin–spin experiments

Geomagnetic field induces electron polarization within the Earth. Hunter *et al.*[132] estimated that there are on the order of 10^{42} polarized electrons in the Earth compared to $\sim 10^{25}$ polarized electrons in a typical laboratory. For spin–spin interaction, from their results there is an improvement in constraining the coupling strength of the intermediate vector boson in the range greater than about 1 km.[132]

Spin–monopole experiments

In Ref. 61, we have used axion-like interaction Hamiltonian

$$H_{\text{int}} = \left[\frac{\hbar(g_s g_p)}{8\pi mc}\right]\left(\frac{1}{\lambda r} + \frac{1}{r^2}\right)\exp\left(\frac{-r}{\lambda}\right)\boldsymbol{\sigma}\cdot\hat{\mathbf{r}}, \tag{122}$$

to discuss the experimental constraints on the dimensionless coupling $g_s g_p/\hbar$ between polarized (electron) and unpolarized (nucleon) particles. In (122), λ is the range of the interaction, g_s and g_p are the coupling constants of vertices at the polarized and unpolarized particles, m is the mass of the polarized particle and $\boldsymbol{\sigma}$ is Pauli matrix 3-vector. Hoedl *et al.*[133] have pushed the constraint to shorter range by about one order of magnitude since our last review.[61] In this update, we see also good progress in the measurement of spin–monopole coupling between polarized neutrons and unpolarized nucleons.[134–136] Tullney *et al.*[136] obtained the best limit on this coupling for force ranges between 3×10^{-4} m and 0.1 m. Regards to a recent analysis of a direct spin-axion momentum interaction and its empirical constraints, see Ref. 137.

Spin–cosmos experiments

For the analysis of spin–cosmos experiments for elementary particles, one usually uses the following Hamiltonian:

$$H_{\text{cosmic}} = C_1\sigma_1 + C_2\sigma_2 + C_3\sigma_3, \tag{123}$$

in the cosmic frame of reference for spin half particle with C's constants and σ's the Pauli spin matrices (see, e.g. Ref. 138 or 61). The best constraint now is on bound neutron from a free-spin-precession ^{3}He–^{129}Xe comagnetometer experiment performed by Allmendinger *et al.*[130] The experiment measured the free precession of nuclear spin polarized ^{3}He and ^{129}Xe atoms in a homogeneous magnetic guiding field of about 400 nT. As the laboratory rotates with respect to distant stars, Allmendinger *et al.* looked for a sidereal modulation of the Larmor frequencies of the collocated spin samples due to (123) and obtained an upper limit of 8.4×10^{-34} GeV (68% c.l.) on the equatorial component $C_\perp^n$ for neutron. This constraint is more stringent by 3.7×10^4 fold than the limit on that for electron.[139] Using a ^{3}He–K co-magnetometer, Brown *et al.*[140] constrained $C_\perp^p$ for the proton to be less than 6×10^{-32} GeV. Recently Stadnika and Flambaum[141] analyzed the nuclear spin contents of ^{3}He and ^{129}Xe together with a re-analysis of the data of Ref. 130 to give the following improved limit on $C_\perp^p$: $C_\perp^p < 7.6 \times 10^{-33}$ GeV.

8. Prospects

After the cosmological electroweak (vacuum) phase transition around 100 ps from the Big Bang, high energy photons came out. At this time it is difficult to do measurement, although things may still evolve according to precise physical law — notably quantum electrodynamics and classical electrodynamics. When our universe cooled down, precision metrology became possible. Metrological standards could be defined and implemented according to the fundamental physical laws. The cosmic propagation according to Galileo's Weak Equivalence Principle for photons (nonbirefringence) in the framework of premetric classical electrodynamics of continuous media dictates that the spacetime constitutive tensor must be of core metric form with an axion (pseudoscalar) degree of freedom and a dilaton (scalar) degree of freedom. Propagation of pulsar pulses, radio galaxy signals and cosmological GRBs has verified this conclusion empirically down to 10^{-38}, i.e. to $10^{-4} \times O([M_{\text{Higgs}}/M_{\text{Planck}}]^2)$. This is also the order that the generalized closure relations of electrodynamics are verified empirically. The axion and dilaton degrees of freedom are further constrained empirically in the present phase of the cosmos (Table 1). However, we should give a different thought to the axion and dilaton degrees of freedom in exploring spacetime and gravitation in the very early universe within 100 ps from the "Big Bang"; we could look for imprints of new physics and new principles.

On the other hand, experiments with spin are important in verifying Galileo Equivalence Principle and Einstein Equivalence Principle which are important cornerstones of spacetime structure and gravitation. It is not surprising that cosmological observations on polarization phenomena become the ultimate test ground of the equivalence principles, especially for the photon sector. Some of the dispersion relation tests are reaching second-order in the ratio of Higgs boson mass and Planck mass. Ultra-precise laboratory experiments are reaching ground in advancing constraints on various (semi-)long-range spin interactions. Sooner or later, experimental efforts will reach the precision of measuring the gyrogravitational ratios of elementary particles. All these developments may facilitate ways to explore the origins of gravity.

Acknowledgments

I would like to thank Sperello di Serego Alighieri for helpful comments on the manuscript.

References

1. W.-T. Ni, Genesis of general relativity: A concise exposition, in *One Hundred Years of General Relativity: From Genesis and Foundations to Gravitational Waves, Cosmology and Quantum Gravity*, Chap. 2, ed. W.-T. Ni (World Scientific, Singapore, 2016).
2. J. D. Jackson, *Classical Electrodynamics*, 2nd edn. (Wiley, Hoboken, 1975).
3. H. Minkowski, Die Grundgleichungen für die elektromagnetischen Vorgänge in bewegten Körpern, *Königliche Gesellschaft der Wissenschaften zu Göttingen. Mathematisch–Physikalische Klasse. Nachrichten* (1908), pp. 53–111; (English translation) The fundamental equations for electromagnetic processes in moving bodies, translated from German by M. Saha and Wikisource, http://en.wikisource.org/wiki/Translation:The_Fundamental_Equations_for_Electro. . . .
4. F. W. Hehl, *Ann. Phys. (Berlin)* **17** (2008) 691.
5. H. Bateman, The transformation of the electrodynamical equations, *Proc. Camb. Math. Soc., Ser. 2* **8** (1910) 223.
6. F. W. Hehl and Yu. N. Obukhov, *Foundations of Classical Electrodynamics: Charge, Flux, and Metric* (Birkhäuser, Boston, MA, 2003).
7. I. E. Tamm, Electrodynamics of an anisotropic medium in special relativity theory, *Zhurn. Ross. Fiz.-Khim. Ob.* **56** (1924) 248–262 (in Russian); Reprinted in I. E. Tamm, *Collected Papers*, Vol. 1 (Nauka, Moscow, 1975), pp. 19–32 (in Russian).
8. L. Mandelstam and J. Tamm, Elektrodynamik der anisotropen Medien in der speziellen Relativitätstheorie, *Math. Ann.* **95** (1926) 154–160 [Erratum *ibid* **96** (1927) 600]; Reprinted in I. E. Tamm, *Collected Papers*, Vol. 1 (Nauka, Moscow, 1975), pp. 62–67 (in Russian).
9. M. V. Laue, *Die Relativitätstheorie*, Vol. 1: Die spezielle Relativittstheorie, 5th rev. edition (Vieweg, Braunschweig, 1952).

10. E. J. Post, *Formal Structure of Electromagnetics — General Covariance and Electromagnetics* (North Holland, Amsterdam, 1962; Dover, Mineola, New York, 1997).
11. A. Einstein and M. Grossmann, Entwerf einer ver-allgemeinerten relativitätstheorie und einer theorie der gravitation, *Z. Math. Phys.* **63** (1913) 215–225; Outline of a generalized theory of relativity and of a theory of gravitation, in *The Collected Papers of Albert Einstein*, Vol. 4 (Princeton University Press, Princeton, NJ, 1995).
12. A. Einstein, Die formale Grundlage der allgemeinen Relätivitatstheorie, *Königlich Preußische Akademie der Wissenschaften (Berlin). Sitzungsberichte* (1914) 1030–1083; The formal foundation of the general theory of relativity, in *The Collected Papers of Albert Einstein*, Vol. 4 (Princeton University Press, Princeton, NJ, 1995).
13. A. Einstein, Eine Neue Formale Deutung der Maxwellschen Feldgleichungen der Elektrodynamik, *Königlich Preußische Akademie der Wissenschaften* (Berlin) (1916), pp. 184–188, A new formal interpretation of Maxwell's field equations of electrodynamics, in *The Collected Papers of Albert Einstein*, Vol. 6 (Princeton University Press, Princeton, NJ, 1997).
14. W.-T. Ni, Equivalence principles and precision experiments, in *Precision Measurement and Fundamental Constants II*, eds. B. N. Taylor and W. D. Phillips, US National Bureau of Standards Special Publication, Vol. 617 (NBS, Gaithersburg, MD, USA, 1984), p. 647.
15. W.-T. Ni, Timing observations of the pulsar propagations in the galactic gravitational field as precision tests of the Einstein equivalence principle, in *Proc. Second Asian-Pacific Regional Meeting of the Int. Astronomical Union on Astronomy*, Bandung, Indonesia, August, 24–29 1981, eds. B. Hidayat and M. W. Feast (Tira Pustaka, Jakarta, Indonesia, 1984), pp. 441–448.
16. W.-T. Ni, Equivalence principles, their empirical foundations, and the role of precision experiments to test them, in *Proc. 1983 Int. School and Symp. Precision Measurement and Gravity Experiment*, Taipei, Republic of China, January 24–February 2, 1983, ed. W.-T. Ni (National Tsing Hua University, Hsinchu, Taiwan, Republic of China, 1983), pp. 491–517, http://astrod.wikispaces.com/.
17. W.-T. Ni, Spacetime structure and asymmetric metric from the premetric formulation of electromagnetism, *Phys. Lett. A* **379** (2015) 1297–1303, arXiv:1411.0460; and references therein.
18. 1. I. Newton, Philosophiae Naturalis Principia Mathematica, London (1687).
19. G. Galilei, *Discorsi e Dimostrazioni Matematiche Intorno a Due Nuove Scienze* (Elzevir, Leiden, 1638). English translation by H. Crew and A. de Salvio, *Dialogues Concerning Two New Sciences* (Macmillan, New York, 1914); reprinted by (Dover, New York, 1954).
20. W.-T. Ni, *Phys. Rev. Lett.* **38** (1977) 301.
21. W.-T. Ni, *Bull. Amer. Phys. Soc.* **19** (1974) 655.
22. A. Einstein, Ist die Trägheit eines Körpers won seinem Energieinhalt abhängig? *Ann. Phys.* **18** (1905) 639.
23. R. V. Eötvös, *Math. Naturwiss. Ber. Ungarn* **8** (1891) 65.
24. M. Planck, *Berl. Sitz.* 13 June 1907, p. 542, specially at p. 544.
25. A. Einstein, *Jahrb. Radioakt. Elektronik* **4**, 411 (1907); Corrections by A. Einstein in *Jahrb. Radioakt. Elektronik* **5** (1908) 98; English translations by H. M. Schwartz in *Amer. J. Phys.* **45** (1977) 512, 811, 899.
26. C. W. Misner, K. S. Thorne and J. A. Wheeler, *Gravitation* (Freeman, 1973).
27. A. S. Eddington, A generalization of Weyl's theory of the electromagnetic and gravitational fields, *Proc. R. Soc. Lond. A* **99** (1921) 104.
28. É. Cartan, Sur une généralisation de la notion de courbure de Riemann et les espaces à torsion. *C. R. Acad. Sci. (Paris)* **174** (1922) 593.

29. É. Cartan, Sur les variétés à Connexion affine et la théorie de la relativitée généralisée I, I (suite), II, *Ann. Sci. Ecole. Norm. Sup.* **40** (1923) 325; **41** (1924) 1; **42** (1925) 17.
30. O. Stern, *Z. Phys.* **1** (1921) 249; O. Stern and W. Gerlach, *Z. Phys.* **8** (1922) 110; **9** (1922) 349.
31. G. Uhlenbeck and S. Goudsmit, *Naturwiss* **13** (1925) 953; *Nature* **117** (1926) 264.
32. D. W. Sciama, On the analogy between charge and spin in general relativity, in *Recent Developments in General Relativity* (Pergamon + PWN, Oxford, 1962), p. 415.
33. D. W. Sciama, *Rev. Mod. Phys.* **36** (1964) 463, 1103.
34. T. W. B. Kibble, *J. Math. Phys.* **2** (1961) 212.
35. R. Utiyama, *Phys. Rev.* **101** (1956) 1579.
36. F. W. Hehl, J. Nitsch and P. von der Heyde, Poincaré gauge field theory with quadratic Lagrangian, in *General Relativity and Gravitation — One Hundred Years After the Birth of Albert Einstein*, Vol. 1, ed. A. Held (Plenum, New York, 1980), pp. 329–355.
37. K. Hayashi and T. Shirafuji, *Prog. Theor. Phys.* **61** (1980) 866–882.
38. F. W. Hehl, P. von der Heyde, G. D. Kerlick and J. M. Nester, *Rev. Mod. Phys.* **48** (1976) 393.
39. P. von der Heyde, *Nuovo Cimento Lett.* **14** (1975) 250.
40. W.-T. Ni, Spin, Torsion and polarized test-body experiments, in *Proc.* 1983 *Int. School and Symp. Precision Measurement and Gravity Experiment*, Taipei, Republic of China, January 24–February 2, 1983, ed. W.-T. Ni (National Tsing Hua University, Hsinchu, Taiwan, Republic of China, 1983), pp. 532–540, http://astrod.wikispaces.com/.
41. W.-T. Ni, *Phys. Lett. A* **120** (1986) 174–178.
42. K. Nordtvedt, Jr., *Phys. Rev.* **169** (1968) 1014, 1017; **170** (1968) 1186.
43. R. H. Dicke, *Gravitation and the Universe* (American Philosophical Society, Philadelphia, 1969), pp. 19–24.
44. R. H. Dicke, *The Theoretical Significance of Experimental Relativity* (Gordon and Breach, New York, 1964).
45. C. M. Will and K. Nordtvedt, Jr., *Astrophys. J.* **77** (1972) 757.
46. K. Nordtvedt, Jr. and C. M. Will, *Astrophys. J.* **77** (1972) 775.
47. A. M. Nobili *et al.*, *Amer. J. Phys.* **81** (2013) 527.
48. E. Di Casola, S. Liberati and S. Sonego, *Amer. J. Phys.* **83** (2015) 39–46.
49. L. I. Schiff, *Amer. J. Phys.* **28** (1960) 340.
50. R. H. Dicke, *Amer. J. Phys.* **28** (1960) 344.
51. K. S. Thorne, D. L. Lee and A. P. Lightman, *Phys. Rev. D* **7** (1973) 3563.
52. A. P. Lightman and D. L. Lee, *Phys. Rev. D* **8** (1973) 364.
53. W.-T. Ni, A nonmetric theory of gravity, preprint, Montana State University, Bozeman, Montana, USA (1973), http://astrod.wikispaces.com/.
54. P. Wolf *et al.*, *Nature* **467** (2010) E1.
55. H. Müller, A. Peters and S. Chu, *Nature* **467** (2010) E2.
56. J. M. Gérard, *Class. Quantum Grav.* **24** (2007) 1867.
57. E. Di Casola, S. Liberati and S. Sonego, *Phys. Rev. D* **89** (2014) 084053.
58. M. Haugan and T. Kauffmann, *Phys. Rev. D* **52** (1995) 3168.
59. C. Lämmerzahl and F. W. Hehl, *Phys. Rev. D* **70** (2004) 105022.
60. W.-T. Ni, *Prog. Theor. Phys. Suppl.* **172** (2008) 49.
61. W.-T. Ni, *Rep. Progr. Phys.* **73** (2010) 056901.
62. W.-T. Ni, *Phys. Lett. A* **378** (2014) 1217–1223.
63. W.-T. Ni, *Phys. Lett. A* **378** (2014) 3413.

64. A. Favaro and L. Bergamin, *Ann. Phys.* **523** (2011) 383–401.
65. M. F. Dahl, *J. Phys. A: Math. Theor.* **45** (2012) 405203.
66. F. G. Smith, *Pulsars* (Cambridge University Press, Cambridge, UK, 1977).
67. H.-W. Huang, Pulsar timing and equivalence principle tests, Master thesis, National Tsing Hua University (2002).
68. P. M. McCulloch, P. A. Hamilton, J. G. Ables and A. J. Hunt, IAU Circular (USA), No. 3703, p. 1, 15 June (1982).
69. D. C. Backer, S. R. Kulkarni, C. Heiles, M. M. Davis and W. M. Goss, *Nature* **300** (1982) 615.
70. V. A. Kostelecky and M. Mewes, *Phys. Rev. D* **66** (2002) 056005.
71. J. W. Moffat, in *Gravitation* 1990 *Proc. Banff Summer Institute*, Banff, Canada, eds. R. D. Mann and P. Wesson (World Scientific, Singapore, 1991).
72. N. J. Cornish, J. W. Moffat and D. C. Tatarshi, *Gen. Relativ. Gravit.* **27** (1995) 933.
73. T. P. Krisher, *Phys. Rev. D* **44** (1991) R2211.
74. D. Götz, S. Covino, A. Fernández-Soto, P. Laurent and Ž. Bosnjak, *Mon. Not. R. Astron. Soc.* **431** (2013) 3550.
75. D. Götz *et al.*, *Mon. Not. R. Astron. Soc.* **444** (2014) 2776.
76. P. Laurent, D. Götz, P. Binétruy, S. Covino and A. Fernández-Soto, *Phys. Rev. D* **83** (2011) 12.
77. V. A. Kostelecký and M. Mewes, *Phys. Rev. Lett.* **110** (2013) 201601.
78. W.-T. Ni, *Chin. Phys. Lett.* **22** (2005) 33.
79. Y. N. Obukhov and F. W. Hehl, *Phys. Lett. A* **341** (2005) 357.
80. Y. Itin, *Gen. Relativ. Gravit.* **40** (2008) 1219.
81. D. J. Fixsen, *Astrophys. J.* **707** (2009) 916.
82. S. di Serego Alighieri, W.-T. Ni and W.-P. Pan, *Astrophys. J.* **792** (2014) 35.
83. H.-H. Mei, W.-T. Ni, W.-P. Pan, L. Xu and S. di Serego Alighieri, *Astrophys. J.* **805** (2015) 107.
84. S. di Serego Alighieri, *Int. J. Mod. Phys. D* **24** (2015) 1530006.
85. P. A. R. Ade *et al. Astron. Astrophys.* **571** (2014) A16.
86. Yu. N. Obukhov and F. W. Hehl, *Phys. Rev. D* **70** (2004) 125105.
87. A. S. Eddington, *The Mathematical Theory of Relativity*, 2nd edn. (Cambridge University Press, 1924).
88. A. Einstein and E. G. Straus, *Ann. Math.* **47** (1946) 731.
89. E. Schrödinger, *Proc. R. Ir. Acad. A* **51** (1947) 163.
90. E. Schrödinger, *Space-Time Structure* (Cambridge University Press, 1950).
91. I. V. Lindell and K. H. Wallén, *J. Electromag. Waves Appl.* **18** (2004) 957.
92. A. Favaro, Recent advances in classical electromagnetic theory, Ph.D. thesis, Imperial College London (2012).
93. R. Toupin, Elasticity and electromagnetics, in *Non-linear Continuum Theories, C.I.M.E. Conf.* Bressanone, Italy (1965), eds. C. Truesdell and G. Grioli, pp. 203–342.
94. M. Schönberg, Electromagnetism and Gravitation, *Rev. Bras. Fis.* **1** (1971) 91.
95. A. Jadczyk, Electromagnetic permeability of the vacuum and light-cone structure, *Bull. Acad. Pol. Sci. Sér. Sci. Phys. Astron.* **27** (1979) 91.
96. R. D. Peccei and H. R. Quinn, *Phys. Rev. Lett.* **38** (1977) 1440.
97. S. Weinberg, *Phys. Rev. Lett.* **40** (1978) 233.
98. F. Wilczek, *Phys. Rev. Lett.* **40** (1978) 279.
99. J. Kim, *Phys. Rev. Lett.* **43** (1979) 103.
100. M. A. Shifman, A. I. Vainshtein and V. I. Zakharov, *Nucl. Phys. B* **166** (1980) 493.
101. M. Dine, Fischler and M. Srednicki, *Phys. Lett. B* **104** (1981) 199.

102. S.-L. Cheng, C.-Q. Geng and W.-T. Ni, *Phys. Rev. D* **52** (1995) 3132.
103. M. Yu. Khlopov, *Cosmoparticle Physics* (World Scientific, 1999).
104. W.-T. Ni, *Phys. Lett. A* **120** (1987) 174.
105. W.-T. Ni, Some basic points about metrology, in *Proc.* 1983 *Int. School and Symp. Precision Measurement and Gravity Experiment*, Taipei, Republic of China, January 24–February 2, 1983, ed. by W.-T. Ni (Published by National Tsing Hua University, Hsinchu, Taiwan, Republic of China, 1983), pp. 121–134, http://astrod.wikispaces.com/.
106. B. W. Petley, Fundamental Physical Constants and the Frontier of Measurement (A. Hilger, Bristol and Boston, 1985).
107. T. J. Quinn, *IEEE Trans. Instrum. Meas.* **40** (1991) 81.
108. P. Becker, *Contemp. Phys.* **53** (2012) 461.
109. C. A. Sanchez, B. M. Wood, R. G. Green, J. O. Liard and D. Inglis, *Metrologia* **51** (2014) S5–S14.
110. D. B. Newell, *Phys. Today* **67**(7) (2014) 35.
111. R. D. Vocke, S. A. Raab and G. C. Turk, *Metrologia* **51** (2014) 361.
112. B. Andreas *et al.*, *Metrologia* **48** (2011) S1.
113. L. Yang, Z. Mester, R. E. Sturgeon and J. Meija, *Anal. Chem.* **84** (2012) 2321.
114. P. J. Mohr, B. N. Taylor and D. B. Newell, *Rev. Mod. Phys.* **84** (2012) 1527.
115. A. D. Ludlow, M. M. Boyd, J. Ye, E. Peik and P. O. Schmidt, *Rev. Mod. Phys.* **87** (2015) 637.
116. K. Kuroda, W.-T. Ni and W.-P. Pan, Gravitational waves: Classification, methods of detection, sensitivities, and sources, in *One Hundred Years of General Relativity: From Genesis and Empirical Foundations to Gravitational Waves, Cosmology and Quantum Gravity*, Chap. 10, ed. W.-T. Ni (World Scientific, Singapore, 2016); *Int. J. Mod. Phys. D* **24** (2015) 1530031.
117. I. Ciufolini and E. C. Pavlis, *Nature* **431** (2004) 958.
118. J. Lense and H. Thirring, *Phys. Z.* **19** (1918) 156.
119. C. W. F. Everitt *et al.*, *Phys. Rev. Lett.* **106** (2011) 221101.
120. L. I. Schiff, *Phys. Rev. Lett.* **4** (1960) 215.
121. W.-T. Ni, *Phys. Rev. Lett.* **107** (2011) 051103.
122. I. Ciufolini *et al.*, *Eur. Phys. J. Plus* **127** (2012) 133.
123. F. Bosi *et al.*, *Phys. Rev. D* **84** (2011) 122002.
124. Y.-C. Huang and W.-T. Ni, Propagation of Dirac wave functions in accelerated frames of reference, arXiv:gr-qc/0407115.
125. Y. N. Obukhov, A. J. Silenko and O. V. Teryaev, *Phys. Rev. D* **88** (2013) 084014 and references therein.
126. H.-H. Tseng, On the equation of motion of a Dirac particle in gravitational field and its gyro-gravitational ratio, M. S. Thesis (in Chinese with an English abstract, Advisor: W.-T. Ni), National Tsing Hua University, Hsinchu (2001).
127. A. Randono, *Phys. Rev. D* **81** (2010) 024027.
128. T. Schuldt *et al.*, *Exp. Astron.* **39** (2015) 167.
129. L. Zhou *et al.*, *Phys. Rev. Lett.* **115** (2015) 013004.
130. F. Allmendinger *et al.*, *Phys. Rev. Lett.* **112** (2014) 110801.
131. W.-T. Ni, Searches for the role of spin and polarization in gravity: A five-year update, arXiv:1501.07696.
132. L. Hunter *et al.*, *Science* **339** (2013) 928.
133. S. A. Hoedl *et al.*, *Phys. Rev. Lett.* **106** (2011) 100801.
134. P.-H. Chu *et al.*, *Phys. Rev. D* **87** (2013) 011105(R).
135. M. Bulatowicz *et al.*, *Phys. Rev. Lett.* **111** (2013) 102001.
136. K. Tullney *et al.*, *Phys. Rev. Lett.* **111** (2013) 100801.

137. Y. V. Stadnik and V. V. Flambaum, *Phys. Rev. D* **89** (2014) 043522.
138. P. R. Phillips, *Phys. Rev. B* **139** (1965) 491.
139. B. R. Heckel, E. G. Adelberger, C. E. Cramer, T. S. Cook, S. Schlamminger and U. Schmidt, *Phys. Rev. D* **78** (2008) 092006.
140. J. M. Brown, S. J. Smullin, T. W. Kornack and M. V. Romalis, *Phys. Rev. Lett.* **105** (2010) 151604.
141. Y. V. Stadnik and V. V. Flambaum, *Eur. Phys. J. C* **75** (2015) 110.

Chapter 6

Cosmic polarization rotation: An astrophysical test of fundamental physics*

Sperello di Serego Alighieri
INAF - Osservatorio Astrofisico di Arcetri,
Largo E. Fermi 5, 50125 Firenze, Italy
sperello@arcetri.astro.it

Possible violations of fundamental physical principles, e.g. the Einstein equivalence principle on which all metric theories of gravity are based, including general relativity (GR), would lead to a rotation of the plane of polarization for linearly polarized radiation traveling over cosmological distances, the so-called cosmic polarization rotation (CPR). We review here the astrophysical tests which have been carried out so far to check if CPR exists. These are using the radio and ultraviolet polarization of radio galaxies and the polarization of the cosmic microwave background (both E-mode and B-mode). These tests so far have been negative, leading to upper limits of the order of one degree on any CPR angle, thereby increasing our confidence in those physical principles, including GR. We also discuss future prospects in detecting CPR or improving the constraints on it.

Keywords: Polarization; radio galaxies; cosmic background radiation.

1. Introduction

Linear polarization is a simple phenomenon by which a single photon is able to transmit across the universe the information about the orientation of a plane. The question which we discuss in this paper is whether the orientation of the plane of linear polarization, the so-called position angle (PA[a]), is conserved for electromagnetic radiation traveling long distances, i.e. if there is any cosmic polarization rotation (CPR). Clearly, if the CPR angle α is not zero, symmetry must be broken at some level, since α must be either positive or negative, for a counterclockwise or clockwise rotation. This immediately suggests that CPR should be connected with the violation of fundamental physical principles. Indeed, it is linked also to a possible violation of the Einstein equivalence principle (EEP), which is the foundation of any metric theory of gravity, including general relativity (GR). Therefore it deserves a chapter in this volume.

*This article was also published in *Int. J. Mod. Phys. D* **24**, 1530016 (2015). This version has been updated to include the results of Planck and POLARBEAR.

[a]We adopt the International Astronomical Union (IAU) convention for PA: it increases counterclockwise facing the source, from North through East.[36]

The fundamental principles whose violation would imply CPR are briefly discussed in Sec. 2 (please refer also to other chapters in this volume). For most of them, the CPR angle would be independent of wavelength. However the violation of some principles would imply a wavelength-dependent CPR, not to be confused with the Faraday rotation, which is a well-known effect for radiation passing through a plasma with a magnetic field. CPR, if it exists, would occur in vacuum. CPR has sometimes been inappropriately called "cosmological birefringence." However we follow here the advice of Ni,[60] since birefringence is only appropriate for a medium whose index of refraction depends on the direction of polarization of the incident light beam, which is then split in two. The phenomenon we are considering here is pure rotation of the polarization, without any splitting.

Testing for CPR is simple in principle: it requires a distant source of linearly polarized radiation, for which the orientation PA_{em} of the polarization at the emission can be established. Then CPR is tested by comparing the observed orientation PA_{obs} with PA_{em}:

$$\alpha = PA_{obs} - PA_{em}.$$

In practice, it is not easy to know *a priori* the orientation of the polarization for a distant source: in this respect the fact that scattered radiation is polarized perpendicularly to the plane containing the incident and scattered rays has been of great help, applied both to radio galaxies (RGs) (see Sec. 4) and to the cosmic background (CMB) radiation (see Sec. 5). For those cases in which CPR depends on wavelength, one can also test CPR by simply searching for variation of PA with the wavelength of the radiation, even without knowing PA_{em}. In this paper, we will review the astrophysical methods which have been used to test CPR, we list the results of these test, discuss the advantages and disadvantages of the various methods and suggest future prospects for these tests.

2. Impact of CPR on Fundamental Physics

This possibility of CPR arises in a variety of important contexts, like the presence of a cosmological pseudoscalar condensate, Lorentz invariance violation and charge, parity and time reversal (CPT) violation, neutrino number asymmetry, the EEP violation. In particular, the connection of the latter with CPR is relevant for this GR Centennial year, since all metric theories of gravity, including GR are based on the EEP. Since the weak equivalence principle (WEP) is tested to a much higher accuracy than the EEP, Schiff[68] conjectured that any consistent Lorentz-invariant theory of gravity which obeys the WEP would necessarily also obey the EEP. If these were true, the EEP would tested to the same accuracy as the WEP, increasing our experimental confidence in GR. However, Ni[57,58] found a unique counter example to Schiff's conjecture: a pseudoscalar field which would lead to a violation of the EEP, while obeying the WEP. Such field would produce a CPR. Therefore, testing for the CPR is important for our confidence in GR. For the other theoretical impacts of CPR we refer the reader to Refs. 59, 60 and 62.

3. Constraints from the Radio Polarization of RGs

Already in his seminal paper about the unique counter-example to Schiff's conjecture giving rise to CPR, Ni[57] suggested that observations of polarized astrophysical sources could give constraints on the CPR. However, only in 1990, the polarization at radio wavelengths of RGs and quasars was used for the first astrophysical test of CPR.[12,b] Ref. 12 has used the fact that extended radio sources, in particular, the more strongly polarized ones, tend to have their plane of integrated radio polarization, corrected for Faraday rotation, usually perpendicular and occasionally parallel to the radio source axis,[18] to put a limit of 6° at the 95% confidence level (CL) to any rotation of the plane of polarization for the radiation coming from these sources in the redshift interval $0.4 < z < 1.5$.

Reanalyzing the same data, Nodland and Ralston[63] claimed to have found a rotation of the plane of polarization, independent of the Faraday one, and correlated with the angular positions and distances to the sources. Such rotation would be as much as 3 rad for the most distant sources. However, several authors have independently and convincingly rejected this claim, both for problems with the statistical methods,[13,27,51] and by showing that the claimed rotation is not observed for the optical/ultraviolet (UV) polarization of two RGs (see below) and for the radio polarization of several newly observed RGs and quasars.[72]

In fact, the analysis of Leahy[48] is important also because it introduces a significant improvement to the radio polarization method for the CPR test. The problem with this method is the difficulty in estimating the direction of the polarization at the emission. Since the radio emission in RGs and quasars is due to synchrotron radiation, the alignment of its polarization with the radio axis implies an alignment of the magnetic field, which is not obvious *per se*. In fact, theory and magnetohydrodynamics simulations foresee that the projected magnetic field should be perpendicular to strong gradients in the total radio intensity.[7,66] For example, for a jet of relativistic electrons the magnetic field should be perpendicular to the local jet direction at the edges of the jet and parallel to it where the intensity changes along the jet axis.[10] On the other hand, such alignments are much less clear for the *integrated* polarization, because of bends in the jets and because intensity gradients can have any direction in the radio lobes, which emit a large fraction of the polarized radiation in many sources. In fact, it is well-known that the peaks at 90° and 0° in the distribution of the angle between the direction of the radio polarization and that of the radio axis are very broad and the alignments hold only statistically, but not necessarily for individual sources (see e.g. Fig. 1 of Ref. 12). More stringent tests can be carried out using high angular resolution data on radio polarization and the local magnetic field's alignment for individual sources,[72] although to our knowledge, only once[48] this method has been used to put quantitative limits on

[b]Ref. 9 had earlier claimed a substantial anisotropy in the angle between the direction of the radio axis and the direction of linear radio polarization in a sample of high-luminosity classical double radio sources, but used it to infer rotation of the universe, not to test for CPR.

the polarization rotation. For example, Carroll,[14] using the data on the ten RGs of Leahy,[48] obtains an average constraint on any CPR angle of $\alpha = -0.6° \pm 1.5°$ at the mean redshift $\langle z \rangle = 0.78$. However, the preprint by Leahy[48] remained unpublished and does not explain convincingly how the angle between the direction of the local intensity gradient and that of the polarization is derived. For example, for 3C9, the source with the best accuracy, Leahy[48] refers to Ref. 47, who however, do not give any measurements of local gradients.

4. Constraints from the UV Polarization of RGs

Another method to test for CPR has used the perpendicularity between the direction of the elongated structure in the UV[c] and the direction of linear UV polarization in distant powerful RGs. The test was first performed by Refs. 16 and 24, who obtained that any rotation of the plane of linear polarization for a dozen RGs at $0.5 \leq z \leq 2.63$ is smaller than 10°.

Although this UV test has sometimes been confused with the one at radio wavelengths, probably because they both use RGs polarization, it is a completely different and independent test, which hinges on the well-established unification scheme for powerful radio-loud Active Galatic Nuclei (AGN).[5] This scheme foresees that powerful radio sources do not emit isotropically, but their strong UV radiation is emitted in two opposite cones, because the bright nucleus is surrounded by an obscuring torus: if our line of sight is within the cones, we see a quasar, otherwise we see a RG. Therefore, powerful RGs have a quasar in their nuclei, which can only be seen as light scattered by the interstellar medium of the galaxy. Often, particularly in the UV, this scattered light dominates the extended radiation from RGs, which then appear elongated in the direction of the cones and strongly polarized in the perpendicular direction.[23] The axis of the UV elongation must be perpendicular to the direction of linear polarization, because of the scattering mechanism which produces the polarization. Therefore, in this case it is possible to accurately predict the direction of polarization at the emission and compare it with the observed one. This method of measuring the polarization rotation can be applied to any single case of distant RG, which is strongly polarized in the UV, allowing independent CPR tests in many different directions. Another advantage of this method is that it does not require any correction for Faraday rotation, which is large at radio wavelengths, but negligible in the UV.

In the case of well resolved sources, the method can be applied also to the polarization which is measured locally at any position in the elongated structures around RGs, and which has to be perpendicular to the vector joining the observed position with the nucleus. From the polarization map in the V-band (~3000 Å rest-frame) of 3C 265, a RG at $z = 0.811$,[70] the mean deviation of the 53 polarization

[c]When a distant RG ($z > 0.7$) is observed at optical wavelengths ($\lambda_{\text{obs.}} \sim 5000$ Å), these correspond to the UV in the rest frame ($\lambda_{\text{em.}} \leq 3000$ Å).

vectors plotted in the map from the perpendicular to a line joining each to the nucleus is $-1.4° \pm 1.1°$.[72] However, more distant RGs are so faint that only the integrated polarization can be measured, even with the largest current telescopes: strict perpendicularity is expected also in this case, if the extended emission is dominated by the scattered radiation, as is the case in the UV for the strongly polarized RGs.[71]

Recently, the available data on all RGs with redshift larger than two and with the measured degree of linear polarization larger than 5% in the UV (at ~1300 Å) have been reexamined, and no rotation within a few degrees in the polarization for any of these eight RGs has been found.[25] In addition, assuming that the CPR angle should be the same in every direction, an average constraint on this rotation $\langle\alpha\rangle = -0.8° \pm 2.2°$ (1σ) at the mean redshift $\langle z\rangle = 2.80$ has been obtained.[25] The same data have been used by Ref. 39 to set a CPR constraint in case of a nonuniform polarization rotation, i.e. a rotation which is not the same in every direction: in this case the variance of any rotation must be $\langle\alpha^2\rangle \leq (3.7°)^2$. The CPR test using the UV polarization has advantages over the other tests at radio or CMB wavelengths, if CPR effects grow with photon energy (the contrary of Faraday rotation), as in a formalism where Lorentz invariance is violated but CPT is conserved.[43,44]

5. Constraints from the Polarization of the CMB Radiation

A more recent method to test for the existence of CPR is the one that uses the CMB polarization, which is induced by the last Thomson scattering of decoupling photons at $z \sim 1100$, resulting in a correlation between temperature gradients and polarization.[49] CMB photons are strongly linearly polarized, since they result from scattering. However the high uniformity of CMB produces a very effective averaging of the polarization in any direction. It is only at the CMB temperature disuniformities that the polarization does not average out completely and residual polarization perpendicular to the temperature gradients is expected. Therefore, also for the CMB polarization it is possible to precisely predict the polarization direction at the emission and to test for any CPR. After the first detection of CMB polarization anisotropies by Degree Angular Scale Interferometer (DASI),[46] there have been several CPR tests using the CMB E-mode polarization pattern.

Unfortunately, the scientists working on the CMB polarization have adopted for the polarization angle a convention which is opposite to the IAU one, used for decades by all other astrophysicists and enforced by the IAU[36]: for the CMB polarimetrists, following a software for the data pixelization on a sphere,[30] the polarization angle increases clockwise, instead of counterclockwise, facing the source. This produces an inversion of the U Stokes parameter, corresponding to a change of PA sign. Obviously, these different conventions have to be taken into account, when comparing data obtained with the different methods used for CPR searches. As mentioned in the introduction, all PA in this paper are given in the IAU convention. Independently of the adopted convention, a problem of CMB polarimetry is the calibration of the PA, which is not easy at CMB frequencies. Although different methods are used, like the *a priori* knowledge of the detector's orientation, the use of calibration sources both near the experiment on the ground and on the sky, the current calibration accuracy is of the order of one degree, producing a nonnegligible systematic error β on the measured PA. In order to alleviate the PA calibration problem, Ref. 42 have suggested a self-calibration technique consisting in minimizing EB and TB power spectra with respect to PA offset. Unfortunately, such a calibration technique would eliminate not just the PA calibration offset β, but also $\alpha - \beta$, where α is the uniform CPR angle, if it exists. Therefore, no independent information on the uniform CPR angle can be obtained, if this calibration technique is adopted, like with the Background Imaging of Cosmic Extragalactic Polarization 2 (BICEP2)[2] experiment.

In the following, we summarize the most recent and accurate CPR measurements obtained using the CMB polarization (see Table 1 and Fig. 1). The BOOMERanG collaboration, revisiting the limit set from their 2003 flight, found a CPR angle $\alpha = 4.3° \pm 4.1°$ (68% CL), assuming uniformity over the whole sky.[64] The QUEST at DASI (QUaD) collaboration found $\alpha = -0.64° \pm 0.50°$ (stat.) $\pm 0.50°$ (syst.) (68% CL),[11] while using three years of BICEP1 data one gets $\alpha = 2.77° \pm 0.86°$ (stat.) $\pm 1.3°$ (syst.) (68% CL).[40] Combining nine years of Wilkinson microwave anisotropy probe (WMAP) data and assuming uniformity, a limit to CPR angle $\alpha = 0.36° \pm 1.24°$ (stat.) $\pm 1.5°$ (syst.) (68% CL) has been set, or $-3.53° < \alpha < 4.25°$ (95% CL), adding in quadrature statistical and systematic errors.[33] The POLARBEAR collaboration[3] reports about a difference of 1.08° in the instrument polarization angle obtained at 148 GHz minimizing the EB spectrum and that obtained from their data on the Crab Nebula using the PA measurement at 90 GHz of Ref. 6. This corresponds to a measurement of CPR, performed with the effect of a rotation on the EB spectrum and using the Crab Nebula for the PA calibration, and giving a CPR angle $\alpha = 1.08° \pm 0.2°$(stat.) $\pm 0.5°$(syst.), assuming that the Crab Nebula polarization angle does not change between 90 and 148 GHz. A consistency check with the value of the Cen A polarization angle measured by POLARBEAR confirms this result. Recently the ACTPol (Atacama Cosmology Telescope Polarimeter) team[56] have used their first three months of observations to measure the CMB polarization over four sky regions near the celestial equator. They do not give an explicit value for the CPR, also because they have used the EB and TB[d] power minimization technique of Ref. 42. However it is possible to derive a value of the CPR from their data, since they have measured a PA of $150.9 \pm 0.6°$ for Crab Nebula (Tau A, a polarization standard source), using the EB and TB nulling procedure (Hasselfield, private communication). The most precise fiducial measurement at CMB frequencies of the Crab Nebula polarization angle is a PA of $149.9 \pm 0.2°$ at 90 GHz.[6] If we assume that the Crab Nebula polarization PA would not change

[d]EB and TB are the cross-correlation power spectra between E- and B-modes and between temperature and B-mode.

Table 1. Measurements of CPR with different methods (in chronological order).

Method	CPR angle ± stat. (± syst.)	Frequency or λ	Distance	Direction	Reference
RG radio pol.	$\lvert\alpha\rvert < 6°$	5 GHz	$0.4 < z < 1.5$	All-sky (uniformity ass.)	12
RG UV pol.	$\lvert\alpha\rvert < 10°$	~3000 Å rest-frame	$0.5 < z < 2.63$	All-sky (uniformity ass.)	16,17
RG UV pol.	$\alpha = -1.4° \pm 1.1°$	~3000 Å rest-frame	$z = 0.811$	$RA: 176.4°, Dec: 31.6°$	72
RG radio pol.	$\alpha = -0.6° \pm 1.5°$	3.6 cm	$\langle z\rangle = 0.78$	All-sky (uniformity ass.)	14,48
CMB pol. BOOMERanG	$\alpha = 4.3° \pm 4.1°$	145 GHz	$z \sim 1100$	$RA \sim 82°, Dec \sim 45°$	64
CMB pol. QUAD	$\alpha = -0.64° \pm 0.50° \pm 0.50°$	100–150 GHz	$z \sim 1100$	$RA \sim 82°, Dec \sim 50°$	11
RG UV pol.	$\alpha = -0.8° \pm 2.2°$	~1300 Å rest-frame	$\langle z\rangle = 2.80$	All-sky (uniformity ass.)	25
RG UV pol.	$\langle\delta\alpha^2\rangle \leq (3.7°)^2$	~1300 Å rest-frame	$\langle z\rangle = 2.80$	All-sky (stoch. var.)	25,39
CMB pol. WMAP9	$\alpha = 0.36° \pm 1.24° \pm 1.5°$	23–94 GHz	$z \sim 1100$	All-sky (uniformity ass.)	33
CMB pol. BICEP1	$\alpha = 2.77° \pm 0.86° \pm 1.3°$	100–150 GHz	$z \sim 1100$	$-50° < RA < 50°, -70° < Dec < -45°$	40
CMB pol. POLARBEAR	$\alpha = 1.08° \pm 0.2° \pm 0.5°$	148 GHz	$z \sim 1100$	$RA \sim 70°, 178°, 345°; Dec \sim -45°, 0°, -33°$	3
CMB pol. ACTPol	$\alpha = 1.0° \pm 0.63°$**	146 GHz	$z \sim 1100$	$RA \sim 35°, 150°, 175°, 355°, Dec \sim 50°$	54,56
CMB pol. B-mode	$\langle\delta\alpha^2\rangle \leq (1.36°)^2$	95–150 GHz	$z \sim 1100$	Various sky regions	54
CMB pol. Planck	$\alpha = 0.35° \pm 0.05° \pm 0.28°$	30–353 GHz	$z \sim 1100$	All-sky (uniformity ass.)	79

Note: **A systematic error should be added, equal to the unknown difference of the Crab Nebula polarization PA between 146 GHz and 90 GHz.

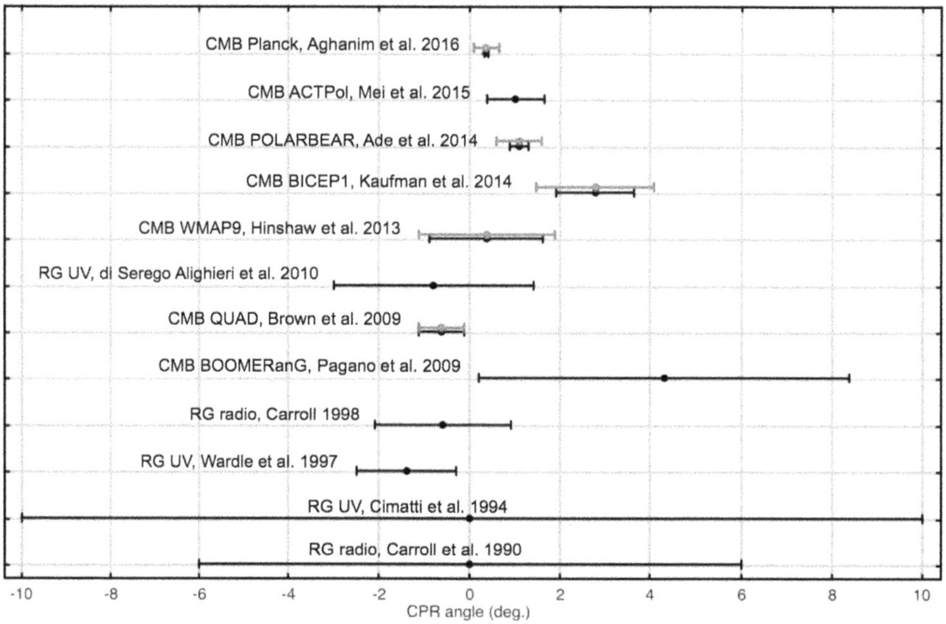

Fig. 1. CPR angle measurements by the various experiments, displayed in chronological order. Black error bars are for the statistical error, while red ones are for the systematic one, if present. A systematic error should be added to the ATCPol measurement, equal to the unknown difference of the Crab Nebula PA between 146 GHz and 90 GHz.[54]

between 90 GHz and 146 GHz (see e.g. the discussion in Sec. 6 of Ref. 6), then the average CPR angle over the ACTPol equatorial regions would be the difference between the above values $\alpha = 1.0° \pm 0.63°$ (see Ref. 54); however the above assumption leaves room for some systematic error. We could instead use the data of Ref. 56 in a different way: since the PA offset angle which they obtain from the EB minimization technique is $-0.2° \pm 0.5°$, i.e. consistent with zero, Ref. 56 suggests that their optical modeling procedure should be free of systematic errors at the 0.5° level or better. If these were true, then $\alpha = 0.22° \pm 0.32° \pm 0.5°$.[54] In summary, for the ACTPol result we prefer the assumption on the constancy of the Crab Nebula polarization angle between 90 GHz and 146 GHz, also because this can be tested *a posteriori* and an eventual correction applied. Recently the results on CPR from the Planck satellite have finally been published giving $\alpha = 0.35° \pm 0.05° \pm 0.28°$ with the stacking analysis.[78] Thanks to the very good quality of the Planck data, they achieve, as expected, a very small statistical uncertainty, considerably lower than any previous measurement. However their accuracy is limited by the uncertainty in the calibration of the position angle: even using the best calibrators, their systematic uncertainty is more than 5 times larger than the statistical one. In fact, most likely their measurement of the CPR angle (see Table 1) is actually a measurement of the Planck polarization angle offset.

In summary, although some have claimed to have detected a rotation,[40,75] the CMB polarization data appear well consistent with a null CPR. In principle the CMB polarization pattern can be used to test CPR in specific directions. However, because of the extremely small anisotropies in the CMB temperature and polarization, these tests have so far used averages over relatively large regions of sky, assuming uniformity.

Recently, Ref. 26 has suggested the possibility of setting constraints on the CPR also using measurements of the B-mode polarization of the CMB, because of the coupling from E-mode to B-mode polarization that any such rotation would produce. This possibility is presently limited by the relatively large systematic errors on the polarization angle still affecting current data. The result is that from the SPTpol (South Pole Telescope polarimeter), POLARBEAR and BICEP2 B-mode polarization data it is only possible to set constraints on the fluctuations $\langle\delta\alpha^2\rangle \leq (1.56°)^2$ of the CPR, not on its mean value. Ref. 54 have similarly obtained an upper limit on the CPR fluctuations $\langle\delta\alpha^2\rangle \leq (1.68°)^2$ from the ACTPol B-mode data of Ref. 56. By considering also SPTPol B-mode polarization data, Ref. 79 have recently improved this upper limit to $\langle\delta\alpha^2\rangle \leq (0.97°)^2$. The one-but-last row of Table 1 reports the combined constraint on the CPR fluctuations obtained from all the B-mode data mentioned above.

6. Other Constraints

Observations of nearby polarized galactic objects could contribute to the CPR test, in particular, for those cases where polarization measurements can be made with high accuracy and at very high frequencies (useful if CPR effects grow with photon energy). Pulsars and supernova remnants emit polarized radiation in a very broad frequency range. For example, hard X-ray polarization observations of the Crab Nebula[22] have been used to set a limit to CPR angle $\alpha = -1° \pm 11°$.[52] However this limit is not particularly stringent, both because of the low accuracy of the X-ray polarization measurement and because of the limited distance to the source. In future, more precise X-ray polarization experiments such as POLARIX,[19] could much improve the situation.

Gamma-ray bursts (GRB) are very distant sources which emit polarized radiation both in the optical afterglow[20,73] and in the prompt gamma-ray emission.[31,38] Nevertheless, they cannot be used for CPR searches, since the orientation of the polarization at the emission is unknown. However, they can be used to test for birefringence effects, i.e. an energy-dependent rotation of the polarization angle, such as those produced by Lorentz invariance violation,[50] since the detection of linear polarization in a gamma-ray band excludes a significant rotation of the polarization within that energy band. In this way Ref. 31 was able to put an upper limit to the dimensionless parameter[e] of this birefringence effect of $\xi < 1 \times 10^{-16}$ from

[e] $\xi \equiv (n_0)^3$, where n_0 is the time component of the Myers–Pospelov four-vector n_α, in a reference frame where $n_\alpha = (n_0, 0, 0, 0)$.[32,55]

the gamma-ray polarization of a GRB at $z = 2.74$. Using the same data for testing the Lorentz symmetry and the equivalence principle, Refs. 61 and 62 provide a birefringence constraint of about 10^{-38}.

For an issue related to CPR, Ref. 34 provides evidence that the directions of linear polarization at optical wavelengths for a sample of 355 quasars ($0 \leq z \leq 2.4$) are nonuniformly distributed, being systematically different near the North and South Galactic Poles, particularly in some redshift ranges. Such behavior could not be caused by uniform CPR, since a rotation of randomly distributed directions of polarization could produce the observed alignments only with a very contrived distribution of rotations as a function of distance and position in the sky. Moreover, the claim by Ref. 34 has not been confirmed by the radio polarization directions on a much larger sample of 4290 flat-spectrum radio sources,[37] and Ref. 35 recently suggested that the alignments could be due to an alignment of quasar's spin axis to the structures to which they belong. The possibility that the quasar's polarization alignments could be due to the mixing of photons with axion-like particles is excluded by the absence of circular polarization.[65]

The rotation of the plane of linear polarization can be seen as different propagation speeds for right and left circularly polarized photons ($\Delta c/c$). The sharpness of the pulses of pulsars in all Stokes parameters can be used to set limits corresponding to $\Delta c/c \leq 10^{-17}$. Similarly the very short duration of GRB gives limits of the order of $\Delta c/c \leq 10^{-21}$. However the lack of linear polarization rotation discussed in the previous sections can be used to set much tighter limits ($\Delta c/c \leq 10^{-32}$).[29]

In a complementary way to the astrophysical tests described in the previous sections, also laboratory experiments can be used to search for CPR. These are outside the scope of this paper and have not obtained significant constraints. For example, the PVLAS (Polarizzazione del Vuoto con LASer) collaboration has found a polarization rotation in the presence of a transverse magnetic field,[76] but later refuted this claim, attributing the rotation to an instrumental artifact.[77] The null result is consistent with the measurement of Ref. 15.

7. Discussion

Table 1 and Fig. 1 summarize the most important limits set on the CPR angle with the various methods examined in the previous sections. Only the best and most recent results obtained with each method are listed. For uniformity, all the results for the CPR angle are listed at the 68% CL (1σ), except for the first one, which is at the 95% CL, as given in the original Ref. 12. In general, all the results are consistent with each other and with a null CPR. Even the CMB measurement by BICEP1, which apparently shows a nonzero rotation at the 2σ level, cannot be taken as a firm CPR detection, since it has not been confirmed by other more accurate measurements.

In practice, all CPR test methods have reached so far an accuracy of the order of 1° and 3σ upper limits to any rotation of a few degrees. It has been however useful to use different methods since they are complementary in many ways. They cover different wavelength ranges and, although most CPR effects are wavelength independent, the methods at shorter wavelength have an advantage, if CPR effects grow with photon energy. They also reach different distances, and the CMB method is unbeatable in this respect. However the relative difference in light travel time between $z = 3$ and $z = 1100$ is only 16%. The radio polarization method, when it uses the integrated polarization, has the disadvantage of not relying on a firm prediction of the polarization orientation at the source, which the other methods have. In addition, the radio method requires correction for Faraday rotation. All methods can potentially test for a rotation which is not uniform in all directions, although this possibility has not yet been exploited by the CMB method, which also cannot see how an eventual rotation would depend on the distance. Reference 28 have recently examined the dependence of CPR on the wavelength and on the distance of the source, and found none, which is not surprising for a null (so far) CPR: in practice, they cannot improve the limit already set on the birefringence parameter ξ in Ref. 31 (see Sec. 6).

8. Outlook

In the future, improvements can be expected for all methods, e.g. by better targeted high resolution radio polarization measurements of RGs and quasars, by more accurate UV polarization measurements of RGs with the coming generation of giant optical telescopes,[8,21,67] and by future CMB polarimeters such as BICEP3[4] and COrE+.[80] In any case, since at the moment the limiting factor for improving the constraints on the CPR angle with the CMB are the systematic uncertainty on the calibration of the polarization angle, it will be necessary to reduce these, which at the moment is at best 0.3° for CMB polarization experiments. The best prospects to achieve this improvement are likely to be more precise measurements of the polarization angle of celestial sources at CMB frequencies, e.g. with the Australia Telescope Compact Array[53] and with Atacama Large Millimeter/Submillimeter Array (ALMA),[69] and a calibration source on a satellite.[41]

Acknowledgments

We would like to thank Matthew Hasselfied, Matteo Galaverni and Wei-Tou Ni for useful discussions.

References

1. Planck Collab. (P. A. R. Ade *et al.*), *Astron. Astrophys.* **571** (2014) A16.
2. BICEP2 Collab. (P. A. R. Ade *et al.*), *Phys. Rev. Lett.* **112** (2014) 241101.
3. POLARBEAR Collab. (P. A. R. Ade *et al.*), *Astrophys. J.* **794** (2014) 171.
4. Z. Ahmed *et al.*, *Proc. SPIE-Int Soc. Opt. Eng.* **9153** (2014) 1.
5. R. Antonucci, *Annu. Rev. Astron. Astrophys.* **31** (1993) 473.
6. J. Aumont *et al.*, *Astron. Astrophys.* **514** (2010) A70.

7. M. C. Begelman, R. D. Blandford and M. J. Rees, *Rev. Mod. Phys.* **56** (1984) 255.
8. R. A. Bernstein *et al.*, *Proc. SPIE-Int Soc. Opt. Eng.* **9145** (2014) 91451C.
9. P. Birch, *Nature* **298** (1982) 451.
10. A. H. Bridle and R. A. Perley, *Annu. Rev. Astron. Astrophys.* **22** (1984) 319.
11. M. L. Brown *et al.*, *Astrophys. J.* **705** (2009) 978.
12. S. M. Carroll, G. B. Field and R. Jackiw, *Phys. Rev. D* **41** (1990) 1231.
13. S. M. Carroll and G. B. Field, *Phys. Rev. Lett.* **79** (1997) 2394.
14. S. M. Carroll, *Phys. Rev. Lett.* **81** (1998) 3067.
15. S.-J. Chen, H.-H. Mei and W.-T. Ni, *Mod. Phys. Lett. A* **22** (2007) 2815.
16. A. Cimatti, S. di Serego Alighieri, R. A. E. Fosbury, M. Salvati and T. Duncan, *Mon. Not. Roy. Astron. Soc.* **264** (1993) 421.
17. A. Cimatti, S. di Serego Alighieri, G. B. Field and R. A. E. Fosbury, *Astrophys. J.* **422** (1994) 562.
18. J. N. Clarke, P. P. Kronberg and M. Simard-Normandin, *Mon. Not. Roy. Astron. Soc.* **190** (1980) 205.
19. E. Costa *et al.*, *Exp. Astron.* **28** (2010) 137.
20. S. Covino *et al.*, *Astron. Astrophys.* **348** (1999) L1.
21. T. de Zeeuw, R. Tamai and J. Liske, *The Messenger* **158** (2014) 3.
22. A. J. Dean *et al.*, *Science* **321** (2008) 1183.
23. S. di Serego Alighieri, A. Cimatti and R. A. E. Fosbury, *Astrophys. J.* **431** (1994) 123.
24. S. di Serego Alighieri, G. B. Field and A. Cimatti, *Astron. Soc. Pac. Conf. Ser.* **80** (1995) 276.
25. S. di Serego Alighieri, F. Finelli and M. Galaverni, *Astrophys. J.* **715** (2010) 33.
26. S. di Serego Alighieri, W.-T. Ni and W.-P. Pan, *Astrophys. J.* **792** (2014) 35.
27. D. J. Eisenstein and E. F. Bunn, *Phys. Rev. Lett.* **79** (1997) 1957.
28. M. Galaverni, G. Gubitosi, F. Paci and F. Finelli, *J. Cosmol. Astropart. Phys.* **8** (2015) 31.
29. M. Goldhaber and V. Trimble, *J. Astrophys. Astron.* **17** (1996) 17.
30. K. M. Gorski *et al.*, *Astrophys. J.* **622** (2005) 759.
31. D. Götz *et al.*, *Mon. Not. Roy. Astron. Soc.* **444** (2014) 2776.
32. G. Gubitosi *et al.*, *J. Cosmol. Astropart. Phys.* **0908** (2009) 021.
33. WMAP Collab. (G. Hinshaw *et al.*), *Astrophys. J. Suppl.* **208** (2013) 19.
34. D. Hutsemekers, R. Cabanac, H. Lamy and D. Sluse, *Astron. Astrophys.* **441** (2005) 915.
35. D. Hutsemekers, L. Braibant, V. Pelgrims and D. Sluse, *Astron. Astrophys.* **572** (2014) A18.
36. IAU Commission 40, *Trans. Int. Astron. Union* **XVB** (1974) 166.
37. S. A. Joshi, R. A. Battye, I. W. A. Browne, N. Jackson, T. W. B. Muxlow and P. N. Wilkinson, *Mon. Not. Roy. Astron. Soc.* **380** (2007) 162.
38. E. Kalemci *et al.*, *Astrophys. J. Suppl.* **169** (2007) 75.
39. M. Kamionkowski, *Phys. Rev. D* **82** (2010) 047302.
40. BICEP1 Collab. (J. P. Kaufman *et al.*), *Phys. Rev. D* **89** (2014) 062006.
41. J. P. Kaufman, B. G. Keating and B. R. Johnson, *Mon. Not. Roy. Astron. Soc.* **455** (2016) 1981.
42. B. G. Keating, M. Shimon and A. P. S. Yadav, *Astrophys. J. Lett.* **762** (2013) L23.
43. V. A. Kostelecký and M. Mewes, *Phys. Rev. Lett.* **87** (2001) 251304.
44. V. A. Kostelecký and M. Mewes, *Phys. Rev. D* **66** (2002) 056005.
45. V. A. Kostelecký and M. Mewes, *Phys. Rev. Lett.* **110** (2013) 201601.
46. J. M. Kovac, *et al.*, *Nature* **420** (2002) 722.

47. P. P. Kronberg, C. C. Dyer and H.-J. Röser, *Astrophys. J.* **472** (1996) 115.
48. J. P. Leahy, astro-ph/9704285.
49. N. F. Lepora, arXiv:gr-qc/9812077.
50. S. Liberati and L. Maccione, *Annu. Rev. Nucl. Part. Sci.* **59** (2009) 245.
51. T. J. Loredo, É. É. Flanagan and I. M. Wasserman, *Phys. Rev. D* **56** (1997) 7507.
52. L. Maccione, S. Liberati, A. Celotti, J. G. Kirk and P. Ubertini, *Phys. Rev. D* **78** (2008) 103003.
53. M. Massardi *et al.*, *Mon. Not. Roy. Astron. Soc.* **436** (2013) 2915.
54. H.-H. Mei, W.-T. Ni, W.-P. Pan, L. Xu and S. di Serego Alighieri, *Astrophys. J.* **805** (2015) 107.
55. R. C. Myers and M. Pospelov, *Phys. Rev. Lett.* **90** (2003) 211601.
56. S. Naess *et al.*, *J. Cosmol. Astropart. Phys.* **10** (2014) 007.
57. W.-T. Ni, A Nonmetric Theory of Gravity, Montana State University (1973), http://astrod.wikispaces.com/file/detail/A Non-metric Theory of Gravity.pdf.
58. W.-T. Ni, *Phys. Rev. Lett.* **38** (1977) 301.
59. W.-T. Ni, *Prog. Theor. Phys. Suppl.* **172** (2008) 49.
60. W.-T. Ni, *Rep. Prog. Phys.* **73** (2010) 056901.
61. W.-T. Ni, *Phys. Lett. A* **379** (2015) 1297.
62. W.-T. Ni, Equivalence Principles, Spacetime Structure and the Cosmic Connection, Chapter 5, this book; *Int. J. Mod. Phys. D* **25** (2016) 1630002.
63. B. Nodland and J. P. Ralston, *Phys. Rev. Lett.* **78** (1997) 3043.
64. L. Pagano *et al.*, *Phys. Rev. D* **80** (2009) 043522.
65. A. Payez, J. R. Cudell and D. Hutsemekers, *Phys. Rev. D* **84** (2011) 085029.
66. D. J. Saikia and C. J. Salter, *Annu. Rev. Astron. Astrophys.* **26** (1988) 93.
67. G. H. Sanders, *J. Astrophys. Astron.* **34** (2013) 81.
68. L. I. Schiff, *Am. J. Phys.* **28** (1960) 340.
69. L. Testi and J. Walsh, *The Messenger* **152** (2013) 2.
70. H. D. Tran, M. H. Cohen, P. M. Ogle, R. W. Goodrich and S. di Serego Alighieri, *Astrophys. J.* **500** (1998) 660.
71. J. Vernet, R. A. E. Fosbury, M. Villar-Martin, M. H. Cohen, A. Cimatti, S. di Serego Alighieri and R. W. Goodrich, *Astron. Astrophys.* **366** (2001) 7.
72. J. F. C. Wardle, R. A. Perley and M. H. Cohen, *Phys. Rev. Lett.* **79** (1997) 1801.
73. K. Wiersema *et al.*, *Nature* **509** (2014) 201.
74. J.-Q. Xia, H. Li, X. Wang and X. Zhang, *Astron. Astrophys.* **483** (2008) 715.
75. J.-Q. Xia, H. Li and X. Zhang, *Phys. Lett. B* **687** (2010) 129.
76. E. Zavattini *et al.*, *Phys. Rev. Lett.* **96** (2006) 110406.
77. E. Zavattini *et al.*, *Phys. Rev. D* **77** (2007) 032006.
78. N. Aghanim *et al.* (Planck Collaboration), submitted to *Astron. Astrophys.* (2016), arXiv:1605.08633.
79. W.-P. Pan, S. di Serego Alighieri, W.-T. Ni. and L. Xu, to be published in *Proceedings of the Second LeCosPA Symposium "Everything about Gravity"*, December 14–18, 2015, Taipei (2016), arXiv:1603.08193.
80. P. de Bernardis and S. Masi, *Int. J. Mod. Phys. D* **25** (2016) 1640012.

Chapter 7

Clock comparison based on laser ranging technologies

Étienne Samain

Géoazur, Université de Nice Sophia-Antipolis,
Observatoire de la Côte d'Azur (OCA), CNRS (UMR 7329),
2130 Route de l'Observatoire, 06460 Caussols, France
etienne.samain@oca.eu

Recent progress in the domain of time and frequency standards has required some important improvements of existing time transfer links. Several time transfer by laser link (T2L2) projects have been carried out since 1972 with numerous scientific or technological objectives. There are two projects currently under exploitation: T2L2 and Lunar Reconnaissance Orbiter (LRO). The former is a dedicated two-way time transfer experiment embedded on the satellite Jason-2 allowing for the synchronization of remote clocks with an uncertainty of 100 ps and the latter is a one-way link devoted for ranging a spacecraft orbiting around the Moon. There is also the Laser Time Transfer (LTT) project, exploited until 2012 and designed in the frame of the Chinese navigation constellation. In the context of future space missions for fundamental physics, solar system science or navigation, laser links are of prime importance and many missions based on that technology have been proposed for these purposes.

Keywords: Clock; time transfer; laser link; laser ranging; event timer.

1. Introduction

Instrumentations allowing for the comparison of distant clocks and for the distribution of time scales have some important applications in metrology, navigation and fundamental physics. Recent progress in the domain of time and frequency standards has required some important improvements of existing time transfer links. The most suitable technique available today to realize the best time transfer in free space is based on the propagation of laser pulses. Such a time transfer can be used to compare some clocks in space or to realize some comparison between ground and space or between several users on ground. These optical time transfers rely on the existing laser ranging network natively developed to measure distances between satellites (or the moon) and the ground. The principle of that technique is based on the two-way time of flight measurement of picosecond laser pulses from ground stations to retro-reflectors (passive) on satellite. The distance between the station and the satellite is computed for every laser pulse emitted by the ground station and reflected by the satellite from the time interval between the start and return epochs (Fig. 1). The accurate distance between the satellite and the reference point of the ground station is obtained by a permanent optical calibration at ground and

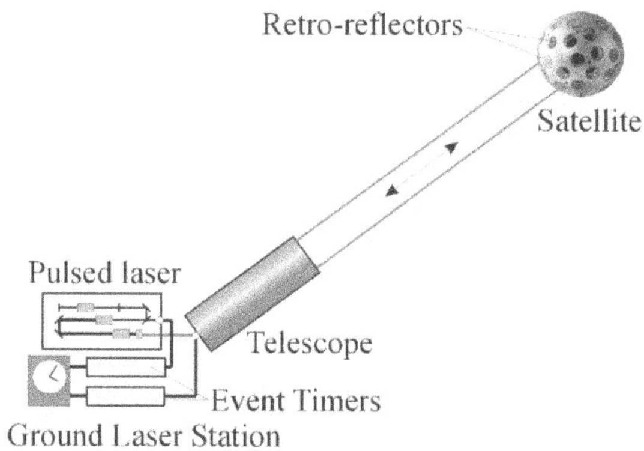

Fig. 1. Laser ranging principle. The laser pulses are emitted and received by the ground laser station. Some retro-reflectors on the satellite reflect a fraction of the incident laser pulse. Usual parameters: Laser rate: 10 Hz (up to 2 kHz); Wavelength: 532 nm; FWHM: 50 ps. (For color version, see page I-CP4.)

knowledge of the geometry of the satellite and of its mass distribution. The location of the laser station on earth and the orbitography of the satellite are computed in the International Terrestrial Reference Frame (ITRF).[1] By using a dedicated active space instrument able to record arrival epochs of laser pulses on the satellite, these satellite laser ranging technologies can be extended to realize a ground to space time transfer.

Two kinds of time transfer can be envisioned: The first, hereafter called the two-way, is based on a transfer with both an uplink and a downlink allowing the measurement, by the process itself, of the propagation delay. The second, called the one-way, is based on a single link with a propagation delay deduced and computed from the distance between the clocks. The two-way can be done either through a simple reflection of the laser beam on the corner cube together with the active space instrument to time tag the laser pulses, or through an active equipment using a synchronous or asynchronous transponder.[2] The passive two-way is well suited for high precision time transfer over distances of a few tens of thousands kilometers while the one-way is more suited for time transfer at very large scale (Solar System). The distance Earth–Moon is typically the maximum distance which can be envisioned in a passive two-way link. Figure 2 is an example of a two-way link in Earth orbit allowing for time transfer between a ground station and a space vehicle.

Based on the scenario depicted in Fig. 2, ground to ground time transfer can be performed with a unique space instrument and several ground laser stations. It can be realized in either a common or a noncommon view mode. In the first case, the satellite is seen during a common period while in the other one, the satellite is observed alternatively (Fig. 3). As a function of the space clock used (quartz, H-Maser, ...), the performance of the ground to ground link in the noncommon view mode can be significantly affected by the noise introduced by the space clock during the nonobserved period τ.

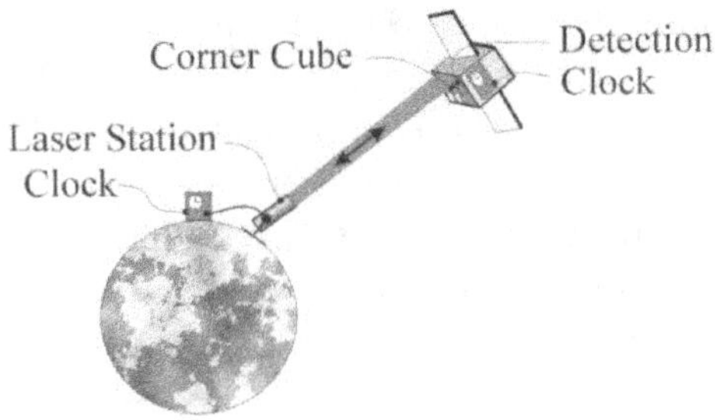

Fig. 2. Ground to space time transfer based on a two-way link in Earth orbit with a laser station on ground and an active space equipment linked with a retro-reflector. (For color version, see page I-CP4.)

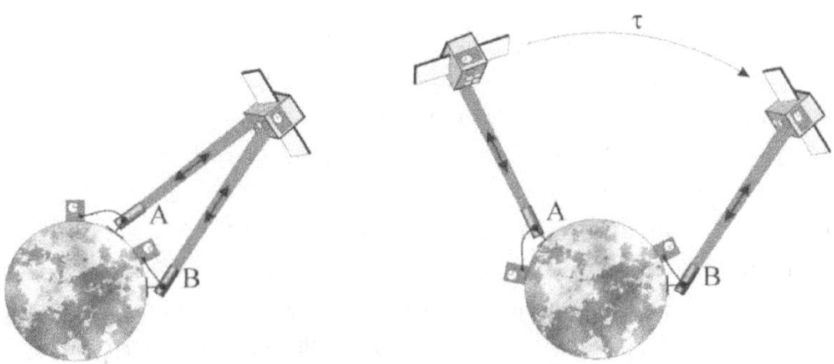

Fig. 3. Ground to ground time transfer in a common view mode (left) and in a noncommon view mode (right). (For color version, see page I-CP4.)

Several time transfer by laser link (T2L2) projects have been tested since 1972 with numerous scientific or technological objectives.

The first two-way optical time transfer experiment was proposed to European Space Agency (ESA) in 1972 with the Laser Synchronization from stationary Orbit (LASSO) project.[3] The mission was a significant success due to some intercontinental ground to ground time transfers with time stabilities in the range of 100 ps over several thousand seconds. The project was embedded on a geostationary satellite (Meteosat P2) launched in 1988. The first common time transfer session between two remote laboratories (Observatoire de la Côte d'Azur (OCA) France and TUG Austria) was carried out in 1989. The primary objective of LASSO was to demonstrate the feasibility of this novel time transfer technique with an uncertainty better than 1 ns.

The following operational laser time transfer was Laser Time Transfer (LTT). The project was developed by the Shanghai Astronomical Observatory (SHAO) in the frame of the Chinese global navigation system Compass.[4] A first LTT equipment was launched in spring 2007 onboard the Compass M1 satellite,[5,6] and two other ones in summer 2010 and spring 2011. The equipment included a very simple and rugged design using a single photon detector, a band pass filter and no other optical component.[7] The detection concept allowed for minimizing the intensity-dependent delay of the detection whatever the energy emitted (single photodetection concept). The first ground to space time transfer based on LTT was done from the Changchun Satellite laser ranging station (China).[8] The time scale synchronization was performed with a data spread of 260 ps and a measurement of the frequency differences between the ground and the space segments with an uncertainty of 3×10^{-14}.[9]

The first active two-way laser ranging experiment at planetary distance[10] was carried out in spring 2005 with the MESSENGER spacecraft by using the embedded laser altimeter MLA.[11] The experiment was based on a double asynchronous one-way transponders operated from the ground and from the spacecraft. The experiment allows for providing range and time transfer between the spacecraft and the Goddard Geophysical and Astrophysical Observatory laser station. The measurements were made at a distance of 24 million km. A second planetary link was done at a distance of 80 million from the altimeter of the Mars Global Surveyor (MGS) spacecraft. This second link was obtained in a one-way uplink configuration.

The third laser link experiment conducted outside the terrestrial orbit was performed with the Lunar Reconnaissance Orbiter (LRO).[12,13] The objective was to obtain routine one-way laser ranging with the Lunar Orbiter Laser Altimeter (LOLA) for demonstrating spacecraft orbit determination. The experiment has been in successful operation since summer 2009. Data acquisition from the laser ranging network was stopped on Oct 1, 2014 but scientific analysis is still ongoing.

In 1994, OCA proposed to build a new generation of optical transfer called T2L2.[14,15] As compared to LASSO, the objective was to improve the performances of both time stability and accuracy by at least two orders of magnitude, and enlarge the number of participating laser stations. After several proposals for the satellites GIOVE (Galileo program), Myriade and the MIR space station,[14] T2L2 was accepted in 2000 in the frame of the Atomic Clock Ensemble in Space (ACES) program[16,17] but unfortunately taken off the mission in 2001 for some problems related to maximum power allowed and mass budget. After T2L2 on ACES was abandoned in 2001, T2L2 was finally accepted in 2005 as a passenger instrument on Jason-2, an altimetry satellite designed to study the internal structure and dynamics of ocean currents.[18–20] T2L2 was launched in June 2008 and has been running without any significant interruption since that date. It was placed by a Delta launcher at an altitude of 1336 km and an inclination of 66°. The T2L2 project is currently supported by both the OCA-GeoAzur and the French space agency CNES.

Section 2 is an overview of the scientific objectives associated to clock comparisons by laser. Sections 3 and 4 are a global description of the two projects currently under exploitation (T2L2 and LRO) and Sec. 5 is a nonexhaustive description of future projects involving a laser link.

2. Scientific Objectives

As compared to classical microwave techniques, laser links have many major advantages for both Earth orbiting and Solar System missions. The most significant benefits are:

- A very high frequency of the carrier (optical) allowing for high bandwidth modulation.
- A good correction of the refraction delay induced by the atmosphere. Ionosphere uncertainty is negligible and the tropospheric correction can be determined through some atmospheric measurements (pressure, temperature, humidity).
- Clear and well-defined reference points for both space and ground segments. A microwave antenna of 70 m used for a mission in the Solar System can be replaced by an optical telescope having an aperture in the range of 1 m.
- A very well-focused beam allowing the use of some compact collectors (relevant for the space segment).

Laser links in space allow for a large number of tests in many fields such as time and frequency metrology, navigation, fundamental physics and Solar System physics. It is clear that the exact science doable with a given mission depends on the precise experimental setup of the considered project (atomic clock, accelerometer, interferometer, or trajectory measurement system in the Solar System), and we cannot give an exhaustive list of what we could measure. This section is a brief overview of the scientific objectives intentionally limited to laser links in space.

2.1. *Time and frequency metrology*

The time and frequency metrology has developed continuously for more than 60 years with regular improvement of both time accuracy and time stability. It has benefited from scientific advances in several domains such as atom laser cooling, optical clocks and frequency comparison with femtosecond laser combs.[21–23]

For instance, the fractional accuracy of Cesium microwaves clocks has been improved by a factor ten every ten years since 1950. Today cold atoms clocks routinely obtain fractional accuracies in the range of a few 10^{-16} and instabilities better than 1.5×10^{-13} over more than ten days.[24]

Progress made in the domain of optical clocks is also very impressive with an improvement of two orders of magnitude every ten years from the 1980's until now. It has been encouraged by the recent advances in atom manipulation and in the optical combs that have allowed establishing a connection between the microwave and optical domains. Today, there are more than ten laboratories working on such optical clocks worldwide. Some of them have a fractional frequency uncertainty of only a few 10^{-18} (Ref. 25) (Fig. 4). The oscillation frequencies of these optical standards are of several hundred terahertz (from IR to UV). Up to now, there is no electronic equipment fast enough to directly use these 100 terahertz signals. The usual solution for establishing the connection between this optical domain and the

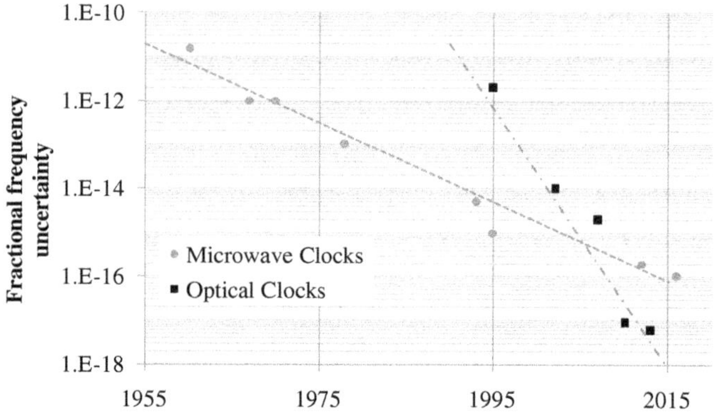

Fig. 4. Evolution of the frequency uncertainty of microwave and optical clocks.

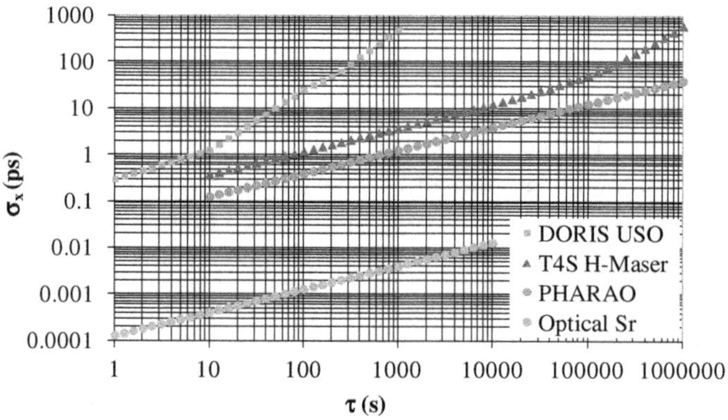

Fig. 5. Time stabilities (root square of the time variance) of some typical clocks. PHARAO will be the first cold atom space clock on ISS in 2017. DORIS USO is an ultrastable quartz oscillator used for both the T2L2 space instrument and the DORIS System. The T4S H-Maser is a commercial clock from T4Science. (For color version, see page I-CP4.)

classical gigahertz domain (electrical signals propagated in some coaxial cables) lies in the use of optical frequency combs generated from a high repetition rate femtosecond laser.[22]

Figure 5 shows the time stability (TDev) of some typical clocks using different technologies. Quartz oscillators are only stable for short time intervals. They are often long-term disciplined on some other sources such as atomic transitions. They can also be used in some applications where short-term stability is required for the detection of rapid changes (Doppler detection). H-Masers are extremely stable over mid-term time intervals (up to 10,000 s). They can be used for instance in some microwave link in the Solar System or for the localization of extra galactic sources (VLBI). Optical clocks are today extremely stable over short- and mid-term time intervals. The very high frequency of the carrier is a major advantage for ultra high time stabilities.

As well, the development of space clocks is a major issue for the field of time and frequency metrology. The uncertainty of gravitational potential at the Earth surface is a limitation to go beyond a fractional frequency uncertainty of 10^{-17}. The access to the space environment allows for decreasing that noise and will become of prime importance for the next future.

Furthermore, all these improvements imply some time and frequency transfer methods well suited to the performances of these clocks. Common View GPS (GPS CV) and Two-Way System Time and Frequency Transfer (TWSTFT) are the most common techniques currently used to achieve comparisons between clocks and distributions of time and frequency references. GPS CV used in the P3 ionosphere free linear combination mode permits to obtain an expanded uncertainty estimated between 3 ns to 7 ns. This result takes into account a numerical factor $k = 2$ used as a multiplier of the standard uncertainty (coverage factor $k = 2$).[26,27] The future European Galileo System will obtain the same kind of performance.

Two-way laser time transfer allows for an enhancement of both the accuracy and time stability of more than one order of magnitude as compared to the classical microwave techniques. Measurements carried out on the T2L2 project for a ground to space link showed a time stability σ_x (root square of the time variance) given by

$$\sigma_x^2(\tau) = (65.10^{-12} \times \tau^{-\frac{1}{2}})^2 + (2.10^{-14} \times \tau^{+\frac{1}{2}})^2, \quad \tau_0 = 1\,\text{s}, \tag{1}$$

where τ_0 is the time interval between consecutive laser pulses. It is the sum of a white phase noise ($\sigma^{-1/2}$) mainly induced by the repeatability error of the optical detectors together with a white frequency noise ($\sigma^{+1/2}$). The typical expended uncertainty between time scale of two distinct clocks A and B $u_E(\Delta_{\text{AB}})$ is (coverage factor $k = 2$)[28]:

$$u_E(\Delta_{\text{AB}}) < 100\,\text{ps}. \tag{2}$$

With these performances, laser time transfer is perfectly well suited to validate the other time transfer techniques such as GPS, TWSTFT and also future technologies currently under studies (optical fiber, MWL (Micro Wave Link) instrument on ACES).

In the time and frequency metrology domain, one relevant objective is to participate in the improvement of time scales. The most crucial of these is the International Atomic Time (TAI) built through an ensemble of ultrastable clocks located all over the world and synchronized to each other with some microwave time transfer techniques. A two-way laser link is a completely independent technique perfectly well suited to calibrate and validate these microwave links.

Ground to space laser time transfer is also essential to validate clocks in space. This is the case for ACES, where European Laser Timing (ELT) will allow to give an independent comparison of both onboard H-Maser–PHARAO clocks, and for T2L2 on Jason-2 equipped with the DORIS quartz oscillator.[29,30] T2L2 is able to measure the performance of the DORIS system quartz oscillator over an integration of roughly 20 s. It gives the opportunity to detect some possible disturbances of the oscillator caused by radiations (South Atlantic Anomaly). Up to now, some effects have been clearly highlighted by T2L2. In the frame of the Jason-2 mission, the perturbations are not high enough to justify a correction model for the DORIS navigation but these effects are of prime importance for quartz technologies in space. Laser time transfer is moreover used to evaluate the performances of the atomic clocks for the GNSS navigation. The LTT project is currently developed in the frame of Compass for this reason and several other proposals (T2L2 and LTT) have been made in the frame of the Galileo program.[6] Some other new developments such as OPTI are currently run for GNSS purposes in general.[31]

2.2. *Fundamental physics*

The possibilities given by an accurate and stable time transfer between remote clocks is also of interest for the domain of fundamental physics. These laser links should contribute to several distinct fields such as the search for a possible anisotropy of the speed of light,[32,33] the measurement of the gravitational redshift,[34] or the measurement of the Eddington's parameter γ.[35]

Laser time transfer on a satellite implies some optical propagation in different directions during the satellite pass. The detection of a possible anisotropy of the speed of light can be performed from the comparison between ground and space clocks as a function of the geometry used during the time transfer with the satellite (Fig. 6).

Two scenarios can be envisioned. The former is based on a given link built between the satellite and a ground station. The location of a station on the ground and the trajectories of the satellite passes are chosen to optimize the global geometry of the test for a given direction. Figure 7 is a geometry example applied to the T2L2 project showing the successive passes of the Jason-2 satellite above Europe.

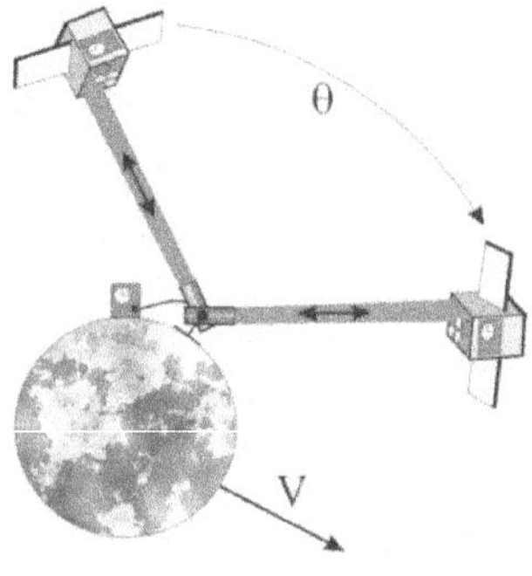

Fig. 6. Delay comparison between the uplink and the downlink for various laser beam orientations.

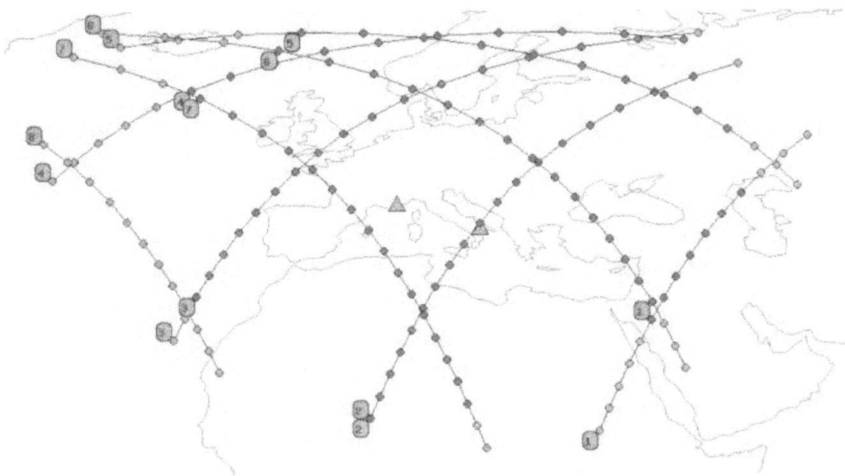

Fig. 7. Successive passes example of the Jason-2 satellite above Europe. Obtained from the T2L2 website and developed by the data mission center.

In cases when the stability of the onboard clock is not high enough, one given or several ground clocks can monitor the space clock from the ground using a H-Maser. The latter is set on numerous ground to space time transfer acquisitions in all possible orientations in order to decrease the uncertainty introduced by both the optical link and the onboard clock, and to provide global orientation coverage. This ground to space time transfer is carried out using several ground stations equipped with ultrastable clocks such as H-Masers. In both scenarios, we get:

$$\frac{\delta c}{c} = \frac{\sigma_x(\tau)}{2T_{\text{Prop}}(1 - \cos(\theta))}, \tag{3}$$

where σ_x is the time deviation of the global link (clocks + laser link), τ is the time integration of the observation, T_{Prop} is the propagation delay and θ is the angle scanned by the laser beam. In cases when only a single measurement is realized for a given orientation, we have to consider in Eq. (3) a time stability σ_{x} of the ground–space link over a time integration τ corresponding to a satellite pass T_{Pass}. In cases when numerous acquisitions N_{Acq} are done over a long time T_{Acq} (for instance 1 year), a part of the uncertainty corresponding to each acquisition can be reduced by averaging. If there are no systematic effects, σ_{x} could be divided by a factor $\sqrt{N_{\text{Acq}}}$. In that case, it is necessary to consider in Eq. (3) the quadratic sum of that term as well as those of the term corresponding to the time stability of the global link over $\tau = T_{\text{Acq}}$. This latter term may be predominant.

When the single acquisition mode is applied to T2L2 on Jason-2, T_{Pass} is roughly 1000 s. In some optimal measurement condition, T2L2 gives a time stability σ_{x} over 1000 s equal to: σ_{x} (1000 s) = 50 ps (see Sec. 4.2). The quantity $\delta c/c$ then can be determined with an uncertainty of roughly 10^{-9}. Regarding the numerous acquisitions mode, implemented with $T_{\rm Acq} = 10$ days, we obtain a time stability of the whole link in the range of 10 ps which could allow for determining $\delta c/c$ at 2×10^{-10} level.

The redshift measurement can be done either with a very eccentric orbit or with some accurate clocks. For instance, in the frame of ACES, by using the high frequency accuracy of the PHARAO space clock the test can be measured with a relative uncertainty of 10^{-16}. In the frame of T2L2 on Jason-2, both the poor accuracy of the embedded quartz oscillator (DORIS) and the quasi-circular orbit of Jason-2 ($< 10^{-3}$) do not allow for measuring any relevant value. The uncertainty on the redshift could be improved by a factor 10^4 for an experiment in the Solar System as compared to the test that would be measured by ACES.

The Eddington parameter γ first introduced by Eddington, Robertson and Schiff, measures the amount of spatial curvature produced by mass. For example, the general relativity predicts a γ value equal to one, and the scalar–tensor theories express this PPN parameter as: $\gamma = (1+\omega)/(2+\omega)$, where ω is an arbitrary coupling function that determines the strength of the scalar field. The ω function could be very large so that the theory's predictions could be almost identical to general relativity as defined today. But ω could take values that would lead to significant differences in cosmological models. Irwin I. Shapiro predicted a relativistic time delay $\Delta t_{\rm Shapiro}$ in the time of flight of photons propagating in a gravity field (Shapiro delay). γ can be deduced from the measurement of that Shapiro delay with the strong gravity field of the Sun.[36] It can be measured with a spacecraft having a solar orbit chosen so that the satellite passes behind the Sun viewed from the Earth. The Shapiro delay is at a maximum near the occultation with a variation as large as 120 μs in a few days during the occultation. It can be enough to only measure a signature of that delay instead of an absolute delay in order to relax the constraint of the long-term time stability of the clocks involved or to use a simple one-way link. The sensitivity of the measurement is at a maximum during the conjunction phase of the spacecraft with the Sun. The relative uncertainty on γ can be evaluated by:

$$\frac{\delta\gamma}{\gamma} = \frac{\sigma_x(\tau)}{\Delta t_{\rm Shapiro}\tau}, \tag{4}$$

where σ_x is the time stability of both the optical link and the clocks and τ the integration duration of the measurement. With a time stability σ_x in the range of $10^{-14} \cdot \tau^{1/2}$ over one day, $(1-\gamma)$ can be assessed at the 10^{-7} level.

2.3. *Solar System science*

The Solar System science relies on the ranging of a given spacecraft in orbit around the Sun or in orbit around a planet. With a spacecraft in orbit around a planet (or a satellite), we can study the gravity field of the planet from the precise trajectory of the vehicle. It is of interest for the mass of the planet, the structure of the gravity field or for studying volcanoes. During occultation phases of the orbiter by the planet, the light beam goes through the atmosphere (if there is an atmosphere) which generates a variation of the time propagation. In the case of the Mars atmosphere, the delays involved can reach a few nanoseconds when the distance between the light beam and the planet surface tends towards zero. If the spacecraft orbit is known during this phase, the analysis of the time propagation variation together with the variation of the laser flux allows for extracting some atmospheric parameters.

Laser ranging in the Solar System allows also for the determination of masses and density of asteroid and the determination of the solar quadruple moment parameter J_2 of the Sun. Here again, ranging can be carried out with a one-way link since most of these parameters can be deduced from a signature on the trajectory of the spacecraft.

2.4. *Solar System navigation based on clock comparison*

Classical navigation in the Solar System is usually done with the microwave links operated from the large antennas of the Deep Space Network (DSN) facility. An interesting alternative is to use laser links. The direct information delivered by a clock comparison is a time flight and a radial distance. This can be achieved at a centimeter level for a time integration of a few thousand seconds. Tracking carried out with several synchronized Earth laser stations allows also for measuring the angular position of the spacecraft (localization in perpendicular plane to the line of sight). This can be done with differential measurements of the arrival time onboard allowing for the geometry of the system to be resolved. Since the measurement done at the spacecraft is differential, no long-term stability is required here for the space instrument. With distances between ground stations in the range of 10,000 km, an uncertainty on the positioning of the station at the centimeter level, and a time synchronization between the stations of 30 ps (obtained from a classical two-way laser like T2L2), an uncertainty on the angular determination of a few nanoradians is doable. Two-way laser time transfer is one unique chance to validate this kind of one-way laser ranging concept by directly comparing the one-way and two-way links together.

3. Time Transfer by Laser Link: T2L2 on Jason-2

3.1. *Principle*

Basically, T2L2 allows for the synchronization between a ground clock linked to a laser station and a clock onboard a satellite. For a ground time transfer between several clocks, several elementary links between ground and space are made and the space segment is only used as a relay between the clocks on ground. To perform a T2L2 time transfer, the laser station emits light pulses (20 ps to 200 ps) toward

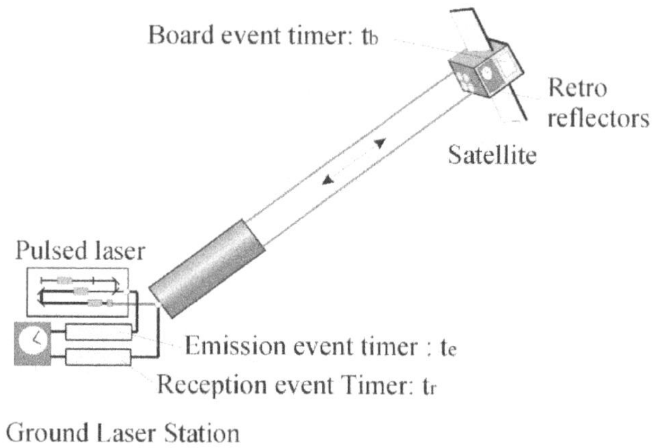

Fig. 8. T2L2 principle. For every laser pulse, the laser station measures the start epoch t_e and the return epoch t_r after reflection from the space. The T2L2 payload records the arrival epoch onboard t_b.

the satellite. A Laser Ranging Array (LRA) onboard the satellite returns a fraction of the received pulses to the ground station. The ground station measures, for each laser pulse, the start epoch t_{e} and the return epoch t_{r} after reflection from the space. The T2L2 payload records, in the time scale of the space oscillator, the arrival epoch onboard t_{b} (Fig. 8).

These data are downloaded to ground with a classical microwave link within 2 h following the record. The differences between the start and return epochs recorded at ground level allow for determining the propagation delay of the transfer.

In the framework of T2L2 on Jason-2, the maximum distance between the stations in common view mode is roughly 6000 km.

3.2. *Laser station ground segment*

The ground segment is based on an international network including more than 40 laser ranging stations.[37] The activities of that network are organized under the International Laser Ranging Service (ILRS)[38] which provides global satellite and lunar laser ranging data to support research in geodesy, geophysics, Lunar science and fundamental physics. That network continuously monitors the distance of more than 40 satellites orbiting around the earth. The distances provided by the station are based on the measurement of time of flight of very short laser pulses. The laser pulses are sent and received by the station and reflected by some corner cubes embedded on the satellite. A typical station (Figs. 9 and 10) is composed of the following elements:

— A pulsed laser to generate pulses between 10 ps to 200 ps (FWHM) at a 10 Hz rate and 532 nm (YAG doubled).
— A start detector to get the start events at the output of the laser.

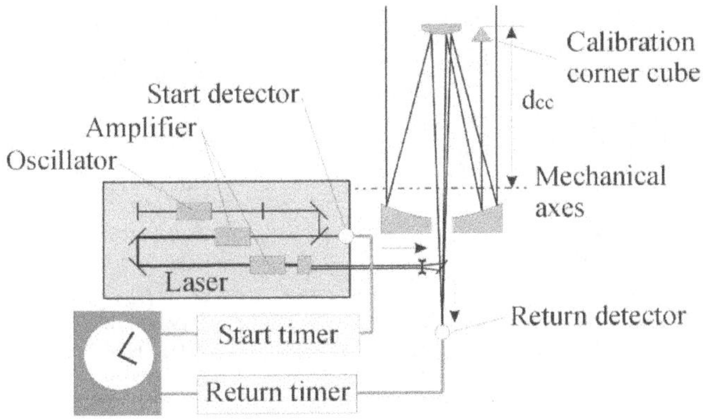

Fig. 9. Typical laser station, using the same telescope for both laser emission and reception. Some stations use separate apertures for emission/reception.

Fig. 10. Photography of the MeO laser station at Grasse (France) built by the end of the seventies for lunar laser ranging and redesigned for satellite and time transfer in 2005. (For color version, see page I-CP5.)

— A telescope to receive the light pulses reflected by the satellite and possibly to emit laser pulses toward the satellite.
— A return detector to detect the pulses reflected by the satellite.
— An event timer to timestamp the events in the time scale of the ground clock to be synchronized.

In a classical laser station, only the time of flight between the ground and the satellite matters. For time transfer purposes, it is also necessary to measure accurately the absolute start time of each laser pulses emitted. This is achieved with some picosecond event timers for the emission epoch and the reception epoch. In order to obtain the picosecond accuracy required for T2L2, laser stations are

Table 1. Main characteristics of a typical laser station.

Subsystem	Characteristics
Telescope diameter	1 m
Telescope slew rate	$5^\circ\ s^{-1}$
Telescope pointing accuracy	5 arcsec
Laser energy	100 mJ
Laser wavelength	532.1 nm
Laser FWHM	100 ps
Laser rate	10 Hz
Event timer standard deviation	5 ps RMS
Photodetection standard deviation	50 ps RMS
Clock stability ADev (H-Maser)	$2 \cdot 10^{-15}$

calibrated with a dedicated calibration station. This calibration is based on some simultaneous measurements done between the usual chronometry of the laser station and the dedicated calibration station specifically installed for that purpose. It allows the measurement of the delay between the optical pulse at the mechanical axis of the telescope and the time reference of the station. The calibration station gathers inside unique equipment for all the metrology required to perform that measurement: a sub-picosecond event timer, an optical module to grab laser pulses from the laser station and an optical fiber.

Main characteristics of a typical laser station are given in Table 1.

Among all laser stations of the international network, 20 are currently contributing actively to T2L2 and ten have the picosecond resolution required for high performance time transfer. Table 2 lists these laser stations together with the type of atomic clock used.

3.3. *Space instrument*

The T2L2 space equipment has been embedded on the satellite Jason-2 as a passenger instrument. Roughly, it is an instrument able to timestamp laser pulses coming from the Earth at the picosecond level. It comprises a photodetection device and an event timer connected to an ultrastable quartz oscillator used as the T2L2 onboard

Table 2. T2L2 participating laser station.

Name	Country	ILRS N°	Clock
Changchung	China	7237	H-Maser
Grasse	France	7845	H-Maser
Herstmonceux	England	7840	H-Maser
Koganei	Japan	7308	Atomic Fountain
Mc Donald	USA	7080	Cesium
Matera	Italy	7941	H-Maser
Mount Stromlo	Australia	7825	Cesium
Potsdam	Germany	7841	Quartz slaved on GPS
Wettzell	Germany	8834	H-Maser
Zimmerwald	Swiss	7810	H-Maser

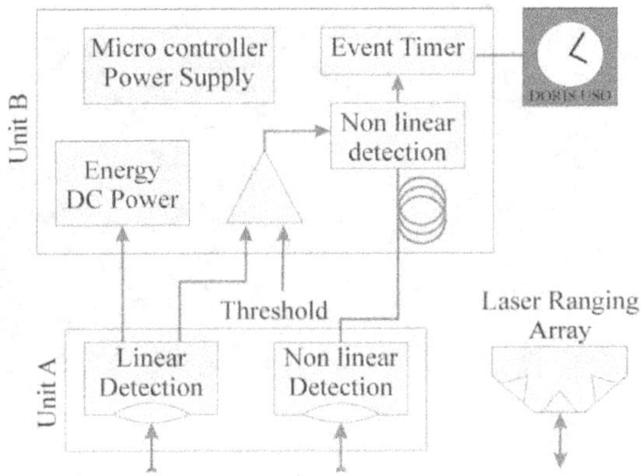

Fig. 11. T2L2 global architecture with unit A outside the satellite oriented toward the Earth and unit B inside the payload. The DORIS USO and the LRA equipment do not belong to the T2L2 assembly. The laser pulse coming from the ground station is on the lower side of the figure.

clock. The reflection of the laser pulse towards the Earth is done with a retro-reflector array. The oscillator and the retro-reflector are used by T2L2 but are not part of the T2L2 package. The oscillator is the frequency reference[39] of the DORIS navigation system, and the LRA is used for the laser ranging satellite positioning system of Jason-2 (Fig. 11). The space instrument is divided into two parts A and B. The A unit includes the photodetection while the B unit contains the event timer, some parts of the detection, the power supply and the microcontroller.

The photodetection device is made with two avalanche photodetectors (unit A in Fig. 11). One of them runs in a nonlinear mode for chronometry,[40] the other in a linear mode to trigger the nonlinear detector and to measure the laser energy. The primary function of the nonlinear photodetection is to generate an electrical pulse from a very weak pulse having a time uncertainty as low as possible. The arrival epoch of the laser pulse is obtained from the time tagging of that electric pulse with the event timer (unit B). The internal delay of the detector has a significant dependency on the energy received. In order to eliminate the temporal noise that would be introduced by some uncontrolled energy variation, this transit delay has to be compensated. This is achieved for each pulse received through the linear photodetector.

The energy of each pulse is recorded together with the arrival epoch measured by the event timer and the compensation is applied through a post treatment on ground. The laser energy received at the satellite plane is sent to the photodetectors through a dedicated optics assembly. These optics allow to limit the field of view of the instrument (Sun protection), to adjust the photon number and to limit the spectral bandwidth. Each photodetection chain (linear and nonlinear) has a distinct optical assembly. The optics of the linear detection is gathered with the linear detector into a unique module. The optics of the nonlinear detection is divided

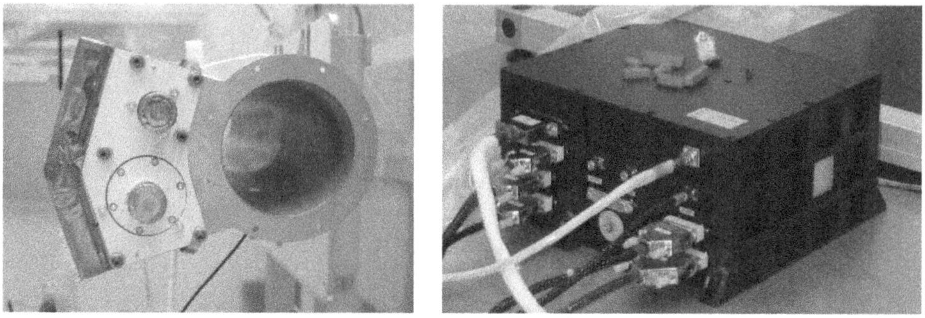

Fig. 12. Photography of units A (right) and B (left) of the T2L2 space instrument. The cylinders on the right are the detection modules (linear and nonlinear). The LRA module is not integrated into the photo. (For color version, see page I-CP5.)

into two subsystems linked together with an optical fiber (between unit A and B). Figure 12 is the photography of both units A and B.

In order to minimize false detections, the nonlinear photodiode is gated a few nanoseconds before the expected arrival of the laser pulses with an external active quenching circuit trigged from the linear avalanche photodiode. The delay between the applied voltage and the arrival time of the laser pulse is made with a few meters optical fiber playing the role of a delay line. The linear detector is used to measure both the laser energy and the DC level produced by the sunlight backscattered by the Earth. These measurements allow to compensate the time walk of the nonlinear photodiode and to adjust automatically the threshold of the trig system (day, night).

The energy density received at the satellite depends on the distance between the ground station and the satellite which also depends on the incident angle ρ between the beam and the optical axes of the instrument through a reversible law. This reversibility allows for compensating the variation of the energy density during the pass of the satellite over the laser station with an optical device generating a variation depending on ρ. This is achieved with a neutral density device having a transmission depending on the incident angle of the laser beam. Among other things, the device allows for minimizing the solar flux backscattered by the Earth and for equalizing the optical flux received onboard whatever the position of the satellite. Each photodetector channel includes an interference filter to improve the signal to noise ratio of the detection. It is tuned to the nominal wavelength of the doubled Nd:YAG laser (532.1 nm). Each channel has also a collimation optic permitting to adjust the field of view to the angular size of the whole Earth seen from the satellite. The reflection function of T2L2 is obtained from the pyramidal LRA unit made with one central corner at the top and eight corner cubes on the periphery. Because the LRA and detection unit are not located at the same place, the reflection and detection points do not coincide. The projected distance between these points on the line of sight is computed through a post treatment on ground.

Table 3. DORIS oscillator characteristics measured before integration.

Characteristics	Measurement
Time stability (TDev)	<1 ps at 1 s 2 ps at 10 s 20 ps at 100 s
Aging	$<1 \times 10^{-11}$ day^{-1}
Thermal sensitivity	6.5×10^{-13} K^{-1}
Acceleration sensitivity	7.6×10^{-10} g^{-1}
Radiation sensitivity	6.7×10^{-12} rad^{-1}

The time reference used onboard is the quartz oscillator of the DORIS equipment. It comprises a dewar to protect both the sensitive electronics and the resonator from temperature fluctuations. Its main characteristics measured before integration, are given in Table 3.

The quartz oscillator is connected to the event timer allowing to get the timestamping of all incoming events. It can be considered as an ultrahigh speed counter made with an analog Vernier and a low frequency digital counter (100 MHz). The Vernier gives the time tag information with a time resolution of 100 fs and a temporal dynamic range of 20 ns while the digital counter has a time resolution of 10 ns and a temporal dynamic range of more than five years. Both Vernier and digital counter are driven by a frequency synthesis module designed to translate the 10 MHz DORIS signal to 100 MHz.

Locations of units A and B on the Jason-2 satellite are depicted in Fig. 13.

Main characteristics of the T2L2 space instrument are given in Table 4.

3.4. *Time equation*

We consider a laser station capable to emit–receive a laser pulse and a satellite capable to reflect the pulse and to measure the arrival time onboard. We note t_e and t_r respectively the emission and reception epochs of laser pulses measured at the laser station and t_b the reception epoch measured at the satellite. Hundred

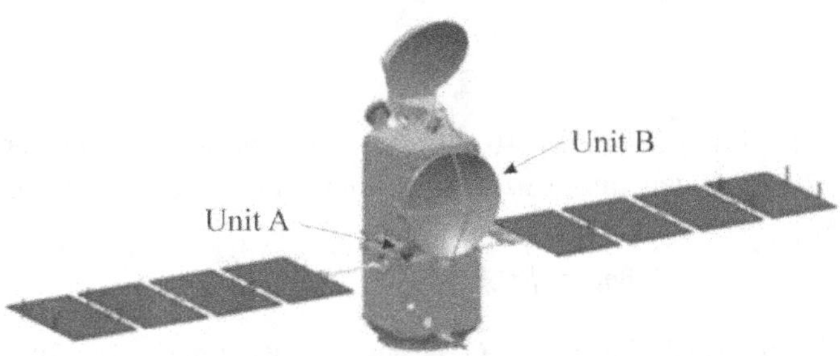

Fig. 13. CAO view of the whole Jason-2 satellite. T2L2 instrumentation is shared into two units A and B respectively outside and inside the satellite. (For color version, see page I-CP5.)

Table 4. Main characteristics of the T2L2 space instrument on Jason-2.

Subsystem	Characteristics
Mass of unit A	1.2 kg
Mass of unit B	8 kg
Volume A	$160 \times 116 \times 103\,\mathrm{mm}^3$
Volume B	$270 \times 280 \times 150\,\mathrm{mm}^3$
Power consumption	42 W
Optical detection wavelength	532.1 nm
Detection threshold	$0.3\,\mu\mathrm{J}\cdot\mathrm{m}^{-2}$
Field of view	55.1°
Photodetection standard deviation	30 ps RMS at mid Energy
Event timer standard deviation	2.8 ps RMS
Quartz Oscillator TDev	20 ps at 100 s

years ago, Poincaré and Einstein defined a procedure to synchronize some clocks in different places. Applied to our ground to space time transfer described in an inertial frame, the satellite clock has to be compared by taking into account the simple propagation delay $\frac{t_r - t_e}{2}$ to synchronize the clocks. Using this procedure in a different inertial frame moving with respect to the first frame, we get a different synchronization. With the assumption of principle of relativity and constant light velocity, Poincaré and Einstein derived the Lorentz transformation. Although this looks very simple from our present point of view, conceptually it was a firm step forward at that time. With acceleration and gravity and other corrections, the time offset Δ_{AS} between a ground clock A and the clock in space can be described by

$$\begin{aligned}\Delta_{\mathrm{AS}} &= t_e + \frac{t_r - t_e}{2} - t_b + C \\ &= \frac{t_e + t_r}{2} - t_b + \frac{C_{\mathrm{Sag}}}{2} + C_{\mathrm{Rel}} + C_{\mathrm{Atm}} + C_{\mathrm{ICal}} + C_{\mathrm{ECal}},\end{aligned} \tag{5}$$

where C_{Sag} is the Sagnac correction, C_{Rel} the relativistic frequency shift, C_{Atm} an atmospheric correction and C_{ICal} and C_{ECal} are some calibration terms for the laser station.[41]

The term C_{Sag} is computed from the coordinates x and X of the satellite and the station for each emission epoch. It is given by

$$C_{\mathrm{Sag}} = \frac{2}{c^2}(x - X)\cdot\dot{X}. \tag{6}$$

C_{Sag} has an amplitude of roughly 10 ns. C_{Rel} represents the relativistic frequency shift of the space oscillator integrated as a function of time. It is computed from

$$C_{\mathrm{Rel}} = \int \frac{1}{c^2}\left(\frac{\dot{x}^2}{2} - \frac{Gm_e}{x}\right)dt, \tag{7}$$

where m_e is the mass of the earth, G is the gravitational constant and $\dot{x}$ is the velocity of the satellite. This relativistic frequency shift includes a periodic part having an amplitude of roughly 100 ps integrated over a whole orbit.

C_{Atm} is a correction introduced by the atmosphere for the optical path difference between the uplink and the downlink. The time delay for the one-way link is roughly 40 ns and the time correction C_{Atm} between the uplink and the downlink is up to 1.5 ps.

C_{ICal} is required to get the absolute time of flight between the reference point of the station and the satellite. It is computed from epochs measured with some laser pulses sent onto the reception chain of the station through a retro-reflector located in the laser beam (Fig. 9) at a given distance d_{cc} to the spatial reference of the station (cross axes of the telescope mount):

$$C_{\text{ICal}} = \langle t_e - t_{\tilde{r}} \rangle_{N_{\text{ICal}}} - \delta_{cc}, \tag{8}$$

where $t_{\tilde{r}}$ is the reception epochs of N_{ICal} laser pulses reflected by the retro-reflector and δ_{cc} is the delay corresponding to the free space propagation between the corner cube and the space reference of the laser station.

The external calibration C_{ECal} allows for setting the absolute epoch of the laser emission. C_{ECal} may be written as

$$C_{\text{ECal}} = \langle t_e - t_{\tilde{e}} \rangle_{N_{\text{ECal}}} - \delta_{\text{ocx}} + \delta_{\text{prg}} \tag{9}$$

where $t_{\tilde{e}}$ is the emission epoch of N_{ECal} events measured by the station calibration, δ_{ocx} is the free space delay propagation between the reference of the station and the optical input of the calibration station and δ_{prg} is the global internal propagation inside the calibration station. By using the same calibration station to calibrate the different laser stations, the delays δ_{prg} in Eq. (9) do not need to be known accurately.

For the ground to ground time transfer in common view mode (distance $<$ 6000 km), the space oscillator is only required over the time interval between laser pulses (typically from 0.1 s to a few seconds). Over longer period, the noise is common for both stations and the difference becomes negligible. In noncommon view mode (distance $>$ 6000 km), the noise of the space oscillator has to be considered over the time interval corresponding to the time delay between the consecutives passes. In that case, the noise of the space oscillator becomes an important source of noise.

A ground to ground time transfer Δ_{AB} between two ground clocks A and B in common view mode is computed from the differences between the individual time transfers x_{AS} and x_{BS}, individually acquired by the stations and corrected with a model C_{osc} illustrating the mid-term behavior of the space oscillator.

3.5. *Error budget*

Tables 5 and 6 summarize the main uncertainties of a typical laser station and those of the space instrument, respectively.

Table 5. Main uncertainties/performances of a typical laser station suited for the T2L2 project.

Subsystem	Characteristics	u (ps)	Comments
Start detector	InGaAs pin photodiode, Bandwidth = 3 GHz	5	$k = 1$
Return detector	Avalanche photodiode, Geiger-mode	50	$k = 1$
Nd:YAG Laser	$\lambda = 532\,\text{nm}$, 10 Hz, $E = 50\,\text{mJ}$, FWHM = 100 ps	25	FWHM variation
Event timer	Two independent channels, resolution 1 ps	5	$k = 1$
Pulse per second (PPS) generator	Numerical division of a frequency source	5	$k = 1$
TF lab to LS cable	$L < 100\,\text{m}$, Thermal $< \pm 5°$ C, $< 0.1\,\text{ps} \cdot \text{K}^{-1} \cdot \text{m}^{-1}$	30	Relative delay
Ref. corner cube	Simple corner cube, Localization $\pm 1\,\text{mm}$	4	Absolute location

Table 6. Main uncertainties/performances of the T2L2 space instrument.

Subsystem	Characteristics	u (ps)	Comments
Board detector	Si avalanche photodiode (Geiger), Bandwidth = 1 GHz	70	$k = 1$, (at min. Energy)
Event timer	Resolution 1 ps, dead time 200 μs	2.8	$k = 1$
Oscillator	DORIS Quartz USO	20	σ_x at 100 s
LRA	Nine Suprasil corner cubes 32 mm Pyramidal 50°	13	Laser signature

Considering that the terms of Eq. (5) are independent, the combined uncertainty of a time transfer u_c (Δ_{AS}) between a ground clock A and the space clock is computed from the quadratic sum of the uncertainty of each term. The combined uncertainty of a ground to ground time transfer in common view mode is therefore the quadratic sum of uncertainties of ground to space time transfers $u_c(\Delta_{\text{AS}}), u_c(\Delta_{\text{BS}})$ and the uncertainty of the onboard model $u_c(C_{\text{Osc}})$. A detailed analysis of the whole experimental setup allows for determining an uncertainty budget for each term. Table 7 gives a summary of these combined uncertainties u_c for a set of data corresponding to a complete acquisition of a satellite pass over a given laser station. Table 8 is the ground to space and ground to ground time transfer overall T2L2 uncertainties.

In the common view mode, because the space oscillator is only required over the time interval between laser pulses coming from each laser station (at a maximum of few seconds), the uncertainty $u(C_{\text{osc}})$ can be neglected as compared to the other noises.

The highest uncertainty terms used in the computation of the error budget are coming from the delay variation of the cable between the time and frequency laboratory and the event timer of the laser station, and also from the laser pulse

Table 7. Combined uncertainty for a typical laser station and for a complete pass acquisition.

Uncertainty source	u (ps)	Comments
Emission epoch $u_c\,(t_E)$	34	Laser station
Reception epoch $u_c\,(t_E)$	17	Laser station
Onboard epoch $u_c\,(t_B)$	16	Laser and space instrument
Internal calibration $u_c\,(C_{\text{CICal}})$	21	Laser station
External calibration $u_c\,(C_{\text{CECal}})$	36	Laser and calibration station
Atmospheric $u_c\,(C_{\text{Atm}})$	1	—
Sagnac $u_c\,(C_{\text{Sag}})$	1	Orbitography
Relativity	1	Orbitography

Table 8. Ground to space and ground to ground uncertainties for a complete satellite pass for the whole project.

Time transfer	u (ps)	Comments
Ground to Space expanded uncertainty $u_E(\Delta_{\text{GS}})$	98	Coverage factor = 2
Ground to Ground expanded uncertainty $u_E(\Delta_{\text{GG}})$	138	Coverage factor = 2

width variations. The first contribution can be reduced by monitoring the delay propagation measured by an event timer through a double propagation of the signal emitted by a distributor and repeated by the user. With such monitoring and with a laser having a pulse width uncertainty of 10 ps, the ground to ground expanded uncertainty becomes better than 100 ps.

3.6. *Link budget*

The energy received at the space segment is deduced from the study of the uplink. It depends on the atmospheric transmission T_{Atm} and on the distance R between the satellite and the laser station. The energy received on the ground is deduced from both the uplink and downlink and also depends on the characteristics of the corner cubes embedded on the satellite and the characteristics of the telescope.

The energy distribution at the output of the telescope can be usually considered as uniform over the whole aperture of the telescope $2 \cdot r_{\text{Tel}}$. At a distance $R \gg r_{\text{Tel}}$, if the beam is diffraction limited, the distribution of such a beam can be approximated by the distribution of a Gaussian beam having a beam waist $\omega_0 = r_{\text{Tel}}$. The atmospheric turbulence modifies significantly the propagation of the beam. It introduces a beam spreading and also creates a speckle pattern in the plane of the satellite by interference. Because the atmosphere changes in time and that the satellite moves along its orbit, this pattern evolves very rapidly and introduces some important variation of energy from one shot to another. To take into account the beam spreading generated by the atmosphere, we must consider an equivalent aperture given by r_0, where r_0 is the Fried parameter which depends on the atmospheric condition. The energy flux D_{E0} (time-averaged) at the center of the beam

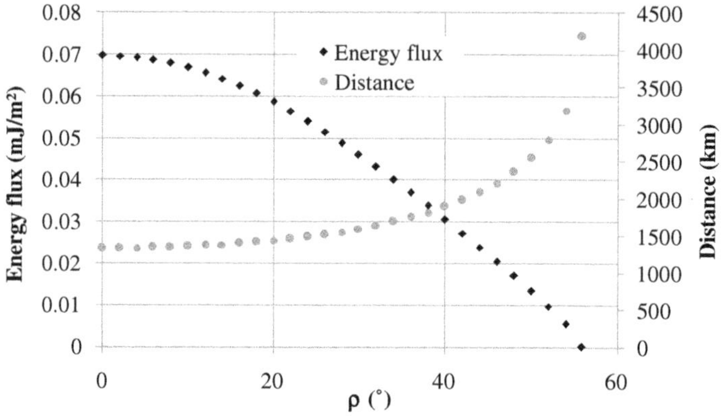

Fig. 14. Energy density and space segment distance as a function of the incident angle ρ.

at the distance R can be approximated by

$$D_{\mathrm{E0}} = E_{\mathrm{Las}} \frac{2T_{\mathrm{Atm}}\eta_{\mathrm{Tel}}\pi r_0^2}{R(\rho)^2\lambda^2}, \tag{10}$$

where E_{Las} is the laser energy by pulse, η_{Tel} is the transmission of the telescope, λ is the wavelength, T_{Atm} is the transmission of the atmosphere and ρ is the incident angle between the laser beam and the axis Earth–satellite. To illustrate the speckle pattern, the maximum and minimum energy densities (in the center of the speckles) have to be considered respectively two times greater than the mean density of the whole beam and close to zero. Figure 14 is an illustration of Eq. (10) as a function of the incident angle ρ of the beam as compared to the optical axis of the instrument, using the following data set: $E_{\mathrm{Las}} = 30\,\mathrm{mJ}$; $r_0 = 24\,\mathrm{mm}$; $T_{\mathrm{Atm}} = 0.81$; $H_{\mathrm{St}} = 1300\,\mathrm{m}$; $H_S = 1330\,\mathrm{km}$; $\eta_{\mathrm{Tel}} = 0.44$; $\lambda = 532\,\mathrm{nm}$; $R_E = 6371\,\mathrm{km}$.

The energy received on the reception channel of the laser station E_{Tel} is given by

$$E_{\mathrm{Tel}} = D_{\mathrm{E0}} \cdot \frac{\sigma_{cc}(\rho)}{4R^2} \cdot T_{\mathrm{Atm}} \cdot \eta_{\mathrm{Tel}} \cdot r_{\mathrm{Tel}}^2, \tag{11}$$

where σ_{cc} is the cross-section of the LRA. σ_{cc} depends on the incident angle ρ because of the geometry of the pyramidal supporting the individual corner cubes, and because of the speed aberration which introduces an angle between the laser beam and the real line of sight between the satellite and the laser station. The typical energy received with a 1.5 m telescope on the ground is in the range of 0.4 fJ (roughly 1000 photons).

3.7. *Exploitation*

Two mission centers operate the exploitation activities of the T2L2 project. The first, operated by the CNES, is the Instrument Mission Center (IMC) which is in charge to gather all the raw data coming from the satellite and generate some preanalyzed products. The second, under the responsibility of GeoAzur-OCA, is the Analysis Mission Center (AMC) which allows to generate the final ground to space and ground to ground time transfers from the preanalyzed product delivered by the IMC. The AMC operates continuously on a daily basis and is able to compute a given link three days after a given laser acquisition.

When T2L2 is operated simultaneously from several laser stations, the arrival epochs of all laser events are mixed together. The first important step carried out by the AMC is to identify each event recorded with the corresponding laser station. This is done by comparing emission epochs of a given laser station with all epochs measured onboard. The next step is to reject the outliers coming from false detections of both the space instrument and laser stations. Data are then corrected to take into account the instrumental model of the space segment[42] including geometry, photodetection time walk versus energy and time walk versus attitude. At this stage, the AMC generates some data files usable by the scientific community comprising schematically emission and reception epochs together with arrival epochs onboard. From these files, it is then possible for everybody to compute the basic ground to space and the ground to ground time transfers. The AMC also computes all available ground to space time transfers, and opens the possibility of computing any ground to ground time transfer on demand. The AMC has developed a dedicated T2L2 website[43] in order to share all these data and computations with the scientific community.

Figure 15 is an example of what we get for a single satellite pass obtained with the MeO Laser Station linked to a H-Maser.

The corresponding time stability computed from the root square of the time variance is illustrated in Fig. 16.

The accumulation of several consecutive passes over the same station allows giving a mid-term time transfer comparison. Figure 17 is an example of such six

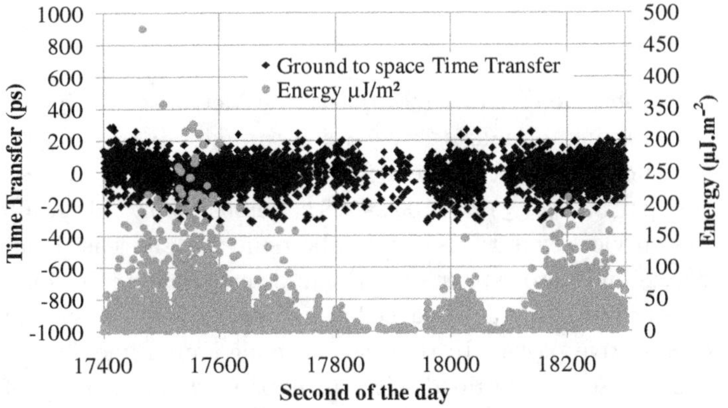

Fig. 15. OCA H-Maser Jason-2 quartz time transfer example acquired on November 2013. A linear regression has been subtracted to the time transfer to take into account the frequency offset between clocks.

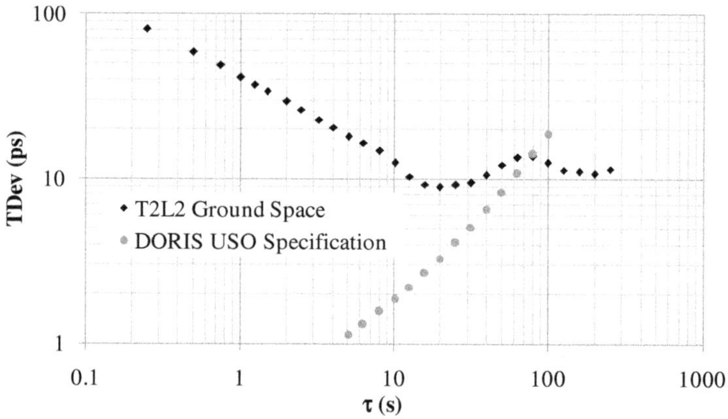

Fig. 16. OCA–DORIS time deviation (black, left scale) and modified Allan deviation (red, right scale). Performances are in accordance with the DORIS USO.

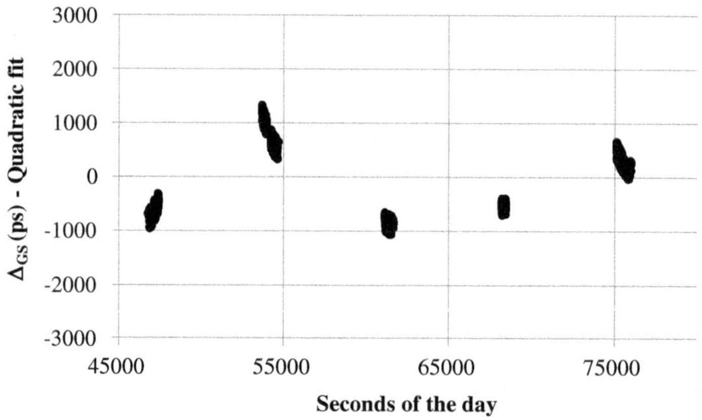

Fig. 17. OCA–DORIS comparison over several consecutive passes. Up to seven passes on Jason-2 can be acquired consecutively.

consecutive passes obtained in spring 2010 with the MeO laser station linked to a H-Maser. After subtracting a quadratic regression to take into account the frequency offset between the clocks and the linear frequency drift, we get over all the passes $u(\Delta_{\mathrm{MeOS}}) = 725\,\mathrm{ps}$ ($u(\Delta_{\mathrm{MeOS}}) = 7100\,\mathrm{ps}$ with a linear fit). This result illustrates that a noncommon view time transfer must be done with at least a subtraction of a quadratic regression in order to remove the mid-term drift of the space oscillator. This can be achieved by using a reference laser station capable to track the satellite over several consecutive passes. In that case, a noncommon view ground to space time transfer Δ_{GS} can be obtained with an expanded uncertainty of two times $u(\Delta_{\mathrm{GS}})$ obtained with the quadratic fit: $u_E(\Delta_{\mathrm{GS}}) = 1450\,\mathrm{ps}$ ($k = 2$). A more sophisticated oscillator model taking into account several external parameters such as thermal changes or radiation is currently under study.

Several co-location time transfers have been made since the beginning of the T2L2 project between the two laser stations belonging to the OCA: MeO and the French transportable laser system (FTLRS).[44,45] Among them, one was conducted in 2010,[46] with an unique H-Maser connected to both MeO and FTLRS to validate the accuracy of the time transfer and another one was made with a high performance time distribution system (Sigma Time STX201) to validate the long-term time stability. The system was able to monitor the absolute time delay variation introduced by coaxial cables used between laser stations and the time and frequency laboratory. The first campaign allowed to validate the expected error budget (Table 8) from nine common satellite passes collected over four days. The time offset measured between MeO and FTLRS was $\Delta_{\mathrm{Meo \cdot FTLRS}} = 37\,\mathrm{ps}$ (filtered at $\pm 3\sigma$) with an uncertainty $u(\Delta_{\mathrm{Meo \cdot FTLRS}}) = 60\,\mathrm{ps}$, $(k = 1)$. The second campaign was based on the same setup except for the distribution of time signals between the time and frequency laboratory and the laser which was made with a continuous monitoring of the delay variation (measured by a dedicated PPS generator including an internal event timer). Figure 18 illustrates the time stability TDev between MeO and FTLRS obtained during that campaign. This long-term time stability result is also in a good agreement with the error budget described in Sec. 3.5.

Some other dedicated experiments were made to validate the time transfer between remote clocks in common view. A first experiment of that kind was performed in 2010 with atomic fountains between two observatories in France: the "Observatoire de Paris" (OP) and OCA.[18]

The mobile laser station FTLRS was installed on a dedicated platform at OP. Some special authorizations were obtained to range with a laser from Paris. A mobile atomic fountain (FOM)[24] designed by OP-SYRTE in the frame of the ACES program, was installed at the OCA during the same period. At the OP, FTLRS, the atomic fountain and both GPS and TWSTFT were connected to the same

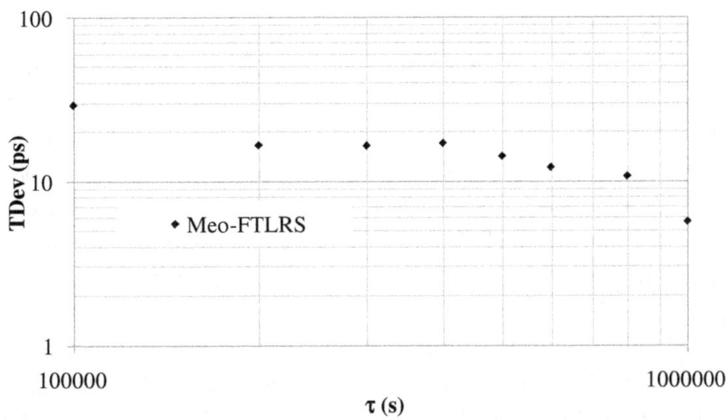

Fig. 18. Estimation of the T2L2 time stability measured between Meo and FTLRS in co-location. The time interval between consecutive acquisitions is not perfectly constant.

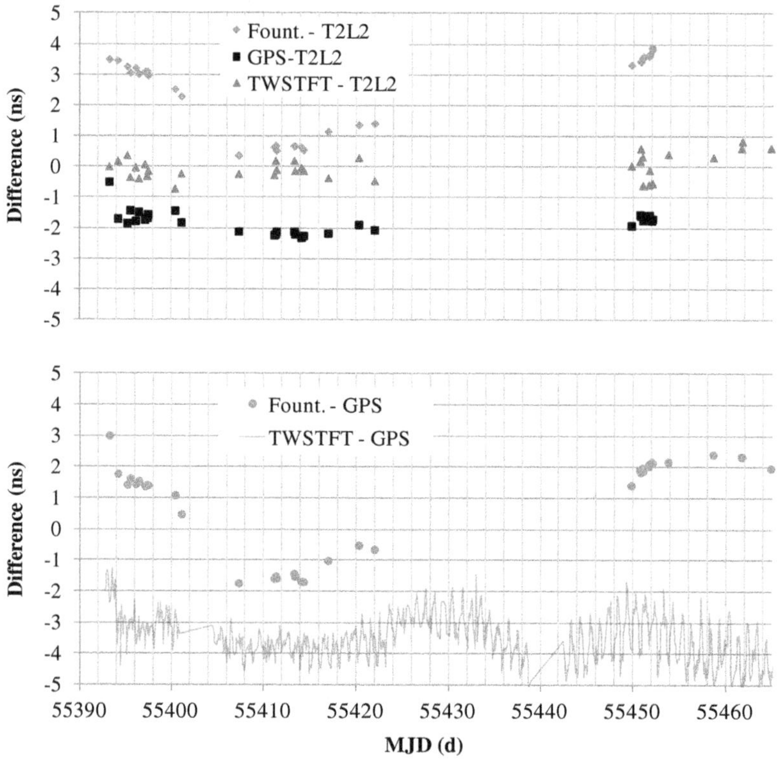

Fig. 19. Microwave and T2L2 time transfer comparison together with atomic fountain differences. A time offset between each plot is voluntarily introduced to facilitate the reading of the graph. Microwave time transfer solutions and fountain differences were computed by SYRTE (D. Rovera).

H-Maser. At OCA, MeO, FOM and FTLRS were also connected to a common H-Maser. About 56 satellite passes were recorded over 114 days.

Figure 19 shows both the time transfer comparison between T2L2–GPS (carrier phase based)–TWSTFT (code) and the differences between atomic fountains. The GPS analysis is based on the PPP NRCan algorithm (carrier phase technique) developed by Natural Resources Canada. This result doesn't take into account calibrations between each of the techniques: An offset for each solution was introduced to facilitate graph reading. Regardless of these absolute aspects, the global drift in the microwave–T2L2 time comparisons is better than 2 ns over two months, which is in accordance with the classical long-term stability of microwave time transfers.

The phase of the atomic fountain comparison is computed from the frequency information between the fountain interrogation and the H-Maser frequency reference. For some technical reasons, the atomic fountains did not operate continuously during the whole campaign. This has probably introduced a significant mid-term noise.

A second campaign between remote clocks in common view was performed by the end of 2013 between four distinct sites in Europe in order to realize a subnanosecond comparison between T2L2 and GPS CV.[47,48] The GPS CV comparisons are done from the differences made individually for all satellites in common view computed from an average over several minutes with some geometrical compensation and some atmospheric corrections. The campaign was jointly conducted in autumn 2013 by

— GRSM 7845 OCA Grasse France.
— FTLRS 7828 OP Paris France.
— HERL 7840 SGF Hertsmonceux United Kingdom.
— WETL 8834 BKG, FESG Wettzell Germany.

For some calibration reasons, only the results between the three first sites were analyzed. During the whole campaign, 28 common passes were obtained by the OCA and Herstmonceux, and 13 by the OCA and OP. Each station was calibrated based on a joint calibration of both laser and GPS. Laser stations were calibrated with the T2L2 calibration station while GPS receivers were calibrated using dedicated equipment moving between stations and conducted by OP-SYRTE.[49] The uncertainty of the GPS CV time transfer was estimated at 2.1 ns for the link OCA–OP, 1.5 ns for SGF–OP and 2.1 ns for SGF–OCA ($k = 2$).[50] The average differences between the calibrated links T2L2 and GPS were below 250 ps with a standard deviation below 500 ps. The very good agreement between GPS CV and T2L2 confirms that the uncertainty budget is consistent with the experimental setup. This is the first validation ever done which results are in agreement at sub-ns level between a GPS CV time transfer and another fully independent technique (T2L2) performed over long distances. This is strong evidence of T2L2's ability to compare other time transfer techniques accurately.

4. One-Way Lunar Laser Link on LRO Spacecraft

LRO is a NASA's mission that aims at exploring the Moon with numerous scientific objectives. LRO is orbiting around the Moon at an altitude of 50 km. Among the on-board instruments, LOLA allows for providing a precise global topographic model of the Moon surface. LOLA emits a single laser pulse divided into five distinct beams. The beams are backscattered from the lunar surface and detected by the instrument. For each beam and for each laser pulse emitted, LOLA measures time of flight, pulse width and energy. The nominal accuracy of the instrument is 10 cm. To realize this task LOLA includes two distinct emission–reception optical devices, a 1064 nm 28 Hz-pulsed laser, a photodetection system and an event timer linked to an ultrastable oscillator. The mid-term time stability of the oscillator is $\sigma_y = 10^{-12}$ over several thousand seconds.

LOLA can also be used to realize a one-way laser ranging from the Earth to the Moon. One of the motives for equipping LRO with the one-way ranging functionality

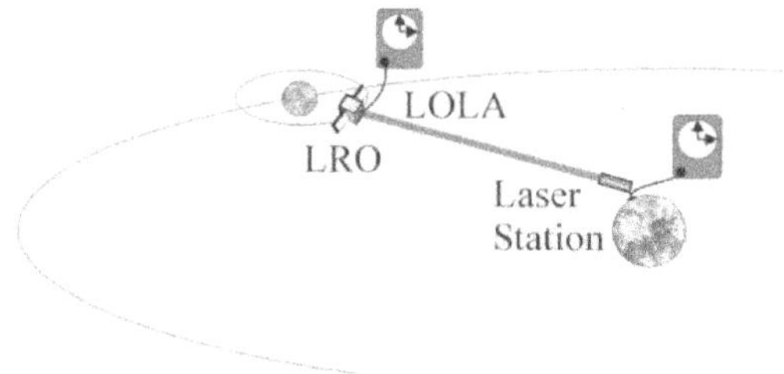

Fig. 20. Earth to Moon one-way laser ranging with LRO through the LOLA of the spacecraft and some laser stations on ground. (For color version, see page I-CP6.)

was the expectation that the orbit determination of LRO could be improved due to a better satellite clock model, leading to a better altimetry profiling of the Moon. The distance is then computed from the differences between the start and the arrival epochs of an ensemble of laser events. The principle of the measurement is depicted in Fig. 20.

As compared to a passive two-way laser ranging configuration, the link budget of such a one-way link varies only with the inverse of the square of the distance R. The signal detected may be written as

$$N_{\mathrm{Le}^-} = \frac{E\lambda}{hc}\frac{4r_{\mathrm{tel}}^2}{\theta^2 R^2}\rho_{\mathrm{det}}\,\rho_{\mathrm{atm}}\,\rho_{\mathrm{opt}}, \tag{12}$$

where E is the energy emitted by the ground station, θ is the divergence of the laser beam, R is the distance between the Earth and LOLA, λ is the wavelength, ρ_{det} is the quantum efficiency of the detector, ρ_{atm} is the atmospheric transmission, ρ_{opt} is the transmission of the optics and r_{tel} is the radius of the detection telescope. The measurement is made with a dedicated $r_{\mathrm{tel}} = 22\,\mathrm{mm}$ receiver telescope pointed toward the Earth[51] and linked to the altimeter using an optical fiber (Fig. 21).

The ground segment of this one-way laser ranging is represented by ten laser stations of the international laser ranging network. The primary ground station is NASA's Next Generation Satellite Laser Ranging (NGSLR) station. LOLA receives the signal from the laser station through a wavelength multiplexer allowing to discriminate the altimeter pulses at 1064 nm from laser ranging station pulses at 532 nm. The collimation telescope on the LRO spacecraft has an aperture of 22 mm and a field of view wide enough to cover the whole Earth. It is mounted on the high gain RF antenna which is pointed to Earth.

For each laser pulse emitted by a given laser station and measured by LOLA, one gets an emission epoch in the time scale of the laser station and a reception epoch in the LOLA time scale. The range is deduced from the differences of these epochs. Data are downloaded in real time (+30 s) with a classical microwave link and preprocessed in order to provide the observer prompt information on the status of the uplink. Figure 22 is an example of what the observer can get in real time.

Since the commissioning of the LRO spacecraft, several thousand hours have been recorded in the frame of this project. The primary orbit determination of the

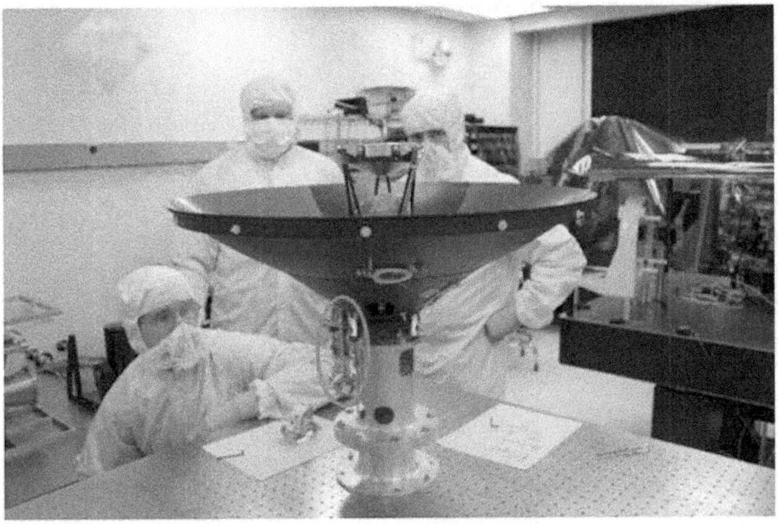

Fig. 21. RF antenna. The Laser ranging receiver telescope is on the left from the center of the main RF antenna (red ellipse). LOLA is coupled with the telescope through an optical fiber. (Courtesy: NASA Goddard Space Flight Center). (For color version, see page I-CP6.)

LRO spacecraft is realized through microwave technology. Besides, one-way laser ranging is used to give independent orbit solutions. Some comparisons between ranges computed from the two techniques have shown an average total RMS difference in the range of 10 m. The optical link between laser ground station and LOLA has been tested for communication transfer[52] at a rate of 300 bits/s.

LOLA allows also performing ground to ground time transfers between laser stations in a common view mode.[53] The major objective is to establish accurate ground station times and improve LRO orbit determination. The time transfer is computed from the onboard epochs recorded by LOLA and from the differences of the times-of-flight from each ground station to the spacecraft. Distances between laser stations being short compared to Earth–Moon distances, much higher uncertainties in times-of-flight from each SLR station to LRO than those required with T2L2 can be tolerated. As a consequence, the times-of-flight estimated from the conventional RF link are sufficient for time transfer at subnanosecond accuracy. For instance, if we consider a distance between laser stations equal to half the planet and a RF link uncertainty of 100 m, the corresponding differential time of flight uncertainty is only 100 ps.[51] The validation of that concept has been undertaken at Goddard Geophysical and Astronomical Observatory between Moblas-7 and NGSLR laser stations. Outputs are currently being analyzed.

It has been proposed to use the LOLA on LRO instrumentation in a differential mode to realize a localization of the spacecraft also in the tangential plane (three-dimensional (3D) localization). Unfortunately, the short-term uncertainty of

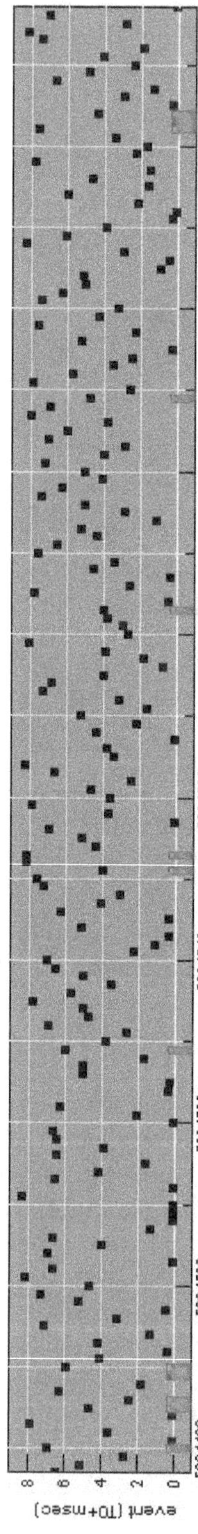

Fig. 22. Acquisition from the MeO station with the LRO-LOLA website. Each dot is an event (noise or laser pulse from laser station) detected by LOLA. The x-axis represents the arrival epoch of the detected event. The y-axis is the arrival epoch in a given temporal gate at the LOLA time scale.

the event timer onboard is not high enough (~ 1 ns, $k = 1$) to get an interesting outcome.

5. Prospective

Numerous missions have been proposed with a laser link. Some of them are based on a coherent laser link well suited for frequency transfer or measurements of position variation, and some other are based on the propagation of laser pulses most suited to meet the needs of absolute localization or time transfer.

Several missions are proposed at the scale of the Solar System with distances in the range of several billion kilometers. Such distances cannot be measured through a classical passive two-way laser ranging scheme: The Earth–Moon distance is now considered as a maximum with a link budget in the ratio of $1/10^{20}$.[54] To go further, it is necessary to use a one-way scheme where the link budget varies only with the inverse of the square of the distance Eq. (12). With a payload instrument based on optics having an aperture of 100 mm, a beacon divergence of 5 arcsec, 300 mJ per pulse and a distance of 400 million km, we get: $N_{\mathrm{Le}^-} \sim 1$ electron.

In the context of future space missions for fundamental physics, Solar System science and navigation, a laser link for clock comparisons, ranging and data transmission is of prime importance. Table 9 gives a nonexhaustive list of some future missions and associated laser links.

ELT is a two-way pulsed laser link currently under study for the ACES mission. The onboard hardware consists of a retro-reflector, a single photon avalanche diode and an event timer connected to the ACES clocks. The ground segment is based on the international network of laser ranging stations. The ELT experiment should allow a space to ground clock comparison with time stability (TDev) of 4 ps over 300 s and 6 ps for the ground to ground time transfer. Thanks to the very good time stability of the ACES clocks, the ground to ground comparison in a noncommon view configuration over one orbit period should be of the same order of magnitude. The time transfer accuracy should be better than 50 ps. Some time transfer comparisons between ELT and T2L2 should be scheduled at the ACES mission start-up

Table 9. Some future space missions and associated laser links.

Mission	Laser link type	Laser link name	Funded	Expected launch
ACES	Two-way laser pulsed	ELT	Yes	2017
GNSS Navigation	Two-way laser pulsed	OPTI	Yes	2017
Tiangong	Two-way laser pulsed	LTT	Yes	—
LATOR	$2\times$ one-way pulsed laser transponder	LATOR	No	—
STE-QUEST	PRN modulation	OPL	No	—
ASTROD I	$2\times$ one-way pulsed laser transponder	ASTROD I	No	
SAGAS	Two-way coherent-modulated	DOLL	No	—
OSS	One-way laser pulsed/coherent	TIPO/DOLL	No	—

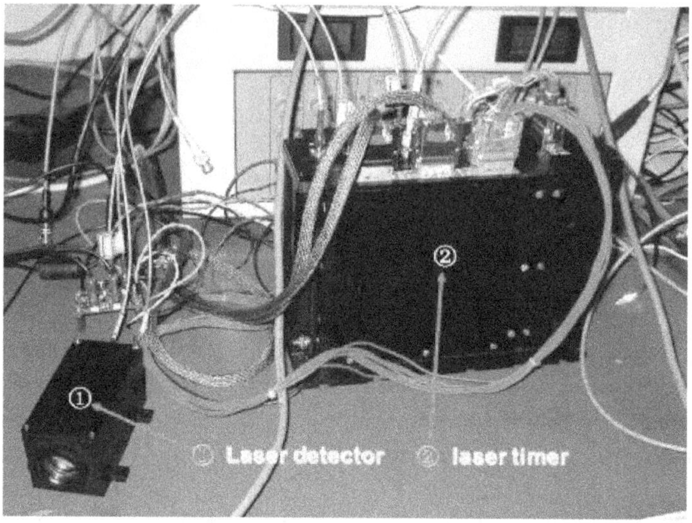

Fig. 23. LTT equipment. On the left is the detector; in the middle lies the main electronic package including the event timer. (Courtesy: Shanghai Astronomical observatory). (For color version, see page I-CP6.)

together with some comparisons with microwave link also embedded on ACES. The project should be launched on the International Space Station (ISS) in 2017.

The OPTI two-way link aims at synchronizing any satellite navigation systems. It is designed as a compact device for real-time ground to space clock corrections, using the existing satellite laser ranging network. The instrument should demonstrate time transfer with an uncertainty of 100 ps. This instrument could be launched in 2017.

The time transfer project proposed on the Chinese Tiangong station will be based upon a LTT instrument designed by SHAO. Figure 23 is photography of the space equipment.

The Laser Astrometric Test of Relativity (LATOR) mission[55,56] architecture is based on a light triangle formed by laser beacons between two spacecrafts placed in heliocentric orbits and a laser terminal on the ISS. LATOR uses both an optical interferometer and classical laser ranging techniques to accurately measure deflection of light in the Solar System. Laser ranging would be done through a double one-way link scheme working in an asynchronous mode. The distance accuracy required for the final scientific objective is 3 mm corresponding to a delay uncertainty measurement of 10 ps.

The STE-QUEST optical link[57,58] is based on an initial design of the TESAT laser communication terminal. Scientific objectives of the STE-QUEST mission require a common view comparison of clocks on ground at the 10^{-18} fractional frequency uncertainty level after a few hours of integration and a space to ground and a ground to ground time transfers with accuracy better than 50 ps. The link includes a space terminal with two optical telescopes allowing for simultaneous bidirectional links and at least two ground stations equipped with an Optical Link Ground Terminal. The laser link is based on an optical carrier modulated with a pseudo-random noise signal. The optical carrier of the link is generated from the optical signal of the onboard atomic clock. The PRN modulation transmitted to the ground station is based on the microwave reference signal. The space optical terminal receives and cross-correlates the PRN optical carriers replicated by the ground from the onboard reference signal. The differential delays of the link have to be calibrated with an uncertainty better than 50 ps.

The objectives of the Astrodynamical Space Test of Relativity using Optical Devices (ASTROD) mission include measurement of some relativistic parameters and measurement of several Solar System parameters. ASTROD is developed in two distinct missions ASTROD I[60] and ASTROD-GW.[59] ASTROD I uses a single spacecraft carrying four lasers, a clock and some laser stations on ground. The distance is measured with a double asynchronous one-way laser ranging system with an uncertainty in the range of 1 mm. ASTROD-GW is based on three spacecrafts with two spacecrafts in separate solar orbits and one near earth. In ASTROD-GW, each payload comprises a proof mass, two lasers, a clock and a drag-free system. The distances between each spacecraft are optically ranged coherently.

The Search for Anomalous Gravitation using Atomic Sensors (SAGAS) mission[61] aims at flying sensitive atomic sensors and a laser link on a Solar System escape trajectory in 2020–2030. SAGAS has several science objectives in fundamental physics and Solar System science. The payload comprises an optimized optical atomic clock, an absolute atomic accelerometer and a laser link (DOLL) for ranging, frequency comparison and communication. The optical link is based on a continuous stabilized laser locked to the optical clock and a coherent heterodyne detection system. The link is based on a 40 cm telescope for the space segment and 1.5 m on ground. Some preliminary projects were initiated in 2009–2012 by the SYRTE institute (Mini-DOLL) in order to define more precisely the design of that laser link.[62,63] The Allan deviation obtained on a fixed target through a turbulent atmosphere (5 km) was $\sigma_x\,(1\,\text{ms}) = 28\,\text{nm}$ and $\sigma_x\,(1\,\text{s}) = 1.4\,\mu\text{m}$.

The aims of the Outer Solar System (OSS) mission[64] are shared by planetary science and fundamental physics. The mission uses a single spacecraft in the Solar System equipped with several specific instruments. In particular, it comprises some instruments (optic and microwave) allowing a precise tracking of payload during the cruise for the measurement of the Eddington parameter γ. Two solutions are envisioned for the measurement in the optical domain, with a one-way pulsed laser system TIPO,[65] or a two-way coherent laser concept DOLL identical to the instrument used in the SAGA mission. In the TIPO concept, the distance is computed from the measurement of the time-of-flight of laser pulses emitted from an Earth laser station and received by the space vehicle carrying a clock, a time tagging unit and a photodetection system. The time-of-flight is deduced from the differences between start and arrival times measured at the respective time scales of the clock on ground and the clock in space. The TIPO instrument is made up of a telescope, a photodetection device and an event timer linked to an ultrastable clock. At the lowest level, the propagation delay and distance between Earth and spacecraft are deduced from the differences between the arrival and departure times. The behavior of the clocks is a major factor in the performances of the experiment. With a compact rubidium clock,[66] having a time stability of $\sigma_x = 2.5 \cdot 10^{-13} \cdot \tau^{1/2}$ over a few thousands of seconds, some interesting measurements like planetary gravity fields or Shapiro signatures can be done. With that kind of rubidium clock, the instrument is able to measure short-term distance variations with an uncertainty of the centimeter level over an integration time of one day.

6. Conclusion and Outlook

Modern laser ranging is today a mature technology able to routinely produce range measurements with a subcentimeter uncertainty. There is a widespread network of laser ranging stations on ground capable to work together in order to meet the classical laser ranging activities but also some specific issues such as time transfer. The first proposal for using the SLR technology in order to realize a time transfer was submitted in the early seventies by the LASSO project. Since that date, many other projects have been led in Earth's orbit as well as at the Solar System scale. The best operational laser time transfer available today is T2L2 on Jason-2. The project has been in successful operation since summer 2008 and extended until 2016. Since 2008, several dedicated campaigns have been carried out to demonstrate the performance of that time transfer technology and realize several scientific objectives. T2L2 has proved its ability to synchronize remote ground clocks with an uncertainty better than 100 ps ($k = 2$). It has routinely established some link from ground to space with a time stability of a few picoseconds over several hundred seconds. Three laser time transfer projects, developed in the frame of the Compass system, have been able to realize a ground to space time scale synchronization with a data spread of a few hundred picoseconds. Laser systems have been also used to establish links in deep space and are able to achieve better performance at lower power with some small apertures as compared to classical microwave systems. Laser ranging technology has demonstrated its capability to realize some one-way laser links beyond the Earth's orbit with the operational mission LRO and two impressive demonstrations in the Solar System at 24 and 80 million km. The next high performance LTT will be supported through the ACES mission with the ELT instrument. Some major projects are envisioned with both T2L2 and ELT to realize picosecond time transfers in noncommon view configuration and to compare microwaves techniques with laser technologies. In the context of fundamental physics, Solar System science or navigation, several other challenging missions, pending for approval, suggest to work at the scale of the Solar System with distances in the range of hundreds of millions of kilometers, such as ASTROD in 2025.

Direct comparison of clocks over short distances can be made through the use of subpicosecond event timers. In that case, the event timer measures the time interval between either the PPS signal or the frequency reference of each clock. Some event timers today are able to obtain a fractional frequency stability of 10^{-16} over an integration period of only a few hundred seconds, or of 10^{-18} over one day.[67] In cases when the distances between clocks are greater than some meters, event timers can also monitor the propagation delay in the coaxial cables to subtract the possible thermal drift. This gives also the possibility to measure the shape of the signals used to synchronize the clocks (PPS) with two event timers running in an oscilloscope mode. One event timer is set as the trigger for the time reference and the other is used to measure the signal for several thresholds. The x-axis is the time difference between the two event timers, and the y-axis is the threshold of the signal. This concept is of crucial importance to compare different time transfer techniques.

Recent advances in optical fiber technology are now allowing the realization of precise frequency and time transfer through the optical fiber network infrastructure used for worldwide communication.[68,69] Preliminary experiments done in several countries such as France, Germany or the USA have shown the possibility to realize some links over distances of up to several hundred kilometers. Some demonstrations have been made over several hundred kilometers with a fractional frequency fluctuation of only a few 10^{-19} over one day. These performances are well suited to distribute the signal generated by the best optical clocks over continental distances over the next ten years. It has been also demonstrated that the Internet traffic could be maintained during the frequency distribution without any notable degradation of the performance. Time transfer accuracy better than 100 ps is likely to be achievable by using some specific modulation of the optical carrier. The simplest way to do such time transfers through optical fibers is to use optical pulses and an asynchronous transponder based on event timers at both ends to measure the two-way propagation. Time transfer can also be performed by using some modulated codes added on the carrier and some specific modems in order to extract a usable synchronization signal. Today, the deployment of these techniques through the infrastructure currently used for Internet communication requires the installation of some specific equipment to meet metrology requirements. We can expect that further progress in the telecommunication technology will enable its development as applied to time and frequency transfer. As well, optical communication in free space turns out to be a very promising technique for very high speed data rate communication. Several proposals are currently under review.[70] We can expect that the development of such free space optical communications will be used in the future for time and frequency transfer. T2L2 and time transfer by optical fiber are complementary techniques that are both extremely promising, especially if developed jointly.

The precise and accurate determination of distances is a critical issue in many fields. Numerous formation flight space missions require significant improvement in the distance measurement between spacecrafts. It is now a general tendency to increase the number of space vehicles to realize some very huge space detectors through a high precision laser metrology. Recent researches in the field of ultrastable event timers have allowed to obtain some repeatability error in the subpicosecond domain and some time stabilities of only a few femtoseconds over some hundred seconds.[71–73] Such femtosecond time stabilities converted into light distance allow for resolving the wavelength of the light. The association of the time-of-flight with an interferometric measurement gives the possibility to obtain length measurements with accuracy much better than a single wavelength. Combined with an optical interferometer, these femtosecond measurements could allow to combine the very high resolution given by the interferometer together with the absolute measurement of the time-of-flight measured by the event timer. Several experimental variations of that concept were proposed in the framework of the ILIADE project.[74] One of these concepts was based on a femtosecond laser frequency comb associated with some interferometric Fabry–Perot filters, a high speed optical modulator and some event timers. The Fabry–Perot filters were used to isolate a single line from the comb to generate two continuous carriers (respectively on the laser output and the return path of the beam) which were mixed together in order to generate the interferometric measurement of the system. The high speed modulator was used to subtract some pulses from the continuous laser train to create a low frequency coded modulation. A high speed detection linked to the event timers was used to detect this modulation and to compute the absolute distance.

Acknowledgments

E. Samain would like to thank the AstroGeo-GeoAzur team for all regular observations of Jason-2 and all dedicated campaigns that were carried out within the T2L2 project, and also for the development of the T2L2 Analysis Center. He would like to thank the laser ranging community for observation campaigns dedicated to time transfer and the French Centre National d'Etudes Spatiales for funding the T2L2 project and developing the T2L2 space instrument. E. Samain is also grateful to Ulrich Schreiber for the very fruitful review of the whole manuscript.

References

1. Z. Altamimi, X. Collilieux and L. Métivier, *J. Geod.* **85** (2011) 457.
2. J. J. Degnan, Laser transponders for high-accuracy interplanetary laser ranging and time transfer in *Lasers, Clocks and Drag-Free Control: Exploration of Relativistic Gravity*, eds. H. Dittus, C. Lämmerzahl and S. G. Turyshev (Springer, 2009), pp. 231–242.
3. P. Fridelance and C. Veillet, *Metrologia* **32** (1995) 27.
4. F. M. Yang and X. Li, *Prog. Astron.* **22** (2004) 10.
5. W. Meng *et al.*, *Adv. Space Res.* **51** (2013) 951.
6. I. Prochazka, U. Schreiber and W. Schafer, *Adv. Space Res.* **47** (2011) 239.
7. I. Prochazka, K. Hamal and K. Kral, *J. Mod. Opt.* **54** (2007) 151.
8. I. Prochazka and Y. Fumin, *J. Mod. Opt.* **56** (2009) 253.
9. Y. Fumin *et al.*, Preliminary results of the laser time transfer (LTT) experiment, in *Proc. of the 16th Int. Workshop on Laser Ranging*, Poznan, Poland, October 12–17, 2008 (Polish Academy of Sciences, 2008), pp. 648–652.

10. M. T. Zuber, *Photonics Spectra* **40** (2006) 56.
11. J. F. Cavanaugh *et al.*, *Space Sci. Rev.* **131** (2007) 451.
12. M. T. Zuber *et al.*, *Space Sci. Rev.* **150** (2009) 63.
13. D. E. Smith *et al.*, *Space Sci. Rev.* **150** (2010) 209.
14. E. Samain and P. Fridelance, *Metrologia* **35** (1998) 151.
15. P. Fridelance, E. Samain and C. Veillet, *Exp. Astron.* **7** (1997) 191.
16. L. Cacciapuoti and C. Salomon, *J. Phys. Conf. Ser.* **327** (2011) 012049.
17. U. Schreiber, I. Prochazka, P. Lauber, U. Hugentobler, W. Schafer, L. Cacciapuoti and R. Nasca, The European laser timing (ELT) experiment on-board ACES, in *Proc. Frequency Control Symp. Joint with the 22nd European Frequency and Time Forum* (IEEE International, 2009), pp. 594–599.
18. E. Samain *et al.*, Time Transfer by Laser Link — T2L2: Current Status and Future Experiments, in *Proc. of the European Frequency and Time Forum*, San Francisco, California, USA (IEEE, 2011).
19. P. Exertier, E. Samain, P. Bonnefond and P. Guillemot, *Adv. Space Res., DORIS Spec. Issue* **46** (2010) 1559.
20. E. Samain, J. Weick, P. Vrancken, F. Para, D. Albanese, J. Paris, J. M. Torre, C. Zhao, P. Guillemot and I. Petitbon, *Int. J. Mod. Phys. D* **17** (2008) 1043.
21. P. Lemonde *et al.*, Cold-Atom Clocks on earth and in space, in *Frequency Measurement and Control*, ed. A. N. Luiten, Topics in Applied Physics, Vol. 79 (Springer, 2001), pp. 131–153.
22. N. Poli, C. W. Oates, P. Gill and G. M. Tino, arXiv:1401.2378 [physics.atom-ph].
23. S. A. Diddams, J. Ye and L. Hollberg, Femtosecond lasers for optical clocks and low noise frequency synthesis, in *Femtosecond Optical Frequency Comb: Principle, Operation, and Application* (Springer, 2005), pp. 225–262.
24. J. Guéna, M. Abgrall, D. Rovera, P. Laurent, B. Chupin, M. Lours, G. Santarelli, P. Rosenbusch, M. E. Tobar, R. Li, K. Gibble, A. Clairon and S. Bize, *IEEE Trans. Ultrason. Ferroelectr. Freq. Control* **59** (2012) 391.
25. B. J. Bloom, T. L. Nicholson, J. R. Williams, S. L. Campbell, M. Bishof, X. Zhang, W. Zhang, S. L. Bromley and J. Ye, *Nature* **506** (2014) 71.
26. M. Weiss *et al.*, Coordinating GPS calibrations among NIST, NRL, USNO, PTB and OP, in *Proc. UFFC-FCS EFTF Joint Meeting* (IEEE, 2011), pp. 1070–1075.
27. G. Petit, P. Defraigne, B. Warrington and P. Uhrich, Calibration of dual frequency GPS receivers for TAI, in *Proc. of 20th EFTF* (BIPM, 2006), pp. 455–459.
28. E. Samain, P. Exertier, C. Courde, P. Fridelance, P. Guillemot, M. Laas-Bourez and J.-M. Torre, *Metrologia* **52** (2014) 423.
29. A. Debaisieux, J. P. Aubry, E. Gerard and M. Brunet, A satellite oscillator for very precise orbitography: The DORIS Program, in *IEEE Proc. of 39th Annual Symp. on Frequency Control* (IEEE, 1985), pp. 202–211.
30. A. Auriol and C. Tourain, *Adv. Space Res.* **46** (2010) 1484.
31. J. W. Conklin, N. Barnwell, L. Caro, M. Carrascilla, O. Formoso, S. Nydam and P. Serra, Optical time transfer for future disaggregated small satellite navigation systems, in *Proc. of 28th Annual AIAA/USU Conf. on Small Satellites* (Utah State University, 2014).
32. C. M. Will, *Phys. Rev. D* **45** (1992) 403.
33. P. Wolf and G. Petit, *Phys. Rev. A* **56** (1997) 4405.
34. L. Cacciapuoti *et al.*, Atomic clock ensemble in space: Scientific objectives and mission status, in *Proc. of the Third Int. Conf. on Particle and Fundamental Physics in Space* Beijing, Vol. 166 (Elsevier, 2007), pp. 303–306.

35. C. Veillet, ESA proposal for M3 mission Séminaire de prospective du CNES, CNES, St. Malo (1993).
36. W.-T. Ni, A. M. Wu and J. T. Shy, A mission concept to measure second-order relativistic effect, solar angular momentum and low-frequency gravitational waves, in *Proc. of the Seventh Marcel Grossmann Meeting on General Relativity* (Stanford, California, 1994).
37. J. J. Degnan, *AGU Geodyn. Ser.* **25** (1993) 133.
38. M. R. Pearlman, J. J. Degnan and J. M. Bosworth, *Adv. Space Res.* **30** (2002) 135.
39. V. Candelier *et al.*, Recent progress in ultra stable oscillators for space onboard and ground applications, in *Proc. of the 15th EFTF* (BIPM, 2001).
40. E. Samain, *Appl. Opt.* **37** (1998) 502.
41. P. Exertier, E. Samain, N. Martin, C. Courde, M. Laas-Bourez and C. Foussard, *Adv. Space Res.* **54** (2014) 2371.
42. E. Samain, P. Vrancken, P. Guillemot, P. Fridelance and P. Exertier, *Metrologia* **51** (2014) 503.
43. P. Exertier *et al.*, T2L2 time transfer bf laser link (2014), http://www.geoazur.fr/t2l2/en/data/v4.
44. E. Samain, P. Exertier, P. Guillemot, F. Pierron, D. Albanese, J. Paris, J. M. Torre and S. Leon, Time transfer by laser link-T2L2: Current status of the validation program, in *Proc. of the 24th EFTF* (IEEE, 2010), pp. 1–8.
45. E. Samain, M. Lass-Bourez, C. Courde, P. Exertier, J. M. Torre, N. Martin, J. L. Oneto, M. Aimar, F. Pierron, P. Guillemot and S. Leon, T2L2: Ground to ground time transfer, in *Proc. of the 26th EFTF* (OP-SYRTE, 2012), pp. 36–40.
46. P. Guillemot, P. Exertier, E. Samain, F. Pierron, P. Laurent, M. Abgrall, J. Achkar, D. Rovera, K. Djeroud and S. Leon, in *Proc. of 42nd Annual Precise Time and Time Interval (PTTI)* Meeting (ION, 2010), pp. 397–412.
47. G. D. Rovera, M. Abgrall, G. Appleby, C. Courde, P. Exertier, P. Fridelance, Ph. Guillemot, M. Laas-Bourez, S. Leon, N. Martin, E. Samain, R. Sherwood, J.-M. Torre and P. Uhrich, A direct comparison between two independently calibrated time transfer techniques: T2L2 and GPS common views, in *Proc. CPEM*, Rio de Janeiro (OP-SYRTE, 2014).
48. E. Samain, M. Laas-Bourez, C. Courde, P. Exertier, N. Martin, J.-M. Torre, G. D. Rovera, M. Abgrall, P. Uhrich, Ph. Guillemot, R. Sherwood, G. Appleby and P. Fridelance, A sub-ns comparison between GPS common view and T2L2, in *Proc. of 28th EFTF* (OP-SYRTE, 2014).
49. P. Uhrich and D. Valat, GPS receiver relative calibration campaign preparation for Galileo in-orbit validation, in *Proc. of 24th EFTF* (IEEE, 2010).
50. G. D. Rovera, J.-M Torre, R. Sherwood, M. Abgrall, C. Courde, M. Laas-Bourez and P. Uhrich, *Metrologia* **51** (2014) 476.
51. D. Mao, X. Sun, J. McGarry, M. Torrence, E. Mazarico, G. Neumann, D. Smith and M. Zuber, One-way laser ranging to LRO, in *Proc. of Frascati ILRS Workshop*, Frascati, Italy (INFN-LNF, 2012).
52. X. Sun *et al.*, *Opt. Express* **21** (2013) 1865.
53. X. Sun, D. R. Skillman, J. McGarry, G. Neumann, D. Mao, M. Torrence and E. Hoffman, Time transfer between satellite laser ranging stations via simultaneous laser ranging to the lunar reconnaissance orbiter, in *Proc. of 19th Int. Laser Ranging Workshop*, Maryland (NASA, 2014).
54. E. Samain *et al.*, *Astron. Astrophys. Suppl. Ser.* **130** (1998) 235.
55. S. G. Turyshev *et al.*, *Class. Quantum Grav.* **21** (2004) 2773.
56. S. G. Turyshev *et al.*, *Exp. Astron.* **27** (2009) 27.

57. B. Altschul *et al.*, *Adv. Space Res.* **55** (2014) 501.
58. STE-QUEST Study Team, STE-QUEST payload definition document, SRE-PA/2011-075/TN/PW, European Space Agency (2012).
59. W.-T. Ni, *Int. J. Mod. Phys. D* **22** (2013) 1341004.
60. C. Braxmaier *et al.*, *Exp. Astron.* **34** (2012) 181.
61. P. Wolf *et al.*, *Exp. Astron.* **23** (2009) 651.
62. K. Djerroud, O. Acef, A. Clairon, P. Lemonde, C. N. Man, E. Samain and P. Wolf, *Opt. Lett.* **35** (2010) 1479.
63. N. Chiodo, K. Djerroud, O. Acef, A. Clairon and P. Wolf, *Appl. Opt.* **52** (2013) 7342.
64. B. Christophe *et al.*, *Exp. Astron.* **34** (2012) 203.
65. E. Samain, P. Bonnefond and J. Nicolas, One way laser ranging on the solar system — The TIPO project on Mars, in *Evolving Space Geodesy Techniques*, eds. R. Weber, W. Schlüter, U. Schreiber and O. Titov (Vienna University of Technology, 2004), p. 80.
66. J. Delporte, M. Brunet and T. Tournier, Complete evaluation of a Perkin Elmer RAFS in the Galileo context, in *Proc of 14th EFTFS*, Torino, Italy (Istituto Elettrotecnico Nazionale Galileo Ferraris, 2000).
67. P. Fridelance, SigmaTime (2015), http://sigmatime.fr.
68. S. M. Foreman, K. W. Holman, D. D. Hudson, D. J. Jones and J. Ye, *Rev. Sci. Instrum.* **78** (2007) 021101.
69. A. Bercy, F. Stefani, O. Lopez, C. Chardonet, P. E. Pottie and A. Amy-Klein, *Phys. Rev. A* **90** (2014) 061802(R).
70. D. H. Phung, E. Samain, N. Maurice, H. Mariey, C. Courde, G. Artaud and J. L. Issler, DOMINO — Laser Communication between SOTA, onboard SOCRATES satellite, and MEO Optical Ground, in *Proc. of the 19th Int. Workshop on Laser Ranging* (ILRS, 2014).
71. E. Samain, J. M. Torre, D. Albanese, Ph. Guillemot, F. Para, J. Paris, I. Petitbon, P. Vrancken and J. Weick, OCA Event timer, in *Proc. of the 15th Int. Workshop on Laser Ranging*, Canberra, Australia (NASA, 2006).
72. E. Samain, P. Fridelance and P. Guillemot, An ultrastable event timer designed for T2L2, in *Proc. of the 24th EFTF* (IEEE, 2010).
73. P. Panek, J. Kodet and I. Prochazka, *Metrologia* **50** (2013) 60.
74. M. Lintz, E. Samain and S. Pitois, Blanc (BLANC) 2007, Projet ILIADE, ANR-07-BLAN-0309, Agence Nationale de la Recherche, Paris (2007).

Chapter 8

Solar-system tests of the relativistic gravity

Wei-Tou Ni

School of Optical-Electrical and Computer Engineering,
University of Shanghai for Science and Technology,
516, Jun Gong Rd., Shanghai 200093, P. R. China

weitou@gmail.com

In 1859, Le Verrier discovered the Mercury perihelion advance anomaly. This anomaly turned out to be the first relativistic gravity effect observed. During the 157 years to 2016, the precisions and accuracies of laboratory and space experiments, and of astrophysical and cosmological observations on relativistic gravity have been improved by 3–4 orders of magnitude. The improvements have been mainly from optical observations at first followed by radio observations. The achievements for the past 50 years are from radio Doppler tracking and radio ranging together with Lunar Laser Ranging (LLR). At present, the radio observations and LLR experiments are similar in the accuracy of testing relativistic gravity. We review and summarize the present status of solar system tests of relativistic gravity. With planetary laser ranging, spacecraft laser ranging and interferometric laser ranging (laser Doppler ranging) together with the development of drag-free technology, the optical observations will improve the accuracies by another 3–4 orders of magnitude in both the equivalence principle tests and solar system dynamics tests of relativistic gravity. Clock tests and atomic interferometry tests of relativistic gravity will reach an ever-increasing precision. These will give crucial clues in both experimental and theoretical aspects of gravity, and may lead to answers to some profound issues in gravity and cosmology.

Keywords: General relativity; experimental tests of relativistic gravity; solar system dynamics; ephemerides.

PACS Number(s): 04.80.Cc, 04.80.−y, 95.10.−a

1. Introduction and Summary

The development of gravity theory stems from the experiments. Newton's theory of gravity[1] is empirically based on Kepler's laws[2] (which are based on Brahe's observations) and Galileo's law of free-falls[3] (which is based on Galileo's experiment of motions on inclined planes). Towards the middle of the 19th century, astronomical observations accumulated a precision which enabled Le Verrier[4] to discover the Mercury perihelion advance anomaly in 1859. This anomaly is the first relativistic gravity effect observed. Michelson–Morley experiment,[5] via various developments,[6] prompted the final establishment of the special relativity theory in 1905.[7,8] Motivation for putting electromagnetism and gravity into the same theoretical framework,[7] the precision of Eötvös experiment[9] on the equivalence, the formulation of Einstein Equivalence Principle (EEP)[10] together with the perihelion advance anomaly led to the road for the final genesis of General Relativity (GR) theory[11–15] in 1915.

As we discussed in Refs. 16 and 17, Einstein proposed the mass-energy equivalence using the formula $E = mc^2$ in 1905[18]; Planck reasoned that all energy must gravitate in 1907.[19] To characterize the strength of a gravitational source, it would then be natural to compare magnitude of the gravitational energy $mU(\boldsymbol{x}, t)$ of a test particle in the gravitational potential $U(\boldsymbol{x}, t)$ of a gravitating source to the total mass-energy mc^2 of the test particle and define this ratio $\xi(\boldsymbol{x}, t)$ as the dimensionless gravitational strength of the source at a spacetime point $\wp$ [with coordinates $(\boldsymbol{x}, t)$]:

$$\xi(\boldsymbol{x}, t) = \frac{U(\boldsymbol{x}, t)}{c^2}. \tag{1}$$

GR gives strong-field corrections to the Newtonian gravity. The first-order correction is proportional to this strength $\xi(\boldsymbol{x}, t)$.

For a point source with mass M in Newtonian gravity,

$$\xi(\boldsymbol{x}, t) = \frac{GM}{Rc^2}, \tag{2}$$

where R is the distance to the source. For a nearly Newtonian system, we can use Newtonian potential for U. The strength of gravity for various configurations is tabulated in Table 1.

From Table 1, it is clear that in the solar system, Mercury has the largest solar system gravitational potential among all planets and satellites, and hence the largest general-relativistic solar system gravitational correction. This is why the general-relativistic deviation of the Mercury orbit from Newtonian theory — the Mercury perihelion advance anomaly of about 40″ per century was first observed. When the observations reached an accuracy of the order of 1″ per century (transit observa-

Table 1. The strength of gravity for various configurations.

Source	Field position	Strength of gravity ξ
Sun	Solar surface	2.1×10^{-6}
Sun	Mercury orbit	2.5×10^{-8}
Sun	Earth orbit	1.0×10^{-8}
Sun	Jupiter orbit	1.9×10^{-9}
Earth	Earth surface	0.7×10^{-9}
Earth	Moon's orbit	1.2×10^{-11}
Galaxy	Solar system	10^{-5}–10^{-6}
Significant part of observed universe	Our galaxy	1–10^{-2}

tions) in the 19th century, a discrepancy from Newtonian gravity would be seen. In a century, Mercury orbits around the Sun 400 times, amounting to a total angle of 5×10^8 arc sec. The fractional relativistic correction (perihelion advance anomaly) of Mercury's orbit is of order $\lambda GM_{\text{Sun}}/dc^2$ with $\lambda = 3$ for GR (i.e. about 8×10^{-8}) and d the distance of Mercury to the Sun. Therefore, the relativistic correction for perihelion advance is about 40 arc sec per century. As the orbit determination of Mercury reached an accuracy better than 10^{-8} (about 1 arc sec for solar transit observations in 100 years), the relativistic corrections to Newtonian gravity became manifest. Le Verrier discovered this perihelion advance anomaly (anomaly to Newton's theory) and measured it to be 38 arc sec per century.[4] In 1881, Newcomb obtained a more precise value (43 arcsecond per century) of Mercury perihelion advance anomaly.[20]

In 1907, Einstein proposed his equivalence principle and derived the gravitational redshift[10]; in 1911, Einstein derived the light deflection in the solar gravitational field.[21,a] In 1913, Besso and Einstein[27] worked out a Mercury perihelion advance formula in the "Einstein–Grossmann Entwert" theory,[28] but the calculation contained an error and did not agree with the experimental value. During the final genesis of GR,[11–15] Einstein[13] corrected their 1913 error and obtained a Mercury perihelion advance value in agreement with the observation.[20] Apparently, this correct calculation played a significant role in the final genesis of GR.

Gravitational redshift, gravitational light deflection and relativistic perihelion advance are called three classical tests of GR in Einstein's "The Foundation of the General Theory of Relativity."[29]

Towards the end of the 19th century, there were studies whether the solar spectra were displaced from the Doppler spectra.[30,31] Various causes (such as pressure effect, pole effect, asymmetrical broadening) were found and investigated before 1910.[32–35] In 1911, Einstein[21] noticed the work of Buisson and Fabry,[34] and explicitly proposed that gravitational redshift might be tested by the examination of the solar spectra. From 1914 to 1919, re-analysis of previous solar spectra together with a number of new measurements were made. However, the outcome is controversial and inconclusive. Earman and Glymour[36] gave a detailed account of this history.

Before Einstein's proposal of relativistic solar deflection of light in 1911, there were photographs taken for studying the solar corona and to find a sub-Mercurial planet of solar neighborhood during total eclipses. These photographs were considered unsatisfactory to study the deflection of light by Perrine upon a question

[a]Newton in his Opticks[22] of 1704 proposed the following query for further research: "Do not Bodies act upon Light at a distance, and by their action bend its Rays, and is not this action strongest at the least distance?" In 1801, Soldner[23] derived the gravitational bending of light from corpuscular nature of light and Newton's universal gravitation. Soldner[23,b] calculated the deflection angle for light grazing the Sun to be 0.84 arcsec remarkably close to Einstein's 1911 value of 0.83 arcsec.[21] Cavendish's work on the gravitational bending of light (probably around 1784[25]) was published posthumously in 1921.[26]

[b]For an English translation and a historical discussion of Soldner's paper [23]. See Ref. 24.

from Freundlich late 1911 either because of small field and brief exposure time, or because of eccentric position of the Sun on the plates (see e.g. p. 61 of Ref. 37). Before 1919, there were four expeditions intent to measure the gravitational deflection of starlight (in 1912, 1914, 1916 and 1918); because of bad weather or war, the first three expeditions failed to obtain any results, the results of 1918 expedition was never published.[37] In 1919, the observation of gravitational deflection of light passing near the Sun during a solar eclipse[38] confirmed the relativistic deflection of light and made GR famous and popular.

The success of Pound and Rebka[39] in using Mössbauer effect to verify the gravitational redshift in earth-bound laboratory in 1960 marked the beginning of a new era for testing relativistic gravity. At the same time, a careful and more precise test of the equivalence principle was performed in Princeton.[40] With the development of technology and advent of space era, Shapiro[41] proposed a fourth test — the time delay of radar echoes in gravitational field. Since the beginning of this era, we have seen 3–4 orders of improvements for the three classical tests together with many new tests. The current technological development is ripe that we are now in a position to discern another 3–4 orders of improvements further in testing relativistic gravity in the coming 25 years (2016–2040). This will enable us to test the second-order relativistic gravity effects. A road map of experimental progress in gravity together with its theoretical implication has been shown in Table 2 of Ref. 42.

The present review updates the solar system test part of a previous review on "Empirical Foundations of the Relativistic Gravity"[42] (which is a five-year update of the 1999–2000 review[43]). A companion review on equivalence principles and the foundation of metric theories of gravity has been already given in Ref. 17. Recently Manchester[44] has reviewed the pulsar tests of relativistic gravity. A previous review on the solar system tests of relativistic gravity is from Reynaud and Jaekel.[45] A good general review on experimental tests of GR is from Will.[46]

In Sec. 2, we review the post-Newtonian approximation of GR, the Parametrized Post-Newtonian (PPN) framework, and derive the Shapiro time delay and the first-order relativistic light deflection as examples. In Sec. 3, we review and discuss the solar system ephemerides. In Sec. 4, we update the solar system tests since our last review in 2005. In Sec. 5, we discuss ongoing and next generation solar system experiments related to testing relativistic gravity with an outlook.

2. Post-Newtonian Approximation, PPN Framework, Shapiro Time Delay and Light Deflection

The equations of motion of GR, i.e. the Einstein equation is

$$G_{\mu\nu} = \kappa T_{\mu\nu}, \tag{3}$$

where $T_{\mu\nu}$ is the stress–energy tensor and $\kappa = 8\pi G/c^4$ (see e.g. Ref. 47). We use the MTW[47] conventions with signature -2; This is also the conventions used in Refs. 16, 17; Greek indices run from 0 to 3; Latin indices run from 1 to 3; the cosmological constant is negligible for solar system dynamics and solar system ephemeris, and is neglected in this review. Contracting the equations of motion (3), we have

$$R = -\left(\frac{8\pi}{c^4}\right) GT, \tag{4}$$

where $T \equiv T^{\mu}_{\mu}$. Substituting (4) into (3), we obtain the following equivalent equations of motion as originally proposed by Einstein[15]

$$R_{\mu\nu} = \left(\frac{8\pi}{c^4}\right)\left[T_{\mu\nu} - \left(\frac{1}{2}\right)(g_{\mu\nu}T)\right]. \tag{5}$$

For weak field in the quasi-Minkowskian coordinates, we express the metric $g_{\alpha\beta}$ as

$$g_{\alpha\beta} = \eta_{\alpha\beta} + h_{\alpha\beta}, \quad h_{\alpha\beta} \ll 1. \tag{6}$$

Since $h_{\alpha\beta}$ is a small quantity ($< 4\times 10^{-6}$) in the solar system gravitational field, we expand everything in $h_{\alpha\beta}$ and linearize the results to obtain the linear (weak-field) approximation. With the harmonic gauge (coordinate) condition for $h_{\alpha\beta}$,

$$\left[h_{\alpha\beta} - \left(\frac{1}{2}\right)\eta_{\alpha\beta}(\mathrm{Tr}\, h)\right]_{,}{}^{\beta} = 0 + \mathrm{O}(h^2), \quad \text{i.e. } h_{\alpha\beta,}{}^{\beta} = \left(\frac{1}{2}\right)(\mathrm{Tr}\, h)_{,\alpha} + \mathrm{O}(h^2), \tag{7}$$

the linearized Einstein equation is:

$$h_{\mu\nu,\beta}{}^{\beta} = -\left(\frac{16\pi G}{c^4}\right)\left[T_{\mu\nu} - \left(\frac{1}{2}\right)(\eta_{\mu\nu}T)\right] + \mathrm{O}(h^2), \tag{8}$$

where $\mathrm{Tr}(h)$ is defined as the trace of $h_{\alpha}{}^{\beta}$, i.e. $\mathrm{Tr}(h) \equiv h_{\alpha}{}^{\alpha}$, and $\mathrm{O}(h^2)$ denotes terms of order of $h_{\alpha\beta}h_{\mu\nu}$ or smaller (see e.g. Refs. 47 and 48). Analogous to classical electrodynamics, the solution of this equation for GR is

$$h_{\mu\nu} = -\left[\frac{(4G)}{(c^4)}\right]\int \left\{\frac{\left[T_{\mu\nu} - \left(\frac{1}{2}\right)g_{\mu\nu}T\right]}{r}\right\}_{\text{retarded}} (d^3x') + \mathrm{O}(h^2). \tag{9}$$

2.1. *Post-Newtonian approximation*

For solar dynamics and solar system ephemerides, we can impose slow motion condition, in addition to weak field condition, i.e.

$$\frac{U}{c^2} = \mathrm{O}\left(\frac{\nu^2}{c^2}\right); \quad \frac{U_{,ij}}{c^2} = \left(\frac{1}{L^2}\right)\mathrm{O}\left(\frac{\nu^2}{c^2}\right); \quad \frac{U_{,0i}}{c^2} = \left(\frac{1}{L^2}\right)\mathrm{O}\left(\frac{\nu^3}{c^3}\right);$$

$$\frac{U_{,00}}{c^2} = \left(\frac{1}{L^2}\right)\mathrm{O}\left(\frac{\nu^4}{c^4}\right), \tag{10}$$

where L is a typical length scale and ν is a typical velocity of the system (see e.g. Ref. 47). The solution $h_{\mu\nu}$ of (9) and $h_\alpha{}^\alpha$ in this approximation then becomes

$$h_{\mu\nu} = -2\left(\frac{U}{c^2}\right)\delta_{\mu\nu} + \mathrm{O}\left(\frac{\nu^3}{c^3}\right); \quad h \equiv h_\mu{}^\mu = 4\left(\frac{U}{c^2}\right) + \mathrm{O}\left(\frac{\nu^3}{c^3}\right), \tag{11}$$

where U is the Newtonian potential which normally contains multipole terms outside a gravitating body. For point mass or outside spherical Sun,

$$U = \left(\frac{GM}{c^2}\right)\left(\frac{1}{r}\right); \quad r = (x^2 + y^2 + z^2)^{1/2}. \tag{12}$$

With the metric (11), one can already derive the solar deflection of light and the Shapiro time delay. For a derivation of relativistic precession of Mercury's orbit, one needs a full post-Newtonian approximation of GR and needs to calculate h_{00} to $\mathrm{O}(\nu^4/c^4)$ order and h_{0i} to $\mathrm{O}(\nu^3/c^3)$ order. The post-Newtonian approximation for perfect fluid in GR is obtained by Chandrasekhar.[49] The metric $g_{\alpha\beta}(= \eta_{\alpha\beta} + h_{\alpha\beta})$ is given by

$$\begin{aligned} g_{00} &= 1 - 2\frac{U}{c^2} + 2\frac{U^2}{c^4} + 4\Psi + \mathrm{O}\left(\frac{\nu^5}{c^5}\right), \\ g_{0i} &= \left(\frac{7}{2}\right)V_i + \left(\frac{1}{2}\right)W_i + \mathrm{O}\left(\frac{\nu^5}{c^5}\right), \\ g_{ij} &= -\left(1 + 2\frac{U}{c^2}\right)\delta_{ij} + \mathrm{O}\left(\frac{\nu^4}{c^4}\right), \end{aligned} \tag{13}$$

where

$$U(\boldsymbol{x}, t) = \int \left[\frac{\rho_0(\boldsymbol{x}', t)}{|\boldsymbol{x} - \boldsymbol{x}'|}\right] dx', \tag{14}$$

$$\Psi(\boldsymbol{x}, t) = \frac{1}{c^4}\int \left[\frac{\rho_0(\boldsymbol{x}', t)\psi(\boldsymbol{x}', t)}{|\boldsymbol{x} - \boldsymbol{x}'|}\right] dx', \tag{15}$$

$$\psi = \boldsymbol{\nu}^2 + U + \left(\frac{1}{2}\right)\Pi + \left(\frac{3}{2}\right)\frac{p}{\rho_0}, \tag{15a}$$

$$V_i(\boldsymbol{x}, t) = \frac{1}{c^3}\int \left[\frac{\rho_0(\boldsymbol{x}', t)\nu_i(\boldsymbol{x}', t)}{|\boldsymbol{x} - \boldsymbol{x}'|}\right] dx', \tag{16}$$

$$W_i(\boldsymbol{x}, t) = \frac{1}{c^3}\int \left\{\frac{\rho_0(\boldsymbol{x}', t)[(\boldsymbol{x} - \boldsymbol{x}') \cdot \boldsymbol{\nu}(\boldsymbol{x}', t)](\boldsymbol{x} - \boldsymbol{x}')_i}{|\boldsymbol{x} - \boldsymbol{x}'|^3}\right\} dx', \tag{17}$$

with $\rho_0(\boldsymbol{x}, t)$ the rest mass density, $\boldsymbol{\nu}(\boldsymbol{x}, t)[= (\nu_1, \nu_2, \nu_3)]$ the 3-velocity, $U(\boldsymbol{x}, t)$ the Newtonian potential, $\Pi(\boldsymbol{x}, t)$ the internal energy and $p(\boldsymbol{x}, t)$ the pressure of the fluid.

2.2. *PPN framework*

For different theories of relativistic gravity, the post-Newtonian metrics are different. However, the post-Newtonian metric of many relativistic gravity theories can be encompassed in the PPN framework with nine post-Newtonian parameters β, β_1, β_2, β_3, β_4, γ, ζ, Δ_1 and Δ_2[50–54]:

$$
\begin{aligned}
g_{00} &= 1 - 2\frac{U}{c^2} + 2\beta\frac{U^2}{c^4} - 4\underline{\Psi} + \zeta\mathscr{A} + \mathrm{O}\left(\frac{\nu^5}{c^5}\right), \\
g_{0i} &= \left(\frac{7}{2}\right)\Delta_1 V_i + \left(\frac{1}{2}\right)\Delta_2 W_i + \mathrm{O}\left(\frac{\nu^5}{c^5}\right), \\
g_{ij} &= -\left(1 + 2\gamma\frac{U}{c^2}\right)\delta_{ij} + \mathrm{O}\left(\frac{\nu^4}{c^4}\right),
\end{aligned} \tag{18}
$$

where

$$\underline{\Psi}(\boldsymbol{x}, t) = \frac{1}{c^4}\int\left[\frac{\rho_0(\boldsymbol{x}', t)\underline{\psi}(\boldsymbol{x}', t)}{|\boldsymbol{x} - \boldsymbol{x}'|}\right]dx', \tag{19}$$

$$\underline{\psi} = \beta_1\boldsymbol{\nu}^2 + \beta_2 U + \left(\frac{1}{2}\right)\beta_3\Pi + \left(\frac{3}{2}\right)\beta_4\frac{p}{\rho_0}, \tag{20}$$

$$\mathscr{A}(\boldsymbol{x}, t) = \frac{1}{c^4}\int\left\{\frac{\rho_0(\boldsymbol{x}', t)[(\boldsymbol{x} - \boldsymbol{x}')\cdot\boldsymbol{\nu}(\boldsymbol{x}', t)]^2}{|\boldsymbol{x} - \boldsymbol{x}'|^3}\right\}dx'. \tag{21}$$

Each gravity theory has a specific set of values for these PPN parameters if it can be encompassed in the framework. GR has the PPN parameters $\beta = \gamma = 1, \beta_1 = \beta_2 = \beta_3 = \beta_4 = \Delta_1 = \Delta_2 = 1$, and $\zeta = 0$. Brans–Dicke–Jordan theory has the PPN parameters $\beta = 1, \gamma = (1 + \omega)/(2 + \omega)$, $\beta_1 = (3 + 2\omega)/(4 + 2\omega)$, $\beta_2 = (1+2\omega)/(4+2\omega)$, $\beta_3 = 1$, $\beta_4 = (1+\omega)/(2+\omega)$, $\zeta = 0$, $\Delta_1 = (10+7\omega)/(14+7\omega)$ and $\Delta_2 = 1$ with ω the Brans–Dicke parameter. Brans–Dicke–Jordan theory is a scalar–tensor theory. For general scalar–tensor theories without mass terms, their PPN parameters are $\beta = 1 + \Lambda$, $\gamma = (1+\omega)/(2+\omega)$, $\beta_1 = (3+2\omega)/(4+2\omega)$, $\beta_2 = (1+2\omega)/(4+2\omega) - \Lambda$, $\beta_3 = 1$, $\beta_4 = (1+\omega)/(2+\omega)$, $\zeta = 0$, $\Delta_1 = (10+7\omega)/(14+7\omega)$ and $\Delta_2 = 1$ with Λ a second parameter in addition to ω. Both GR and scalar–tensor theories are conservative and nonpreferred-frame theories. For them, it would be more convenient to re-define the following linear combinations of parameters as the new PPN parameters[55,56]:

$$
\begin{aligned}
\alpha_1 &\equiv 7\Delta_1 + \Delta_2 - 4\gamma - 4, \\
\alpha_2 &\equiv \Delta_2 + \zeta - 1, \\
\alpha_3 &\equiv 4\beta_1 - 2\gamma - 2 - \zeta, \\
\zeta_1 &\equiv \zeta,
\end{aligned}
$$

$$\zeta_2 \equiv 2\beta + 2\beta_2 - 3\gamma - 1,$$
$$\zeta_3 \equiv \beta_3 - 1,$$
$$\zeta_4 \equiv \beta_4 - \gamma. \tag{22}$$

In terms of nine parameters β, γ, α_1, α_2, α_3, ζ_1, ζ_2, ζ_3, and ζ_4, the only parameters which are not vanishing for GR and for scalar–tensor theories are the two parameters β and γ. Indeed, Will and Nordtvedt[55,56] showed that a theory of gravity which can be encompassed in the PPN framework at the post-Newtonian level and which possesses all 10 global conservation laws (four for energy–momentum and six for angular momentum) if and only if

$$\zeta_1 = \zeta_2 = \zeta_3 = \zeta_4 = \alpha_3 = 0. \tag{23}$$

α_1, α_2, and α_3 measure the extent and nature of preferred-frame effects[53,56,46]; any gravity theory with at least one of $\alpha_i's$ nonzero is called a preferred-frame theory. In the PPN framework (18), conservative nonpreferred-frame theories can have only two independent parameters β and γ. General scalar–tensor theories without mass term span the whole class of conservative theories fitted in the PPN framework.

Empirically, the preferred-frame and nonconservative parameters α_1, α_2, α_3, ζ_1, ζ_2, and ζ_3 are constrained as follows:

$|\alpha_1| < 3.4 \times 10^{-5}$(limit[57] from the orbit dynamics of the binary pulsar PSR J1738 + 0333[58]),

$|\alpha_2| < 1.6 \times 10^{-9}$(limit from millisecond pulsars PSR B1937 +21 and PSR J1744–1134[59]),

$|\alpha_3| < 4.0 \times 10^{-20}$(limit from the orbital dynamics of the statistical combination of a set of binary pulsars[60]),

$|\zeta_1| < 1.5 \times 10^{-3}$(limit calculated from the constraints on the Nordtvedt parameter $\eta[= 4\beta - \gamma - 3 - \alpha_1 + (2/3)\alpha_2 - (2/3)\zeta_1 - (1/3)\zeta_2]$ and other parameters (Table 2)), (24)

$|\zeta_2| < 4 \times 10^{-5}$(limit from binary pulsar PSR 1913+16 acceleration[61]),

$|\zeta_3| < 1.5 \times 10^{-3}$(limit from confirmation of Newton's third law by lunar acceleration[62–64]),

As to ζ_4, according to Will,[46,65] there is a theoretical relation $6\zeta_4 = 3\alpha_3 + 2\zeta_1 - 3\zeta_3$ for gravity theories whose perfect-fluid equations are blind to different forms of internal energy and pressure in the fluid so that ζ_4 becomes redundant.

Although PPN framework (18) encompasses a large class of gravity theories, there are still many gravity theories outside its scope. One notable example is Whitehead theory as completed by imposing the EEP. Its post-Newtonian metric contains additional terms which has to be parametrized by an additional parameter

ξ (or ξ_W) called Whitehead parameter. These additional terms with parameter ξ can be included in an extended PPN framework.[55,56] For Whitehead theory, $\xi = 1$ (by definition). Solar system tests and constraints on the Whitehead terms have been studied in Refs. 66–69 with $|\xi|$ constrained to order of 10^{-3}. The constraint from millisecond pulsars[70] gives $|\xi| < 3.9 \times 10^{-9}$. Also, the PPN framework (18) does not contain the intermediate-range gravity terms (Yukawa terms). These terms can be included in a separate treatment. Misner *et al.*[47] have treated the case with the anisotropic stresses. For this case, there is a post-Newtonian term in g_{00} with an extra parameter in addition to parameter β_4. However, the anisotropic stresses are much smaller than the isotropic stresses or pressures in the solar system. For solar system dynamics consideration, they are negligible up to now.

Historically, Eddington[71] first used the parametrization of metric for discussing the classical tests of relativistic gravity based on the isotropic post-Newtonian expansion of the Schwarzschild metric with the following line element:

$$ds^2 = \left[1 - 2\alpha\left(\frac{GM}{r}\right) + 2\beta\left(\frac{GM}{r}\right)^2 + \cdots\right] dt^2 - \left[1 + 2\gamma\left(\frac{GM}{r}\right) + \cdots\right](dr^2 + r^2 d\theta^2 + r^2 \sin^2\theta \, d\varphi^2), \tag{25}$$

where α, β, γ are called the Eddington parameters. For metric theories, EEP is assumed already. The Eddington parameters should not depend on the mass-energy contents. To have the correct Newtonian limit, α can be absorbed into a re-definition of Newtonian gravitational constant G (i.e. $\alpha G \to G$). Hence as a parametrized post-Newtonian framework, there are only two effective Eddington parameters β and γ. In 1968, Nordtvedt[72] developed the first modern version of PPN framework for a system of two gravitating point masses with later generalization to more particles. It contains seven parameters in addition to α. In 1971, Will[52] extended the PPN framework to perfect fluid with two additional parameters β_3 and β_4 (or ζ_3 and ζ_4 in the new combination of parameters) for terms on internal energy and pressure. This framework does not contain the parameter α. As we noticed in the comment after (24), ζ_4 (or β_4) would be a redundant parameter if EEP is assumed. So is ζ_3 (or β_3). They are really parameters for testing EEP. As we discussed at the beginning of this paragraph, α parameter tests EEP too. In a metric framework like PPN framework (18), seven parameters can be explored. This means the 1968-Nordtvedt and 1971-Will framework are effectively equivalent.

Moreover, the preferred-frame parameters and the conservation-law parameters α_1, α_2, ζ_1, ζ_2, and α_3 are essentially test parameters for Strong Equivalence Principles (SEP). They are constrained to observe SEP quite well [see e.g. (24)]. In rest of this paper, we concentrate mainly on the experiments to test the two parameters β and γ. In the next two sections, we illustrate the PPN effects with two simple calculations of Shapiro time delay and the gravitational deflection of light passing near the Sun.

2.3. *Shapiro time delay*

One can derive the light propagation equation in the weak field limit for the physical metric (6). Let $\underline{\mathbf{r}} = \underline{\mathbf{r}}(t)$ be the light trajectory where $\underline{\mathbf{r}}(t) = (x(t), y(t), z(t))$ is a 3-vector. Light propagation follows null geodesics of the metric $g_{\alpha\beta}$; its trajectory $\underline{\mathbf{r}}(t)$ satisfies

$$0 = ds^2 = g_{\alpha\beta}dx^\alpha dx^\beta = (1 + h_{00})c^2dt^2 + 2h_{0i}cdx^idt + (\eta_{ij} + h_{ij})dx^idx^j, \quad (26)$$

using (6).

In the Minkowski approximation, the light trajectory can be approximated by

$$\frac{dx^i}{dt} = \left(\frac{dx^i}{dt}\right)^{(0)i} + \mathrm{O}(h) = cn^{(0)i} + \mathrm{O}(h), \quad \text{with} \sum_i (n^{(0)i})^2 = 1, \qquad (27)$$

where $n^{(0)i}$ are constants. In the Post-Minkowski approximation, we express dx^i/dt as

$$\frac{dx^i}{dt} = cn^{(0)i} + cn^{(1)i} + \mathrm{O}(h^2), \qquad (28)$$

where $n^{(1)i}$ is a function of trajectory and of the order of $\mathrm{O}(h)$. Substituting (28) into (26) and dividing by dt^2, we have

$$0 = (1 + h_{00})c^2 + 2h_{0i}c\left(\frac{dx^i}{dt}\right) + \left|\frac{d\underline{\mathbf{r}}}{dt}\right|^2 - h_{ij}\left[\left(\frac{dx^i}{dt}\right)\left(\frac{dx^j}{dt}\right)\right] \qquad (29a)$$

$$= (1 + h_{00})c^2 + 2h_{0i}c(cn^{(0)i} + cn^{(1)i}) - c^2\sum_{i=1}^{3}(n^{(0)i} + n^{(1)^i})^2$$

$$+ h_{ij}n^{(0)i}n^{(0)j}c^2 + \mathrm{O}(h^2). \qquad (29b)$$

Simplifying (29b), we have

$$\sum_{i=1}^{3} n^{(0)i}n^{(1)i} = \left(\frac{1}{2}\right)(h_{00} + 2h_{0i}n^{(0)i} + h_{ij}n^{(0)i}n^{(0)j}) + \mathrm{O}(h^2), \qquad (30)$$

and solving for $|d\underline{\mathbf{r}}/dt|$ in (29a), we obtain the light propagation equation to $\mathrm{O}(h)$ order:

$$\left|\frac{d\underline{\mathbf{r}}}{dt}\right| = c[(1 + h_{00} + 2h_{0i}n^{(0)i} + h_{ij}n^{(0)i}n^{(0)j}) + \mathrm{O}(h^2)]^{1/2}$$

$$= c\left[1 + \left(\frac{1}{2}\right)h_{00} + h_{0i}n^{(0)i} + \left(\frac{1}{2}\right)h_{ij}n^{(0)i}n^{(0)j} + \mathrm{O}(h^2)\right]. \qquad (31)$$

From (31), we calculate the light travel time Δt_{TT} between two observers (time delay)[41] as

$$\Delta t_{\mathrm{TT}} = \left(\frac{1}{c}\right)\int |d\underline{\mathbf{r}}|\left[1 - \left(\frac{1}{2}\right)h_{00} - h_{0i}n^{(0)i} - \left(\frac{1}{2}\right)h_{ij}n^{(0)i}n^{(0)j} + \mathrm{O}(h^2)\right]. \quad (32)$$

Choosing the z-axis along the initial light propagation direction, i.e. $n^{(0)i} = (0, 0, 1)$, and using (18) or (25) for a slow-motion observer and Sun (or a central mass), we have

$$\Delta t_{\mathrm{TT}} = \int dt = \left(\frac{1}{c}\right) \int dz[1 + (1+\gamma)U + \mathrm{O}(h^2)] = \Delta t^{\mathrm{N}} + \left[\frac{(1+\gamma)}{2}\right] \Delta t_S^{\mathrm{GR}}$$
$$= \left(\frac{1}{c}\right)(z_2 - z_1) + (1+\gamma)\left(\frac{GM}{c^3}\right)$$
$$\times \ln\left\{\frac{[(z_2^2 + b^2)^{1/2} + z_2]}{[(z_1^2 + b^2)^{1/2} + z_1]}\right\} + \mathrm{O}(h^2), \quad (z_1 < 0, z_2 > 0) \tag{33}$$

where the first term is the Newtonian travel time Δt^{N} (Römer delay), the second term is the relativistic Shapiro time delay[41] with Δt_S^{GR} the general relativistic Shapiro time delay, and b is the impact parameter of light propagation to the Sun.

2.4. *Light deflection*

The geodesic equation for light and for test particle in GR and in the metric theories of gravity

$$d^2 \frac{x^\mu}{d\lambda^2} + \Gamma^\mu_{\sigma\rho}\left(\frac{dx^\sigma}{d\lambda}\right)\left(\frac{dx^\rho}{d\lambda}\right) = 0, \quad \lambda : \text{affine parameter} \tag{34}$$

can be cast in the form

$$d\frac{\left(\dfrac{g_{\mu\nu}dx^\nu}{d\lambda}\right)}{d\lambda} = \left(\frac{1}{2}\right) g_{\sigma\rho},_\mu \left(\frac{dx^\sigma}{d\lambda}\right)\left(\frac{dx^\rho}{d\lambda}\right). \tag{35}$$

Integrating, we obtain

$$\left(\frac{g_{\mu\nu}dx^\nu}{d\lambda}\right)\Bigg|_{x0}^{x1} = \left(\frac{1}{2}\right)\int_{x0}^{x1}\left[g_{\sigma\rho},_\mu \left(\frac{dx^\sigma}{d\lambda}\right)\left(\frac{dx^\rho}{d\lambda}\right)\right] d\lambda. \tag{36}$$

To obtain light deflection angle in a weak gravitational field of the Sun or other source, we choose x-axis in the initial light (photon) propagation direction, y-axis in the plane spanned by the Sun or other gravitational source and the light trajectory, and the sense of the x-axis is in the direction of the trajectory. From the $\mu = y$ component of (36), we obtain

$$\left(g_{0y} + g_{xy} - \left(\frac{1}{c}\right)\frac{dy}{dt}\right)\Bigg|_{x0}^{x1} = \left(\frac{1}{2}\right)\int_{x0}^{x1}(h_{00,y} + h_{xx,y} + 2h_{0x,y})cdt + \mathrm{O}(h^2). \tag{37}$$

Solving for dy/dt in (37), substituting (18) or (25) in and simplifying, we obtain

$$\Delta\varphi_{\text{deflection}} = \left(\frac{1}{c}\right)\left(\frac{dy}{dt}\right)\Bigg|_{x0}^{x1} = (g_{0y} + g_{xy})|_{x0}{}^{x1}$$
$$- \left(\frac{1}{2}\right)\int_{x0}^{x1}(h_{00,y} + h_{xx,y} + 2h_{0x,y})cdt + \mathrm{O}(h^2)$$

$$= \left\{ -\frac{\left[\frac{(1+\gamma)}{c^2 b}\right] GMx}{(x^2+b^2)^{1/2}} \right\}\Bigg|_{x=x1} + \left\{ \frac{\left[\frac{(1+\gamma)}{c^2 b}\right] GMx}{(x^2+b^2)^{1/2}} \right\}\Bigg|_{x=x0}$$

$$= -(1+\gamma)\left(\frac{G_N M}{c^2 b}\right)(\cos\theta_1 - \cos\theta_0). \tag{38}$$

for the deflection angle $\Delta\varphi_{\text{deflection}}$, where θ_0 (θ_1) is the angle between the position vector of the light emitter (observer) and x-axis. For star light and close impact, i.e. $b \ll r_1$, we have

$$\Delta\varphi_{\text{deflection}} = -(1+\gamma)\left(\frac{G_N M}{c^2 b}\right)(\cos\theta_1 + 1). \tag{39}$$

If cos $\theta_1 \approx 1$ and $\gamma = 1$, we obtain the usual formula of GR, i.e. $\Delta\varphi_{\text{deflection}} = -4(G_N M/c^2 b)$.

3. Solar System Ephemerides

Planetary ephemerides are a must for precision tests of relativistic gravity in the solar system and for the orbit design of spacecraft and missions. Before the advent of space age, the analytical theories developed by Le Verrier, Hill, Newcomb, and Clemens on planetary motion had sufficient accuracy to account for ongoing optical observations. With the Doppler radio tracking of spacecraft and the radio/laser ranging to planets/Moon, the required accuracy of planetary ephemerides increased tremendously. The accuracy of analytical theories became inadequate. The usage and development of numerical methods started.

Since the motion of planets and the moon are influenced by other planets/moon, to test relativistic gravity, one needs a complete solar system ephemeris. To do this, one would start with the PPN equations of motion in an appropriate gauge for celestial bodies. Because the separation of planets/moon are large compared with their sizes, one could treat the planets/moon as point particles with suitable multipole moments. Such a set of PPN equations of motion is the post-Newtonian barycentric equations of motion as derived in Brumberg[73] from the post-Newtonian barycentric metric with PPN parameters β and γ for solar system bodies. The metric with the gauge parameter α (not to be confused with Eddington parameter α in the last section) and ν set to zero corresponding to a harmonic gauge adopted by the 2000 IAU resolution[74] is

$$ds^2 = \left[1 - 2\sum_i \frac{m_i}{r_i} + 2\beta\left(\sum_i \frac{m_i}{r_i}\right)^2 + (4\beta-2)\sum_i \frac{m_i}{r_i}\sum_{j\neq i}\frac{m_j}{r_{ij}}\right.$$
$$- c^{-2}\sum_i \frac{m_i}{r_i}\left(2(\gamma+1)\dot{\boldsymbol{x}}_i^2 - \boldsymbol{r}_i\cdot\ddot{\boldsymbol{x}}_i - \frac{1}{r_i^2}(\boldsymbol{r}_i\cdot\dot{\boldsymbol{x}}_i)^2\right)$$

$$+\frac{m_1 R_1^2}{r_1^3} J_2 \left(3\left(\frac{\boldsymbol{r}_1\cdot\hat{\mathbf{z}}}{r_1}\right)^2 - 1\right)\Bigg] c^2 dt^2$$

$$+ 2c^{-1}\sum_i \frac{m_i}{r_i}((2\gamma+2)\dot{\boldsymbol{x}}_i)\cdot d\boldsymbol{x} c dt - \left[1 + 2\gamma\sum_i \frac{m_i}{r_i}\right](d\boldsymbol{x})^2 \tag{40}$$

where $\boldsymbol{r}_i = \boldsymbol{x} - \boldsymbol{x}_i, \boldsymbol{r}_{ij} = \boldsymbol{x}_i - \boldsymbol{x}_j, r_i = |\boldsymbol{r}_i|, r_{ij} = |\boldsymbol{x}_i - \boldsymbol{x}_j|, m_i = GM_i/c^2$, and M_i's are the masses of the celestial bodies with M_1 the solar mass.[73] J_2 is the quadrupole moment parameter of the Sun. $\hat{\mathbf{z}}$ is the unit vector in the direction of solar angular momentum. The associated equations of motion of N-mass problem derived from the geodesic variational principle of this metric (the effect of solar quadrupole moment not yet added) are

$$\ddot{\boldsymbol{x}}_i = -\sum_{j\neq i}\frac{GM_j}{\boldsymbol{r}_{ij}^3}\boldsymbol{r}_{ij} + \sum_{j\neq i} m_j(A_{ij}\boldsymbol{r}_{ij} + B_{ij}\dot{\boldsymbol{r}}_{ij}),$$

$$A_{ij} = \frac{\dot{\boldsymbol{x}}_i^2}{\boldsymbol{r}_{ij}^3} - (\gamma+1)\frac{\dot{\boldsymbol{r}}_{ij}^2}{r_{ij}^3} + \frac{3}{2r_{ij}^5}(\boldsymbol{r}_{ij}\dot{\boldsymbol{x}}_j)^2 + G[(2\gamma+2\beta+1)M_i + (2\gamma+2\beta)M_j]\frac{1}{r_{ij}^4}$$

$$+ \sum_{k\neq i,j} GM_k\Bigg[(2\gamma+2\beta)\frac{1}{r_{ij}^3 r_{ik}} + (2\beta-1)\frac{1}{r_{ij}^3 r_{jk}} + \frac{2(\gamma+1)}{r_{ij} r_{jk}^3} \tag{41}$$

$$- \left(2\gamma+\frac{3}{2}\right)\frac{1}{r_{ik} r_{jk}^3} - \frac{1}{2r_{jk}^3}\frac{r_{ij} r_{ik}}{r_{ij}^3}\Bigg],$$

$$B_{ij} = \frac{1}{r_{ij}^3}[(2\gamma+2)(\boldsymbol{r}_{ij}\dot{\boldsymbol{r}}_{ij}) + (\boldsymbol{r}_{ij}\dot{\boldsymbol{x}}_j)].$$

These equations can be used to build a computer-integrated planetary ephemeris framework. For a complete ephemeris, one needs to fit observational data to obtain solar and planetary parameters together with a set of initial conditions at a specific epoch; for a working ephemeris, one could simply adopt the parameters including planetary positions and velocities at some epoch from a complete fundamental ephemeris. For example, in our working CGC 1 ephemeris (CGC: Center for Gravitation and Cosmology),[75] we computer-integrated equations (41) with (40) (setting $\beta = \gamma = 1$ for GR, and $J_2 = 2\times 10^{-7}$ for the Sun) for eight-planets plus Pluto, the Moon, the Sun and the three big asteroids — Ceres, Pallas and Vesta (14-body evolution); the positions and velocities at the epoch 2005.6.10 0:00 are taken from the DE403 ephemeris.[76] The evolution (can go forward or backward in time) is solved by using the fourth-order Runge–Kutta method with the step size $h = 0.01$ day. Since tilt of the axis of the solar quadrupole moment to the perpendicular of the elliptical plane is small (7°), in CGC 1 ephemeris, we have neglected this tilt. In CGC 2 ephemeris,[77] we have added the perturbations of additional 489 asteroids. Such ephemerides can be used for mission orbit design/optimization and mission simulation. We used CGC 1 for orbit simulation and parameter determination for Astrodynamical Space Test of Relativity using Optical Devices (ASTROD). Using this ephemeris as a deterministic model and adding stochastic terms to simulate noise, we generated simulated ranging data and used Kalman filtering to determine the accuracies of fitted relativistic and solar system parameters for 1050 days of the ASTROD mission.[75] This way, we simulated the accuracy achievable for the ASTROD mission concept. For a better evaluation of the accuracy of $\dot{G}/G$, we need also to monitor the masses of other asteroids. For this, we considered all known 492 asteroids with diameter greater than 65 km to obtain an improved ephemeris framework — CGC 2, and calculated the perturbations due to these 492 asteroids on the ASTROD spacecraft.[77] More recently, we apply different variants of CGC 2 ephemeris framework to study the ASTROD I orbit design/optimization/simulation,[78] the ASTROD-GW orbit design/optimization,[79] and the numerical Time Delay Interferometry (TDI) for space gravitational-wave mission concepts ASTROD-GW,[80] LISA[81] and eLISA.[82]

At present, there are three series of complete fundamental ephemerides for the solar system — Development Ephemerides (DE),[83] Intégrateur Numérique Planétaire de l'Observatoire de Paris (INPOP)[84] and Ephemerides of Planets and Moon (EPM).[85] The major common feature of these three series of ephemerides are the simultaneous numerical integration of the equations of motion of the eight planets plus Pluto, the Sun, the Moon, and the lunar physical libration using the post-Newtonian approximation of GR in a harmonic coordinate system. In addition, they take different number of asteroids and Trans-Neptunian objects (TNOs) in the integration of ephemerides.

Let us illustrate with the EPM ephemerides.[85] Specifically, the basic dynamical model of EPM2011 is the post-Newtonian equations of motion of the Sun, the Moon, the eight planets plus Pluto (now a TNO), and five largest asteroids with the following relatively weak gravitational effects taken into account:

(a) perturbations from the known 301 of the most massive asteroids;
(b) perturbations from the other minor planets in the main asteroid belt, modeled by a homogeneous ring;
(c) perturbations from the known 21 largest TNOs;
(d) perturbations from the other trans-Neptunian planets, modeled by a homogeneous ring at a mean distance of 43 AU;
(e) perturbations from the solar oblateness (2×10^{-7}).

The main data set for the three current ephemerides to fit comes from the astrometric observations of planets and spacecraft. For EPM2011,[85] it includes (i) optical observations of the outer planets and their satellites made from 1913 to 2011 (57560 data points); (ii) radar observations of Mercury, Venus, and Mars from 1961 to 1997 (58112 data points); (iii) radio data provided by spacecraft from 1971 to 2010 (561998 data points).

From (41) with (40) or equivalent formulas and observations to obtain a complete ephemeris, one needs to fit for the mass parameters, relevant multipole moments, initial positions and initial velocities of the planets, the Moon and the Sun together with some other solar system bodies like the three largest asteroids — Ceres, Pallas and Vesta. Before 2009, in fitting the data for ephemerides, instead of the mass parameter GM_{Sun} of the Sun, one could use the following relation to fit or adjust the astronomical unit (au):

$$GM_{\text{Sun}}(\text{m}^3\text{s}^{-2}) = k^2\text{au}(\text{m}^3)/86400^2(\text{s}^2) \tag{42}$$

with $k = 0.017\ 202\ 098\ 95$, the Gaussian gravitational constant. The astronomical unit is a basic unit in astronomy and was supposed to be close to the mean Earth distance to the Sun. With the development of ranging observations in the solar system, it could be related precisely to the SI meter through the ephemeris fitting. Standish determined the au to be 149 597 870 697.4 m when worked out DE410 in 2003.[86] Pitjeva[87] determined the au to be 149 597 870 696.0 m when worked out EPM2004. The difference of 1.4 m represents the realistic error in the determination of the au. In 2009, Pitjeva and Standish[88] proposed to the IAU Working group on Numerical Standards for fundamental Astronomy (NSFA) the masses of three largest asteroids (Ceres, Pallas and Vesta), the ratio of Moon's mass to Earth's mass, and the au from the fitting/adjustment of DE421[89] and EPM2009.[90] In this determination from ephemerides, the DE421[89] value of au is 149 597 870 699.6.±0.15 m and the EPM 2009[90] value of au is 149 597 870 696.6 ± 0.1 m with the quoted uncertainty formal uncertainty. They estimated the realistic error to be 3 m. From this result, they proposed to adopt the numerical value of the au in meter to be 149 597 870 700 (3) m. They also concluded that the numerical value of the au in meters is identical in both the TDB-based Barycentric Dynamical Time (TDB) and the TCB-based Barycentric Coordinate Time (TCB) systems of units if one uses the conversion proposed by Irwin and Fukushima,[91] Brumberg and Groten,[92] and Brumberg and Simon.[93] This value of au was accepted and included by the XXVIIth IAU General Assembly at Rio de Janeiro in 2009 as part of the IAU (2009) System of Astronomical Constants (Resolution B2).[94]

In the XXVIIIth IAU General Assembly at Beijing in 2012, the astronomical unit in meters is changed from a fitted value to a defining constant similar to speed of light: 1 au = 149 597 870 700 m. The abbreviation of the astronomical unit should be au (lower case). In the fitting of ephemeris, GM_{Sun} should then be used instead of au. The current values from the ephemeris fitting are: DE430,[83] $GM_{\text{Sun}} = 132\ 712\ 440\ 042(10)\ \text{km}^3/\text{s}^2$; INPOP13b,c,[95] $GM_{\text{Sun}} = 132\ 712\ 440\ 044.5(0.2)\ \text{km}^3/\text{s}^2$; EPM2014,[96] $GM_{\text{Sun}} = 132\ 712\ 440\ 053(1)\ \text{km}^3/\text{s}^2$.

In the actual testing of relativistic theories of gravity, one fits additional PPN or relativistic parameters. In the next section, we compile these ephemeris tests together with other solar system tests.

4. Solar System Tests

For last 50 years, we have seen great advances in the dynamical testing of relativistic gravity. This is largely due to interplanetary radio ranging/tracking and

Lunar Laser Ranging (LLR). Interplanetary radio ranging and tracking provided more stimuli and progresses at first. However, with improved accuracy of 2 mm from 20 to 30 cm and long-accumulation of observation data, LLR reaches similar accuracy in determining relativistic parameters as compared to interplanetary radio ranging despite that in LLR the relativistic effects are weaker. Table 2 gives such a comparison.

Relativistic perihelion advance and the solar quadrupole moment. In the PPN equations of motion (41) with two PPN parameters β and γ and with the effect of solar quadrupole moment added, the solar contribution to the secular planetary perihelion advances is given by the well-known formula (see e.g. Ref. 47, p. 1116):

$$\Delta\phi_0 = \left[\frac{(2-\beta+2\gamma)}{3}\right]\left[\frac{6\pi GM}{(c^2a(1-e^2))}\right] + J_2\left[\frac{3\pi R^2}{(a^2(1-e^2)^2)}\right], \tag{43}$$

where a and e are the semimajor axis and the eccentricity of the planet orbit. If Sun is uniformly rotating throughout, J_2 would be about 2×10^{-7} and its magnitude amounts to about 0.05% of the general relativistic perihelion advance for Mercury. To measure or separate the relativistic term (the first term), one needs to know or measure the solar quadrupole parameter. There are three ways to measure the solar quadrupole parameter: (i) through the solar oblateness measurement based on brightness of solar surface; (ii) through the helioseismology measurement; (iii) through the measurement of perihelion advance of different planets and asteroids. Although the solar oblateness measurement up to 1980's might imply large solar quadrupole parameter,[97] the determination of internal rotation of the Sun through measurement of rotation-induced frequency splitting in the observed solar surface acoustic power spectrum about the same period of time gave the value $J_2 = (1.7 \pm 0.4) \times 10^{-7}$, rather close to the value for a uniformly rotating Sun.[98]

In the 12th International Conference on GR and Gravitation at Boulder in 1989, Shapiro[99] reported the cumulative measurement of the relativistic Mercury perihelion advance to be $(42''.98 \pm 0.04)$/century assuming the solar quadrupole parameter J_2 to be about 2×10^{-7}; this gives $\beta = 1.000 \pm 0.003$.

In the 1990's, the solar quadrupole moment issue basically settled: (i) The solar oblateness ε measured in a balloon flight of the Solar Disk Sextant (SDS) on 1992 September 30[100] is $(8.63 \pm 0.88) \times 10^{-6}$ with the inferred $J_2 = 3 \pm 6 \times 10^{-7}$ (in agreement with the measured and inferred values $\varepsilon = (9.6 \pm 6.5) \times 10^{-6}$ and $J_2 = 10 \pm 43 \times 10^{-7}$ of Hill and Stebbins[101] in 1975). A subsequent analysis[102] based on SDS balloon flight data both in 1992 and 1994 combined with solar surface angular rotation data gives the solar quadrupole moment parameter $J_2 = 1.8 \times 10^{-7}$ and the solar octopole moment parameter $J_4 = 9.8 \times 10^{-7}$. (ii) The high quality helioseismological data obtained from the Solar Heliospheric Satellite (SoHO) and from the Global Oscillations Network Group (GONG) had made a much better determination of solar internal structure and solar differential rotation possible; this in turn led to a good determination of solar quadrupole moment and solar angular momentum. Pijpers[103] did an analysis and obtained $J_2 = (2.14 \pm 0.09) \times 10^{-7}$ from the GONG data, and $J_2 = (2.23 \pm 0.09) \times 10^{-7}$ from the SoHO/MDI data, with an error-weighted mean $J_2 = (2.18 \pm 0.06) \times 10^{-7}$. Godier and Roselot[104] used a differential rotation model established from helioseismological data and integrated J_2 from core to the surface to obtain a slightly lower value of $J_2 = 1.60 \times 10^{-7}$.

The results of space-borne measurements of solar oblateness from 1997 to 2011 are basically giving consistent numbers as summarized by Meftah *et al.* in 2015[105]: SoHO/MDI by Emilio *et al.* in 2007[106] with the oblateness (the solar equator-to-pole radius difference) $\Delta r = 8.7 \pm 2.8$ mas using 676.78 nm (λ) observation (2007), RHESSI/SAS by Fivian *et al.* in 2008[107] with $\Delta r = 8.01 \pm 0.14$ mas using 670.0 nm observation in 2004, SDO/HMI by Kuhn *et al.* (2012)[108] with $\Delta r = 7.2 \pm 0.49$ mas using 617.3 nm observation (2011–2012),*Picard*/SODISM by Irbah *et al.* (2014)[109] with $\Delta r = 8.4 \pm 0.3$ mas using 535.7 nm observation (2011), *Picard*/SODISM by Meftah *et al.* (2015)[105] $\Delta r = 7.86 \pm 0.32$ mas using 782.2 nm observation (2010–2011). It is to be noted that $\Delta r = 8$ mas corresponds to $J_2 = 1.60 \times 10^{-7}$.

The third way (iii) to measure the solar quadrupole moment is through its gravitational field generated. This will be discussed in the following together with the ephemeris fitting.

Test of relativistic gravity through ephemeris fitting. As planetary ephemerides became more and more precise, Anderson *et al.*[110] in 2002, used JPL archive of planetary positional data and DE ephemerides fitting method to solve all the conventional parameters in the DE ephemeris, plus four more parameters β, γ, J_2 and $\dot{G}/G$ specific to test relativistic gravity. In fitting the data, they weighted the separate data sets, except four data sets for the Mars, such that the assumed standard error for each data set is equal to the RMS residual for that particular set after the fit. For Mars, they used a standard error equal to five times the RMS residual for each of the four data sets — orbit data from Mariner 9 (1971–1972), lander data from Viking (1976–1982), orbit data from Mars global Surveyor (1998–2000) and Lander data from Pathfinder (1997), to compensate for systematic error from asteroids perturbations. This way they interpreted their resulting parameter values after fit as realistic errors instead of formal errors. The results of their fitting[110] are $\beta = 0.9990 \pm 0.0012, \gamma = 0.9985 \pm 0.0021$, $J_2 = (2.3 \pm 5.2) \times 10^{-7}$, and $\dot{G}/G = \pm(1.1$–$1.8) \times 10^{-12}$/yr (the $\dot{G}/G$ value is the same as their previous result in Ref. 111).

As an application of the developing EPM ephemerides, Pitjeva[87] (EPM2004 fitting) in 2004 obtained a determination of β and γ simultaneously with estimations for the solar quadrupole parameter and the possible variability of the gravitational constant: $\beta = 1.0000 \pm 0.0001, \gamma = 0.9999 \pm 0.0001, J_2 = (1.9 \pm 0.3) \times 10^{-7}$ and $\dot{G}/G = (1 \pm 5) \times 10^{-14}$/yr. In working out INPOP2010a planetary ephemeris, Fienga *et al.*[112] tested relativistic gravity by fitting β or γ, and obtained: $\beta = 0.999959 \pm 0.000078$; $\gamma = 1.000038 \pm 0.000081$; $J_2 = (2.4 \pm 0.25) \times 10^{-7}$. Pitjeva in working out EPM2011 ephemerides in 2013[85] obtained $\beta = 0.99998 \pm 0.00003$, $\gamma = 1.00004 \pm 0.00006$, $J_2 = (2.0 \pm 0.2) \times 10^{-7}$ and $[d(GM_{\rm Sun})/dt]/GM_{\rm Sun} = (-5.0 \pm 4.1) \times 10^{-14}$/yr. Verma *et al.*[113] included the radio ranging observations of MESSENGER, improved our knowledge of the orbit of Mercury, obtained INPOP13a ephemeris, and used it to perform tests of relativistic gravity. Their estimations of parameters are: $\beta = 1.000002 \pm 0.000025$; $\gamma = 0.999997 \pm 0.000025$ ($\beta = 1$ fixed); $J_2 = (2.4 \pm 0.2) \times 10^{-7}$.

Fienga *et al.*[114] added supplementary range tracking data obtained from the analysis of the MESSENGER spacecraft from 2011 to 2014 and included in their INPOP15a planetary ephemerides the new JPL datasets[115] obtained after the new analysis of Cassini tracking data from 2004 to 2014. They use INPOP15a to estimate possible supplementary advances of perihelia for Mercury and Saturn to test GR and presented their results in the 14th Marcel Grossmann meeting.[115] The results are basically consistent with the previous analysis; no violations of GR are found.

Using analytic and numerical methods, Anderson *et al.*[116] demonstrated that Earth–Mars ranging could provide a useful estimate of the SEP parameter η. For Mars ranging measurements with an accuracy of σ meters for 10 years, the expected accuracy for the Nordtvedt SEP parameter η would be of order $(1 - 12) \times 10^{-4}\sigma$ according to Ref. 116. The SEP for the Earth–Mars–Sun–Jupiter system is probably already tested implicitly in the ephemeris fit. It remains to separate the effect of SEP violation in the fit.

Time variability of the gravitational constant and mass loss from the Sun. The solar system dynamics could measure the possible time variability of the gravitational constant and the mass change of the Sun when the precision becomes good. If the gravitational constant does not change or its change is measured in another way, the mass loss (change) from the Sun can be measured dynamically. We advocate this potential during the 1990's when we propose the concept of ASTROD.[117] The electromagnetic radiation of the Sun carries 6.8×10^{-14} fractional mass from the Sun each year. This is the largest mass change of the Sun. Other mass change mechanisms give similar fractional change but of smaller magnitude. In order to separate the time variability of the gravitational constant, estimation of the mass loss from the Sun and the mass accretion into the Sun is needed. Pitjeva and Pitjev[118] made an estimate of the mass of celestial bodies falling into the Sun (mainly comets) and gave the following annual upper limit:

$$\frac{M_{\rm comet}}{M_{\rm Sun}} < 3.2 \times 10^{-14}. \tag{44}$$

Combined with the estimate of annual solar wind loss of $(2\text{–}3)\times10^{-14} M_{\rm Sun}$ per year (See Ref. 118 for references), Pitjeva and Pitjev gave the following bounds on the annual mass loss $M_{\rm loss}$ of Sun:

$$-9.8 \times 10^{-14} < \frac{M_{\rm loss}}{M_{\rm Sun}} < -3.6 \times 10^{-14}. \tag{45}$$

The fitted value of

$$\frac{\left[\dfrac{d(GM_{\rm Sun})}{dt}\right]}{GM_{\rm Sun}} = (-5.0 \pm 4.1) \times 10^{-14}/\text{yr}, \quad (3\sigma) \tag{46}$$

from EPM[85,118] led to the following relation[118] with 95% confidence (2σ) level

$$-7.8 \times 10^{-14}/\text{yr} < \left(\frac{\dot{G}}{G} + \dot{M}_{\text{Sun}}/M_{\text{Sun}}\right) < -2.3 \times 10^{-14}/\text{yr}, \quad (2\sigma). \qquad (47)$$

Equation (47) together with (45) gave bound on $\dot{G}/G$[118] as

$$-4.2 \times 10^{-14}/\text{yr} < \frac{\dot{G}}{G} < +7.5 \times 10^{-14}/\text{yr}. \qquad (48)$$

More recently, Fienga *et al.*[119] used Monte Carlo simulations to find constraints on the possible variation of the gravitational constant. They deduced the values of $\dot{G}/G$ considering a fixed value for annual mass loss of the Sun (including radiation and solar winds):

$$\frac{M_{\text{loss}}}{M_{\text{Sun}}} = (5.5 \pm 1.5) \times 10^{-14}, \qquad (49)$$

extracted from solar physics measurements and variations of $M_{\text{loss}}/M_{\text{Sun}}$ during the 11-year solar cycle of Pinto *et al.*[120] The values of $\dot{G}/G$ are typically within $\pm 10 \times 10^{-14}$.

Solar system dynamics also constrains dark energy models. For interested readers, please see Refs. 121 and 122.

Light/radio wave deflection, Shapiro time delay and constraint on the Eddington parameter γ. As we have seen in Sec. 1, gravitational light deflection is one of the three classical tests of GR. Before the ephemeris determination of the Eddington (light deflection) parameter γ, the Very Long Baseline Interferometry (VLBI) measurement of the gravitational deflection of radio waves by the Sun from astrophysical radio sources had been an important method. The accuracy of the observation had been improved to 1.7×10^{-3} for γ (Robertson *et al.*,[123] Lebach *et al.*[124]) in 1995. Analysis using VLBI data from 1979–1999 improved the result by about four times to 0.99983 ± 0.00045 (Shapiro *et al.*[125]). Fomalont *et al.*[126] used the Very Long Baseline Array (VLBA) at 43, 23 and 15 GHz to measure the solar gravitational deflection of radio waves among four radio sources during an 18-day period in October 2005 and determined the Eddington parameter γ to be 0.9998 ± 0.0003. Fomalont and Kopeikin[127,128] measured the effect of retardation of gravity by the field of moving Jupiter via VLBI observation of light bending from a quasar.

In 2003, Bertotti *et al.*[129] reported a measurement of the frequency shift of radio photons due to relativistic Shapiro time delay effect from the Cassini spacecraft as they passed near the Sun during the June 2002 solar conjunction. From this measurement, they determined γ to be 1.000021 ± 0.000023.

With the Hipparcos mission, very accurate measurements of star positions at various elongations from the Sun were accumulated.[130] Most of the measurements were at elongations greater than 47° from the Sun. At these angles, the relativistic light deflections are typically a few mas; it is 4.07 mas at right angles to the solar direction for an observer at 1 AU from the Sun according to GR. In the Hipparcos

Table 2. Relativity-parameter determination from interplanetary radio ranging/tracking and from lunar/satellite laser ranging.

Parameter	Meaning	Value from solar system determinations and from gravity probe B	Value from lunar/satellite laser ranging
β	PPN[55] Nonlinear gravity	1.000 ± 0.003[99] (Perihelion shift with $J_2(\mathrm{Sun}) = 10^{-7}$ assumed) 0.9990 ± 0.0012[110] (solar system tests with $J_2(\mathrm{Sun}) = (2.3 \pm 5.2) \times 10^{-7}$ fitted) 1.0000 ± 0.0001[87] (EPM2004 fit) 0.999959 ± 0.000078[112] (INPOP10a fit) 0.99998 ± 0.00003[85] (EPM2011 fit)	1.003 ± 0.005[135] 1.00012 ± 0.0011[136,129] (with Cassini γ) 1.00017 ± 0.00015[138] 1.00006 ± 0.00011[138] (from η)
γ	PPN[55] Space curvature	1.000 ± 0.002[99] (Viking ranging time delay) 0.9985 ± 0.0021[110] (solar system tests) 1.000021 ± 0.000023[129] (Cassini S/C ranging) 0.9999 ± 0.0001[87] (EPM2004 fit) 0.9998 ± 0.0003[126] (VLBI deflection) 1.000038 ± 0.000081[112] (INPOP10a fit) 1.00004 ± 0.00006[85] (EPM2011 fit)	1.000 ± 0.005[135]
K_{gp}	Geodetic precession	0.99935 ± 0.0028[149] (gravity probe B)	0.997 ± 0.007[135] 0.9981 ± 0.0064[136] 0.997 ± 0.005[138]
K_{L-T}	Lense–Thirring effect	0.95 ± 0.19[149] (gravity probe B)	$0.994 \pm (0.1\text{–}0.3)$[146] (LAGEOS) $0.994 \pm 0.002 \pm 0.05$[150] (LAGEOS & LARES)

Table 2. (*Continued*)

Parameter	Meaning	Value from solar system determinations and from gravity probe B	Value from lunar/satellite laser ranging
E (η)	SEP (Nordtvedt parameter)	The SEP for the Earth–Mars–Sun–Jupiter system[116] is probably already tested implicitly in the ephemeris fit. It remains to separate the effect of SEP violation.	$(3.2 \pm 4.6) \times 10^{-13}$ [135] $(-2.0 \pm 2.0) \times 10^{-13}$ [136,143] $(-0.8 \pm 1.3) \times 10^{-13}$ [137] $(0.9 \pm 1.9) \times 10^{-13}$ [138] ($\eta = (1 \pm 3) \times 10^{-4}$ [139])
$\dot{G}/G$	Temporal change in G	$\pm(1.1 - 1.8) \times 10^{-12}/\mathrm{yr}$ [111] (Solar System Tests) $(1 \pm 5) \times 10^{-14}/\mathrm{yr}$.[87] (EPM2004 fitting) $(-4.2 \text{ to } 7.5)\times 10^{-14}/\mathrm{yr}$ [118] (Planets & S/C observations with solar mass loss estimate) $\pm 10 \times 10^{-14}/\mathrm{yr}$ [119] (INPOP & Monte Carlo with solar mass loss estimate)	$(1 \pm 8) \times 10^{-12}/\mathrm{yr}$ [135] $(4 \pm 9) \times 10^{-13}/\mathrm{yr}$ [136] $(14 \pm 15) \times 10^{-14}/\mathrm{yr}$ [138]
$\ddot{G}/G$	Temporal change in $\dot{G}$		$(4 \pm 5) \times 10^{-15}\mathrm{yr}^{-2}$ [140]
$\alpha_{\mathbf{Y}\mathrm{ukawa}}$	Intermediate range force	$\alpha_{\lambda=1.5\,\mathrm{au}} = (2 \pm 13) \times 10^{-10}$ [141] (perihelion of Mars)	$\alpha_{\lambda=380\,000\,\mathrm{km}} = (-0.6 \pm 1.8) \times 10^{-11}$ [138]

measurements, each abscissa on a reference great-circle has a typical precision of 3 mas for a star with 8–9 mag. There are about 3.5 million abscissae generated, and the precision in angle or similar parameter determination is in the range. Frœschlé *et al.*[131] analyzed these Hipparcos data and determined the light deflection parameter γ to be 0.997 ± 0.003. This result demonstrated the power of precision optical astrometry.

Global Astrometric Interferometer for Astrophysics (Gaia)[132] is an ambitious astrometric mission aiming at the broadest possible astrophysical exploitation of optical interferometry using a modest baseline length ($\sim$ 3 m). Gaia, launched on 19 December 2013 by Arianespace using a Soyuz ST-B/Fregat-MT rocket flying from Kourou in French Guiana in a Lissajous orbit around the Sun–Earth Lagrangian point L_2, is charting a three-dimensional map of our Galaxy, the Milky Way, in the process revealing the composition, formation and evolution of the Galaxy. Operating in the depths of space, far beyond the Moon's orbit, ESA's Gaia spacecraft had completed two years of a planned five-year survey of the sky on 16 August 2016. Data Release 1 (Gaia DR1)[133] was already public and contained astrometric results for more than 1 billion stars brighter than magnitude 20.7 based on observations collected by the Gaia satellite during the first 14 months of its operational phase. Gaia has already provided unprecedented positional and radial velocity measurements with the accuracies needed to produce a stereoscopic and kinematic census of about one billion stars in our Galaxy and throughout the Local Group. This amounts to about 1% of the Galactic stellar population. To increase the weight of measuring the relativistic light deflection parameter γ, Gaia observes at elongations greater than 35° (as compared to essentially 47° for Hipparcos) from the Sun. A simulation shows that GAIA could measure γ to 1×10^{-5}–2×10^{-7} accuracy.[133]

LLR Tests of relativistic gravity. In the last column of Table 2, the values come from LLR observations.[135–140] Reference 135 gave the results as of 1996. In Ref. 136, Williams *et al.* used a total of 15 553 LLR normal-point data in the period of March 1970 to April 2004 from Observatoire de la Côte d'Azur, McDonald Observatory and Haleakala Observatory in their determination. Each normal point comprises from 3 to about 100 photons. The weighted rms scatter after their fits for the ten-year ranges from 1994 to 2004 is about 2 cm (about 5×10^{-11} of range). Müller *et al.* wrote a comprehensive chapter on "Lunar Laser Ranging and Relativity" and summarized their work on the LLR tests of the relativistic gravity.[138] From Table 2, we can see clearly that the LLR tests of relativistic gravity have the same level of precision as the radio solar system tests. Constraints on intermediate range force is from LLR[138] and from the Mars perihelion precession (Iorio[141]) are compiled in the last row.

LLR also constrains dark energy models. For interested readers, please see Ref. 142 and references therein.

Frame Dragging Effects. In 1918, Lense and Thirring[144] predicted that a rotating body drags the local inertial frames of reference around it in GR. In 1960,

Schiff[145] showed that in GR, the spin axis of a gyroscope orbiting around Earth would undergo both geodetic drift in the orbit plane due to motion through the spacetime curved by the Earth's mass and frame-dragging due to the Earth's rotation with respect to a distant inertial frame. The dragging of gyro's spin axis is sometimes called the Schiff effect while both spin axis dragging and orbiting axis dragging can be grouped as Lense–Thirring frame-dragging effects. In 2004, Ciufolini and Pavlis[146] reported a measurement of the Lense–Thirring effect on the two Earth satellites, LAGEOS and LAGEOS2; it is 0.99 ± 0.10 of the value predicted by GR. In the same year, Gravity Probe B (GP-B, a space mission to test GR using cryogenic gyroscopes in orbit)[147] was launched in April[148]; their final results are a geodetic drift rate of $-6,601.8 \pm 18.3$ mas/yr and a frame-dragging drift rate of -37.2 ± 7.2 mas/yr, to be compared with the GR predictions of $-6,606.1$ mas/yr and -39.2 mas/yr, respectively; i.e. GP-B[148] provides independent measurements of the geodetic and frame-dragging effects at an accuracy of 0.28% and 19%, respectively. GP-B experiment has also verified the weak equivalence principle for macroscopic rotating bodies to ultra-precision.[149] Recently, Ciufolini *et al.*[150] have used about 3.5 years of laser-range observations of the LARES, LAGEOS, and LAGEOS2 satellites together with the Earth gravity field model GGM05S produced by the space geodesy mission GRACE to measure the Earth's dragging of inertial frames to be $0.994 \pm 0.002 \pm 0.05$ of the GR value with 0.002 the 1-σ formal error and 0.05 their preliminary estimate of systematic error.

5. Outlook — On Going and Next-Generation Tests

In the early days, astronomical observations of the solar system provided the basis for developing gravitation theories. Gravitation theories provide the scientific basis of space exploration of Earth and the entire solar system. The advent of space age and solar system exploration required the range measurements in the solar system that made possible the creation of high-accuracy planetary and lunar ephemerides. These ephemerides in turn provide dynamical positioning atlases for the solar system exploration and the precision tests of relativistic gravitational theories. As we have seen in the last section, ephemeris fitting for gravitational parameters in relativistic gravitational theories is playing more and more important role in the experimental tests. Experimentally, the improvement depends on the technological advance of radio ranging/Doppler tracking and laser ranging/tracking of spacecraft and celestial bodies in the solar system.

In Table 2, we have seen that LLR reaches similar accuracy in determining relativistic parameters as compared to interplanetary radio ranging despite that in LLR, the relativistic effects are weaker. The main reason is that the resolution depends on wavelength. Optical wavelength is four orders of magnitude shorter than microwave wavelength. The most precise radio Doppler tracking experiment is Cassini radio wave retardation measurement.[129] Cassini multilink radio system consists of a sophisticated multilink radio system that simultaneously receives two uplink signals at frequencies of X and Ka bands and transmits three downlink signals with X-band coherent with the X-band uplink, Ka-band coherent with the X-band uplink, and Ka-band coherent with the Ka-band uplink. X-band is a standard deep space communication frequency band about 8.4 GHz; Ka-band is another deep space communication frequency band about 32 GHz. The wavelength of Ka-band microwave is about 1 cm. The reason to use multilink system is to measure and subtract the plasma dispersion which is proportional to the wavelength square. For laser optical ranging, a typical wavelength is about 1 μm. There is a four-order difference in wavelength. For laser ranging, the plasma effect is eight-order smaller; in the interplanetary space the subtraction is not needed. If one link is on Earth, subtraction of extra optical path length by two-wavelength observation or other means is still needed. With four-order improvement in ranging, monitoring, the noninertial spacecraft motion is required. One way is to use drag-free technology. LISA Pathfinder launched on 3 December 2015 has successfully tested and demonstrated the drag-free technology to satisfy not just the requirement of LISA Pathfinder, but also basically the drag-free requirement of LISA gravitational-wave space mission concept.[151] The drag-free technology is ripe for relativistic missions in the solar system. Hence, we envisage a 3–4 orders of improvement in testing the relativistic gravity and the solar system dynamics, say, in the next 25 years or so. This improvement is for all relativistic parameters. In the following, we give an outlook of improvements on the Eddington parameter γ for various ongoing/proposed experiments. Table 3 lists the aimed accuracy of such experiments. Some motivations for determining γ precisely to 10^{-5}–10^{-9} are given in Refs. 152 and 153.

First, as we have discussed in Sec. 4, Gaia Data Release 1 (Gaia DR1)[133] has already become public and contained astrometric results for more than 1 billion stars brighter than magnitude 20.7 based on observations collected by the Gaia satellite during the first 14 months of its operational phase. With expected 4-year observation period, a simulation shows that GAIA could measure γ to 1×10^{-5}–2×10^{-7} accuracy.[133] This is listed as the second row in Table 3.

BepiColombo is a joint mission to Mercury[154] between ESA and the Japan Aerospace Exploration Agency (JAXA), executed under ESA leadership. The mission comprises two spacecrafts: The Mercury Planetary Orbiter (MPO) and the

Table 3. Aimed accuracy of PPN space parameter γ for various ongoing/proposed experiments.

Ongoing/Proposed experiment	Aimed accuracy of γ	Type of experiment
GAIA[132–134]	1×10^{-5}–2×10^{-7}	Deflection
Bepi-Colombo[154,155]	2×10^{-6}	Retardation
ASTROD I[115]	3×10^{-8}	Retardation
ASTROD[117]	1×10^{-9}	Retardation
Super-ASTROD[161]	1×10^{-8}	Retardation
Odyssey[162]	1×10^{-7}	Retardation
SAGAS[163]	1×10^{-7}	Retardation
OSS[164]	1×10^{-7}	Retardation

Mercury Magnetospheric Orbiter (MMO). It will set off in 2018 on a journey to the smallest and least explored terrestrial planet in our Solar System. When it arrives at Mercury in late 2024, it will endure temperatures in excess of 350°C and gather data during its 1-year nominal mission, with a possible 1-year extension. Milani *et al.*[155] have simulated the radio science of this mission: "While determining its orbit around Mercury, it will be possible to indirectly observe the motion of its center-of-mass, with an accuracy of several orders of magnitude better than what is possible by radar ranging to the planet's surface. This is an opportunity to conduct a relativity experiment which will be a modern version of the traditional tests of GR, based upon Mercury's perihelion advance and the relativistic light propagation near the Sun." They predict that the determination of γ can reach 2×10^{-6}.

ASTROD I is envisaged as the first in a series of ASTROD missions.[78,156–159] ASTROD I mission concept is to use one spacecraft carrying a telescope, four lasers, two event timers and a clock with a Venus swing-by orbit. Two-way, two-wavelength laser pulse ranging will be used between the spacecraft in a solar orbit and deep space laser stations on Earth, to achieve the ASTROD I goals of testing GR with an improvement in sensitivity of over three orders of magnitude, improving our understanding of gravity and aiding the development of a new quantum gravity theory; to measure key solar system parameters with increased accuracy; and to measure the time rate of change of the gravitational constant with improvement. Using the achieved accuracy of 3 ps in laser pulse timing and the demonstrated LISA Pathfinder drag-free capability, a simulation showed that accuracy of the determination of γ will reach 3×10^{-8}.

The general concept of ASTROD is to have a constellation of drag-free spacecraft navigate through the solar system and range with one another using optical devices to map the solar system gravitational field, to measure related solar system parameters, to test relativistic gravity, to observe solar g-mode oscillations, and to detect gravitational waves. A baseline implementation of ASTROD, also called ASTROD, is to have two spacecraft in separate solar orbits (one in inner solar orbit, the other in outer solar orbit), each carrying a payload of a proof mass, two telescopes, two 1–2 W lasers, a clock and a drag-free system, together with a similar spacecraft near Earth around one of the Lagrange points L1/L2. The three spacecraft range coherently with one another using lasers to map solar system gravity, to test relativistic gravity, to observe solar g-mode oscillations, and to detect gravitational waves. Since it will be after ASTROD I, we assume 1 ps timing accuracy and 10 times better drag-free performance than what LISA Pathfinder achieved. With these requirements, the accuracy of the determination of γ will reach 1×10^{-9} in 3.5 years.[160]

Super-ASTROD,[161] Odyssey,[162] SAGAS (Search for Anomalous Gravitation using Atomic Sensors)[163] and OSS (Outer Solar System)[164] are four mission concepts to test fundamental physics and to explore outer solar system.

Solar System Odyssey[162] is designed to perform a comprehensive set of gravitational tests in the Solar System. The mission has four major scientific objectives: (1)

significantly improve the accuracy of deep space gravity test; (2) investigate planetary flybys; (3) improve the current accuracy of the measurements of the Eddington parameter; (4) map the gravity field in outer regions of the Solar System. For improving the current accuracy of the measurement of the Eddington parameter γ, Odyssey proposes to use an improved multi-frequency radio link of the Cassini type together with a precision accelerometer and a possible laser tracking option and aims at measuring γ at an accuracy of 10^{-7}.

SAGAS[163] aims at flying highly sensitive atomic sensors (optical clock, cold atom accelerometer, optical link) on a Solar System escape trajectory. It also aims at measuring γ at an accuracy of $1 - 2 \times 10^{-7}$.

OSS[164] is an outer solar system exploration mission concept. The OSS probe would carry instruments allowing precise tracking of the spacecraft during the cruise. It would facilitate improved tests of the laws of gravity in deep space. A largely improved accuracy can be attained with the up-scaling option of a laser ranging equipment onboard to measure the parameter γ at the 10^{-7} level.

Super-ASTROD[161] is a mission concept with four spacecraft in 5 AU orbits together with an Earth-Sun L1/L2 spacecraft ranging optically with one another to probe primordial gravitational waves with frequencies 0.1 μHz–1 mHz, to test fundamental laws of spacetime and to map the outer-solar-system mass distribution and dynamics. With larger orbits, the main goal of Super-ASTROD in test relativistic gravity is not to improve on Parametrized Post-Newtonian (PPN) parameters over ASTROD I / ASTROD, instead it is to test cosmological theories which give larger modifications from GR for larger orbits. However, with same or better ranging capability than ASTROD I / ASTROD, the accuracy of its determination of γ will be better than 1×10^{-8}.

All four mission concepts explore gravity at deep space to bridge the gap between inner solar system tests and cosmological tests. They are most relevant to the detection/constraint of dark matter and dark energy, and to the tests of MOND models and the dark energy dynamical models.

Since we had another review on "Equivalence principles, spacetime structure and the cosmic connection"[17] early this year, we did not discuss space missions for testing (weak) equivalence principles. Here we just mention in passing that Microscope (MICROSCOPE: MICRO-Satellite à trainée Compensée pour l'Observation du Principle d'Équivalence)[165,166] has been in orbit since 26 April 2016 with the aim of improving the test accuracy by two orders of magnitude than any of the ground-based weak-equivalence-principle experiment, and is performing functional tests successfully.[166]

With increasing outreach and precision of observations, astrophysics and cosmology became increasingly important for developing gravitation theories; notably the precise timing of pulses from pulsars[44] and various cosmological tests.[17]

During last 157 years, the precisions of laboratory and space experiments, and the precisions of astrophysical and cosmological observations on the tests of relativistic gravity have improved by 3–4 orders of magnitude. The advent of space age has stimulated the development of numerical ephemerides. Doppler and ranging observations from various space missions drive the ephemerides to ever increasing precision. For the last decade, we have seen great progress in various aspects of testing relativistic gravity in the solar system. Systematic modeling and ephemeris fitting of all the observational data becomes standard. The pending testing of relativistic gravity better than 10^{-5}–10^{-6} precision requires the development of 2PN (post–post-Newtonian) numerical ephemerides. In the next 25 years, we envisage another 3–4 order improvement in all directions of tests of relativistic gravity. These will give enhanced interest and development both in experimental and theoretical aspects of gravity, and may lead to answers to some profound questions of gravity and the cosmos.

Gravitation is clearly empirical. As precision is increased by orders of magnitude, we are in a position to explore deeper into the origin of gravitation. The current and coming generations hold such promises.

Acknowledgments

I am grateful to Jürgen Müller and Franz Hofmann for helpful discussions on the LLR tests of relativistic gravity. I would also like to thank Science and Technology Commission of Shanghai Municipality (STCSM-14140502500) and Ministry of Science and Technology of China (MOST-2013YQ150829, MOST-2016YFF0101900) for supporting this work in part.

This review is dedicated to my mother Suh-Ling Huang (1923.12.13–2016.01.24) on her obituary 2016.01.29. Her amity, competence and diligence as mother fostered my confidence and perseverance to continue.

References

1. I. Newton, *Philosophiae Naturalis Principia Mathematica* (Benjamin Motte, London, 1687).
2. J. Kepler, *Astronomia Nova de Motibus Stellae Martis* (Prague, 1609); *Harmonice Mundi* (Linz, (Austria) 1619).
3. G. Galilei, *Discorsi e Dimostrazioni Matematiche Intorno a Due Nuove Scienze* (Elzevir, Leiden, 1638).
4. U. J. J. Le Verrier, Theorie du mouvement de mercure, *Ann. Observ. Imp. Paris (Mém.)* **5** (1859) 1.
5. A. A. Michelson and E. W. Morley, On the relative motion of the Earth and the luminiferous ether, *Am. J. Sci.* **34** (1887) 333.
6. H. A. Lorentz, *Kon. Neder. Akad. Wet. Amsterdam. Versl. Gewone Vergad. Wisen Natuurkd. Afd.* **6** (1904) 809.
7. H. Poincaré, *C. R. Acad. Sci.* **140** (1905) 1504.
8. A. Einstein, Zur elektrodynamik bewegter Körper [On the Electrodynamics of Moving Bodies], *Ann. Phys. (Germany)* **17** (1905) 891.
9. R. V. Eötvös, *Math. Naturwiss. Ber. Ungarn* **8** (1889) 65.
10. A. Einstein, Über das Relativitätprinzip und die aus demselben gezogenen Folgerungen, *Jahrb. Radioakt. Elektronik* **4** (1907) 411; Corrections by Einstein in *Jahrb.*

Radioakt. Elektronik **5** (1908) 98 (in German); H. M. Schwartz, *Am. J. Phys.* **45** (1977) 512, 811, 899.

11. A. Einstein, Zür allgemeinen Relativitätstheorie, *Preuss. Akad. Wiss. Berlin, Sitzber.* **47** (1915) 778–786 (presented November 4, published November 11).
12. A. Einstein, Zür allgemeinen Relativitätstheorie (Nachtrag), *Preuss. Akad. Wiss. Berlin, Sitzber.* **47** (1915) 799–801 (presented November 11, published November 18).
13. A. Einstein, Erklärung der Perihelbewegung des Merkur aus allgemeinen Relativitätstheorie, *Preuss. Akad. Wiss. Berlin, Sitzber.* **47** (1915) 831–839 (presented November 18, published November 25).
14. D. Hilbert, Die Grundlagen der Physik, *Konigl. Gesell. d. Wiss. Göttingen, Nachr., Math.-Phys. KL.* **47** (1915) 395–407 (presented November 20).
15. A. Einstein, Die Feldgleichungen der Gravitation, *Preuss. Akad. Wiss. Berlin, Sitzber.* **47** (1915) 844–847 (presented November 25, published December 2).
16. W.-T. Ni, Genesis of general relativity: A concise exposition, in *One Hundred Years of General Relativity: From Genesis and Foundations to Gravitational Waves, Cosmology and Quantum Gravity*, ed. W.-T. Ni, Chap. 2 (World Scientific, Singapore, 2016); *Int. J. Mod. Phys. D* **25** (2016) 1530004.
17. W.-T. Ni, Equivalence principles, spacetime structure and the cosmic connection, in *One Hundred Years of General Relativity: From Genesis and Foundations to Gravitational Waves, Cosmology and Quantum Gravity*, ed. W.-T. Ni, Chap. 5 (World Scientific, Singapore, 2016); *Int. J. Mod. Phys. D* **25** (2016) 1530002.
18. A. Einstein, Ist die Trägheit eines Körpers von seinem Energieihalt abhängig? *Ann. Phys.* **18** (1905) 639.
19. M. Planck, Zur Dynamik bewegter Systeme, *Sitzungsber. Königlich-Preuss. Akad. Wiss. Berlin* 13, June 1907, 542–570 (in German); On the Dynamics of Moving Systems (1907), https://en.wikisource.org/wiki/Translation:On_the_Dynamics_of_Moving_Systems.
20. S. Newcomb, *Discussion and Results of Observations on Transits of Mercury from 1677 to 1881* (U.S. Govt. Printing Office, Washington, D.C., 1882), pp. 367–487.
21. A. Einstein, Über den Einfluß der Schwerkraft auf die Ausbreitung des Lichtes, *Ann. Phys.* **35** (1911) 898.
22. I. Newton, *Opticks* (Dover, New York, 1952), http://find.galegroup.com.nthulib-oc.nthu.edu.tw/ecco/infomark.do?&source=gale&prodId=ECCO&userGroupName=twnsc070&tabID=T001&docId=CW3309834823&type=multipage&contentSet=ECCOArticles&version=1.0&docLevel=FASCIMILE.
23. J. G. von Soldner, On the deviation of a light ray from its motion along a straight line through the attraction of a celestial body which it passes close by, *Astronomishes Jahrbuch für das Jahr 1804* (C. F. E. Späthen, Berlin, 1801), pp. 161–172.
24. S. L. Jaki, Johann Georg von Soldner and the Gravitational Bending of Light, with an English Translation of His Essay on It Published in 1801, *Found. Phys.* **8** (1978) 927.
25. C. M. Will, Henry Cavendish, Johann von Soldner, and the deflection of light, *Am. J. Phys.* **56** (1988) 413.
26. H. Cavendish, *The Scientific Papers of the Honourable Henry Cavendish, F. R. S., Vol. II: Chemical and Dynamical*, ed. E. Thorpe (Cambridge U. P., London, 1921), pp. 433–437.
27. A. Einstein and M. Besso, Manuscript on the Motion of the Perihelion of Mercury, June 1913, pp. 360–473 in *The Collected Papers of Albert Einstein, Volume 4: The Swiss Years: Writings*, 1912–1914 *Albert Einstein*, eds. M. J. Klein, A. J. Kox,

J. Renn and R. Schulmann (Princeton University Press, US, 1995), p. 344, http://press.princeton.edu/einstein/digital/.
28. A. Einstein and M. Grossmann, Entwurf einer verallgemeinerten Relativitätstheorie und einer Theorie der Gravitation [Outline of a Generalized Theory of Relativity and of a Theory of Gravitation], *Z. Math. Phys.* **62** (1913) 225.
29. A. Einstein, Die Grundlage der allgemeinen Relätivitatstheorie, *Ann. Phys.* **49** (1916) 769.
30. H. Rowland, *Preliminary Table of Solar-Spectrum Wave Lengths* (Carnegie Institute of Washington, 1895).
31. L. Jewell, The coincidence of solar and metallic lines, *Astrophys. J.* **3** (1896) 89.
32. W. J. Humphreys and J. F. Mohler, Effect of pressure on the wave-lengths of lines in the arc-spectra of certain elements, *Astrophys. J.* **3** (1896) 114.
33. L. Jewell, J. Mohler and W. J. Humphreys, Note on the pressure of the "Reversing Layer" of the solar atmosphere, *Astrophys. J.* **3** (1896) 138.
34. H. Buisson and C. Fabry, Mesures de Petites Variations de Longueurs d'Onde par la Méthode Interférentielle, *J. Phys. Théor. Appl.* **9** (1910) 298.
35. J. Evershed, Pressure in the reversing layer, *Bull. Kadaikánal Obs.* **18** (1909); A new interpretation of the general displacement of the lines of the solar spectrum towards the Red, *Bull. Kadaikánal Obs.* **36** (1913).
36. J. Earman and C. Glymour, The gravitational red shift as a test of general relativity: History and analysis, *Stud. Hist. Philos. Sci.* **11** (1980) 175.
37. J. Earman and C. Glymour, Relativity and eclipses: The British eclipse expeditions of 1919 and their predecessors, *Hist. Stud. Phys. Sci.* **11** (1980) 49.
38. F. Dyson, A. Eddington and C. Davidson, A determination of the deflection of light by the sun's gravitational field, from observations made at the total eclipse of May 29, 1919, *Philos. Trans. R. Soc.* **220A** (1920) 291.
39. R. V. Pound and G. A. Rebka, Apparent weight of photons, *Phys. Rev. Lett.* **4** (1960) 337; R. V. Pound and J. L. Snider, Effect of gravity on nuclear resonance, *Phys. Rev. Lett.* **13** (1964) 539.
40. P. G. Roll, R. Krotkov and R. H. Dicke, The equivalence of inertial and passive gravitational mass, *Ann. Phys.* (U. S. A) **26** (1964) 442.
41. I. I. Shapiro, Fourth test of general relativity, *Phys. Rev. Lett.* **13** (1964) 789.
42. W.-T. Ni, Empirical foundations of the relativistic gravity, *Int. J. Mod. Phys. D* **14** (2005) 901.
43. W.-T. Ni, Empirical tests of the relativistic gravity: The past, the present and the future, in *Recent Advances and Cross-Century Outlooks in Physics: Interplay Between Theory and Experiment: Proc. Conf.* March 18–20, (Atlanta, Georgia, 1999), eds. P. Chen and C.-Y. Wong (World Scientific, Singapore, 2000); in *Gravitation and Astrophysics*, eds. L. Liu, J. Luo, X.-Z. Li and J.-P. Hsu (World Scientific, Singapore, 2000), pp. 1–19.
44. R. N. Manchester, Pulsars and gravity, in *One Hundred Years of General Relativity: From Genesis and Empirical Foundations to Gravitational Waves, Cosmology and Quantum Gravity*, ed. W.-T. Ni, Chap. 9 (World Scientific, Singapore, 2016); *Int. J. Mod. Phys. D* **24** (2015) 1530018.
45. S. Reynaud and M.-T. Jaekel, Tests of general relativity in the solar system, in *Series Proc. International School of Physics "Enrico Fermi", Atom Optics and Space Physics*, Vol. 168 (Societa Italiana de Fisica & IOS Press, 2009), pp. 203–217.
46. C. M. Will, The confrontation between general relativity and experiment, *Living Rev. Rel.* **17** (2014) 4.

47. C. W. Misner, K. S. Thorne and J. A. Wheeler (MTW), *Gravitation* (Freeman, San Francisco, 1973).
48. K. Kuroda, W.-T. Ni and W.-P. Pan, Gravitational waves: Classification, methods of detection, sensitivities, and sources, in *One Hundred Years of General Relativity: From Genesis and Empirical Foundations to Gravitational Waves, Cosmology and Quantum Gravity*, ed. W.-T. Ni, Chap. 10 (World Scientific, Singapore, 2016); *Int. J. Mod. Phys. D* **24** (2015) 1530031.
49. S. Chandrasekhar, The post-Newtonian equations of hydrodynamics in general relativity, *Astrophys. J.* **142** (1965) 1488.
50. K. S. Thorne, C. M. Will and W.-T. Ni, Theoretical frameworks for testing relativistic gravity: A review, in ed. R. W. Davis, *Proc. Conf. Experimental Tests of Gravitational Theories*, November 11–13, 1970, California Institute of Technology, JPL Technical Memorandum 33-499 (1971), pp. 10–31.
51. K. S. Thorne and C. M. Will, Theoretical frameworks for testing relativistic gravity, I: Foundations, *Astrophys. J.* **163** (1971) 595.
52. C. M. Will, Theoretical frameworks for testing relativistic gravity, II: Parametrized Post-Newtonian hydrodynamics and the Nordtvedt effect, *Astrophys. J.* **163** (1971) 611.
53. C. M. Will, Theoretical frameworks for testing relativistic gravity, III: Conservation laws, Lorentz invariance and the values of the PPN parameters, *Astrophys. J.* **169** (1971) 125.
54. W.-T. Ni, Theoretical frameworks for testing relativistic gravity, IV: A compendium of metric theories of gravity and their post-Newtonian limits, *Astrophys. J.* **176** (1972) 769.
55. C. M. Will and K. Nordtvedt, Jr., Conservation laws and preferred frames in relativistic gravity, I: Preferred frame theories and an extended PPN formalism, *Astrophys. J.* **177** (1972) 757.
56. K. Nordtvedt, Jr. and C. M. Will, Conservation laws and preferred frames in relativistic gravity, II: Experimental evidence to rule out preferred frame theories of gravity, *Astrophys. J.* **177** (1972) 775.
57. L. Shao and N. Wex, New tests of local Lorentz invariance of gravity with small-eccentricity binary pulsars, *Class. Quantum Grav.* **29** (2012) 215018.
58. P. C. C. Freire *et al.*, The relativistic pulsar-white dwarf binary PSR J1738 + 0333 – II. The most stringent test of scalar–tensor gravity, *Mon. Not. R. Astron. Soc.* **423** (2012) 3328.
59. L. Shao, R. N. Caballero, M. Kramer, N. Wex, D. J. Champion and A. Jessner, A new limit on local Lorentz invariance violation of gravity from solitary pulsars, *Class. Quantum Grav.* **30** (2013) 165019.
60. I. H. Stairs *et al.*, Discovery of three wide-orbit binary pulsars: Implications for binary evolution and equivalence principle, *Astrophys. J.* **632** (2005) 1060.
61. C. M. Will, Is momentum conserved? A test in the binary system PSR 1913+16, *Astrophys. J. Lett.* **393** (1992) L59.
62. D. F. Bartlett and D. van Buren, Equivalence of active and passive gravitational mass using the moon, *Phys. Rev. Lett.* **57** (1986) 21.
63. C. M. Will, *Theory and Experiment in Gravitational Physics*, 2nd edn. (Cambridge University Press, Cambridge, UK, New York, USA, 1993).
64. K. Nordtvedt, Testing Newton's third law using lunar laser ranging, *Class. Quantum Grav.* **18** (2001) L133.
65. C. M. Will, Active mass in relativistic gravity: Theoretical interpretation of the Kreuzer experiment, *Astrophys. J.* **204** (1976) 224.

66. C. M. Will, Relativistic gravity in the solar system. III. Experimental disproof of a class of linear theories of gravitation, *Astrophys. J.* **185** (1973) 31.
67. R. J. Warburton and J. M. Goodkind, Search for evidence of a preferred reference frame, *Astrophys. J.* **208** (1976) 881.
68. J. M. Goodkind, The superconducting gravimeter, *Rev. Sci. Instrum.* **70** (1999) 4131.
69. S. Shiomi, Testing gravitational physics with superconducting gravimeters, *Prog. Theor. Phys. Suppl.* **172** (2008) 61.
70. L. Shao and N. Wex, New limits on the violation of local position invariance of gravity, *Class. Quantum Grav.* **30** (2013) 165020.
71. A. S. Eddington, *The Mathematical Theory of Relativity* (Cambridge University Press, 1922).
72. K. Nordtvedt, Jr., Equivalence principle for massive bodies, II.Theory, *Phys. Rev.* **169** (1968) 1017.
73. V. A. Brumberg, *Essential Relativistic Celestial Mechanics* (Adam Hilger, Bristol, 1991), pp. 176–177.
74. M. Soffel *et al.*, The IAU 2000 resolutions for astrometry, celestial mechanics and metrology in the relativistic framework: Explanatory supplement, *Astron. J.* **126** (2003) 2687.
75. D.-W. Chiou and W.-T. Ni, Orbit simulation for the determination of relativistic and solar-system parameters for the ASTROD space mission, *presented to 33rd COSPAR Scientific Assembly*, Warsaw, 16–23 July (2000), arXiv:astro-ph/0407570.
76. E. M. Standish, X. X. Newhall, J. G. Williams and W. M. Folkner, *JPL Planetary and Lunar Ephemerides*, DE403/LE403, JPL Interoffice Memorandum IOM 314.10–127, 27 pages (1995), ftp://ssd.jpl.nasa.gov/pub/eph/planets/ioms/de403.iom.pdf.
77. C.-J. Tang and W.-T. Ni, CGC 2 ephemeris framework (in Chinese with an English abstract), *Publ. Yunnan Obs.* **2002** (2002) 21.
78. C. Braxmaier and the ASTROD I team, Astrodynamical space test of relativity using optical devices I (ASTROD I) — a class-M fundamental physics mission proposal for cosmic vision 2015–2025: 2010 Update, *Exp. Astron.* **34** (2012) 181.
79. G. Wang and W.-T. Ni, Time-delay interferometry for ASTROD-GW, *Chin. Astron. Astrophys.* **36** (2012) 211.
80. G. Wang and W.-T Ni, Orbit optimization for ASTROD-GW and its time delay interferometry with two arms using CGC ephemeris, *Chin. Phys. B* **22** (2013) 049501.
81. S. V. Dhurandhar, W.-T. Ni and G. Wang, Numerical simulation of time delay interferometry for a LISA-like mission with the simplification of having only one interferometer, *Adv. Space Res.* **51** (2013) 198.
82. G. Wang and W.-T. Ni, Numerical simulation of time delay interferometry for eLISA/NGO, *Class. Quantum Grav.* **30** (2013) 065011.
83. W. M. Folkner, J. G. Williams, D. H. Boggs, R. S. Park and P. Kuchynka, The planetary and lunar ephemerides DE430 and DE431, Interplanetary Network Progress Report **196** (2014) C1.
84. A. Fienga, H. Manche, J. Laskar and M. Gastineau, INPOP06: A new numerical planetary ephemeris, *Astron. Astrophys.* **477** (2008) 315.
85. E. V. Pitjeva, Updated IAA RAS planetary ephemerides-EPM2011 and their use in scientific research, *Sol. Syst. Res.* **47** (2013) 386.
86. E. M. Standish, *JPL Planetary Ephemerides*, DE410, JPL Interoffice Memorandum IOM 312.N — 03–009, pp. 16 (2003), ftp://ssd.jpl.nasa.gov/pub/eph/planets/ioms/de410.iom.pdf.

87. E. Pitjeva, Precise determination of the motion of planets and some astronomical constants from modern observations, in *IAU Coll. N 196 / Transit of Venus: New Views of the Solar System and Galaxy*, ed. D. W. Kurtz (Cambridge University Press, Cambridge, 2005), pp. 230–241.
88. E. V. Pitjeva and E. M. Standish, Proposals for the masses of the three largest asteroids, the Moon-Earth mass ratio and the Astronomical Unit, *Celest. Mech. Dyn. Astron.* **103** (2009) 365.
89. W. M. Folkner, J. G. Williams and D. H. Boggs, The planetary and lunar ephemerides DE421, Interoffice Memorandum 343.R-08-003, pp. 31 (2008); published as *IPN Progress Report* 42–178, August 15 (2009), 34 pages.
90. E. V. Pitjeva, Ephemerides EPM2008: The updated model, constants, data, *Proc. Journées 2008 "Systèmes de référence spatio-temporels" & X. Lohrmann-Kolloquium: Astrometry, Geodynamics and Astronomical Reference Systems*, TU Dresden, Germany, 22–24 September 2008, eds. M. Soffel and N. Capitaine, Lohrmann-Observatorium and Observatoire de Paris, ISBN 978-2-901057-63-5, pp. 57–60.
91. A. W. Irwin and T. Fukushima, A numerical time ephemeris of the Earth, *Astron. Astrophys.* **348** (1999) 642.
92. V. A. Brumberg and E. Groten, IAU resolutions on reference systems and time scales in practice, *Astron. Astrophys.* **367** (2001) 1070.
93. V. A. Brumberg and J.-L. Simon, Relativistic indirect third-body perturbations in the SMART Earth's rotation theory, in *Astrometry, Geodynamics and Solar System Dynamics: From Milliarcsecond to Microarcseconds/Journees-2003*, eds. A. Finkelstein and N. Capitaine, IAA RAS, St.Petersburg (2004), pp. 302–313.
94. B. Luzum, N. Capitaine, A. Fienga, W. Folkner, T. Fukushima, J. Hilton, C. Hohenkerk, G. Krasinsky, G. Petit, E. Pitjeva, M. Soffel and P. Wallace, The IAU 2009 system of astronomical constants: The report of the IAU working group on numerical standards for fundamental astronomy, *Celest. Mech. Dyn. Astron.* **110** (2011) 293. The Working Group's web pages are at http://maia.usno.navy.mil/NSFA/.
95. A. Fienga, H. Manche, J. Laskar, M. Gastineau and A. Verma, INPOP new release: INPOP13b/INPOP13c, arXiv:1405.0484v1/v2.
96. E. V. Pitjeva and N. P. Pitjev, Development of planetary ephemerides EPM and their applications, *Celest. Mech. Dyn. Astron.* **119** (2014) 237.
97. R. H. Dicke, J. R. Kuhn and K. G. Libbrecht, The variable oblateness of the Sun — Measurements of 1984, *Astrophys. J.* **311** (1986) 1025.
98. T. L. Duvall, Jr. *et al.*, Internal rotation of the Sun, *Nature* **310** (1984) 22.
99. I. Shapiro, Solar system tests of GR: Recent results and present plans, *General Relativity and Gravitation: Proc. 12th Int. Conf. General Relativity and Gravitation, University of Colorado at Boulder*, July 2–8, 1989, eds. N. Ashby, D. F. Bartlett and W. Wyss (Cambridge University Press, Cambridge, 1990), pp. 313–330.
100. S. Sofia, W. Heaps and L. W. Twigg, The solar diameter and oblateness measured by the solar disk sextant on the 1992 September 30 balloon flight, *Astrophys. J.* **427** (1994) 1048.
101. H. A. Hill and R. T. Stebbins, The intrinsic visual oblateness of the Sun, *Astrophys. J.* **200** (1975) 471.
102. T. J. Lydon and S. Sofia, A measurement of the shape of the solar disk: The solar quadrupole moment, the solar octopole moment, and the advance of perihelion of the planet Mercury, *Phys. Rev. Lett.* **76** (1996) 177.
103. F. P. Pijpers, Helioseismic determination of the solar gravitational quadrupole moment, *Mon. Not. R. Astron. Soc.* **297** (1998) L76.

104. S. Godier and J.-P. Roselot, Quadrupole moment of the Sun. Gravitational and rotational potentials, *Astron. Astrophys.* **350** (1999) 310.
105. M. Meftah, A. Irbah, A. Hauchecorne, T. Corbard, S. Turck-Chièze, J.-F. Hochedez, P. Boumier, A. Chevalier, S. Dewitte, S. Mekaoui and D. Salabert, On the determination and constancy of the solar oblateness, *Sol. Phys.* **290** (2015) 673.
106. M. Emilio, R. I. Bush, J. Kuhn and P. Scherrer, A changing solar shape, *Astrophys. J. Lett.* **660** (2007) L161.
107. M. D. Fivian, H. S. Hudson, R. P. Lin and H. J. Zahid, A large excess in apparent solar oblateness due to surface magnetism, *Science* **322** (2008) 560.
108. J. R. Kuhn, R. Bush, M. Emilio and I. F. Scholl, The precise solar shape and its variability, *Science* **337** (2012) 1638.
109. A. Irbah, M. Meftah, A. Hauchecorne, D. Djafer, T. Corbard, M. Bocquier and E. Momar Cisse, New space value of the solar oblateness obtained with Picard, *Astrophys. J.* **785** (2014) 89.
110. J. D. Anderson, E. L. Lau, S. Turyshev, J. G. Williams and M. M. Nieto, Recent results for solar-system tests of general relativity, *Bull. Am. Astron. Soc.* **34** (2002) 660.
111. J. G. Williams, J. D. Anderson, D. H. Boggs, E. L. Lau and J. O. Dickey, Solar system tests for changing gravity, *Bull. Am. Astron. Soc.* **33** (2001) 836.
112. A. Fienga *et al.*, The INPOP10a planetary ephemeris and its applications in fundamental physics, *Celest. Mech. Dyn. Astron.* **111** (2011) 363.
113. A. K. Verma, A. Fienga, J. Laskar, H. Manche and M. Gastineau, Use of MESSENGER radioscience data to improve planetary ephemeris and to test general relativity, *Astron. Astrophys.* **561** (2014) A115.
114. A. Fienga, J. Laskar, H. Manche and M. Gastineau, Tests of GR with INPOP15a planetary ephemerides: Estimations of possible supplementary advances of perihelia for Mercury and Saturn, to be published in *Proc. 14th Marcel Grossmann Meeting*, arXiv:1601.00947.
115. W. M. Folkner, J. G. Williams, D. H. Boggs, R. S. Park and P. Kuchynka, *JPL IOM* 42–196 (2014).
116. J. D. Anderson, M. Gross, K. L. Nordtvedt and S. G. Turyshev, The solar test of the equivalence principle, *Astrophys. J.* **459** (1996) 365–370.
117. W.-T. Ni, ASTROD mission concept and measurement of the temporal variation in the gravitational constant, pp. 309–320 in *Proc. Pacific Conf. Gravitation and Cosmology*, February 1–6, 1996, Seoul, Korea, eds. Y. M. Cho, C. H. Lee and S.-W. Kim (World Scientific, Singapore, 1996).
118. E. V. Pitjeva and N. P. Pitjev, Changes in the Sun's mass and gravitational constant estimated using modern observations of planets and spacecraft, *Sol. Syst. Res.* **46** (2012) 78.
119. A. Fienga, J. Laskar, P. Exertier, H. Manche and M. Gastineau, Tests of general relativity with planetary orbits and Monte Carlo simulations, arXiv:1409.4932.
120. R. F. Pinto, A. S. Brun, L. Jouve and R. Grappin, Coupling the solar dynamo and the corona: Wind properties, mass, and momentum losses during an activity cycle, *Astrophys. J.* **737** (2011) 72.
121. E. V. Pitjeva and N. P. Pitjev, Relativistic effects and dark matter in the solar system from observations of planets and spacecraft, *Mon. Not. R. Astron. Soc.* **432** (2013) 3431.
122. E. V. Pitjev and N. P. Pitjeva, Constraints on dark matter in the solar system, *Astron. Lett.* **39** (2013) 141.

123. D. S. Robertson, W. E. Carter and W. H. Dillinger, New measurement of solar gravitational deflection of radio signals using VLBI, *Nature* **349** (1991) 768.
124. D. E. Lebach *et al.*, Measurement of the solar gravitational deflection of radio waves using very-long-baseline interferometry, *Phys. Rev. Lett.* **75** (1995) 1439.
125. S. S. Shapiro, J. L. Davis, D. E. Lebach and J. S. Gregory, Measurement of the solar gravitational deflection of radio waves using geodetic very-long-baseline interferometry data, 1979–1999, *Phys. Rev. Lett.* **92** (2004) 121101(4).
126. E. Fomalont, S. Kopeikin, G. Lanyi and J. Benson, Progress in measurements of the gravitational bending of radio waves using the VLBA, *Astrophys. J.* **699** (2009) 1395.
127. E. B. Fomalont and S. M. Kopeikin, The measurement of the light deflection from Jupiter: Experimental results, *Astrophys. J.* **598** (2003) 704.
128. E. Fomalont and S. Kopeikin, Radio interferometric tests of general relativity, *IAU Symp.* **248** (2008) 383.
129. B. Bertotti, L. Iess and P. Tortora, A test of general relativity using radio links with the Cassini spacecraft, *Nature* **425** (2003) 374.
130. ESA, ed., ESA Special Publication, Vol. 1200, The HIPPARCOS and TYCHO catalogues. Astrometric and photometric star catalogues derived from the ESA HIPPARCOS Space Astrometry Mission (1997).
131. M. Frœschlé, F. Mignard and F. Arenou, in *Proc. ESA Symp. "Hipparcos-Venice'97"*, 13–16 May, Venice, Italy, ESA SP-402 (July 1997), p. 49.
132. Gaia Team, http://sci.esa.int/gaia/.
133. L. Lindegren *et al.*, Gaia data release 1. Astrometry: One billion positions, two million proper motions, and parallaxes, *Astron. Astrophys.*, arXiv:1609.04303.
134. A. Vecchiato *et al.*, Testing general relativity by micro-arcsecond global astrometry, *Astron. Astrophys.* **399** (2003) 337.
135. J. G. Williams, X. X. Newhall and J. O. Dickey, Relativity parameters determined from lunar laser ranging, *Phys. Rev. D* **53** (1996) 6730.
136. J. G. Williams, S. G. Turyshev and D. H. Boggs, Progress in lunar laser ranging tests of relativistic gravity, *Phys. Rev. Lett.* **93** (2004) 261101(4).
137. J. G. Williams, S. G. Turyshev and D. H. Boggs, Lunar laser ranging tests of the equivalence principle, *Class. Quantum Grav.* **29** (2012) 184004(11).
138. J. Müller, L. Biskupek, F. Hofmann and E. Mai, Lunar laser ranging and relativity, *Frontiers in Relativistic Celestial Mechanics — Volume 2 : Applications and Experiments*, ed. S. Kopeikin (deGruyter, Berlin, Boston, 2014), pp. 103–156.
139. J. Müller, F. Hofmann, X. Fang and L. Biskupek, Lunar Laser Ranging: Recent Results Based on Refined Modelling, eds. C. Rizos and P. Willis, *Earth on the Edge: Science for a Sustainable Planet, Int. Association of Geodesy Symp. 139*, doi: 10.1007/978-3-642-37222-3_59 (Springer-Verlag, Berlin, Heidelberg, 2014).
140. J. Müller and L. Biskupek, Variations of the gravitational constant from lunar laser ranging data, *Class. Quantum Grav.* **24** (2007) 4533.
141. L. Iorio, Constraints on the range λ of Yukawa-like modifications to the Newtonian inverse-square law of gravitation from solar system planetary motions, *J. High Energy Phys.* **2007** (2007) 10.
142. F. Hofmann, J. Müller, L. Biskupek, E. Mai and J. M. Torre, Lunar laser ranging — What is it good for? *Proc. 18th Int. Workshop on Laser Ranging*, 13-0402 (2013).
143. Y. Su *et al.*, New tests of the universality of free fall, *Phys. Rev. D* **50** (1994) 3614; S. Baessler, *et al.*, Improved test of the equivalence principle for gravitational self-energy, *Phys. Rev. Lett.* **83** (1999) 3585; E. G. Adelberger, New tests of Einstein's

equivalence principle and Newton's inverse-square law, *Class. Quantum Grav.* **18** (2001) 2397.
144. J. Lense and H. Thirring, *Phys. Zeits.* **19** (1918) 156; *Gen. Relativ. Gravit.* **16** (1984) 712.
145. L. I. Schiff, Possible new experimental test of general relativity theory, *Phys. Rev. Lett.* **4** (1960) 215.
146. I. Ciufonlini and E. C. Pavlis, A confirmation of the general relativistic prediction of the Lense–Thirring effect, *Nature* **431** (2004) 958; N. Ashby, General relativity: Frame-dragging confirmed, *Nature* **431** (2004) 918.
147. GP-B (Gravity Probe B) http://einstein.stanford.edu/; C. W. F. Everitt, S. Buchman, D. B. DeBra, G. M. Keiser, J. M. Lockhart, B. Muhlfelder, B. W. Parkinson, J. P. Turneaure and other members of the Gravity Probe B team, "Gravity Probe B: Countdown to Launch", in *Gyros, Clocks, Interferometers ... : Testing Relativistic Gravity in Space*, eds. C. Lämmerzahl, C. W. F. Everitt and F. W. Hehl (Berlin, Springer-Verlag, 2001), pp. 52–82.
148. C. W. F. Everitt *et al.*, Gravity probe B: Final results of a space experiment to test general relativity, *Phys. Rev. Lett.* **106** (2011) 221101.
149. W.-T. Ni, Rotation, the equivalence principle, and the gravity probe B experiment, *Phys. Rev. Lett.* **107** (2011) 051103.
150. I. Ciufolini *et al.*, A test of general relativity using the LARES and LAGEOS satellites and a GRACE earth's gravity model, *Eur. Phys. J. C* **76** (2016) 120.
151. M. Armano *et al.*, Sub-femto-g free fall for space-based gravitational wave observatories: LISA Pathfinder results, *Phys. Rev. Lett.* **116** (2016) 231101.
152. T. Damour and K. Nordtvedt, Jr., General relativity as a cosmological attractor of tensor-scalar theories, *Phys. Rev. Lett.* **70** (1993) 2217.
153. T. Damour, F. Piazza and G. Veneziano, Violations of the equivalence principle in a dilaton-runaway scenario, *Phys. Rev. D* **66** (2002) 046007.
154. BepiColombo Collab., http://sci.esa.int/bepicolombo/.
155. A. Milani, D. Vokrouhlicky, D. Villani, C. Bonanno and A. Rossi, Testing general relativity with the BepiColombo radio science experiment, *Phys. Rev. D* **66** (2004) 082001.
156. H. Selig, C. Lämmerzahl and W.-T. Ni, Astrodynamical space test of relativity using optical devices I (ASTROD I) — Mission Overview, *Int. J. Mod. Phys. D* **22** (2013) 1341003.
157. T. Appouchaux *et al.*, Astrodynamical space test of relativity using optical devices I (ASTROD I) — A class-M fundamental physics mission proposal for Cosmic Vision 2015-2025, *Exp. Astron.* **23** (2009) 491.
158. W.-T. Ni *et al.*, ASTROD I: Mission concept and Venus flybys, in *Proc. 5th IAA Int. Conf. Low-Cost Planetary Missions, ESTEC, Noordwijk*, The Netherlands, 24–26 September 2003, ESA SP-542, November 2003, 79–86.
159. W.-T. Ni *et al.*, ASTROD I: Mission concept and Venus flybys, *Acta Astron.* **59** (2006) 598.
160. W.-T. Ni, ASTROD and ASTROD I — Overview and Progress, *Int. J. Mod. Phys. D* **22** (2008) 921.
161. W.-T. Ni, Super-ASTROD: Probing primordial gravitational waves and mapping the outer solar system, *Class. Quantum Grav.* **26** (2009) 075021.
162. B. Christophe *et al.*, Odyssey: A solar system mission, *Exp. Astron.* **23** (2009) 529.
163. P. Wolf *et al.*, Quantum physics exploring gravity in the outer solar system: The SAGAS project, *Exp. Astron.* **23** (2009) 651.

164. B. Christophe *et al.*, OSS (Outer Solar System): A fundamental and planetary physics mission to Neptune, Triton and the Kuiper Belt, *Exp. Astron.* **34** (2012) 203.
165. P. Touboul, M. Rodrigues, G. Métris and B. Tatry, MICROSCOPE, testing the equivalence principle in space, *C. R. Acad. Sci. Ser.* IV **2** (2001) 1271.
166. MICROSCOPE Collab., https://presse.cnes.fr/en/cnes-onera-cooperation-first-ultra-precise-measurements-microscope.

Chapter 9

Pulsars and gravity

R. N. Manchester
CSIRO Astronomy and Space Science,
Epping NSW 1710, Australia
dick.manchester@csiro.au

Pulsars are wonderful gravitational probes. Their tiny size and stellar mass give their rotation periods a stability comparable to that of atomic frequency standards. This is especially true of the rapidly rotating "millisecond pulsars" (MSPs). Many of these rapidly rotating pulsars are in orbit with another star, allowing pulsar timing to probe relativistic perturbations to the orbital motion. Pulsars have provided the most stringent tests of theories of relativistic gravitation, especially in the strong-field regime, and have shown that Einstein's general theory of relativity is an accurate description of the observed motions. Many other gravitational theories are effectively ruled out or at least severely constrained by these results. MSPs can also be used to form a "Pulsar Timing Array" (PTA). PTAs are Galactic-scale interferometers that have the potential to directly detect nanohertz gravitational waves from astrophysical sources. Orbiting super-massive black holes in the cores of distant galaxies are the sources most likely to be detectable. Although no evidence for gravitational waves has yet been found in PTA data sets, the latest limits are seriously constraining current ideas on galaxy and black-hole evolution in the early universe.

Keywords: Gravity; gravitational waves; pulsars; pulsar timing; general relativity.

1. Introduction

Pulsars are rotating neutron stars that emit beams of radiation which sweep across the sky as the star rotates. A beam sweeping across the Earth may be detected as a pulse that repeats with a periodicity equal to the rotation period of the star. Because of the large mass of neutron stars, $\sim 1.4\,\mathrm{M}_\odot$, and their tiny size, radii $\sim 15\,\mathrm{km}$, (see Ref. 76 for a review of neutron-star properties) the rotation period of neutron stars is incredibly stable, with a stability comparable to that of the best atomic clocks on Earth. This great period stability, combined with the fact that pulsars are often in a binary orbit with another star, makes them wonderful probes of relativistic gravity. Tiny perturbations to their period resulting from, for example, relativistic effects in a binary orbit, may be detected and compared with the predictions of a gravitational theory. Pulsars may also be used as detectors for gravitational waves passing through the Galaxy. To separate the effects of gravitational waves from other perturbations, signals from pulsars in different directions on the sky must be compared — exactly analogous to the way laser-interferometer gravitational-wave detectors compare laser phases in perpendicular arms.

More than 2400 pulsars are now known.[a] The vast majority of them reside in our Galaxy, typically at distances of a few thousand light-years from the Sun. Their pulse periods, P, range between 1.4 milliseconds and 12 s and fall into two main groups. The so-called "normal" pulsars have periods longer than about 30 milliseconds and the "millisecond" pulsars (MSPs) have shorter periods. MSPs comprise about 15% of the known pulsar population.

Although pulsar periods are very stable, they are not constant. In their own reference frame, all pulsars are slowing down, albeit very slowly. Pulsars are powered by their rotational energy. They have extremely strong magnetic fields and, as they spin, they emit streams of relativistic particles and so-called "dipole radiation", electromagnetic waves with period equal to the rotation period of the star. These carry energy away from the star resulting in a steady increase in the pulse period. Typical rates of period increase, $\dot{P}$, are a part in 10^{15} for normal pulsars and much less for MSPs. Assuming that the surrounding magnetic field is predominantly dipole, the characteristic age τ_c of a pulsar is given by

$$\tau_c = \frac{P}{(2\dot{P})}, \tag{1}$$

and their surface dipole magnetic field strength (in gauss[b]) is

$$B_s \approx 3.2 \times 10^{19} (P\dot{P})^{1/2}\,\mathrm{G}. \tag{2}$$

Figure 1 shows the distribution of pulsars on the $P - \dot{P}$ plane, with several different types of pulsars indicated. For most normal pulsars, τ_c is between 10^3 and 10^7 years and B_s is between 10^{11} and 10^{13} G. For MSPs, the corresponding ranges are 10^9 to 10^{11} yrs and 10^8 to 10^{10} G.

Normal pulsars and MSPs have quite different evolutionary histories. Most if not all normal pulsars are formed in supernova explosions at the death of a massive star. They age with relatively constant B_s until the pulse emission mechanism begins to fail when τ_c reaches about 10^6 yrs. Many young pulsars ($\tau_c \lesssim 10^4$ yrs) are located within supernova remnants, with the most famous example being the Crab pulsar, PSR B0531+21, located near the center of the Crab Nebula. Most young pulsars lie relatively close to the Galactic Plane, consistent with the idea that they are formed from massive stars. MSPs on the other hand are much more widely distributed in the Galactic halo. They are believed to originate from old, slowly rotating and probably dead neutron stars that accrete matter and angular momentum from an evolving binary companion. This "recycling" process increases their spin rate so that they have periods in the millisecond range and re-energizes the beamed emission (see, e.g. Ref. 17).

Figure 1 shows that the majority of MSPs are members of a binary system, consistent with this formation scenario. The accretion also suppresses their apparent

[a]See the ATNF Pulsar Catalogue: http://www.atnf.csiro.au/research/pulsar/psrcat.
[b]1 gauss (G) = 10^{-4} tesla.

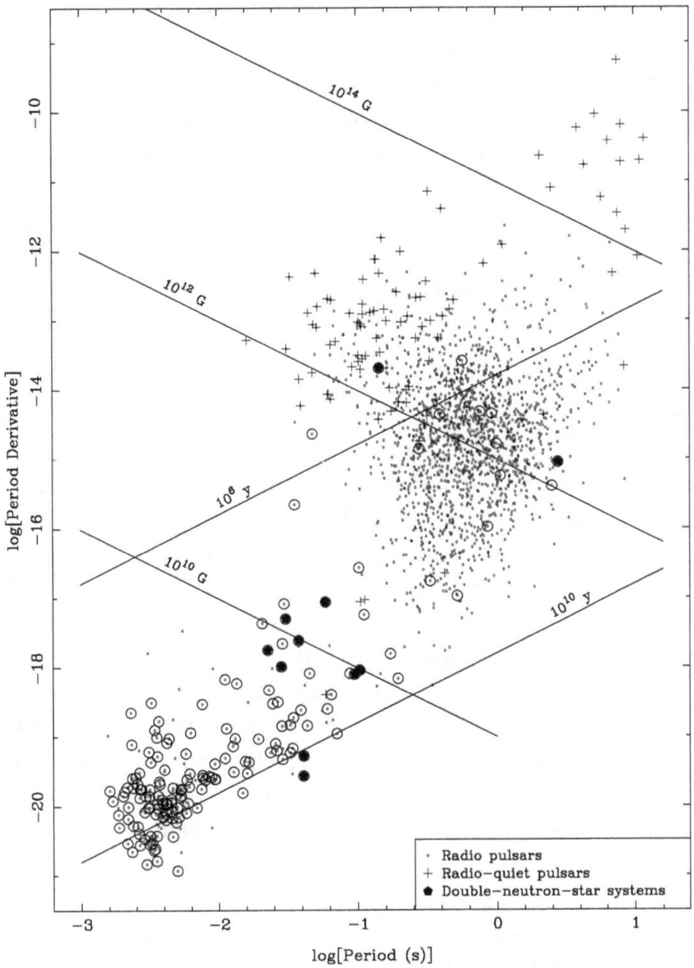

Fig. 1. Distribution of pulsars on the $P - \dot{P}$ plane with radio-loud and radio-quiet pulsars indicated. Binary pulsars are indicated by a circle around the point and, for those with a neutron-star companion, the circle is filled in. Lines of constant characteristic age (τ_c) and surface-dipole magnetic field (B_s) are shown.

magnetic field, so that MSPs have $\dot{P}$ values about five orders of magnitude less than normal pulsars. Since the level of intrinsic period irregularities is related to $\dot{P}$,[117] it is this low B_s that makes MSPs extremely stable clocks and suitable tools for the study of relativistic gravitation. Figure 1 also indicates the class of radio-quiet pulsars. Essentially all of the 72 known radio-quiet pulsars are young and solitary. They include the so-called "magnetars" which lie in the upper right side of the $P - \dot{P}$ diagram where dipole magnetic fields are strongest.

Most double-neutron-star systems lie in the zone between the MSPs and the normal pulsars. These systems are believed to have been partially recycled prior to the formation of the second-born neutron star. In almost every case, the observed pulsar is the recycled one since it has a much longer active lifetime than the newly formed young pulsar. A famous exception to this is the Double Pulsar (PSR J0737−3039A/B) in which the second-born star (B) is (or was) still visible.[83] In Fig. 1, the B pulsar is the solitary double-neutron-star system on the right-hand side of the plot. The double-neutron-star system identified near the middle of the plot contains the relatively young pulsar, PSR J1906+0746. There is some doubt about the nature of the companion in this system — it could be a heavy white dwarf.[139] These double-neutron-star systems and their use as probes of relativistic gravity are discussed in some detail in Sec. 2.1.

1.1. *Pulsar timing*

The most important contributions of pulsars to investigations of gravitational theories and gravitational waves rely on precision pulsar timing observations. These allow both the relativistic perturbations to binary orbits to be studied in detail and the potential detection of the tiny period fluctuations generated by gravitational waves passing through our Galaxy. Because of the importance of pulsar timing to these studies, we give here a brief review of its basic principles.

The basic observable in pulsar timing is the time of arrival (ToA) of a pulse at an observatory. In fact, because of signal/noise limitations and the intrinsic fluctuation of individual pulse shapes, ToA measurements are based on mean pulse profiles formed by synchronously averaging the data, typically for times between several minutes and an hour. The time at which a fiducial pulse phase (usually near the pulse peak) arrives at the telescope is determined by cross-correlating the observed mean pulse profile with a standard pulse template. A series of these ToAs is measured over many days, months, years and even decades for the pulsar of interest.

These observatory ToAs are affected by the rotational and orbital motion of the Earth (and for satellite observatories, the orbital motion of the satellite). To remove these effects, the observed ToAs are referred to the solar-system barycenter (center of mass) which is assumed to be inertial (unaccelerated) with respect to the distant universe.[c] This correction makes use of a solar-system ephemeris giving predictions of the position of the center of the Earth with respect to the solar-system barycenter. Such ephemerides, for example, the Jet Propulsion Laboratory ephemeris DE 421,[48] are generated by fitting relativistic models to planetary and spacecraft data. The correction also takes into account the relativistic variations in terrestrial time resulting from the Earth's motion.

The resulting barycentric ToAs are then compared with predicted pulse ToAs based on a model for the pulsar. The pulsar model can have 20 or more parameters; generally included are the pulse frequency ($\nu = 1/P$), frequency time derivative

[c]This neglects any acceleration of the solar-system barycenter resulting from, for example, Galactic rotation. For some precision timing experiments, such effects are taken into account at a later stage of the analysis.

($\dot{\nu}$), the pulsar position and the five Keplerian binary parameters if the pulsar is a member of a binary system. If the data are recorded at different radio frequencies, it is also necessary to include the dispersion measure (DM) which quantifies the frequency-dependent delay suffered by the pulses as they propagate through the interstellar medium.

The differences between the observed and predicted ToAs are known as "timing residuals". Errors in any of the model parameters result in systematic variations in the timing residuals as a function of time. For example, if the model pulse frequency is too small, the residuals will grow linearly as illustrated in Fig. 2. The required correction to the pulse frequency is given by the slope of this variation. An error in the pulsar position results in an annual sine curve which arises from the barycentric correction. The phase and amplitude of this curve give the corrections to the two position coordinates. Similarly, a pulsar proper motion results in a linearly growing sine curve (away from the reference epoch). For a sufficiently

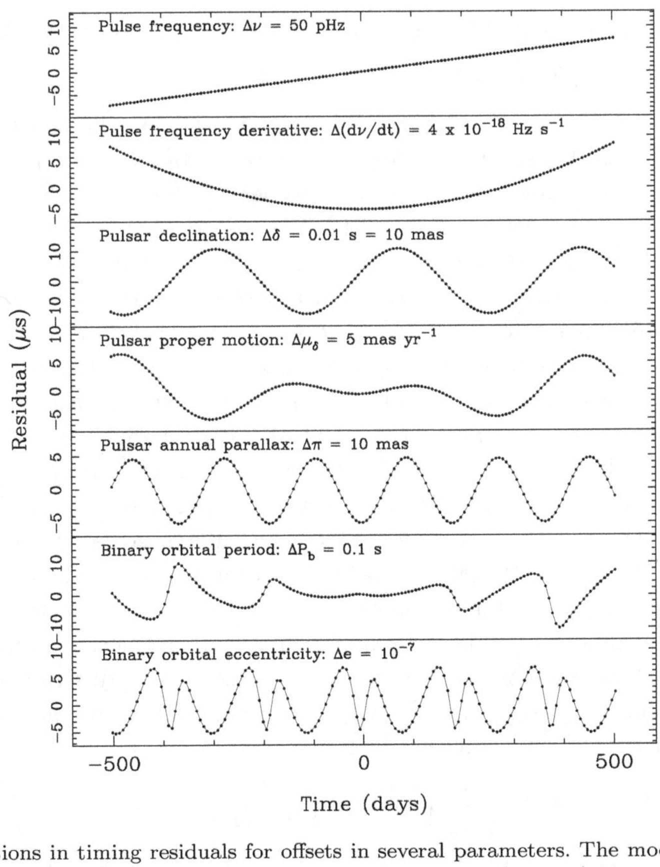

Fig. 2. Variations in timing residuals for offsets in several parameters. The model pulsar has a pulse frequency of 300 Hz (3.3 ms period) and is in an eccentric ($e = 0.4$) binary orbit of period 190 days. The reference epochs for period, position and binary phase are at the middle of the plotted range.

close pulsar, the curvature of the wavefront results in a biannual term in the residuals, and offsets in binary parameters result in terms which vary with the orbital periodicity.

Since offsets in different parameters in general result in different systematic residual variations, it is possible to do a least-squares fit to the observed residuals for the correction to any desired parameter.[d] The precision of the fitted parameters is often amazingly high. For example, the relative precision of the pulse frequency determination is $\sim \delta t/T$, where δt is the typical uncertainty in the ToAs and T is the data span. Since, for MSPs at least, δt is often $< 1\mu$s and T is often many years, the relative precision of the measured ν can easily exceed 1:10^{14}. Similarly, pulsar positions can be measured to micro-arcseconds and binary eccentricities measured to $1 : 10^8$. These very high precisions often allow higher-order terms, for example, resulting from relativistic perturbations to the binary orbit, to be measured.

2. Tests of Relativistic Gravity

2.1. *Tests of general relativity with double-neutron-star systems*

2.1.1. *The Hulse–Taylor binary, PSR B*1913 + 16

The discovery at Arecibo Observatory in 1974 of the first-known binary pulsar, PSR B1913+16, by Hulse and Taylor[61] was remarkable in a number of respects. First, it showed that pulsars with short pulse periods but large characteristic ages (59 ms and 10^8 yr, respectively, for PSR B1913+16) could exist. The period of PSR B1913+16 was second only to the Crab pulsar, but its age was enormously greater than that of other short-period pulsars known at the time. Second, it was in a binary orbit with a relatively massive star, very likely another neutron star, showing that an evolutionary pathway to such systems existed. Thirdly, its orbital period was extraordinarily short, only 7.75 h, its eccentricity large, ~ 0.617, and, as shown in Fig. 3, its maximum orbital velocity very high, $\sim 300\,\mathrm{km\,s^{-1}}$ or 0.1% of the velocity of light. As was immediately recognized by Hulse and Taylor, these properties opened up the possibility that relativistic perturbations to the orbit were potentially measurable. Lowest-order relativistic effects go as $(v/c)^2$, and so the variations are of order 1:10^6, enormous by the standards of pulsar measurements.

Relativistic effects in binary motion can be expressed in terms of "post-Keplerian" parameters that describe departures from Keplerian motion (see, e.g. Ref. 129). The first such parameter to be observed was periastron precession.[135] In Einstein's general theory of relativity (GR) the rate of periastron precession (averaged over the binary orbit) is given by:

$$\dot{\omega} = 3\left(\frac{P_b}{2\pi}\right)^{-5/3}(T_\odot M)^{2/3}(1-e^2)^{-1}, \tag{3}$$

[d]The data span must be sufficiently long to avoid excessive covariance between the variations for different fitted parameters.

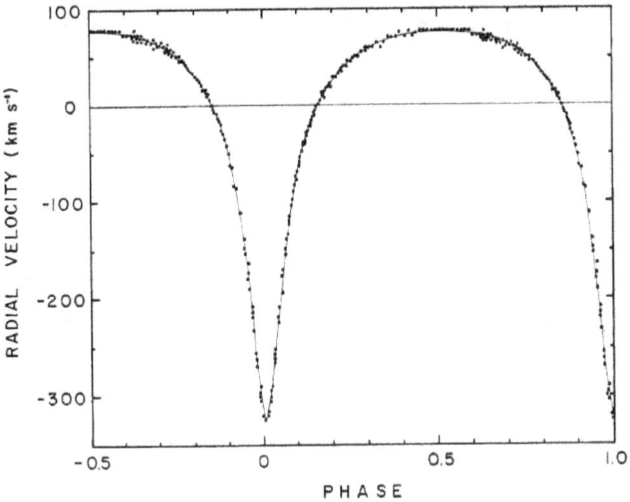

Fig. 3. Velocity curve for the Hulse–Taylor binary pulsar, PSR B1913+16 (Ref. 61).

where ω is the longitude of periastron measured from the ascending node, P_b is the orbital period, $T_\odot \equiv GM_\odot/c^3 = 4.925490947\mu s$, G is the Gravitational Constant, $M_\odot$ is the mass of the Sun, e is the orbital eccentricity and $M = m_1 + m_2$, the sum of the pulsar mass m_1 and the companion mass m_2 in solar units. It is worth noting that this is the same effect as the excess perihelion advance of Mercury that was used by Einstein[43] as an observational verification of GR. The relativistic effect for Mercury is just 43 arcsec per century, minuscule compared to the $4.^\circ 22$ per year predicted and observed for PSR B1913+16.

The next most significant parameter, normally labeled γ, describes the combination of gravitational redshift and 2nd-order (or transverse) Doppler shift, both of which have the same dependence on orbital phase. In GR

$$\gamma = e \left(\frac{P_b}{2\pi}\right)^{1/3} T_\odot^{2/3} M^{-4/3} m_2(m_1 + 2m_2). \tag{4}$$

Since the Keplerian parameters are very well determined, measurement of $\dot{\omega}$ and γ gives two equations in two unknowns, m_1 and m_2, and so the two stellar masses can be determined. For PSR B1913+16, these are both close to $1.4\,M_\odot$, confirming the double-neutron-star nature of the system. An important consequence of this was that the two stars could safely be treated as point masses in GR, thereby allowing precise tests of the theory.

Given the Keplerian parameters and the two masses, the next post-Keplerian parameter, orbital decay due to the emission of gravitational waves from the system, given in GR by

$$\dot{P}_b = -\frac{192\pi}{5}\left(\frac{P_b}{2\pi}\right)^{-5/3}\left(1 + \frac{73}{24}e^2 + \frac{37}{96}e^4\right)(1-e^2)^{-7/2}\, T_\odot^{5/3}\, m_1 m_2 M^{-1/3}, \tag{5}$$

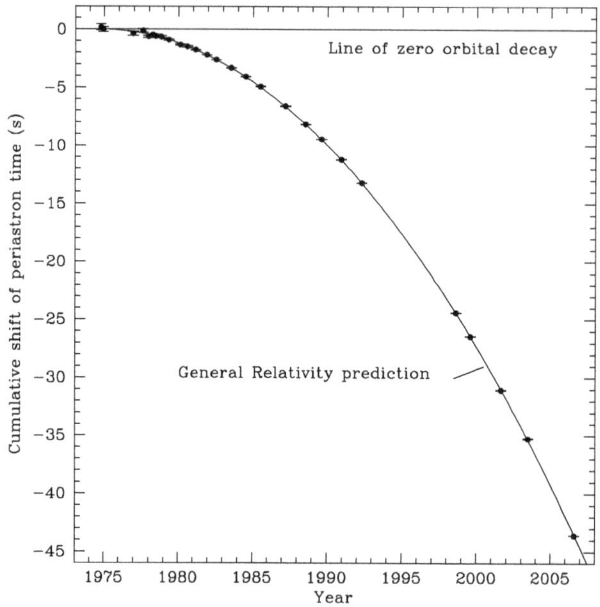

Fig. 4. Comparison of the observed and predicted orbital period decay in PSR B1913+16. The orbital decay is quantified by the shift in the time of periastron passage with respect to a nondecaying orbit. The parabolic curve is the predicted decay from GR (Ref. 143).

could be predicted and compared with observation. This therefore constituted a unique test of GR in strong-field gravity. After just four years, the $\dot{P}_b$ term was well measured and shown to be consistent with the GR prediction.[134] The orbital decay, including more recent results, is illustrated in Fig. 4; the ratio of observed to predicted orbit decay is 0.997 ± 0.002.[143] It should be mentioned that this near-perfect agreement relies on correcting the observed $\dot{P}_b$ for the differential acceleration of the solar system and the pulsar system in the gravitational field of the Galaxy. This correction is $-0.027 \pm 0.005 \times 10^{-12}$ or about 1% of the observed value, and the uncertainty in the final result is dominated by the uncertainty in this correction. Unfortunately, it is unlikely that this uncertainty will be significantly reduced in the near future since it mainly depends on the poorly known distance to the binary system.

Two more post-Keplerian parameters, denoted by r and s, relate to the Shapiro delay suffered by the pulsar signal while passing through the curved spacetime surrounding the companion star.[123] The relations for them in GR are as follows:

$$r = T_\odot m_2, \tag{6}$$

$$s \equiv \sin i = \left(\frac{a_1}{c}\right) \sin i \left(\frac{P_b}{2\pi}\right)^{-2/3} T_\odot^{-1/3} M^{2/3} m_2^{-1}, \tag{7}$$

where $(a_1/c)\sin i$ is the projected semi-major axis of the pulsar orbit (in time units), a Keplerian parameter. The Shapiro delay is generally only observable when the orbital inclination is relatively close to 90°, that is, the orbit is seen close to edge-on. For the Hulse–Taylor binary pulsar the orbital inclination is about 47° and the Shapiro delay is small and covariant with Keplerian parameters. However it has been observed in a number of other neutron-star binary systems as will be discussed below.

Another relativistic effect, geodetic precession, has observable consequences for PSR B1913+16 and several other binary pulsars. A neutron star formed in a supernova explosion receives a "kick" during or shortly after the explosion which typically gives the star a velocity of several hundred $\mathrm{km\,s^{-1}}$.[59] If the pulsar is a member of a binary system which is not disrupted by the kick, the orbital axis is changed so that pulsar spin axis, which was most likely aligned with the orbital axis prior to the explosion, is no longer aligned. Since kick velocities are often comparable to or even larger than the orbital velocities, the misalignment angle can be large. Precession of the spin axis will therefore alter the aspect of the radio beam seen by an observer and may even move the beam out of the line of sight.

In GR, the precessional angular frequency is given by

$$\Omega_p = \frac{1}{2}\left(\frac{P_b}{2\pi}\right)^{-5/3} T_\odot^{2/3} \frac{m_2(4m_1 + 3m_2)}{(1 - e^2)M^{4/3}}. \tag{8}$$

For PSR B1913+16, the corresponding precessional period is about 297 years, so the aspect changes over observational data spans are small. Never-the-less changes in the relative amplitude of the two peaks in the PSR B1913+16 profile were reported by Weisberg *et al.*[144] and attributed to the effects of geodetic precession with a "patchy" beam. For a basically conal beam geometry, the separation of the two profile components would be expected to change and evidence for this was first found by Kramer[72] leading to an estimate of the misalignment angle of about 22°. A data set extending to 2001 was analyzed by Weisberg and Taylor,[145] obtaining results similar to those of Kramer.[72] Their best-fitting model gives the "peanut" shaped beam shown in Fig. 5. Clifton and Weisberg[28] have shown that a set of circular nested emission cones can also give apparent pulse-width variations similar to those observed. Over the 20-year interval covered by the data set, the "impact parameter" (minimum angle between the beam symmetry axis and the observer's line of sight) has changed by a rather small amount, about 3°, from $-3\overset{\circ}{.}5$ to $-6\overset{\circ}{.}5$. Extrapolation of this model suggests that the pulsar will become unobservable in about 2025. While this result is compatible with relativistic precession, it is not possible to derive an independent measure of the precessional rate from these data.

2.1.2. *PSR B*1534 + 12

PSR B1534+12, discovered by Wolszczan in 1990,[152] is a binary system with parameters quite similar to those of B1913+16, notably a short orbital period (~10.1 hours), relatively high eccentricity (~0.27) and a compact orbit about 60% larger than that of PSR B1913+16. Analysis of less than a year's data already

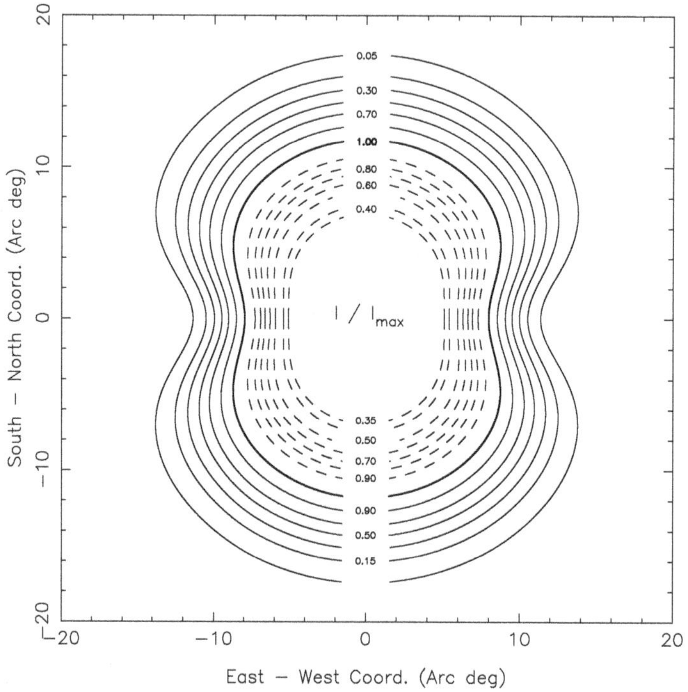

Fig. 5. Contours of the symmetric part of the radio emission beam for PSR B1913+16, obtained by fitting to the variation in pulse width over the interval 1981 to 2001 (Ref. 145).

allowed measurement of two post-Keplerian parameters, $\dot{\omega}$ and γ, thereby determining the masses of the two stars and confirming that they are both neutron stars. These results also showed that the orbit was more edge-on than that of PSR B1913+16, with an implied inclination angle of about 77°. Analysis of timing data extending over 22 years by Fonseca *et al.*[49] built on earlier results by Stairs *et al.*,[131] with significant detections of r and s, the Shapiro-delay parameters, and the orbital decay, $\dot{P}_b$, giving five post-Keplerian parameters and three independent tests of GR. A fourth test of GR, albeit less precise, was provided by an analysis of the evolution of the profile shape and polarization, yielding a rate for the geodetic precession of the pulsar spin axis $\Omega_p = 0\overset{\circ}{.}59^{+0.12}_{0.08}\ \mathrm{yr}^{-1}$ which is consistent with the value predicted by GR. Figure 6 shows the so-called "mass–mass" diagram for PSR B1534+12, illustrating these constraints.

If GR gives a correct description of the post-Keplerian parameters, all of these constraints should be consistent with an allowed range of m_1 and m_2. For PSR B1534+12, the masses are most accurately constrained by $\dot{\omega}$ and γ, but the predicted constraint on $\dot{P}_b$ appears to be inconsistent. As for PSR B1913+16, the measured value of $\dot{P}_b$ must be corrected for kinematic effects resulting from the differential acceleration of the binary and solar system in the Galaxy. PSR B1534+12 is much closer to the Sun that PSR B1913+16 and the so-called "Shklovskii" effect,[125]

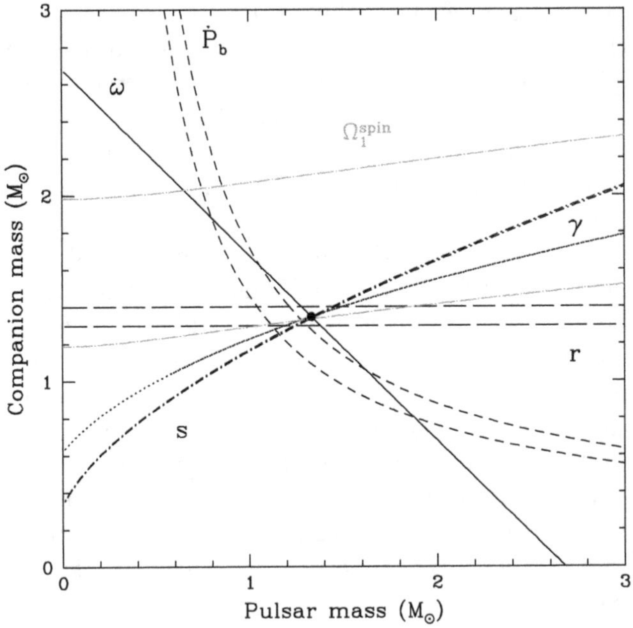

Fig. 6. Plot of companion mass (m_2) versus pulsar mass (m_1) for PSR B1534+12. Constraints derived from the measured post-Keplerian parameters assuming GR are shown pairs of lines with separation indicating the uncertainty range. For $\dot{\omega}$, γ and s, the uncertainty ranges are too small to see on the plot. In addition, a constraint from the measured spin precession rate is shown. (Ref. 49).

an apparent radial acceleration resulting from transverse motion of the binary system ($\dot{P}/P = \mu^2 d/c$, where μ is the pulsar proper motion and d is the pulsar distance), is also important. The distance estimate is based on the pulsar DM and a model for the free electron density in the Galaxy and is quite uncertain. Stairs and her colleagues[49,131] inverted the problem, assuming that the GR prediction for $\dot{P}_b$ is correct, thereby deriving an improved value for the pulsar distance, a technique first suggested by Bell and Bailes.[12]

2.1.3. *The double pulsar, PSR J0737 − 3039A/B*

The discovery of the double pulsar system[21,83] heralded a remarkable era for investigation of relativistic effects in double-neutron-star systems. In this system, the A pulsar was formed first and subsequently spun up to approximately its current period of 23 ms by accretion from its evolving binary companion. The companion then imploded to form the B pulsar which has since spun down to its current period of about 2.8 s. The orbital period is only 2.5 h, less than a third of that for PSR B1913+16, and the projected semi-major axis $a_1 \sin i$ is about 60% that of PSR B1913+16. These parameters imply relativistic effects much larger than those seen for PSR B1319+16. For example, the predicted relativistic periastron advance is

$16\overset{\circ}{.}9\,\mathrm{yr}^{-1}$, more than four times the value for PSR B1913+16. Added to that, the orbit is seen within a degree or so of edge-on, not only allowing detailed measurement of the Shapiro delay, but also resulting in eclipses of the A pulsar emission by the magnetosphere of the B pulsar. Finally, the still-unique detection of the second neutron star as a pulsar allows a direct measurement of the mass ratio of the two stars from the ratio of the two nonrelativistic Roemer delays ($a_1 \sin i$).[e]

Four post-Keplerian parameters ($\dot{\omega}$, γ, r and s) were detected in just seven months of timing data from the Parkes 64-m and Lovell 76-m Telescopes.[83] Further observations, including data from the Green Bank 100 m Telescope, with a 2.5-year data span give the currently most stringent test of GR in strong-field conditions.[74,75] Figure 8 shows the mass–mass diagram based on these results together with the measurement of geodetic spin precession for the B pulsar described below. A total of six post-Keplerian parameters together with the mass ratio R gives five independent tests of GR. As well as the three post-Keplerian parameters, $\dot{\omega}$, γ and $\dot{P}_b$ observed for PSRs B1913+16 and PSR B1534+12 (Fig. 6), for the double pulsar we have the mass ratio R, the Shapiro delay parameters r and s and a measurement of the rate of geodetic precession Ω_p. The observed Shapiro delay (Fig. 7) shows that the J0737−3039A/B orbit inclination angle is $88\overset{\circ}{.}7 \pm 0\overset{\circ}{.}7$.

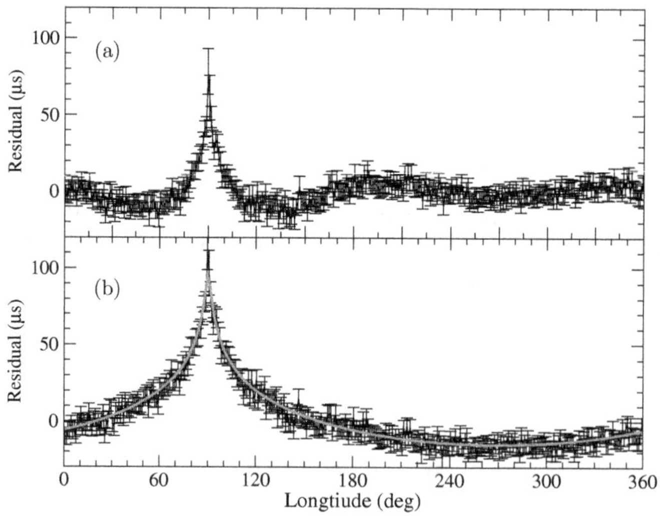

Fig. 7. Observed Shapiro delay as a function of orbital phase for PSR J0737−3039A. Upper panel: Timing residuals after fitting for all parameters except the Shapiro-delay terms, r and s, with these set to zero. Lower panel: The full Shapiro delay obtained by taking the best-fit values from a full solution, but setting r and s to zero. The grey line is the prediction based on GR (Ref. 74).

[e]The B pulsar became undetectable as a radio pulsar in 2008, most probably because the beam precessed out of our line of sight.[101] Because of uncertainty about the beam shape, the date of its return to visibility is very uncertain, but it should be before 2035.

As mentioned above, this nearly edge-on view of the orbit results in eclipses of the A-pulsar emission by the magnetosphere of the B pulsar. These eclipses last only about 30 s, showing that the B-pulsar magnetosphere is highly modified by the relativistic wind from pulsar A.[83] Remarkably, observations with high time resolution made with the Green Bank Telescope showed that the eclipse is modulated at the spin period of pulsar B.[88] Modeling of the detailed eclipse profile by absorption in the doughnut-shaped closed-field-line region of the magnetosphere by Lyutikov and Thompson[84] allowed determination of the geometry of the binary system including the offset between the B-pulsar spin axis and the orbital angular momentum axis, the so-called "misalignment angle" which they estimate to be about 60°. Even more remarkably, detailed measurements of the eclipse profile over about four years enabled Breton *et al.*[20] to directly estimate the rate of geodetic precession as $4^\circ\!.77 \pm 0^\circ\!.66\,\mathrm{yr}^{-1}$, consistent with the predicted precessional period of 70.95 years based on GR [Eq. (8)]. It is notable that no secular profile evolution has been observed for PSR J0737−3039A, implying that, unlike for the B pulsar, the misalignment angle for pulsar A is very small.[45]

The most precisely measured post-Keplerian parameter is $\dot{\omega}$. This constraint is nearly orthogonal to that from the mass ratio R (Fig. 8) giving values for the two neutron-star masses of $m_1 = 1.3381 \pm 0.0007\,\mathrm{M}_\odot$ and $m_2 = 1.2849 \pm 0.0007\,\mathrm{M}_\odot$.[74] Together with the accurately measured Keplerian parameters, these two masses can

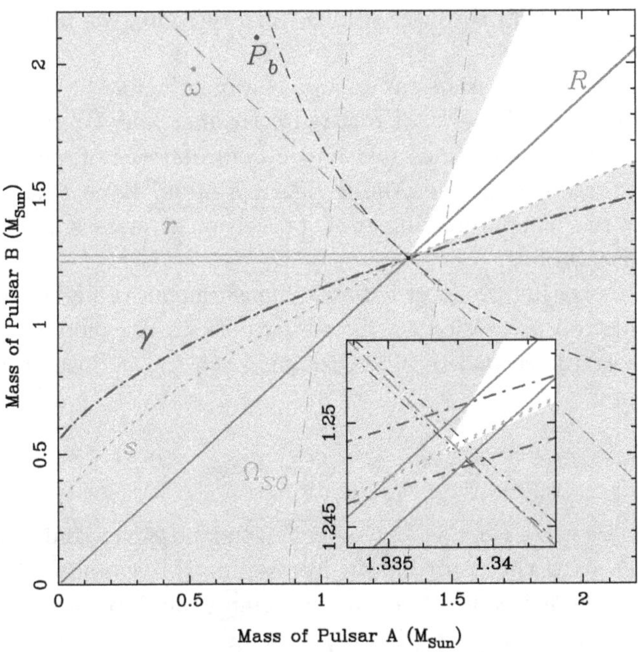

Fig. 8. Plot of companion mass (m_2) versus pulsar mass (m_1) for PSR J0737−3039 with observed constraints interpreted in the framework of GR. The inset shows the central region at an expanded scale, illustrating that GR is consistent with all constraints (M. Kramer, private communication). (For color version, see page I-CP7.)

be used to predict the remaining five post-Keplerian parameters using Eqs. (4)–(8). As Fig. 8 shows, GR gives a self-consistent description of the orbital motions, with all five independent constraints consistent with the masses derived from $\dot{\omega}$ and R.

The most stringent test comes from the derived value of s, 0.99974 ± 0.00039. The ratio of the observed and predicted values of s is 0.99987 ± 0.0005, by far the most constraining test of GR in the strong-field regime. Furthermore, this test is qualitatively different to that for PSR B1913+16 in that it is based on a nonradiative prediction. With just 2.5 years of data analyzed, the orbital decay term for the double pulsar is not measured as precisely as that for PSR B1913+16. However, since the phase offset grows quadratically (Fig. 4) and the effect of receiver noise decreases as $T^{1/2}$ where T is the data span (assuming approximately uniform sampling), the precision of the $\dot{P}_b$ measurement should improve as $T^{5/2}$. Furthermore, J0737−3039A/B is much closer to the Sun than PSR B1913+16, thus reducing the magnitude and uncertainty of the correction due to differential Galactic acceleration. Also, using very long-baseline interferometry (VLBI), Deller *et al.*[38] have shown that the transverse space velocity of the binary system is small, just 24^{+9}_{-6} km s^{-1}, and so the Shklovskii correction to the observed $\dot{P}_b$ is also small and accurately known. For these reasons, the orbit-decay test with the Double Pulsar will not be limited by these corrections, at least for another decade or so. Beyond that, improved models for the Galactic gravitational potential can be expected from analysis of GAIA data[108] and improved parallax and proper motion measurements may come from further VLBI observations, both reducing the uncertainty in these kinematic corrections.

Several other post-Keplerian parameters can in principle be observed in binary systems,[35] testing other aspects of relativity. Kramer and Wex[75] show that only one of these, δ_θ, which quantifies relativistic deformations of the binary orbit, is potentially measurable with the double pulsar system. Even this will take more than a decade to reach an interesting level of precision given reasonable expectations about future observations.

A potentially exciting prospect is to use measurement of higher-order terms for relativistic periastron precession to put constraints on the moment of inertia of a neutron star.[32] At the second post-Newtonian level, the periastron precession has three components:

$$\dot{\omega} = \dot{\omega}_{1pN} + \dot{\omega}_{2pN} + \dot{\omega}_{SO}, \tag{9}$$

where $\dot{\omega}_{1pN}$ is given by Eq. (3), $\dot{\omega}_{2pN}$ is the 2nd post-Newtonian $(v/c)^4$ contribution and $\dot{\omega}_{SO}$ gives the contribution from spin-orbit coupling.[16,32] This latter term depends on the moment of inertia and spin of the A pulsar (the spin angular momentum of the B pulsar is negligible) and can in principle be measured if the two leading components in Eq. (9) can be measured to sufficient precision. Kramer and Wex[75] show that, given expected advances in precision timing and astrometry, a significant result could be obtained in 20 years or so.

Spin-orbit coupling can also lead to a nonlinear variation in ω as a function of time and a time variation in the projected semi-major axis of pulsar A, $a_1 \sin i$.[75] However, the nonlinear terms depend on $\sin\theta$, where θ is the misalignment angle. As mentioned above, it seems as though θ_A is very small, so unfortunately these terms may be difficult to detect.

2.1.4. *Measured post-Keplerian parameters*

Table 1 summarizes the measured post-Keplerian parameters for the eleven pulsar binary systems where three or more post-Keplerian parameters have been measured. Since only two measurements are required to determine the stellar masses, these systems provide the opportunity for tests of theories of relativistic gravity. Of the eleven systems, in six cases the companion star is believed to be a neutron star, in a further four cases a white dwarf companion is more likely and in one case the companion is probably a main sequence star. This latter system, PSR J1903+0327, evidently has had an unusual evolutionary history as a triple system in which one component was ejected.[50] It is currently not clear if there is a significant contribution to the observed $\dot{\omega}$ from kinematic effects and/or spin-orbit coupling, so the utility of this system for tests of relativistic gravity is limited. There a further ten pulsar systems in the literature where just two post-Keplerian parameters have been measured, enabling estimates of the stellar masses, but no tests of relativistic gravity.

Although an independent measurement of relativistic spin precession has been possible for just two systems so far, as shown in Table 1, secular changes in observed pulse profiles have been observed for four other pulsars and attributed to geodetic precession of the pulsar spin axis.

2.2. *Tests of equivalence principles and alternative theories of gravitation*

Pulsars and especially binary pulsars have unique advantages in testing theories of relativistic gravitation as a result of their often rapid spin, short orbital periods and the ultra-high density of the under-lying neutron stars. As we have shown above, GR has been amazingly successful in describing all measurements to date. Never-the-less, investigations of quantum gravity and cosmology suggest that, in some regimes, extensions or modifications of GR may be required. This strongly motivates a search for departures from GR within existing experimental capabilities.

Gravitational theories have equivalence principles at their heart. The weak equivalence principle (WEP) is basic to Newtonian gravity, stating that acceleration in a gravitational field is independent of mass or composition. The strong equivalence principle (SEP) adds Lorentz invariance (no preferred reference frame) and position invariance (no preferred location) for both gravitational and non-gravitational interactions. GR satisfies the SEP whereas other theories may violate the SEP or even the WEP in one or more respects.

Table 1. Binary pulsars with three or more significant measured post-Keplerian parameters.

Pulsar/Parameter	J0437−4715	J0737−3039A/B	J0751+1807	J1141-6545	B1534+12	J1756−2251
Peri. advance $\dot{\omega}$ (°yr^{-1})	0.016(8)	16.8995(7)	—	5.3096(4)	1.7557950(19)	2.58240(4)
Time dilation γ (ms)	—	0.3856(26)	—	0.773(11)	2.0708(5)	0.001148(9)
Orb.P deriv. $\dot{P}_b$ (10^{-12})	3.73(6)[b]	−1.252(17)	−0.031(5)	−0.403(25)	−0.1366(5)	−0.229(5)
$s \equiv \sin i$	0.6746(28)[b]	0.99974(39)	0.90(5)	—	0.9772(16)	0.93(4)
Comp. mass m_2 ($M_\odot$)	0.254(18)	1.2489(7)	0.191(15)	—	1.35(5)	1.6(6)
Geod. prec. Ω_p (°yr^{-1})	—	4.77(66)[c]	—	Note d	0.59(10)	—
Binary companion[a]	He WD	NS	He WD	CO WD	NS	NS
References	141	74, 20	96, 97	15	49	46
Pulsar/Parameter	**J1807−2459B**	**J1903+0327**	**J1906+0746**	**B1913+16**	**B2127+11C**	
Peri. advance $\dot{\omega}$ (°yr^{-1})	0.018339(4)	0.0002400(2)[b]	7.5841(5)	4.226598(5)	4.4644(1)	
Time dilation γ (ms)	26(14)	—	0.470(5)	4.2992(8)	4.78(4)	
Orb.P deriv. $\dot{P}_b$ (10^{-12})	—	—	−0.56(3)	−2.423(1)[b]	−3.96(5)	
$s \equiv \sin i$	0.99715(20)	0.9759(16)	—	—	—	
Comp. mass m_2 ($M_\odot$)	1.02(17)	1.03(3)	—	—	—	
Geod. prec. Ω_p (°yr^{-1})	—	–	Note d	Note d	Note d	
Binary companion[a]	CO WD(?)	MS	NS(?)	NS	NS	
References	82	50	139, 41	145, 143	62	

Note: [a]Binary companion types: CO WD: Carbon-Oxygen White Dwarf; He WD: Helium White Dwarf; MS: Main-sequence star; NS: Neutron star.

[b]Dominated or significantly biased by kinematic effects.

[c]For PSR J0737−3039B.

[d]Effects of precession observed, but no independent determination of $\dot{\Omega}_p$.

Comparison of different gravitational theories has been greatly facilitated by the "parametrized post-Newtonian" (PPN) formalism which describes observable or potentially observable phenomena in a theory-independent way. This formalism was first developed by Will and Nordtvedt[150] for "weak-field" situations, that is, where $\epsilon \sim GM/Rc^2 \ll 1$, where G is the Newtonian gravitational constant, M and R characterize the size and mass of the object and c is the velocity of light. Many gravitational tests are performed within the solar system where $\epsilon \lesssim 10^{-5}$, firmly in the weak-field regime. However, in the vicinity of a neutron star $\epsilon \sim 0.2$ and so strong-field effects are potentially important. A number of authors have considered the generalization of the PPN formalism to strong-field situations (see, e.g. Refs. 31, 148 and 35) allowing investigation of these effects.

Some Lorentz-violating theories predict a dependence of the velocity of light on photon energy or polarization.[149] Pulsar observations can be used to limit these theories, but potentially stronger limits come from observations of gamma-ray bursts, polarized extra-galactic radio sources and the cosmic microwave background.[95]

A recent comprehensive review of observational limits on theories of gravitation by Will can be found in Ref. 149. Pulsar tests of relativistic gravitation have been reviewed by Stairs[129] and, more recently, by Wex.[146] Further details on many of the topics discussed here may be found in these reviews.

2.2.1. *Limits on PPN parameters*

The standard PPN formalism has ten parameters: $\gamma_{\rm PPN}$, $\beta_{\rm PPN}$, ξ, α_1, α_2, α_3, ζ_1, ζ_2, ζ_3 and ζ_4. In GR $\gamma_{\rm PPN}$, describing space curvature per unit mass, and $\beta_{\rm PPN}$, describing superposition of gravitational fields, are unity and all others are zero. ξ describes preferred location effects, the α_n preferred frame effects and the others describe violations of momentum conservation. (α_3 also may be nonzero in this case.) Pulsar observations do not directly constrain $\gamma_{\rm PPN}$ or $\beta_{\rm PPN}$ but are important in constraining many of the remaining PPN parameters and in fact currently place the strongest constraints on several parameters.

Damour and Schäfer[33] recognized that wide-orbit low-eccentricity binary pulsars, which generally have a white-dwarf companion, could provide a valuable test of the SEP through a strong-field extension of the solar-system tests pioneered by Nordtvedt.[98] A violation of the SEP would cause bodies with different gravitational self-energy to fall at different rates in an external gravitational field. In a pulsar–white-dwarf binary system, this results in a forced eccentricity in the direction of the gravitational field, that of the Galaxy in this case. This eccentricity is given by

$$\mathbf{e}_{\rm F} = \frac{3}{2} \frac{\Delta \mathbf{g}_\perp}{\dot{\omega} a \left(\dfrac{2\pi}{P_b} \right)}, \tag{10}$$

where $\mathbf{g}_\perp$ is the projection of the Galactic gravitational field on to the orbital plane, $\dot{\omega}$ is the relativistic periastron advance, a is the orbit semi-major axis and P_b the orbital period. The ratio of the gravitational mass m_g and the inertial mass m (which is exactly one if the SEP is obeyed) for body i is described by

$$\left(\frac{m_g}{m}\right)_i = 1 + \Delta_i = 1 + \eta_N \left(\frac{E_g}{mc^2}\right)_i + \eta'_N \left(\frac{E_g}{mc^2}\right)_i^2 + \cdots, \tag{11}$$

where η_N is the Nordtvedt parameter, a function of several PPN parameters (see, e.g. Ref. 149) and $\Delta = \Delta_1 - \Delta_2$. Violations of the SEP will result in nonzero Δ.

Since Damour and Schäfer[33] first proposed this method of testing SEP violations, the number of suitable pulsar–white-dwarf binary systems has increased greatly. Gonzalez *et al.*[53] combined data for 27 systems to place a 95% confidence upper limit on $|\Delta|$ of 4.6×10^{-3}.[f] Since $E_g/mc^2 \sim 0.1$ for a neutron star, this limit is not as strong as the weak-field limit on $\eta_N \lesssim 3 \times 10^{-4}$ from lunar-laser ranging.[151] However, it does enter the strong-field regime and test possible violations of the SEP that solar-system tests cannot reach.

The recent discovery by Ransom *et al.*[104] of a remarkable triple system containing a pulsar, PSR J0337+1715, and two white-dwarf stars in essentially coplanar orbits, one in a relatively tight 1.6-day orbit with the pulsar and the other in a much wider 327-day orbit around the inner system, opens up the possibility of a much more sensitive test of the SEP. Precise timing observations of the pulsar have already shown that the motion of the inner system is strongly affected by the gravitational field of the outer white dwarf. The gravitational field of this star, which has a mass of about $0.41\,\mathrm{M}_\odot$, at the inner system is at least six orders of magnitude larger than the Galactic gravitational field at the position of a typical pulsar and so the effect of SEP violations may be expected to be correspondingly larger.[51] Observations over several orbital periods of the outer star will almost certainly be necessary to isolate any SEP-related effects from other orbit perturbations.

The wide-orbit low-eccentricity binary pulsars can also be used test for violations of local Lorentz invariance (LLI) of the gravitational interaction and momentum conservation that are described by the parameter α_3 and its strong-field generalization $\hat{\alpha}_3$. Bell and Damour[13] showed that such violations produce a forced eccentricity analogous to that produced by the Nordtvedt effect given by

$$|\mathbf{e}_F| = \hat{\alpha}_3 \frac{c_p|\mathbf{w}|}{24\pi} \frac{P_b^2}{P} \frac{c^2}{G(m_1 + m_2)} \sin\beta, \tag{12}$$

where c_p is a dimensionless "compactness" parameter, for neutron stars about 0.2,[31] and β is the angle between $\mathbf{w}$, the velocity of the system with respect to a reference frame defined (for example) by the cosmic microwave background, and the pulsar's spin axis. A similar analysis to that for the generalized Nordtvedt effect resulted in a 95% confidence limit of $|\hat{\alpha}_3| < 4.0 \times 10^{-20}$, some 13 orders of magnitude lower than the best solar-system test.[130] It is worth noting that the observed small scatter

[f]This limit may be slightly under-estimated — see Wex.[146]

in the period derivatives of MSPs had already been used by Bell[11] to limit $|\alpha_3|$ to $< 5 \times 10^{-16}$.

Strong-field limits on the other two PPN parameters describing preferred frame effects, specifically LLI, $\hat{\alpha}_1$ and $\hat{\alpha}_2$, can also be obtained from low-eccentricity binary pulsar observations.[31] Nonzero $\hat{\alpha}_1$ induces a forced eccentricity in the direction of motion of the binary system velocity $\mathbf{w}$, analogous to Nordtvedt $\hat{\alpha}_3$ tests discussed above, whereas nonzero $\hat{\alpha}_2$ induces a precession of the orbital angular momentum about $\mathbf{w}$. The best current limits come from observations of the binary pulsar systems PSRs J1012+5307 and J1738+0333, both of which have short orbital periods, ~ 0.60 and ~ 0.35 days respectively, and extremely low orbital eccentricities $e \lesssim 10^{-7}$.[121] Furthermore, both of these pulsars have optically identified binary companions which, together with proper motion measurements from the timing observations, allow the three-dimensional space velocity of the binary system to be determined. For these pulsars, the observational data span is sufficiently long ($\gtrsim 10$ years) that relativistic periastron advance has significantly changed the orientation of the intrinsic eccentricity vector relative to the direction of any forced eccentricity, resulting in a potentially detectable change in the total eccentricity. The strongest limit on $\hat{\alpha}_1$ comes from observations of PSR J1738+0333 as shown in Fig. 9, conservatively $\hat{\alpha}_1 < 4 \times 10^{-5}$. This is not only better than the best weak-field limit of $\alpha_1 < 2 \times 10^{-4}$ from lunar laser ranging,[93] but also constrains strong-field effects as well.

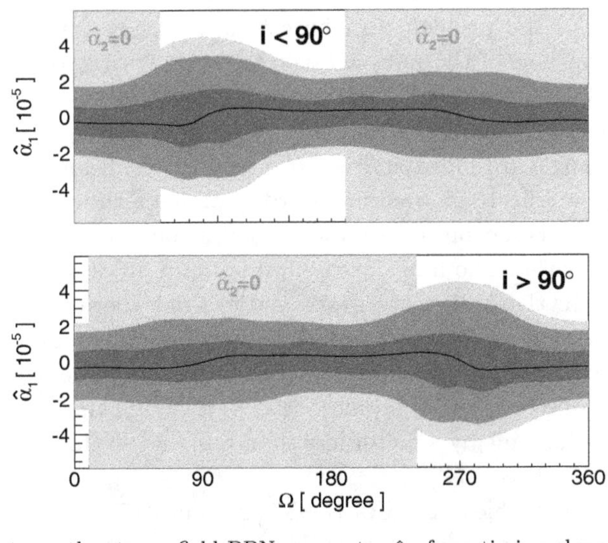

Fig. 9. Constraints on the strong-field PPN parameter $\hat{\alpha}_1$ from timing observations of the low-eccentricity binary pulsar PSR J1738+0333. The constraints are a function of the unknown orientation of the binary system on the sky (described by the longitude of the ascending node Ω) and the unknown sign of $\sin i$. If the PPN parameter $\hat{\alpha}_2$ is assumed to be zero, then only certain ranges of Ω are permitted. The shading corresponds to 68%, 95% and 99.7% confidence limits on $\hat{\alpha}_1$ (Ref. 121).

A precession of the orbital axis about the system velocity vector $\mathbf{w}$ induced by a nonzero value of $\hat{\alpha}_2$ is potentially observable as a time variation $\dot{x}$ in the projected semi-major axis of the pulsar orbit $x \equiv a_1 \sin i$. $\dot{x}$ is one of the possible post-Keplerian parameters in standard timing solutions and significant values have been measured for both PSRs J1012+5307 and J1738+0333.[121] There are several possible contributions to the observed $\dot{x}$ but in Ref. 121 it is shown that all of these are negligible in these systems except that due to the changing orbit inclination i resulting from proper motion of the system. This is a function of the unknown Ω values and for certain Ω values there is no constraint. Consequently only a probabilistic limit on $\hat{\alpha}_2$ can be obtained. By combining results for the two pulsars, a 95% confidence limit of $|\hat{\alpha}_2| < 1.8 \times 10^{-4}$ is obtained.

This is not as constraining as a solar-system limit $|\alpha_2| < 2.4\times10^{-7}$ obtained from the present deviation of the Sun's spin axis from the solar-system orbit normal,[99] a limit that rests on the assumption that the two axes were aligned at the time of formation of the solar system.

However, pulsars provide an even stronger constraint based on the stability of the spin axis of isolated pulsars. Any precession of the spin axis of a pulsar is likely to result in changes in the observed pulse profile (as observed with geodetic precession in binary pulsars as discussed in Sec. 2). Shao *et al.*[120] compared mean pulse profiles for the isolated MSPs B1937+21 and J1744−1134 taken 10–12 years apart with the same observing system and found no perceptible change in the pulse width at 50% of the peak amplitude. They interpreted these results by assuming a circular beam profile for the main pulse in PSR B1937+21 and for PSR J1744−1134. All of the angles in the problem can be estimated from modeling of radio and gamma-ray observations of the pulsar, taken together with known direction of the system velocity $\mathbf{w}$ with respect to the cosmic microwave background, except the angle of the projected spin axis on the sky. Probability histograms for $\hat{\alpha}_2$ allowing for this unknown angle are shown in Fig. 10. The final result for the 95% confidence upper limit is $|\hat{\alpha}_2| < 1.6\times10^{-9}$, about four orders of magnitude better than the limit described above based on orbital precession in pulsar binary systems and two orders of magnitude better than the limit based on solar spin precession. The assumption of circular beams in these MSPs is problematic, since there is good evidence for caustic enhancement in MSP pulse profiles, both radio and gamma-ray,[54,105] which would tend to elongate the beam in the latitude direction, reducing the effect of precession on the observed pulse profiles. However, taking this into account would probably increase the limit by a factor less than ten, and so this limit would remain the best available.

The stable pulse profiles of PSRs B1937+21 and J1744−1134 have also been used to limit the PPN parameter ξ describing local position invariance (LPI), also known as the Whitehead parameter, and its strong-field counterpart $\hat{\xi}$ (Ref. 122). The centripetal acceleration of Galactic rotation results in an anisotropy in the local gravitational field at a pulsar resulting in a precession of the pulsar spin

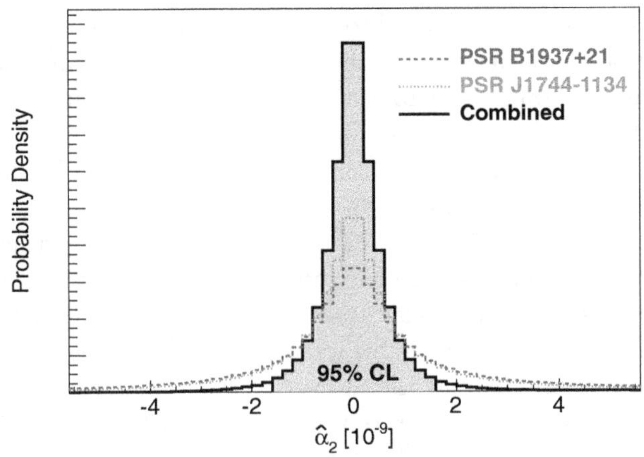

Fig. 10. Constraints on the strong-field PPN parameter $\hat{\alpha}_2$ based on the long-term stability of the mean pulse profiles for the isolated MSPs B1937+21 and J1744−1134 (Ref. 120).

vector around the direction of the Galactic acceleration with period

$$\Omega_p = \xi \left(\frac{2\pi}{P}\right) \left(\frac{v_G}{c}\right)^2 \cos\psi, \tag{13}$$

where v_G is the velocity of Galactic rotation at the pulsar and ψ is the (unknown) angle between the pulsar spin vector and the Galactic acceleration. Combining the results for the two pulsars in a way analogous to that for the α_2 test described above, Shao and Wex[122] obtain a 95% confidence limit $|\hat{\xi}| < 3.9 \times 10^{-9}$. Even with the qualification about beam shapes mentioned above, this limit is at least two orders of magnitude better than the next best limit obtained from considering of the evolution of the solar spin-misalignment angle over the lifetime of the solar system.[122]

The PPN parameter ζ_2 is one of a number of parameters that may be nonzero in gravitational theories that violate conservation of total momentum.[149] A nonzero ζ_2 would result in an acceleration of the center of mass of a binary system that changes direction with periastron precession. This is best measured by looking for a change in the rate of orbital decay in a binary system with large periastron precession and a long data span. Will[147] used observations of PSR B1913+16 with a 15-year data span to obtain a limit $|\zeta_2| < 4 \times 10^{-5}$. Obviously, this test could be made more stringent using the long data spans now available for PSR B1913+16 and other double-neutron-star systems with high $\dot{\omega}$.

2.2.2. *Dipolar gravitational waves and the constancy of G*

As described above, because of its tiny eccentricity, the pulsar–white-dwarf binary system PSR J1738+0333 has played a significant role in limiting the PPN parameters describing preferred frame effects. This system is composed of two very different stars, the neutron star we see as a pulsar and the companion which we know to be a white dwarf of mass 0.181 $M_\odot$ through its optical identification.[7] This and its short orbital period (~ 8.5 h) make it an ideal system for testing theories of gravity that predict a dipolar component to gravitational-wave damping as well as general scalar-tensor theories.[52]

From analysis of about 10 years of timing data obtained at Parkes and Arecibo observatories, Freire *et al.*[52] found an observed rate of orbital decay for PSR J1738+0333 of $\dot{P}_b = (-17.0 \pm 3.1) \times 10^{-15}$. To obtain the intrinsic rate of orbital decay, kinematic contributions from the differential acceleration of the binary system and the solar system in the Galactic gravitational field and the Shklovskii effect due to transverse motion must be subtracted, giving $\dot{P}_b^{\rm Int} = (-25.9 \pm 3.2) \times 10^{-15}$. Since the orbital parameters and the masses of the two stars are well known, the orbit decay due to GR can be accurately determined: $\dot{P}_b^{\rm GR} = (-27.7^{+1.5}_{-1.9}) \times 10^{-15}$, leaving a residual orbit decay of $\dot{P}_b^{\rm Res} = (2.0^{+3.7}_{-3.6}) \times 10^{-15}$.

This residual orbit decay is consistent with zero, which can be interpreted as a further confirmation of the accuracy of GR. However, because of the very different nature of the two stars in this binary system, this result also places strong constraints on theories of gravity that predict a dipolar component to gravitational-wave emission. Besides a dipolar component $\dot{P}_b^{\rm D}$, there are several other possible contributions to $\dot{P}_b^{\rm Int}$:

$$\dot{P}_b^{\rm Int} = \dot{P}_b^{\dot{M}} + \dot{P}_b^{\rm T} + \dot{P}_b^{\rm D} + \dot{P}_b^{\dot{G}}, \tag{14}$$

where $\dot{P}_b^{\dot{M}}$ is due to mass loss from the binary system, $\dot{P}_b^{\rm T}$ is a term resulting from tidal effects on the white dwarf (tidal effects on the neutron star are negligible) and $\dot{P}_b^{\dot{G}}$ is decay resulting from a possible variation in the gravitational "constant" G.[34] Freire *et al.*[52] showed that the likely $\dot{M}$ and tidal terms are small for this system, $\lesssim 10^{-15}$, and so the limit on $\dot{P}_b^{\rm Int}$ is effectively a limit on $\dot{P}_b^{\rm D} + \dot{P}_b^{\dot{G}}$.

Within certain restrictions on strong-field effects,[52] the dipole term is given by

$$\dot{P}_b^{\rm D} = -\frac{4\pi^2}{P_b} T_\odot m_c \frac{q}{q+1} \kappa_D \mathcal{S}^2 + \mathcal{O}(s^3), \tag{15}$$

where $q = m_1/m_2$ is the mass ratio (with subscript 1 referring to the neutron star), $\mathcal{S} = s_1 - s_2$ is the difference in "sensitivity" s of the mass of each body to a scalar field ϕ, where

$$s \equiv \left(\frac{d \ln m(\phi)}{d \ln \phi} \right), \tag{16}$$

and κ_D is a body-independent constant that describes the dipole self-gravity contribution in a given theory of gravity (see, e.g. Ref. 149). The sensitivity s_i depends on the stellar equation of state and, for neutron stars, is typically about 0.15, whereas for a white dwarf it is $\sim 10^{-4}$. Therefore, if κ_D is nonzero, dipole radiation will contribute to the orbit decay.

The remaining term in the residual orbit decay is that due to possible variations in G. In weak gravity, $\dot{G}/G$ has been constrained to be less than $4 \times 10^{-13}\,{\rm yr}^{-1}$

from lunar laser ranging experiments,[60] giving

$$\dot{P}_b^{\dot{G}} = -2\frac{\dot{G}}{G}P_b < 0.8 \times 10^{-15}, \tag{17}$$

and hence $|\kappa_D| < 2 \times 10^{-4}$. However, it is also possible to obtain independent estimates of the effects of dipole radiation and $\dot{G}$ by combining the J1738+0333 results with those for PSR J0437−4715.[39] This southern pulsar is in a wider orbit than PSR J1738+0333 and hence has a different mix of the dipole and $\dot{G}$ components, allowing them to be separated. After accounting for the fact that a changing G will also change the stellar masses,[100] the formal results are $\dot{G}/G = (-0.6 \pm 1.6) \times 10^{-12}\,\mathrm{yr}^{-1}$ and $\kappa_D = (-0.3 \pm 2.0) \times 10^{-4}$, both effectively upper limits. While the limit on κ_D is the best available, the derived limit on $\dot{G}/G$ is about an order of magnitude weaker than the result (actually a limit on the variation of $GM_\odot$) from the Mars Reconnaissance Orbiter[71] and from lunar laser ranging.[60]

Interestingly, pulsars have provided two other independent limits on $\dot{G}/G$. Thorsett[136] used determinations of neutron star masses from timing observations of double-neutron-star systems that formed many gigayears ago. In standard formation scenarios, the mass of a neutron star depends on the Chandrasekhar mass, the maximum possible mass of a white dwarf star, just prior to the collapse to a neutron star. The Chandrasekhar mass is proportional to $G^{-3/2}$ and so the observed small range of neutron star masses implies that $\dot{G}/G < 4 \times 10^{-12}$. This limit has been somewhat weakened by recent discoveries of both less massive and more massive neutron stars in pulsar binary systems (see Ref. 70 for a recent review).

The very small observed rate of change of pulsar period $\dot{P}$ observed in some pulsars (after correction for kinematic effects) may be used to set a further independent limit.[153] A variation in G will result in an inverse variation in the stellar moment of inertia, with the exact relation depending on the neutron-star structure. If the observed (intrinsic) $\dot{P}$ is entirely attributed to this effect, a limit of $\dot{G}/G \lesssim 2 \times 10^{-11}$ is obtained.

2.2.3. *General scalar–tensor and scalar–vector–tensor theories*

Many alternate theories of gravity can be expressed in a "tensor–scalar" framework in which a scalar field ϕ contributes to the "physical metric" $\tilde{g}_{\mu\nu}$ through a coupling function $A(\phi)$

$$\tilde{g}_{\mu\nu} \equiv A^2(\phi) g_{\mu\nu}, \tag{18}$$

where $g_{\mu\nu}$ is the usual tensor metric. The coupling constant may be expressed in different ways,[149] one of which is as an expansion around the asymptotic value of the scalar field ϕ_0

$$\ln A(\phi) = \ln A(\phi_0) + \alpha_0(\phi - \phi_0) + \beta_0(\phi - \phi_0)^2 + \cdots \tag{19}$$

(Ref. 31). For GR, $\alpha_0 = \beta_0 = 0$. In the well-known example of a tensor–scalar theory, the Brans–Dicke theory, the scalar coupling is described by a single parameter ω_{BD}. For $\omega_{BD} \to \infty$, the Brans–Dicke theory approaches GR. For this theory, $\alpha_0 = 1/(2\omega_{BD} + 3)$ and $\beta_0 = 0$. In other theories, both the linear and quadratic terms in Eq. (19) (and higher-order terms) may be nonzero.

The various observational constraints on PPN and post-Keplerian parameters can be expressed as limits in the (α_0,β_0) space for tensor–scalar theories as shown in Fig. 11.[52] The Cassini experiment[14] placed a strong limit on the PPN parameter $\gamma_{PPN} = 1+(2.1\pm2.3)\times10^{-5}$ which translates to a limit on $|\alpha_0|$ of about 0.003. Only the limits on dipole gravitational radiation from the asymmetric binary systems PSR J1141−6545[15] and PSR J1738+0333[52] rival the Cassini limit over most of the space. Since the precision of these measurements will increase with time, it seems likely that ultimately binary systems such as these will provide the strongest constraints on tensor–scalar theories.

PSR J0348+0432 and its white-dwarf companion form another asymmetric binary system, one that is distinguished by its very short orbital period (2.46 h) and

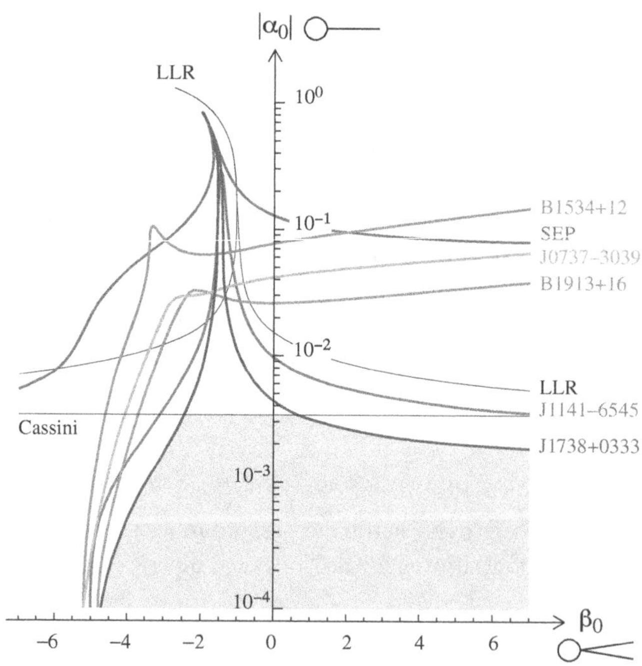

Fig. 11. Constraints on the scalar-field parameters in the $(|\alpha_0|,\beta_0)$ plane from various observational tests. Only the region below each line is allowed by the corresponding test. "SEP" refers to the test of the strong equivalence principle based on low-eccentricity pulsar–white-dwarf binary systems,[53] "LLR" refers to lunar laser ranging results[151] and Cassini to the "Shapiro delay" experienced by signals to and from the Cassini spacecraft (on its way to Saturn) as its line of sight passed close to the Sun.[14] Other binary-system tests are labeled according to the pulsar concerned. In GR, both α_0 and β_0 are zero (Ref. 52).

massive neutron star ($2.01 \pm 0.04\,\mathrm{M}_\odot$).[6] The high neutron-star mass is of interest since some scalar–tensor theories predict a strongly nonlinear relationship between the strength of dipole GW emission and neutron-star mass or self-gravity [Eq. (15)]. For most of the parameter space of the class of theories discussed by Freire *et al.*,[52] the limits on effective scalar coupling from PSR J0348+0432 are currently not as strong as those from PSR J1738+0333 but, because of the high neutron-star mass, they place stronger limits on some other theories with a greater degree of nonlinear coupling.[146]

Bekenstein[10] has proposed a relativistic generalization of the so-called "MOND" theory of gravity[91] that seeks to avoid the need for dark matter in galactic dynamics. The generalization invokes an additional vector field and hence is known as a tensor–vector–scalar theory. Such theories relax some of the constraints on tensor–scalar theories, in particular, the dipole radiation constraints, and allow significantly larger values of α_0. However, as shown by Freire *et al.*,[52] the binary pulsar results still significantly constrain theories of this type and in fact are more constraining than solar-system tests. With future observations, binary pulsar tests have the potential to make this class of theories untenable.

In another example of the use of pulsar observations, especially the limits on dipolar-GW radiation, to place limits on gravitational theories, Yagi *et al.*[155] have strongly limited the allowed parameter space for the LLI-violating "Einstein–Æther" and the "Khronometric" theories.

2.3. *Future prospects*

As described above, GR has provided an accurate description of all pulsar timing results obtained so far. However, continued refinement of existing methods and development of new tests is highly desirable. Continued pulsar timing measurements, especially with the advent of new and more sensitive observing facilities such as the 500-m Arecibo-type *FAST* radio telescope in China[94] and the Square Kilometre Array (*SKA*) in South Africa and Australia[24] will certainly improve on existing limits and enable new tests of gravitational theories. They may even demonstrate a failure of GR and hence a need for a modified or conceptually different theory of gravity. Conversely, if GR is assumed to be valid, results of astrophysical significance can be deduced from the observations. For example, as described in Sec. 2.1.3, observations of higher-order terms in the relativistic perturbations may enable a measurement of the moment of inertia of a neutron star.

These more sensitive radio telescopes can also be used to search for previously unknown pulsars and binary systems that are suitable for tests of gravitational theories. Past experience has shown that pulsar searches repeatedly turn up new classes of object. This potential is wonderfully illustrated by the recent discovery of the pulsar triple system PSR J0337+1715 which promises to provide a strong limit on violations of the strong equivalence principle. A dream for such searches is the discovery of a pulsar in a close orbit around a black hole as this would offer much more stringent tests of gravity in the strong-field regime.[80] Discovery of a pulsar orbiting the black hole at the center of our Galaxy with an orbital period of a few months or less could even allow a test of the so-called "no-hair" theorem for black holes.[81]

3. The Quest for Gravitational-Wave Detection

The direct detection of the gravitational waves (GWs) predicted by Einstein's general theory of relativity and other relativistic theories of gravity is one of the major goals of current astrophysics. As described in Sec. 2, we have excellent evidence from the orbital decay of binary systems for the existence of GWs at the level predicted by GR, but up to now there has been no direct detection of the changing curvature of spacetime induced by a passing GW. This changing curvature induces a change in the proper distance between two test masses, described by the gravitational strain $h = \delta L/L$. The problem is that, for any likely source, h is tiny. For example, the *LIGO* gravitational-wave detector[2] hopes to detect the merger of two neutron stars at a distance of 100 Mpc for which $h \sim 10^{-22}$, a change in the length of its 4 km arms of 10^{-18} m or 10^{-3} of the diameter of a proton.

Pulsar timing can measure a change in the proper distance between the pulsar and the telescope. Systematic changes in timing residuals for a given pulsar reflect unmodeled changes in the effective time of emission, the pulsar position, the propagation path or the position of the telescope. With care, and with observations of multiple pulsars, residual delays due to changing proper distances can be isolated, effectively giving us a set of interferometers with baselines of $\gtrsim 10^{16}$ km. However, even in the best cases, we can only measure the interferometer "phase" to about 100 ns, so that the limiting strain is about 10^{-18}. Unlike ground-based laser-interferometer systems, which are most sensitive to GW signals with frequencies around 100 Hz, pulsar timing systems are most sensitive to signals with frequencies around the inverse of the data span, typically a few nanohertz. Potential sources of detectable GWs in this low-frequency band include super-massive black-hole (SMBH) binary systems in distant galaxies and cosmic strings in the early universe.

3.1. *Pulsar timing arrays*

The effect of GWs on pulsar timing signals was first considered by Sazhin[112] and Detweiler,[42] with the latter being the first to consider the effect of a GW from a distant source passing over a pulsar and the Earth. For this case, it can be shown that the net effect on the observed pulsar arrival times is simply the difference between the effect of the GW passing over the pulsar and the effect of the GW passing over the Earth. For a GW travelling in the $\hat{z}$-direction, the redshift z of the pulse frequency ν for a pulsar at distance d with direction cosines (α, β, γ) is given by

$$z(t) = \frac{\nu_0 - \nu(t)}{\nu_0} = \frac{\alpha^2 - \beta^2}{2(1+\gamma)}\Delta h_+ + \frac{\alpha\beta}{1+\gamma}\Delta h_\times, \tag{20}$$

where $\Delta h_A = h_A^p - h_A^E$, with A representing the two possible wave polarization states $(+,\times)$ in GR, and $h_A^p(t - d/c)$ and $h_A^E(t)$ the gravitational strain at the pulsar and the Earth, respectively.[g] The observed timing residuals are then given by the integral of the redshift,

$$R(t) = \int_0^t z(t')dt'. \tag{21}$$

The coefficients multiplying the Δh_+ and $\Delta h_\times$ terms are the "antenna patterns" for the two polarizations as illustrated in Fig. 12. A pulsar located in the GW propagation direction has zero sensitivity to the GW since its direction cosines α and β are zero. Furthermore, despite the $(1+\gamma)$ term in the denominator of Eq. (20), the response for a pulsar exactly in the $-z$ direction (the same direction as the GW source) is also zero. This comes about because of a cancellation of the $(1+\gamma)$ term by the expansion of Δh_A when $1 + \gamma \ll 1$ (Ref. 78).

A pulsar timing array (PTA) consists of a set of pulsars spread across the sky which have precise timing measurements over a long data span. The detection of

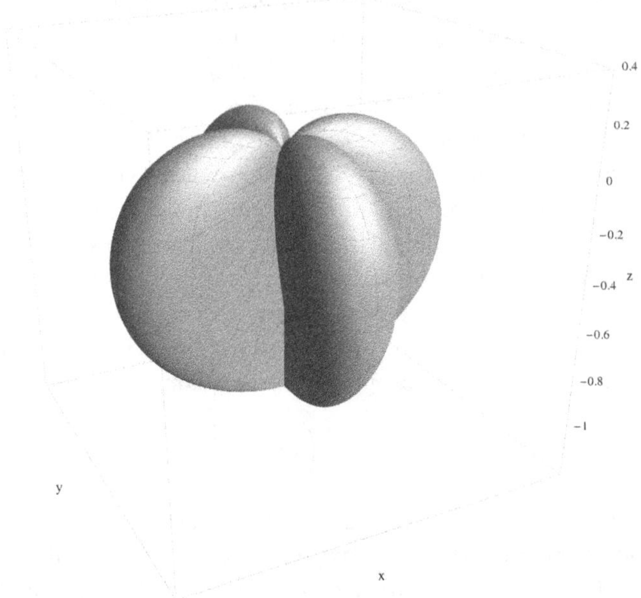

Fig. 12. Effective "antenna pattern" for detection of a GW with pulsar timing. The wave is propagating in the $+z$ direction and is assumed to have the $+$ polarization. The pattern for the $\times$ polarization is the same but rotated by 45° about the z-axis (Ref. 25).

[g]See Ref. 5 for a rederivation of the Detweiler result.

GWs by PTAs depends on the correlated timing residuals for different pulsars given by the Earth term $h_A^E(t)$ in Eq. (20). GWs passing over the pulsars produce uncorrelated residuals because of both the retarded time and the different GW environment for each pulsar. Also the pulsars themselves have uncorrelated timing noise at some level, either intrinsic or resulting from uncorrected variations in interstellar delays. Because the expected GW strain is so weak, only MSPs have sufficient timing precision to make GW detection with PTAs feasible.

For an isolated source of continuous GWs, say an SMBH binary system in a nearby galaxy, in principle, both the Earth term and the pulsar term in Eq. (20) could be detected. For a rapidly evolving source, the pulsar term and the Earth term may be at different frequencies because of the retarded time of the pulsar term (see, e.g. Ref. 65). They can then be added incoherently to increase the detection sensitivity. For a nonevolving source, i.e. $\delta f_b \ll 1/T$ where δf_b is the change in binary orbital frequency over the pulsar timing data span T, in principle the pulsar term and the Earth term could be summed coherently for optimal sensitivity. As is discussed further in Sec. 3.4, unfortunately we currently do not know enough pulsar distances to sufficient accuracy to make this coherent addition possible. In a PTA, the pulsar terms therefore add with random phase, washing out the fringes in the antenna pattern (see also Ref. 78) and adding "self-noise" to the signal from the Earth term. Since the antenna pattern (Fig. 12) has a maximum for pulsars roughly in the same direction as the GW source, the maximum response of a PTA is toward the greatest concentration of pulsars in the array.

A stochastic background of nanohertz GW from many SMBH binary systems in distant galaxies is likely to be the signal first detected by PTAs. To a first approximation, this background is also likely to be statistically isotropic, i.e. the expectation value $\langle h^2 \rangle$ is independent of direction when averaged over typical data spans. Hellings and Downs[55] were the first to show that, in this case, the correlation between GW-induced timing residuals for two pulsars separated by an angle θ on the sky is dependent only on θ and not on the sky positions of the two pulsars. The zero-lag correlation function, commonly known as the Hellings and Downs curve and obtained by integrating the product of the antenna patterns (Fig. 12) for the two pulsars over all possible GW propagation directions, is given by:

$$c_{\mathrm{HD}} = \frac{1}{2} + \frac{3x}{2}\left(\ln x - \frac{1}{6}\right), \tag{22}$$

where $x = (1 - \cos\theta)/2$, and is plotted in Fig. 13. c_{HD} goes negative for angular separations around 90° and then positive again for pulsars that are more-or-less opposite on the sky — this is a direct consequence of the quadrupolar nature of GWs. It is also important to note that the limiting value as $\theta \to 0$ is 0.5, not 1.0. This is a consequence of the fact that the pulsar terms in Eq. (20) are uncorrelated and, on average, of equal amplitude to the Earth term. The scatter in the simulated correlations results from the random phases of the pulsar terms and illustrates the "self-noise" that limits the sensitivity of PTA experiments in the strong-signal limit.

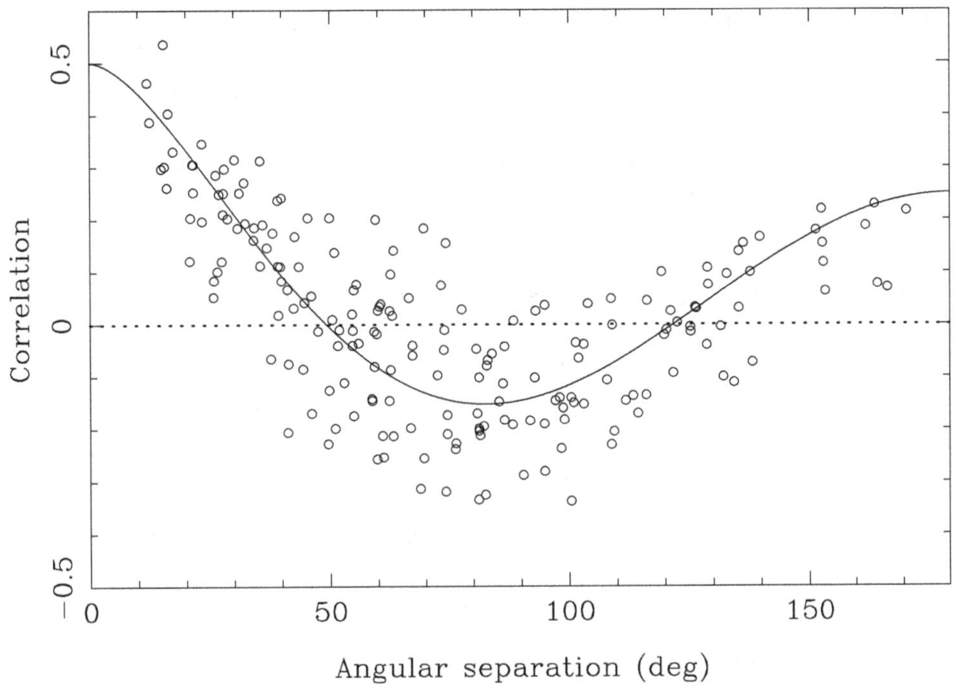

Fig. 13. The Hellings and Downs correlation function, i.e. the correlation between timing residuals for pairs of pulsars as a function of their angular separation for an isotropic stochastic background of GWs. Also shown are simulated correlations between the 20 pulsars of the Parkes PTA for a single realization of a strong GW signal that dominates all other noise contributions (Ref. 58).

3.2. *Nanohertz gravitational-wave sources*

3.2.1. *Massive black-hole binary systems*

There is good evidence that massive black holes form in the center of galaxies at very early times (see, e.g. Ref. 140) and also that merger events play a major role in galaxy growth (see, e.g. Ref. 132). When two galaxies, each containing a central massive black hole, merge, dynamical friction will result in the two black holes migrating to the center of the merged galaxy to form a binary system, with an estimated timescale for the migration of the order of giga-years (see, e.g. Ref. 68). When the binary separation is less than about 1 pc, loss of energy to GWs becomes the dominant orbital decay mechanism and the binary system will ultimately coalesce, becoming a strong GW source as it spirals in. There remains much controversy about the efficiency of orbital decay mechanisms as the binary separation approaches a parsec — known as the "last parsec problem". Some (see, e.g. Refs. 69 and 107) argue that dissipation mechanisms will quickly move the binary system through this phase, whereas others (see, e.g. Ref. 27) argue that the binary is likely to stall at separations where the gravitational decay is ineffective. Detection of a stochastic GW background (GWB) would resolve this issue.

Since large numbers of binary systems with different orbital periods contribute to the GWB, it is a broadband signal which is best described in the spectral domain. It is convenient to express the amplitude of the GW signal in terms of the dimensionless "characteristic strain", defined by

$$h_c = 2f|\tilde{h}(f)|, \tag{23}$$

where $\tilde{h}(f)$ is the Fourier transform of $h(t)$ and f is the GW frequency. (Note that, for a circular binary system, the frequency of the emitted GW is twice the binary orbital frequency $f = 2f_b$.) Two other quantities that are often used to parametrize GW spectra and detector spectral sensitivities are the square root of the one-sided strain power spectral density

$$S_h^{1/2}(f) = h_c f^{-1/2}, \tag{24}$$

and the GW energy density as a fraction of the closure energy density of the universe

$$\Omega_{GW} = \frac{2\pi^2}{H_0^2} f^2 h_c^2(f), \tag{25}$$

where H_0 is the Hubble constant.

In order to understand the astrophysical implications of results obtained from PTA experiments, it is necessary to have estimates of the likely strength of signals from potential sources of nanohertz GW. For a cosmological population of SMBH binary systems at luminosity distance D_L and redshift z, the local energy density in GW at frequency f is given by:

$$fS_E(f) = \int_0^\infty dz \int_0^\infty dM_c \frac{d^2n}{dzdM_c} \frac{1}{(1+z)} \frac{1}{D_L^2} \frac{dE_g}{d\ln f_r}, \tag{26}$$

where $M_c = (M_1 M_2)^{3/5}(M_1 + M_2)^{-1/5}$ is the binary chirp mass and M_1 and M_2 are the masses of the binary components, $d^2n/(dzdM_c)$ is the comoving density of binary systems with redshift and chirp mass between z and $z + dz$ and M_c and $M_c + dM_c$, respectively, and $dE_g/d\ln f_r$ is the total energy emitted by a single binary system in the logarithmic frequency interval $d\ln f_r$, where $f_r = f(1+z)$.[92,103,115] The local GW energy density is related to the local characteristic strain by

$$fS_E(f) = \frac{\pi c^2}{4G} f^2 h_c^2(f) \tag{27}$$

and for a circular binary system

$$\frac{dE_g}{d\ln f_r} = \frac{G^{2/3}\pi^{2/3}}{3} M_c^{5/3} f_r^{2/3}. \tag{28}$$

Therefore, we have

$$h_c^2(f) = \frac{4G^{5/3}}{3\pi^{1/3}c^2} f^{-4/3} \int_0^\infty dz \int_0^\infty dM_c \frac{d^2n}{dzdM_c} \frac{1}{(1+z)^{1/3}} \frac{1}{D_L^2} M_c^{5/3}. \tag{29}$$

As Phinney[103] has emphasized, the result that $h_c \propto f^{-2/3}$ for a cosmological population of circular binary systems decaying through GW emission is quite general and independent of any particular cosmology, black-hole mass function or galaxy merger scenario. Consequently the spectrum of the GWB is often parametrized as follows:

$$h_c(f) = A_{1\mathrm{yr}} \left(\frac{f}{f_{1\mathrm{yr}}} \right)^{\alpha}, \tag{30}$$

where $f_{1\mathrm{yr}} = (1\mathrm{yr})^{-1}$, $A_{1\mathrm{yr}}$ is the characteristic strain at $f_{1\mathrm{yr}}$ and $\alpha = -2/3$ for the case described above. For pulsar timing experiments, the one-sided power spectrum of the timing residuals is given by

$$P(f) = \frac{1}{12\pi^2} \frac{1}{f^3} h_c^2(f). \tag{31}$$

Consequently, a GWB produces a very "red" modulation of the timing residuals with a spectral index of $-13/3$ for $\alpha = -2/3$.

In order to estimate the likely strength of this modulation, the factor $d^2n/(dzdM_c)$ in Eq. (29) must be evaluated. This requires a prescription for the cosmological evolution of massive black-hole binary systems in galaxies. Different approaches to this problem have been taken by different authors. An early paper by Jaffe and Backer[63] used observational constraints on close galaxy pairs coupled with a black-hole mass function, whereas another early paper by Wyithe and Loeb[154] used a prescription for merger of dark-matter halos coupled with different scenarios for growth of massive black holes in galaxies. The latter approach was developed further by Sesana *et al.*[115] who showed that the GWB spectrum steepens at frequencies above about 10^{-8} Hz since the number of binary systems contributing to the background at these frequencies becomes small. This is illustrated in the left panel of Fig. 14 which shows that binary systems at $z \lesssim 2$ contribute most of the strain to the GWB. The right panel shows that massive binary systems with $M_c \gtrsim 10^8\,\mathrm{M}_\odot$ contribute most of the low-frequency GWB but that only systems with $M_c \lesssim 10^8\,\mathrm{M}_\odot$ contribute to the high-frequency end. Importantly, at the high-mass end, only a few systems contribute to the GWB. As shown in Fig. 15, this leads to a steepening and large uncertainties in the expected GWB spectrum at frequencies $\gtrsim 10^{-8}$ Hz. These results were confirmed and extended by Sesana *et al.*[116] and Ravi *et al.*[106] using the *Millenium* simulation of the cosmological evolution of dark matter structures[128] to define a merger history together with various prescriptions for galaxy and black-hole formation and growth.

While there is some consensus about the form of the GWB spectrum, there remain significant uncertainties. For example, Ravi *et al.*[107] consider the effects of the stellar environment on the late evolution of massive black-hole binary systems in the cores of galaxies and conclude that the effect of dynamical friction is important, both in extracting energy from the binary system and inducing an eccentricity. Both

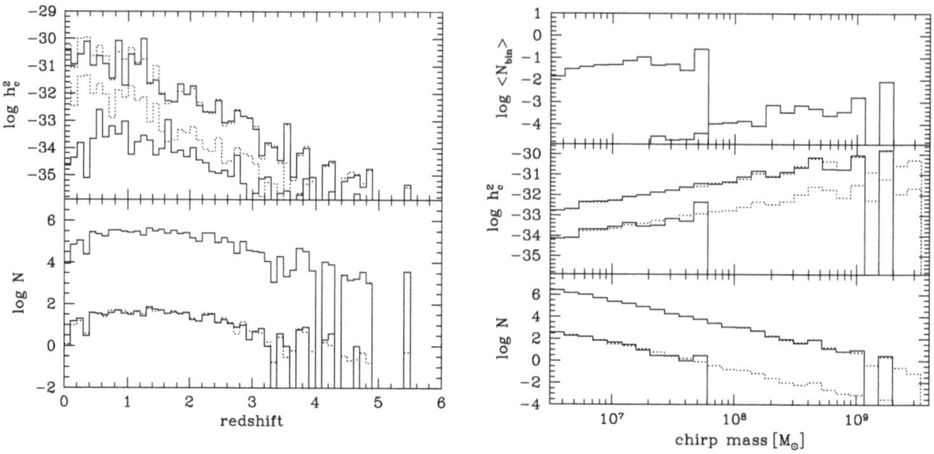

Fig. 14. Left panel: Number of binary black-hole systems and their contribution to the characteristic strain h_c of the GWB as function of source redshift z. The solid lines are results from a Monte Carlo approach and the dotted lines are from a semi-analytic analysis. In both panels, the upper histograms are for a GW frequency $f = 8 \times 10^{-9}$ Hz and the lower histograms for $f = 10^{-7}$ Hz. Right panel: The lower two panels are as for the left panel but as a function of source chirp mass M_c. The upper panel shows the number of frequency bins spanned by the chirp over the 5-year span of the simulation (Ref. 115).

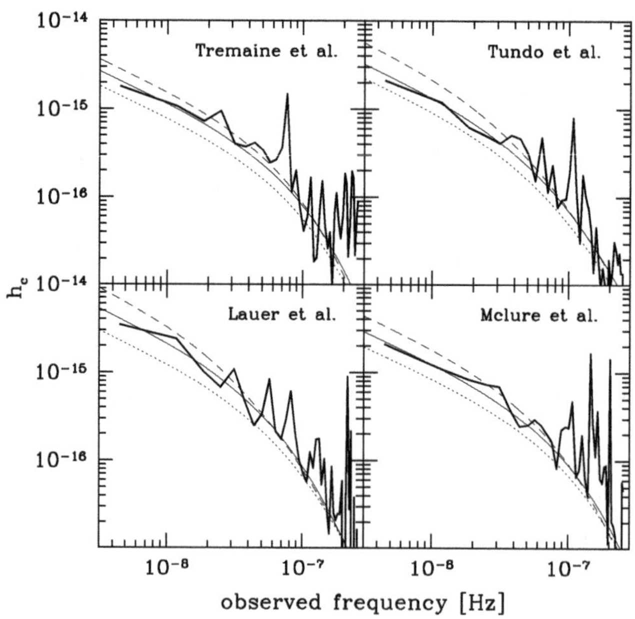

Fig. 15. Characteristic strain spectrum for the GWB based on different prescriptions for black-hole growth by accretion between mergers (solid, dashed and dotted lines) and different black-hole mass functions (different panels) (Ref. 116).

of these have the effect of reducing the strength of the predicted GWB, especially at the low-frequency end, consequently making its detection by PTAs more difficult. On the other hand, McWilliams *et al.*[89] argue for a model in which all black-hole growth is by merger rather than by accretion after coalescence, which is the main contributor to black-hole growth in the models discussed above. This leads to predictions of a significantly larger GWB characteristic strain compared to previous predictions and, hence, the imminent detection of the GWB by PTAs.

As Fig. 15 indicates, there is a possibility that the nearby universe could contain a massive black-hole binary system with an orbital period of the order of a few years that actually dominates the nanohertz GW spectrum. This opens up the exciting prospect of the GW detection and study of an isolated supermassive black-hole binary system using pulsar timing and even the possibility of detection and study of the system in the electromagnetic bands — so-called "multi-messenger" astronomy. Searches for binary GW sources will be described in Sec. 3.3 and the prospects for their detailed study will be discussed in Sec. 3.4.

3.2.2. *Cosmic strings and the early universe*

Cosmic strings and the related cosmic super-strings are one-dimensional topological defects which may have formed in phase transitions in the early universe. Cosmic strings occur in standard field-theory inflation models, whereas superstrings are found in brane inflationary models. The idea that such strings will oscillate and hence emit GWs was first proposed by Vilenkin.[142] Such oscillations may contribute to the stochastic GWB (see, e.g. Ref. 23) or generate bursts of GW radiation from string cusps and kinks (see, e.g. Ref. 36). The amplitude of GWs from cosmic strings is dependent on a large number of poorly known (or unknown) parameters and hence is very uncertain (see, e.g. Ref. 111). Key parameters are the string tension μ, usually parametrized by the dimensionless quantity $G\mu/c^2$, and the size α of string loops relative to the horizon radius at the time of birth. Other significant parameters for the GWB are the intrinsic spectral index q of the GW emission, a characteristic node number n_* for the high-frequency cutoff in the emission spectrum and the probability p of "intercommutation", that is, intersecting strings dividing and the two parts exchanging. Such intercommutation can, for example, form two smaller loops from an intersecting twist in a larger loop. For standard strings, $p = 1$ but it may be less for superstrings.

Vibrating cosmic strings are likely to decay by emission of GWs in a series of harmonics with fundamental frequency $2c/l$, where l is the length of the loop, with an initial value αD_H, where D_H is the horizon distance at the time of loop creation. The rate of energy loss for vibration mode n of a given loop is:

$$\frac{dE_{GW}}{dt} = \Gamma \frac{n^{-q}}{\sum\limits_{m=1}^{\infty} m^{-q}} G\mu^2 c, \tag{32}$$

where Γ is factor depending on the shape of the loop, typically about 50.[111] As the loop loses energy, it shrinks and eventually disappears. The creation of loops through intercommutation and their decay through GW emission sets up an equilibrium distribution of loop sizes. Sanidas *et al.*[111] compute the number density of loops as a function of loop length and time and hence, using Eq. (32), the predicted spectrum of the GWB from string loops as a function of the various parameters. As Fig. 16 shows, the spectrum is very broad, extending all the way from nanohertz to Megahertz. Since the lowest ($n = 1$) frequency is proportional to the string length, the weaker and smaller loops do not contribute to the nanohertz background.

3.2.3. *Transient or burst GW sources*

The prime target of ground-based laser-interferometer GW detectors is the burst emitted at the coalescence of a double-neutron-star system. Such a burst is intense for just a few milliseconds and clearly cannot be detected by PTA experiments which typically are sensitive to signals of duration between a few weeks and a few years. Possible sources of longer-duration bursts are coalescence of SMBH binary systems, highly eccentric massive black-hole binary systems, and formation and decay of cusps and kinks in cosmic strings.

A major difference between detection of GW bursts and continuous GW sources is that there is generally no interference between the Earth term and the pulsar terms in the signal detected by PTAs [Eq. (20)]. This is because the duration of the burst (by definition, less than the observational data span) is much less than the light-time to the pulsars and so, except for a source in the same direction as a pulsar, the burst will occur at very different times in the Earth and pulsar terms.

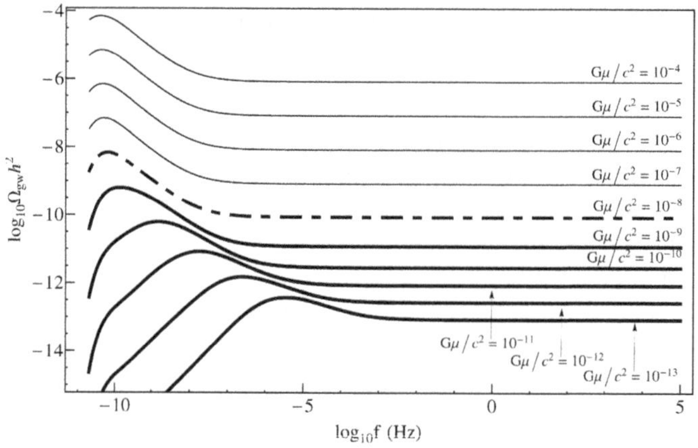

Fig. 16. Energy density spectrum for GWs emitted by cosmic strings as a function of the dimensionless string tension $G\mu/c^2$. Other parameters are held fixed at values $\alpha = 10^{-7}$, $q = 4/3$, $n_* = 1$ and $p = 1$. The dashed spectrum is for the critical point where $\Gamma G\mu/c^2 \approx \alpha$; spectra above this are for large loops and spectra below are for small loops (Ref. 111).

The pulsar term reflects the effect of the burst on the pulsar at a time d/c *before* the burst arrives at the Earth, but the pulsar term is detected at a time $(d/c)(1+\cos\theta)$ *later* than the Earth term, where d is the pulsar distance and θ is the angle between the pulsar and the GW propagation direction as seen from the Earth.

Figure 17 shows the GW waveform produced by the coalescence of two black holes in a coordinate system where all the signal is in h_+. The maximum amplitude of the waveform is about 0.1 in the time units of Fig. 17, or about $0.1\,cMT_\odot/D$ or $\sim 10^{-14}M_9/D_{Gpc}$ where $M_9 \equiv 10^{-9}M$ is the total system mass in solar units and D_{Gpc} is the (comoving) distance in Gpc. Although this is comparable to the strain sensitivities achieved by current PTAs for continuous GW signals (see Sec. 3.3 below), even for the largest SMBHs, the timescale of the burst, $\sim 200\,MT_\odot$ is only of order 10 days and the period of the oscillation is about an order of magnitude less than that. Not only is this too short to be resolved by any existing PTA, but the sensitivity over this short interval would be much less than that achieved for CW signals integrated over the entire data span. Space-based laser interferometer systems such as the planned *eLISA*[3] have the potential to directly detect these bursts.

However, Fig. 17 also shows another effect known as GW "memory" which is potentially detectable with pulsar timing. During the coalescence event, a non-oscillatory component to the gravitational strain builds up, so that at the end of the "ring-down" phase, the strain has a permanent offset from the pre-coalescence

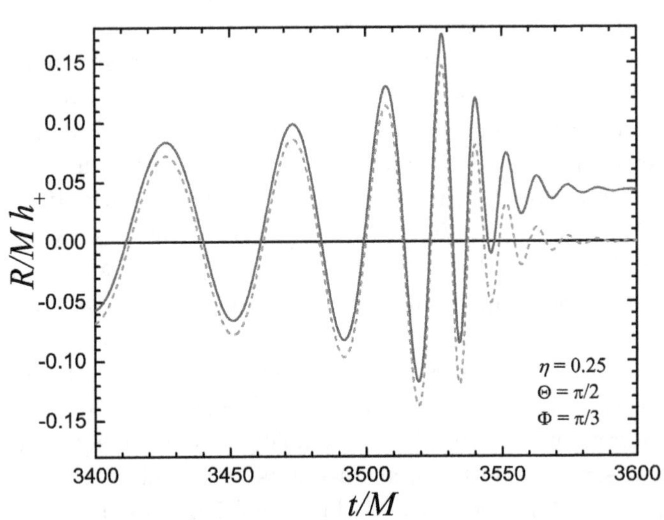

Fig. 17. Gravitational waveform resulting from the coalescence of two equal-mass black holes (reduced mass ratio $\eta = 0.25$) with total mass M at distance R. Both M and R are expressed in time units; the conversions to conventional units are $M \to MT_\odot$ and $R \to R/c$. Θ and Φ are assumed source directions. The dashed line is the predicted waveform if the gravitational memory effect is ignored (Ref. 44).

value. The amplitude of the memory effect is

$$h_m = \frac{\epsilon\eta M T_\odot}{\frac{D}{c}} \approx 10^{-15}\frac{M_9}{D_9}, \tag{33}$$

where $\epsilon \approx 0.07$ is the mass fraction contributing to the memory effect and η is the reduced mass fraction, 0.25 for equal-mass binary components.[30,44]

This step change in h produces a step change in the observed pulse frequency $\Delta\nu/\nu = h_m$, i.e. a "glitch". This glitch persists until it is reversed by the pulsar term at a time $(d/c)(1+\cos\theta)$ later. In a PTA, these reversals will occur at different times for different pulsars. Of course, it is also possible that a GW memory jump could be detected in the pulsar terms, but there it may be confused with a real glitch in the intrinsic pulse frequency, whereas in the Earth term there is a correlation in the effect on different pulsars. However, glitches in MSPs are rare (only one very small glitch detected so far: Ref. 29). Also, real pulsar glitches are generally spin-ups, whereas a GW-memory jump may be of either sign, depending on pulsar-source angle. Therefore, as Cordes and Jenet[30] have discussed, the pulsar terms may give an improved probability of detection.

Black-hole binary systems with circular orbits emit GW at the second harmonic of the orbital frequency, i.e. $f = 2f_b$. For eccentric orbits, the GW emission becomes more burst-like as the accelerations and hence GW power are greatest around periastron when the two black holes are closest together.[102] In the spectral domain, power spreads to higher harmonics and also to the fundamental frequency f_b. In the gravitational-decay phase of evolution, when energy loss is dominated by GW emission, the orbit tends to circularise.[8] However, at earlier phases of the orbital decay when three-body stellar interactions or interaction with a gaseous disk surrounding the binary system are important, the eccentricity may grow. Stellar three-body interactions can result in orbital decay through dynamical friction, but probably result in a modest increase in the eccentricity of the black-hole binary system (e.g. Ref. 90). However, Roedig *et al.*[110] find that initially mildly eccentric binary systems decaying through interaction with a gaseous disk evolve toward a limiting eccentricity in the 0.6–0.8 range. In a regime of frequent mergers it is even possible that a black-hole triplet could form[4] and, in this case, eccentricities as high as 0.99 could exist.

Finn and Lommen[47] have investigated the GW emission and the resulting timing residuals from a close parabolic encounter of two massive black holes. As Fig. 18 shows, an encounter of two $10^9 M_\odot$ black holes located at a distance of 15 Mpc with a minimum separation of 0.02 pc produces a GW burst of duration about 1 year and maximum GW strain $\sim 10^{-13}$. This results in potentially detectable PTA timing residuals of about 1μs amplitude with the same timescale. Unfortunately, the probability of having such a close encounter of two SMBHs in the local universe within PTA observational data spans is not high.

Cosmic strings are another potential source of GW burst emission. They can radiate over a wide frequency range with a huge range of possible amplitudes and

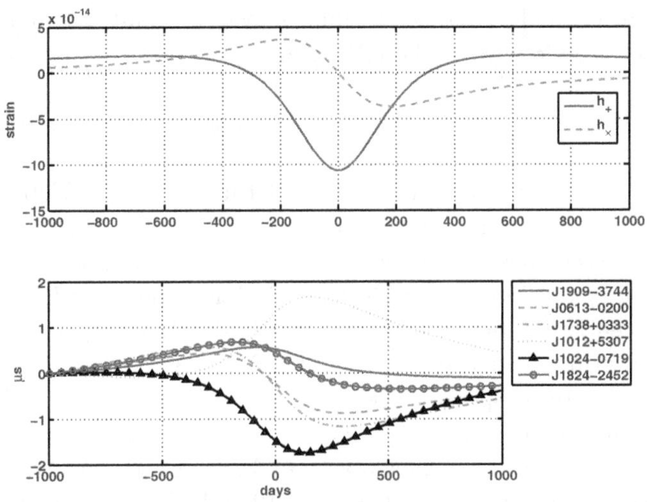

Fig. 18. The upper plot shows the gravitational waveforms in the h_+ and h_- polarizations resulting from a parabolic encounter with an impact parameter of 0.02 pc of two $10^9 M_\odot$ black holes located at a distance of 15 Mpc in the direction of the Virgo Cluster. The lower plot shows the resulting timing residuals for several PTA pulsars (Ref. 47).

timescales (Fig. 16) depending on the detailed mechanism invoked (loops, cusps, short strings, etc.) and the very wide (virtually unlimited) parameter space.[36,77,127] Short bursts radiating in the *LIGO* and *eLISA* bands may be frequent and unresolved, producing a GWB at these frequencies. However, bursts with longer timescales, producing radiation in the nanohertz band, are also possible but are likely to be extremely rare.[36] Consequently, while in principle such bursts could be detected, in practice it is unlikely that PTAs will be able to significantly constrain models for GW burst emission from cosmic strings.

3.3. *Pulsar timing arrays and current results*

In this section, we first describe the three main PTAs currently operating worldwide: the European Pulsar Timing Array (*EPTA*), the North American pulsar timing array (*NANOGrav*) and the Parkes Pulsar Timing Array (*PPTA*), and the collaboration between them, the International Pulsar Timing Array (*IPTA*). PTAs have many possible applications such as establishing a pulsar-based timescale,[57] investigating the accuracy of solar-system ephemerides,[26] and investigating the properties of the pulsars themselves (e.g. Refs. 156 and 118) and of the intervening interstellar medium (e.g. Ref. 67). However, here we concentrate on what is undoubtedly their primary scientific goal, the direct detection of gravitational waves. Unfortunately, in common with other GW detection efforts around the world, PTAs have so far only been able to place limits on the strength of signals from potential GW sources. However, these limits are now beginning to seriously constrain the astrophysical source models and the assumptions that go into them and hence have implications that go far beyond the GW studies themselves.

3.3.1. *Existing PTAs*

The *EPTA* uses five large radio-telescopes in Europe, the Effelsberg 100 m telescope in Germany, the Nançay Radio Telescope in France (95 m equivalent area), the Westerbork Synthesis Radio Telescope in the Netherlands (similar effective area to the Nançay telescope), the 76 m Lovell Telescope at Jodrell Bank in England and the recently completed 64 m Sardinia Radio Telescope in Italy, to observe about 40 MSPs with a cadence of between a few days and 30 days for different pulsars.[73] Different telescopes observe at different frequencies in the range 0.3–2.6 GHz, but all are instrumented at 1.4 GHz. Normally the five telescopes observe independently, but in a project known as the "Large European Array for Pulsars" (*LEAP*) 1.4 GHz signals over a bandwidth of 128 MHz from the five telescopes can be summed coherently to form a 194 m equivalent diameter radio telescope. The different telescopes use different signal-processing systems, either digital filterbanks or coherent dedisperion systems or both.

NANOGrav makes use of the 300 m Arecibo radio telescope in Puerto Rico and the 100 m Green Bank Telescope (*GBT*) in West Virginia.[87] A sample of about 36 pulsars is observed, typically at 3 week intervals. At Arecibo, observations are made in bands centered at 430 MHz and 1410 MHz, whereas at the *GBT*, the observed bands are centered at 820 MHz and 1500 MHz. Currently both radio telescopes use coherent dedispersion systems with bandwidths up to 800 MHz, but in the past a range of filterbank and coherent dedispersion systems with more limited bandwidths have been used.

As the name suggests, the *PPTA* uses the Parkes 64 m radio telescope located in New South Wales, Australia. A sample of 22 pulsars is currently being observed with regular observations at 2–3 week intervals in three bands around 730 MHz, 1400 MHz and 3100 MHz respectively.[56,86] Coherent dedispersion systems are used at 730 MHz and 1400 MHz with bandwidths up to 310 MHz and digital filterbanks at 1400 MHz and 3100 MHz with bandwidths of 256 MHz and 1024 MHz, respectively.

Data sets from all three PTAs have spans ranging from a few years up to about 20 years for different pulsars; three (including the original MSP, PSR B1937+21) have Arecibo data spans of nearly 30 years. The three PTAs together observe about 50 pulsars with some being observed by two or even all three of the PTAs. Their distribution on the sky is shown in Fig. 19.

Given that the combined data set of the three PTAs contains a larger number of pulsars, improved observation cadence and greater frequency diversity than the data set of any one PTA, there is a strong motivation to combine all the available data sets to obtain maximum sensitivity for PTA scientific objectives. The *IPTA* consortium was set up to facilitate progress toward this goal.[85] The *IPTA* also

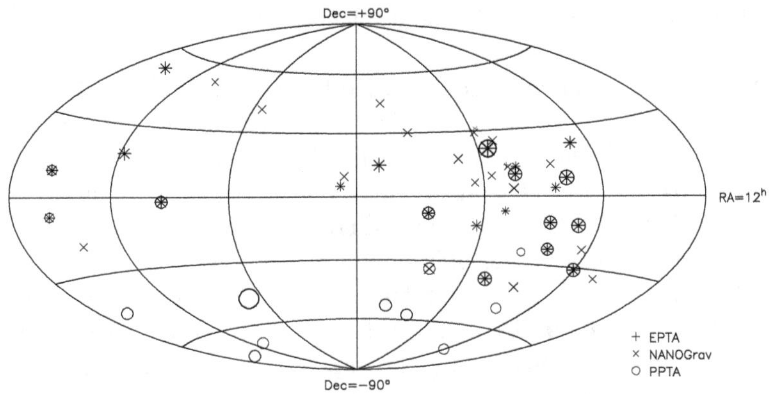

Fig. 19. Distribution on the sky of MSPs being timed by the three PTAs, with different symbols for each PTA. Right ascension increases to the left with 0^h at the plot center. The symbol size is related to the ratio S_{1400}/P, where S_{1400} and P are the pulsar 1400 MHz flux density and pulse period respectively (Ref. 85).

arranges annual science meetings and student workshops and provides a forum for outreach programs and other activities related to PTA research.

3.3.2. *Limits on the nanohertz GW background*

As discussed in Sec. 3.2.1, GWs from a cosmological distribution of SMBH binary systems are expected to contribute a very "red" signal to the spectrum of pulsar timing residuals. The expected signal from other GWB sources is similar. Consequently, long-term observations of a single pulsar with little or no detectable intrinsic timing irregularities can be used to place a limit on the strength of the GWB in the Galaxy. Of course, statistical limits can be improved by using data from several such pulsars. An early limit on the GWB at a frequency of 4.5 nHz was set by Kaspi *et al.*[66] using Arecibo observations of two MSPs, PSR B1855+09 and PSR B1937+21, with a 95% confidence limit on $\Omega_{GW}h^2$ of 6×10^{-8}, where $h = H_0/100\,\mathrm{km\,s^{-1}}$.

Since the advent of the various PTA projects, both the quality and quantity of timing data sets has improved and a variety of analysis techniques have been employed to extract increasingly restrictive limits. Based on early *PPTA* data on seven pulsars combined with the Kaspi *et al.* PSR B1855+09 data set, Jenet *et al.*[64] used a "frequentist" approach with a statistic based on the amplitude of the low-frequency components in the power spectrum of the timing residuals to set a 95% confidence limit of about 2×10^{-8} on Ω_{GW} at a GW frequency of 1/8 yr or 4 nHz. From Eqs. (25) and (30), this result is equivalent to a characteristic strain at frequency 1/1 yr, $A_{1\mathrm{yr}} \approx 1.1 \times 10^{-14}$.

van Haasteren *et al.*[138] analyzed *EPTA* 1400 MHz data sets for five MSPs with spans of 5–8 years using a Bayesian analysis to place limits on the GWB amplitude as a function of its spectral index α. For $\alpha = -2/3$, the derived limit at the 95%

confidence level is $A_{1\mathrm{yr}} \approx 6\times 10^{-15}$, about a factor 1.8 better than the Jenet *et al.*[64] limit.

NANOGrav multi-band data sets recorded between 2005 and 2010 for 17 MSPs were analyzed by Demorest *et al.*[40] Timing analyses taking into account time-varying dispersion delays and frequency-dependent pulse profiles were carried out to form sets of post-fit residuals and the corresponding covariance matrices for each pulsar. For most MSPs in the sample, no red noise signal was detectable in the post-fit residuals. Considering just the pulsar with the smallest post-fit residuals, PSR J1713+0747, and taking into account absorption of red noise by the timing fit, Demorest *et al.* obtained a 95% confidence limit for $A_{1\mathrm{yr}}$ of 1.1×10^{-14}. A separate cross-correlation analysis weighted by the expected Hellings and Downs function [Eq. (22)] across all the pulsars in the sample resulted in a somewhat better limit $A_{1\mathrm{yr}} \approx 7 \times 10^{-15}$, although this limit was dominated by correlations with the two best pulsars in the sample, PSRs J1713+0747 and J1909−3744.

Based on *PPTA* and earlier Parkes timing observations made in three observing bands centered near 700 MHz, 1400 MHz and 3100 MHz respectively with data spans of up to 17 years,[86] together with the PSR B1855+09 archival Arecibo data,[66] Shannon *et al.*[119] placed a limit of 1.3×10^{-9} on Ω_{GW} at a GW frequency of 2.8 nHz. The corresponding limit on $A_{1\mathrm{yr}}$ assuming a GW spectral index of $-2/3$ is 2.4×10^{-15}. This analysis, which included dispersion correction for the *PPTA* data sets and was based on the six best *PPTA* pulsars, used a statistical method similar to that of Jenet *et al.*[64] but included modeling of the red noise in the timing residuals. As shown in Fig. 20, this limit rules out a model in which the growth of SMBH in galaxies is dominated by mergers[89] at the 91% confidence level, but is consistent with other models for galaxy and SMBH evolution where much of the SMBH growth is by accretion.

As discussed in Sec. 3.2.2, topological defects in the early universe are another potential source for the GWB. Figure 21 shows limits on the dimensionless string tension $G\mu/c^2$ as a function of loop size for various sets of other relevant parameters.[111] The middle solid curves are limits based on the current *EPTA* data sets and the lower dashed curves are projections for LEAP data sets that coherently combine data from the *EPTA* telescopes. The upper dot-dashed line is a limit based on LIGO data.[1] The current *EPTA* results give a conservative upper limit on the string tension of 5.3×10^{-7}. Lower limits can be obtained with more restricted assumptions about the string parameters (e.g. Refs. 37, 64 and 138).

3.3.3. *Limits on GW emission from individual black-hole binary systems*

For an isolated binary system at a luminosity distance d_L, the intrinsic GW strain amplitude is given by

$$h_0 = \frac{(GM_c)^{5/3}}{c^4}\frac{(\pi f)^{2/3}}{d_L}, \tag{34}$$

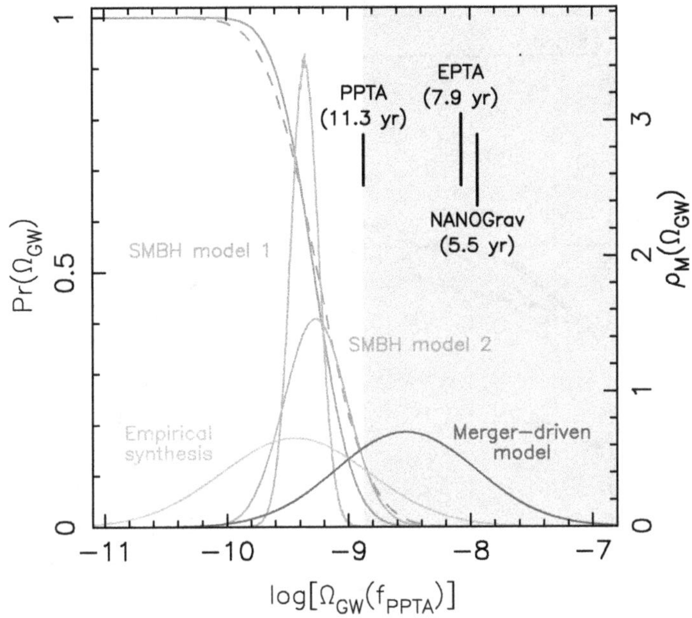

Fig. 20. Limits on the relative energy density of the GWB, Ω_{GW} at a GW frequency of 2.8 nHz based on the *PPTA* data sets, together with predictions for Ω_{GW} based on several different models for the GWB.[119] The solid and dashed lines that are asymptotic to 1.0 at low Ω_{GW} show the probability Pr that a GWB signal of energy density Ω_{GW} can exist in the *PPTA* data sets, based on Gaussian and non-Gaussian GWB statistics respectively. The shaded region is ruled out with 95% confidence by the *PPTA* data. Corresponding limits from analysis of *EPTA*[138] and *NANOGrav*[40] data sets, scaled to $f_{GW} = 2.8\,\text{nHz}$, are also shown. The Gaussian curves show the probability density functions ρ_M for the existence of a GWB with energy density Ω_{GW} based on a merger-driven model for growth of SMBHs in galaxies,[89] an empirical synthesis of observational constraints on SMBHs in galaxies,[113] and based on the Millennium dark matter simulations[18] together with semi-analytic models for growth of SMBHs in galaxies (see Ref. 119 for more details). (For color version, see page I-CP7.)

where M_c is the binary chirp mass and $f = 2f_b$ is the GW frequency. The actual observed signal depends on the orbital orientation and phase as well as the GW polarization angle. By averaging over these quantities, PTAs can set probabilistic limits for the strain amplitude as a function of f, both in a given direction and averaged over all directions (see, e.g. Refs. 9 and 158). In these analyses, there is assumed to be negligible evolution of f over the data span and only the Earth term [Eq. (20)] is considered, because of uncertainties in the pulsar distances, the pulsar terms cannot be added coherently and just contribute noise.

Figure 22 shows both sky-averaged upper limits and detection sensitivity for continuous-wave GW signals as a function of GW frequency based on the *PPTA* data set and using a frequentist analysis method.[158] The best limits and sensitivity are obtained for GW frequencies around 10^{-8} Hz, where the upper limit on h_0 is about 1.1×10^{-14}. A similar analysis of the five-year *NANOGrav* data set by Arzoumanian *et al.*[9] using both frequentist and Bayesian analysis methods gave a

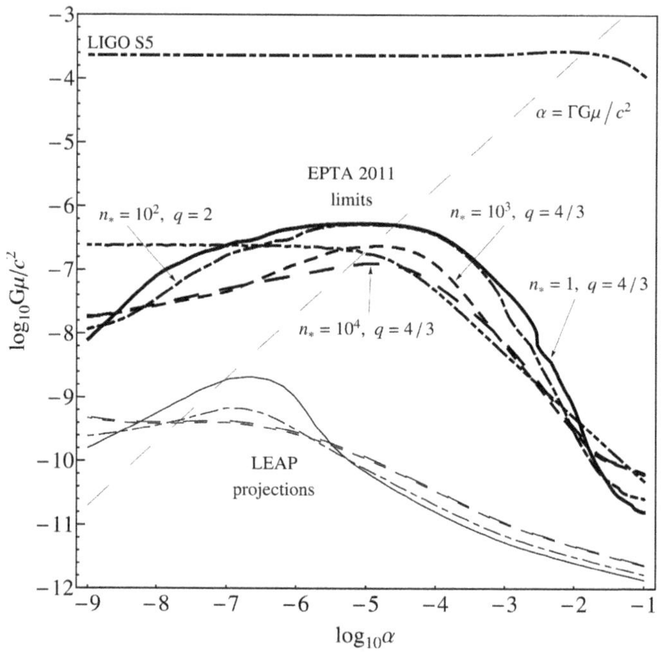

Fig. 21. Limits on cosmic string tension as a function of the loop size scale parameter α for different cutoff node numbers n_* and intrinsic spectral slopes q for the current *EPTA* limit on the energy density of the GWB (thick lines) and the projected sensitivity of the *LEAP* PTA (corresponding thin lines). The line with one long dash and three short dashes is an analytic approximation which is valid for large loops. The uppermost line is the *LIGO* limit at $f = 1\,\mathrm{kHz}$ (Refs. 111 and 1).

somewhat higher sky-averaged upper limit of about 5×10^{-14} at 10^{-8} Hz. Because of the uneven sky distribution of PTA pulsars, there is quite a strong dependence of sensitivity on source direction. This is illustrated in Fig. 23 which shows that sensitivity is greater toward the greatest concentration of PTA pulsars, roughly toward the Galactic Center.

Figure 22 also shows that we can effectively rule out the existence of SMBH binary systems with $M_c = 10^9\,\mathrm{M}_\odot$ and orbital frequencies around 10^{-8} Hz at distances closer than 30 Mpc. Similarly, a system with $M_c = 10^{10}\,\mathrm{M}_\odot$ at a distance of 400 Mpc should be detectable. Unfortunately, as Fig. 23 shows, the nearby galaxy clusters such as Virgo, Coma and Fornax are all in regions of relatively low sensitivity for the *PPTA* (and other PTAs), so the effective limits for these clusters are a factor of a few higher. It is unlikely that such massive binary systems exist in these clusters. More generally, limits can also be placed on the SMBH binary coalescence rate in the nearby universe ($z \lesssim 0.1$). Based on the *PPTA* results, Zhu *et al.*[158] place a 95% confidence limit of $4 \times 10^{-3}(10^{10}\mathrm{M}_\odot/M_c)^{10/3}\,\mathrm{Mpc}^{-3}\,\mathrm{Gyr}^{-1}$ on the coalescence rate. This limit is about two orders of magnitude above current estimates of the galaxy merger rate in the local universe (Ref. 9).

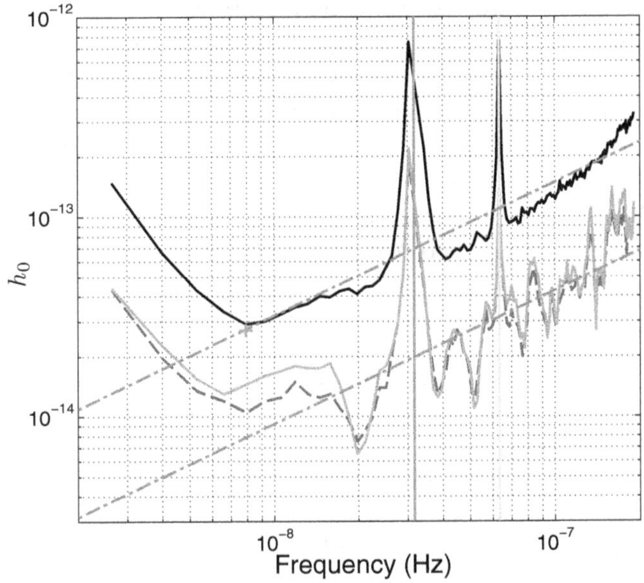

Fig. 22. Sky-averaged limits on the intrinsic GW strain amplitude h_0 as a function of GW frequency f based on the *PPTA* data sets. The lower curves represent the largest GW signal (with a false-alarm probability of 1%) that could be present in the real *PPTA* data (dashed line) and a simulated data set (solid line). The upper solid line gives the sensitivity of the *PPTA* to a continuous-wave source, i.e. the minimum signal that could be detected with 95% probability. The upper limits and sensitivities are higher at frequencies of 1/1 yr and 1/6 months as these frequencies are absorbed by the timing fits for position and parallax respectively. The sloping dot-dashed lines are the expected signal levels for a SMBH binary systems with $M_c = 10^{10}\,\mathrm{M}_\odot$ and distance 400 Mpc (upper line) and $M_c = 10^9\,\mathrm{M}_\odot$ and distance 30 Mpc (lower line) (Ref. 158).

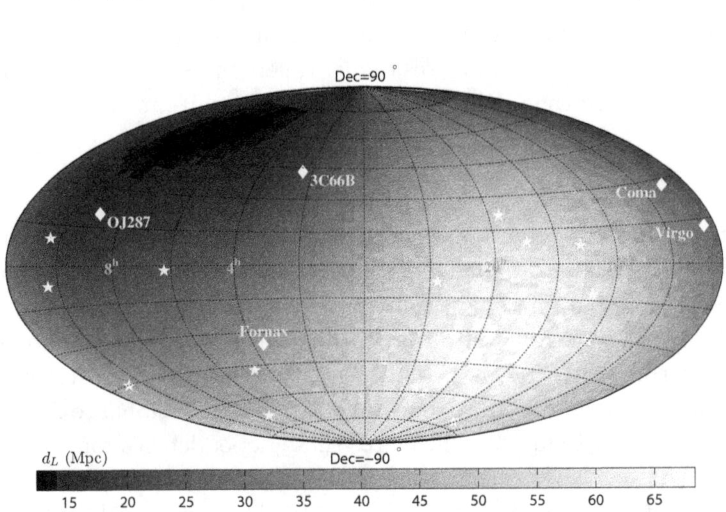

Fig. 23. Sky distribution of the luminosity distance d_L to which a binary system with chirp mass $M_c = 10^9\,\mathrm{M}_\odot$ radiating at 10^{-8} Hz could be detected. The stars indicate the positions of the 20 *PPTA* pulsars and the diamonds are potential sources of GW continuous-wave emission (Ref. 158).

These rate estimates are based on the orientation-averaged and sky-averaged amplitudes. It is of course possible that a favorably oriented and located SMBH binary system in the late stages of coalescence could exist. On the other hand, the estimates are based on circular binary orbits and, as discussed in Secs. 3.2.1 and 3.2.3, short-period SMBH binaries may have significant eccentricity which reduces the GW power at the fundamental frequency $f = 2f_b$ and hence the detectability of such systems. On balance, it seems unlikely that GW from an individual coalescing SMBH binary system will be detected with the current generation of PTAs.

3.4. *Future prospects*

PTAs have now achieved data spans and ToA precisions that would allow detection of the GWB predicted by some models for the evolution of galaxies and the SMBHs at their core (see, e.g. Ref. 119). Up to now, no detections have been made. While this is disappointing from the point of view of GW astrophysics, it is starting to have important implications for galaxy and SMBH evolution models and to rule out some scenarios. It also implies that PTAs are close to detecting the GWB if current predictions for its amplitude are correct.

Siemens *et al.*[126] have considered the sensitivity of an idealized PTA to a GWB. At low signal levels, when the lowest signal frequencies are below the white noise level, the detection signal-to-noise ratio (S/N) is

$$\langle \rho \rangle \propto Mc\frac{A_{1\mathrm{yr}}^2}{\sigma^2}T^{\beta}, \tag{35}$$

where M is the number of pulsars in the array, c is the observing cadence (frequency of observations), $A_{1\mathrm{yr}}$ is the GWB amplitude [Eq. (30)], σ is the rms level of the white timing noise, T is the observing data span and β is the inverse spectral index of the GWB signal in the timing residuals, taken to be 13/3 [Eq. (31)]. In the detection regime where the GWB signal exceeds the white noise level, the S/N is

$$\langle \rho \rangle \propto M\left(\frac{\sqrt{c}\,A_{1\mathrm{yr}}}{\sigma}\right)^{1/\beta}T^{1/2}. \tag{36}$$

Consequently, in the pre-detection regime, the S/N increases rapidly with increased observing cadence and data span and decreased timing noise, but has a much weaker dependence on these parameters in the strong signal regime. The reason for this is that noise from the uncorrelated pulsar term, which is also proportional to $A_{1\mathrm{yr}}$, dominates over the white "receiver" noise, greatly modifying the statistical behavior. Importantly though, in both regimes, the S/N is proportional to M, the number of pulsars in the PTA. Figure 24 illustrates these dependencies for a range of plausible future PTAs.

As Siemens *et al.*[126] point out, if $A_{1\mathrm{yr}} \sim 10^{-15}$, current PTAs are already in the "strong signal" regime. This means that increasing the observing data spans and cadence or decreasing ToA uncertainties has limited effect on the S/N of a potential detection. Increasing the number of pulsars in the PTA is a much more

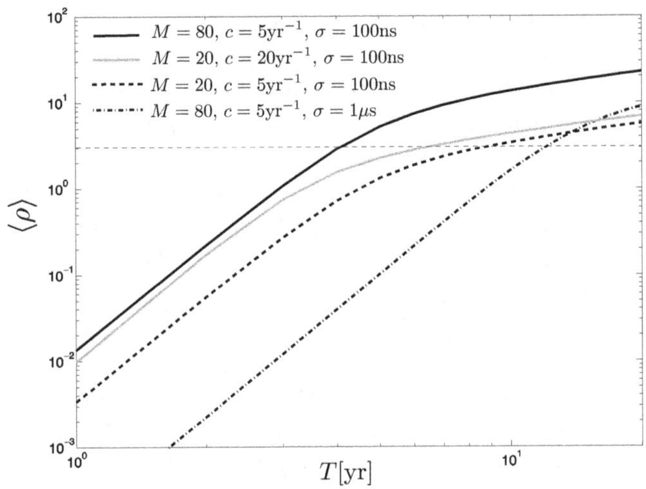

Fig. 24. Detection S/N for a GWB as a function of PTA data span for four different, but plausible, future PTAs. See text for the meaning of the PTA parameters (Ref. 126).

cost-effective way to increase detection sensitivity. This fact provides much of the motivation to combine data from existing PTAs to form the *IPTA*. In the future, the Chinese *FAST* radio telescope[94] and the *SKA*[24] will provide a large increase in radiometer sensitivity compared to existing instruments. The discussion above shows that this increased sensitivity will be best employed in increasing the number of (weaker) pulsars that are timed, rather than improving the ToA precision on the stronger existing PTA pulsars. Considerations of "jitter noise" in ToAs resulting from shape variations in individual pulses[118] lead to the same conclusion.

While direct detection of the GWB would be enormously exciting and significant, there is no doubt that direct detection of GW from individual SMBH binary systems is potentially much more interesting from an astrophysical perspective. It opens up the possibility of identifying a GW source with source or region identified through electromagnetic-wave (radio, optical, X-ray or γ-ray) emission and the advent of "multi-messenger" astronomy (e.g. Refs. 78, 114 and 22). For example, many active galactic nuclei (AGNs) show evidence of a close binary SMBH at their core. Examples of this include X-shaped radio lobes (e.g. Ref. 157), double-peaked or variable emission lines from the core region (e.g. Ref. 124), quasi-periodic modulation of core radio (e.g. Ref. 137) or X-ray emission (e.g. Refs. 114 and 79), and direct imaging of double AGN (e.g. Ref. 109).

With one exception, existing PTA systems do not have sufficient sensitivity to detect these potential GW sources. The exception is the claimed SMBH binary system identified by VLBI astrometry of the nearby quasar 3C 66B[133] which was effectively ruled out by pulsar timing observations.[65] Future PTAs including *FAST* and the *SKA* will have much increased sensitivity, making searches for other GW candidate sources potentially more productive.

Similarly, although blind searches for continuous-wave GW signals in current PTA data sets have a low probability of successful detection, in future this should not be the case. The possibility of identification of a source galaxy or AGN then depends critically on the accuracy of the position determination for the GW source. When only detection of the "Earth term" is considered, the accuracy is at best many tens of square degrees (e.g. Ref. 158) containing thousands if not millions of galaxies. Only a correlation of the GW signal with a modulation of some property of a galaxy core (e.g. intensity or velocity) would establish an identification.

The situation changes dramatically if the pulsar terms can be added coherently with the Earth term [Eq. (20)]. Each Earth–pulsar system then forms an interferometer with baseline d and fringe spacing $\sim \lambda_{GW}/[2\pi d(1+\gamma)]$, where d is the pulsar distance, λ_{GW} is the GW wavelength and γ is the direction cosine between the pulsar direction and the GW propagation direction.[19,78] Positional accuracies are roughly the fringe spacing divided by the S/N of the GW detection. Since d is typically $> 10^3$ light-years and λ_{GW} is a few light-years, sub-arc-minute positional accuracies are possible for the stronger sources. However, to achieve the coherent summation, the distance to the pulsars must be known to better than $\lambda_{GW}/(1+\gamma)$, i.e. about 1 pc unless the GW source is nearly aligned with the pulsar. Currently, only one PTA pulsar, PSR J0437−4715, has a distance known to this accuracy, measured through VLBI astrometry,[39] but in the *SKA* era this will change. As Boyle and Pen[19] pointed out, with a high density of PTA pulsars on the sky, advantage can be taken of the $(1+\gamma)$ factor and so pulsars with less precisely determined distances located in the general direction of the GW source could be useful. In this situation, PTAs would not be confusion-limited and, in principle, many individual GW sources could be identified. With a high enough number of PTA pulsars (say 1000 or more) it may be possible to localize a binary GW source by the quadrupolar pattern of timing residuals in pulsars surrounding the source, even if the pulsar distances are poorly determined.

4. Summary and Conclusion

Nature has been very kind in providing us with a set of near-perfect celestial clocks, many in situations of rapidly varying gravitational accelerations. Not only are these celestial clocks, known as pulsars, precise time-keepers, they are also exceedingly compact. This enables them to be treated as point masses in theoretical analyses of their motion and also permits tests in the regime of strong gravitational fields. These qualities result in a very wide range of applications for pulsar time-keeping, most importantly, at least in the context of this review, to investigations of relativistic gravitation.

Observations of double-neutron-star systems, wide circular pulsar–white-dwarf systems and even isolated pulsars have been used to test the accuracy of gravitational theories. Remarkably, GR is unscathed by all of these tests and hence remains the most viable theory of gravitation. Pulsar timing has provided the strongest available limits on at least six parameters describing deviations from GR. Continued and improved pulsar timing measurements, especially with new and highly sensitive radio telescopes such as *FAST* and the *SKA*, will both improve on these limits and enable new and different tests of relativistic gravity. They may even demonstrate a failure of GR to adequately account for the observations, leading to new or modified theories of gravitation.

Continuing and new searches for previously unknown pulsars, especially with *FAST* and the *SKA*, will not only increase the number of pulsars that can be used in tests of relativistic gravity. They will also turn up new and exciting classes of object such as the recently discovered triple system, PSR J0337+1715. Such discoveries enrich the investigations of relativistic astrophysics that can be undertaken with pulsars.

One of the outstanding goals of current astronomy and astrophysics is the direct detection and study of GWs. PTAs provide a viable mechanism for detection of GWs with frequencies in the nanohertz range. They therefore complement other existing or planned instruments such as the laser-interferometer systems *LIGO* and *eLISA* which are sensitive to GWs at frequencies of around 100 Hz and millihertz respectively. The most probable sources for GW detection by PTAs are binary super-massive black holes in the cores of distant galaxies. These produce an unresolved background of GWs that is potentially detectable, but there may also be individual binary systems that could be detected by PTAs.

There are currently three major PTAs operating, one each in Europe (*EPTA*), North America (*NANOGrav*) and Australia (*PPTA*). Up to now, no GWs have been detected by PTAs (or other GW detection systems) so the direct detection of GWs remains a goal. However, recent limits on the GW background are placing significant constraints on existing models for galaxy mergers over cosmological time and the formation and evolution of super-massive black holes in the cores of these galaxies. For example, a model in which black-hole growth is dominated by mergers is essentially ruled out.

The sensitivity of PTAs to GWs is a function of several factors including the precision of the pulse arrival-time measurements, the data span of the PTA observations and the cadence or frequency of observations within this data span. However the most important single factor is the number of pulsars in the PTA. Of course, these pulsars must meet certain timing-precision and period-stability criteria to usefully contribute to a PTA. There are two main approaches to increasing the number of pulsars. First, existing data sets can be combined to form a single PTA — this is the goal of the IPTA project. Second, searches can be undertaken to increase the number of known pulsars suitable for PTA projects. With *FAST* and the *SKA* it is possible that hundreds of MSPs that are suitable for PTA projects will be both discovered and subsequently timed to high precision. This will surely lead to the detection of GWs and to detailed investigations of both the GWs themselves and the sources that generate them.

Acknowledgments

I thank my colleagues in the pulsar community for their insights and hard work that have led to the results discussed in this review. In particular, I thank Paulo Freire, Matthew Kerr and Norbert Wex for their helpful comments on earlier versions of the manuscript and Sydney Chamberlin, Sotìris Sanidas and Xingjiang Zhu for producing revised versions of figures from their papers. I also thank the CSIRO and the Australian Research Council for supporting my research over the years and, in particular, CSIRO Astronomy and Space Science for their continued support.

References

1. B. P. Abbott *et al.*, *Nature* **460** (2009) 990.
2. B. P. Abbott *et al.*, *Rep. Progr. Phys.* **72** (2009) 076901.
3. P. Amaro-Seoane *et al.*, *Class. Quantum Grav.* **29**(12) (2012) 124016.
4. P. Amaro-Seoane, A. Sesana, L. Hoffman, M. Benacquista, C. Eichhorn, J. Makino and R. Spurzem, *Mon. Not. R. Astron. Soc.* **402** (2010) 2308.
5. M. Anholm, S. Ballmer, J. D. E. Creighton, L. R. Price and X. Siemens, *Phys. Rev. D* **79**(8) (2009) 084030.
6. J. Antoniadis *et al.*, *Science* **340** (2013) 448.
7. J. Antoniadis, M. H. van Kerkwijk, D. Koester, P. C. C. Freire, N. Wex, T. M. Tauris, M. Kramer and C. G. Bassa, *Mon. Not. R. Astron. Soc.* **423** (2012) 3316.
8. P. J. Armitage and P. Natarajan, *Astrophys. J.* **634** (2005) 921.
9. Z. Arzoumanian *et al.*, *Astrophys. J.* **794** (2014) 141.
10. J. D. Bekenstein, *Phys. Rev. D* **70**(8) (2004) 083509.
11. J. F. Bell, *Astrophys. J.* **462** (1996) 287.
12. J. F. Bell and M. Bailes, *Astrophys. J.* **456** (1996) L33.
13. J. F. Bell and T. Damour, *Class. Quantum Grav.* **13** (1996) 3121.
14. B. Bertotti, L. Iess and P. Tortora, *Nature* **425** (2003) 374.
15. N. D. R. Bhat, M. Bailes and J. P. W. Verbiest, *Phys. Rev. D* **77**(12) (2008) 124017.
16. B. M. Barker and R. F. O'Connell, *Phys. Rev. D* **12** (1975) 329.
17. D. Bhattacharya and E. P. J. van den Heuvel, *Phys. Rep.* **203** (1991) 1.
18. M. Boylan-Kolchin, V. Springel, S. D. M. White, A. Jenkins and G. Lemson, *Mon. Not. R. Astron. Soc.* **398** (2009) 1150.
19. L. Boyle and U.-L. Pen, *Phys. Rev. D* **86**(12) (2012) 124028.
20. R. P. Breton, V. M. Kaspi, M. Kramer, M. A. McLaughlin, M. Lyutikov, S. M. Ransom, I. H. Stairs, R. D. Ferdman, F. Camilo and A. Possenti, *Science* **321** (2008) 104.
21. M. Burgay *et al.*, *Nature* **426** (2003) 531.
22. S. Burke-Spolaor, *Class. Quantum Grav.* **30**(22) (2013) 224013.
23. R. R. Caldwell and B. Allen, *Phys. Rev. D* **45** (1992) 3447.
24. C. Carilli and S. Rawlings, *New Astron. Rev.* **48** (2004) 11.
25. S. J. Chamberlin and X. Siemens, *Phys. Rev. D* **85**(8) (2012) 082001.
26. D. J. Champion *et al.*, *Astrophys. J.* **720** (2010) L201.
27. D. Chapon, L. Mayer and R. Teyssier, *Mon. Not. R. Astron. Soc.* **429** (2013) 3114.
28. T. Clifton and J. M. Weisberg, *Astrophys. J.* **679** (2008) 687.
29. I. Cognard and D. C. Backer, *Astrophys. J.* **612** (2004) L125.
30. J. M. Cordes and F. A. Jenet, *Astrophys. J.* **752** (2012) 54.
31. T. Damour and G. Esposito-Farèse, *Class. Quantum Grav.* **9** (1992) 2093.

32. T. Damour and G. Schäfer, *Nuovo Cim.* **101** (1988) 127.
33. T. Damour and G. Schäfer, *Phys. Rev. Lett.* **66** (1991) 2549.
34. T. Damour and J. H. Taylor, *Astrophys. J.* **366** (1991) 501.
35. T. Damour and J. H. Taylor, *Phys. Rev. D* **45** (1992) 1840.
36. T. Damour and A. Vilenkin, *Phys. Rev. D* **64**(6) (2001) 064008.
37. T. Damour and A. Vilenkin, *Phys. Rev. D* **71**(6) (2005) 063510.
38. A. T. Deller, M. Bailes and S. J. Tingay, *Science* **323** (2009) 1327.
39. A. T. Deller, J. P. W. Verbiest, S. J. Tingay and M. Bailes, *Astrophys. J.* **685** (2008) L67.
40. P. B. Demorest *et al.*, *Astrophys. J.* **762** (2013) 94.
41. G. Desvignes, M. Kramer, I. Cognard, L. Kasian, J. van Leeuwen, I. Stairs and G. Theureau, PSR J1906+0746: From relativistic spin-precession to beam modeling, in *Neutron Stars and Pulsars: Challenges and Opportunities After 80 Years, IAU Symposium 291*, ed. J. van Leeuwen (Cambridge University Press, Cambridge, 2013), pp. 199.
42. S. Detweiler, *Astrophys. J.* **234** (1979) 1100.
43. A. Einstein, Erklärung der Perihelbewegung des Merkur aus der allgemeinen Relativitätstheorie, *Sitzungsberichte der Königlich Preußischen Akademie der Wissenschaften (Berlin)* (1915), pp. 831.
44. M. Favata, *Phys. Rev. D* **80**(2) (2009) 024002.
45. R. D. Ferdman *et al.*, *Astrophys. J.* **767** (2013) 85.
46. R. D. Ferdman *et al.*, *Mon. Not. R. Astron. Soc.* **443** (2014) 2183.
47. L. S. Finn and A. N. Lommen, *Astrophys. J.* **718** (2010) 1400.
48. W. M. Folkner, J. G. Williams and D. H. Boggs, IPN Progress Report 42-178, NASA Jet Propulsion Laboratory (2009).
49. E. Fonseca, I. H. Stairs and S. E. Thorsett, *Astrophys. J.* **787** (2014) 82.
50. P. C. C. Freire *et al.*, *Mon. Not. R. Astron. Soc.* **412** (2011) 2763.
51. P. C. C. Freire, M. Kramer and N. Wex, *Class. Quantum Grav.* **29**(18) (2012) 184007.
52. P. C. C. Freire *et al.*, *Mon. Not. R. Astron. Soc.* **423** (2012) 3328.
53. M. E. Gonzalez *et al.*, *Astrophys. J.* **743** (2011) 102.
54. L. Guillemot *et al.*, *Astrophys. J.* **744** (2012) 33.
55. R. W. Hellings and G. S. Downs, *Astrophys. J.* **265** (1983) L39.
56. G. Hobbs, *Class. Quantum Grav.* **30**(22) (2013) 224007.
57. G. Hobbs *et al.*, *Mon. Not. R. Astron. Soc.* **427** (2012) 2780.
58. G. Hobbs *et al.*, *Mon. Not. R. Astron. Soc.* **394** (2009) 1945.
59. G. Hobbs, D. R. Lorimer, A. G. Lyne and M. Kramer, *Mon. Not. R. Astron. Soc.* **360** (2005) 974.
60. F. Hofmann, J. Müller and L. Biskupek, *Astron. Astrophys.* **522** (2010) L5.
61. R. A. Hulse and J. H. Taylor, *Astrophys. J.* **195** (1975) L51.
62. B. A. Jacoby, P. B. Cameron, F. A. Jenet, S. B. Anderson, R. N. Murty and S. R. Kulkarni, *Astrophys. J.* **644** (2006) L113.
63. A. H. Jaffe and D. C. Backer, *Astrophys. J.* **583** (2003) 616.
64. F. A. Jenet *et al.*, *Astrophys. J.* **653** (2006) 1571.
65. F. A. Jenet, A. Lommen, S. L. Larson and L. Wen, *Astrophys. J.* **606** (2004) 799.
66. V. M. Kaspi, J. H. Taylor and M. Ryba, *Astrophys. J.* **428** (1994) 713.
67. M. J. Keith *et al.*, *Mon. Not. R. Astron. Soc.* **429** (2013) 2161.
68. F. M. Khan, I. Berentzen, P. Berczik, A. Just, L. Mayer, K. Nitadori and S. Callegari, *Astrophys. J.* **756** (2012) 30.
69. F. M. Khan, K. Holley-Bockelmann, P. Berczik and A. Just, *Astrophys. J.* **773** (2013) 100.

70. B. Kiziltan, A. Kottas, M. De Yoreo and S. E. Thorsett, *Astrophys. J.* **778** (2013) 66.
71. A. S. Konopliv, S. W. Asmar, W. M. Folkner, Ö. Karatekin, D. C. Nunes, S. E. Smrekar, C. F. Yoder and M. T. Zuber, *Mars High Resolution Gravity Fields from MRO, Mars Seasonal Gravity, and Other Dynamical Parameters*, Vol. 211 (Elsevier, 2011), pp. 401.
72. M. Kramer, *Astrophys. J.* **509** (1998) 856.
73. M. Kramer and D. J. Champion, *Class. Quantum Grav.* **30**(22) (2013) 224009.
74. M. Kramer *et al.*, *Science* **314** (2006) 97.
75. M. Kramer and N. Wex, *Class. Quantum Grav.* **26**(7) (2009) 073001.
76. J. M. Lattimer, *Ann. Rev. Nucl. Part. Sci.* **62** (2012) 485.
77. L. Leblond, B. Shlaer and X. Siemens, *Phys. Rev. D* **79**(12) (2009) 123519.
78. K. J. Lee, N. Wex, M. Kramer, B. W. Stappers, C. G. Bassa, G. H. Janssen, R. Karuppusamy and R. Smits, *Mon. Not. R. Astron. Soc.* **414** (2011) 3251.
79. F. K. Liu, S. Li and S. Komossa, *Astrophys. J.* **786** (2014) 103.
80. K. Liu, R. P. Eatough, N. Wex and M. Kramer, *Mon. Not. R. Astron. Soc.* **445** (2014) 3115.
81. K. Liu, N. Wex, M. Kramer, J. M. Cordes and T. J. W. Lazio, *Astrophys. J.* **747** (2012) 1.
82. R. S. Lynch, P. C. C. Freire, S. M. Ransom and B. A. Jacoby, *Astrophys. J.* **745** (2012) 109.
83. A. G. Lyne *et al.*, *Science* **303** (2004) 1153.
84. M. Lyutikov and C. Thompson, *Astrophys. J.* **634** (2005) 1223.
85. R. N. Manchester, *Class. Quantum Grav.* **30**(22) (2013) 224010.
86. R. N. Manchester *et al.*, *PASA* **30** (2013) e017.
87. M. A. McLaughlin, *Class. Quantum Grav.* **30**(22) (2013) 224008.
88. M. A. McLaughlin *et al.*, *Astrophys. J.* **616** (2004) L131.
89. S. T. McWilliams, J. P. Ostriker and F. Pretorius, *Astrophys. J.* **789** (2014) 156.
90. D. Merritt and M. Milosavljević, *Liv. Rev. Relativ.* **8** (2005) 8.
91. M. Milgrom, *Astrophys. J.* **270** (1983) 365.
92. C. J. Moore, R. H. Cole and C. P. L. Berry, *Class. Quantum Grav.* **32**(1) (2015) 015014.
93. J. Müller, J. G. Williams and S. G. Turyshev, Lunar laser ranging contributions to relativity and geodesy, in *Lasers, Clocks and Drag-Free Control: Exploration of Relativistic Gravity in Space*, *Astrophysics and Space Science Library*, eds. H. Dittus, C. Lammerzahl and S. G. Turyshev, Vol. 349 (Springer, Berlin, 2008), p. 457.
94. R. Nan, D. Li, C. Jin, Q. Wang, L. Zhu, W. Zhu, H. Zhang, Y. Yue and L. Qian, *Int. J. Mod. Phys. D* **20** (2011) 989.
95. W.-T. Ni, *Rep. Progr. Phys.* **73**(5) (2010) 056901.
96. D. J. Nice, E. M. Splaver, I. H. Stairs, O. Löhmer, A. Jessner, M. Kramer and J. M. Cordes, *Astrophys. J.* **634** (2005) 1242.
97. D. J. Nice, I. H. Stairs and L. E. Kasian, Masses of neutron stars in binary pulsar systems, in 40 *Years of Pulsars*: *Millisecond Pulsars, Magnetars and More*, eds. C. Bassa, Z. Wang, A. Cumming and V. M. Kaspi, Vol. 983 (American Institute of Physics, New York, 2008), pp. 453.
98. K. Nordtvedt, *Phys. Rev.* **170** (1968) 1186.
99. K. Nordtvedt, *Astrophys. J.* **320** (1987) 871.
100. K. Nordtvedt, *Phys. Rev. Lett.* **65** (1990) 953.
101. B. B. P. Perera *et al.*, *Astrophys. J.* **721** (2010) 1193.
102. P. C. Peters and J. Mathews, *Phys. Rev.* **131** (1963) 435.

103. E. S. Phinney, arXiv:astro-ph/0108028.
104. S. M. Ransom *et al.*, *Nature* **505** (2014) 520.
105. V. Ravi, R. N. Manchester and G. Hobbs, *Astrophys. J.* **716** (2010) L85.
106. V. Ravi, J. S. B. Wyithe, G. Hobbs, R. M. Shannon, R. N. Manchester, D. R. B. Yardley and M. J. Keith, *Astrophys. J.* **761** (2012) 84.
107. V. Ravi, J. S. B. Wyithe, R. M. Shannon, G. Hobbs and R. N. Manchester, *Mon. Not. R. Astron. Soc.* **442** (2014) 56.
108. H. W. Rix and J. Bovy, *Astron. Asrophys. Rev.* **21** (2013) 61.
109. C. Rodriguez, G. B. Taylor, R. T. Zavala, A. B. Peck, L. K. Pollack and R. W. Romani, *Astrophys. J.* **646** (2006) 49.
110. C. Roedig, M. Dotti, A. Sesana, J. Cuadra and M. Colpi, *Mon. Not. R. Astron. Soc.* **415** (2011) 3033.
111. S. A. Sanidas, R. A. Battye and B. W. Stappers, *Phys. Rev. D* **85**(12) (2012) 122003.
112. M. V. Sazhin, *Sov. Astron.* **22** (1978) 36.
113. A. Sesana, *Mon. Not. R. Astron. Soc.* **433** (2013) L1.
114. A. Sesana, C. Roedig, M. T. Reynolds and M. Dotti, *Mon. Not. R. Astron. Soc.* **420** (2012) 860.
115. A. Sesana, A. Vecchio and C. N. Colacino, *Mon. Not. R. Astron. Soc.* **390** (2008) 192.
116. A. Sesana, A. Vecchio and M. Volonteri, *Mon. Not. R. Astron. Soc.* **394** (2009) 2255.
117. R. M. Shannon and J. M. Cordes, *Astrophys. J.* **725** (2010) 1607.
118. R. M. Shannon *et al.*, *Mon. Not. R. Astron. Soc.* **443** (2014) 1463.
119. R. M. Shannon *et al.*, *Science* **342** (2013) 334.
120. L. Shao, R. N. Caballero, M. Kramer, N. Wex, D. J. Champion and A. Jessner, *Class. Quantum Grav.* **30**(16) (2013) 165019.
121. L. Shao and N. Wex, *Class. Quantum Grav.* **29**(21) (2012) 215018.
122. L. Shao and N. Wex, *Class. Quantum Grav.* **30**(16) (2013) 165020.
123. I. I. Shapiro, *Phys. Rev. Lett.* **13** (1964) 789.
124. Y. Shen, X. Liu, A. Loeb and S. Tremaine, *Astrophys. J.* **775** (2013) 49.
125. I. S. Shklovskii, *Sov. Astron.* **13** (1970) 562.
126. X. Siemens, J. Ellis, F. Jenet and J. D. Romano, *Class. Quantum Grav.* **30**(22) (2013) 224015.
127. X. Siemens, V. Mandic and J. Creighton, *Phys. Rev. D* **98**(11) (2007) 111101.
128. V. Springel *et al.*, *Nature* **435** (2005) 629.
129. I. H. Stairs, *Living Rev. Relat.* **6** (2003).
130. I. H. Stairs *et al.*, *Astrophys. J.* **632** (2005) 1060.
131. I. H. Stairs, S. E. Thorsett, J. H. Taylor and A. Wolszczan, *Astrophys. J.* **581** (2002) 501.
132. K. R. Stewart, J. S. Bullock, E. J. Barton and R. H. Wechsler, *Astrophys. J.* **702** (2009) 1005.
133. H. Sudou, S. Iguchi, Y. Murata and Y. Taniguchi, *Science* **300** (2003) 1263.
134. J. H. Taylor, L. A. Fowler and P. M. McCulloch, *Nature* **277** (1979) 437.
135. J. H. Taylor, R. A. Hulse, L. A. Fowler, G. E. Gullahorn and J. M. Rankin, *Astrophys. J.* **206** (1976) L53.
136. S. E. Thorsett, *Phys. Rev. Lett.* **77** (1996) 1432.
137. M. J. Valtonen, H. J. Lehto, L. O. Takalo and A. Sillanpää, *Astrophys. J.* **729** (2011) 33.
138. R. van Haasteren *et al.*, *Mon. Not. R. Astron. Soc.* **414** (2011) 3117.
139. J. van Leeuwen *et al.*, *Astrophys. J.* **798** (2015) 118.

140. B. P. Venemans, J. R. Findlay, W. J. Sutherland, G. De Rosa, R. G. McMahon, R. Simcoe, E. A. González-Solares, K. Kuijken and J. R. Lewis, *Astrophys. J.* **779** (2013) 24.
141. J. P. W. Verbiest *et al.*, *Astrophys. J.* **679** (2008) 675.
142. A. Vilenkin, *Phys. Lett. B* **107** (1981) 47.
143. J. M. Weisberg, D. J. Nice and J. H. Taylor, *Astrophys. J.* **722** (2010) 1030.
144. J. M. Weisberg, R. W. Romani and J. H. Taylor, *Astrophys. J.* **347** (1989) 1030.
145. J. M. Weisberg and J. H. Taylor, *Astrophys. J.* **576** (2002) 942.
146. N. Wex, in *Brumberg Festschrift*, ed. S. M. Kopeikein (de Gruyter, Berlin, 2014).
147. C. M. Will, *Astrophys. J.* **393** (1992) L59.
148. C. M. Will, *Theory and Experiment in Gravitational Physics* (Cambridge University Press, Cambridge, 1993).
149. C. M. Will, *Living Rev. Relat.* **17** (2014) 4.
150. C. M. Will and K. J. Nordtvedt, *Astrophys. J.* **177** (1972) 757.
151. J. G. Williams, S. G. Turyshev and D. H. Boggs, *Class. Quantum Grav.* **29**(18) (2012) 184004.
152. A. Wolszczan, *Nature* **350** (1991) 688.
153. A. Wolszczan, O. Doroshenko, M. Konacki, M. Kramer, A. Jessner, R. Wielebinski, F. Camilo, D. J. Nice and J. H. Taylor, *Astrophys. J.* **528** (2000) 907.
154. J. S. B. Wyithe and A. Loeb, *Astrophys. J.* **590** (2003) 691.
155. K. Yagi, D. Blas, E. Barausse and N. Yunes, *Phys. Rev. D* **89**(8) (2014) 084067.
156. W. M. Yan *et al.*, *Mon. Not. R. Astron. Soc.* **414** (2011) 2087.
157. X.-G. Zhang, D. Dultzin-Hacyan and T.-G. Wang, *Mon. Not. R. Astron. Soc.* **377** (2007) 1215.
158. X.-J. Zhu *et al.*, *Mon. Not. R. Astron. Soc.* **444** (2014) 3709.

Part III. Gravitational Waves

Chapter 10

Gravitational waves: Classification, methods of detection, sensitivities and sources

Kazuaki Kuroda[†,§], Wei-Tou Ni[‡,¶] and Wei-Ping Pan[‡,‖]

[†]*Institute for Cosmic Ray Research, The University of Tokyo, 5-1-5 Kashiwanoha, Kashiwa, Chiba 277-8582, Japan*

[‡]*Center for Gravitation and Cosmology (CGC), Department of Physics, National Tsing Hua University, Hsinchu, Taiwan 300, ROC*

[§]*kuroda@icrr.u-tokyo.ac.jp*

[¶]*weitou@gmail.com*

[‖]*d9722518@oz.nthu.edu.tw*

After giving a brief introduction and presenting a complete classification of gravitational waves (GWs) according to their frequencies, we review and summarize the detection methods, the sensitivities and the sources. We notice that real-time detections are possible above 300 pHz. Below 300 pHz, the detections are possible on GW imprints or indirectly. We are on the verge of detection. The progress in this field will be promising and thriving. We will see improvement of a few orders to several orders of magnitude in the GW detection sensitivities over all frequency bands in the next hundred years.

Keywords: Gravitational waves (GWs); GW spectrum classification; GW sources; methods of detecting GWs; GW sensitivities.

1. Introduction and Classification

Soon after the proposal of general relativity (GR), Einstein predicted the existence of gravitational waves (GWs) and estimated its strength from the wave equation he obtained in his 1916 paper on "Approximative Integration of the Field Equations of Gravitation".[1] Toward the end of his paper, he obtained the expression of the radiation A of the system per unit time in GR as (Eq. (23) in his paper) $A = (\kappa/24\pi)\Sigma_{\alpha\beta}(\partial^3 J_{\alpha\beta}/\partial t^3)^2$ with $J_{\alpha\beta}$ defined as the time-variable components of moment of inertia of the radiating system ($\kappa = 8\pi G_{\rm N}$ in terms of Newtonian gravitational constant $G_{\rm N}$).[a] He then continued that "This expression (for the radiation A) would get an additional factor $1/c^4$ if we would measure time in seconds and energy in Erg (erg). Considering $\kappa = 1.87 \cdot 10^{-27}$ (in units of cm and gm), it is

[a]This radiation formula is corrected with the trace contribution of the moment of inertia subtracted and the overall factor replaced by $\kappa/80\pi$ [a factor 2 off compared with (37)] in Einstein's next paper on GWs.[2] With his correction, Einstein noted that "This result shows that a mechanical system which permanently retains spherical symmetry cannot radiate. ..."

obvious that A has, in all imaginable cases, a practically vanishing value." Indeed at that time, possible expected source strengths and the detection capability had a huge gap. However, with the great strides in the advances of astronomy and astrophysics and in the development of technology, this gap is largely bridged. White dwarf was discovered in 1910 with its density soon estimated. Now we understand that GWs from white dwarf binaries in our Galaxy form a stochastic GW background ("confusion limit")[3] for space (low frequency) GW detection in GR. The first artificial satellite Sputnik was launched in 1957. However, at present the space GW missions are only expected to be launched in about 19 years later (~2034).[4]

The existence of GWs is the direct consequence of GR and unavoidable consequence of all relativistic gravity theories with finite velocity of propagation. Maxwell's electromagnetic theory predicted electromagnetic waves. Einstein's GR and other relativistic gravity theories predict the existence of GWs. GWs propagate in spacetime forming ripples of spacetime geometry.

The role of GW in gravity physics is like the role of electromagnetic wave in electromagnetic physics. The importance of GW detection is two-fold: (i) as probes to explore fundamental physics and cosmology, especially black hole physics and early cosmology and (ii) as a tool in astronomy and astrophysics to study compact objects and to count them, complement to electromagnetic astronomy and cosmic ray (including neutrino) astronomy.

The existence of gravitational radiation is demonstrated by binary pulsar orbit evolution.[5,6] In GR, a binary star system would emit energy in the form of GWs. The loss of energy results in the shrinkage of the orbit and shortening of orbital period. Based on more than 32 years (from 1974 through 2006) of timing observations of the relativistic binary pulsar B1913 + 16, the cumulative shift of peri-astron time is over 43 s. The calculated orbital decay rate in GR using parameters determined from pulsar timing observations agreed with the observed decay rates. From this and a relative acceleration correction due to solar system and pulsar system motion, Weisberg, Nice and Taylor[6] concluded that the measured orbital decay to the GR predicted value from the emission of gravitational radiation is 0.997 ± 0.002 providing conclusive evidence for the existence of gravitational radiation as their previous papers. Kramer *et al.*[7] did an orbit analysis of the double pulsar system PSR J0737-3039A/B from 2.5 years of pulse timing observations and found that the orbit period shortening rate 1.252(17) agreed with the GR prediction of 1.24787(13) to 1.003(14) fraction. Freire *et al.*[8] analyzed about 10 years of timing data of the binary pulsar J1738 + 0333 and obtained the intrinsic orbital decay rate to be $(-25.9 \pm 3.2) \times 10^{-15}$, agreed well with the calculated GR value $(-27.7^{+1.5}_{-1.9}) \times 10^{-15}$ using the determined orbital parameters. Further precision and many more systems are expected in the future for observable GW radiation reaction imprint on the orbital motion.

The usual way of detection of GW is by measuring the strain $\Delta l/l$ induced by it. Hence, GW detectors are usually amplitude sensors, not energy sensors.

The detection of GWs can be resolved into characteristic frequencies. The conventional classification of GW frequency bands, as given by Thorne[9] in 1995, was into (i) High-frequency band (1 Hz–10 kHz); (ii) Low-frequency band (100 μHz–1 Hz); (iii) Very-low-frequency band (1 nHz–100 nHz); (iv) Extremely-low-frequency band (1 aHz–1 fHz). This classification was mainly according to frequency ranges of corresponding types of detectors/detection methods: (i) ground GW detectors; (ii) space GW detectors; (iii) pulsar timing method and (iv) cosmic microwave background (CMB) methods. In 1997, we followed Ref. 9 and extended the band ranges to give the following classification[10,11]:

(i) High-frequency band (1–10 kHz).
(ii) Low-frequency band (100 nHz–1 Hz).
(iii) Very-low-frequency band (300 pHz–100 nHz).
(iv) Extremely-low-frequency band (1 aHz–10 fHz).

Subsequently, we added the very-high-frequency band and the middle-frequency band for there were enhanced interests and activities in these bands. Recently, we added the missing band (10 fHz–300 pHz) and the two bands beyond to give a complete frequency classification of GWs as compiled in Table 1.[12–17]

In Sec. 2, we give a brief introduction to GWs in GR. In Sec. 3, we review various methods of detection together with their typical/aimed sensitivities. In Sec. 4, we review various astrophysical and cosmological sources. In Sec. 5, we present an outlook.

Table 1. Frequency classification of GWs.[16–17]

Frequency band	Detection method
Ultra-high frequency band: above 1 THz	Terahertz resonators, optical resonators and magnetic conversion detectors.
Very-high-frequency band: 100 kHz–1 THz	Microwave resonator/wave guide detectors, laser interferometers and Gaussian beam detectors.
High-frequency band (audio band)*: 10 Hz–100 kHz	Low-temperature resonators and ground-based laser-interferometric detectors.
Middle frequency band: 0.1 Hz–10 Hz	Space laser-interferometric detectors of arm length 1,000 km–60,000 km.
Low-frequency band (milli-Hz band)†: 100 nHz–0.1 Hz	Space laser-interferometric detectors of arm length longer than 60,000 km.
Very-low-frequency band (nano-Hz band): 300 pHz–100 nHz	Pulsar timing arrays (PTAs).
Ultra low-frequency band: 10 fHz–300 pHz	Astrometry of quasar proper motions.
Extremely-low (Hubble)-frequency band (cosmological band): 1 aHz–10 fHz	CMB experiments.
Beyond Hubble-frequency band: below 1 aHz	Through the verifications of inflationary/ primordial cosmological models.

Notes: *The range of audio band normally goes only to 10 kHz.
†The range of milli-Hz band is 0.1 mHz–100 mHz.

2. GWs in GR

The equations of motion of GR, i.e. the Einstein equation is

$$G_{\mu\nu} = \kappa T_{\mu\nu}, \tag{1}$$

where $T_{\mu\nu}$ is the stress–energy tensor and $\kappa = 8\pi G_{\rm N}$. (We use the MTW[18] conventions with signature −2; this is also the convention used in Ref. 19; Greek indices run from 0 to 3; Latin indices run from 1 to 3; the cosmological constant is negligible for treating the methods of GW detection and for evaluating GW sources at the aimed accuracy of this paper except in the Hubble frequency band and beyond the Hubble frequency band and will be neglected in this treatment except in association with cosmological models.). Contracting the equations of motion (1), we have

$$R = -8\pi G_{\rm N} T, \tag{2}$$

where $T \equiv T_\mu{}^\mu$. Substituting (2) into (1), we obtain the following equivalent equations of motion

$$R_{\mu\nu} = 8\pi G_{\rm N}\left[T_{\mu\nu} - \frac{1}{2}g_{\mu\nu}T\right]. \tag{3}$$

For weak field in the quasi-Minkowskian coordinates, we express the metric $g_{\alpha\beta}$ as

$$g_{\alpha\beta} = \eta_{\alpha\beta} + h_{\alpha\beta}, \quad h_{\alpha\beta} \ll 1. \tag{4}$$

Since $h_{\alpha\beta}$ is a small quantity, we expand everything in $h_{\alpha\beta}$ and linearize the results to obtain the linear approximation. For linearized quantities, we use the Minkowski metric $\eta_{\alpha\beta}$ to raise and lower indices without affecting the linearized results. The Riemann curvature tensor can be expressed as

$$R_{\alpha\beta\gamma\delta} = \frac{1}{2}(g_{\alpha\delta},_{\beta\gamma} + g_{\beta\gamma},_{\alpha\delta} - g_{\alpha\gamma},_{\beta\delta} - g_{\beta\delta},_{\alpha\gamma}) + g_{\mu\nu}(\Gamma^\mu{}_{\beta\gamma}\Gamma^\nu{}_{\alpha\delta} - \Gamma^\mu{}_{\beta\delta}\Gamma^\nu{}_{\alpha\gamma}). \tag{5}$$

After linearization, we have

$$R_{\alpha\beta\gamma\delta} = \frac{1}{2}(h_{\alpha\delta},_{\beta\gamma} + h_{\beta\gamma},_{\alpha\delta} - h_{\alpha\gamma},_{\beta\delta} - h_{\beta\delta},_{\alpha\gamma}) + O(h^2), \tag{6}$$

$$R_{\alpha\gamma} = \frac{1}{2}(h_{\alpha\beta},_\gamma{}^\beta + h_{\beta\gamma},_\alpha{}^\beta - h_{\alpha\gamma},_\beta{}^\beta - h_\beta{}^\beta,_{\alpha\gamma}) + O(h^2), \tag{7}$$

$$R = h_{\alpha\beta},^{\alpha\beta} - h_\beta{}^\beta,_\alpha{}^\alpha + O(h^2), \tag{8}$$

where $O(h^2)$ denotes terms of order of $h_{\alpha\beta}h_{\mu\nu}$ or smaller. Now we choose the harmonic gauge condition for $h_{\alpha\beta}$,

$$\left[h_{\alpha\beta} - \frac{1}{2}\eta_{\alpha\beta}(\mathrm{Tr}\,h)\right],^\beta = 0 + O(h^2), \quad \text{i.e. } h_{\alpha\beta},^\beta = \frac{1}{2}(\mathrm{Tr}\,h),_\alpha + O(h^2), \tag{9}$$

where $\mathrm{Tr}(h)$ is defined as the trace of $h_\alpha{}^\beta$, i.e. $\mathrm{Tr}(h) \equiv h_\alpha{}^\alpha$. Now the linearized Einstein equation can be derived from (3), (7) and (9) and written in the form:

$$h_{\mu\nu},_\beta{}^\beta = -16\pi G_{\rm N}\left[T_{\mu\nu} - \frac{1}{2}\eta_{\mu\nu}T\right] + O(h^2). \tag{10}$$

This is the linearized wave equation for GR. The corresponding equation for electromagnetism is

$$A_{\mu},_{\beta}{}^{\beta} = 4\pi J_{\mu} \tag{11}$$

with gauge condition

$$A_{\alpha},^{\alpha} = 0. \tag{12}$$

The retarded solution of Eq. (12) is

$$A_{\mu} = \int \frac{J_{\mu}}{r_{\text{retarded}}} d^3x'. \tag{13}$$

Analogously, the solution of equation for GR in the harmonic gauge is

$$h_{\mu\nu} = -\frac{4G_{\text{N}}}{c^4} \int \left\{ \frac{T_{\mu\nu} - \frac{1}{2} g_{\mu\nu} T}{r} \right\}_{\text{retarded}} d^3x' + O(h^2). \tag{14}$$

In the linearized scheme, it is useful to represent the solution $h_{\mu\nu}(x, y, z, t)$ *outside of the source* region by its spectral components with wave vector (k_x, k_y, k_z) and frequency f. First, find the Fourier transform $h_{\mu\nu}(k_x, k_y, k_z)$ of $h_{\mu\nu}(x, y, z, t)|_{t=0}$:

$$^{(\text{k})}h_{\mu\nu}(k_x, k_y, k_z) \equiv c^{-3} \int h_{\mu\nu}(x, y, z, 0) \exp(-\text{i}k_x x - \text{i}k_y y - \text{i}k_z z)\, dxdydz. \tag{15}$$

The integration is from $-\infty$ to ∞ for each integration variable. From Eq. (10), for each spectral components, the frequency f is given by the dispersion relation

$$f = \frac{c}{2\pi}(k_x^2 + k_y^2 + k_z^2)^{1/2} \equiv \frac{c}{2\pi} k. \tag{16}$$

Hence the solution is

$$h_{\mu\nu}(x, y, z, t) = \left(\frac{c}{2\pi}\right)^3 \int {}^{(\text{k})}h_{\mu\nu}(k_x, k_y, k_z) \times \exp(\text{i}k_x x + \text{i}k_y y + \text{i}k_z z - 2\pi \text{i} f t)\, dk_x dk_y dk_z, \tag{17}$$

with f given by (16).

For plane GW $h_{\mu\nu}(n_x x + n_y y + n_z z - ct)$ propagating in the (n_x, n_y, n_z) direction with $n_x^2 + n_y^2 + n_z^2 = 1$, letting

$$U \equiv u - ct \equiv n_x x + n_y y + n_z z - ct, \tag{18}$$

we can resolve the plane GW into the following spectral representation:

$$\begin{aligned} h_{\mu\nu}(u, t) &\equiv h_{\mu\nu}(u - ct) = h_{\mu\nu}(U) \\ &= \frac{c}{2\pi} \int_{-\infty}^{\infty} {}^{(\text{k})}h_{\mu\nu}(k) \exp(\text{i}ku) \exp(-2\pi \text{i} f t)\, dk \end{aligned} \tag{19}$$

with

$$^{(k)}h_{\mu\nu}(k) \equiv c^{-1} \int_{-\infty}^{\infty} h_{\mu\nu}(U) \exp(-\mathrm{i}kU) dU. \tag{20}$$

The plane wave (19) and (20) can also be written as

$$h_{\mu\nu}(u,t) \equiv h_{\mu\nu}(u-ct) = h_{\mu\nu}(U) = \int_{-\infty}^{\infty} {}^{(f)}h_{\mu\nu}(f) \exp\left(\frac{2\pi \mathrm{i} fU}{c}\right) df \tag{21}$$

with

$$^{(f)}h_{\mu\nu}(f) \equiv {}^{(k)}h_{\mu\nu}\left(k = \frac{2\pi f}{c}\right) = \int_{-\infty}^{\infty} h_{\mu\nu}(u-ct)|_{u=0} \exp(2\pi \mathrm{i} ft) dt. \tag{22}$$

We note that since $h_{\mu\nu}(t)$ is real, $^{(f)}h_{\mu\nu}(-f) = {}^{(f)}h_{\mu\nu}{}^*(f)$ with * the complex conjugation

$$\begin{aligned} h_{\mu\nu}(U) &= \int_0^{\infty} 2|^{(f)}h_{\mu\nu}(f)| \cos\left(\frac{2\pi fU}{c}\right) df \\ &= \int_0^{\infty} 2f|^{(f)}h_{\mu\nu}(f)| \cos\left(\frac{2\pi fU}{c}\right) d(\ln f). \end{aligned} \tag{23}$$

From (21, 22) and the Parseval's equality, we have

$$\int_{-\infty}^{\infty} |h_{\mu\nu}(t)|^2 dt = \int_{-\infty}^{\infty} |^{(f)}h_{\mu\nu}(f)|^2 df$$

$$\text{(No summations in the indices } \mu \text{ and } \nu). \tag{24}$$

The squared-amplitude integral is equal to its squared-spectral-amplitude integral. One can also obtain a similar identity relating the integral on the (absolute) square of $h_{\mu\nu}(x,y,z,t)$ over (x,y,z) and the integral on the absolute square of $^{(k)}h_{\mu\nu}(k_x,k_y,k_z)$ over (k_x,k_y,k_z) using (15), (17) and the Parseval's equality in three dimensions.

For weak GW $h_{\mu\nu}$ propagating in the spacetime background $g_{\mu\nu}$ (i.e. the total spacetime metric is $g_{\mu\nu} + h_{\mu\nu}$.), Isaacson[20,21] showed that the GW stress–energy averaged over several wavelength is

$$t_{\mu\nu} = \frac{c^4}{32\pi G_{\mathrm{N}}} \langle \partial_\mu h^{\mathrm{TT}}{}_{\alpha\beta} \partial_\nu h^{\mathrm{TT}\alpha\beta} \rangle. \tag{25}$$

Here, $h^{\mathrm{TT}}{}_{\mu\nu}$ is the transverse traceless part of $h_{\mu\nu}$. *In the special harmonic gauge called radiation gauge (similar to radiation gauge in electrodynamics)*, $h^{\mathrm{TT}}{}_{\mu\nu} = h_{\mu\nu}$. Far from the sources, the GW can be approximated by plane waves. For a wave propagating in the z-direction, the only nonvanishing components of $h_{\mu\nu}$ in radiation gauge are $h_{11}, h_{22}(= -h_{11}), h_{12}$ and $h_{21}(= h_{12})$. The mass-energy density

t_{00} and mass-energy flux ct_{03} are given by

$$\rho c^2 = t_{00} = t_{03} = \frac{c^4}{16\pi G_{\mathrm{N}}} \left\langle \left[\left(\frac{1}{4}\right) (\partial_0 h_{11} - \partial_0 h_{22})^2 + (\partial_0 h_{12})^2 \right] \right\rangle$$
$$= \frac{c^2}{16\pi G_{\mathrm{N}}} \langle (\partial_0 h_+^2 + \partial_0 h_\times^2) \rangle \quad (26)$$

in agreement with Ref. 22. Here, $h_+ (\equiv (1/2)(h_{11} - h_{22}) = h_{11} = -h_{22})$ is the amplitude of e_1-polarization (+-polarization); $h_\times (\equiv h_{12} = h_{21})$ is the amplitude of e_2-polarization ($\times$-polarization). Due to gauge (coordinate) invariance from the linearized wave equation (10) in GR, for plane GW waves in the direction of z-axis, there are two polarizations $\boldsymbol{e}_+$ and $\boldsymbol{e}_\times$:

$$\boldsymbol{e}_+ = \underline{\boldsymbol{xx}} - \underline{\boldsymbol{yy}}, \quad \boldsymbol{e}_\times = \underline{\boldsymbol{xy}} + \underline{\boldsymbol{yx}} \quad (27)$$

with $\underline{\boldsymbol{x}}$ and $\underline{\boldsymbol{y}}$ the unit vectors in the directions of x-axis and y-axis. The product $\underline{\boldsymbol{xx}}$ is tensor product. The metric tensor of e_+-polarization GW is $h_+ \boldsymbol{e}_+$; that of $e_\times$-polarization GW is $h_\times \boldsymbol{e}_\times$. The total GW metric $\boldsymbol{h}$ is

$$\boldsymbol{h} = h_+ \boldsymbol{e}_+ + h_\times \boldsymbol{e}_\times; \quad \text{in component form,} \quad h_{\mu\nu} = h_+ e_{+\mu\nu} + h_\times e_{\times\mu\nu}. \quad (28)$$

GW with e_+-polarization contributes to the first term of the energy density formula (26) with squared amplitude $(\partial_0 h_+)^2$; GW with $e_\times$-polarization contributes to the second term of the energy density formula (26) with squared amplitude $(\partial_0 h_\times)^2$.

Far from the GW sources as it is in the present experimental/observational situations, the plane wave approximation is valid. Space averages can be replaced with time averages. For orthogonal modes, the energy can be added in quadrature. For multi-frequency plane GW, the total energy density in the spectral representation of (26) becomes then

$$\rho c^2 = t_{00} = \frac{c^2}{16\pi G_{\mathrm{N}}} \int_{-\infty}^{\infty} (2\pi)^2 f^2 [|^{(\mathrm{f})}h_+(f)|^2 + |^{(\mathrm{f})}h_\times(f)|^2] df$$
$$\equiv \int_0^\infty {}^{(\mathrm{E})}S_h(f) df. \quad (29)$$

${}^{(\mathrm{E})}S_h(f)$ is defined as the (one-sided) energy spectral density of h and is given by

$${}^{(\mathrm{E})}S_h(f) = \frac{\pi c^2}{2G_{\mathrm{N}}} f^2 [|^{(\mathrm{f})}h_+(f)|^2 + |^{(\mathrm{f})}h_\times(f)|^2] = \frac{\pi c^2}{8G_{\mathrm{N}}} f S_h(f) \quad (30)$$

with

$$S_h(f) = 4f(|^{(\mathrm{f})}h_+(f)|^2 + |^{(\mathrm{f})}h_\times(f)|^2) = f^{-1}(h_{\mathrm{c}}(f))^2;$$
$$S_{h\mathrm{A}}(f) = 4f|^{(\mathrm{f})}h_{\mathrm{A}}(f)|^2 = f^{-1}(h_{\mathrm{cA}}(f))^2 \text{ for a single polarization A } (\mathrm{A} = +, \times), \quad (31)$$

the spectral power density of h and h_{A}, respectively and

$$h_{\mathrm{c}}(f) \equiv 2f[(|^{(\mathrm{f})}h_+(f)|^2 + |^{(\mathrm{f})}h_\times(f)|^2)]^{1/2}; \quad h_{\mathrm{cA}}(f) \equiv 2f|^{(\mathrm{f})}h_{\mathrm{A}}(f)| \quad (31a)$$

the characteristic strains [comparing with Eq. (23)]. For unpolarized GWs, $|^{(\mathrm{f})}h_+(f)|^2 = |^{(\mathrm{f})}h_\times(f)|^2$ and we have

$$\begin{aligned} {}^{(\mathrm{E})}S_h(f) &= \frac{\pi c^2}{G_\mathrm{N}} f^2 |^{(\mathrm{f})}h_+(f)|^2 \\ &= \frac{\pi c^2}{G_\mathrm{N}} f^2 |^{(\mathrm{f})}h_\times(f)|^2 \\ &= \frac{\pi c^2}{4G_\mathrm{N}} f S_{h\mathrm{A}}(f). \end{aligned} \tag{32}$$

From (30), the energy density is proportional to h_A^2 for a particular polarization. General GW can be resolved into superposition of plane GWs, the formula (29)–(32) are still applicable. For an early motivation and an in-step mathematical derivation, see, e.g. Refs. 23 and 24, respectively.

For background or foreground stochastic GWs, it is common to use the critical density ρ_c for closing the universe as fiducial:

$$\rho_\mathrm{c} = \frac{3H_0^2}{8\pi G_\mathrm{N}} = 1.878 \times 10^{-29}\,\mathrm{g/cm}^3, \tag{33}$$

where H_0 is the Hubble constant at present. Throughout this article, we use the Planck 2015 value 67.8 (± 0.9) km s^{-1}Mpc^{-1} for H_0.[25] In the cosmological context, it is more convenient to define a *normalized GW spectral energy density* $\Omega_\mathrm{g}(f)$ and express the GW spectral energy density in terms of *the energy density per logarithmic frequency interval divided by the cosmic closure density* ρ_c for a cosmic GW sources or background, i.e.

$$\Omega_\mathrm{gw}(f) = \frac{f}{\rho_\mathrm{c}}\frac{\mathrm{d}\rho(f)}{\mathrm{d}f} = \frac{\pi}{8G_\mathrm{N}} f^3 \frac{S_h(f)}{\left(\dfrac{3H_0^2}{8\pi G_\mathrm{N}}\right)} = \frac{\pi^2}{3H_0^2} f^3 S_h(f)$$

$$\left(= \frac{2\pi^2}{3H_0^2} f^3 S_{h\mathrm{A}}(f) \text{ for unpolarized GW}\right). \tag{34}$$

For the very-low-frequency band, the ultra-low-frequency band and the extremely-low-frequency band, this is a common choice.

From Eq. (14), one can derive the quadrupole formulas of the gravitational radiation metric and the radiated power at the lowest approximation[22]:

$$h_{ij}(t,x,y,z) = -\left[\frac{2G_\mathrm{N}}{c^6 r}\frac{d^2 Q_{ij}}{dt^2}\right]_\mathrm{retarded}, \tag{35}$$

$$\frac{dP}{d\Omega} = \frac{G_\mathrm{N}}{8\pi c^5}\left[\frac{d^3 Q_{ij}}{dt^3} e_{ij}\right]^2. \tag{36}$$

Here, $dP/d\Omega$ is the power radiated into the solid angle $d\Omega$ in the polarization e_{ij}, $Q_{ij}(= \int \rho x_i x_j d^3x)$ is the moment of inertia of the radiating system and e_{ij} is the polarization of the emitted GW. Summed over two polarizations and integrated over solid angles, the total power emitted is

$$P = \frac{G_N}{5c^5}\left[\frac{d^3Q_{ij}}{dt^3}\frac{d^3Q_{ij}}{dt^3} - \frac{1}{3}\frac{d^3Q_{ii}}{dt^3}\frac{d^3Q_{jj}}{dt^3}\right]. \tag{37}$$

Inserting the moment of inertia of the binary Keplerian orbit motion into Eq. (37) and average over one orbit period, Peters and Mathews[26] obtained the following formula for the gravitational radiation loss:

$$\langle P\rangle = \frac{32G_N^4}{5c^5}\frac{M_1^2M_2^2(M_1+M_2)}{a^5(1-e)^{7/2}}\left[1+\frac{73}{24}e^2+\frac{37}{96}e^4\right]. \tag{38}$$

Here, M_1 and M_2 are the two masses of the binary, e is the eccentricity of the elliptic orbit, and a the semi-major axis. Peters[27] further obtained the average angular momentum emission rate:

$$\left\langle\frac{dL}{dt}\right\rangle = -\frac{32G_N^{7/2}}{5c^5}\frac{M_1^2M_2^2(M_1+M_2)^{1/2}}{a^{7/2}(1-e^2)^2}\left[1+\frac{7}{8}e^2\right]. \tag{39}$$

From the Peters–Mathews radiation formula (38) and Peters' angular momenta radiation formula (39), the orbital period P_b decay rate can be calculated as[27]

$$\frac{dP_b}{dt} = -\frac{192\pi}{5}\left(\frac{P_b}{2\pi}\right)^{-5/3} \times\left[1+\frac{73}{24}e^2+\frac{37}{96}e^4\right](1-e)^{7/2}M_1M_2(M_1+M_2)^{-1/3}. \tag{40}$$

From (39) and (40) Peters obtained the time evolution equations for $\langle da/dt\rangle$ and $\langle de/dt\rangle$, and found the time dependence of the semi-major axis $a(t)$ and the merging time $T_c(a_0)$ for circular orbits starting from initial semi-major axis $a = a_0$:

$$a(t) = (a_0^4 - 4\beta t)^{1/4};\quad T_c(a_0) = \frac{a_0^4}{4\beta},\quad \text{with } \beta \equiv \frac{64G_N^3}{5c^5}M_1M_2(M_1+M_2) \tag{41}$$

in reasonable agreement with estimates from higher-order approximations and results from numerical relativity.

For a binary system of masses M_1 and M_2 with Schwarzschild radius R_1 and R_2, the strain h calculated from (35) of its emitted gravitational radiation is of the order of

$$h \approx \frac{R_1R_2}{Dd}, \tag{42}$$

where d is the distance between M_1 and M_2, D the distance to the observer. For neutron star or black hole, d can be of the order of Schwarzschild radius and the estimation can be simplified:

$$h \le \frac{R}{D}. \tag{43}$$

For black hole of solar masses, $R = 3\,\text{km}$ and $d = 10^8$ l.y., $h \le 3\times 10^{-21}$; for inspiral of neutron star binaries, the GW strain generated is smaller.

GWs in GR have two independent polarizations. GWs in a general metric theory of gravity can have up to six independent polarizations according to the Riemann tensor classification of Eardley *et al.*[28]; in terms of helicity, there are two scalar modes, one helicity +2 mode, one helicity +1 mode, one helicity −1 mode and one helicity −2 mode.[29] Therefore, by measuring the GW polarizations, different theories can be distinguished and tested. For a general metric theory with additional fields (scalar, vector, etc.), there are monopole and/or dipole contributions to the quadrupole radiation formula (36). However, due to conservation of mass and conservation of linear momentum in the Newtonian order, the leading order of monopole and dipole contributions are of the same order or less compared with the quardrupole contribution in GR.[30] Nevertheless, experiments/observations do distinguish them. For example, pulse timing observations on the relativistic pulsar-white dwarf binary PSR J1738 + 0333 have given stringent tests on some of these theories already.[8]

3. Methods of GW Detection, and Their Sensitivities

Similar to the frequency classification of electromagnetic waves to radio wave, millimeter wave, infrared, optical, ultraviolet, X-ray and γ-ray etc., in Table 1, we have compiled a complete frequency classification of GWs. This classification together with the current and aimed sensitivities of various detection methods plus predicted GW source strengths are plotted in Figs. 1–4. Figure 1 shows the spectrum classification together with detection methods and projects. Figs. 2–4 show respectively the characteristic strain h_c versus frequency plot, the strain power spectral density (psd) amplitude $[S_h(f)]^{1/2}$ versus frequency plot and the normalized GW spectral energy density Ω_g versus frequency plots for various GW detectors and sources. Detailed accounts and explanations of Figs. 2–4 are given in the following subsections and in Sec. 4.

For the methods of detecting GWs, we first classify them into real-time detection and imprint detection. For real-time detection, we use the time scale of 100 year — the life span of a human being. Although this scale could be extended, it is at least good for next few 100 years. Above 300 pHz [$\sim (100\,\mathrm{yr})^{-1}$], real-time detections are possible. These detections include using resonators, interferometers and pulsar timing for detection in the first six GW bands in Table 1. Below 300 pHz, the detections are possible on GW imprints. Imprint (or snapshot) detections include (i) using the method of quasar astrometry for detection in the ultra-low-frequency GW band, (ii) using CMB observations for detection in Hubble frequency GW band and (iii) using indirect verifications of primordial (inflationary or noninflationary) cosmological models beyond the Hubble frequency band.

There are basically two kinds of GW detectors for real-time detection — (i) the resonant type: GW induces resonances in detectors (metallic bars, metallic spheres, resonant cavities...) to enhance sensitivities; (ii) detectors measuring distance change using microwave/laser/X-ray/atom/molecule... between/among

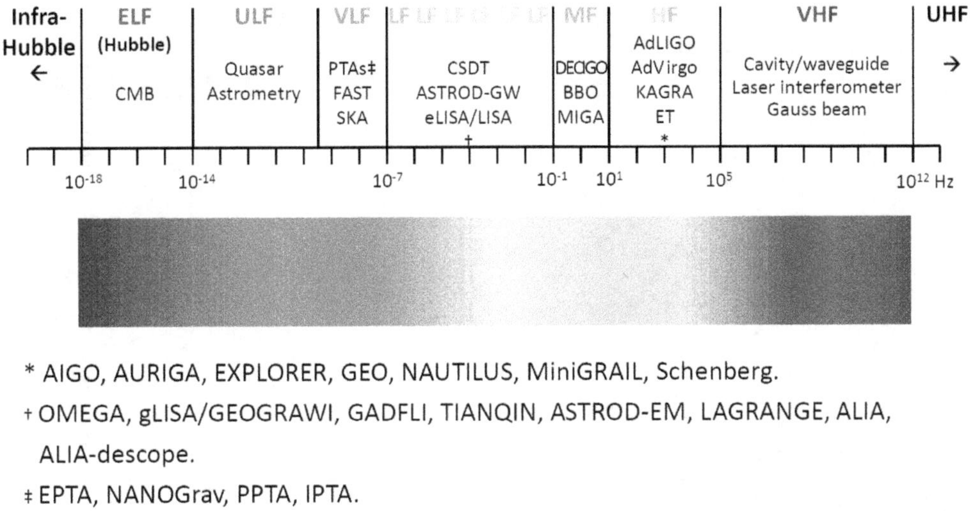

Fig. 1. The GW Spectrum Classification (updated from Refs. 16 and 17).

suspended/floating test bodies. In the case of PTAs for detection in the very-low-frequency GW band, the floating test bodies are the pulsars and observatories while the relative distance change are through pulsar timing variations. Two crucial issues in real-time GW detection are (i) to lower disturbance effects and/or to model the residuals: suspension isolation, drag-free to decrease the effects of surrounding disturbances and appropriate modeling of the motion and the disturbances to reduce the uncertainties in the measurement; (ii) to increase measurement sensitivity: capacitive sensing, microwave sensing, SQUID transducing, optical sensing, X-ray sensing, atom sensing, molecule sensing and timing. . . .

3.1. *Sensitivities*

The input and output of a detector are scalar quantities. The input of a GW detector is a time series $h(t)$ of GW signals which can be written as a functional of the GW metric $h_{\alpha\beta}(\boldsymbol{x}', t')$. For weak GW as in most situations, this functional can be linearized and approximated by a linear functional D of $h_{\alpha\beta}(\boldsymbol{x}', t')$:

$$h(t) = D(h_{\alpha\beta}(\boldsymbol{x}', t')). \tag{44}$$

For a stationary local detector, D may further be reduced to a constant tensor $D^{\alpha\beta}$ such that

$$h(t) = D^{\alpha\beta} h_{\alpha\beta}(\boldsymbol{x}, t). \tag{45}$$

In a transverse, traceless coordinate gauge, $h(t)$ is further reduced to

$$h(t) = D^{ij} h^{\mathrm{TT}}{}_{ij}(\boldsymbol{x}, t). \tag{46}$$

D^{ij} (or $D^{\alpha\beta}$) is called the detector tensor which depends on detection geometry. As an example, for a GW interferometer oriented with two arms on the x-axis and

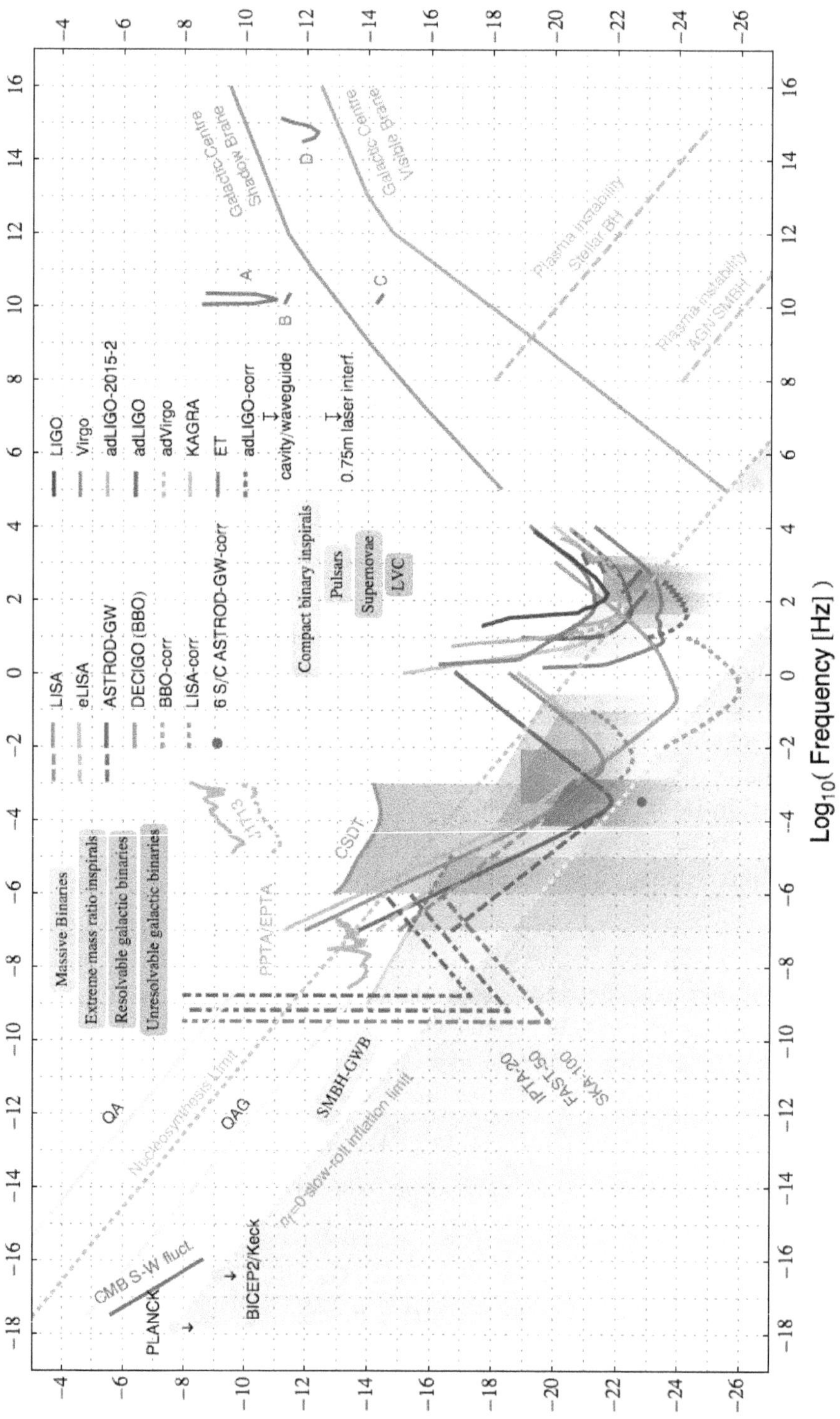

Fig. 2. Characteristic strain h_c versus frequency for various GW detectors and sources. [QA: Quasar Astrometry; QAG: Quasar Astrometry Goal; LVC: LIGO-Virgo Constraints; CSDT: Cassini Spacecraft Doppler Tracking; SMBH-GWB: Supermassive Black Hole-GW Background.] (For color version, see page I-CP8.)

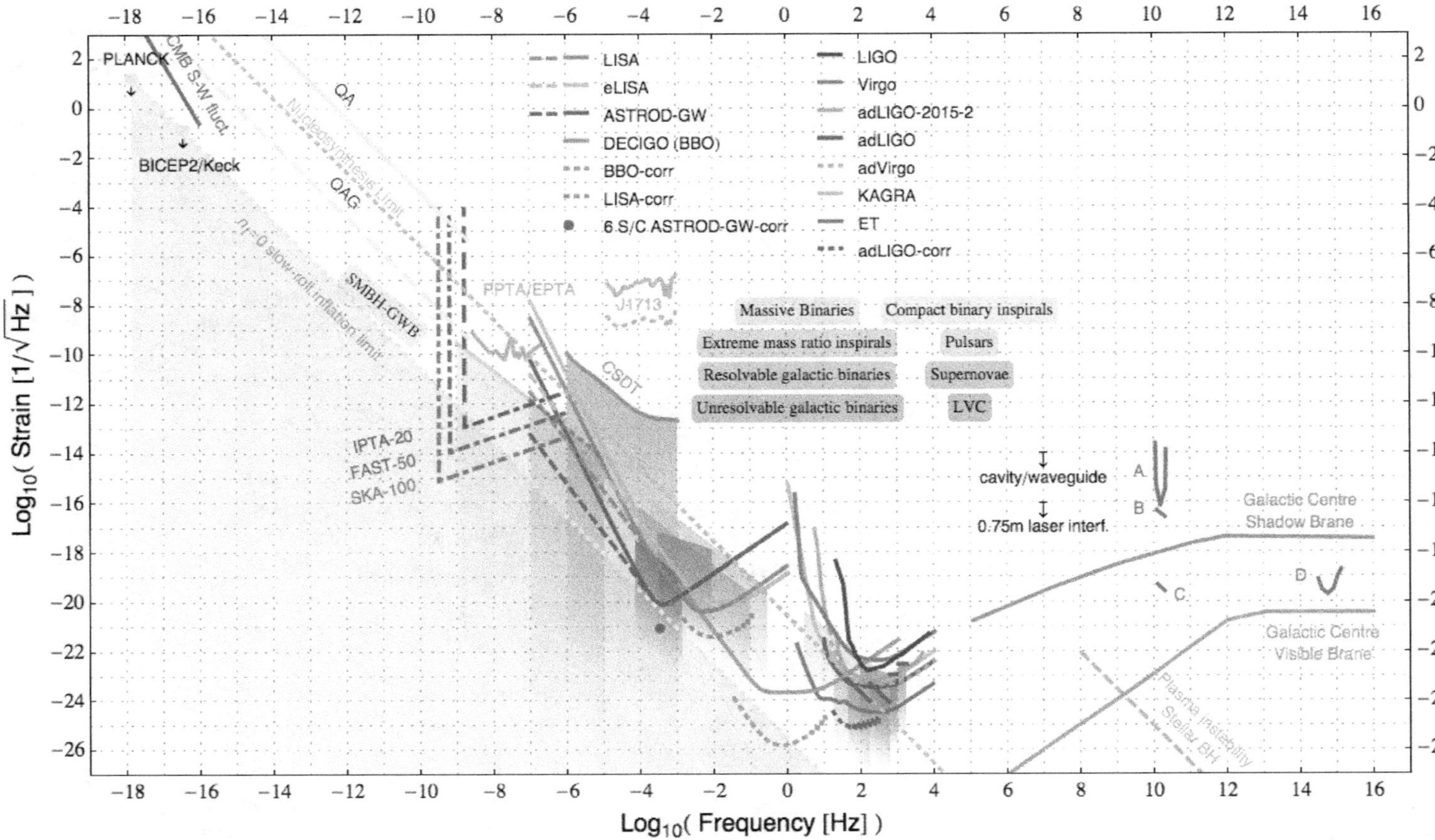

Fig. 3. Strain psd amplitude versus frequency for various GW detectors and GW sources. See Fig. 2 caption for the meaning of various acronyms. (For color version, see page I-CP9.)

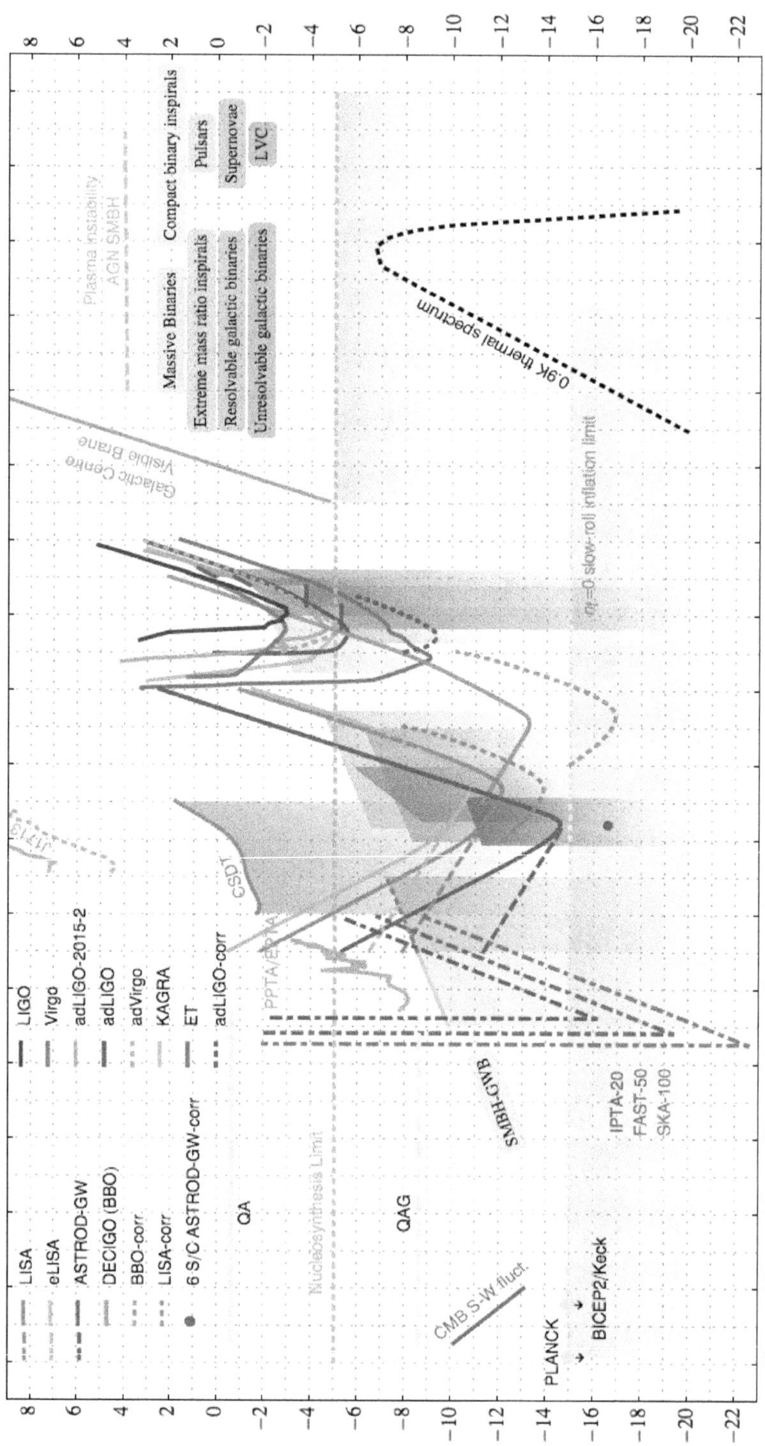

Fig. 4. Normalized GW spectral energy density Ω_{gw} versus frequency for GW detector sensitivities and GW sources. See Fig. 2 caption for the meaning of various acronyms. (For color version, see page I-CP10.)

y-axis with nearly equal arm lengths in the long wavelength limit, the detector tensor has $D^{11} = 1/2, D^{22} = -1/2$ and all other components vanishing; we have $h(t) = h_{11}(t), ^{(\mathrm{f})}h(f) = {}^{(\mathrm{f})}h_{11}(f) = {}^{(\mathrm{f})}h_+(f)$ and $h_{c+}(f) \equiv 2f|^{(\mathrm{f})}h_+(f)|$.

In the case of linear response, the detector output ${}^{(\mathrm{f})}h^{(\mathrm{out})}(f)$ is related to the input by

$$ {}^{(\mathrm{f})}h^{(\mathrm{out})}(f) = T(f) \times {}^{(\mathrm{f})}h(f), $$
$$ \text{(or simply } h^{(\mathrm{out})}(f) = T(f) \times h(f) \text{ without heavy notations)}, \tag{47} $$

where $T(f)$ is the transfer function or the response function of the detector. In the output of any detector there will be noise also. The total output $s^{(\mathrm{out})}(t)$ is the addition of the GW signal output $h^{(\mathrm{out})}(t)$ and the noise output $n^{(\mathrm{out})}(t)$:

$$ s^{(\mathrm{out})}(t) = h^{(\mathrm{out})}(t) + n^{(\mathrm{out})}(t); \tag{48} $$

in frequency space the total output ${}^{(\mathrm{f})}s^{(\mathrm{out})}(f)$ is

$$ {}^{(\mathrm{f})}s^{(\mathrm{out})}(f) = {}^{(\mathrm{f})}h^{(\mathrm{out})}(f) + {}^{(\mathrm{f})}n^{(\mathrm{out})}(f). \tag{49} $$

Habitually, it is convenient to refer and compare noise at the input port by defining

$$ {}^{(\mathrm{f})}n(f) \equiv [T(f)]^{-1} \times {}^{(\mathrm{f})}n^{(\mathrm{out})}(f). \tag{50} $$

From (50), we have

$$ {}^{(\mathrm{f})}s^{(\mathrm{out})}(f) = [T(f)] \times [{}^{(\mathrm{f})}h(f) + {}^{(\mathrm{f})}n(f)], \tag{51} $$

and we can define

$$ {}^{(\mathrm{f})}s(f) \equiv [T(f)]^{-1} \times {}^{(\mathrm{f})}s^{(\mathrm{out})}(f) = {}^{(\mathrm{f})}h(f) + {}^{(\mathrm{f})}n(f) \tag{52} $$

to be the total input signal. In time domain, we have then

$$ s(t) = h(t) + n(t). \tag{53} $$

It is convenient to take $n(t)$ as the detector noise.

It is also convenient and practical to assume that the detector noise is stationary and Gaussian, and the different Fourier components are independent. We then have

$$ \langle {}^{(\mathrm{f})}n^*(f)\; {}^{(\mathrm{f})}n(f') \rangle = \left(\frac{1}{2}\right) \delta(f - f') S_n(f), \tag{54} $$

where $S_n(f)$ is defined by the equation. From this equation, one can derive

$$ \langle n^2(t) \rangle = \langle n^2(t=0) \rangle = \int_{-\infty}^{\infty} df df'\; \langle {}^{(\mathrm{f})}n * (f) {}^{(\mathrm{f})}n(f') \rangle = \int_0^{\infty} df\; S_n(f). \tag{55} $$

Hence, $S_n(f)$ is called the noise power spectrum, the noise power spectrum density (noise psd), the noise spectral density, or the noise spectral sensitivity. It is one-sided since the integration only takes on the positive axis. For a more detailed derivation of Eq. (55), we refer the reader to Refs. 24 and 31. For a dimensionless description of noise power at a particular frequency, one usually use noise amplitude $h_n(f)$ which is defined as

$$h_n(f) \equiv [f S_n(f)]^{1/2}. \tag{56}$$

For comparison, we note that for a GW interferometer oriented with two arms oriented on the x-axis and y-axis with nearly equal arm lengths in the long wavelength limit, the GW signal $h(t) = h_{11}(t) = h_+(t)({}^{(\mathrm{f})}h(f) = {}^{(\mathrm{f})}h_{11}(f) = {}^{(\mathrm{f})}h_+(f))$ with GW propagating perpendicular to xy-plane corresponds to detector geometry with $D_{\mu\nu} = e_{+\mu\nu}$. Its associated strain psd $S_h(f)$ is

$$\begin{aligned} S_h(f) &= 4f|{}^{(\mathrm{f})}h(f)|^2 = |2f^{1/2(\mathrm{f})}h_{11}(f)|^2 \\ &= |2f^{1/2(\mathrm{f})}h_+(f)|^2 = |f^{-1/2(\mathrm{f})}h_{\mathrm{c}+}(f)|^2 = |f^{-1/2(\mathrm{f})}h_{\mathrm{c}}(f)|^2. \end{aligned} \tag{57}$$

In general, GW detector has different geometric sensitivity to monochromatic GW coming from different directions and with different polarization. Hence each detector has its own pattern function of directions and polarizations. In plotting GW sensitivities, one usually takes average over directions and polarizations for a detector.

In the discussion of GW sensitivities and GW signal strengths, there are three customary ways to plot: characteristic strain $h_{\mathrm{c}}(f)$ versus frequency f, square-root psd $[S_h(f)]^{1/2}$ versus frequency f and the normalized GW spectral energy density Ω_{gw} versus frequency f. From (57), for this case, we define the (dimensionless) characteristic strain for the singal ${}^{(\mathrm{f})}h(f)$ as in (31a)

$$h_{\mathrm{c}}(f) \equiv 2f|{}^{(\mathrm{f})}h_{11}(f)| = 2f|{}^{(\mathrm{f})}h_+(f)|. \tag{58}$$

With this definition, we have from (57) and (58)

$$S_h(f) = f^{-1}|h_{\mathrm{c}}(f)|^2 = f|2^{(\mathrm{f})}h_{11}(f)|^2 = f|2^{(\mathrm{f})}h_+(f)|^2. \tag{59}$$

Now we relate the three quantities $h_{\mathrm{c}}(f)$, $[S_h(f)]^{1/2}$ and $\Omega_{\mathrm{gw}}(f)$ using Eqs. (34) and (58):

$$\Omega_{\mathrm{gw}}(f) = \frac{2\pi^2}{3H_0^2} f^3 S_h(f); \quad h_{\mathrm{c}}(f) = f^{1/2}[S_h(f)]^{1/2}. \tag{60}$$

Table 2 compiles the conversion factors among the characteristic strain $h_{\mathrm{c}}(f)$, the strain psd $[S_h(f)]^{1/2}$ and the normalized spectral energy density $\Omega_{\mathrm{gw}}(f)$. In using (60) and Table 2, especially the conversion to $\Omega_{\mathrm{gw}}(f)$, we assume a baseline detector and source configuration just mentioned. For other configuration, its specific detector geometry, source geometry and GW polarization need to be taken care of.

Table 2. Conversion factors among the characteristic strain $h_c(f)$, the strain psd $[S_h(f)]^{1/2}$ and the normalized spectral energy density $\Omega_{gw}(f)$.

	Characteristic strain $h_c(f)$	Strain psd $[S_h(f)]^{1/2}$	Normalized spectral energy density $\Omega_{gw}(f)$
$h_c(f)$	$h_c(f)$	$f^{1/2}[S_h(f)]^{1/2}$	$[(3H_0^2/2\pi^2 f^2)\Omega_{gw}(f)]^{1/2}$
Strain psd $[S_h(f)]^{1/2}$	$f^{-1/2}h_c(f)$	$[S_h(f)]^{1/2}$	$[(3H_0^2/2\pi^2 f^3)\Omega_{gw}(f)]^{1/2}$
$\Omega_{gw}(f)$	$(2\pi^2/3H_0^2)f^2 h_c^2(f)$	$(2\pi^2/3H_0^2)f^3 S_h(f)$	$\Omega_{gw}(f)$

In data analysis, the optimal signal-to-noise ratio η that can be obtained is by using Wiener matched filter ${}^{(f)}h(f)/S_n(f)$:

$$\eta^2 = \int_0^\infty df \frac{4|{}^{(f)}h(f)|^2}{S_n(f)}. \tag{61}$$

Using Eqs. (58) and (56), (61) can be written as

$$\eta^2 = \int_0^\infty df\, f^{-1}\left[\frac{h_c(f)}{h_n(f)}\right]^2 = \int_{-\infty}^\infty d(\log f)\left[\frac{h_c(f)}{h_n(f)}\right]^2. \tag{62}$$

Hence in the log–log plot of characteristic strain versus frequency, the square of signal-to-noise ratio is equal to the integral of the square of the ratio of characteristic strain of source over the characteristic noise strain. For large signal-to-noise ratio, it is approximately equal to the area between the characteristic strain curve and the characteristic noise strain curve in the detection bandwidth.

For the parameter fitting, the more discernable the structure, the better are the parameters fitted. Please see Refs. 32–35 for good accounts.

3.2. *Very high frequency band* (100 kHz–1 THz) *and ultrahigh frequency band* (*above* 1 THz)

In the very high frequency band (100 kHz–1 THz), there are two experiments completed. A cavity/waveguide detector, where the polarization of electromagnetic wave changes its direction under incoming GW, was operated at 100 MHz and gave upper limit of the background GW radiation of around $10^{-14}\,\text{Hz}^{-1/2}$.[36] And a 0.75 m arm length laser interferometer, where synchronous amplification of the phase shift due to GW occurs, achieved a noise level limiting the existence of 100 MHz background GW down to $10^{-16}\,\text{Hz}^{-1/2}$.[37,38] These two upper limits are marked on Fig. 3 and the corresponding limits on Figs. 2 and 4.

Cruise has described two types of magnetic conversion prototype detectors A and D being commissioned at Birmingham in Sec. 9 of Ref. 39. The basic principle of a magnetic conversion detector is to convert GWs in a laboratory magnetic field to electromagnetic waves which are then focused or concentrated on one detector element to be measured. Detector A has a room temperature microwave receiver sensing the waveguide conversion volume with a magnetic field 0.2 T. The expected sensitivity curve of prototype detector A for one year integration with 1 MHz bandwidth is shown in Fig. 1 of Ref. 39 as curve A. In the same figure, curve B shows the expected sensitivity for a larger detector, $60 \times 500 \times 800\,\mathrm{mm}^3$ having a 20 K noise temperature amplifier and curve C shows that for a pair of such detectors in correlation over a one year period. Curve D is for the detector with the cooled CCD sensing the same waveguide and field as A, also for a one year integration.

We put curves A, B, C and D from Fig. 1 of Ref. 39 on Fig. 3 in this section. The definition of $\Omega_{\mathrm{gw}}(f)$ in Eq. (1) of Ref. 39 is the same as ours; while their conversion to their h Eq. (2) of Ref. 39 is

$$h = 5.8 \times 10^{-22} \left(\frac{100}{f}\right)^{3/2} (\Omega_{\mathrm{gw}}(f))^{1/2}. \tag{63}$$

Our strain psd $[S_h(f)]^{1/2}$ using Planck H_0[25] is

$$[S_h(f)]^{1/2} = \left[\frac{3H_0^2}{2\pi^2 f^3}\Omega_{\mathrm{gw}}(f)\right]^{1/2} = 8.57 \times 10^{-22} \left(\frac{100}{f}\right)^{3/2} (\Omega_{\mathrm{gw}}(f))^{1/2}. \tag{64}$$

Therefore, their h is basically our $[S_h(f)]^{1/2}$ with a multiplicative factor 0.677. Hence, Fig. 1 of Ref. 39 corresponds to our Fig. 3 basically. We adjust the factor 0.677 for the nucleosynthesis limit in Fig. 4 and corresponding places in Figs. 2 and 3. We have not adjusted other parts of Fig. 1 of Ref. 39 while transport to our figures since the multiplicative factor is not large in our log-log plots.

Curve C and curve D have sensitivities in strain psd $[S_h(f)]^{1/2}$ close to 10^{-20} for frequencies around 10^{10} Hz in the very high frequency band and 10^{15} Hz in the ultrahigh frequency band, respectively. The corresponding curves are also shown in Figs. 2 and 4. Possible sensitivity enhancements have been suggested by generating electromagnetic power depending linearly on the GW amplitudes;[40] however, the associated noise issues are still pending on solutions.[41,39] Signal amplitudes from various GW sources are summarized in Sec. 4.6.

3.3. *High frequency band* (10 Hz–100 kHz)

Most of the current activities of GW detection on the ground or in the underground are in the high frequency band. In the following, we summarize the activities and sensitivities. For a detailed exposition, we refer to Ref. 42.

In this band, the cryogenic resonant bar detectors have already reached a strain spectral sensitivity of $10^{-21}\,\mathrm{Hz}^{-1/2}$ in the kHz region. NAUTILUS put an upper limit on periodic sources ranging from 3.4×10^{-23} to 1.3×10^{-22} depending on frequency in their all-sky search.[43] The AURIGA-EXPLORER-NAUTILUS-Virgo Collaboration applied a methodology to the search for coincident burst excitations over a 24 h long joint data set.[44] The MiniGRAIL[45] and Schenberg[46] cryogenic spherical GW detectors are for omnidirectional GW detection.

Major detection efforts in the high frequency band are in the long arm laser interferometers. The TAMA 300 m arm length interferometer,[47] the GEO 600 m interferometer,[48] and the kilometer size laser-interferometric GW detectors —

LIGO[49] (two 4 km arm length, one 2 km arm length) and VIRGO[50] all achieved their original sensitivity goals basically. Around the frequency 100 Hz, the LIGO and Virgo sensitivities are both in the level of 10^{-23} $(\text{Hz})^{-1/2}$. The LIGO and Virgo achieved sensitivity curves are shown in Figs. 2–4.[49,50] Interference spikes are taken out for clarity in the presentation in these figures. Various limits on the GW strains for different sources become significant. For example, analyses of data from S6 (sixth science run) of LIGO and GEO600 GW detectors and VSR 2 and VSR 4 (Virgo science runs) of Virgo detector set strain upper limits on the GW emission from 195 radio pulsars; specifically, the strain upper limit on the Vela pulsar is comparable to the spin-down limit and that on the Crab pulsar is about a factor of 2 below the spin-down limit.[51] The 2009 analysis of the data from a LIGO two-year science run constrained the normalized spectral energy density $\Omega_{\text{gw}}(f)$ of the stochastic GW background in the frequency band around 100 Hz, to be 6.9×10^{-6} at 95% confidence.[52] This search for the stochastic background improved on the indirect limit from the Big Bang nucleosynthesis at 100 Hz. In 2014 further improvement and refinement on the limit of the stochastic GW background were obtained from the analysis of the 2009–2010 LIGO and Virgo Data.[53] Assuming a stochastic GW spectrum of

$$\Omega_{\text{gw}}(f) = \Omega_\alpha \left(\frac{f}{f_{\text{ref}}}\right)^\alpha, \tag{65}$$

LIGO and Virgo collaboration placed 95% confidence level upper limits on the normalized spectral energy density of the background in each of four frequency bands spanning 41.5–1726 Hz:

$$\begin{aligned}
&\Omega_{\text{gw}}(f) < 5.6 \times 10^{-6}, && \text{for the frequency band 41.5–169.25 Hz;}\\
&\Omega_{\text{gw}}(f) < 1.8 \times 10^{-4}, && \text{for the frequency band 170–600 Hz;}\\
&\Omega_{\text{gw}}(f) < 0.14 \left(\frac{f}{900\ \text{Hz}}\right)^3, && \text{for the frequency band 600–1000 Hz;}\\
&\Omega_{\text{gw}}(f) < 1.0 \left(\frac{f}{900\ \text{Hz}}\right)^3, && \text{for the frequency band 1000–1726 Hz.}
\end{aligned} \tag{66}$$

These constraints [LVC: LIGO-Virgo Constraints] are plotted on Fig. 4 and the corresponding constraints on Figs. 2 and 3.

Also in the analysis of jointly conducted science runs (LIGO S6 and Virgo VSR 2 and VSR 3), two kinds of search were done for possible GWs associated with 154 gamma ray bursts that were observed by satellite experiments in 2009–2010: the first search is for a signal from coalescence of two neutron stars or a neutron star and a black hole and the second search is for a burst-like GW signal from the collapse of a massive star. No signals were detected. This results places limits of 17 Mpc for no collapsing star, 16 Mpc for nonexistence of the coalescence of binary neutron stars and 28 Mpc for that of a neutron star and a black hole associated with the observed 154 gamma ray bursts.[54]

Observations by all above long baseline laser interferometers have finished their first phase operation by 2010. Sensitivity improvement of one order of magnitude is underway by upgrading LIGO and Virgo as advanced interferometers, adLIGO[55] and adVirgo[56] and by initiating a new project, KAGRA/LCGT.[57] This improvement will increase the detection volume by three orders of magnitudes. These GW detectors are the second-generation interferometers. The advanced LIGO is the earliest started and has achieved 3.5 fold better sensitivity improvement already; it began its first observing run (O1) on September 18, 2015 searching for GWs. We plot the February-2015 achieved adLIGO sensitivity together with the planned strain sensitivities of adLIGO,[55] adVirgo[56] and KAGRA[57] on Figs. 2–4. KAGRA will be a cryogenic underground interferometer with 3 km arm length; it will already have some features of the third generation GW interferometers. ET (Einstein Telescope)[58] is a third generation GW interferometer. It will be a cryogenic underground interferometer with 10 km arm length. Its goal sensitivity is also plotted on Figs. 2–4.

As to the upper range of this band, it is noticed that every free spectral range (FSR) relative to the lock point, there would be good sensitivity. The FSRs of LIGO and VIRGO/KAGRA are 37.5 MHz and 50 MHz. LIGO is considering/discussing this frequency. Although digitation under 100 kHz is not a technological feasibility problem, it is a practical problem in sampling/digitizing the data at these high frequencies. Nevertheless, the upper range of the high frequency band is accessible to the km-sized GW interferometers.

3.4. *Doppler tracking of spacecraft* (1 μHz–1 mHz *in the low-frequency band*)

Doppler tracking of spacecraft can be used to constrain (or detect) the level of low-frequency GWs.[59] The separated test masses of this GW detector are the Doppler tracking radio antenna on Earth and a distant spacecraft. Doppler tracking measures relative distance-change. Estabrook and Walquist derived[59] the effect of GWs passing through the line of sight of spacecraft on the Doppler tracking frequency measurements (see also Ref. 60). From these measurements, GWs can be detected or constrained. The most recent measurements came from the Cassini spacecraft Doppler tracking (CSDT). Armstrong *et al.*[61] used precision Doppler tracking of the Cassini spacecraft during its 2001–2002 solar opposition to derive improved observational limits on an isotropic background of low-frequency GW. They used the Cassini multilink radio system and an advanced tropospheric calibration system to remove the effects of leading noises — plasma and tropospheric scintillation to a level below the other noises. The resulting data were used to construct upper limits on the strength of an isotropic background in the 1 μHz–1 mHz band.[61] The characteristic strain upper limit curve labelled CSDT in Fig. 2 is a smoothed version of the curve in Fig. 4 of Ref. 61. The corresponding CSDT curves on the strain psd amplitude in Fig. 3 and the normalized spectral energy density in Fig. 4 are calculated using Table 2 for conversion. The minimal points on these curves are

$$h_c(f) < 2 \times 10^{-15}, \qquad \text{at frequency about } 0.3\,\text{mHz};$$

$$[S_h(f)]^{1/2} < 8 \times 10^{-13}\,\text{Hz}^{-1/2}, \qquad \text{at several frequencies in the } 0.2\text{–}0.7\,\text{mHz band};$$

$$\Omega_{gw}(f) < 0.03, \qquad \text{at frequency } 1.2\,\mu\text{Hz}. \tag{67}$$

The GW sensitivity of spacecraft Doppler tracking could still be improved by 1–2 order of magnitude with a space borne optical clock on board.[62]

The basic principle of spacecraft Doppler tracking, of spacecraft laser ranging, of space laser interferometer and of PTAs for GW detection are similar. In the development of further GW detection methods, spacecraft Doppler tracking method has stimulated significant inspirations. ASTROD I (Astrodynamical Space Test of Relativity Using Optical Devices I)[63] using a space borne precision clock has included as one of its goals GW sensitivity improvement of the CSDT by one order of magnitude. The methods using space laser interferometers and using PTAs are two important methods of detecting GWs; their sensitivities will be discussed in Secs. 3.5 and 3.6, respectively.

3.5. *Space interferometers (low-frequency band,* 100 nHz–100 mHz; *middle-frequency band,* 100 mHz–10 Hz)

Space laser interferometers for GW detection (*eLISA/LISA,*[64,65] *ASTROD,*[66,67] *ASTROD-GW,*[13–16,68,69] *ASTROD-EM,*[69,70] *Super-ASTROD,*[71] *DECIGO,*[72] *Big Bang Observer,*[73] *ALIA,*[74] *ALIA-descope,*[75] *gLISA/GEOGRAWI,*[76–78] *GADFLI,*[79] *LAGRANGE,*[80] *OMEGA,*[81] *and TIANQIN*[82]) hold the most promise with high signal-to-noise ratio. Laser Interferometer Space Antenna (LISA)[65] is aimed at detection of 10^{-4}–1 Hz GWs with a strain sensitivity of $4 \times 10^{-21}/(\text{Hz})^{1/2}$ at 1 mHz. There are abundant sources for eLISA/LISA, ASTROD, ASTROD-GW and Earth-orbiting missions: (i) In our Galaxy: galactic binaries (neutron stars, white dwarfs, etc.); (ii) Extra-galactic targets: supermassive black hole binaries, supermassive black hole formation and (iii) Cosmic GW background. A date of launch of eLISA or substitute mission is set around 2034.[4]

Early in 2009, responding to the call for GW mission studies of Chinese Academy of Sciences (CAS), a dedicated GW mission concept ASTROD-GW with 3 S/C (spacecraft) orbiting near Sun–Earth Lagrange points L3, L4 and L5, respectively was proposed and studied. Before that, Super-ASTROD which was proposed in 1997[10] with S/C in Jupiter-like orbits was studied as a dual mission for GW measurement and for cosmological model/relativistic gravity test.[71] With the proposal of ASTROD-GW, the baseline GW configuration of Super-ASTROD takes 3 S/C orbiting respectively near Sun–Jupiter Lagrange points L3, L4 and L5. For the possibility of a down scaled version of ASTROD-GW mission, the ASTROD-EM with the orbits of 3 S/C near Earth–Moon Lagrange points L3, L4 and L5, respectively has been under study.[70]

DECi-hertz Interferometer GW Observatory (DECIGO)[72] was proposed in 2001 with the aim of detecting GWs from early universe in the observation band (the middle frequency band) between the terrestrial band and the band of low-frequency space GW detectors. It will use a Fabry–Perot method (instead of a delay line method) as in the ground interferometers but with a 1000 km arm length. As a LISA follow-on, BBO (Big Bang Observer)[73] was proposed in the United States with a similar goal. A likely version of DECIGO/BBO is to have 12 S/C in LISA-like orbits with correlated detection. They will be used for the direct measurement of the stochastic GW background by correlation analysis.[83] 6S/C-ASTROD-GW has also been considered to possibly explore the relic GWs in the lower part of the low-frequency band. ALIA[74] was proposed as a less-ambitious LISA follow-on. A de-scoped ALIA[75] has also been proposed and under study.

After the end in 2011 of ESA-NASA partnership for flying LISA, NASA solicited "Concepts for the NASA Gravitational Wave Mission" proposals on September 27, 2011 for study of low cost GW missions (http://nspires.nasaprs.com/external/). gLISA/GEOGRAWI[76–78] (geosynchronous LISA/GEOstationary GRAvitational Wave Interferometer), GADFLI (Geostationary Antenna for Disturbance-Free Laser Interferometer), and LAGRANGE[80] (Laser Gravitational-wave Antenna at Geo-lunar Lagrange points) were proposed; OMEGA[81] (Orbiting Medium Explorer for Gravitational Astronomy) re-emerged. OMEGA was first proposed as a low-cost alternative to LISA in the 1990s. In China, a GW mission in Earth orbit called TIANQIN[82] has been proposed and under study.

Table 3 lists the orbit configuration, arm length, orbit period and S/C number of various GW space mission proposals.

Typical frequency sensitivity spectrum of strain for space GW detection consists of three regions (Fig. 3), the acceleration noise region, the shot noise (flat for current space detector projects like LISA in strain psd) region, if any, and the antenna response region. The lower frequency region for the detector sensitivity is dominated by vibration, acceleration noise or gravity-gradient noise. The higher frequency part of the detector sensitivity is restricted by antenna response (or storage time). In a power-limited design, sometimes there is a middle flat region in which the sensitivity is limited by the photon shot noise.[10,65,84]

The shot noise sensitivity limit in the strain for GW detection is inversely proportional to $P^{1/2}l$ with P the received power and l the distance. Since P is inversely proportional to l^2 and $P^{1/2}l$ is constant, this sensitivity limit is independent of the distance. For 1–2 W emitting power, the limit is around $10^{-21}\,\text{Hz}^{-1/2}$. As noted in the LISA study,[65] making the arms longer shifts the time-integrated sensitivity curve to lower frequencies while leaving the bottom of the curve at the same level. Hence, ASTROD-GW with longer arm length has better sensitivity at lower frequency. eLISA and GW interferometers in Earth orbit have shorter arms and therefore have better sensitivities at higher frequency.

Table 3. A compilation of GW mission proposals.

Mission concept	S/C configuration	Arm length	Orbit period	S/C #
Solar-Orbit GW Mission Proposals				
LISA[65]	Earth-like solar orbits with 20° lag	5 Gm	1 year	3
eLISA[64]	Earth-like solar orbits with 10° lag	1 Gm	1 year	3
ASTROD-GW[68]	Near Sun–Earth L3, L4, L5 points	260 Gm	1 year	3
Big Bang Observer[73]	Earth-like solar orbits	0.05 Gm	1 year	12
DECIGO[72]	Earth-like solar orbits	0.001 Gm	1 year	12
ALIA[74]	Earth-like solar orbits	0.5 Gm	1 year	3
ALIA-descope[75]	Earth-like solar orbits	3 Gm	1 year	3
Super-ASTROD[71]	Near Sun–Jupiter L3, L4, L5 points (3 S/C), Jupiter-like solar orbit(s)(1-2 S/C)	1300 Gm	11 year	4 or 5
Earth-Orbit GW Mission Proposals				
OMEGA[81]	0.6 Gm height orbit	1 Gm	53.2 days	6
gLISA/GEOGRAWI[76–78]	Geostationary orbit	0.073 Gm	24 hours	3
GADFLI[79]	Geostationary orbit	0.073 Gm	24 hours	3
TIANQIN[82]	0.057 Gm height orbit	0.11 Gm	44 hours	3
ASTROD-EM[69,70]	Near Earth–Moon L3, L4, L5 points	0.66 Gm	27.3 days	3
LAGRANGE[80]	Near Earth–Moon L3, L4, L5 points	0.66 Gm	27.3 days	3

In Figs. 2–4, we plot sensitivity curves for LISA, eLISA and ASTROD-GW for the low-frequency GW band. In the Mock LISA Data Challenge (MLDC) program, the consensus goal for the LISA instrumental noise density amplitude ${}^{(\mathrm{MDLC})}S_{\mathrm{Ln}}^{1/2}(f)$ is

$$ {}^{(\mathrm{MDLC})}S_{\mathrm{Ln}}^{1/2}(f) = \frac{1}{L_{\mathrm{L}}} \times \left\{\left(1+0.5\left(\frac{f}{f_{\mathrm{L}}}\right)^2\right) \times S_{\mathrm{Lp}} + [1+(10^{-4}\,\mathrm{Hz}/f)^2]\frac{4S_{\mathrm{a}}}{(2\pi f)^4}\right\}^{1/2} \mathrm{Hz}^{-1/2}, \tag{68} $$

where $L_{\mathrm{L}} = 5\times 10^9$ m is the LISA arm length, $f_{\mathrm{L}} = c/(2\pi L_{\mathrm{L}})$ is the LISA arm transfer frequency, $S_{\mathrm{Lp}} = 4\times 10^{-22}\,\mathrm{m}^2\,\mathrm{Hz}^{-1}$ is the LISA (white) position noise level due to photon shot noise and $S_{\mathrm{a}} = 9\times 10^{-30}\ \mathrm{m}^2 s^{-4}\,\mathrm{Hz}^{-1}$ is the LISA white acceleration noise (power) level.[85] Note that (68) contains the "reddening" factor $[1+(10^{-4}\,\mathrm{Hz}/f)^2]$ in the acceleration noise term.

If we drop the "reddening factor", the enhanced LISA instrumental noise density amplitude ${}^{(\mathrm{Enhanced})}S_{\mathrm{Ln}}^{1/2}(f)$ becomes

$$ {}^{(\mathrm{Enhanced})}S_{\mathrm{Ln}}^{1/2}(f) = \frac{1}{L_{\mathrm{L}}} \times \left\{\left(1+0.5\left(\frac{f}{f_{\mathrm{L}}}\right)^2\right) \times S_{\mathrm{Lp}} + \frac{4S_{\mathrm{a}}}{(2\pi f)^4}\right\}^{1/2} \mathrm{Hz}^{-1/2}. \tag{69} $$

The eLISA arm length $L_{e\mathrm{L}}$ is five times shorter. Its instrumental noise density amplitude ${}^{(\mathrm{MDLC})}S_{e\mathrm{Ln}}^{1/2}(f)$ is

$$ {}^{(\mathrm{MDLC})}S_{e\mathrm{Ln}}^{1/2}(f) = \frac{1}{L_{e\mathrm{L}}} \times \left\{\left(1+0.5\left(\frac{f}{f_{e\mathrm{L}}}\right)^2\right) \times S_{e\mathrm{Lp}} + [1+\left(10^{-4}\,\mathrm{Hz}/f\right)^2]\frac{4S_{\mathrm{a}}}{(2\pi f)^4}\right\}^{1/2} \mathrm{Hz}^{-1/2}, \tag{70} $$

where $L_{e\mathrm{L}} = 10^9\,\mathrm{m}$ is the eLISA arm length, $f_{e\mathrm{L}} = c/(2\pi L_{e\mathrm{L}})$ is the eLISA arm transfer frequency, $S_{e\mathrm{Lp}} = 1\times 10^{-22}\,\mathrm{m}^2\,\mathrm{Hz}^{-1}$ is the eLISA (white) position noise level due to photon shot noise assuming that the telescope diameter is 25 cm (compared with 40 cm for that of LISA) and that the laser power is the same as LISA. The corresponding enhanced eLISA instrumental noise density amplitude ${}^{(\mathrm{Enhanced})}S_{e\mathrm{Ln}}^{1/2}(f)$ is

$$ {}^{(\mathrm{MDLC})}S_{e\mathrm{Ln}}^{1/2}(f) = \frac{1}{L_{e\mathrm{L}}} \times \left\{\left(1+0.5\left(\frac{f}{f_{e\mathrm{L}}}\right)^2\right) \times S_{e\mathrm{Lp}} + \frac{4S_{\mathrm{a}}}{(2\pi f)^4}\right\}^{1/2} \mathrm{Hz}^{-1/2}. \tag{71} $$

For ASTROD-GW, our goal on the instrumental strain noise density amplitude is

$$ S_{\mathrm{An}}^{1/2}(f) = \frac{1}{L_{\mathrm{A}}} \times \left\{\left(1+0.5\left(\frac{f}{f_{\mathrm{A}}}\right)^2\right) \times S_{\mathrm{Ap}} + \frac{4S_{\mathrm{a}}}{(2\pi f)^4}\right\}^{1/2} \mathrm{Hz}^{-1/2}, \tag{72} $$

over the frequency range of $100\,\text{nHz} < f < 1\,\text{Hz}$. Here, $L_A = 260 \times 10^9\,\text{m}$ is the ASTROD-GW arm length, $f_A = c/(2\pi L_A)$ is the ASTROD-GW arm transfer frequency, $S_a = 9 \times 10^{-30}\,\text{m}^2\,\text{s}^{-4}\,\text{Hz}^{-1}$ is the white acceleration noise level (the same as that for LISA) and $S_{Ap} = 10816 \times 10^{-22}\,\text{m}^2\,\text{Hz}^{-1}$ is the (white) position noise level due to laser shot noise which is 2704 ($= 52^2$) times that for LISA.[13,14,16,68,86] The corresponding noise curve for the ASTROD-GW instrumental noise density amplitude ${}^{(\text{MDLC})}S_{An}^{1/2}(f)$ with the same "reddening" factor as specified in MLDC program is

$$S_{An}^{1/2}(f) = \frac{1}{L_A} \times \left\{ \left(1 + 0.5\left(\frac{f}{f_A}\right)^2\right) \times S_{Ap} + [1 + (10^{-4}/f)^2]\,\frac{4S_a}{(2\pi f)^4} \right\}^{1/2} \text{Hz}^{-1/2} \tag{73}$$

over the frequency range of $100\,\text{nHz} < f < 1\,\text{Hz}$. The sensitivity curves from the six formulas (68)–(73) are shown in Fig. 3. The corresponding sensitivity curves in terms of $h_c(f)$ and $\Omega_{gw}(f)$ are shown in Figs. 2 and 4, respectively.

The LISA Pathfinder Mission has been shipped to Kourou and is scheduled for launch from Kourou Spaceport on Arianespace Flight VV06 on 2nd December 2015. It is a technology demonstration mission. Its success will pave the road for future space GW missions. (Note added in proof: LISA Pathfinder was successfully launched on 3rd December 2015.)

The sensitivity curve of a single DECIGO interferometer as shown in Fig. 3 is from Ref. 87. BBO has a similar single-interferometer sensitivity curve. One-sigma, power-law integrated sensitivity curve for BBO (BBO-corr) as shown in Fig. 3 is obtained by Thrane and Romano.[88] That of DECIGO is similar. We also put in the plot their LISA autocorrelation measurement sensitivity curve (LISA-corr) in a single detector assuming perfect subtraction of instrumental noise and/or any unwanted astrophysical foreground.[88] The minimum autocorrelation sensitivity using the same method for ASTROD-GW is also estimated and plotted in Fig. 3; this would also be the level that 6 S/C ASTROD-GW[68] (6 S/C ASTROD-GW-corr) could reach. For comparison, the one-sigma, power-law integrated sensitivity curve for the adLIGO H1L1 (adLIGO-corr) from Ref. 88 is also plotted in Fig. 3. All of the corresponding curves are plotted in Figs. 2 and 4.

The development in atom interferometry is fast and promising. It already contributes to precision measurement and fundamental physics. A proposal using atom interferometry to detect GWs has been raised at Stanford University as an alternate method to LISA on the LISA bandwidth.[89,90] Issues have arisen on its realization of LISA sensitivity.[91,92] In Observatoire de Paris, SYRTE has started the first stage of its project — MIGA (Matter-wave laser Interferometric Gravitation Antenna)[93] of building a 300 m long optical cavity to interrogate atom interferometers at the underground laboratory LSBB in Rustrel. In the second stage of the project (2018–2023), MIGA will be dedicated to science runs and data analyses in order to probe the spatio-temporal structure of the local field of the LSBB region. In the meantime,

MIGA will assess future potential applications of atom interferometry to GW detection in the middle-frequency band (0.1–10 Hz).

3.6. *Very-low-frequency band* (300 pHz–100 nHz)

When GWs are propagating across the line of sight of pulsar observations, the pulse arrival times are affected. This effect can be used to observe the GWs. For isotropic stochastic GW background, Hellings and Downs derived a formula on the correlation in the timing residuals as a function of pairs of pulsars and used it to constrain the energy density in GWs of frequency between 4–10 nHz to be less than 1.4×10^{-4} of the cosmic critical density in 1983.[23] In 1996 and 2002, the upper limits from pulsar timing observations on a GW background derived by McHugh *et al.*[94] and by Lommen[95] are $\Omega_{\rm gw} \le 10^{-7}$ in the frequency range 4–40 nHz, and $\Omega_{\rm gw} \le 4 \times 10^{-9}$ at 6×10^{-8} Hz, respectively.

Now there are 4 major PTAs: the European PTA (EPTA),[96] the NANOGrav,[97] the Parks PTA (PPTA)[98] and the International (EPTA, NANOGrav and PPTA combined) PTA (IPTA).[99] For recent reviews on pulsar timing and PTAs for GW detection, please see Refs. 100 and 101. These 4 PTAs have improved greatly on the sensitivity for GW detection recently.[102–104] Upper limits on the isotropic stochastic background from EPTA, PPTA and NANOGrav are listed in Table 4. These limits assumes that the GW background has the following frequency dependence with $\alpha = -(2/3)$:

$$h_c(f) = A_{\rm yr}\left[f/(1\,{\rm yr}^{-1})\right]^{\alpha}. \tag{74}$$

The most stringent limit is from Shannon *et al.*[103] using observations of millisecond pulsars from the Parks telescope to constrain $A_{\rm yr}$ to less than 1.0×10^{-15} with 95% confidence. This limit already excludes present and most recent model predictions with 91–99.7% probability.[103] The three experiments form a robust upper limit of 1×10^{-15} on $A_{\rm yr}$ at 95% confidence level ruling out most models of supermassive black hole formation. The limit is shown as constraint on the Supermassive Black Hole Binary GW Background (SBHB-GWB) in Fig. 2; the corresponding constraints are also shown in Figs. 3 and 4. Since more energy of GWs might be emitted with higher frequency in the hierarchy of supermassive black hole formation, we extrapolate this constraint linearly instead with a knee using dotted line to 1×10^{-5} Hz with some confidence. Constraints with other α values have similar order of magnitudes.

To have an outlook of sensitivity of PTAs for the next hundred years, we adopt and extrapolate the estimates of Moore, Taylor and Gair.[105] The sensitivity for a monochromatic GW of a PTA is mainly dependent on the timing accuracies including timing residuals after modelling (rms deviations in timing residuals). The bandwidth depends on the sampling frequencies, i.e. cadences and the duration of the data span. For observations every Δt of time and an observation span of T the

Table 4. Upper limits on the isotropic stochastic background from three PTAs.

	No. of pulsars included	No. of years observed	Observation radio band [MHz]	Constraint on characteristic strain $h_c(f)[= A_{yr}[f/(1yr^{-1})]^{-(2/3)}$, $(f = 10^{-9} - 10^{-7}$ Hz)]
EPTA[102]	6	18	120–3000	$A_{yr} < 3 \times 10^{-15}$
PPTA[103]	4	11	3100	$A_{yr} < 1 \times 10^{-15}$
NANOGrav[104]	27	9	327–2100	$A_{yr} < 1.5 \times 10^{-15}$

bandwidth f is $[1/T, 1/\Delta t]$. The frequency dependence of the sensitivity in $h_c(f)$ is linear in f:

$$h_c(f) = B_{yr}\,(f/\mathrm{yr}^{-1}), \quad \frac{1}{T} < f < \frac{1}{\Delta t}. \tag{75}$$

We assume B_{yr} is proportional to the timing accuracy, inversely proportional to the observation time span and inversely proportional to the number of pulsars in the PTA. In Moore *et al.*,[105] a canonical PTA (MTG canonical PTA) with 36 pulsars randomly distributed on the sky, observed every two weeks with a precision of 100 ns over five years was assumed; this canonical PTA has a sensitivity (75) with $B_{yr} = 4 \times 10^{-16}$ and is roughly equivalent to OPEN 1 mock dataset in the IPTA data challenge.[106] In Table 5, we compile the projected sensitivities for IPTA, FAST[107] and SKA[108] for an observation span of 20 years, 50 years and 100 years, respectively. To obtain a fiducial sensitivity of IPTA, we take the MTG canonical PTA,[105] but extend the observation time span to 20 years. The sensitivity is 1×10^{-16} at $f = \mathrm{yr}^{-1} = 3.17 \times 10^{-8}$. With the advent of new and more sensitive observing facilities, PTA sensitivity will be improved. The 500 m Aperture Spherical Radio Telescope (FAST)[107] is under construction in Guei-Zhou, China. Since the 305 m radio telescope of Arecibo Observatory has been working for 52 years, we expect that FAST will work for more than 50 years also. In obtaining a fiducial sensitivity, we assume the FAST PTA to observe 50 pulsars with 50 ns timing accuracy for a 50 year time span. The Square Kilometre Array (SKA)[108] in South Africa and Australia will certainly improve on existing limits and we assume pulsar timing measurements every two weeks for 100 pulsars with 20 ns timing accuracies for 100 years. Table 5 lists the basic assumptions for IPTA, FAST and SKA and their projected sensitivities in B_{yr} on the characteristic strain.

Table 5. Sensitivities of IPTA, FAST and SKA to monochromatic GWs.

	No. of pulsars	No. of years of observation	Timing accuracy (ns)	Sensitivity in characteristic strain $h_c(f)[= B_{yr}(f/\mathrm{yr}^{-1})]$ for monochromatic GWs
IPTA[106]	36	20	100	$B_{yr} = 1 \times 10^{-16}$
FAST[107]	50	50	50	$B_{yr} = 1.5 \times 10^{-17}$
SKA[108]	100	100	20	$B_{yr} = 1.5 \times 10^{-18}$

The sensitivity curves of IPTA, FAST and SKA:

$$h_c(f) = 1 \times 10^{-16}(f/\text{yr}^{-1}),$$
$$1.58 \times 10^{-9}\,\text{Hz} < f < 8.27 \times 10^{-7}\,\text{Hz}, \quad \text{for IPTA-20},$$
$$h_c(f) = 1.5 \times 10^{-17}(f/\text{yr}^{-1}),$$
$$6.34 \times 10^{-10}\,\text{Hz} < f < 8.27 \times 10^{-7}\,\text{Hz}, \quad \text{for FAST-50},$$
$$h_c(f) = 1.5 \times 10^{-18}(f/\text{yr}^{-1}),$$
$$3.17 \times 10^{-10}\,\text{Hz} < f < 8.27 \times 10^{-7}\,\text{Hz}, \quad \text{for SKA-100}$$

are plotted on Fig. 2. The corresponding sensitivity curves in terms of $[S_h(f)]^{1/2}$ and $\Omega_g(f)$ are plotted in Figs. 3 and 4, respectively. We note that the SKA sensitivity for monochromatic GWs reaches 10^{-22} in $\Omega_g(f)$ at frequency around 3.17×10^{-10} Hz. The acronyms for these curves are IPTA-20, FAST-50 and SKA-100.

As to the single source GW limits. The bounds from PPTA[109] and EPTA[110] are in the order of 10^{-14} for h_c in the frequency range 5×10^{-9} to 2×10^{-7}. They are drawn on Fig. 2 with the corresponding curves on Figs. 3 and 4. A 24-Hour Global Campaign for GW from J1713+0747 gives upper limits in the frequency range 10^{-5}–10^{-3} Hz; the solid line shows the upper limit in random direction while the dotted line show the upper limit in the direction of pulsars.[111]

3.7. *Ultra-low-frequency band* (10 fHz–300 pHz)

GWs with periods longer than the time span of observations produce a simple pattern of apparent proper motions over the sky.[112] Therefore, precise measurement of proper motion of quasars would be a method to detect ultra-low-frequency (10 fHz–300 pHz) GWs. Gwinn *et al.*[113] used this method to constrain the normalized spectral energy density of stochastic GWs with frequencies less than 2×10^{-9} Hz and greater than 3×10^{-18} Hz (including frequencies in the ultra-low-frequency band) to less than 0.11 h^{-2} (95 % confidence) times the critical closure density of our universe. In Fig. 4, we use the Planck 2015 value[25] of Hubble constant $H_0 = (67.8 \pm 0.9)$ km s^{-1} Mpc^{-1} to set $h = 0.678$ in their original plot and obtain a bound of 0.24 in terms of the critical density (the bound is labelled Quasar Astrometry (QA) in Fig. 4). Long baseline optical interferometer with sub-micro-arcsecond and nano-arcsecond (nas) astrometry is technologically feasible.[114] With this kind of interferometer implemented, precision astrometry of quasar proper motions may possibly be improved by four orders of magnitude and reach nas yr^{-1}. In terms of energy, the precision of determining/constraining $\Omega_{gw}(f)$ could reach a sensitivity of 2.4×10^{-9} or better (Fig. 4; the curve is labelled QAG [Quasar Astrometry Goal]).

Using (60) or Table 2, we have the bound on characteristic strain $h_c(f)$:

$$h_c(f) < 4.2 \times 10^{-19}\,(\mathrm{Hz}/f) \quad \text{for } 3 \times 10^{-18}\,\mathrm{Hz} < f < 2 \times 10^{-9}\,\mathrm{Hz}. \tag{76}$$

When the angle resolution is improved by four orders of magnitude, the sensitivity will reach

$$h_c(f) = 4.2 \times 10^{-23}\,(\mathrm{Hz}/f) \quad \text{for } 3 \times 10^{-18}\,\mathrm{Hz} < f < 2 \times 10^{-9}\,\mathrm{Hz}. \tag{77}$$

Both the bound (76) and the curve (77) are plotted on Fig. 2. They are labelled QA and QAG. Using (60), we also convert $\Omega_{\mathrm{gw}}(f)$ to $[S_h(f)]^{1/2}$ and plot the results on Fig. 3.

3.8. *Extremely-low (Hubble)-frequency band* (1 aHz–10 fHz)

The successful prediction of nucleosynthesis of primordial abundances of ^{3}He, ^{4}He, ^{7}Li and deuterium put a constraint on integrated tensor perturbations $\int d(\log f)\Omega_{\mathrm{gw}}(f)$ of 10^{-5}.[115] This is plotted on Fig. 4 as the $\Omega_{\mathrm{gw}}(f) = 10^{-5}$ line. CMB experiments are most sensitive to the extremely-low (Hubble)-frequency band (1 aHz–10 fHz). First, a strong GW background at extremely-low-frequency produces stochastic redshift of CMB (Sachs–Wolfe effect; S-W effect).[116,117] The COBE observation gives CMB S-W redshift fluctuation bound which was plotted on Figs. 2–4 as CMB S-W fluct. The COBE microwave-background quadrupole anisotropy measurement[118,119] gives a limit $\Omega_{\mathrm{gw}}(1\,\mathrm{aHz}) \sim 10^{-9}$ on the extremely-low-frequency GW background.[120,121] WMAP[122–124] improves on the COBE constraints; the constraint on Ω_{gw} for the higher frequency end of this band is better than 10^{-14}. Planck Surveyor space mission has recently probed anisotropies with l up to 2000 and with higher sensitivity. Ground and balloon experiments probe smaller-angle anisotropies and, hence, higher-frequency background. ACTpol has probed anisotropies with l from 225 up to 8725.[125] These CMB experiments probe the 1 aHz–10 fHz extremely-low (Hubble) frequency band GWs. In inflationary cosmology these GWs give the tensor mode density and temperature perturbations (imprints) on CMB.

Inflation postulates a rapid accelerated expansion which set the initial moments of the Big Bang Cosmology.[126–131] Expansion drives the universe towards a homogeneous and spatially flat geometry that accurately describes the average state of the universe. The quantum fluctuations in this era grow into the galaxies, clusters of galaxies and temperature anisotropies of the CMB.[132–137] Modern inflation has been originated from efforts of unification, but its mechanism still remains unclear. The quantum fluctuations in the spacetime geometry in the inflationary era generate GWs which would have imprinted tensor perturbations on the CMB anisotropy. There is no confirmed discovery of these tensor perturbations yet (Refs. 138 and 139 and references therein). The analysis of *Planck*, SPT and ACT temperature data together with WMAP polarization did not discover these tensor perturbations and showed that the scalar index is $n_s = 0.959 \pm 0.007$ and the tensor-to-scalar perturbation ratio r is less than 0.11 (95% CL; no running).[140] The pivot scale of this constraint is 0.002 Mpc^{-1}, corresponding to GW frequency $f \sim 1.5 \times 10^{-18}$ Hz at present. From Refs. 141 and 117, this constraint corresponds to $\Omega_{\mathrm{gw}} < 10^{-15}$ and $h_c < 2.34 \times 10^{-9}$; it is plotted on Figs. 2–4 with label *Planck*. In March 2014, the announcement of BICEP2 of the detection of B-mode polarization excess and their interpretation of this excess as imprint from primordial tensor waves immediately attracted the imagination of the scientific community and the public.[142] The September 2014 announcement of dust measurement including the BICEP2 observation region from the Planck team convinced the physics community that the excess is consistent with dust emission.[143] The new *Keck Array* data[138] confirmed the BICEP2 B-mode polarization excess. The combined analysis of BICEP2/*Keck Array* and *Planck* Collaboration[138] convincingly showed that this excess is consistent with the *Planck* dust measurement and that the tensor-to-scalar perturbation ratio r is constrained to less than 0.12 (95% CL; no running). The pivot scale of this constraint is 0.05 Mpc^{-1}, corresponding to GW frequency $f \sim 3.8 \times 10^{-17}$ Hz at present. From Refs. 141 and 117, this constraint corresponds $\Omega_{\mathrm{gw}} < 1.1 \times 10^{-15}$ and $h_c < 5.57 \times 10^{-8}$; it is plotted on Figs. 2–4 with label BICEP2/*Keck*.

The sources for B-mode polarization in CMB could come from (i) GW imprints on CMB; (ii) gravitational lensing during the CMB propagation and (iii) pseudo-scalar-photon interaction during the CMB propagation. In the BICEP2 analysis,[142] gravitational lensing effect is subtracted. Einstein equivalence principle dictates that the propagation of electromagnetic waves (photons) observes Maxwell equations locally and there is no rotation of polarization plane during propagation [i.e. no cosmic polarization rotation (CPR)]. However, this is exactly a soft spot in the empirical foundation of EEP.[144] For a survey of constraints on CPR from astrophysical and cosmological observations, see di Serego Alighieri's review.[145] Basically, both the mean CPR and the CPR fluctuation magnitude are constrained to a couple of degrees. For example, from the ACTpol CMB polarization data fitting,[146] the mean CPR angle α is constrained from the EB correlation power spectra to be less than about 1° and the fluctuation (rms) is constrained from the BB correlation power spectra to $\langle\delta\alpha^2\rangle^{1/2} < 1.68°$. Including CPR effect together with the Planck dust measurement in a joint fitting of ACTpol, BICEP2, and POLARBEAR gives the values of the mean squares of the CPR fluctuation $\langle\delta\alpha^2\rangle = 41 \pm 522\,[\mathrm{mrad}^2]$ and the tensor-to-scalar ratio $r = -0.012 \pm 0.109$; this in turn gives a 1 σ bound on the rms of the CPR magnitude $\langle\delta\alpha^2\rangle^{1/2} < 23.7\,\mathrm{mrad}\,(1°.36)$ and that of $r < 0.097$. This result not only gives the best constraint on the CPR fluctuation magnitude, but also is consistent with the Planck and the joint BICEP2/*Keck Array*-Planck bound.

The ongoing situation as said in view given by Halverson[139] is "The competition is fierce, with at least six funded ground-based experiments underway (including the third version of BICEP), several balloon-borne experiments, and a number of proposed space missions. Finally, thanks to the new data, galactic foreground contaminants — and strategies for removing them — are now better understood."

The present consensus is that when the present ground-based and balloon-borne experiments are performed, the accuracy in determine r will have one order of magnitude improvement to 0.01; when the proposed space missions are flown and completed, the accuracy will have another order of magnitude improvement to 0.001. This means that the sensitivity in the Ω_{gw}-f plot will be improved to 10^{-17}.

4. Sources of GWs

In this section, we discuss sources of GWs concisely while refer to various references for more complete treatment.

4.1. *GWs from compact binaries*

Binary neutron stars coalesce by losing kinetic energy of orbital motion due to the emission of GW. When the orbital radius is much larger than the radius of stars, the radiation of GW is described by the quadrupole approximation reviewed in Sec. 2 until merging starts where two stars are deformed by each other's tidal forces.[147] The wave signal chirps according to the advancement of time and the frequency ranges from the low-frequency orbital motion period to high frequency merger characteristic frequency ($\sim 1\,\mathrm{kHz}$). Since the amplitude of the signal increases till the merger (inspiral phase), the signal of this inspiral phase is the most probable target of all ground-based laser interferometers with km-scale baseline length.

There are several neutron star binaries in our Galaxy (in the case of J1906+0746 the companion star may be a white dwarf). In Table 6, all that may coalesce due to the emission of GW within the age of the universe are listed. After the merger, coalesced neutron stars form a black hole and it oscillates due to dynamical energy of the coalescence just after the merger, which is known as quasi-normal mode oscillation of black hole. Since its typical frequency is several kHz for black hole with a few $M_\odot$, the oscillation (ring down) is the GW target of detectors that have sensitivity at high frequencies such as GEO-HF detector or cryogenic mechanical detectors.

Table 6. Neutron star binaries in our galaxy that may coalesce within the age of the universe (Companion star of J1906 + 0746 may be a white dwarf). P_s is the pulse period, P_b the orbital period of the binary system, e the eccentricity of the orbit and τ_{life} the life time of the binary system.

	P_s (ms)	P_b (hr)	e	τ_{life} (Gyr)
B1913+16[6]	59.03	7.75	0.62	0.37
B1534+12[148]	37.40	10.10	0.27	2.93
J0737-3039A[149]	22.70	2.45	0.088	0.23
J1756-2251[150]	28.46	7.67	0.18	2.03
J1906+0746[151]	144.14	3.98	0.085	0.082
J2127+11C[152]	32.76	8.04	0.68	0.32

The coalescence rate of binary neutron stars is estimated by knowing both the distribution of binaries and the life time of the binary systems. Due to the small number statistics and to the uncertainty biases of pulsar observation (e.g. dissipation of electromagnetic waves in our Galaxy, beaming angle, faint pulsars, etc.), the estimated event rate ranges more than two orders of magnitude, where typical value is once per 100,000 years in such matured galaxy as ours.[153] Since the population of such matured galaxy is roughly 0.01 per cubic Mpc, at least one event per year can be detected if the sensitivity to catch events occurring at 130 Mpc is achieved by ground-based detector. Advanced LIGO has initiated observation run 1 (O1) starting 18th September 2015 with sensitivity reaching 70 Mpc for the coalescence of nominal binary neutron stars.

In the coalescence of binary black holes, the frequency of the chirping signal shifts down to lower frequencies. If their initial mass ranges around $10\,\mathrm{M}_\odot$, the merger may occur at around 200 Hz. The signal is in the most sensitive frequencies of the second generation ground-based interferometers.

Dominick estimated that the population and the coalescence rate of binary black holes is smaller than that of binary neutron stars.[154] However, a theoretical study shows that merger rate of black holes based on ejections from globular clusters is larger than that of neutron star binaries.[155] This is still an issue of different opinions. Moreover, since the amplitude of GWs from the coalescence of black holes is larger, possible detection rate will be larger if the detector has sensitivity at lower frequencies ($<\sim 10$ Hz), which will be realized by the third-generation detectors.

We plot the source strengths of compact binary inspirals, pulsars, resolvable galactic binaries and unresolvable galactic binaries [confusion background][3,156] in Figs. 2–4 by adopting those of Moore *et al.*[35]

4.2. *GWs from supernovae*

Massive stars heavier than 8 $\mathrm{M}_\odot$ collapse due to gravity after burning out and a neutron star may be born. This collapse produces burst GW. Taking the second derivative of the quadrupole moment of the star and using (35), the maximum amplitude $h_{\max}$ is estimated to be $\varsigma MR^2(2\pi f)^2$, where ς [of the order of $G_{\mathrm{N}}/(c^4r)$ times nonsphericity of the explosion] is a calculable numerical factor, M is the mass of the initial neutron star born just after the collapse, R is the radius of the star and r is the observation distance to the star. If the collapse occurs in the center of our galaxy in a favorable condition, the burst wave signal may be detected by resonant antennas. Also it is a target source of GW of ground-based interferometric detectors.[157]

The GW form information is useful to enhance the signal-to-noise ratio of the detector. Since stellar core collapse is a complex physical phenomena that involves GR, hydrodynamics, and neutrino transport with thermonuclear kinetics in short time duration, it is not easy to conduct a full numerical simulation to obtain the GW form. In 2002 Dimmelmeier *et al.*[158] first performed axisymmetric hydrodynamic simulations of rotational core collapse and its associated GW emission in 26 general-relativistic and Newtonian models. The total energy of GWs emitted is only about 10^{-7}–$10^{-8}\,M_\odot c^2$. Recent development in three-dimensional numerical simulation which requires longer computing time shows that strong burst GW of total luminosity of $0.01\,M_\odot c^2$ can be produced by an initially nonrotating star due to standing accretion shock instability (SASI).[159,160] For a review on this subject, see Ref. 161. These may be the plausible candidates for the second-generation ground-based detectors. The event rate of supernova explosions in our Galaxy is estimated as once per 40 ± 10 yr.[162] According to Abadie *et al.*,[163] supernova explosion with GW energy $0.056\,M_\odot c^2$ could be detected at 16 Mpc with LIGO-Virgo achieved sensitivity; hence supernova explosion with GW energy $\sim 0.01\,M_\odot c^2$ should be detected up to 6.8 Mpc with the LIGO-Virgo achieved sensitivity; if supernova explosion is always accompanied with the emission of GW energy of $\sim 0.01 M_\odot c^2$, the detection rate on the Earth would be 0.04/yr assuming uniform distribution of such galaxies as ours to be 0.01 Mpc^{-3}. This rate would nominally be improved to 1.7 yr^{-1} by adLIGO at present sensitivity (3.5 fold improvement compared with Ref. 163). However, since there is a large uncertainty in the distribution of GW energy strength in the supernova explosion, we just adopting the strength as given by Moore *et al.*[35] for plotting in Figs. 2–4.

4.3. *GWs from massive black holes and their coevolution with galaxies*

Observational evidences indicate that massive black holes (MBHs) residing in most local galaxies. Relations have been discovered between the MBH mass and the mass of host galaxy bulge, and between the MBH mass and the velocity-dispersion. These relations indicate that the central MBHs are linked to the evolution of galactic structure. Newly fueled quasar may come from the gas-rich major merger of two massive galaxies. Recent astrophysical evidences linked together these major galaxy mergers and the growth of supermassive black holes in quasars.[164,165] Distant quasar observations indicate that MBH of billions of solar masses already existed less than a billion years after the Big Bang. At present, there are different theoretical proposals for scenarios of the initial conditions and formations of black holes. These scenarios include BH seeds from inflationary universe and/or from the collapse of Population III stars, different accretion models and binary formation rates. All of these models generate MBH merging scenarios in galaxy co-evolution with GW radiations. Measurement of amplitude and spectrum of these GWs will tell us the cosmic history of MBH formation.

The standard theory of MBH formation is the merger-tree theory with various MBH Binary (MBHB) inspirals acting. The GWs from these MBHB inspirals can be detected and explored to cosmological distances using space GW detectors. Although there are different merger-tree models and models with BH seeds, they all give significant detection rates for space GW detectors and PTAs.[100,166–168]

GW observation in the 300 pHz–0.1 Hz frequency band will be a major observation tool to study the coevolution of galaxy with BHs. This frequency band covers the low frequency band (100 nHz–100 mHz) and very-low-frequency band (300 pHz–100 nHz) GWs and is in the detection range of PTAs, eLISA/LISA-like/Earth-orbiting-missions and ASTROD-GW. PTAs are most sensitive in the frequency range 300 pHz–100 nHz, eLISA/LISA-like/Earth-orbiting space GW detector is most sensitive in the frequency range 2 mHz–0.1 Hz, while ASTROD-GW is most sensitive in the frequency range 500 nHz–2 mHz (Figs. 2–4).

PTAs have been collecting data for decades for detection of stochastic GW background from MBH binary mergers. They already constrain A_{yr} in Eq. (74) to less than 1.0×10^{-15} with 95% confidence.[102–104] This limit excludes present and most recent model predictions of supermassive black hole formation with 91-99.7% probability.[103] This means the detection could be anytime near. Since we know that SMBHs are already formed, it also means that the backgrounds in the higher frequency/shorter wavelength band are higher than original predicted. For most models there is a knee around $f \sim 100$ nHz, now we straighten the knee and extend Eq. (74) to $f \sim 10\,\mu$Hz with dashed line in Fig. 2. Below this 1.0×10^{-15} limit, we plot pink colored region to show possible background source region. Corresponding line and colored region are also shown in Figs. 3 and 4.

eLISA and ASTROD-GW will be able to directly observe how MBHs form, grow and interact over the entire history of galaxy formation. ASTROD-GW will detect stochastic GW background from MBH binary mergers in the frequency range 500 nHz–100 μHz. These observations are significant and important to the study of co-evolution of galaxies with MBHs. The expected rate of MBHB sources is 10 yr^{-1}–100 yr^{-1} for eLISA and 10 yr^{-1}–1000 yr^{-1} for LISA.[64] For ASTROD-GW, similar number of sources as that of LISA is expected with better angular resolution.[68] For a more detailed account, see Ref. 169.

At present, there are different theoretical scenarios for the initial conditions and formations of black holes, e.g. primordial MBH clouds as seeds, direct formation of supermassive black hole via multi-scale gas inflows in galaxy mergers, direct collapse into a supermassive black hole from mergers between massive protogalaxies with no need to suppress cooling and star formation, etc. The mass range and maximum mass of Population III stars is also a relevant issue for seed BHs. With the PPTA constraint, there should be more backgrounds in the μHz region. ASTROD-GW with good sensitivity in the μHz band will contribute to detect or constrain GW background to distinguish various scenarios for finding the history of BH and galaxy coevolution.

With the detection of MBH merger events and background, the properties and distribution of MBHs could be deduced and underlying population models could be tested. Sesana *et al.*[170] consider and compare ten specific models of MBH formation. These models are chosen to probe four important and largely unconstrained aspects of input physics used in the structure formation simulations, i.e. seed formation, metallicity feedback, accretion efficiency and accretion geometry. With Bayesian analyses to recover posterior probability distribution, they show that LISA has enormous potential to probe the underlying physics of structure formation. With better sensitivity in the frequency range 100 nHz–1 mHz, ASTROD-GW will be able to probe the underlying physics of structure formation further. With the detection of the GW background of the MBH mergers, PTAs and ASTROD-GW will add to our understanding of the structure formation.

We plot the source strengths of massive binaries in Figs. 2–4 adopting those of Moore *et al.*[35]

4.4. *GWs from extreme mass ratio inspirals (EMRIs)*

EMRIs are GW sources for space GW detectors. The eLISA sensitive range for central MBH masses is $10^4 - 10^7$ $M_\odot$. The expected number of eLISA detections over two years is 10–20;[64] for LISA, a few tens;[64] for ASTROD-GW, similar or more with sensitivity toward larger central BH's and with better angular resolution.[68] For a more detailed account, we refer to Ref. 169. We plot the source strengths of EMRIs in Figs. 2–4 adopting those in Moore *et al.*[35]

4.5. *Primordial/inflationary/relic GWs*

Relic GWs from inflationary or noninflationary period are commonly called primordial GWs. Relative to primordial GWs, all the GW sources we have discussed are foregrounds. Assuming the primordial GW spectrum is flat in the $\Omega_{\rm gw}(f)$ versus f diagram, i.e. the tensor index $n_{\rm t}$ is 0, we draw an upper bound of inflationary spectrum to saturate the constraints given in Sec. 3.8; it is the flat line (the tensor index $n_{\rm t}$ is 0) about 10^{-15} level in the $\Omega_{\rm gw}(f)$ versus f diagram (Fig. 4) with the very high frequency part dropping steeply above 10^{10} Hz. For comparison, the black dotted curve shows the corresponding $\Omega_{\rm gw}(f)$ for a 0.9 K blackbody radiation. If the GW perturbations had been in equilibrium with the matter fields, it is an expected GW background. We refer the readers to the recent review by Sato and Yokoyama on "Inflationary cosmology: First 30+ years"[171] for a detailed account of the inflationary scenario.

As expected in Sec. 3.8, the present consensus on the CMB B-polarization measurements is that when the present ground-based and balloon-borne experiments are performed the sensitivity in the $\Omega_{\rm gw}$-f plot will have a one-order improvement to 10^{-16} and when the proposed space missions are flown and completed the sensitivity will have another order of magnitude improvement to 10^{-17}.

The instrument sensitivity goals of DECIGO,[72] Big Bang Observer[73] and 6-S/C ASTROD-GW[68] all reach the 10^{-17}-level in terms of $\Omega_{\rm gw}$ (Fig. 4). The sensitivities of IPTA, FAST and SKA also reach the 10^{-17}-level or beyond in terms of $\Omega_{\rm gw}$ (Fig. 4). These instrument sensitivities are good enough to probe primordial GWs down to the 10^{-17}-level or beyond in terms of $\Omega_{\rm gw}$ at frequencies around 1 nHz, 10–300 μHz and 0.1–1 Hz to search and test inflationary/noninflationary physics. The main issue is the level of foreground and whether foreground could be separated.

4.6. *Very-high-frequency and ultra-high-frequency GW sources*

There are four kind of potential GW sources in the very-high-frequency and ultra-high-frequency bands[39]:

(i) Discrete sources.[172]
(ii) Cosmological sources.[173]
(iii) Braneworld Kaluza–Klein (KK) mode radiation.[175,176]
(iv) Plasma instabilities.[176]

In general, objects do not radiate efficiently at wavelengths very different from their size. This implies objects radiate at these bands need to be very small and yet have a very large energy concentration to induce significantly large curvature fluctuations. Grishchuk[173] estimated the GWs generated from the amplification of quantum fluctuations by inflation. GWs in these bands with current wavelengths would had very short wavelengths that new physics might be working in the period of generation. However, the nucleosysthesis bound of $\Omega_{\rm gw}(f) \approx 10^{-5}$ must be satisfied by the spectrum of any GW background.[24] h_c at 100 MHz, 10 GHz and 1 THz would need to be less than $9.5 \times 10^{-29}, 9.5 \times 10^{-31}$ and 9.5×10^{-33}, respectively. The actual signals may be much lower. Various theoretical models[177–184] predict GWs at levels from $\Omega_{\rm gw}(f) \sim 10^{-8}$ to below $\sim 10^{-18}$. See Ref. 39 and references therein for more details.

To close this subsection, we quote from Ref. 39: "Even assuming the most optimistic noise temperatures and the highest magnet strengths, detection of the cosmological signals look beyond reasonable extrapolation of current performance whereas very-high-frequency GWs from braneworld scenarios may be within range of current technology. The most optimistic plasma instability signals from our galaxy if they occur at the low-frequency end of the range could also be above the sensitivity of future microwave detectors. ... There may also be astrophysical processes that convert violent electromagnetic events into very-high-frequency gravitational sources that could be detected but more targeted modelling is needed to identify candidate astronomical objects. The technology for detectors which convert the GW directly to an electromagnetic signal is currently available and builds on decades of development for other applications."

4.7. *Other possible sources*

Cosmic strings are popular GW sources in many theoretical investigations. For possible GW magnitudes in various bands of cosmic-string contribution, please see Ref. 185 and references therein. Recently, Geng *et al.*[186] proposed the coalescence of strange-quark planets with strange stars as a new kind of GW burst sources for ground-based interferometers. As GW astronomy and GW physics progress, there could be detected GW sources of various different origins. This is open until the experiments and observations are performed.

5. Discussion and Outlook

In spite of tremendous efforts in the high frequency band and some efforts in the very high frequency band experiments, GW has not been directly detected yet. This is due to the weakness in the strength of GWs in the present epoch.

The first generation of km-sized arm length interferometers reach the sensitivity of detecting binary neutron star inspirals up to the Virgo cluster distance. From the statistics of astrophysical binary neutron star distribution, the rate of detection is about 0.05 events per year with a large uncertainty. However, with a ten-fold increase of strain sensitivity, the reach in distance increases by ten-fold and the reach in astrophysical volume increases by one thousand fold. Hence, the rate of detection is about 50 events per year. This is the goal of Advanced LIGO,[55] Advanced Virgo[56] and KAGRA/LCGT[57] under construction. Advanced LIGO has achieved 3.5 fold better sensitivities with a reach to neutron star binary merging event at 70 Mpc and began its first observing run (O1) on September 18, 2015 searching for GWs. We could expect detection of GWs anytime. We will see a global network of second generation km-size interferometers for GW detection soon.

Another avenue for real-time direct detection is from the PTAs. The PTA bound on stochastic GW background already excludes most theoretical models; this may mean we could detect very-low-frequency GWs anytime too with a longer time scale.

We have presented a complete frequency classification of GWs according to their detection methods. Although there is no direct real-time detection of GWs yet, several bands are amenable to direct detection. Real-time direct detection may first come in the high frequency band or in the very-low-frequency band. Although the prospect of a launch of space GW is only expected in about 20 years, the detection in the low-frequency band may have the largest signal-to-noise ratios. This will enable the detailed study of black hole coevolution with galaxies and of the dark energy issue. Foreground separation and correlation detection method need to be investigated to achieve the sensitivities 10^{-16}–10^{-17} or beyond in $\Omega_{\rm gw}$ to study the primordial GW background for exploring very early universe and possibly quantum gravity regimes.

When we look back at the theoretical and experimental development of GW physics and astronomy over the last 100 years, there are many challenges, some pitfalls and during last 50 years close interactions among theorists and experimentalists. The subject and community have become clearly multidisciplinary. One example is the interaction of the GW community and the Quantum Optics community in the last 40 years to identify standard quantum uncertainties in measurement, to realize that this is not an obstacle of measurement in principle and to find ways to overcome it. Another example is the interaction of the physics community and the astronomy community to understand and to identify detectable and potentially detectable GW sources. With current technology development and astrophysical understanding, we are in a position using GWs to study more thoroughly galaxies, supermassive black holes and clusters together with cosmology and to explore deeper into the origin of gravitation and our universe. Next 100 years will be the golden age of GW astronomy and GW physics. The current and coming generations are holding such promises.

Note added in proof: After this review appeared in arXiv, Refs. 187–189 have been brought to our attention that the pulsar timing method can also detect the imprint of a stochastic GW background on pulsar timing parameters in the ultralow frequency range down to $f \sim c/r$ where r is the distance to the pulsar. We thank Maxim Pshirkov for his helpful communication.

Acknowledgments

This review extends and updates the former review.[16] It includes significant amount of materials from Refs. 42, 68 and 69. We would like to thank M. Bucher, A. Di Virgilio, N. Kanda, L. Lentati, R. N. Manchester, D. H. Reitze and L. Wen for their help and comments on various stages of writing. This work was supported in part by the National Science Council (Grant No. NSC102-2112-M-007-019) and by the MEXT (JSPS Leading-edge Research Infrastructure Program, JSPS Grant-in-Aid for Specially Promoted Research 26000005, MEXT Grant-in-Aid for Scientific Research on Innovative Areas 24103005, JSPS Core-to-Core Program, A. Advanced Research Networks and the joint research program of the Institute for Cosmic Ray Research, University of Tokyo).

References

1. A. Einstein, *Sitz.ber. Preussischen Akad. Wiss.* **6** (1916) 688; English translation (translated by Alfred Engel) in The Collected Papers of Albert Einstein, Volume 6: The Berlin Years: Writings, 1914–1917 (English translation supplement, Doc. 32, pp. 201–210, Princeton University Press, http://einsteinpapers.press.princeton.edu/vol6-trans).
2. A. Einstein, *Sitz.ber. Königlich Preussischen Akad. Wiss.* **7** (1918) 154; English translation (translated by Alfred Engel) in The Collected Papers of Albert Einstein, Volume 7: The Berlin Years: Writings, 1918–1921 (English translation supplement, Doc. 1 "On gravitational waves", pp. 9–27, Princeton University Press, http://einsteinpapers.press.princeton.edu/vol7-trans).
3. P. L. Bender and D. Hils, *Class. Quantum Grav.* **14** (1997) 1439.
4. EAS's New Vision to Study the Invisible Universe (28 November 2013), http://www.esa.int/Our_Activities/Space_Science/ESA_s_new_vision_to_study_the_invisible_Universe.
5. J. M. Weisberg and J. H. Taylor, Relativistic binary pulsar B1913+16: Thirty years of observations and analysis, in *Proc. Aspen Conf. Binary Radio Pulsars, ASP Conf. Series*, eds. F. A. Rasio and I. H. Stairs (2004), arXiv:astro-ph/0407149.
6. J. M. Weisberg, D. J. Nice and J. H. Taylor, *Astrophys. J.* **722** (2010) 1030.
7. M. Kramer *et al. Science* **314** (2006) 97.
8. P. C. C. Freire, N. Wex, G. Esposito-Farèse, J. P. W. Verbiest, M. Bailes, B. A. Jacoby, M. Kramer, I. H. Stairs, J. Antoniadis and G. H. Janssen, *Mon. Not. R. Astron. Soc.* **423** (2012) 3328.

9. K. S. Thorne, Gravitational waves, in *Particle and Nuclear Astrophysics and Cosmology in the Next Millennium*, eds. E. W. Kolb and R. D. Peccei (World Scientific, Singapore, 1995), p. 160.
10. W.-T. Ni, ASTROD and gravitational waves, in *Gravitational Wave Detection*, eds. K. Tsubono, M.-K. Fujimoto and K. Kuroda (Universal Academy Press, Tokyo, Japan, 1997), pp. 117–129.
11. W.-T. Ni, *Int. J. Mod. Phys. D* **14** (2005) 901.
12. W.-T. Ni, Gravitational waves, in 2007–2008 Report on the Development of Astronomical Sciences, Chinese Astronomical Society Report (Science and Technology Press, Beijing, China, 2008), pp. 100–104 (in Chinese).
13. W.-T. Ni, ASTROD optimized for gravitational-wave detection: ASTROD-GW — A pre-Phase A study proposal, *Chinese Academy of Sciences* (26 February 2009).
14. W.-T. Ni, Testing relativistic gravity and detecting gravitational waves in space, in *Proc. 9th Asia-Pacific International Conf. Gravitation and Astrophysics* (ICGA9), Wuhan, June 28–July 3, 2009, eds. J. Luo, Z. B. Zhou, H.-C. Yeh and J.-P. Hsu (World Scientific, 2010), arXiv:1001.0213, pp. 40–47.
15. W.-T. Ni *et al.*, ASTROD optimized for gravitational wave detection: ASTROD-GW, in *Proc. 6th Deep Space Exploration Technology Symp.*, Sanya, Hainan, China December 3–6, 2009.
16. W.-T. Ni, *Mod. Phys. Lett. A* **25** (2010) 922.
17. W.-T. Ni, Classification of gravitational waves, http://astrod.wikispaces.com/file/view/GW-classification.pdf.
18. C. W. Misner, K. S. Thorne and J. A. Wheeler, *Gravitation* (Freeman, San Francisco, 1973).
19. W.-T. Ni, Genesis of general relativity: A concise exposition, in *One Hundred Years of General Relativity: From Genesis and Empirical Foundations to Gravitational Waves, Cosmology and Quantum Gravity*, ed. W.-T. Ni (World Scientific, Singapore, 2015); *Int. J. Mod. Phys. D* **25** (2016) 1630004.
20. R. A. Isaacson, *Phys. Rev.* **166** (1968) 1263.
21. R. A. Isaacson, *Phys. Rev.* **166** (1968) 1272.
22. L. Landau and E. Lifshitz, *The Classical Theory of Fields*, Chap. 11 (Addison-Wesley, Reading, Massachusetts, 1959).
23. R. Hellings and G. Downs, *Astrophys. J.* **265** (1983) L39.
24. M. Maggiore, *Phys. Rep.* **331** (2000) 283.
25. Planck Collab. (P. A. R. Ade *et al.*), Planck 2015 results. XIII. Cosmological parameters, arXiv:1502.01589.
26. P. C. Peters and J. Mathews, *Phys. Rev.* **131** (1963) 435.
27. P. C. Peters, *Phys. Rev.* **136** (1964) B1224.
28. D. M. Eardley, D. L. Lee, A. P. Lightman, R. V. Wagoner and C. M. Will, *Phys. Rev. Lett.* **30** (1973) 884.
29. D. M. Eardley, D. L. Lee and A. P. Lightman, *Phys. Rev. D* **8** (1973) 3308.
30. W.-T. Ni, A conjecture about gravitational radiation in all viable gravitational theories, *Chin. J. Phys.* **13** (1975) 84, http://psroc.phys.ntu.edu.tw/cjp/download.php?type=paper&vol=13&num=1&page=84.
31. M. Maggiore, *Gravitational Waves* (Oxford University Press, 2008).
32. L. S. Finn, *Phys. Rev. D* **46** (1992) 5236.
33. C. Cutler and É. É. Flanagan, *Phys. Rev. D* **49** (1994) 2658.
34. E. Thrane and J. D. Romano, *Phys. Rev. D* **88** (2013) 124032.
35. C. J. Moore, R. H. Cole and C. P. L. Berry, *Class. Quantum Grav.* **32** (2015) 015014.
36. A. M. Cruise and R. M. J. Ingley, *Class. Quantum Grav.* **23** (2006) 6185.

37. T. Akutsu *et al.*, *Phys. Rev. Lett.* **101** (2008) 101101.
38. A. Nishizawa *et al.*, *Phys. Rev. D* **77** (2008) 022002.
39. A. M. Cruise, *Class. Quantum Grav.* **29** (2012) 095003.
40. F. Li, N. Yang, Z. Fang, R. M. L. Baker, G. V. Stephenson and H. Wen, *Phys. Rev. D* **80** (2009) 064013.
41. L. P. Grishchuk, Electromagnetic generators and detectors of gravitational waves, arXiv:gr-qc/0306013.
42. K. Kuroda, Ground based gravitational wave detectors, in *One Hundred Years of General Relativity: From Genesis and Empirical Foundations to Gravitational Waves, Cosmology and Quantum Gravity*, Chap. 11, ed. W.-T. Ni (World Scientific, Singapore, 2015); *Int. J. Mod. Phys. D* **24** (2015) 1530032.
43. P. Astone *et al.*, *Class. Quantum Grav.* **25** (2008) 184012.
44. F. Acernese *et al.*, *Class. Quantum Grav.* **25** (2008) 205007.
45. L. Gottardi *et al.*, *Phys. Rev. D* **76** (2007) 102005.
46. O. D. Aguiar *et al.*, *Class. Quantum Grav.* **25** (2008) 114042.
47. TAMA300 Collab., http://tamago.mtk.nao.ac.jp/.
48. The GEO600 Team, http://geo600.aei.mpg.de.
49. The LIGO Scientific Collab., http://www.ligo.caltech.edu/.
50. The Virgo Collaboration, http://www.virgo.infn.it/.
51. J. Aasi *et al.*, *Phys. Rev. D* **91** (2015) 022004.
52. The LIGO Scientific Collab. and the Virgo Collab., *Nature* **460** (2009) 990.
53. LIGO and Virgo Collab., (J. Aasi *et al.*) *Phys. Rev. Lett.* **113** (2014) 231101.
54. J. Abadie *et al.*, *Astrophys. J.* **760** (2012) 12.
55. The Advanced LIGO Team, http://www.ligo.caltech.edu/advLIGO/.
56. The Advanced Virgo Team, http://wwwcascina.virgo.infn.it/advirgo/; http://www cascina.virgo.infn.it/advirgo/docs/whitepaper.pdf.
57. KAGRA/LCGT Team, http://gwcenter.icrr.u-tokyo.ac.jp/en/.
58. Einstein Telescope, http://www.et-gw.eu/.
59. F. B. Estabrook and H. D. Wahlquist, *Gen. Relativ. Gravit.* **6** (1975) 439.
60. H. D. Wahlquist, *Gen. Relativ. Gravit.* **19** (1987) 1101.
61. J. W. Armstrong, L. Iess, P. Tortora and B. Bertotti, *Astrophys. J.* **599** (2003) 806.
62. M. Tinto, G. J. Dick, J. D. Prestage and J. W. Armstrong, *Phys. Rev. D* **79** (2009) 102003.
63. C. Braxmaier *et al.*, *Exp. Astron.* **34** (2012) 181.
64. O. Jennrich *et al.*, NGO (New Gravitational wave Observatory) Assessment Study Report, ESA/SRE(2011)19.
65. LISA Study Team, LISA (Laser Interferometer Space Antenna): A cornerstone mission for the observation of gravitational waves, ESA System and Technology Study Report, ESA-SCI 11 (2000).
66. W.-T. Ni, *Int. J. Mod. Phys. D* **17** (2008) 921.
67. A. Bec-Borsenberger *et al.*, Astrodynamical Space Test of Relativity using Optical Devices ASTROD — A Proposal, submitted to ESA in Response to Call for Mission Proposals for Two Flexi-Mission F2/F3 (31 January 2000).
68. W.-T. Ni, *Int. J. Mod. Phys. D* **22** (2013) 1431004.
69. W.-T. Ni, Gravitational wave detection in space, in *One Hundred Years of General Relativity: From Genesis and Empirical Foundations to Gravitational Waves, Cosmology and Quantum Gravity*, Chap. 12, ed. W.-T. Ni (World Scientific, Singapore, 2015); *Int. J. Mod. Phys. D* **25** (2016) 1630001.
70. W.-T. Ni and A.-M. Wu, Orbit design of ASTROD-EM, in preparation.
71. W.-T. Ni, *Class. Quantum Grav.* **26** (2009) 075021.

72. S. Kawamura *et al.*, *Class. Quantum Grav.* **23** (2006) S125.
73. J. Crowder and N. J. Cornish, *Phys. Rev. D* **72** (2005) 083005.
74. P. L. Bender, *Class. Quantum Grav.* **21** (2004) S1203.
75. X. Gong *et al.*, Descope of the ALIA mission, *J. Phys.: Conf. Ser.* **610** (2015) 012011; arXiv:1410.7296.
76. M. Tinto, J. C. N. de Araujo, O. D. Aguiar and M. E. S. Alves, A geostationary gravitational wave interferometer (GEOGRAWI), arXiv:1111.2576.
77. M. Tinto, J. C. N. de Araujo, O. D. Aguiar and M. E. S. Alves, *Astropart. Phys.* **48** (2013) 50.
78. M. Tinto, D. Debra. S. Buchman and S. Tilley, *Rev. Sci. Instrum.* **86** (2015) 014501.
79. S. T. McWilliams, Geostationary Antenna for Disturbance-Free Laser Interferometry (GADFLI), arXiv:1111.3708v1.
80. J. W. Conklin *et al.*, LAGRANGE: Laser Gravitational-wave Antenna at Geo-lunar Lagrange points, arXiv:1111.5264v2.
81. R. Hellings, S. L. Larson, S. Jensen, C. Fish, M. Benacquista, N. Cornish and R. Lang, A low-cost, high-performance space gravitational astronomy mission: A mission-concept white paper submitted to NASA (2011).
82. J. Luo *et al.*, TianQin: A space-borne gravitational wave detector, submitted to *Class. Quantum Grav.* (2015); arXiv:1512.02076.
83. N. Seto, *Phys. Rev. D***73** (2006) 063001.
84. W.-T. Ni, *Int. J. Mod. Phys. D* **11** (2002) 947.
85. K. G. Arun *et al.*, *Class. Quantum Grav.* **26** (2009) 094027.
86. W.-T. Ni, Dark energy, co-evolution of massive black holes with galaxies, and ASTROD-GW, Paper (COSPAR paper number H05-0017-10) presented in the 38th COSPAR Scientific Assembly, 18–25 July 2010, Bremen, Germany (2010); *Adv. Space Res.* **51** (2013) 525, arXiv:1104.5049.
87. M. Ando, presented original DECIGO target sensitivity in the form of numerical data, private communication.
88. E. Thrane and J. D. Romano, *Phys. Rev. D* **88** (2013) 124032.
89. J. M. Hogan *et al.*, *Gen. Relativ. Gravit.* **43** (2011) 1953.
90. J. M. Hogan and M. A. Kasevich, Atom interferometric gravitational wave detection using heterodyne laser links, arXiv:1501.06797.
91. P. L. Bender, *Phys. Rev. D* **84** (2011) 028101.
92. S. Dimopoulos *et al.*, *Phys. Rev. D* **84** (2011) 028102.
93. R. Geiger *et al.*, Matter-wave laser Interferometric Gravitation Antenna (MIGA): New perspectives for fundamental physics and geosciences, in *Proc. 50th Rencontres de Moriond "100 years after GR"*, La Thuile, Italy, 21–28 March, 2015, arXiv:1505.07137.
94. M. P. McHugh, G. Zalamansky, F. Vernotte and E. Lantz, *Phys. Rev. D* **54** (1996) 5993.
95. A. N. Lommen, in *Proc. 270th WE-Heraeus Seminar on: "Neutron Stars, Pulsars and Supernova Remnants"*, Physikzentrum, Bad Honnef, Germany, Jan. 21–25, 2002, eds. W. Becker, H. Lesch and J. Trümper, MPE Report 278, pp. 114–125; arXiv:astro-ph/0208572.
96. http://www.epta.eu.org/.
97. http://nanograv.org/.
98. http://www.atnf.csiro.au/research/pulsar/ppta/.
99. http://www.ipta4gw.org/.
100. R. N. Manchester, Pulsars and gravity, in *One Hundred Years of General Relativity: From Genesis and Empirical Foundations to Gravitational Waves, Cosmology and*

Quantum Gravity, Chap. 9, ed. W.-T. Ni (World Scientific, Singapore, 2015); *Int. J. Mod. Phys. D* **24** (2015) 1530018.
101. X.-J. Zhu, L. Wen, G. Hobbs, R. N. Manchester and R. M. Shannon, Detecting nanohertz gravitational waves with pulsar timing arrays, arXiv:1509.06438v1 [astro-ph.IM].
102. L. Lentati *et al.*, *Mon. Not. R. Astron. Soc.* **453** (2015) 2576.
103. R. M. Shannon *et al.*, *Science* **349** (2015) 1522.
104. Z. Arzoumanian *et al.*, The NANOGrav nine-year data set: Limits on the isotropic stochastic gravitational wave background, arXiv:1508.03024 [astro-ph.GA].
105. C. J. Moore, S. R. Taylor and J. R. Gair, *Class. Quantum Grav.* **32** (2015) 055004.
106. http://ipta4gw.org/?page_id=89.
107. R. Nan *et al.*, *Int. J. Mod. Phys. D* **20** (2011) 989.
108. C. Carilli and S. Rawlings, *New Astron. Rev.* **48** (2004) 979.
109. X.-J. Zhu *et al.*, *Mon. Not. Roy. Astron. Soc.* **444** (2014) 3709.
110. S. Babak *et al.*, European pulsar timing array limits on continuous gravitational waves from individual supermassive black hole binaries, arXiv:1509.02165.
111. NANOGrav Collab. (T. Dolch *et al.*), Single-source gravitational wave limits from the J1713+0747 24-hr global campaign, arXiv:1509.05446.
112. T. Pyne, C. R. Gwinn, M. Birkinshaw, T. M. Eubanks and D. N. Matsakis, *Astrophys. J.* **465** (1996) 566.
113. C. R. Gwinn, T. M. Eubanks, T. Pyne, M. Birkinshaw and D. N. Matsakis, *Astrophys. J.* **485** (1997) 87.
114. S. C. Unwin *et al.*, *Publ. Astron. Soc. Pac.* **120** (2008) 38.
115. C. J. Copi, D. N. Schramm and M. S. Turner, *Phys. Rev. D* **55** (1997) 3389.
116. R. Sachs and A.Wolfe, *Astrophys. J.* **147** (1967) 73.
117. B. Allen, "The stochastic gravity-wave background: sources and detection", in *Relativistic Gravitation and Gravitational Radiation*, eds. J.-A. Marck and J.-P. Lasota (Cambridge University Press, Cambridge, 1997), p. 373.
118. G. F. Smoot *et al.*, *Astrophys. J.* **396** (1992) L1.
119. C. L. Bennett *et al.*, *Astrophys. J.* **464** (1996) L1.
120. L. M. Krauss and M. White, *Phys. Rev. Lett.* **69** (1992) 969.
121. R. L. Davis, H. M. Hodges, G. F. Smoot, P. J. Steinhardt and M. S. Turner, *Phys. Rev. Lett.* **69** (1992) 1856.
122. G. Hinshaw *et al.*, *Astrophys. J. Suppl.* **180** (2009) 225.
123. J. Dunkley *et al.*, *Astrophys. J. Suppl.* **180** (2009) 306.
124. E. Komatsu *et al.*, *Astrophys. J. Suppl.* **180** (2009) 330.
125. S. Naess *et al.*, *J. Cosmol Astropart. Phys.* **10** (2014) 007, arXiv:1405.5524.
126. A. A. Starobinsky, *Phys. Lett. B* **91** (1980) 99.
127. A. H. Guth, *Phys. Rev. D* **23** (1981) 347.
128. K. Sato, *Mon. Not. R. Astron. Soc.* **195** (1981) 467.
129. A. Albrecht and P. J. Steinhardt, *Phys. Rev. Lett.***48** (1982) 1220.
130. A. D. Linde, *Phys. Lett. B***108** (1982) 389.
131. A. A. Starobinsky, *JETP Lett.* **30** (1979) 682.
132. V. F. Mukhanov and G. V. Chibisov, *JETP Lett.* **33** (1981) 532.
133. S. W. Hawking, *Phys. Lett. B***115** (1982) 295.
134. A. A. Starobinsky, *Phys. Lett. B***117** (1982) 175.
135. J. M. Bardeen, P. J. Steinhardt and M. S. Turner, *Phys. Rev. D***28** (1983) 679.
136. A. H. Guth and S.-Y. Pi, *Phys. Rev. D***32** (1985) 1899.
137. V. F. Mukhanov, *JETP Lett.* **41** (1985) 493.
138. BICEP2/Keck and Planck Collab., *Phys. Rev. Lett.* **114** (2015) 101301.

139. N. W. Halverson, *Physics* **8** (2015) 21.
140. Planck Collab. XVI, *Astron. Astrophys.* **571** (2014) A16.
141. M. S. Turner, *Phys. Rev. D* **55** (1997) R435.
142. BICEP2 Collab. (P. A. R. Ade *et al.*), *Phys. Rev. Lett.* **112** (2014) 241101.
143. Planck Collab. (R. Adam *et al.*), Planck intermediate results. XXX. The angular power spectrum of polarized dust emission at intermediate and high Galactic latitudes, accepted for publication in *Astron. Astrophys.* (2014), http://dx.doi.org/10.1051/0004-6361/201425034; arXiv:1409.5738.
144. W.-T. Ni, Equivalence principles, spacetime structure and the cosmic connection, in *One Hundred Years of General Relativity: From Genesis and Empirical Foundations to Gravitational Waves, Cosmology and Quantum Gravity*, Chap. 5, ed. W.-T. Ni (World Scientific, Singapore, 2015); *Int. J. Mod. Phys. D* **25** (2016) 1630002.
145. S. di Serego Alighieri, Cosmic polarization rotation: An astrophysical test of fundamental physics, in *One Hundred Years of General Relativity: From Genesis and Empirical Foundations to Gravitational Waves, Cosmology and Quantum Gravity*, Chap. 16, ed. W.-T. Ni (World Scientific, Singapore, 2015); *Int. J. Mod. Phys. D* **24** (2015) 1530016.
146. H.-H. Mei, W.-T. Ni, W.-P. Pan, L. Xu and S. di Serego Alighieri, *Astrophys. J.* **805** (2015) 107.
147. K. S. Thorne, Gravitational radiation, in *300 Years of Gravitation*, eds. S. W. Hawking and W. Israel (Cambridge University Press, 1987), pp. 330–458.
148. E. Fonseca *et al.*, *Astrophys. J.* **787** (2014) 82.
149. M. Kramers *et al.*, *Science* **314** (2006) 97.
150. R. D. Ferdman *et al.*, *Mon. Not. R. Astron. Soc.* **443** (2014) 2183.
151. J. van Leeuwen *et al.*, arXiv:1411.1518v1[astro-ph.SR].
152. S. B. Anderson *et al.*, *Nature* **346** (1992) 42.
153. C. Kim *et al.*, *New Astron. Rev.* **54** (2010) 148.
154. M. Dominik, *Astrophys J.* **759** (2012) 52.
155. Y.-B. Bae *et al.*, *Mon. Not. R. Astron. Soc.* **440** (2014) 2714.
156. A. J. Farmer and E. S. Phinney, *Mon. Not. Roy. Astron. Soc.* **346** (2003) 1197.
157. J. Abadie *et al.*, *Class. Quantum Grav.* **27** (2010) 173001.
158. H. Dimmelmeier, J. A. Font and E. Müller, *Astron. Astrophys.* **388** (2002) 917; **393** (2002) 523.
159. K. Kotake, W. Iwakami, N. Ohnishi and S. Yamada, *Astrophys J.* **697** (2009) L133.
160. K. Kotake, *J. Phys., Conf. Ser.* **229** (2010) 012011.
161. K. Kotake, *Comptes Rendus Phys.* **14** (2013) 318.
162. S. Ando, J. F. Beacom and H. Yüksel, *Phys. Rev. Lett.* **95** (2005) 171101.
163. J. Abadie *et al.*, *Phys. Rev. D* **85** (2012) 122007.
164. E. Treister *et al.*, *Science* **328** (2010) 600.
165. J. Primack, *Science* **328** (2010) 576.
166. A. Sesana, A. Vecchio and C. N. Colacino, *Mon. Not. R. Astron. Soc.* **390** (2008) 192.
167. A. Sesana, A. Vecchio and M. Volonteri, *Mon. Not. R. Astron. Soc.* **394** (2009) 2255.
168. R. Manchester, *Int. J. Mod. Phys. D* **22** (2013) 1341007.
169. P. Amaro-Seoane *et al.*, *Class. Quantum Grav.* **29** (2012) 124016.
170. A. Sesana, J. R. Gair, E. Berti and M. Volonteri, *Phys. Rev. D* **83** (2011) 044036.
171. K. Sato and J. Yokoyama, Inflationary cosmology: First 30+ years, in *One Hundred Years of General Relativity: From Genesis and Empirical Foundations to Gravitational Waves, Cosmology and Quantum Gravity*, Chap. 18, ed. W.-T. Ni (World Scientific, Singapore, 2015); *Int. J. Mod. Phys. D* **24** (2015) 1530025.

172. G. S. Bisnovatyi-Kogan and V. N. Rudenko, *Class. Quantum Grav.* **21** (2004) 3347.
173. L. P. Grishchuk, *JETP Lett.* **23** (1976) 293.
174. S. S. Seahra, C. Clarkson and R. Maartens, *Phys. Rev. Lett.* **94** (2005) 121302.
175. C. Clarkson and S. S. Seahra, *Class. Quantum Grav.* **24** (2007) F33.
176. M. Servin and G. Brodin, *Phys. Rev. D* **68** (2003) 044017.
177. M. Giovannini, *Phys. Rev. D* **60** (1999) 123511.
178. M. Gasperini and G. Veneziano, CERN-TH/2002-104 (2002).
179. J. Garcia-Bellido and D. G. Figueroa, *Phys. Rev. Lett.* **98** (2007) 061302.
180. R. Easther, J. T. Giblin and E. A. Lim, *Phys. Rev. Lett.* **99** (2007) 2213013.
181. C. Caprini, R. Durrer, T. Konstandin and G. Servant, *Phys. Rev. D* **79** (2009) 083519.
182. E. J. Copeland, D. J. Mulryne, N. J. Nunes and M. Shaeri, *Phys. Rev. D* **79** (2009) 023508.
183. R. R. Caldwell, R. A. Battye and E. P. S. Shellard, *Phys. Rev. D* **54** (1996) 7146.
184. L. Leblond, B. Shlaer and X. Siemens, *Phys. Rev. D* **79** (2009) 123519.
185. D. Chernoff and H. Tye, Inflation, string theory and cosmic strings, in *One Hundred Years of General Relativity: From Genesis and Empirical Foundations to Gravitational Waves, Cosmology and Quantum Gravity*, Chap. 19, ed. W.-T. Ni (World Scientific, Singapore, 2015); *Int. J. Mod. Phys. D* **24** (2015) 1530010.
186. J. J. Geng, Y. F. Huang and T. Lu, *Astrophys. J.* **804** (2015) 21.
187. S. M. Kopeikin, *Phys. Rev. D* **56** (1997) 4455.
188. M. S. Pshirkov, *Mon. Not. R. Astron. Soc.* **398** (2009) 1932.
189. J. A. Ellis, M. A. McLaughlin and J. P. W. Verbiest, *Mon. Not. R. Astron. Soc.* **417** (2011) 2318.

Chapter 11

Ground-based gravitational-wave detectors

Kazuaki Kuroda*

Institute for Cosmic Ray Research, The University of Tokyo, 5-1-5 Kashiwanoha, Kashiwa, Chiba 277-8582, Japan

kuroda@icrr.u-tokyo.ac.jp

Gravitational wave is predicted by Einstein's general relativity, which conveys the information of source objects in the universe. The detection of the gravitational wave is the direct test of the theory and will be used as new tool to investigate dynamical nature of the universe. However, the effect of the gravitational wave is too tiny to be easily detected. From the first attempt utilizing resonant antenna in the 1960s, efforts of improving antenna sensitivity were continued by applying cryogenic techniques until approaching the quantum limit of sensitivity. However, by the year 2000, resonant antenna had given the way to interferometers. Large projects involving interferometers started in the 1990s, and achieved successful operations by 2010 with an accumulated extensive number of technical inventions and improvements. In this memorial year 2015, we enter the new phase of gravitational-wave detection by the forthcoming operation of the second-generation interferometers. The main focus in this paper is on how advanced techniques have been developed step by step according to scaling the arm length of the interferometer up and the history of fighting against technical noise, thermal noise, and quantum noise is presented along with the current projects, LIGO, Virgo, GEO-HF and KAGRA.

Keywords: Gravitational wave; detection; ground based.

PACS Number: 04.80.Nn

1. Introduction to Ground-Based Gravitational-Wave Detectors

The gravitational wave is predicted by Einstein's general relativity, which is produced by dynamic acceleration of celestial objects. The gravitational wave propagates in a speed of light with inducing spacetime distortion around. The detection of gravitational wave is the direct test of the theory of general relativity and becomes a new tool of astronomy to observe dynamic nature of the universe. However, the effect of gravitational wave is too tiny to be easily detected. The first experiment was attempted by J. Weber with his resonant-type antennas placed at Argonne National Laboratory and at the University of Maryland. He declared

*Physics and Engineering Research Company, 3-6-12 Komatsu, Tsuchiura 300-0823, Japan.

that he had succeeded in the detection of gravitational wave coming from the center of our Galaxy.[1] Although the detection was discredited by several independent experiments based on a pair of antennas,[2–6] this opened an era of experimental astrophysics of gravitational waves. Another type of gravitational-wave detection on Earth is the laser interferometric detector.

Full expertise in several discipline of experimental physics is required in order to realize an advanced interferometers. In order to give the reader an idea of the deep knowledge and variety of arguments required, the author tried to include selected R&D items that were confirmed by experimental facts.

In this paper, the detection of gravitational waves is reviewed for both a resonant antenna and a laser interferometer after a brief summary of current status of gravitational-wave detection.

1.1. *Gravitational-wave sources*

Gravitational-wave sources evolves along with the advancement of R&D of gravitational-wave detector. In the era of resonant antennas, a typical source of gravitational wave is supernova explosion and the signal amplitude was roughly estimated from the energy balance between released gravitational-wave energy and the total energy of the gravitational wave propagating in the universe. In the era of laser interferometer that has wider frequency-band of observation, main target of the gravitational-wave source is the coalescence of compact binary stars. We are now in the stage where the second-generation detectors initiate their operations and are making designs for third-generation interferometers that have sensitivity better by one order than the second-generation ones, which can explore much more abundant gravitational-wave sources, which are summarized in Ref. 7.

Currently achieved sensitivities of large projects for gravitational-wave detection are introduced and gravitational-wave sources are presented in this subsection.

1.1.1. *Achieved sensitivities of large projects*

The first-generation laser interferometers are realized in the world as LIGO Hanford (1) and (2), LIGO Livingston,[8] Virgo,[9] GEO,[10] and TAMA.[11] They have achieved design sensitivities by the mid of 2000s. Depending on their baseline length, the highest sensitivity is attained by LIGO 4 km interferometer and the most remote target of neutron star binaries is located at 50 Mpc away. The operation of these first-generation detectors ended by cooperative observation in 2010.[12] We have no report of the detection of gravitational waves by this observation. Considering the event rate of neutron star binaries, we have to let these detectors run for more than 100 years on average. If we assume the coalescence of black hole binaries of masses of $10M_\odot$, we can limit its birth rate down to 6×10^{-6}/year by this observation.

The second-generation detectors are designed to achieve more sensitivity and they are being constructed. They are advanced LIGO (Hanford and Livingston),[13] advanced Virgo,[14] GEO HF,[15] and KAGRA.[16] The interferometer of LIGO-India may be initiated to be constructed soon. If we achieve the first detection of gravitational wave, the astronomy of gravitational waves will start. At this phase, we have to make the frequency band wider for more detections of various types of gravitational-wave sources. Especially, increased the sensitivity of lower frequency will be achieved by a cryogenic interferometer placed underground. This kind of detectors is called the third-generation interferometer. KAGRA will be positioned between the second and the third-generation due to the adoption of cryogenic mirror and underground location.

European scientists plan to construct more sensitive gravitational-wave detector, Einstein Telescope, the design study of which was finished in 2011.[17]

The achieved sensitivities of laser interferometers used for observation are plotted by broken line curves and the sensitivities of the second- and third-generation detectors are shown by solid line curves in Fig. 1. Note that curves show the value of noise spectrum multiplied by square root of frequency, which makes easy to compare the characteristic amplitudes of expected gravitational wave signals from various sources except continuous gravitational waves.

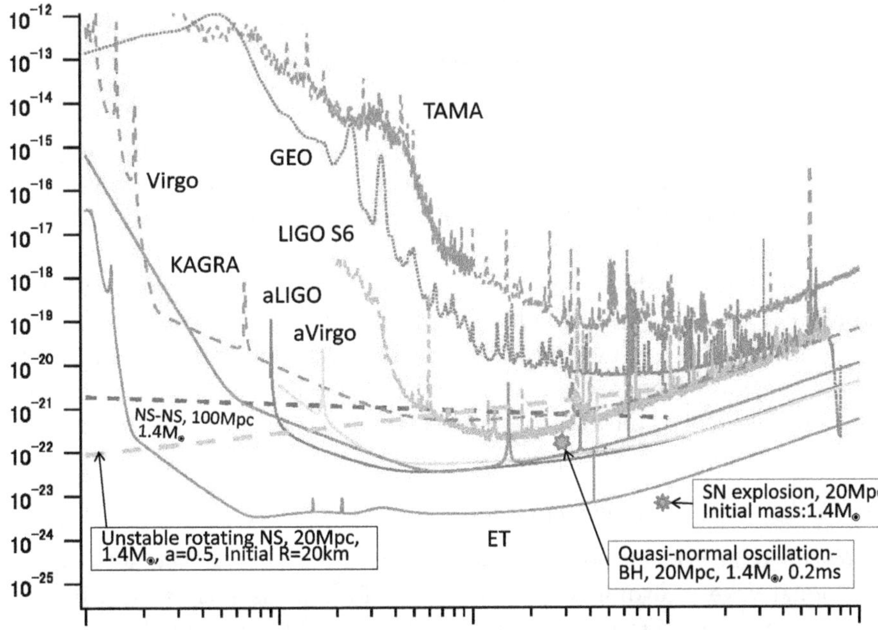

Fig. 1. The achieved sensitivities of laser interferometers, operated for observation, are plotted by broken line curves and the sensitivities of the second- and third-generation detectors are shown by solid curves (KAGRA: Broadband RSE, aLIGO:ZERO_DET_high_P, aVirgo:Wideband SR-tuning). Typical gravitational wave sources are also shown in the figure.

In this figure, examples of gravitational-wave sources, which can be possibly detected by the ground based interferometers, are shown. They are described as follows.

1.1.2. *Coalescences of binary neutron stars*

There are several neutron star binaries in our Galaxy and some of them coalesce due to the emission of gravitational wave within the age of the universe.[18] In a few minutes before the coalescence, chirping gravitational wave is emitted and its signal is the most probable target of ground based laser interferometers. The radiation of gravitational wave by the orbital motion of binary pulsars is described by the quadrupole approximation until merging start where two stars may be deformed by each tidal forces.[19] Although its orbital radius is difficult to be known due to the unknown merging radius of neutron stars because the state equation of neutron is not well known for extensively high density state of neutron stars. Therefore, the analytical calculation is done until the orbital radius approaches 6 times of the neutron-star radius (inspiral phase). The radiation at approaching closely and at merging (merger phase) can be only handled only by numerical calculation.[20]

$$h_c = 2\sqrt{f}|h(f)|$$
$$h(f) = \sqrt{\frac{5}{24\pi^3}\frac{Qc}{D}} f^{-\frac{7}{6}} \left(\frac{G\pi M_{\rm ch}}{c^3}\right)^{\frac{5}{6}}, \tag{1}$$

where Q is a factor representing the dependence of both inclination angles to the source position and the inclination angle to the direction of the orbital plane of the binary stars, D is the distance from the source to the Earth, and $M_{\rm ch}$ is the chirp mass calculated by $\mu^{\frac{3}{5}}M^{\frac{2}{5}}$, where μ is the reduce mass of the binary, $\frac{m_1 m_2}{M}$, M is the total mass of the binary, $M = m_1 + m_2$. G is the Newtonian gravitational constant, and c is the speed of light. The case satisfying $D = 100\,{\rm Mpc}$, $Q = 1$ and for $m_1 = m_2 = 1.4\,M_\odot$, where $M_\odot$ is the mass of the sun, is plotted in Fig. 1.

The estimated event rate ranges more than 2 orders of magnitude and typical value is once per 100 thousands years in such matured galaxy as ours.[21,22]

1.1.3. *Coalescences of binary black holes*

If there is a binary black holes that are rotating each other, the system radiates gravitational wave according to the quadrupole formula as long as the orbital radius is fairly larger than the radius of each event horizon of the black hole. As in the case of a binary neutron stars, the system gradually loses its dynamical energy due to the radiation of gravitational wave and its orbit radius decreases until its merger. A numerical analysis of coalescence of binary black holes with spin is calculated by numerical method.[23] Since the population of binary black holes is estimated to be smaller than that of neutron star binaries, the coalescence rate of the coalescence of black hole binaries is smaller than that of binary neutron stars.[24] However, a theoretical study shows that merger rate of black holes ejected from globular clusters is larger than that of neutron star binaries.[25] Moreover, since the amplitude of gravitational waves from the coalescence of black holes is larger, possible detection rate will be larger if the detector has sensitivity at lower frequencies less than 10 Hz, which will be realized by the third-generation detectors.

1.1.4. *Supernova explosion*

Massive stars heavier than $8\,M_\odot$ collapse due to gravity after burning out and a neutron star may be born. This collapse produces burst gravitational wave. Since supernova explosion is a complex physical phenomena involving general relativity, hydrodynamics of nuclear density, neutrino transport, and thermonuclear kinetics, there is no established scenario widely accepted. The magnitude of the gravitational wave from stellar core collapse was estimated in order by second time-derivative of the quadrupole moment of the core. However, the wave form of the gravitational wave is useful to enhance the signal-to-noise ratio of the detection. By numerical simulation, first wave form catalogue was calculated on rapidly rotating star considering general relativity effect in the collapse.[26] Recent development in three-dimensional numerical simulation which requires higher computing power shows that stronger burst wave by one order than that in the first catalogue is produced due to nonaxisymmetric dynamical instabilities such as rapidly spinning bar-like core[27] and standing accretion shock instability.[28] In any case, expected maximum amplitude $h_{\max}$[29] is calculated by assuming that the magnitude of second derivative of quadrupole moment, $\kappa M R^2 (2\pi f)^2$, where κ is 0.1 for interesting case, M is the mass of the initial neutron star born just after the collapse, $1.4 M_\odot$, and R is the radius of the star, 20 km. The source distance from the Earth is taken as 20 Mpc. Also, the factor 0.2 is assumed to show the departure from the symmetric figure of a sphere represented by δ_I which is 0 for spherically symmetric collapse.

$$h_{\max} \sim 1 \times 10^{-23} \left(\frac{20\,\mathrm{Mpc}}{D}\right)\left(\frac{\kappa}{0.1}\right)\left(\frac{\delta_I}{0.2}\right)\left(\frac{M}{1.4\,M_\odot}\right)\left(\frac{R}{20\,\mathrm{km}}\right)^2\left(\frac{f}{1\,\mathrm{kHz}}\right)^2 . \tag{2}$$

The numerical value is plotted in Fig. 1.

This may be the plausible candidate for the second-generation ground based detectors. The events rate of supernova explosions are estimated as once per a few ten years in our Galaxy.

1.1.5. *Quasi-normal mode oscillation at the birth of black hole*

After the merger of compact star binary, born black hole vibrates and its vibration decays due to energy release by gravitational wave, which is called as ring-down phase. In this phase, expected maximum amplitude assuming the energy released $\Delta E = \epsilon M_\odot c^2$ with $\epsilon = 10^{-6}$ during a time scale of $t_d = 0.2\,\mathrm{ms}$ at $D = 20\,\mathrm{Mpc}$

away is[29]

$$h_{\rm max} \sim 9.4 \times 10^{-23} \left(\frac{20\,{\rm Mpc}}{D}\right) \left(\frac{\epsilon}{10^{-6}}\right)^{\frac{1}{2}} \left(\frac{M}{1.4 M_\odot}\right)^{\frac{1}{2}} \left(\frac{f}{1\,{\rm kHz}}\right)^{-1} \left(\frac{f_d}{1{\rm ms}}\right)^{\frac{1}{2}}. \tag{3}$$

The observation of this oscillation also gives us the understanding about Kerr space-time geometry, if the initial state of the binary has a large angular momentum.

1.1.6. *Unstable fast rotating neutron star*

Angular momentum of the initial massive stars is taken over by new born neutron star, which is possibly a rapidly rotating spheroid with differential rotation speeds between inner sphere and outer sphere. This is unstable system and effectively radiate gravitational wave that damps the rotation speed. The emission of gravitational wave is described by the following formula[29] that is plotted in Fig. 1:

$$h_{\rm eff} \sim \frac{R}{D}\sqrt{\frac{GMf}{c^3}} \sim 1 \times 10^{-21} \left(\frac{20\,{\rm Mpc}}{D}\right) \left(\frac{R}{20\,{\rm km}}\right) \left(\frac{M}{1.4\,M_\odot}\right)^{\frac{1}{2}} \left(\frac{f}{100\,{\rm Hz}}\right)^{\frac{1}{2}}. \tag{4}$$

More detailed description about the inspiral, merger and ring-down of a coalescence of compact binary systems and references for other gravitational-wave sources are found in relevant chapters in the series of this publication.

1.2. *Acceleration due to a gravitational wave*

The detection principle of a gravitational wave is based on excitation due to the tidal force induced by a gravitational wave represented by a metric perturbation. In resonant antennae, the resonant vibration modes of an elastic body will be excited, but in a laser interferometer, the geodesic motions of freely suspended test masses are deviated.[30]

The metric perturbation by a gravitational wave propagating along the z-direction is described in TT gauge by

$$ds^2 = -c^2dt^2 + (1 + h_{xx}^{\rm TT})dx^2 + 2h_{xy}^{\rm TT}dxdy + (1 + h_{yy}^{\rm TT})dy^2 + dz^2, \tag{5}$$

where

$$h_{xx}^{\rm TT} = -h_{yy}^{\rm TT} = A_+(\omega t - kz), \tag{6}$$

$$h_{xy}^{\rm TT} = h_{yx}^{\rm TT} = A_\times(\omega t - kz), \tag{7}$$

are used, and ω is the angular frequency and k is the wave number of the wave. By using these metric perturbations, we can calculate the component of the Riemann tensor as

$$R_{x0x0} = -R_{y0y0} = -\frac{1}{2c^2}\ddot{A}_+(\omega t - kz), \tag{8}$$

$$R_{x0y0} = R_{y0x0} = -\frac{1}{2c^2}\ddot{A}_\times(\omega t - kz), \tag{9}$$

where double dots stands for second derivative with respect to t. In the detector coordinates system, which is the proper reference frame of the experimental room, the wave induces the acceleration of a unit mass relative to the center of mass of the detector as

$$\begin{aligned}\left(\frac{d^2x}{dt^2}\right) &= -R_{x0x0}x - R_{x0y0}y,\\ &= -\frac{1}{2c^2}(\ddot{A}_+x + \ddot{A}_\times y),\end{aligned} \tag{10}$$

$$\begin{aligned}\left(\frac{d^2y}{dt^2}\right) &= -R_{y0y0}y - R_{y0x0}x,\\ &= -\frac{1}{2c^2}(-\ddot{A}_+y + \ddot{A}_\times x),\end{aligned} \tag{11}$$

$$\left(\frac{d^2z}{dt^2}\right) = 0. \tag{12}$$

The energy carried by the wave can be represented by the tensor of the energy–momentum

$$T_{00} = T_{zz} = -T_{0z} = \frac{c^2}{16\pi G}(\dot{A}_+^2 + \dot{A}_\times^2). \tag{13}$$

The force field produced by a gravitational wave is sketched by the force lines as in Fig. 2.

The force caused by the acceleration is normal to the wave propagation direction. Let us assume for simplicity that there are two point masses. If these masses are bound by some elastic body, the acceleration between two masses causes stress inside this body. However, if two masses are free, the acceleration changes the distance between them. Contrary to this picture involving the proper reference frame of the detector, the acceleration in the TT gauge is given by a first-order approximation

$$\left(\frac{d^2x^\alpha}{dt^2}\right) = -c^2\Gamma^\alpha_{00} = 0, \tag{14}$$

where Γ represents the Christoffel symbol. The coordinate value of the mass does not change in the TT gauge even if a gravitational wave passes. Indeed, the proper length in the TT gauge, which is converted from the baseline length between two point masses in the experimental room, does not change due to a gravitational wave in the first-order approximation of the metric perturbation.

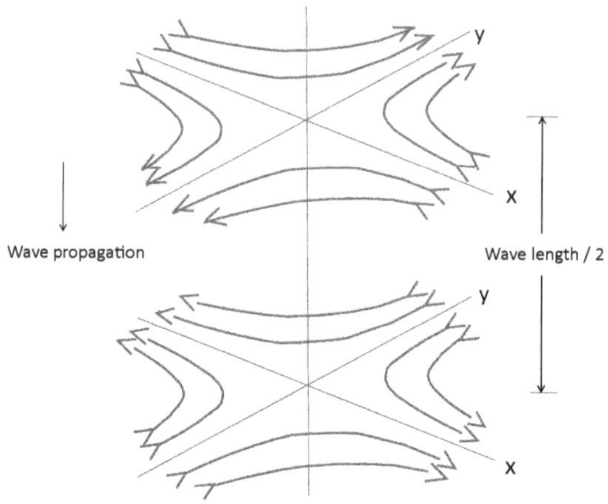

Fig. 2. Gravitational wave propagating along the z-axis which causes acceleration of a mass relative to the center of mass of the detector. The picture shows the force field by "+" polarization. The force field by "×" polarization is obtained by rotating the force field by 45 degrees around the z-axis.

1.3. *Response of a resonant antenna*

A resonant antenna is an elastic body whose mechanical resonance is excited by the stress due to gravitational waves. The response of the detector depends on the shape, rigidity and material density of the antenna.[31] The shape of the antenna developed by J. Weber was a cylindrical bar.[32] Cryogenic bar antennae were developed by various research groups around the world. An analytical calculation for the bar-type antenna was given in the form of the absorption cross-section by Paik and Wagoner.[33] Here, the author introduces an analysis by Hirakawa *et al.*[34] The strain of the antenna body is described by a field of a displacement vector, u_α, where α specifies one of components in three-dimensional coordinates. If a gravitational wave impinges on the antenna, the equation of motion of the antenna is represented by

$$\rho\ddot{u}_\alpha - \mu\Delta u_\alpha - (\lambda+\mu)(\nabla\cdot\mathbf{u})_{,\alpha} = \frac{1}{2}\sum_\beta \ddot{h}_{\alpha\beta}x_\beta, \tag{15}$$

where ρ is the density, and both μ and λ are the Lame's elasticity coefficients of the antenna material. We assume that the material is isotropic. In this case, Young's modulus is given by $\mu(3\lambda+2\mu)/(\lambda+\mu)$ and the Poisson ratio is by $\lambda/[2(\lambda+\mu)]$. According to the above equation, since $u(x,y,z,t)$ can be expanded by $\sum_n c_n(t)w_n(x,y,z)$, the amplitude, c_n, of the nth mode, satisfies the following equation considering the internal mechanical loss, $1/Q_n$:

$$\ddot{c}_n + \frac{\omega_n}{Q_n}\dot{c}_n + \omega^2 c_n = \frac{\sum_{\alpha\beta}\ddot{h}_{\alpha\beta}q_{n\alpha\beta}}{\left(4\int\rho\sum_\alpha w_{n\alpha}^2 dV\right)}, \tag{16}$$

$$q_{n\alpha\beta} = \int \rho \left(w_{n\alpha} x_\beta + w_{n\beta} x_\alpha - \frac{2}{3} \delta_{\alpha\beta} \sum_\gamma w_{n\gamma} x_\gamma \right) dV, \tag{17}$$

where ω_n is the nth eigen-mode angular frequency. The equilibrium energy of the antenna at the nth resonance is given by

$$E = \frac{\frac{Q_n^2 \omega_n^2}{16} \left\langle \left(\sum_{\alpha\beta} h_{\alpha\beta} q_{n\alpha\beta} \right)^2 \right\rangle}{\int \rho \sum_\alpha w_{n\alpha}^2 dV}. \tag{18}$$

This is the saturated vibration energy due to a monochromatic continuous gravitational wave. For burst wave signals, the energy deposited in the antenna is effectively described by the antenna cross-section and the antenna directivity pattern.[35] When a burst of un-polarized wave propagating in the direction $\mathbf{n}(n_x, n_y, n_z)$ impinges on the antenna, the deposited energy is

$$\begin{aligned} E_{\text{deposited}} &= \frac{1}{2} \int \rho (|\dot{\mathbf{u}}|^2 + \omega_0^2 |\mathbf{u}|^2) dV, \\ &= \frac{\pi^3 G}{5c^3} M \nu^2 A_G \Theta(n_x, n_y, n_z) F(\nu), \end{aligned} \tag{19}$$

where M is the mass of the antenna, ν is the frequency of the gravitational-wave, $F(\nu)$ is the energy spectrum density of the burst wave, and A_G is the antenna cross section, given by

$$A_G = \frac{2 \sum q_{\alpha\beta}^2}{M \int \rho |\mathbf{w}|^2 dV}. \tag{20}$$

Also, Θ is the antenna directivity pattern, calculated by

$$\Theta(n_x, n_y, n_z) = \frac{\frac{1}{4} \left(\sum q_{\alpha\beta} n_\alpha n_\beta \right)^2 + \frac{1}{2} \sum q_{\alpha\beta}^2 - \sum q_{\alpha\beta} q_{\alpha\gamma} n_\beta n_\gamma}{\frac{1}{5} \sum q_{\alpha\beta}^2}. \tag{21}$$

Narihara[35] gave numerical values concerning a square antenna, as sketched in Fig. 3, where the mass is M=1400 kg, the length is equal to the width, 1.65 m, and the thickness is 0.14 m. A_G is 0.77 m^2 for an isotropic source

$$E_{\text{deposited}} = 3.5 \times 10^{-28} F(\nu_0). \tag{22}$$

The directivity pattern of the antenna is calculated in polar coordinates

$$\Theta(\theta, \phi) = \frac{5}{2} - \frac{5}{2} \sin^2 \theta + \frac{5}{8} \sin^4 \theta \cos^2 \phi. \tag{23}$$

The response by a disk antenna was similarly given by Paik.[36]

In the late 1990s, the merit of pursuing a spherical antenna was emphasized and practically analyzed by Johnson and Merkowitz,[37] although the spherical antenna's

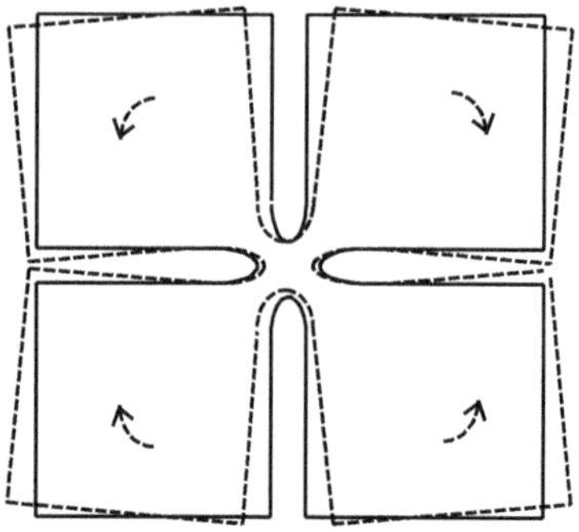

Fig. 3. An deformation image of the square-type antenna that was developed by Hirakawa *et al.*

enhanced gravitational cross-section was recognized in the early 1970s with its ability to measure signal direction and the polarization.[38] Although the bar antenna can detect only one quadrupole mode, the sphere has five degenerate ones that interact strongly with a gravitational wave. Each modes act as a separate antenna, being oriented toward a different polarization or direction. This merit had already been mentioned by Wagoner and Paik,[39] where the improvement in the cross-section is about a factor of 60 compared to a bar with the same quadrupole mode frequency and a typical length/diameter of 4.2.

Prototypes of this sphere detector were made at both LSU and Leiden University (MiniGRAIL as shown in Fig. 4),[40] and its development by Schenberg produced a

Fig. 4. Prototype sphere detector, MiniGRAIL in Leiden University.
Source: Photo is taken from http://www.minigrail.nl/.

spherical detector in Brazil.[41] All of the above resonant detectors were expected to catch burst waves.

1.4. *Response of an interferometer*

It is easier to consider metric perturbation when a laser interferometer responds to a gravitational wave. The basic interferometric detector consists of three main optical components that form the so-called Michelson interferometer, as shown in Fig. 5.

The beam-splitter splits the light beam to both "arm" directions of the x-axis and y-axis. The reflected beams are also split at a beam splitter, and the beam combined toward the output photo-detector experiences interference reflecting the phase shift due to the path difference. For simplicity, a wave of "+" polarization is considered here. During passing of the gravitational wave, the metric perturbation causes a change in the speed of light travelling from the beam splitter to the mirror, and the returning process. Since the light propagation is represented by $ds^2 = 0$, the velocity, c_x, along the x-axis is calculated from Eq. (5) as

$$c_x = c(1 + h_{xx}^{\mathrm{TT}})^{-\frac{1}{2}}. \tag{24}$$

In a similar manner the velocity, c_y, along y-axis is given by

$$c_y = c(1 + h_{yy}^{\mathrm{TT}})^{-\frac{1}{2}}. \tag{25}$$

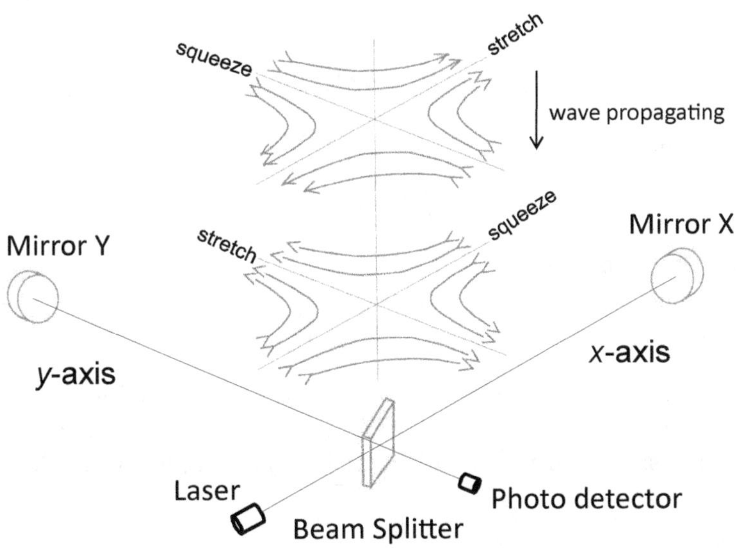

Fig. 5. Michelson interferometer consists of three main components; beam splitter, a mirror on the x-axis and a mirror on the y-axis. Gravitational wave comes from the zenith with "+" polarization in accordance with x- and y-axis directions. In TT-gauge description, the apparent speeds of light in the interferometer arms alternatively change according to the wave propagation, which creates interfered phase shift at the combined beams at the beam splitter.

If the arm lengths (distances between the beam splitter and the x-axis mirror, or the y-axis mirror) are the same in the proper reference frame of the Michelson interferometer, the coordinate positions in the TT gauge corresponding to these positions do not change due to a gravitational wave. This means that the velocity difference experiences a phase shift, $\Delta\phi_{GR}$, at the point of the interfered light at the beam splitter, which is easily obtained by

$$\Delta\phi_{GR} = \frac{2\pi}{\lambda}(2L_0 h), \tag{26}$$

where λ is the wave length of the light, L_0 is the arm length, and h is the amplitude of the gravitational wave ($h = h_{xx}^{\mathrm{TT}} = -h_{yy}^{\mathrm{TT}}$). This phase shift is detected by a photo detector catching the output beam from the beam splitter.

The response due to a gravitational wave was first evaluated by Russian scientists[42] and was experimentally tested by Forward.[43] Initially, there has been a debate about how interferometers respond to gravitational waves.[44]

The above signal output of the interferometer is only valid when the frequency change of the gravitational wave is slower than travel time of the light inside the arms. In order to correctly handle the frequency spectrum of h, we assume that $h(t) = \int h(\omega)e^{i\omega t}d\omega$. Using Eq. (26)

$$\Delta\phi_{GR}(t) = \frac{\Omega}{2}\int\left(h(\omega)\int_{t-2\ell/c}^{t} e^{i\omega t'}dt'\right)d\omega \tag{27}$$

$$= \frac{\Omega}{2}\int h(\omega)\frac{1}{i\omega}[e^{i\omega t} - e^{i\omega(t-2\ell/c)}]d\omega \tag{28}$$

$$= \Omega\int h(\omega)e^{i\omega t}\frac{1-e^{-2i\omega\ell/c}}{2i\omega}d\omega, \tag{29}$$

which is represented by

$$\Delta\phi_{GR}(t) = \int h(\omega)e^{i\omega t}H_M(\omega)d\omega. \tag{30}$$

The response function becomes

$$H_M(\omega) = \frac{\Omega}{\omega}\sin\left(\frac{\ell\omega}{c}\right)e^{-i\omega\ell/c}, \tag{31}$$

which takes maximum value for $\frac{\ell\omega_0}{c} = \pi/2$. For a gravitational wave of 1 kHz, the optimum is obtained by $\ell = 75$ km.[43] The frequency response of a simple Michelson interferometer is shown for the two different baseline lengths in Fig. 6.

1.4.1. *Directivity*

In the previous section, we obtained the response of a Michelson interferometer to an incoming gravitational-wave travelling as shown in Fig. 5, the direction of which gives the maximum signal. In general, the direction of the incoming gravitational-wave is not necessarily aligned along the optimum direction of the interferometer.

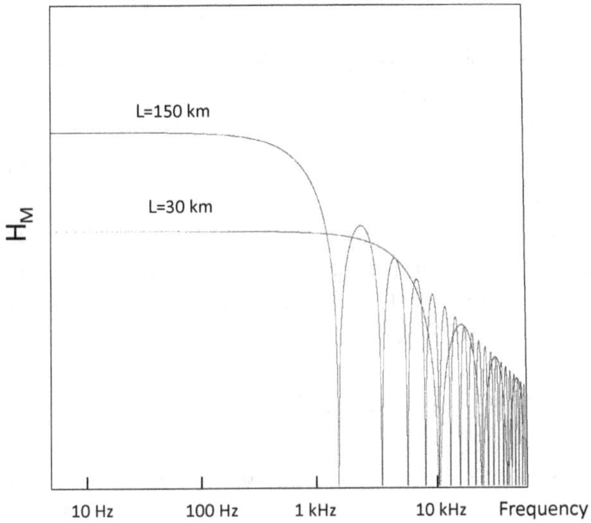

Fig. 6. Frequency response of a simple Michelson interferometer.

The directivity sensitivity of the interferometer is obtained by

$$2h_{11}\left[\frac{1+\cos^2\theta}{2}\cos 2\phi\cos 2\psi - \cos\theta\sin 2\phi\sin 2\psi\right]$$
$$-2h_{12}\left[\frac{1+\cos^2\theta}{2}\cos 2\phi\sin 2\psi + \cos\theta\sin 2\phi\cos 2\psi\right], \tag{32}$$

where θ and ϕ are the direction angles toward the source of gravitational wave and ψ represents the angle of the polarization of the gravitational wave. h_{ij} is defined by the following equations:

$$h_{ij} = \begin{pmatrix} 0 & 0 & 0 & 0 \\ 0 & h_{11} & h_{12} & 0 \\ 0 & h_{21} & h_{22} & 0 \\ 0 & 0 & 0 & 0 \end{pmatrix} = \begin{pmatrix} 0 & 0 & 0 & 0 \\ 0 & h_{11} & h_{12} & 0 \\ 0 & h_{21} & -h_{11} & 0 \\ 0 & 0 & 0 & 0 \end{pmatrix}. \tag{33}$$

If gravitational-wave events occur uniformly in the universe, and there is no bias of the orbital direction, we can calculate the average detection rate by an interferometer located on the Earth using the above result. Since the squared average of both ϕ and θ becomes $\frac{2}{5}$, the average in total is $\frac{1}{\sqrt{5}}$ considering the squared average of the polarizations ($\frac{1}{2}$). Hence, if gravitational-wave sources are binary coalescences, we have to consider the direction of the binary orbit, which is practically calculated by allocating a random direction of the orbit in a computer simulation. Considering all the above, the sensitivity becomes 0.44-times the optimal one.

1.4.2. *Positioning*

Since gravitational waves pass across the Earth, the detectors sense from both directions, from the sky and the underground. The wave propagation line is determined by knowing the exact arrival times of three detectors and the direction is fixed by the fourth detector. However, since we can utilize information from the wave polarization, we need only three detectors in practice.

Three interferometers on Earth are assumed to be directed as $\mathbf{n}_1, \mathbf{n}_2$ and $\mathbf{n}_3$. Let us take a coordinate system x, y, z, where a gravitational wave is passing Earth along the z-axis. $\mathbf{n}_i$ coincides with the z'-axis of a coordinate system, where its x'-axis and y'-axis are along both arms of the detector i, respectively. The coordinate transformation of L, where rotating the coordinate system along the z'-axis in order to make x'-axis parallel to the xy-plane by ϕ_1, and rotating in order to make the z'-axis to be in accord with the z-axis along its rotated x'-axis by θ_1, and finally rotating by ψ_1 along the z-axis in order to make the x'-axis accord with the x-axis, becomes the inverse transformation from vector basis of the detector coordinate to the vector basis of the wave, where the wave propagation direction coincides with the z-axis. Shortly, for the space components,

$$h_{\alpha\beta} = \mathbf{h}(L\mathbf{e}_\alpha, L\mathbf{e}_\beta) = L^\dagger h_{\alpha\beta} L, \tag{34}$$

where L is a transformation tensor of the vector basis of the coordinates, the output signal of the interferometer $S_i(t)$ is proportional to $h_{11\text{detector}} - h_{22\text{detector}}$:

$$\begin{aligned} S_i(t) = &\left[\frac{1+\cos^2\theta_i}{2}\cos 2\phi_i \cos 2\psi_i - \cos\theta_i \sin 2\phi_i \sin 2\psi_i\right] h_{11}\left(\frac{t - R\cos\theta_i}{c}\right) \\ &- \left[\frac{1+\cos^2\theta_i}{2}\cos 2\phi_i \sin 2\psi_i + \cos\theta_i \sin 2\phi_i \cos 2\psi_i\right] h_{12}\left(\frac{t - R\cos\theta_i}{c}\right). \end{aligned} \tag{35}$$

Let us assume here that all detectors are correctly calibrated, and that the terms of h_{11} and h_{12} are all the same, except that its phase is delayed. In the above, the gravitational wave was assumed to propagate along the z-axis direction. However, if the direction is different by some amount, the directions of all detectors are different by the same amount. We take this discrepancy as an evaluation parameter. For a given value of this discrepancy, the coefficients of both h_{11} and h_{12} are fixed in the above equations of S_1 and S_2, and we can solve analytically or by fitting those equations of S_1 and S_2 for h_{11} and h_{12}. Also, solved values are set in the third equation of S_3. If the direction differs from the true value, the third equation may not hold. If the difference becomes minimum, the direction of the gravitational wave is the most probable. In order to efficiently obtain the result, we evaluate the equation:

$$\Delta \equiv (S_1 - [*])^2 + (S_2 - [*])^2 + (S_3 - [*])^2, \tag{36}$$

where $[*]$ is the iterated values corresponding to S_i. For example, we can apply a matched filter technique in this analysis. Each interferometer has directions to whom they are not sensitive, so if the source direction is close to this, the accuracy of determination may not be good. In this case, one more detector is desired to augment the observation network.

1.5. *Comparison of a resonant antenna and an interferometer*

The signal frequency bandwidth of resonant antennas is at most few tens of Hz, while even first-generation interferometric detectors is a bandwidth of several hundreds of Hz, which is sufficiently wider to cover the whole final phase of the coalescence of binary neutron stars (from several 10 Hz to several 100 Hz). Therefore, resonant antennae are only applicable to observe burst waves. However, before the 1980s, just a few researchers believed that laser interferometric gravitational-wave detectors could become dominant tool on the Earth, because the author considers that a large frequency bandwidth requires anti-vibration system with excellent performances. A thermal noise-limited resonant antenna was realized at cryogenic temperature in 1980, but the sensitivity approaching thermal noise by interferometers was much later. In 2005, when LIGO reached its design sensitivity, thermal noise of main optics was regarded to limit the sensitivity in mid-frequencies with a combination with other noise sources.[45] The researchers developing resonant antenna reached a deadlock after achieving thermal noise sensitivity, and had to devise both quantum noise evading theory[46] and technology.[47,48] In general the effort spent in developing the detector reliability was significantly higher concerning cryogenic resonant antennae than interferometric detectors until the construction of LIGO[8] started.

A comparison of a resonant antenna and an interferometer helps us to catch physical priority of the interferometer from the point of view of energy amplification.[49]

2. Resonant Antennae

The announcement by Weber was disproved by several groups, as stated in the previous section. All antennae were operated at room temperature, and almost all detectors adopted a piezo-electric transducer. Based on these experiments, Giffard studied the ultimate sensitivity limit of a resonant gravitational-wave antenna using a linear-motion detector[50]; that is, the sensitivity given by the minimum effective temperature is twice as much the noise temperature of the amplifier; back action evading and quantum nondemolition schemes were introduced later. The paper of Giffard accelerated the R&D on cryogenic resonant antennae, called second-generation antennae. Note that researchers who were developing resonant antennae used the minimum effective temperature for presenting the sensitivity. The sensitivity was shown by the unity signal-to-noise ratio. During the development of second-generation detectors, this custom had been widely used. Technically, in order to attain the minimum effective temperature of the antenna, impedance matching is needed between the impedance of the mechanical resonant body and that of the transducer system. Hence, reducing the effective temperature requires both cryogenics and impedance-matching that lead to multi-mode resonant antenna.

The development of this second-generation antennae had been conducted since the 1970s, and was almost completed by the 1990s, when large-scale laser interferometers were planned to be constructed. In the attempt to lower the effective temperature and to enhance their sensitivity, resonant detectors were cooled down to liquid helium temperature (4.2 K) and some of them even below (the super-fluid phase, 2.17 K).[51]

2.1. *Development of resonant antennae*

The cryogenic bar antenna system is sketched in Fig. 7.

First, two-mode cryogenic bar detector was constructed and operated at liquid-helium temperature at Stanford University in 1977. The vibration mode of the 680 kg aluminum bar was amplified by a small mass supported by a niobium diaphragm, the movement of which was sensed by flat superconducting pick-up coils facing its two sides, and converted into magnetic field detected by a magnetometer using an rf-superconducting quantum interference device (SQUID).[52] The average vibrational energy in the lowest longitudinal mode at 1315 Hz was consistent with the level of thermal noise at the antenna temperature. This became the basis of a 4800 kg cryogenic antenna,[53] which started its operation in 1980. Its resonant frequency was designed to be 830 Hz. This detector was damaged by a 1989 Earthquake and was shut down.

A torsional antenna was developed in Tokyo for low-frequency gravitational-waves.[54] High mechanical Q of aluminum alloy (5056) was originally found at cryogenic temperature by this research group.[55] A series of CRAB experiments were

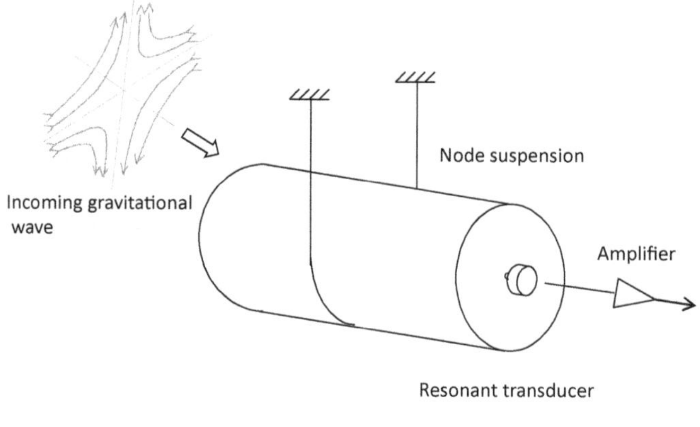

Fig. 7. Sketch representing key features of a resonant antenna system; the bar is suspended through the node of the lowest longitudinal oscillation mode, and a small mass connected to the end of the bar, where the resonant frequency is set to equal the bar frequency. The electromechanical transducer converts the displacement of the small mass to an electric signal. The whole system is in a vacuum chamber, and is cooled down to cryogenic temperature.

conducted at KEK (High Energy Research Organization) in Tsukuba, Japan, for observations of continuous waves at around 60 Hz from the crab pulsar. In the end, a 1.2 tons torsional antenna was developed.[56] Since the resonant frequency is quite lower than any other resonant antennas that were developed to detect burst waves from supernovas, the vibration isolation system required to be more sophisticated.

Explorer was developed by Rome group, and was first operated in 1986 (Fig. 8). The antenna was made of a high-Q alloy, Al-5056, and had a mass of $M = 2270$ kg, 3 m-long. Its fundamental mode frequency was around 900 Hz. A resonant capacitive transducer was mounted, which was followed by a dc-SQUID amplifier. The operation was made at $T \sim 2.6$ K in a cryostat cooled with super-fluid helium, which was able to remove acoustic noise due to the boiling of liquid helium. It was operated for observations for more than 10 years, since 1990 after achieving good duty cycle.[57]

ALLEGRO was a cryogenic bar detector at Louisiana State University.[58] The bar was a cylinder of aluminum alloy (5056; 60 cm in diameter; and 300 cm in length). The mass was 2296 kg and its lowest longitudinal normal mode was at 913 Hz. The longitudinal vibration was amplified by a smaller mushroom type resonator attached to one end of the cylinder, the displacement of which was sensed by an inductor with a persistent current of 10 A. This current was amplified by a dc-SQUID. It was operational from 1991 until 1995.

NIOBE achieved its successful operation in 1995 at University of Western Australia in Perth (Fig. 9).[59] It operated at about 5 K, and consisted of a 1500 kg Nb bar with a fundamental resonant frequency of 710 Hz. A bending flap weighing 450 g, which was attached to the end of the bar, amplified the vibration of the bar. The resonant frequency of the flap was 700 Hz, and the coupled frequencies were

Fig. 8. Explorer resonant antenna in Rome. (For color version, see page I-CP11.)
Source: Photo is reprinted from http://www.roma1.infn.it/rog/explorer/.

Fig. 9. NIOBE resonant antenna at Perth in Western Australia. Antenna body of Nb is housed in the inner most radiation shield.

Source: Photo courtesy of David Blair.

713 Hz and 694 Hz. The vibrational state was monitored by a superconducting re-entrant microwave cavity whose capacitance was modulated by the relative motion of the bar and the bending flap. Thus, this transducer scheme, parametric, was different from linear schemes adopted elsewhere. Possible parametric instability was suppressed by a cold damping technique that controlled the carrier power.

NAUTILUS was designed to achieve mK operation, and built at the Frascati INFN Laboratories (Fig. 10). First cooling was achieved below 0.1 K by the resonant antenna of the Rome Group.[51] In the second observation run from December, 1995,

Fig. 10. NAUTILUS cryogenic resonant antenna at Frascati, Rome.
Source: Photo is reprinted from a presentation file under the permission of *E. Coccia*.

to December, 1996, continuous observation was conducted at $T = 0.1\,\mathrm{K}$ except for maintenance breaks (85% thermal duty cycle achieved).[60]

AURIGA was built as a twin of NAUTILUS and first operated at a cryogenic temperature of several hundred mK since 1995 until 1996. The antenna was located in Legnaro in Italy (Fig. 11) and its body was made of aluminum alloy (5056) and its mass was 2300 kg. It was equipped with a capacitive transducer coupled to an internal SQUID amplifier. The lowest temperature, 140 mK, was achieved by a $^3\mathrm{He} -^4\mathrm{He}$ dilution refrigerator. In the second run, starting in 1997, the operating temperature of the bar and transducer was lowered to about 90 mK, and kept at about 200 mK.[61]

2.2. *Dynamical model of a resonant antenna with two modes*

In order to achieve the ultimate sensitivity of a resonant antenna, impedance matching can be realized by adding a small-mass resonant system to the antenna bar. The

Fig. 11. AURIGA cryogenic resonant antenna is the twin of NAUTILUS, which is placed at Legnaro in Padova. (For color version, see page I-CP11.)

Source: Photo is reprinted from http://www.auriga.lnl.infn.it.

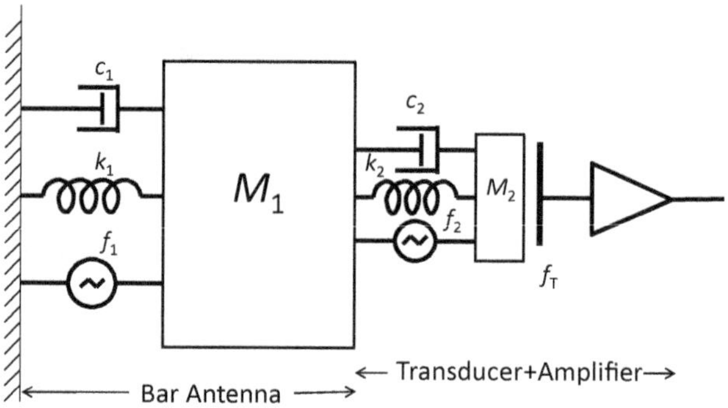

Fig. 12. Dynamical model of a resonant antenna with a resonant transducer. Almost all resonant antennae had this kind of two mode system, which made possible to realize impedance matching between the antenna bar and the transducer. The dynamical response of the antenna has two resonant modes, upper frequency and lower frequency. For example, ALLEGRO had 896.8 Hz and 920.3 Hz, and Explorer had 904.7 Hz and 921.3 Hz.

small-mass system amplifies the displacement of the bar. The dynamical system of the two mode antenna is schematically shown in Fig. 12.

The equations of motion of this model are

$$M_1\ddot{x}_1(t) + \zeta_1\dot{x}_1(t) + k_1x_1(t) - \zeta_2\dot{x}_2(t) - k_2x_2(t)$$
$$= f_1(t) - f_2(t) + f_T(t) + \frac{1}{2}M_1L_1\ddot{h}_{xx}(t), \tag{37}$$
$$M_2\ddot{x}_2(t) + M_1\ddot{x}_1(t) + \zeta_2\dot{x}_2(t) + k_2x_2(t) = f_2(t) - f_T(t). \tag{38}$$

Here, M_1 is the mass of the bar and M_2 is the mass of the mushroom-shaped resonator. L_1 is the effective length of the bar. k_1 and k_2 are the spring constants of the bar and the mushroom resonator, respectively. ζ_1 and ζ_2 are the damping coefficients of those parts. f_1 and f_2 are the Langevin force noises generators associated with the dissipation coefficients of each mass. f_T is the reaction force from the transducer and amplifier due to the fluctuating magnetic pressure in the superconducting pick-up coil. The antenna of Stanford University also adopted a similar system where the change of the magnetic field was directly sensed by a coil and coupled to an ac-SQUID to sense the current change. However, Explorer adopted a capacitance change to extract the current signal with a dc-SQUID. In any case, a back-action force exists, and dictates the performance of the transducer. The output of the transducer is represented by

$$V_{\rm out}(t) = Gx_2(t) + n(t), \tag{39}$$

where G is the gain factor and $n(t)$ is the white noise from the amplifier.

The dynamical response of the antenna has two resonant modes of upper frequency and lower frequency. For example, ALLEGRO had 896.8 Hz and 920.3 Hz. Explorer had 904.7 Hz and 921.3 Hz.

2.3. *Signal-to-noise ratio and noise temperature*

The signal-to-noise ratio (SNR) of a resonant antennae is optimized for burst-wave detection. As long as the signal consists of sufficiently short pulses, the detection procedure does not depend on the exact pulse shape. The purpose of the filtering that is applied to the output of the detector is not to reproduce the signal form, but to make decisions concerning the presence or absence of the signal in a reliable manner, and to determine the strength and arrival time of the signal. This is mostly different from the objective of the data analysis of interferometric detectors.

The voltage of the output of the transducer is sent to a single lock-in amplifier, which demodulates the signal and acts as a low pass filter. The reference frequency is set halfway between the normal-mode frequencies of the antenna, which converts the frequency of the normal mode to low frequency. The demodulated signal is represented in a two-dimensional phase diagram. The state of the signal in terms of energy, $E(t)$, is given by

$$E(t) = x^2(t) + y^2(t). \tag{40}$$

In the absence of any transducer noise, the quantity of $E(t)$ is proportional to the vibrational energy in the antenna mode, which is exp($-E/k_{\rm B}T$), where $k_{\rm B}$ is Boltzmann's constant. At room temperature, the vibration amplitude is dominant, even if the transducer noise is included. If the antenna is cooled down to a cryogenic temperature, the noise of the transducer is not non-negligible. For short pulses of gravitational waves, any change of the signal power, $\Delta E(t)$, is a useful index to measure the sensitivity of the antenna

$$\Delta E(t) = [x(t) - x(t-\tau)]^2 + [y(t) - y(t-\tau)]^2, \tag{41}$$

where τ is the time window sufficient for covering the bandwidth of the target pulse and limiting the noise of the transducer. There is an optimum value of τ, which is determined by a balance between Brownian noise of the bar and the electric noise of the transducer. A practical τ ranges from 0.3–1.0 s. The average $\langle \Delta E \rangle$ represents an impulse energy, and has a mean value corresponding to a temperature that is determined by both the Nyquist noise of the antenna and the transducer noise, which is much smaller than room temperature. This is because the fluctuating excursion of a state in the phase diagram is less than

$$\langle \Delta x^2(\tau) \rangle = \frac{k_{\mathrm{B}} T}{M\omega_0^2} \frac{\tau}{\tau_A}, \tag{42}$$

which is smaller by the extra factor defined by the last fraction τ/τ_A, where τ_A is the relaxation time of the antenna mode vibration. This means that the fluctuation amplitude can be reduced by the low-loss mechanical Q of the antenna material. The reduction factor is affected by the effective noise temperature. The observation of mechanical Nyquist noise in the cryogenic bar antenna was clearly shown in Ref. 62. In order to achieve the optimum sensitivity of a given resonant antenna system, impedance matching of the antenna with the transducer is needed.

It is important to know how reliably the kick amplitudes of an incoming gravitational wave exceed thermal excursion during the time window. Lower the excursion amplitude, the more reliably those kicks from thermal fluctuations are identified.

2.4. *Comparison of five resonant antennae*

Since the first cryogenic operation in 1980, five resonant antennas had been operated until early 2000. Every antenna achieved its own world record during the operation. The performance is described mainly by its strain noise spectral density, its duty cycle, the antenna pattern, and a nonstationary "background" noise in excess of the model. The strain noise spectral density, $S_h(f_0)$, is given by

$$S_h(f_0) = \frac{\pi}{2} \frac{k_{\mathrm{B}} T}{M\nu_s^2 Q f_0}, \tag{43}$$

where ν_s is the elastic sound velocity of the bar material. The frequency bandwidth of the detector is given by the full width at half maximum of the resonance

$$\delta f = \frac{f_0}{Q} \Gamma^{-1/2}, \tag{44}$$

where Γ is

$$\Gamma = \frac{\text{wide band noise in resonance}}{\text{narrow band noise}}. \tag{45}$$

Γ decreases if the noise temperature decreases, and when the efficiency of the transducer becomes large. Using the above strain noise power spectrum, the minimum

Table 1. Summary of resonant antennae.

	ALLEGRO	EXPLORER	NIOBE	NAUTILUS	AURIGA
Bar Working Temp. [K]	4.2	2.6	5	0.13	0.25
Mechanical Q factor	1.5×10^6	2×10^6	20×10^6	0.5×10^6	3×10^6
Strain noise $S_h^{1/2}$ [$\mathrm{Hz}^{-1/2}$]	10×10^{-22}	6×10^{-22}	8×10^{-22}	2×10^{-22}	2×10^{-22}
Eff. Bandwidth Hz	1	0.2	1	0.6	0.5
Eff. noise temp [mK]	10	10	3	2	2
Burst strain sens. h_{min}	8×10^{-19}	8×10^{-19}	10×10^{-19}	4×10^{-19}	4×10^{-19}
Duty cycle	97%	75%	75%	75%	75%
SNR ≥ 5 rate [day^{-1}]	100	150	75	150	200

Note: Since the first cryogenic operation in 1980, five resonant antennae had been operated until the early 2000 with intermittent observations. NAUTILUS and AURIGA are being operated in 2015 for covering the lack of observation by under construction interferometers.

detectable gravitational-wave amplitude (SNR = 1) for short bursts is described by

$$h_0^{\mathrm{min}} = \tau_g^{-1}\left(\frac{S_h(f_0)}{2\pi\delta f}\right), \tag{46}$$

where τ_g is the time duration of the gravitational wave.

Five resonant antennae are summarized in Table 1.[63] At the beginning of the operations by interferometric gravitational-wave detectors, all of the above-mentioned resonant antennae have been shut down, except for NAUTILUS and AURIGA which are both in operation (astro-watch[a]) and their operation is supposed to stop during the current year.

3. Interferometers

The development after Forward was initiated by several prominent researchers: J. Hough in Glasgow University, R. Weiss in Massachusetts Institute of Technology (MIT), Schilling *et al.* in Max Planck Institute in Garching, and R. Drever who moved from the Glasgow University to California Institute of Technology (Caltech).

The development of techniques involving interferometers is characterized by three stages in time. The first stage was the era of prototype interferometers led by the Garching 3 m-long and a 30 m-long delay-line interferometer.[64,65] The Glasgow 10 m[66] and Caltech 40 m[67] ones belong also to this category. ISAS[b] 10 m-interferometers[68] and 100 m one[69] succeeded the technique developed by the Garching 30 m one. At this stage, the basic concept of the interferometer was formulated so as to remove any technical noise sources such as the laser amplitude and frequency noises, the mirror suspension subsystem, and scattering-light noise. The knowledge and experiences accumulated by those R&D efforts were utilized to design the next stage of the first-generation large-scale interferometers

[a]Gravitational Wave International Committee (GWIC) recommends the operation during the absence of operations of large-scale interferometers as astro-watch.

[b]Institute of Space and Astronautical Science, later from University of Tokyo to JAXA, Japan Aerospace Exploration Agency.

with sensitivity limited only by the fundamental noises. Under the condition where technical noises are well suppressed, noises of the interferometers consist of photon-counting noise, thermal noise originating from mirrors, including the suspension system, seismic noise at low frequencies in the first-generation interferometers.[8,9,70,71] At the beginning stage of the installation of the first-generation detectors, there was no standard estimation method to assess the SNR for compact binary coalescence. During the operation of the first-generation detectors, almost all fundamental noise limits were reached, except for a few new noise sources, which were unidentified noise sources, as up-conversion noise and electro-static charge noise affecting the sensitivity at low-frequency region (around 40 Hz).[72] In this second stage, optical-coating thermal noise became close-up and vacuum squeezing injection was tested. At the time of outlining this paper, the construction and installation of the second-generation detectors are ongoing. This is the third stage, where new optical configurations are considered, and techniques to reduce any mechanical loss of the optical coating is urgent concern.[73] Newtonian gravity-gradient noise is expected to affect the second-generation detectors in the low-frequency range, where the first-generation detectors could not achieve good sensitivity.

The above three stages roughly correspond to three stages of fighting against: (i) technical noises, (ii) thermal noises, and (iii) quantum noises. The advancement of techniques is quite rapid, and is drastically changing. Noise handling in the stage where first-generation detectors were designed is not correct any more. For example, thermal noise based on velocity damping is being replaced by structure damping. There have been many good review articles at hand through the Internet concerning interferometric gravitational-wave detectors since the 1980s.[74] The most recent one is by R. Adhikari.[75] In this section, the kinds of noise limits to the sensitivity and related techniques to reduce them are described.

3.1. *First stage against technical noises in prototype interferometers*

3.1.1. 3 *m-Garching interferometer*

In the earlier stage of the laser-interferometer development,[64] the optical configuration was that of Michelson interferometer with multi-bounce delay-line arms. This kind of delay-line Michelson interferometer was developed at Max-Planck-Institute fur Quantenoptik at Garching near Munchen, the Massachusetts Institute of Technology (MIT) and at the Institute of Space and Astronautical Science (ISAS) in Tokyo. The achievement of this prototype interferometer was to successfully show that the shot-noise level at frequency higher than 1 kHz, and to study the feasibility of a laser interferometer that can attain the sensitivity required to detect gravitational-wave events occurring at the distance of Virgo cluster, which was estimated at that time to be $\delta L/L = 10^{-21}$. The baseline length of the interferometer was only 3.2 m, and the reflections were 46 between mirrors. The laser source was an argon-ion laser of 1.5 W, the efficiency of which was not so good as that presently used, a Nd:YAG laser of infra-red light (1 μm).

Photon shot noise arising from the quantum nature of photon limits the sensitivity of laser interferometers. For catching a concrete image of the noise, the author would like to take an example of fluctuation being accompanied with the $I = 1$ pA current. This is a typical amount of gate leakage current in a low-noise Field Effect Transistor (FET). The magnitude of the noise current is represented by its power spectrum,

$$\sqrt{\langle i_n^2 \rangle} = \sqrt{2eI}, \tag{47}$$

which is $\sim 5\times10^{-16}\,\mathrm{A}/\sqrt{\mathrm{Hz}}$, where e is the electron charge. If this current, $I = 1\,\mathrm{pA}$, is created in a photo-detector by a photon, a photon quanta of 6×10^6 per second should be put into the photo-detector, 6×10^6 per second should be read-out, assuming 100% efficiency. This is equivalent to a power of $1.8 \times 10^{-18}\,\mathrm{pW}$ if the light frequency is 500 THz. According to Poisson's statistics, N quanta fluctuate by $\sqrt{N}$, and if the quanta consists of electrons, the fluctuation produces a noise current.

The minimum phase observed by the Michelson interferometer that creates the photon current with a fluctuating noise current is given by assuming $\phi_1 - \phi_2 \equiv \phi_0 + \delta\phi$,

$$\delta I = -\frac{I_0}{2} \sin\phi_0 \cdot \delta\phi, \tag{48}$$

where $I_0 = I_{\mathrm{max}} - I_{\mathrm{min}}$. Equating this with the above noise current,

$$\delta\phi_{\mathrm{min}} = \frac{\sqrt{2eI}}{\dfrac{I}{2}\sin\phi_0} \geq \sqrt{\frac{2e}{I_{\mathrm{max}}}} \frac{1}{\sin\left(\dfrac{\phi_0}{2}\right)}, \tag{49}$$

where $I = \frac{1}{2}(I_{\mathrm{max}} + I_{\mathrm{min}} + I_0 \cos\phi_0)$, and the equality holds at the ideal condition of $I_{\mathrm{min}} = 0$. The minimum depends on the value of ϕ_0, and is obtained at the condition that $\phi_0 = \pi$, where the beams destructively interfere, and the output of the photo-detector is null. In this case, the output signal cannot be extracted without a modulation method (see Appendix B). Even if the modulation method is applied to extract the signal, the minimum condition of noise becomes the same as that given above.

The minimum detectable phase of the Michelson interferometer is given by

$$\delta\phi_{\mathrm{min}} = \sqrt{\frac{2\hbar\Omega}{\eta P}}, \tag{50}$$

where $I_{\mathrm{max}} = e\frac{\eta P}{\hbar\Omega}$. Since the output of the signal is proportional to P, the signal-to-noise ratio is in proportion to $1/\sqrt{P}$, which means that a high-power laser is required to attain high sensitivity.

In the prototype experiment, the quality of the laser source (frequency & amplitude noise and beam-jitter noise) was studied and its remedy was presented.[64] The concerning factors turned out to be:

(i) Intensity variation that induces noise with the fluctuating deviation from the operating point,
(ii) Frequency variation that produces the absolute path difference,
(iii) Lateral jitter that couples to the mirror tilts, which produces a false displacement,
(iv) Pulsation in width that creates a curved wave front difference.

Remedies to reduce the above noises led to the achievement of the sensitivity of the photon shot noise level, and it is impressive to realize how a study concerning the stray-light problem in this rather earlier prototype interferometer helped in designing second-generation detectors.

3.1.2. 30 *m-Garching interferometer*

Based on the previous development of the 3 m-interferometer, a longer scale delay-line interferometer was developed in Garching,[65] which formed the basis of GEO 600.[70] It was a 30 m baseline interferometer with delay-lines along the arms using green light from an argon-ion laser (Fig. 13). The operating point was set at the dark fringe using internal modulation scheme (refer to Appendix B).

The laser frequency was stabilized both by a reference cavity and the arm length change. The end mirrors were suspended, and the central part of the beam splitter and input mirrors were fixed onto a suspended mass together. Figure 14 shows the

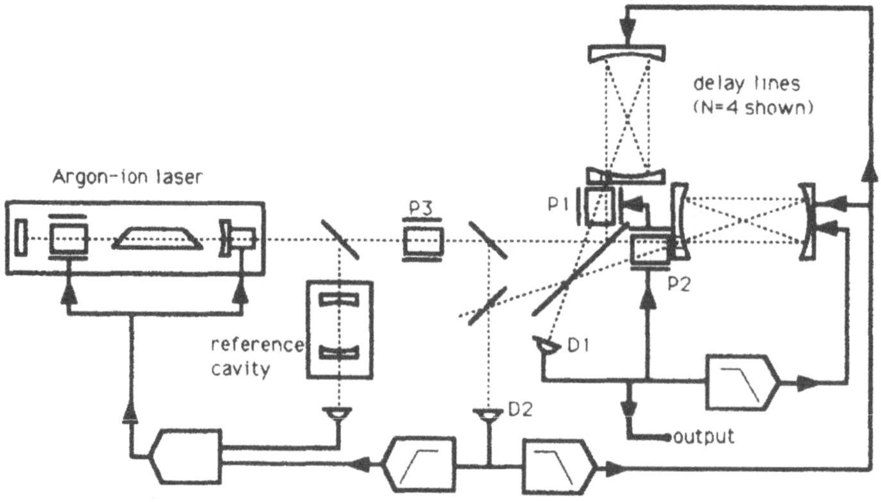

Fig. 13. Schematic view of Garching 30 m interferometer.

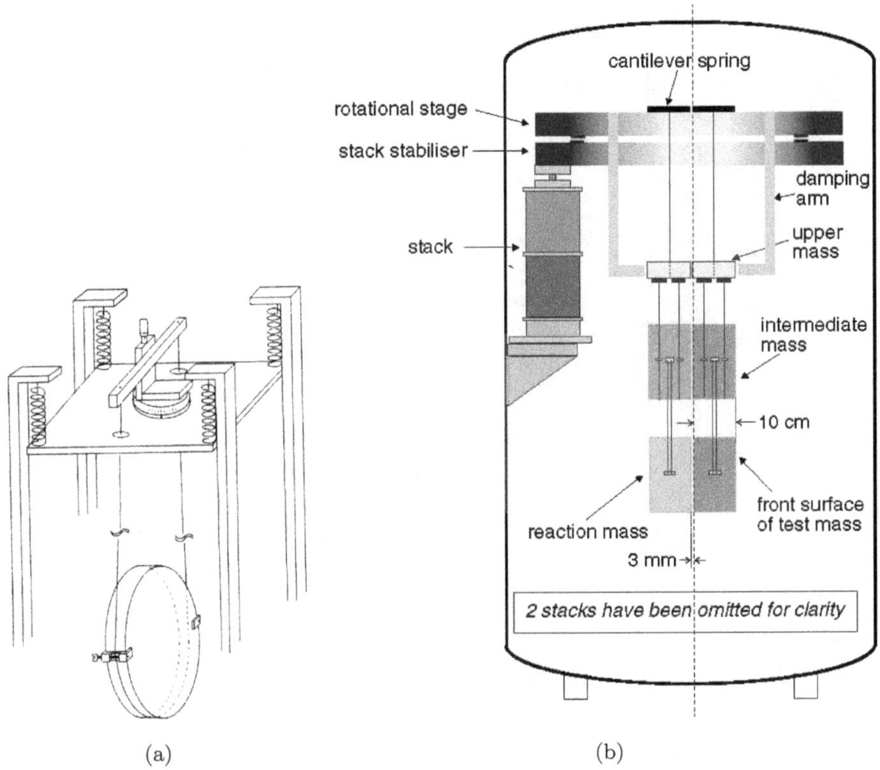

Fig. 14. Development of the suspension system from the Garching 30 m prototype (a) to the triple suspension system of the GEO interferometer (b). The advanced structure looks complicated, but it is simpler from the point of view of thermal noise.

development of the suspension system from the Garching 30 m prototype to the main suspension of the GEO interferometer,[76] where a triple pendulum structure is used in place of a single one. The advanced structure looks complicated, but it is simpler from the point of view of thermal noise. The important point is that the Garching 30 m prototype established the suspension, the heart of which prevails all over the world. In this prototype, both the thermal noise and the transfer function of the mirror suspension system were studied. However, only the thermally excited peak corresponding to the lowest modal oscillation of the mirror was observed, and the thermal noise of the slope was far less than the shot noise and seismic noise.

The behavior of seismic noise is determined by the ground continuous vibration. The amplitude of the seismic noise usually decreases with the frequency and its power spectrum density looks like as shown in Fig. 15. It is not easy to find the quiet place being represented by the low-noise model in this figure. The amplitude below 1 Hz largely depends on how strong the crust of the Earth is excited by external forces, due to wind and sea, atmospheric and ground surface waves.

A survey, recently performed, in the context of site selection for third-generation interferometric detectors provides an overview on this subject as shown in Fig. 16.

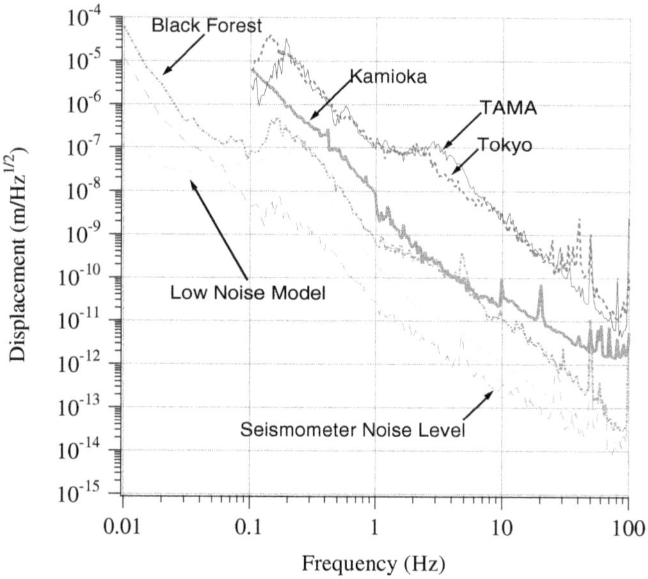

Fig. 15. Square root of the power spectrum of typical seismic noise that affects the sensitivity of interferometers.

Source: This figure is reproduced from Ref. 77.

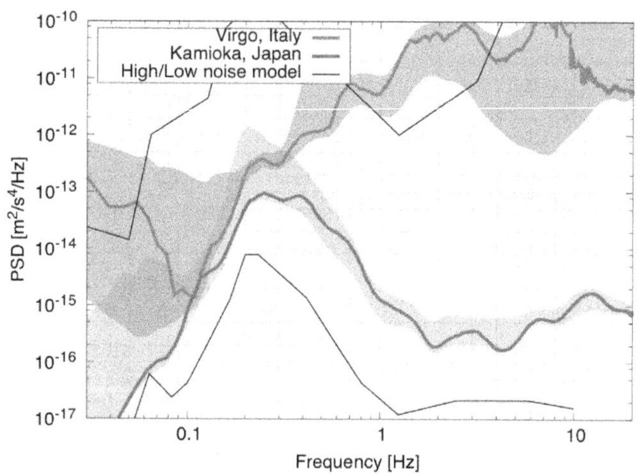

Fig. 16. A survey, recently performed, in the context of site selection for third-generation interferometric detectors. This figure shows the power spectrum of high/low change at the KAGRA and Virgo sites.

Source: The figure was taken from the doctoral thesis of M. Beker (see Ref. 78).

The square root of the power spectrum of the seismic vibration of a ground floor, G_s, is expressed by a widely known formula

$$\sqrt{G_s} = 10^{-7}/f^2[m/\sqrt{\mathrm{Hz}}]. \tag{51}$$

This formula can be applied to both directions of horizontal and vertical. It is natural that the amplitude changes both place by place and time by time, where a micro-seismic noise peak is seen at between 0.1 Hz and 1 Hz at any place on Earth.

The basic principle of the vibration isolation is based on a mechanical pendulum or an oscillator. For example, a single pendulum of 25 cm long (1 s of pendulum period) can achieve a suppression of 10^{-4} at 100 Hz under the assumption of the ideal point mass pendulum. The suspension system of the Garching 30 m-interferometer was a simple pendulum, which is the basis of the suspension system, as its development shows in Fig. 14.

The noise source limiting the sensitivity in the observed power spectrum was explained by the estimated photon shot noise while considering the modulation and transfer function of the suspension system, which was developed in this interferometer. The best sensitivity was a strain h of 3×10^{-18} in a 1-kHz bandwidth. Although scattered multi-beam light pushed the noise level up, the fundamental technique to realize a highly sensitive laser interferometer was acquired at this stage.[74]

3.1.3. *Glasgow 10 m-Fabry–Perot Michelson interferometer*

The basic idea of utilizing Fabry–Perot cavities arose from a proposal and experiment by Drever.[79] The Fabry–Perot cavity can enclose light between two facing high-reflectivity mirrors (refer to Appendix C). A Michelson interferometer was developed with a Fabry–Perot cavity in each arm at Glasgow. The baseline length was 10 m, using a cw argon-ion green laser. Figure 17 shows a schematic view of the Glasgow 10 m Fabry–Perot Michelson interferometer.[66] Different from the simple Michelson interferometer, returning beams from both cavities were not directly recombined, but electrically subtracted after being converted to electrical signals through photo detectors. This optical configuration arose in order to stabilize the laser source frequency by using one of Fabry–Perot cavities, while other cavity responses to incoming gravitational waves. Also, it could reduce the number of feedback systems. Successful direct recombination of the Fabry–Perot Michelson interferometer was first demonstrated in mid of the 1990s.[80] The output of the primary arm was fed back to adjust the laser frequency so as to keep the resonance of the cavity, the subtracted output was used to actuate the mirror of the secondary cavity to keep its resonance. The feedback signal had an information about any possible gravitational wave. This optical configuration is called a locked Fabry–Perot cavity interferometer. A low-loss optical coating was applied to obtain the mirror of the Fabry–Perot cavity. A multi-loop feedback system was applied to stabilize the frequency of the laser; the amplitude was controlled by another feedback system. The beam-splitter was mounted on a center plate that was suspended by a single

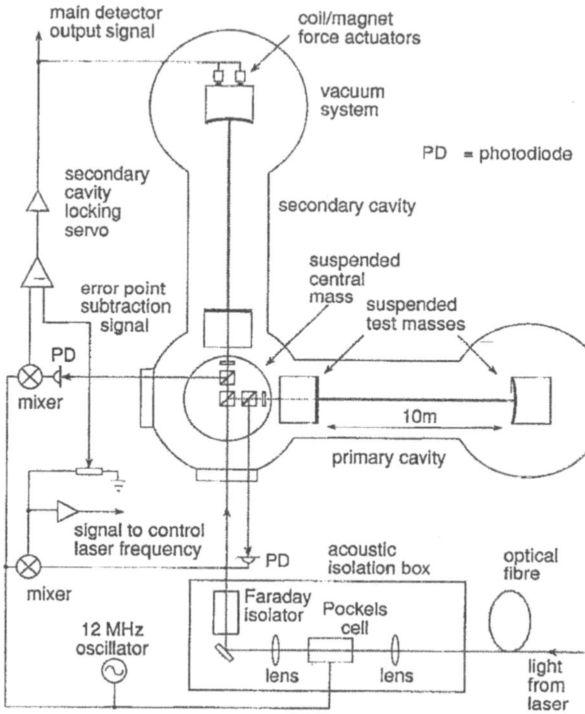

Fig. 17. A Fabry–Perot Michelson interferometer at Glasgow University. The primary arm was set in resonance by frequency feedback to the laser, and the secondary arm was kept in resonance by fed-backed actuation of the rear mirror, the actuation signal of which reflects the gravitational-wave signal. It was called a locked Fabry–Perot cavity interferometer.

pendulum, and input mirrors were also suspended by single pendulum systems. The end mirrors were applied by double pendulum suspension systems so as to improve the low frequency noise performance.

As deduced in Sec. 3.1.1, the noise of a Fabry–Perot Michelson interferometer is obtained by equating the photo-current corresponding to a gravitational wave and the shot noise,[81] if the effect of modulation is neglected. Taking η as the detection efficiency of the photo-detector,

$$|H_{\mathrm{FP}}|\delta h = \sqrt{\frac{2\hbar\Omega}{\eta P_0}}. \tag{52}$$

From this formula, h in units of the band frequency is represented by

$$h_{\min} = \frac{\omega(1 - r_1 r_2)}{\alpha_c \left|\sin\left(\frac{\omega\ell}{c}\right)\right|}\sqrt{\frac{\hbar\left[1 + F\sin^2\left(\frac{\omega\ell}{c}\right)\right]}{2\Omega\eta P_0}}, \tag{53}$$

which reduces to

$$h_{\min} \approx \sqrt{\frac{\hbar[1+(\tau_s\omega)^2]}{2\Omega\eta P_0\tau_s^2}} \tag{54}$$

if $\omega\ell/c \ll 1$ and $r_1 < r_2$ or $1-r_1 \ll 1$ holds, where the light travel time is $\tau_s = \frac{2\ell}{\pi c}\mathcal{F}$. A formula considering modulation is given in Ref. 66.

The achieved sensitivity by Glasgow 10 m-interferometer was $\sim 7\times 10^{-20}/\sqrt{\text{Hz}}$ from 500 Hz to 3 kHz, which is better than Garching 30 m-interferometer.

3.1.4. *Caltech* 40 *m-Fabry–Perot Michelson interferometer*

The Caltech 40 m prototype was a locked Fabry–Perot cavity interferometer succeeding the configuration of Glasgow 10 m.[67] In the initial operation, which ended in 1992 (Mark I), the shot-noise level was achieved at higher frequencies. On the other hand, seismic noise leaking through isolation system dictated a steep increase of the noise spectrum at lower frequencies. In the mid-frequency region between these, unidentified noise prevented from reaching thermal-noise level, except for peaks of the violin modes. By a revision of the isolation stacks and the test mass, the noise spectrum of the mid-frequency region was improved as shown (Mark II) in Fig. 18, which compares with the sensitivity of the initial LIGO design in terms of not by strain sensitivity but by displacement one. The test mass was made of fused silica housing a mirror with optical contact. The basic difference between the initial LIGO and the Mark II stays in this point. Without a complete understanding of unidentified noise of the Mark II, the initial LIGO was constructed and operated.

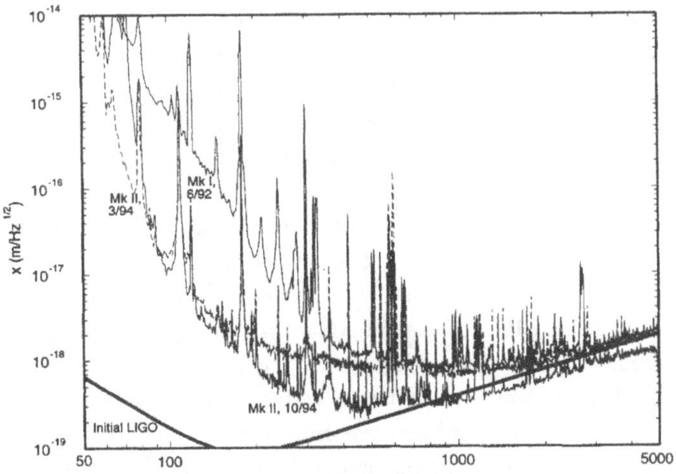

Fig. 18. Improvement of sensitivity in Caltech 40 m prototype interferometer. In order to compare LIGO and Mark II, the displacement sensitivity is shown in place of the strain one. Photon shot noise level was achieved at higher frequencies, and mirror thermal noise level was considered to be achieved, which made the start of the initial LIGO possible.

Many new unidentified noises were found during operation of the first-generation detectors.

3.1.5. *ISAS* 10 *m and* 100 *m delay-line interferometer*

Before TAMA[71] that was started in 1995, the Japanese research community conducted to choose which type of interferometer should be adopted for a future large-scale detector in Japan. The ISAS group introduced a delay-line interferometer and constructed a 10 m baseline detector[68] and a 100 m one.[69] The NAOJ[c] group constructed a 20 m Fabry–Perot Michelson interferometer.[80] Although the delay-line one had a simpler optical configuration, the main mirror should be larger, which means that the elastic resonant mode frequencies are lower and thermal noise in the observation frequency band was larger in the end. As stray-light noise had been confirmed by the 30 m-Garhing interferometer, noise hunting on the 100 m-ISAS interferometer was braked by up-converted stray-light noise.[82] On the other hand, the mirror quality of the Fabry–Perot cavity needed to be higher. Considering both practice and design, TAMA decided to adopt the same Fabry–Perot Michelson interferometer as LIGO[8] and Virgo.[9] In parallel with this determination, TAMA took a diode pumped NPRO YAG laser (in the next section) as the light source, which had been already tested by the 20 m Fabry–Perot prototype at NAOJ. When this was decided, LIGO kept the original plan of using an argon-ion laser. The 100 m-ISAS detector adopted an argon-ion laser.

3.2. *Further R&D efforts in the first-generation detectors*

In order to increase the sensitivity, a higher power of laser is required as discussed in Sec. 3.1.1. For example, we need a laser power of 10 MW level to obtain a sensitivity of $h \sim 10^{-23}$ using 534 nm light in the 500 Hz bandwidth. The leading technology for this objective is the invention of nonplaner ring oscillator (NPRO) by Byer and Kane[83] in 1985. For the first-generation interferometers, a stable and single-mode laser source was developed by a laser-diode pumped Nd:YAG laser, which produced a few W power level. Using this laser-diode pumped Nd:YAG laser, a consistent R&D effort has achieved 100 W level high-power lasers, so far.[84] In order to increase the power up to 100 W, an injection-locking system[85] or a master oscillator with a power amplifier system (MOPA) is needed, where a single-mode 100 W power level has been achieved.[86] Under this high-power optical situation, all optical parts need to endure such a high power level without reducing their performances.[87] It is not an exaggeration to describe that the laser-interferometric detector became possible after establishing the manufacturing of highly low-loss mirror with the measurement.[88] In the second-generation detectors, the power inside the cavity reaches the 1 MW level by utilizing high-power laser and a power recycling technique.

[c]National Astronomical Observatory, Japan.

In place of increasing the power, an alternative method, power recycling technique, was developed. Along with this power recycling, another idea, that of signal recycling, arose, which induced the resonant side-band extraction method.

3.2.1. *Power recycling*

Since the sensitivity of a laser interferometer is limited by the shot noise, being represented in phase noise as

$$\phi_{\text{noise}} = \sqrt{\frac{2\hbar\Omega}{\eta P}}, \tag{55}$$

the output signal is proportional to the optical power in the interferometer for a given phase shift. The phase shift produced by a given mirror displacement is increased by taking the multi-path of the beam or using a resonant Fabry–Perot cavity. The power recycling technique, invented by Drever[89] and confirmed by several experiments[90–94] became a standard techniques to further enhance the power stored in the interferometers and to lower the impact of shot noise on the sensitivity. This idea arose from the condition of the dark-fringe operation of a Michelson interferometer, which produces the minimum shot noise at the output port of the interferometer. In this condition, if power loss inside the interferometer is negligible, all laser power returns back to the laser source, which can be reflected by a mirror in phase with the original input light. This situation forms a new Fabry–Perot cavity including both optical arms.

Let us consider that the arms are made of Fabry–Perot cavities. The ratio of input power to internal one in a simple Fabry–Perot cavity is given by

$$\frac{P_{\text{int}}}{P_0} = \frac{T_1}{[1-\sqrt{R_1R_2}]^2} = \frac{T_1}{[1-\sqrt{(1-A_1-T_1)(1-A_2)}]^2}, \tag{56}$$

where T_i, R_i and A_i are the power transmittance, power reflectivity, and power loss in each mirror, respectively. The transmittance of the second mirror is included in the power loss. The power recycling mirror is regarded as being the first mirror, and the mirrors in the interferometer arms are regarded as being a combined second mirror. Under this condition, the power ratio in the above equation, P_{int}/P_0, is called the recycling gain G_{rec}. If the reflectivity of the second mirror and the loss of the first one are given by R_2 and A_1, respectively, the power ratio reaches the maximum as

$$T_1 = (1-A_1)[1-R_2(1-A_1)]. \tag{57}$$

Utilizing this simple model, the maximum power of the interferometer can be obtained as

$$G_{\text{rec}}^{\max} = \frac{P_{\text{int}}^{\max}}{P_0} = \frac{T_1}{A_1+A_2-A_1A_2} \approx \frac{1}{A_1+A_2}, \tag{58}$$

which shows that the maximum gain is 1/(all loss).

Thanks to recycling, the shot noise can be further reduced by $\sqrt{G_{\rm rec}}$. The gravitational-wave signal arising from the phase difference of two arms goes out to the photo-detector port without any affect of the power recycling, except for being affected by the storage time of the Fabry–Perot cavity, which can be remedied by the resonant side-band extraction described in the next subsection.

Note here that: (i) two stages of the Fabry–Perot cavities create couplings of the optical modes and (ii) the transmittance of the input mirror (first mirror) should be much larger than its optical loss.

In applications to practical interferometers, plural modulation frequencies need to be adopted, and a control signal extraction scheme must be developed. For the initial LIGO, a frontal modulation was applied to a table-top Michelson interferometer with Fabry–Perot cavities.[95] Virgo project also adopted the frontal modulation method.[96] From TAMA project, practical results of that technique are referred in Refs. 97 and 98.

3.2.2. *Signal recycling and resonant side-band extraction*

Adding another mirror, M_3 to a power-recycled Fabry–Perot Michelson interferometer, between the beam splitter and the signal output port as shown in Fig. 19, creates a possible increase in the sensitivity within specific bandwidth of interferometric detectors. This signal recycling mirror allows us to realize standard, detuned, dual or resonant recycling.[81,99] Meers neatly analyzed the frequency response of interferometric gravitational-wave detectors[100] that can be applied to all optical systems with slight extensions, where the interferometer can contain delay-lines or Fabry–Perot cavities, whether or not power recycling is used and whether the recycling scheme is standard, detuned or dual. The interferometer is arranged so

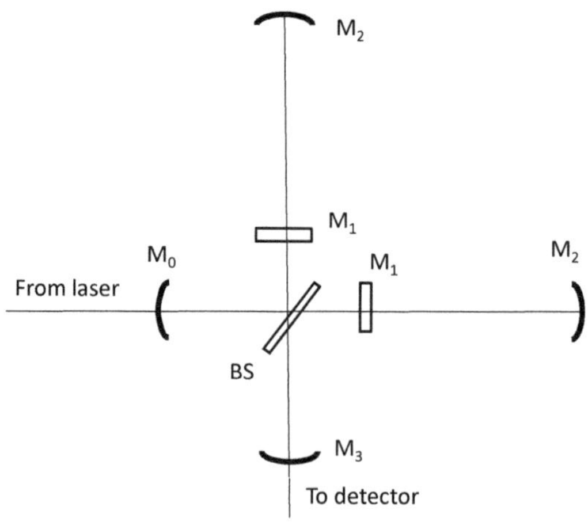

Fig. 19. Adding another mirror, M_3, to a power-recycled Fabry–Perot Michelson interferometer creates a possible increase of the sensitivity and bandwidth of interferometric detectors. M_0 is the power recycling mirror and mirror M_1 and M_2 form a Fabry–Perot cavity.

that, when light beams from the two arms meet at the beam-splitter, the original laser frequency heads back to the mirror M_0. On the other hand, any side-bands produced by differential phase modulation travel towards mirror M_3 and the output of the interferometer. The mirrors M_0 and M_3 will have different relative positions and reflectivity, which should be taken into account when calculating how both the laser light and side-bands resonate, which means that the laser frequency and the side-band frequency experience different reflectivity and optical lengths.

Dual recycling was experimentally demonstrated with fixed mirrors[101] and with suspended mirrors later.[102]

In this signal recycling, the bandwidth for gravitational waves is limited by the photon storage time in the combined cavity. To escape from this limit, a resonant side-band extraction (RSE) configuration is proposed in the interferometer adopting Fabry–Perot cavites.[103] This configuration puts a signal-extraction cavity that is combined cavities formed by the mirror M_3 and the arm cavity to be in resonance for side-band frequencies (different from signal recycling where the signal recycling cavity does not resonate). The bandwidth can be adjusted so as to remain wide even if the finesse of the Fabry–Perot cavity is increased to improve the sensitivity. Since the signal recycling mirror introduces another degree of freedom, the length control of the interferometer becomes more complicated, which required more sophisticated sensing and control scheme.[104] The resonant side-band extraction configuration is applied to second-generation interferometers.

3.3. *Fighting with thermal noise of the second stage*

In designing the first-generation interferometers, thermal noise is considered to dictate the mechanical vibration of the optical mirror and suspension system. Thermal noise is one of such fundamental noises. These fundamental noise sources are schematically shown in Fig. 20 with an inset figure showing how much frequency band each noise dominates. They are:

(i) seismic noise
(ii) thermal noise
(iii) shot noise.

Newtonian gravity gradient noise, coating thermal noise and radiation pressure noise will be considered when we discuss about the sensitivity limit of the second-generation detectors. At the early stage of the first-generation detector era, coating thermal noise was considered to be smaller than the that of mirror substrate. However, it was recognized that it was not true by simulations and measurements.[105] Needless to say, the sensitivity of an interferometer is determined not only by these noise sources, but also by noise sources arising from cross-coupling between unsuppressed technical noise with any imperfection of the interferometer optical system. For example, noise is induced by laser power fluctuations via any absorption asymmetry,[106] and higher frequency noise is induced by any relatively large vibration of

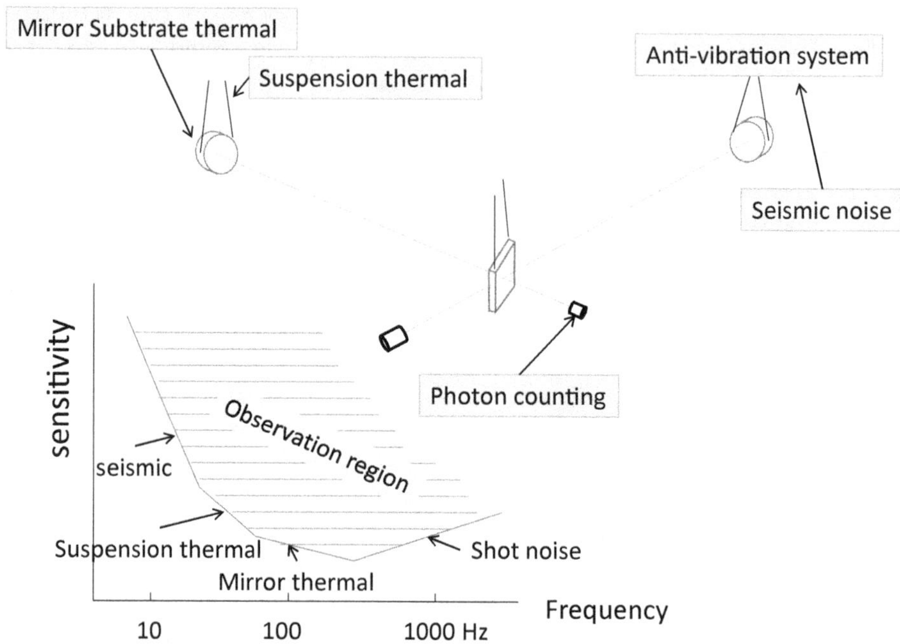

Fig. 20. Summary of fundamental noises that limit the sensitivity of a first-generation interferometer. Newtonian gravity gradient noise, coating thermal noise and radiation pressure noise will appear in the discussion of second-generation detectors.

the suspension pendulum due to seismic noise.[107] Also, mirror tilts may couple with Earth's gravitational field, which may induce fluctuations of the baseline of the arm length.[108] Also, many more noise sources of this kind exist, and unidentified ones will be born according to removing and/or reducing those noises in the future.

3.3.1. *Mirror and suspension thermal noise*

Thermal fluctuation arises from the dissipative mechanism of a material,[109] which limits the sensitivity of interferometers. Thermal vibrations that affect the optical path length of the laser beam are:

(i) internal elastic vibration mode
(ii) pendulum swing mode
(iii) violin mode of the suspension wire.

The thermal noise of a coating material is considered in the next subsection. The effect of the third item is reduced in terms of the optical path direction by the ratio of the "mass of the wire" (at most 0.1 g) and the "mass of the mirror" (more than 10 kg).

The thermal noise power spectrum due to mechanical modes of mirrors can be approximated by summing the contribution of all the internal modes of each test mass as

$$\sqrt{\langle |x(\omega)|^2 \rangle} = \sqrt{\frac{4k_{\rm B}T}{m\omega} \sum_{n=1}^{\infty} \frac{\omega_n^2 \phi_n}{|-\omega^2 + [1 + i\phi_n]\omega_n^2|^2}}. \tag{59}$$

This is correct in the case where the damping ϕ_n of each mode does not depend on frequency and assuming the modes are normal.[110] Several experiments gave null-dependence, even over a wide range of frequency, but in general the general frequency independence has not been demonstrated. Indeed, thermal noise caused by some inhomogeneously distributed loss should be considered to evaluate the overall effect of a compound system such as a mirror with suspension and an additive actuator part.[111]

In place of the mode-expansion method, a direct method was applied using Green's function by Nakagawa,[112] and also presented by Levin.[113] This method is a standard one for thermal noise estimation.

Let us consider, as an example of the lowest mechanical internal mode term in Eq. (59), coupled to the beam, of a test mass with the size of TAMA mirrors ($\omega_0/2\pi = 51\,\text{kHz}$, $\phi = 10^{-8}$ if cooled down to 20 K, one has

$$\sqrt{\langle |x(\omega)|^2 \rangle} = 6.9 \times 10^{-21} \left(\frac{T}{20\,\text{K}}\right)^{1/2} \left(\frac{\phi}{10^{-8}}\right)^{1/2} \left(\frac{3.6\,\text{kg}}{m}\right)^{1/2}$$
$$\times \left(\frac{2\pi \times 100\,Hz}{\omega}\right)^{1/2} \left(\frac{2\pi \times 51\,\text{kHz}}{\omega_0}\right) \text{m}/\sqrt{\text{Hz}}. \tag{60}$$

Higher modes of the cylindrical body were analysed and confirmed by experiment,[114] from which non-negligible contribution to the thermal noise was shown.[115] If all higher-mode contributions are added, the amplitude becomes three-times larger than the lowest-order value, which was estimated for TAMA sized mirror.[116] There are two mirrors in each arm of the Fabry–Perot cavity, and there is no correlation among those mirrors. Therefore, the total thermal noise is obtained by taking the squared sum, which is about 6-times larger in amplitude than the above value, when considering higher mode contributions.

In respect with the thermal noise of the pendulum mode, the damping factor that has a relation with dissipation needs to be found. Considering the pendulum frequency, since that is typically rather low (e.g. 0.6–0.8 Hz), it is difficult to experimentally find the factor ϕ of the pendulum. It can be estimated by a measurement of the factor of the violin mode of the wire.[117]

The structure damping is given by $F = -k[1 + i\phi]x$, where k is the spring constant and x is the displacement. If $\phi \ll 1$, the above equation is represented by $F = -ke^{i\phi}x$. This suggests that the displacement is retarded with respect to the force by the phase ϕ on an imaginary space. If $x = x_0 e^{i\omega_0 t}$ holds, the time average of $F\dot{x}$ becomes

$$\langle F \cdot \dot{x} \rangle = 2\pi\phi f k x_0^2 = 2\pi\phi f E. \tag{61}$$

This means that the energy relatively dissipates the vibration energy by $2\pi\phi$ in one cycle of the vibration. In the case of the pendulum, since the recovering force has no dissipation owing to gravity, dissipation occurs only in the deformation of the wire. Accordingly, the damping of the pendulum system is given using the wire damping, ϕ_{w}, by[118]

$$2\pi\phi_{\mathrm{pend}}(E_{\mathrm{grav}} + E_{\mathrm{wire}}) = 2\pi\phi_{\mathrm{w}}E_{\mathrm{wire}}. \tag{62}$$

That is

$$\phi_{\mathrm{pend}} = \phi_{\mathrm{w}}\frac{E_{\mathrm{wire}}}{E_{\mathrm{grav}} + E_{\mathrm{wire}}} \sim \phi_{\mathrm{w}}\frac{k_{\mathrm{el}}}{k_{\mathrm{grav}}}, \tag{63}$$

where $k_{el} = n\sqrt{TEI}/2\ell^2$ with a number of wires, n, and the wire tension, T, that suspends the mirror. Here, E is Young's modulus and I is the second moment of the cross area of the wire. The damping factor is rewritten by

$$\phi_{\mathrm{pend}} = \phi_{\mathrm{w}}\frac{n\sqrt{TEI}}{2\,mg\ell}. \tag{64}$$

Since $T \sim mg/n$, the coefficient of ϕ_w, called dilution factor is of the order of a few thousands.

Finally, the thermal noise of the pendulum is calculated[119] and the magnitude is in a TAMA-size sapphire mirror suspension at a temperature of 20 K,

$$\sqrt{\langle x(\omega)^2\rangle} = 3.5\times10^{-22}\left(\frac{T}{20\,\mathrm{K}}\right)^{1/2}\left(\frac{\omega_0}{2\pi\times1\,\mathrm{Hz}}\right)\left(\frac{\phi_{\mathrm{pend}}}{10^{-9}}\right)^{1/2}$$
$$\times\left(\frac{3.6\,\mathrm{kg}}{m}\right)^{1/2}\left(\frac{2\pi\times100\,\mathrm{Hz}}{\omega}\right)^{5/2}\ \mathrm{m}/\sqrt{\mathrm{Hz}}, \tag{65}$$

where the damping factor of the wire is the measured value of the violin mode, assuming no dependence of the frequency.

3.3.2. *Thermal noise of optical coating*

Mirrors have optical coatings for controlling their reflectivity and transmittance. Commonly, the dielectric coating consists of alternating layers of SiO_2 (silica) and Ta_2O_5 (tantala), the latter of which is substantially larger.[105] Although the mechanical loss of the substrate is quite low ($\sim 10^{-8}$), that of the coating is relatively high, such as 10^{-3}–10^{-4}. However, it had been considered that since the thickness of the coating was quite thinner than that of the substrate, the thermal noise arising from this coating was negligible for the first-generation detectors, which is not correct as stated in the beginning of this section. The mechanical loss of the coating partly arises from thermoelastic damping, where the coating and the substrate have different thermal properties.[120]

The mechanical loss, ϕ, is related to the quality factor, Q, which is $\phi = 1/Q$. If all other noises than that of the substrate and coating can be neglected, ϕ_{total} is equal to the sum of the intrinsic loss of the substrate plus any loss associated with the coating,

$$\phi_{\rm total} = \phi_{\rm s} + \frac{U_{\rm c}}{U_{\rm s}}\phi_{\rm c}, \tag{66}$$

where subscripts "c" and "s" denote the coating and substrate, respectively. $U_{\rm c}$ is the energy stored in the coating and $U_{\rm s}$ is that in the substrate. The term $\phi_{\rm c}$ includes losses in the coating materials, in the coating interfaces and in the coating-substrate interface. $\phi_{\rm c}$ is the sum of the residual loss of the coating and a thermoelastic term, $\phi_{\rm c} = \phi_{\rm residual} + \phi_{\rm th}$. $\phi_{\rm th}$ is given by[120]

$$\phi_{\rm th} = \frac{2C_F T}{\left(\dfrac{E}{1-\nu}\right)_{\rm avg}} \left[\frac{1}{C_F}\left(\frac{E\alpha}{1-\nu}\right)_{\rm avg} - \frac{1}{C_{\rm s}}\frac{E_{\rm s}\alpha_{\rm s}}{1-\nu_{\rm s}}\right]^2 g(f), \tag{67}$$

where $g(f)$ is a frequency dependent term; E, C are the Young's modulus and heat capacity, respectively. α is coefficient of thermal expansion, and ν is Poisson's ratio.

The coating thermal-noise is the most dangerous noise in both advanced LIGO and Virgo detectors. A study to replace the coating material by lower loss material is urgent. A recent discovery[121,122] is titania-doped Ta_2O_5, which produced a mechanical loss of 2×10^{-4} with optical absorption less than 0.5 ppm. This is promising. As for a competitive replacement, Nb_2O_5 may be promising due to its low mechanical loss, but higher optical absorption. Another approach in Italy gave a promising result using a nano-layer coating with appropriate annealing,[123] which agrees well with the mixture formulas for the complex bulk and shear-loss angles in connection with a new coating noise model.[124]

Since KAGRA interferometer adopts a cryogenic mirror, the coating thermal noise is less effective compared with a room-temperature interferometer. In order to confirm this point, the mechanical loss of a silica/tantala coating on a sapphire substrate was measured at cryogenic temperature, and was found to be a temperature independent mechanical loss, which supports the KAGRA design.[125] However, other measurements at Glasgow University gave a conflicting result.[126,127] This apparent conflicting results needs to be clarified, being pursued by an experiment, but not yet be clearly solved.[128]

3.4. *Fighting against quantum noises and squeezing*

Radiation-pressure noise and Newtonian gravity gradient noise appear in the second-generation interferometers.[129–131] In this section, quantum noise-related topics are described.

3.4.1. *Radiation pressure noise*

Radiation-pressure noise arises from random hitting of the test mass mirror by amplitude fluctuations on the laser, which behaves as carrier light to readout the displacement of the mirror. This is a manifestation of quantum back action. Higher power makes the noise larger. This noise was irrelevant to the first-generation interferometer; however, it constrains the sensitivity at lower frequencies of the second-generation ones in future. So far, we have seen no experimental report presenting any successful observation of the radiation pressure noise by interferometers, including prototype ones.

3.4.2. *Squeezing*

Remarkably, although radiation pressure noise has not been directly observed, its studies toward squeezing and standard-quantum-limit related issues brought to scheduling the implementation of squeezing strategies in advanced detectors (e.g. upgrade of second-generation detectors and third-generation ones). Squeezing related to a Fabry–Perot Michelson interferometer is described in this subsection. The first motivation of squeezing arises from the necessity of higher light power to suppress the photon shot noise. The cavity power of the first-generation detectors was at the 10 kW level, which needs to be increased up to the 1 MW level in a second-generation detector. Even for the first-generation detectors, the Fabry–Perot cavity should be equipped with a compensation system to cancel out the thermal-lensing effect on input mirrors. In addition to this problem, all optical parts need to maintain their performances under high-power heat production. Reducing the requirements by squeezing will improve as well the sensitivity and the events rate.

Quantum squeezing and quantum noise belong to quantum optics, and were originally formalized by Glauber.[132] A quantized single-mode electromagnetic (light) field is represented by phase quadrature and amplitude quadrature. These are non-commuting Hermitian operators that obey the uncertainty principle of Heisenberg.[133] By applying this uncertain principle to the momentum measurement of the test mass, the standard quantum-limit sensitivity, h_{SQL}, is obtained as

$$h_{\mathrm{SQL}} = \sqrt{\frac{8\hbar}{ML^2\Omega^2}}, \tag{68}$$

where L, M and Ω are the baseline length of the interferometer, the mass of the test mass and the angular frequency of the observation band, respectively. Note that test-mass quantization has no influence on the output noise.[134] Shot noise is associated with the phase quadrature of the input vacuum field, while radiation pressure noise is associated with the input amplitude quadrature. The power spectrum of the radiation noise, S_h^{rp}, is proportional to the cavity power, I_0, whereas that of the photon shot noise, S_h^{shot} changes according with $1/I_0$. The total noise of the power spectrum, S_h, is the sum of these noises,

$$S_h = S_h^{\mathrm{rp}} + S_h^{\mathrm{shot}}. \tag{69}$$

At the frequency of the optimum sensitivity in the first-generation interferometer, where the input laser power is at the few W level, the radiation pressure noise is lower. If the power is increased, S_h^{rp} increases with S_h^{shot} decreasing and at some level, $S_h^{\rm rp}$ becomes equal to $S_h^{\rm shot}$, that is, $S_h = 2S_h^{\rm shot}$, which cannot be less than the noise given by Eq. (68).[135] At this optimization, the input laser power, I_0, is estimated by $I_{\rm SQL} = mL^2\gamma^4/(4\omega_0)$ ~10 kW, where ω_0 is the cavity resonance angular frequency, m the mirror mass, L the cavity length, and γ the cavity resonance width. The power spectrum of the quantum noises is given at this optimized condition ($I_0 = I_{SQL}$) by

$$S_h = \frac{h_{\rm SQL}^2}{2}\left[\mathcal{K} + \frac{1}{\mathcal{K}}\right], \tag{70}$$

where $\mathcal{K} = 2\gamma^4/\Omega^2(\gamma^2 + \Omega^2)$. Ω is the angular frequency of the gravitational-wave.

Squeezing is a technique used to reduce the fluctuation of one of the quadratures at the expense of increasing fluctuation of the other quadrature. The first proposal was presented by Braginsky, who called it quantum nondemolishing (QND).[136] Kimble *et al.* proposed to convert conventional interferometric detectors to quantum nondemolition interferometers by modifying their input and/or output optics with the analysis of three kinds of squeezing for designing the future LIGO interferometer in 2001.[135] They were

(i) squeezed-input interferometer: squeezed vacuum with frequency dependent squeeze angle is injected to the interferometer's dark port,
(ii) variational output interferometer: homodyne detection with frequency dependent homodyne phase is performed on the output light,
(iii) squeezed variational interferometer: squeezed input and frequency dependent homodyne output.

Here, the noise spectrum evaluation was conducted assuming a lossless cavity performance of the interferometer.

The techniques for preparing vacuum squeezing needed to be suitably adapted for practical implementation in gravitational-wave interferometers. In the first demonstration of the squeezing injection effective in the gravitational-wave frequency band conducted by a bench-top apparatus, the $\xi^{(2)}$ nonlinearity in optical media was used to create a squeezed vacuum.[137] This technique was also used to produce a squeezed vacuum for the Caltech 40 m prototype interferometer, and achieved a squeezing enhancement of 3 dB at shot-noise-limited frequencies above 42 kHz.[138] Squeezing light was also injected into GEO 600, and achieved 3.7 dB squeezing,[139] described in the next sub-section. The strain sensitivity of LIGO was improved up to 2.15 dB by injecting squeezed light at the Hanford site.[140] Apart from the technique utilizing nonlinearity of optical media, Corbit succeeded to extract the radiaton-induced squeezing, called "ponderomotive", generated inside an interferometer, which is a result of the coupling between the optical field and the mechanical motion of the mirror in a table-top experiment.[141]

Optical loss essentially dictates that the squeezed vacuum and the control of losses is a key to achieve good performance of squeezing. This situation is reviewed in Ref. 142.

4. Large Scale Projects

The beginning of constructing large-scale interferometers opened the world of interferometric gravitational-wave detectors in the 1990s. The sensitivity improvement of cryogenic resonant antennas was blocked by quantum limit, which was hard to be escaped from. Contrary to cryogenic bar antennas, the sensitivity of interferometers can be simply increased by lengthening the baseline with the help of known and available technical improvement. Based on experimental results owing to prototype interferometers, practical designs of large scale interferometers were conducted and their construction have been started. The first-generation detectors have finished their role and renovation to second-generation ones is ongoing, which will end in this year. It is suitable for initiating operations of those detectors for the centennial anniversary of Einstein's general relativity. In this section, large-scale projects around the world are introduced.

4.1. *LIGO project*

LIGO project started in 1994 in order to practically catch gravitational-wave events by constructing a pair of 4 km baseline-length scale facilities separated by 3030 km, in Livingston, Louisiana and in Hanford, Washington,[8] as shown in Fig. 21. In Hanford, two parallel interferometers were installed: a 4 km baseline one (H1) and a half-sized (2 km) one (H2). The design target was to observe neutron star coalescence occurring at Virgo cluster, 20 Mpc away, where the theoretical event rate is typically 4×10^{-3} to 4×10^{-4} per year. The first observation, called science run #1 (S1), was conducted in 2002, and the observation was repeated 6 times until 2010 (S1–S6), where the project conducted in 2009–2010 was called enhanced LIGO due to advanced technologies being applied and tested for the advanced LIGO.[13]

The optical configuration of LIGO was a power-recycled Fabry–Perot Michelson interferometer adopting a nonplanar ring oscillator of a Nd:YAG 10 W laser source.

Fig. 21. LIGO project started in 1994 to construct a pair of 4 km baseline length scale facilities for laser interferometers separated by 3030 km, which were in Livingston, Louisiana and in Hanford, Washington. (For color version, see page I-CP11.)

Source: These pictures are taken from Ref. 72.

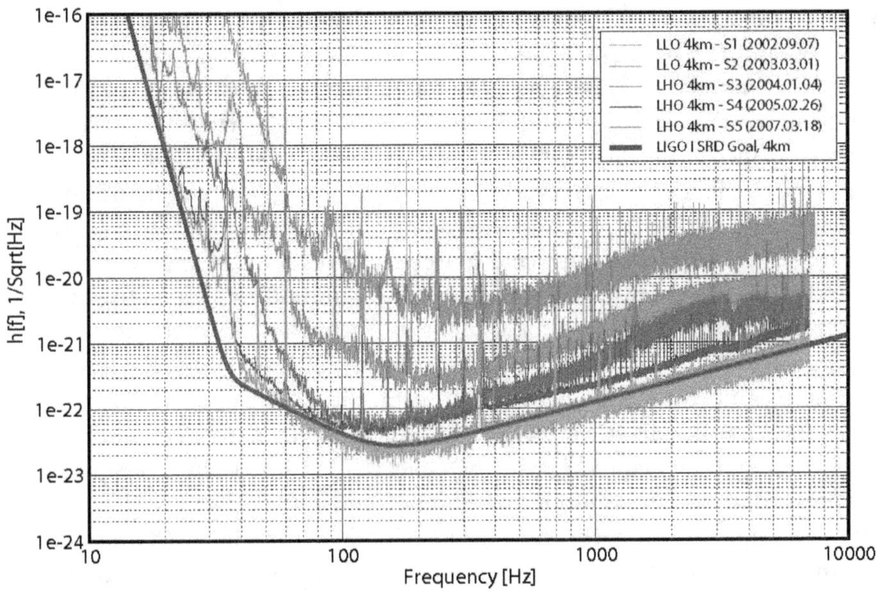

Fig. 22. During the initial LIGO project, it took about five years to attain the target design sensitivity after the installation. However, much more short period is expected to achieve the sensitivity of the advanced LIGO, the installation of which is finished in 2015. (For color version, see page I-CP12.)

It took roughly 5 years to reach the target sensitivity of the project design by 2006, which was 33 Mpc for neutron-star binary coalescence in its optimum of both the direction[d] and orbital axis configuration (in Fig. 22). A summary of the detector performance along with scientific results up to S5 can be found in Ref. 72. Over the period of S5, individual duty cycles, which involve the statistical figure-of-merit representing the detector's operation time during the observation period, were 78%, 79%, and 67% for H1, H2, and L1, respectively; for double coincidence between L1 and H1 or H2, the duty cycle was 60%; for triple coincidence of all three detectors it was 54%.[72] Due to the published result of a data analysis applied to data collected in those scientific runs, there is no report concerning the detection of gravitational-wave events, so far. Since 2009, the modification of interferometers was initiated through the enhanced LIGO for the advanced LIGO project so as to assure detection,[13] which is completed in 2015, and commissioning accordingly starts.

The strain sensitivity of the advanced LIGO is better by about a factor of 10 than the initial LIGO over a broad band while lowering the lowest observation frequency down to 10 Hz, which is attained by new seismic isolation consisting of three-stage anti-vibration, four-stage pendulum design, 40 kg test-mass optics, and

[d]13 Mpc in average distance, which is all sky and inclination average for coalescence of 1.4 $M_\odot$ mass neutron star binary in SNR=8.

180 W laser subsystems. The input laser power has been increased to 125 W from 7 W to 8 W, and the light power in the cavity is 800 kW.

In order to reduce thermal lensing in the input test mass, very low absorption coatings and a substrate are being pursued and designed. However, compensation is required as in the initial LIGO,[143] the experience of which is being applied to develop a more efficient compensation system. In its optical configuration, power recycling and signal recycling are applied along with changing the shape of the sensitivity curves for various astrophysical sources. DC-readout technique can eliminate electric noise during both modulation and demodulation. It was tested at the Caltech 40 m prototype interferometer which confirmed the expected performance.[144] For coping with thermal noise, a larger beam size was designed, and a low-loss optical coating was developed. A parametric instability[145,146] that occurs from mode coupling between any acoustic oscillation of the mirror substrate and the optical cavity electromagnetic field is one of the concerns in a high-power resonant cavity, which can be damped by appropriate optical feedback.[147,148]

By these improvements, the advanced LIGO detector can catch an event occurring at more than 300 Mpc in the optimum configuration.[e] The interferometer is operated under the standard quantum limit at the most sensitive frequency. As described in the previous section, vacuum squeezing light was introduced into the anti-symmetric port, which obtained a sensitivity improvement.[140]

Figure 23 shows the progress of the strain sensitivity at Livingston, which surpasses that of any interferometers in the initial LIGO[149,f]

LIGO plans to export one of the interferometers in Hanford to India for LIGO-India observations.[150]

4.2. *Virgo project*

Under French and Italian collaboration, the construction of the 3 km baseline length interferometer of the Virgo project[9] was completed in 2003 at Cascina near Pisa, Italy (Fig. 24). The scientific objective of the Virgo project is to detect gravitational-wave events as LIGO project, and the target sensitivity was similar to that of LIGO. By 2005, its final configuration was settled, and a scientific data-taking run was started in 2007,[151] which ended in 2008 with LIGO after conducting collaborative observations (named as VSR1).

In coincidence with the enhanced LIGO, the upgraded configuration of Virgo+ was used for a cooperative observation run, VSR2, with LIGO S6. The optical configuration of the Virgo interferometer was the same as that of LIGO, and was characterized by its seismic-noise attenuation system (SAS) to achieve attenuation of seismic noise at low frequencies, the design of which is shown in Fig. 25.[152] Larger sensitivity in the low frequency band gives the possibility to set upper limits for

[e] 120 Mpc in average distance.

[f] The advanced LIGO has initiated its observation run (O1) in September, 2015 and has improved the sensitivity by three times compared with S6.

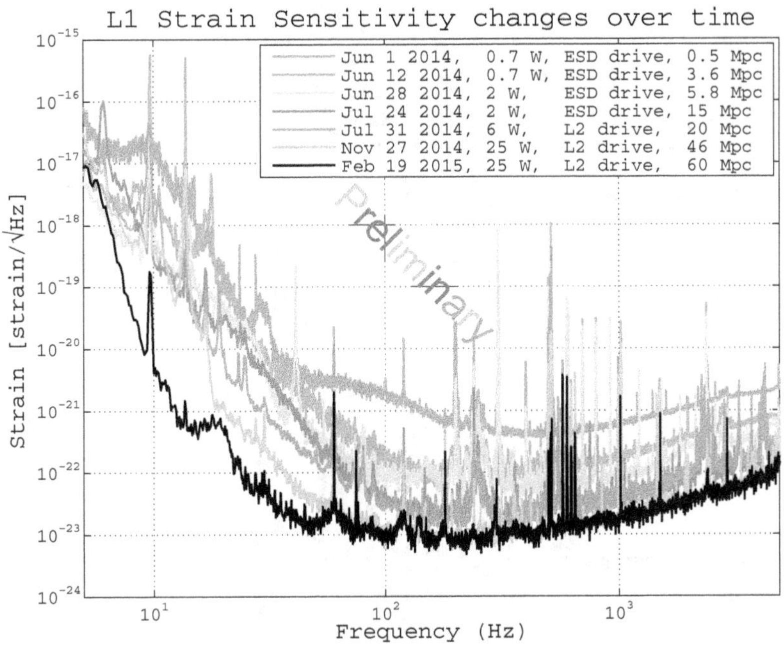

Fig. 23. Progress of the strain sensitivity of one of the advanced LIGO interferometer at Livingston.

Figure credit: LIGO Laboratory/LIGO Scientific Collaboration.

Fig. 24. Virgo interferometer is constructed in the suburb of Pisa, Italy.

Photo credit: European Gravitational Observatory.

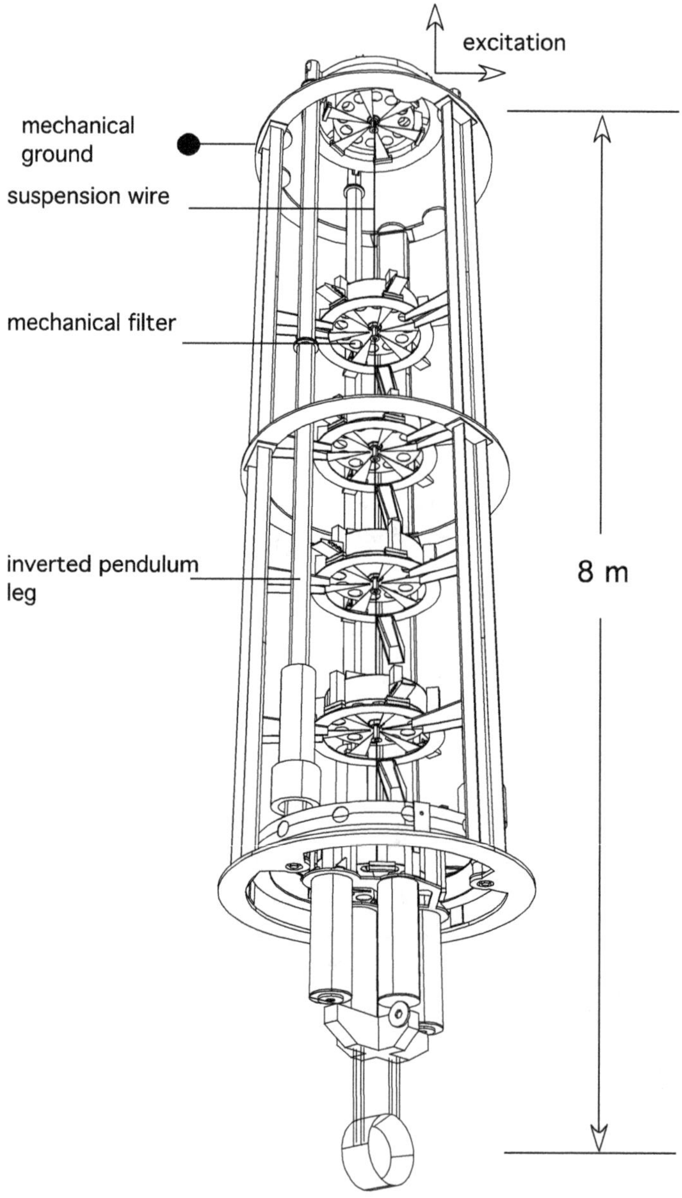

Fig. 25. Virgo detector, characterized by its seismic noise attenuation system (SAS) to achieve attenuation at low frequencies, especially for continuous waves.

Source: The figure is taken from Ref. 152.

signals coming from continuous sources as the Vela and Crab pulsars. The sensitivity improvement was similar to that of LIGO, and it took about four years to reach the level of the design sensitivity, and further improvement was applied during as shown in Fig. 26. In 2007, collaborations of LIGO and Virgo exchanged a memorandum

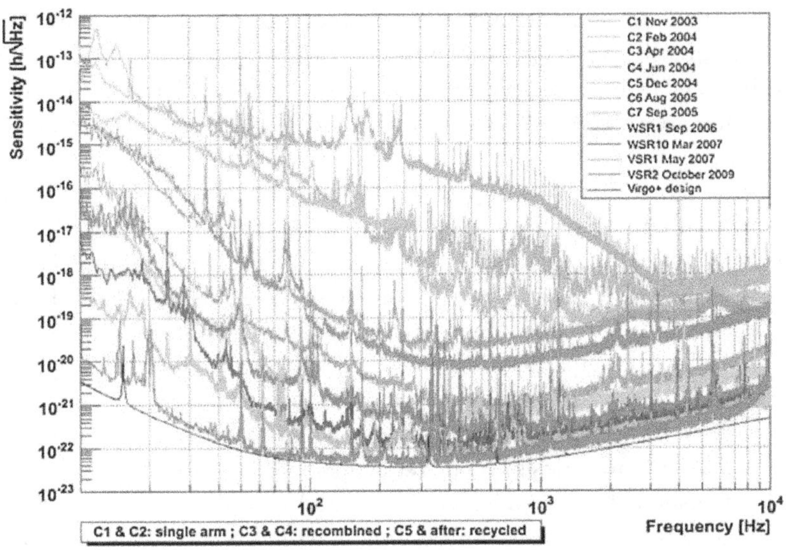

Fig. 26. Sensitivity improvement of Virgo. The sensitivity is inferior to that of LIGO at around mid frequencies. However, it is much better than LIGO at lower frequencies. (For color version, see page I-CP12.)

Source: The figure is taken from a paper after VSR2 (see Ref. 153).

of agreement to conduct a cooperative observation run, which will be effective to promote the research of gravitational-wave astronomy.[9] During this cooperative observation (S6 in LIGO and VSR1 and VSR2 in Virgo), the goal sensitivity was mostly achieved as shown in Fig. 26.[153] The sensitivity is inferior to that of LIGO at around the mid frequencies, while it was significantly better than LIGO at lower frequencies of less than 40 Hz, the cross frequency of which increased up to 70 Hz by the end of VSR2(S6) in 2010. The observable optimum distance was 15 Mpc for neutron-star binary coalescence during VSR4 (2011), where an analysis of fast spin-down young pulsars was published.[154]

If an earthquake greater than magnitude 6 occurred on Earth, the interferometer lost the state of locking, irrelevantly to where it occurred. The performance of this SAS system was enhanced during VSR1 from several times of losing locks per week to less than one per week by applying adaptive control. The Virgo interferometer was heavily affected by the environment, such as strong winds, sea waves, and earthquakes. Actually, this effect was operated at low frequency, even if this is implicitly assessed, from the text it appears that the cavity-lock and the operation were fragile, while, with respect to LIGO, that was exactly the opposite. Virgo was incredibly more robust. Clearly, at low frequency there was a significant influence of wind sources and micro-seismic noise as the weather conditions were not good, but those effects were unobservable with LIGO. Nowadays, it is going to be different because the advanced LIGO has a very good attenuation system. The earthquake magnitude 6 occurring on the other side of the world was effective to unlock cavities.

Virgo loosed the lock just once per week in average due to earthquake. That was actually an impressive stability, at the times with respect to LIGO.

Reflecting relatively large seismic noise, scatter light, which is regarded as being the major source of noise at mid frequencies, had to be reduced to reach the design sensitivity.

As in the advanced LIGO, an upgrade to the advanced Virgo began in December, 2011, and is now ongoing.[131] It adopts a dual recycling scheme: ordinary power recycling and signal recycling along with a parameter modification. The test mass is increased to 42 kg. A laser beam is supplied by utilizing a fiber laser with a power of 200 W level in its final stage. At the beginning (first year), the Virgo laser will be used, which is capable of providing up to 60 W. Since the optical power is greater, the optical parts have to be compliant with a 10-fold increase in the optical power. A DC-readout scheme is applied as the advanced LIGO. A larger spot size on the mirror is designed to reduce the thermal noise. Any thermal-lens effect is compensated by a sophisticated thermal compensation system, as in the advanced LIGO, which is based on off-axis Hartmann wave-front sensors[155] and a phase camera. New diaphragm baffles are installed to suppress any stray light merging into the main beams of the interferometer. The payloads and vibration isolation will be upgraded. Advanced Virgo is scheduled to have three different operation modes:

(i) power recycling, 25 W,
(ii) dual recycled, 125 W, tuned signal recycling,
(iii) dual recycled, 125 W, detuned signal recycling,

where the detuned signal recycling is chosen to optimize the BNS inspiral range. They can be considered as benchmark configurations for a reasonable step-by-step approach when facing increasing complexity.

The installation began by the end of 2013, and the revised interferometer will be handed to researchers within 2015.[14]

4.3. *GEO project*

GEO 600 is a project under collaboration between German (Albert Einstein Institute, AEI) and British (University of Glasgow, Cardiff University, etc.) researchers. The interferometer is placed near Hannover, Germany (Fig. 27).[10] The optical layout of GEO 600 is a simple Micheleson interferometer of baseline length, 600 m, with a folded optical path for each arm and equipped with two additional mirrors: a power recycling mirror and a signal recycling mirror, which form a dual recycled optical scheme to enhance the sensitivity.[156]

The GEO detector combines a feature of an observing instrument and that of prototype for new technologies "Advanced techniques" of GEO, which are:

- Monolithic suspensions (Refs. 76 and 157)
- Electro-static actuators

Fig. 27. GEO 600 interferometer is placed near Hannover, which surrounded by vine field. Photo credit: AEI/GEO600.

- Signal recycling
- Squeezing (since 2010).

All the above upper three items are applied to the design of the second-generation detectors that are now under construction.

Figure 28 shows the noise projection of the interferometer[158] to find each noise source linearly coupled to the observed noise spectrum, first proposed by Hild *et al.*[159] The plots in the figure correspond to data taken at the S5 LSC science run.

With respect to squeezing, squeezed vacuum-state light was introduced from its signal port so as to reduce the shot noise of GEO 600. 3.5 dB improvement was achieved in 2010 in the shot-noise limited frequency band. This is equivalent to about 3-times enhancement of detectable sources. This technique was applied for a long duration of time; a 2.0 dB improvement on the time average was obtained (Fig. 29).[139] In 2013, 3.7 dB improvement has been achieved, which is 3.7 times enhancement from the point of view of event rate.[160]

GEO 600 partially joined the collaborative observation with LIGO and Virgo during both S6 and VSR4. Researchers of GEO closely collaborate with LIGO people. Glasgow researchers developed a monolithic suspension system for the advanced LIGO and Virgo researchers developed a variant of these for their advanced detector. AEI supported by supplying the advanced LIGO with a high-power laser system.

GEO was upgraded to GEO-HF (optimized in the high frequency range). Its commissioning was conducted in 2011 by utilizing a squeezing technique through an automatic alignment control with achieving 4 + dB, where the noise was reduced

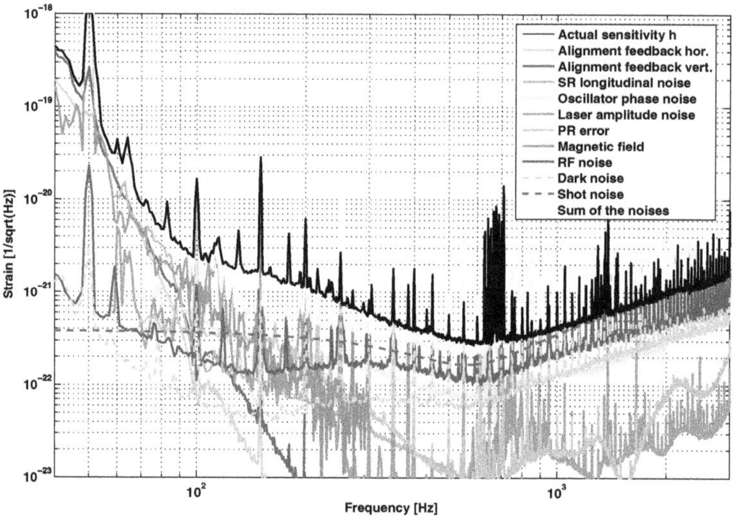

Fig. 28. Noise projection of various noise sources of GEO 600 taken at the S5 LSC science run. Shot noise, feedback noise, magnetic noise, laser amplitude noise, oscillator phase noise, RF noise and so on are plotted in order to find unidentified noise sources. There is a discrepancy between the un-correlated sum of all noise projections and the observed sensitivity curve.

Source: This figure is taken from Ref. 158.

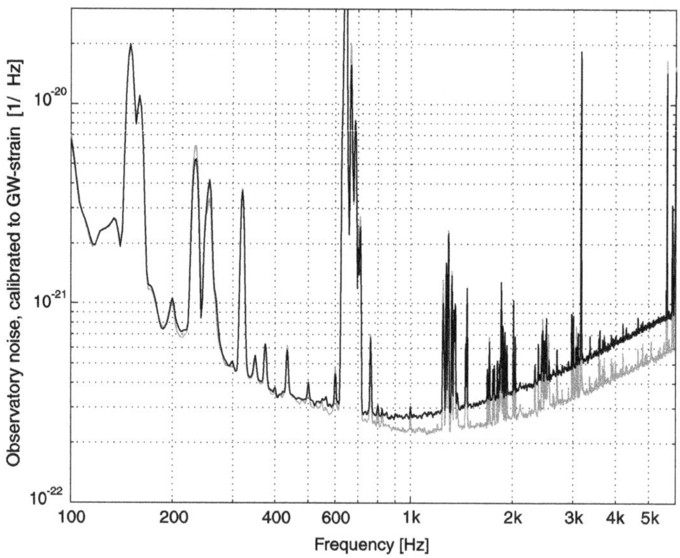

Fig. 29. Squeezed vacuum state light is introduced from its signal port in order to reduce the shot noise of GEO 600. A 3.7 dB improvement was achieved in the shot-noise limited frequency band in 2013.

in intermediate frequencies. The power of the main laser was increased to 30 W; further, a new inner mirror compensation system was installed and also a thermal compensation scheme was applied to the beam-splitter. The operation of GEO-HF is during the term of installing both the advanced LIGO and the advanced Virgo (probably until 2017). The duty cycle of the interferometer, which was 80–90%, so far will be improved by the advancing GEO-HF project.

4.4. *TAMA/CLIO/LCGT(KAGRA) project*

In the early 1990s, researchers in Japan had been separately conducting R&D effort by utilizing the 20 m-Fabry–Perot interferometer of NAOJ and the 100 m-delay-line interferometer of ISAS, as discussed in Sec. 3.1.5. After this period, researchers formed a kind of consortium in order to develop a larger scale interferometric gravitational-wave detector in Japan. This was not easy from the point of view of funding to obtain a large amount to construct such a km-scale interferometer in Japan. In Japan, a conservative approach was adopted: acquiring and testing the know-how on a given scaled prototype and rescale the design through steps by one order of magnitude in arm length. In spite of the quite effective approach, driving the Japanese researchers at the top of the expertise in this field, it was not straightforward to promote at national funding agencies the poject of a km-scale interferometer. Since a 10 m-scale prototype had been tested, the next one should be a 300 m scale one. Therefore, researchers took TAMA project as the next step. When TAMA began to be constructed in 1995, the site construction of the initial LIGO had already started. Researchers had to conduct both the construction of TAMA and the design of the km-scale interferometer project. It was not easy to persuade the funding agency to approve big funding without any fruitful result of TAMA. Large seismic noise did not permit to reach the design sensitivity within the funded term (5 years + extended 2 years). TAMA introduced the so-called TAMA-SAS to reduce any large ground vibration at low frequencies, which improved the low frequency performance, but, also, enlightened the need of more crucial choices to be considered for the km-scale interferometer. In these struggles, researchers modified the design of km-scale that would be more appealing, and would accentuate its distinctive features among gravitational-wave projects in the world and proposed the Large scale Cryogenic Gravitational wave Telescope (LCGT), which adopted an underground facility and cryogenics. Utilizing both the TAMA and CLIO interferometers, various R&D projects were conducted, leading to the second-generation interferometer techniques necessary for LCGT(now KAGRA).

4.4.1. *TAMA*

TAMA has a 300 m baseline length Fabry–Perot Michelson interferometer placed at the Mitaka campus of the National Astronomical Observatory, Japan (NAOJ) with power recycling. It achieved the best sensitivity and longest observation run earlier than any other long-baseline interferometers by 2000.[11] The optical configuration

Fig. 30. TAMA vacuum tubes are placed half underground (upper left). An input mode cleaner vacuum tube, 10 m long, is used (lower left). Test mass is suspended by double pendulum system on an isolation stack (right).

of TAMA is a power-recycled Michelson interferometer with a Fabry–Perot cavity in each arm that is similar to those of LIGO and Virgo. The laser beam produced by a 10 W Nd:YAG laser was fed through a 10 m length ring mode-cleaner cavity. The vacuum tubes, 40 cm in diameter, are placed half underground (3 m in depth) as shown in the upper left of Fig. 30. A test mass mirror, 10 cm in diameter, 6 cm in length and weighing 1 kg, was suspended from an intermediate mass that was suspended by a control platform fixed through four stems on the last stage of the vibration isolation stack, consisting of three stages in the vacuum chamber. The vibration isolation was augmented by an additional active isolator between the legs and the floor under the vacuum chamber. It took two years to encounter the deadlock of the noise spectrum, as shown in Fig. 31, which was the world-best sensitivity, realized by an interferometric gravitational-wave detector.[11]

The lower-frequency noise arises from the fact that Mitaka is in the Kanto area, a large part of which was formed during ancient times by volcano ash from Mount Fuji. TAMA researchers tried to reduce the effect of the large seismic noise by installing SAS, which was originally developed at the Virgo project.[161,162] The improvement is shown in Fig. 32.

Shot-noise sensitivity in higher frequencies was achieved immediately after its construction was finished, but the sensitivity at lower frequencies was largely

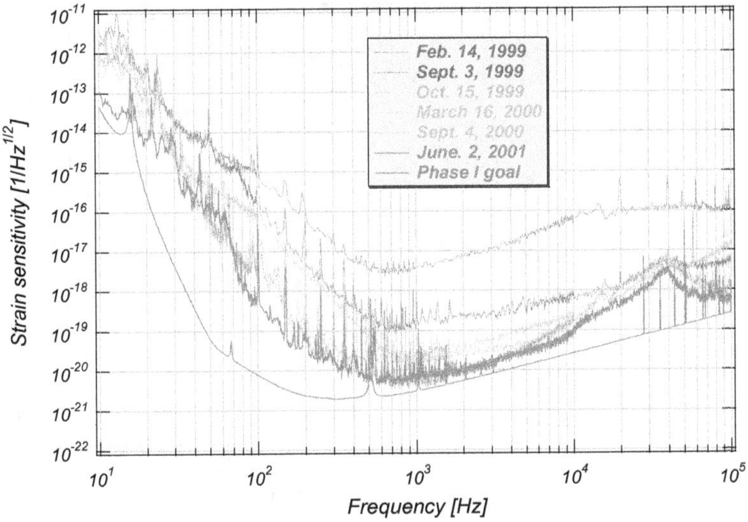

Fig. 31. Sensitivity improvement of TAMA just after installation until achieving the world record at that time, where the initial LIGO began its commissioning one year later.

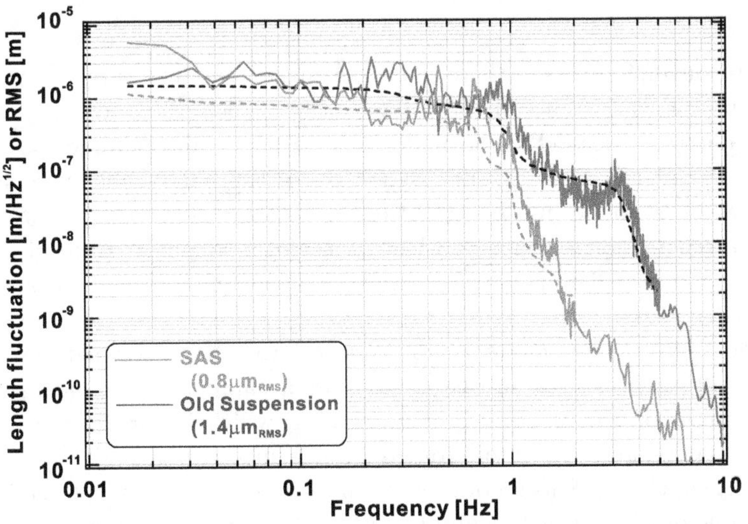

Fig. 32. Vibration noise at low frequencies was effectively reduced by the installation of TAMA-SAS.

Source: This figure is taken from Ref. 161.

dictated by mechanical noises originating from seismic noise, such as Barkhausen noise[107] in the actuator bar magnet and/or stray light-scattered noise possibly due to the large amplitude of the mirror suspension pendulum. Figure 33 shows the achieved sensitivity of TAMA and also shows that the major limit arose from a

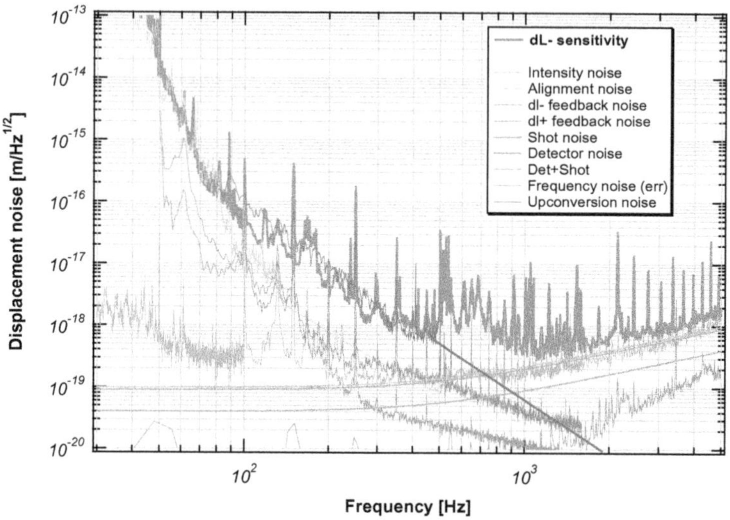

Fig. 33. Achieved sensitivity spectrum of TAMA at the end of the term under the TAMA project. The design sensitivity was attained at higher than 800 Hz, where the shot noise limited the spectrum, and obtained noise spectrum had a discrepancy at lower frequencies. The curve of upconversion was empirically determined by knowing the linear dependence of the noise spectrum against the current of the coil for test-mass actuation.

possible up-conversion noise, where the noise curves were measured under the conditions of increased currents of the coil actuator for length control and estimated by a typical current needed for normal operation of the interferometer.

The noise caused by the Barkhausen effect in the bar magnet used for mirror actuation was first found in the initial LIGO; the large amplitude of the pendulum motion induced up-converted broad-band noise at frequencies from of 10 Hz to a few 100 Hz, which still remained in the noise spectrum of the final stage of the initial LIGO.[72]

The effect of large seismic noise on the interferometer was experimentally clarified by the 20 m prototype moved from Mitaka campus to Kamioka mine. The sensitivity improvement is compared in Fig. 34, which is surprisingly large. The optical configuration of the interferometer was a locked Fabry–Perot cavity interferometer. Also, the suspension of the test mass was simply a single pendulum, the support frame of which was fixed on an optical table in the vacuum chamber. The stability of the interferometer placed underground was reported to be good.[163]

4.4.2. *CLIO*

CLIO is a 100 m baseline-length cryogenic locked Fabry–Perot interferometer placed underground at Kamioka mine (Fig. 35). A thermally limited sensitivity was achieved in 2009 by cooling the mirrors down to 10 K.[164] Until reaching this result, a series of key technical developments are described in this subsection.

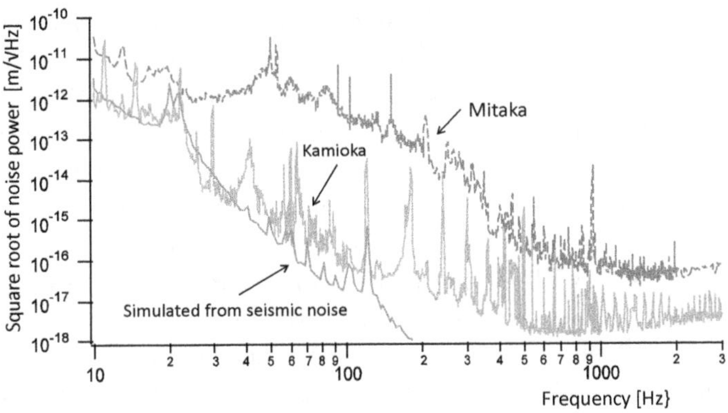

Fig. 34. Effect of a large seismic noise on the interferometer, experimentally clarified by the 20 m prototype moved from Mitaka campus to Kamioka mine. The optical configuration of the interferometer was a locked Fabry–Perot type and the suspension of the test mass was simply a single pendulum, the support frame of which was fixed on an optical table inside the vacuum chamber.

Fig. 35. An end cryostat of CLIO placed underground at Kamioka mine. Thermal noise at cryogenic temperature, 10 K, was achieved in 2009. (For color version, see page I-CP13.)

Considering the sensitivity improvement of the first-generation laser interferometer, both LIGO and Virgo chose the way not to adopt any cryogenic mirror. The financial reason why LCGT had to choose cryogenics was touched upon at the beginning of this section. Using cooling mirrors is a direct way to reduce thermal noise, unless its mechanical loss increases at cryogenic temperature. Since fused silica, which is widely used in an interferometer operating at room temperature, has a higher mechanical loss if it is cooled, a sapphire crystal is chosen in place of fused silica for the substrate of the cryogenic mirror. Sapphire crystal has high mechanical Q and extremely high thermal conductivity at cryogenic temperature. Since it has greater optical loss, how to extract heat produced inside the substrate is serious concern. In the designing process of the whole payload structure, we recognized no-need to worry about thermal lensing efefct at early stage, which is the main concern of room temperature operated system.

Since it is not realistic to cool down the whole interferometer, including the beam tubes extending to km-scale, only the test-mass mirror is cooled, as is schematically shown with a suspension subsystem in Fig. 36. The problem of how to cool it is easy to be answered. Since the mirror produced high power, even if a low-loss absorption substrate and optical coating were developed, the heat must be efficiently extracted. We have no way other than using a heat conductor attached directly to the mirror. It is the suspension fiber. The second question is how much of an improvement is expected by cooling the mirror, which depends on the mechanical loss change according to lowering the temperature. The third one is how to block thermal radiation incoming from a beam tube that is placed at room temperature.

Cryogenic cooling tests of sapphire mirrors were conducted to address the first question. They confirmed that suspension fibers with reasonable thickness were able

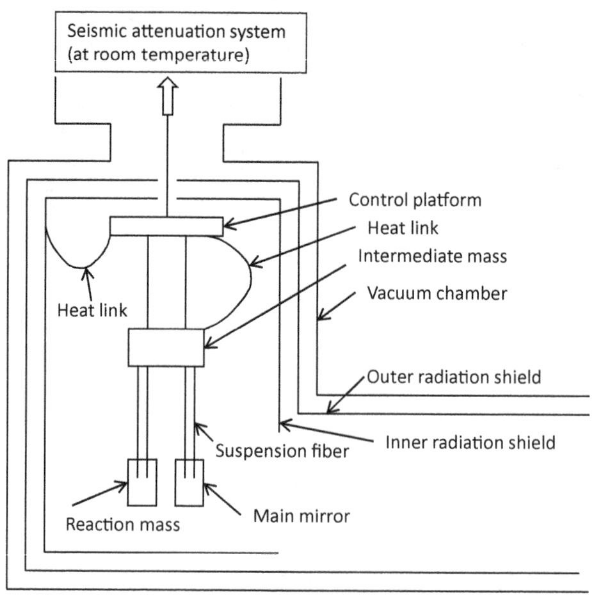

Fig. 36. Schematic structure of the cryogenic mirror with the suspension system. The heat produced in the mirror is extracted through heat conductors. The reaction mass is not for CLIO, but for LCGT(KAGRA).

to extract enough heat.[165] This is due to a rapid increase of thermal conductivity by four orders at the cryogenic temperature, where the fiber is made of pure crystal sapphire. However, the thermal conductivity is affected by the diameter of the fiber due to phonon scattering.[166] Regarding the second question, the mechanical quality factors of the mirror substrate, suspension fiber, and optical coating were measured, and confirmed the improvement by cooling the mirror with the suspension system.[125,167,168] In relation to the third question, a conduction effect of thermal radiation in a metal shield pipe in a cryostat was studied, and it was significant reduction of the heat flow by radiation baffles with their appropriate arrangement was discovered.[169] Even if heat is suppressed, contamination residual gas may degrade the mirror quality. The contamination speed was measured simulating practical arrangement of the high finesse cavity mirror and found that the effect can be controlled.[170]

There was no measurement concerning the optical absorption in the sapphire material around a wavelength of 1 μm, which is expected to be used. The measurement was made to check the optical quality of the sapphire substrate at cryogenic temperature. Sapphire crystal was one of the candidates of the advanced LIGO optics; however, it was dropped due to an unreliable production quality.[13] The sapphire substrates were all produced by Crystal System Ltd. in 2001. The measured absorption was 90 ppm/cm,[171] which was higher than expected. Significant quality control is necessary for practical usage at the cryogenic temperature. The worst situation will arise due to the absorption of laser power inside the mirror substrate, which is thermal lensing. This effect was harmful in both the initial LIGO and Virgo, where TCS was applied to compensate for the optical deformation. However, the effect is greatly reduced in the cryogenic sapphire substrate owing to the huge thermal conductivity.[172]

By utilizing those experiences and knowledge, the CLIO interferometer was designed and developed.[173] By this cryogenic interferometer, direct measurement of the thermal fluctuation of high-Q pendulum was obtained.[174]

Since CLIO was placed underground, valuable knowledge and experiences were obtained for the km-scale detector. They include knowledge about the cryo-cooler system,[175] the maintenance of mechanical devices, dust preventing techniques, and so on.

4.4.3. *LCGT(KAGRA)*

Although LCGT (now, KAGRA) was originally planned in 1999,[130] its funding was approved as one of national scientific projects in 2010[176] and its construction started. As all other national projects, a nickname of LCGT project, KAGRA, was chosen from submissions from the public in 2012. It is a 3 km baseline length power-recycled Fabry–Perot Michelson interferometer having the RSE configuration with cryogenic mirrors, and is placed underground at Kamioka in Gifu prefecture, which is perspectively shown in Fig. 37.

Fig. 37. KAGRA is a 3 km baseline length power-recycled Fabry–Perot Michelson interferometer having RSE configuration with cryogenic mirrors, and is placed underground at Kamioka in Gifu prefecture. (For color version, see page I-CP13.)

The KAGRA interferometer is placed underground deeper than the surface of the mountain by more than 200 m. Also, a 500 m long horizontal access tunnel to the center area is dug from the surface entrance. The design of the center area is shown in Fig. 38.

The rock of the mountain is utilized to form a 2-layer structure for a tall vibration isolation (SAS) system above the cryostat housing cryogenic mirror as shown in Fig. 39. The top of the isolation system is in 14 m high above the floor shown in Fig. 40. The design of SAS is based on knowledge of TAMA-SAS.

The optical configuration of KAGRA is a power-recycled Fabry–Perot Michelson interferometer utilizing the RSE scheme, as shown in Fig. 41. The sensitivity design was based on the signal recycling.[16] A variable RSE with DC read out is planed to be installed and its control method is investigated.[177] The limit beyond classical noises is quantum noise, which can be conquered, later, by adopting QND squeezing strategies. First, a homodyne phase is determined to cancel any photon shot noise and radiation–pressure noise. Second, an optical spring effect will be utilized by a detuning technique. Figure 42 shows the designed noise power spectrum. DRSE in the right-hand side figure is more sensitive at frequencies less than 500 Hz, while BRSE in the left-hand side figure is better in higher frequencies. The first detection of gravitational wave can be achieved by DRSE, and the details of a merger can be detected by BRSE. The most relevant characteristic features are the adoption of a cryogenic mirror and that the location is placed underground. The third-generation

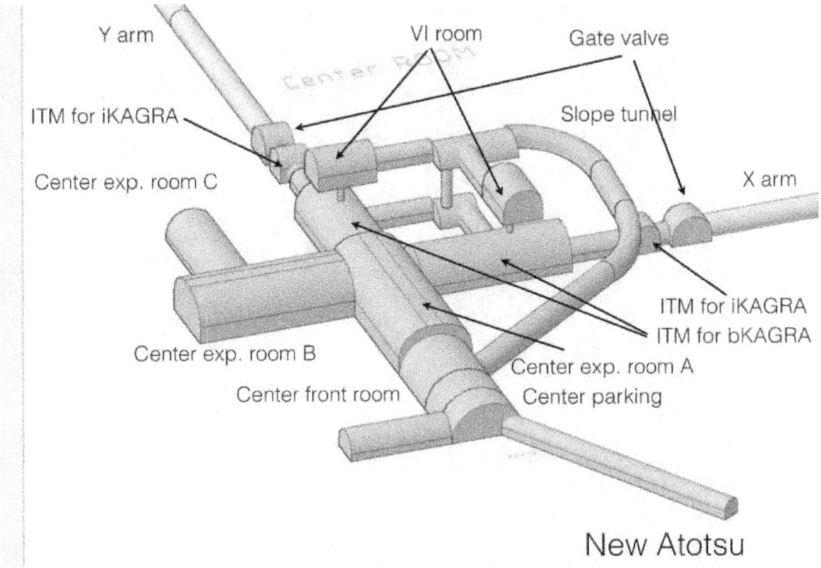

Fig. 38. The KAGRA cavern is designed to keep more than 200 m depth from the mountain surface at any places of the interferometer. The figure schematically shows the center area. A 500 m long horizontal access tunnel to the center area is dug from the surface entrance.

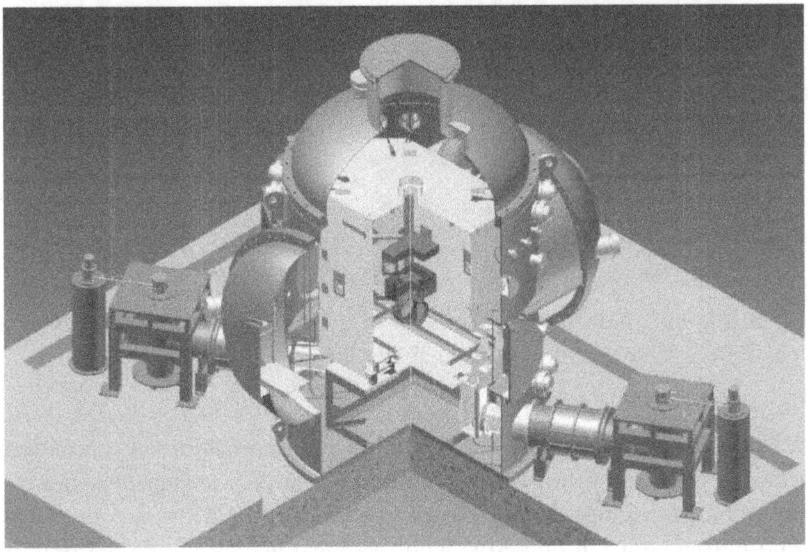

Fig. 39. Cryogenic mirror suspended inside the cryostat. The concept of cooling system is described in relation to Fig. 36.

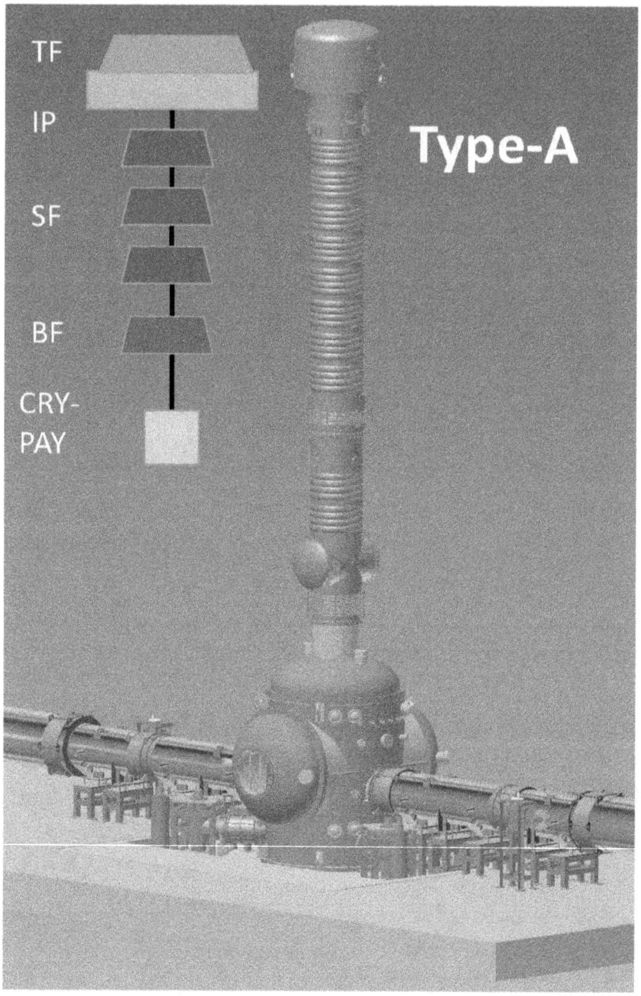

Fig. 40. Design of SAS, based on the knowledge of TAMA-SAS. A series of GAS filter stages is supported from the top housing, which is fixed on the second floor caved in the mountain rock.

detector, Einstein telescope (ET), planed in EU countries, adopts both underground location and cryogenics. Since other second-generation interferometers are placed on the ground surface; and do not use of cryogenics, KAGRA is sometimes called as a second-half generation detector. The technical achievements detained for this KAGRA project are based on cryogenic mirror development conducted in the CLIO project and cryogenic experiments done for CLIO. The practical design of the cryogenic payload is being conducted under EU-Japan research collaboration supported by ELiTES.[178] The techniques developed in KAGRA will contribute to the advancement of gravitational-wave physics.

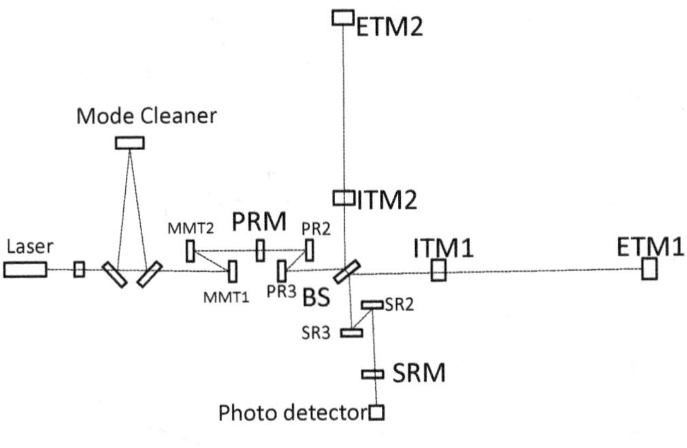

Fig. 41. Optical configuration of KAGRA, which is a 3 km baseline length Fabry–Perot Michelson interferometer with power recycling utilizing the RSE scheme.

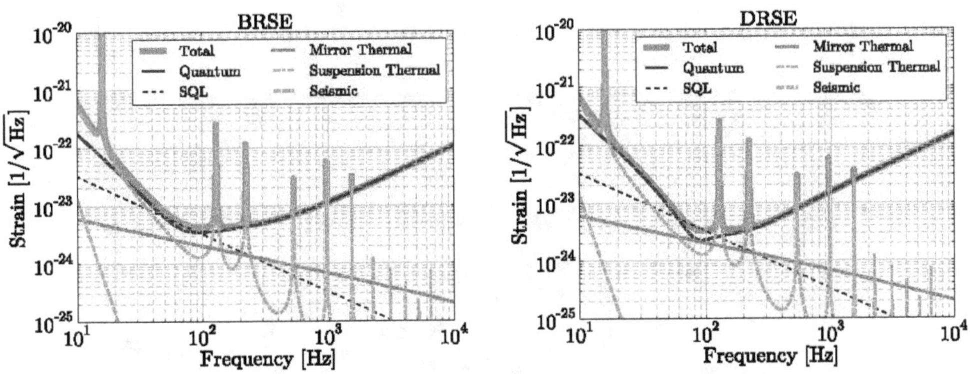

Fig. 42. Design sensitivity of KAGRA. DRSE, shown in the right-hand side figure is more sensitive at frequencies of less than 500 Hz, while BRSE in the left-hand side figure is better at higher frequencies.

4.4.4. *Einstein telescope*

Advanced gravitational-wave detectors will soon succeed in the first detection of gravitational wave. However, since the detection rate will be a few in a year, it is not sufficient to open the era of precision gravitational-wave astronomy. Higher detection rate can be achieved by enhancing the sensitivity, which makes SNR better for closer gravitational-wave sources. ET was planned to achieve this requirement of 10-fold sensitivity improvement (in Fig. 43). This will be realized by an interferometer of 10 km baseline length with cryogenic mirrors, placed underground as schematically shown in Fig. 44),[17] the design of which was disclosed in 2011 being financially supported as design study by EU committee.

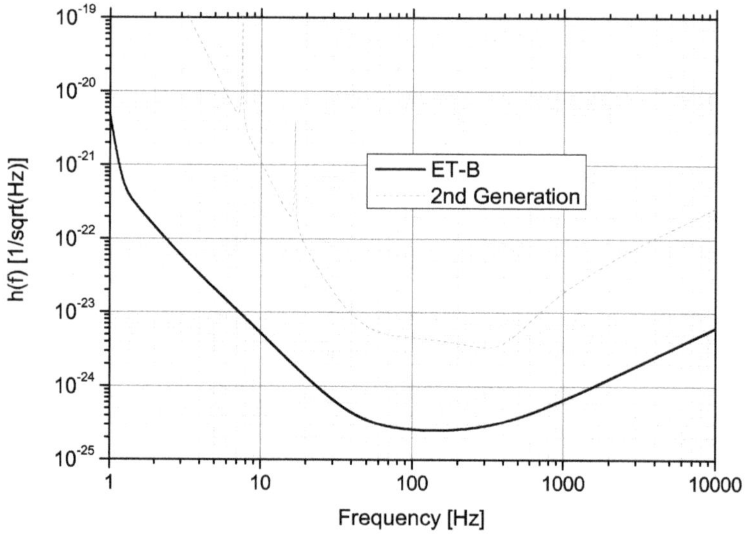

Fig. 43. The sensitivity of ET is designed to achieve sensitivity improvement by a factor of 10.

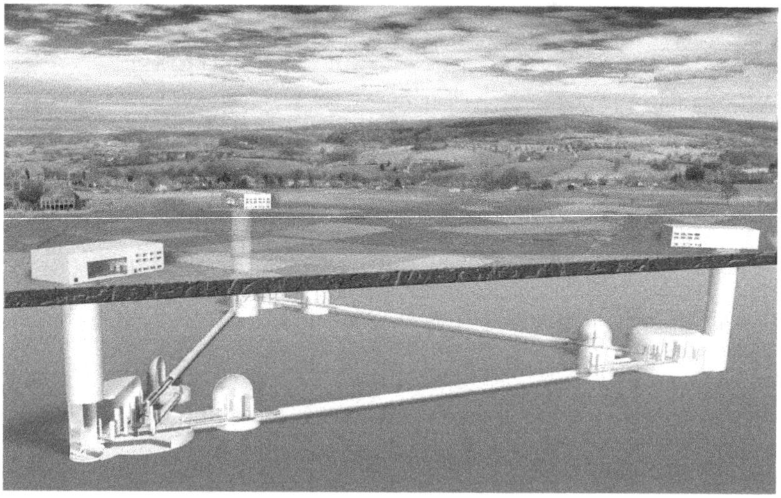

Fig. 44. Einstein telescope planed in EU countries. 10 km long triangular shaped interferometer arms are placed underground and main mirrors are cooled down to cryogenic temperature. The artistic view is taken from the ET design document.

5. Summary

Regardless of a 60-years effort to improve the sensitivity of gravitational-wave detectors, the detection of a gravitational-wave event has not yet succeeded. The author had hopefully expected to have a report of the detection of a gravitational-wave when cryogenic resonant antennae achieved their design sensitivities in the late 1980s. However, no detection had been achieved and the theoretical bound receded further. Since then almost 30 years have passed. In this centennial anniversary year of the birth of Einstein's general relativity, the advanced LIGO should start operation, and in a few years all second-generation interferometers should begin observations for the practical detection of gravitational-waves possibly under a world-wide observation network. The author strongly believe that one can hear the report of the first detection of a gravitational-wave in a few years.

Acknowledgments

The author thanks J. Hough, E. Majorana, and A. Virgilio for their refereeing and valuable comments, which corrected flaws of several essential points and made originally poor sentences to be readable. This work was supported by MEXT, Leading-edge Research Infrastructure Program, JSPS Grant-in-Aid for Specially Promoted Research 26000005, and JSPS Core-to-Core Program, A. Advanced Research Networks.

Appendix A. Thermal Noise

After J. Weber it is not an exaggeration to describe that thermal noise has been the most difficult phenomenon to be conquered among experimentalists for aiming to detect gravitational-waves. Thermal noise prevails everywhere; not only in the antenna body itself, but also in the transducer in the case of resonant antennae, and not only in the mirror, itself, but also in the optical coating material in the case of laser interferometers. Thermal noise dictates the ultimate performance of all electronic devices.

A.1. *Nyquist theorem*

A fluctuating voltage appears between two poles of an electric resistance, R, which is thermally in an equilibrium state. Assuming temperature T, the mean square of voltage V is given using the Boltzmann constant, k_{B}:

$$\bar{V}^2 = 4Rk_{\mathrm{B}}T\Delta f, \tag{A.1}$$

where Δf is the frequency bandwidth where the voltage is measured.

The power spectrum of random processes is described by the power spectrum density, $G(f)$, which is defined by an ensemble average of the time average of the power consumption occurring in a unit resistance per unit frequency bandwidth. Taking the frequency bandwidth between adjacent frequencies as

$$\Delta f_n = f_{n+1} - f_n = \frac{n+1}{T_p} - \frac{n}{T_p} = \frac{1}{T_p}, \tag{A.2}$$

an ensemble average of the time average of the power consumption, $\langle \mathcal{P}_n \rangle$, is represented by

$$G(f_n)\Delta f_n = \langle \mathcal{P}_n \rangle = V_n^2, \tag{A.3}$$

where V_n is the voltage in the frequency bandwidth and $G(f) = 4Rk_{\mathrm{B}}T$.

Let us consider a function system, where an output O is produced responding to an input I. The input is something like a force, voltage and/or current, and the corresponding output is a displacement, velocity and/or current. The function is a linear system characterized by an imaginary response function, $Z(f)$, that produces an output of $Z(f)A(f)e^{i\omega t}$ for an input of $A(f)e^{i\omega t}$. Assuming that the input fluctuates in time T, as given by

$$I(t) = \int A(f)e^{2\pi ft}df, \tag{A.4}$$

the fluctuation of the output is calculated by

$$O(t) = \int Z(f)A(f)e^{2\pi ift}df. \tag{A.5}$$

Since the power spectrum of the input $G_I(f)$ is given by

$$\lim_{T\to\infty}\frac{2}{T}|A(f)|^2, \tag{A.6}$$

the power spectrum of the output $G_O(f)$ becomes

$$\lim_{T\to\infty}\frac{2}{T}|Z(f)A(f)|^2. \tag{A.7}$$

That is,

$$G_\mathrm{O}(f) = |Z(f)|^2 G_\mathrm{I}(f), \tag{A.8}$$

$$\bar{O}^2 = \int_0^\infty G_\mathrm{O}(f)df = \int_0^\infty |Z(f)|^2 G_\mathrm{I}(f)df. \tag{A.9}$$

This formula gives a method to evaluate the response of the fluctuated quantity in a linear system that has been affected by a noise source.

A.2. *Thermal noise of a harmonic oscillator*

Let us consider a harmonic oscillator of mass m and spring constant $k_{sp} = m\omega_0^2$ with damping $\beta = m/\tau_0$. To the oscillator is applied a fluctuating force with a power spectrum irrelevant to the eigen-mode frequency, ω_0. From the applied force, $Ae^{2\pi ift}$, the behavior of the oscillator is determined by

$$m\ddot{x} + \beta\dot{x} + k_{sp}x = Ae^{2\pi ift}. \tag{A.10}$$

Since the stationary solution of the equation is given by

$$x = \frac{Ae^{2\pi ift}}{-4\pi^2 mf^2 + 2\pi i\beta f + m\omega_0^2}, \tag{A.11}$$

the response function of this system is

$$Z(f) = \frac{1}{-4\pi^2 mf^2 + 2\pi i\beta f + m\omega_0^2}. \tag{A.12}$$

Denoting the input spectrum as G_N, the output mean square is calculated by

$$\bar{x}^2 = G_{\rm N}\int_0^\infty \frac{1}{(c-4\pi^2 m f^2)^2 + 4\pi^2\beta^2 f^2}df,$$
$$= G_{\rm N}\frac{1}{4\beta k_{\rm sp}}. \tag{A.13}$$

We assume here that the damping given by β arises due to a statistical-fluctuating force independent from any special system. If the system is in a thermal-equilibrium state of temperature T, $k_{\rm sp}\bar{x}^2 = k_{\rm B}T$ holds. By combining the above equations, we obtain

$$G_{\rm N}(f) = 4\beta k_{\rm B}T. \tag{A.14}$$

In general, if a dynamical system described by a generalized coordinate, q, has a damping of $-\beta\dot{q}$, a fluctuating force, $4\beta k_{\rm B}T$, always exists.

The damping that dictates the behavior of a dynamical system is not velocity, but structure damping in the frequency region where mechanical vibration dominates.[118] If this is correct, the response function deduced in Eq. (A.12) needs to be modified by replacing the spring constant, $k_{\rm sp}$, by an imaginary spring constant, $k_{\rm sp}(1+{\rm i}\phi)$. The spectrum of the fluctuation force is not white, but becomes $G_{\rm N} = 4(k_{\rm sp}\phi/\omega)k_{\rm B}T$ according to the fluctuation–dissipation theorem. The output response to this force is in frequency spectrum,

$$|x(\omega)|^2 = \frac{4k_{\rm B}T}{m\omega}\frac{\omega_0^2\phi}{|-\omega^2 + [1+{\rm i}\phi]\omega_0^2|^2}. \tag{A.15}$$

A large difference from the case of velocity damping is that the magnitude of the spectrum at lowering frequencies increases in proportion to f^{-1}, and at increasing frequencies decreases more rapidly. That is, for $\omega \ll \omega_0$:

$$|x(\omega)|^2 \sim \frac{4k_{\rm B}T\phi}{m\omega_0^2\omega}, \tag{A.16}$$

which is useful to calculate the noise power spectrum arising from the thermal noise of the mirror substrate in observation–frequency band and for $\omega \gg \omega_0$,

$$|x(\omega)|^2 \sim \frac{4k_{\rm B}T\omega_0^2\phi}{m\omega^5}, \tag{A.17}$$

which is useful to evaluate the noise power spectrum from pendulum thermal noise in observation–frequency band.

Appendix B. Modulation

In order to make the condition of minimum noise in a Michelson interferometer, the output signal becomes null. Even in this condition, a modulation technique is applied to extract a non-null signal by placing an optical phase modulator in each beam path, as shown in Fig. 45.

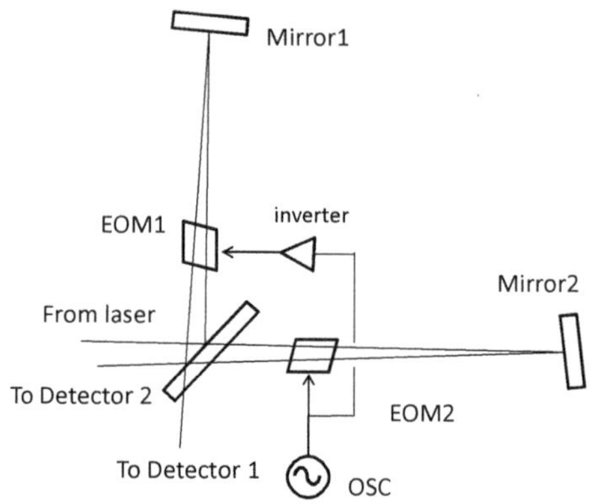

Fig. 45. Optical phase modulators are equipped to extract signal under the minimum shot-noise condition. Electro-optical modulators (EOMs) are set in the beam paths, and applied modulation voltages in an alternative phase. This is called as the internal modulation technique. Since the optical phase is distorted through the modulator imperfection and there is a limit of light power, the internal modulation is not applied to present practical interferometers.

EOM_1 and EOM_2 are driven by counter phases, and the modulation index is $\frac{m}{2}$. The electric fields of light beams that enter the beam-splitter are combined as:

$$E = \frac{E_0}{2} e^{-i(\phi_1 + \frac{m}{2}\sin\omega_m t)} - \frac{E_0}{2} e^{-i(\phi_2 - \frac{m}{2}\sin\omega_m t)}, \tag{B.1}$$

where $\phi_1 = \phi_2$ is assumed. The photo-current is described using I_0 to be 2-times the current in the case of $m = 0$ as:

$$\frac{I}{I_0} = \frac{1}{2} - \frac{1}{2}\cos\Delta\phi\cos(m\sin\omega_m t) + \frac{1}{2}\sin\Delta\phi\sin(m\sin\omega_m t), \tag{B.2}$$

which is expanded by a Bessel function as

$$I = \frac{I_0}{2}[1 - J_0(m)] + I_0 J_1(m)\Delta\phi\sin\omega_m t + (\text{higher-harmonics-of} - \omega_m). \tag{B.3}$$

The first term is the DC component, I_{dc}, and the second term is the first-order term of the modulation, I_{ω_m}. This current is fed into a demodulator circuit and the coefficient of $\sin\omega_m t$ is extracted. The noise current is $I_{\text{mod}} = \sqrt{2}\sqrt{2eI_{\text{dc}}}$, where white noise by I_{dc} produces two side-band noises at around $\omega_m/2\pi$. The phase noise equivalent to this current becomes the minimum detectable phase, which is described by

$$\Delta\phi_{\text{min}} = \frac{I_{\text{mod}}}{\frac{I_{\omega_m}}{\Delta\phi}} = \frac{\sqrt{1 - J_0(m)}}{J_1(m)}\sqrt{\frac{2e}{I_0}}. \tag{B.4}$$

Note here that the magnitude tends to be equal to the minimum phase obtained earlier if $m \to \infty$.

On the other hand, if we take the condition $\phi_1 - \phi_2 = \pi + \Delta\phi$, the operating fringe becomes bright, but the signal-to-noise ratio becomes worse when m is taken to be small as described by

$$\Delta\phi_{\min} = \frac{I_{\text{mod}}}{\frac{I_{\omega_m}}{\Delta\phi}} = \frac{\sqrt{1+J_0(m)}}{J_1(m)}\sqrt{\frac{2e}{I_0}}. \tag{B.5}$$

Because the above I_{dc} is $\frac{I_0}{2}[1 + J_0(m)]$, I_{ω_m} becomes $-I_0 J_1(m)\Delta\phi$. The above explanation is based on the book.[179]

Appendix C. Fabry–Perot Interferometer

C.1. *Fabry–Perot cavity*

Here, the response of the Fabry–Perot cavity to phase-modulated light was analyzed and compared with an experiment.[180] The Fabry–Perot cavity consists of two high-reflection mirrors facing each other at a distance of ℓ and can trap light inside, as shown in Fig. 46.

Mirrors have high-reflectivity optical coatings on their facing inner surfaces, and have anti-reflection coatings on the outside surfaces. When light is reflected at the inner surface, the light phase changes due to the reflection by the harder refractive material. The phase advancement due to propagation inside the cavity is $\Delta = \Omega\ell/c$, where Ω is the angular frequency of light. Assume r, t to be the amplitude reflectivity and the amplitude transmittance (differentiating both mirrors by suffix 1 and 2). Denoting the amplitude of the input light beam as A_i, the amplitude of the reflection beam, A_r, is represented by

$$\begin{aligned} A_r &= [(ir_1) + t_1^2(ir_2)e^{-2i\Delta} + t_1^2(ir_1)(ir_2)^2e^{-4i\Delta} + \cdots]A_i, \\ &= \left[ir_1 + t_1^2(ir_2)e^{-2i\Delta}\sum_{n=0}^{\infty}(ir_1)^n(ir_2)^n e^{-2in\Delta}\right]A_i, \\ &= \left[ir_1 + \frac{t_1^2(ir_2)e^{-2i\Delta}}{1+r_1r_2e^{-2i\Delta}}\right]A_i. \end{aligned} \tag{C.1}$$

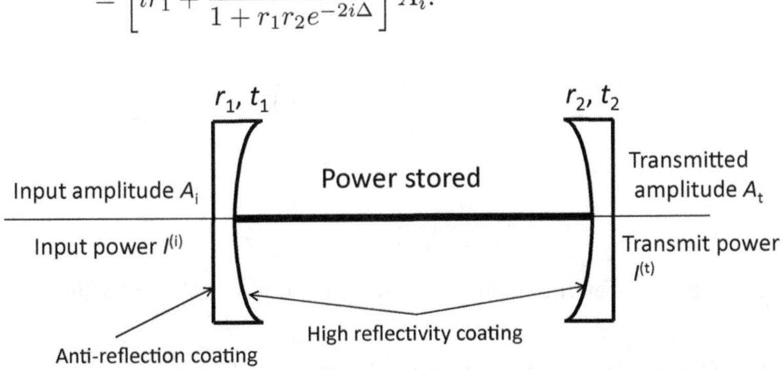

Fig. 46. Light beam introduced from the left hand side. The outer surface of the mirror has an anti-reflection optical coating and the inner surface has high reflectivity due to an optical coating. The mirror of the right-hand side has a similar surface treatment. Light is trapped inside the cavity, and a tiny amount of light power leaks from the right-hand side mirror.

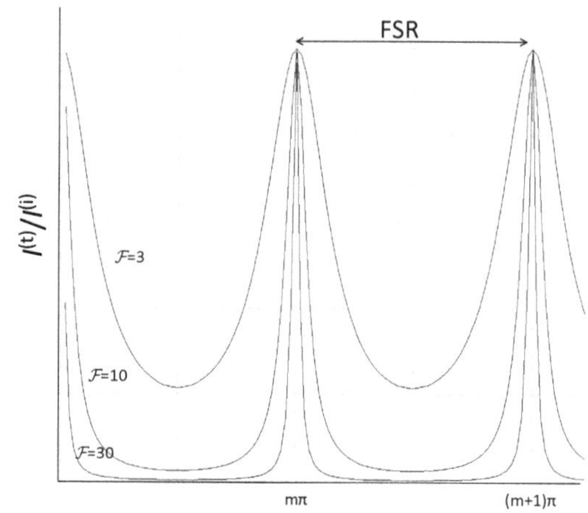

Fig. 47. Transmitted light of the Fabry–Perot cavity in resonance. The frequency gap between adjacent resonances is called FSR, which is given by $\Delta\nu_{\rm FSR} = \frac{c}{2\ell}$. The Finesse, $\mathcal{F}$, is defined by the ratio of $\Delta\nu_{\rm FSR}$ and $\Delta\Omega/2\pi$ (width of resonance).

Also, the amplitude of the transmitted light, A_t, is

$$A_t = [t_1t_2e^{-i\Delta} + t_1t_2(ir_1)(ir_2)e^{-3i\Delta} + t_1t_2(ir_1)^2(ir_2)^2e^{-5i\Delta} + \cdots]A_i,$$
$$= t_1t_2e^{-i\Delta}\sum_{n=0}^{\infty}(ir_1)^n(ir_2)^ne^{-2in\Delta}A_i = \frac{t_1t_2e^{-i\Delta}}{1+r_1r_2e^{-2i\Delta}}A_i. \quad \text{(C.2)}$$

When the condition $e^{-2i\Delta} = -1$ holds, light is resonantly trapped; the transmitted light is drawn in Fig. 47. The gap between the resonances is called FSR, which is given by $\Delta\nu_{FSR} = \frac{c}{2\ell}$; the Finesse, $\mathcal{F}$, is defined by the ratio of $\Delta\nu_{FSR}$ and $\Delta\Omega/2\pi = \Delta\nu$ (width of resonance):

$$\mathcal{F} = \frac{\Delta\nu_{FSR}}{\Delta\nu}. \quad \text{(C.3)}$$

The power of the transmitted light is given by

$$\frac{I_t}{I_i} = \left|\frac{t_1t_2e^{-i(\Delta+\delta)}}{1-r_1r_2e^{-2i\delta}}\right|^2 = \left(\frac{t_1t_2}{1-r_1r_2}\right)^2\frac{1}{1+F\sin^2\delta}, \quad \text{(C.4)}$$

where

$$F = \frac{4r_1r_2}{(1-r_1r_2)^2} = \left(\frac{2\mathcal{F}}{\pi}\right)^2. \quad \text{(C.5)}$$

This represents a transfer function of a bandpass filter with a Q-value of $2\ell\mathcal{F}/\lambda$.

C.2. *Frequency response of a Fabry–Perot Michelson interferometer*

A Fabry–Perot Michelson interferometer has Fabry–Perot cavities in place of arm mirrors on the Michelson interferometer. The frequency response of the Fabry–Perot

Michelson interferometer is obtained by applying repeatedly the calculation used in the simple Michelson interferometer, which considering amplitude reduction by r_1r_2 with a retarded phase of 2Δ in each light return. The response function is

$$H_{FP}(\omega) = \frac{\alpha\Omega}{\omega}\sin\left(\frac{\omega\ell}{c}\right)e^{-i\omega\ell/c}\frac{1}{1-r_1r_2e^{-2i\omega\ell/c}}, \tag{C.6}$$

where $\alpha = \frac{t_1^2r_2}{1-r_1r_2}$. Also,

$$|H_{FP}(\omega)| = \frac{\alpha\Omega}{\omega(1-r_1r_2)}\frac{\left|\sin\frac{\omega\ell}{c}\right|}{\sqrt{1+F\sin^2\frac{\omega\ell}{c}}}, \tag{C.7}$$

where F is given by Eq. (C.5).

Appendix D. Newtonian Noise

If the surrounding gravity gradient changes, suspended test masses experience acceleration due to the changes, which causes the so-called Newtonian noise. The source of the gravity gradient change is any density change due to some elastic deformation of the ground, atmospheric pressure, underground water level and so on. As is easily expected, since the period of the vibration is long compared with other noisy dynamics, it affects the sensitivity at lower than a few Hz in third-generation detectors. The first analytical estimation was made by Saulson,[181] where $3\times10^{-20}\,\mathrm{m}/\sqrt{\mathrm{Hz}}$ was estimated at 10 Hz by seismic noise. Although it is not harmful in second-generation detectors, suppression of the Newtonian noise is a benefit to widen the frequency band towards lower frequency, where astrophysical sources are much more abundant. An experiment to reduce the effect on test masses was conducted by feeding back filtered seismic array data.[182]

References

1. J. Weber, *Phys. Rev. Lett.* **22** (1969) 1320–1324.
2. R. W. P. Drever *et al.*, *Nature* **246** (1973) 340–344.
3. V. B. Braginskii *et al.*, *Sov. Phys.-JETP* **39** (1974) 387–392.
4. H. Billing *et al.*, *Lett. Nuovo Cimento* **12** (1975) 111–116.
5. D. H. Douglass *et al.*, *Phys. Rev. Lett.* **35** (1975) 480–483.
6. H. Hirakawa and K. Narihara, *Phys. Rev. Lett.* **35** (1975) 330–334.
7. K. Kuroda, W.-T. Ni and W.-P. Pan, Gravitational waves; Classification, Methods of detection, Sensitivities, and Sources, Chapter 10, in *One Hundred Years of General Relativity: From Genesis and Empirical Foundations in Gravitational Waves, Cosmology and Quantum Gravity*, ed. W.-T. Ni (World Scientific, Singapore, 2015), *Int. J. Mod. Phys. D* **24** (2015) 1530031.
8. A. Abramovici *et al.*, *Science* **256** (1992) 325–333.
9. F. Acernese *et al.*, *Class. Quantum Grav.* **25** (2008) 184001.
10. H. Lück *et al.*, *Class. Quantum Grav.* **23** (2006) S71–S78.
11. M. Ando *et al.*, *Phys. Rev. Lett.* **86** (2001) 3950–3954.

12. J. Abadie *et al.*, *Phys. Rev. D* **85** (2012) 082002.
13. G. H. Harry, *Class. Quantum Grav.* **27** (2010) 084006 (12 pp).
14. F. Acernese *et al.*, arXiv:1408.3978v3 [gr-qc].
15. H. Lück *et al.*, *J. Phys. Conf. Ser.* **228** (2010) 012012.
16. Y. Aso *et al.*, *Phys. Rev. D* **88** (2013) 043007.
17. M. Punturo *et al.*, *Class. Quantum Grav.* **27** (2010) 194002.
18. I. H. Stairs, *Science* **304** (2004) 547–552.
19. K. S. Thorne, *300 Years of Gravitation, Gravitational Radiation*, eds. S. W. Hawking and W. Israel (Cambridge University Press, 1987), pp. 330–458.
20. M. Shibata *et al.*, *Phys. Rev. D* **71** (2005) 084021.
21. V. Kalogera *et al.*, *Astrophys. J.* **601** (2004) L179–L182.
22. C. Kim *et al.*, *New Astron. Rev.* **54** (2010) 148–151.
23. K. Kyutoku *et al.*, *Phys. Rev. D* **84** (2011) 064018.
24. M. Dominik, *Astrophys. J.* **759** (2012) 52–79.
25. Y.-B. Bae *et al.*, *Mon. Not. R. Astron. Soc.* **440** (2014) 2714–2725.
26. H. Dimmelmeier *et al.*, *Astron. Astrophys.* **388** (2002) 917–935: **393** (2002) 523–542.
27. M. Shibata and Y. Sekiguchi, *Phys. Rev. D* **71** (2005) 024014.
28. K. Kotake, *J. Phys. Conf. Ser.* **229** (2010) 012011.
29. M. Shibata, *Gravitational Wave and Numerical Relativity*, Vol. 3 (University of Tokyo Press, 2007) (in Japanese).
30. C. W. Misner, K. S. Thorne and J. A. Wheeler, *Gravitation*, Chap. 35 (W. H. Freeman and Company San Francisco, 1973).
31. C. W. Misner, K. S. Thorne and J. A. Wheeler, *Gravitation*, Chap. 37 (W. H. Freeman and Company San Francisco, 1973).
32. J. Weber, *Phys. Rev.* **117** (1969) 306–313.
33. H. J. Paik and R. V. Wagoner, *Phys. Rev. D* **13** (1976) 2694–2699.
34. H. Hirakawa *et al.*, *J. Phys. Soc. Jpn.* **41** (1976) 1093–1101.
35. K. Narihara and H. Hirakawa, *J. Appl. Phys.* **15** (1976) 833–842.
36. H. J. Paik, *Phys. Rev. D* **15** (1977) 409–415.
37. W. W. Johnson and S. M. Merkowitz, *Phys. Rev. Lett.* **70** (1993) 2367–2370.
38. N. Ashby and J. Dreitlein, *Phys. Rev. D* **12** (1975) 336–349.
39. R. V. Wagoner and H. J. Paik, in *Proc. Int. Symp. Experimental Gravitation* (RomanAccademia Nazionale dei Linccei, Roma, 1977), pp. 257–265.
40. A. de Waard *et al.*, *Class. Quantum Grav.* **21** (2004) S465–S471.
41. O. D. Aguiar *et al.*, *Class. Quantum Grav.* **25** (2008) 114042.
42. M. E. Gertsenshtein and V. I. Pustovoit, *Sov. Phys.-JETP* **16** (1962) 433–435.
43. R. I. Forward, *Phys. Rev. D* **17** (1978) 379–390.
44. P. R. Saulson, *Am. J. Phys.* **65** (1997) 501–505.
45. LIGO Scientific Collab. (J. R. Smith), arXiv:0902.0381v2 [gr-qc].
46. C. M. Caves *et al.*, *Rev. Mod. Phys.* **52** (1980) 341–392.
47. C. Cinquegrana *et al.*, *Phys. Rev. D* **48** (1993) 448–465.
48. C. Cinquegrana *et al.*, *Phys. Rev. D* **50** (1994) 3596–3607.
49. P. R. Saulson, *Class. Quantum Grav.* **14** (1997) 2435–2454.
50. R. P. Giffard, *Phys. Rev. D* **14** (1976) 2478–2486.
51. P. Astone *et al.*, *Europhys. Lett.* **16** (1991) 231–235.
52. H. J. Paik, *J. Appl. Phys.* **47** (1976) 1168–1178.
53. R. P. Giffard *et al.*, *Physics and Astrophysics of Neutron Stars and Black Holes* (Society Italiana di Fisica, Bologna, 1978), p. 166.
54. S. Kimura *et al.*, *Phys. Lett. A* **81** (1981) 302–304.
55. T. Suzuki *et al.*, *Phys. Lett. A* **67** (1978) 2–4.

56. T. Suzuki, in *Proc. 1st Edoardo Amaldi Conf. Gravitational Wave Experiment*, Villa Tuscolana, Frascati, Rome, 14–17 June 1994.
57. P. Astone *et al.*, *Phys. Rev. D* **47** (1993) 362–375.
58. E. Mauceli *et al.*, *Phys. Rev. D* **54** (1996) 1264–1275.
59. D. G. Blair *et al.*, *Phys. Rev. Lett.* **74** (1995) 1908–1911.
60. P. Astone *et al.*, *Astropart. Phys.* **7** (1997) 231–241.
61. G. A. Prodi *et al.*, in *Proc. 2nd Edoardo Amaldi Conf.*, Geneve, 1–4 July, 1997), pp. 148–158.
62. S. P. Boughn *et al.*, *Phys. Rev. Lett.* **38** (1977) 454–457.
63. M. Cerdonio *et al.*, in *Proc. XXXIth Rencontres De Moriond, Gravitational Waves and Experimental Gravity*, eds. J. T. T. Van, J. D. Dumarchez, S. Reynaud, C. Salomon, S. Thorsett and J. Y. Vinet (World Publishers, Hanoi, 2000), pp. 33–43.
64. K. Maischberger *et al.*, in *Proc. Second Marcel Grossmann Meeting on General Relativity*, ed. R. Ruffini (North-Holland Publishing Company, 1982), pp. 1083–1100.
65. D. Shoemaker *et al.*, *Phys. Rev. D.* **38** (1988) 423–432.
66. D. I. Robertson *et al.*, *Rev. Sci. Instrum.* **66** (1995) 4447–4452.
67. A. Abramovici *et al.*, *Phys. Lett. A* **218** (1996) 157–163.
68. R. Takahashi *et al.*, *Phys. Lett. A* **187** (1994) 157–162.
69. E. G. Heflin and N. Kawashima, *ISAS Research Note* **567** (1995) 1–12.
70. K. Danzmann *et al.*, *Proposal for a 600 m Laser-Interferometric Gravitational Wave Antenna* **M P Q 190** (1994) 1–17.
71. D. Tatsumi and TAMA Collab., *J. Phys. Conf. Ser.* **120** (2008) 032011.
72. B. P. Abbott *et al.*, *Rep. Prog. Phys.* **72** (2009) 076901.
73. G. H. Harry *et al.*, *Optical Coatings and Thermal Noise in Precision Measurement* (Cambridge University Press, 2012).
74. J. Hough and S. Rowan, *J. Opt. A, Pure Appl. Opt.* **7** (2005) S257–S264.
75. R. X. Adhikari, *Rev. Mod. Phys.* **86** (2014) 121–151.
76. S. Gossler *et al.*, *Class. Quantum Grav.* **19** (2002) 1835–1842.
77. M. Ohashi *et al.*, *Class. Quantum Grav.* **20** (2003) S599–S607.
78. M. G. Beker, Doctor Thesis, de Vrije Universiteit Amsterdam (2013), pp. 189–192.
79. R. W. P. Drever *et al.*, in *Proc. Ninth Int. Conf. General Relativity and Gravitation*, ed. E. Schmutzer (Cambridge University Press, 1980), pp. 265–267.
80. M. Ohashi, Doctor Thesis, University of Tokyo (1994).
81. B. Meers, *Phys. Rev. D* **38** (1988) 2317–2326.
82. S. Miyoki, Doctor Thesis, The University of Tokyo (1996).
83. T. J. Kane *et al.*, *IEEE J Quantum Electron.* **21** (1985) 1195–1210.
84. K. Takeno *et al.*, *Opt. Lett.* **30** (2005) 2110–2112.
85. S. T. Yang *et al.*, *Opt. Lett.* **21** (1996) 1676–1678.
86. B. Willke, *Laser Photon. Rev.* **4** (2010) 780–794.
87. C. N. Man, *Class. Quantum Grav.* **20** (2003) S117–S125.
88. N. Uehara *et al.*, *Opt. Lett.* **20** (1995) 530–532.
89. R. W. P. Drever, in *Gravitational Radiation, Les Houches 1982*, eds. N. Deruelle and T. Piran (North Holland, Amsterdam, 1983), pp. 321–338.
90. P. Fritschel *et al.*, *Appl. Opt.* **31** (1992) 1412–1418.
91. M. W. Regehr *et al.*, *Opt. Lett.* **20** (1995) 1507–1509.
92. D. Schnier *et al.*, *Phys. Lett. A* **225** (1997) 210–216.
93. M. Ando *et al.*, *Phys. Lett. A* **248** (1998) 145–150.
94. S. Sato *et al.*, *Appl. Opt.* **39** (2000) 4616–4620.
95. M. W. Regehr, Doctor Thesis (California Institute of Technology, 1995).
96. R. Flaminio and H. Heitmann, *Phys. Lett. A* **214** (1996) 112–122.

97. M. Ando *et al.*, *Phys. Lett. A* **237** (1997) 12–20.
98. K. Arai *et al.*, *Phys. Lett. A* **273** (2000) 15–24.
99. J.-Y. Vinet *et al.*, *Phys. Rev. D* **38** (1988) 433–447.
100. B. J. Meers, *Phys. Lett. A* **142** (1989) 465–470.
101. K. A. Strain and B. J. Meers, *Phys. Rev. Lett.* **66** (1991) 1391–1394.
102. G. Heinzel *et al.*, *Phys. Rev. Lett.* **81** (1998) 5493–5496.
103. J. Mizuno *et al.*, *Phys. Lett. A* **175** (1993) 273–276.
104. K. A. Strain *et al.*, *Appl. Opt.* **42** (2003) 1244–1256.
105. S. D. Penn *et al.*, *Class. Quantum Grav.* **20** (2003) 2917–2928.
106. P. Hello and J.-Y. Vinet, *Phys. Lett. A* **230** (1997) 12–18.
107. P. J. Cote and L. V. Meisel, *Phys. Rev. Lett.* **67** (1991) 1334–1337.
108. S. Kawamura and M. E. Zucker, *Appl. Opt.* **33** (1994) 3912–3918.
109. H. B. Callen and T. A. Welton, *Phys. Rev.* **83** (1951) 34–40.
110. E. Majorana and Y. Ogawa, *Phys. Lett. A* **233** (1997) 162–168.
111. K. Yamamoto, Doctor Thesis, The University of Tokyo (2000).
112. N. Nakagawa *et al.*, *Rev. Sci. Instrum.* **68** (1997) 3553–3556.
113. Y. Levin, *Phys. Rev. D* **57** (1998) 659–663.
114. J. R. Hutchinson, *J. Appl. Mech.* **47** (1980) 901–907.
115. A. Gillespie and F. Raab, *Phys. Rev. D* **52** (1995) 577–585.
116. K. Tsubono, Private Communication.
117. J. E. Logan *et al.*, *Phys. Lett. A* **170** (1992) 352–358.
118. P. R. Saulson, *Pys. Rev. D* **42** (1990) 2437–2445.
119. J. E. Logan *et al.*, *Phys. Lett. A* **183** (1993) 145–152.
120. M. M. Fejer *et al.*, *Phys. Rev. D* **70** (2004) 082003.
121. G. M. Harry *et al.*, *Class. Quantum Grav.* **24** (2007) 405–415.
122. R. Flaminio *et al.*, *Class. Quantum Grav.* **27** (2010) 084030.
123. H.-W. Pan *et al.*, *Opt. Exp.* **22** (2014) 29847.
124. T. Hong *et al.*, *Phys. Rev. D* **87** (2014) 082001.
125. K. Yamamoto *et al.*, *Phys. Rev. D* **74** (2006) 022002.
126. M. Granata *et al.*, *Opt. Lett.* **38** (2013) 5268–5271.
127. I. Martin *et al.*, *Class. Quantum Grav.* **31** (2014) 035019.
128. E. Hirose *et al.*, *Phys. Rev. D* **90** (2014) 102004 (6 pp).
129. D. Sigg *et al.*, *Class. Quantum Grav.* **21** (2004) S409–S415.
130. K. Kuroda *et al.*, *Int. J. Mod. Phys. D* **8** (1999) 557–579.
131. F. Acernese *et al.*, *Class. Quantum Grav.* **32** (2015) 024001.
132. R. J. Glauber, *Phys. Rev.* **131** (1963) 2766–2788.
133. A. Yariv, *Quantum Electronics*, Chap. 17, 3rd edn. (John Wiley & Sons, 2006).
134. V. B. Braginsky and F. Y. Khalili, *Quantum Measurement* (Cambridge University Press, Cambridge, England, 1992).
135. H. J. Kimble *et al.*, *Phys. Rev. D* **65** (2001) 022002.
136. V. B. Braginsky and Y. I. Vorontsov, *Sov. Phys. Usp.* **17** (1975) 644–650.
137. K. McKenzie *et al.*, *Phys. Rev. Lett.* **88** (2002) 231102.
138. K. Goda *et al.*, *Nature Phys.* **4** (2008) 472–476.
139. H. Grote *et al.*, *Phys. Rev. Lett.* **110** (2013) 181101.
140. The LIGO Scientific Collaboration, *Nature Photon.* **7** (2013) 613–619.
141. T. Corbitt *et al.*, *Phys. Rev. A* **73** (2006) 023801.
142. S. S. Y. Chua *et al.*, *Class. Quantum Grav.* **31** (2014) 183001.
143. G. Mueller *et al.*, *Class. Quantum Grav.* **19** (2002) 1793–1801.
144. R. L. Ward *et al.*, *Class. Quantum Grav.* **25** (2008) 114030.
145. V. B. Braginsky *et al.*, *Phys. Lett. A* **287** (2001) 331–338.

146. V. B. Braginsky *et al.*, *Phys. Lett. A* **305** (2002) 111–124.
147. L. Ju *et al.*, *Class. Quantum Grav.* **26** (2009) 015002.
148. M. Evans *et al.*, *Phys. Rev. Lett.* **114** (2015) 161102.
149. Advanced LIGO (2015) LIGO-G1401390-v5.
150. https://dcc.ligo.org/cgi-bin/DocDB/ShowDocument?docid=75988.
151. F. Frasconi *et al.*, *J. Phys., Conf. Ser.* **120** (2008) 032007.
152. S. Braccini *et al.*, *Astropart. Phys.* **23** (2005) 557–565.
153. T. Accadia *et al.*, *Class. Quantum Grav.* **28** (2011) 114002.
154. J. Aasi *et al.*, *Astrophys. J.* **785** (2014) 119.
155. A. Brooks *et al.*, *Gen. Relativ. Gravit.* **37** (2005) 1575–1580.
156. W. Winkler *et al.*, *Opt. Commun.* **280** (2007) 492–499.
157. M. V. Plissi *et al.*, *Rev. Sci. Instrum.* **71** (2000) 2539–2545.
158. H. Grote for LSC, *Class. Quantum Grav.* **25** (2008) 114043.
159. S. Hild *et al.*, *Class. Quantum Grav.* **32** (2006) 66–73.
160. C. Affeldt *et al.*, *Class. Quantum Grav.* **31** (2014) 224002.
161. R. Takahashi *et al.*, *Class. Quantum Grav.* **25** (2008) 114036.
162. K. Agatsuma *et al.*, *J. Phys., Conf. Ser.* **122** (2008) 012013.
163. S. Sato *et al.*, *Phys. Rev. D* **69** (2004) 102005.
164. T. Uchiyama *et al.*, *Phys. Rev. Lett.* **108** (2012) 141101.
165. T. Uchiyama *et al.*, *Phys. Lett. A* **242** (1998) 211–214.
166. T. Tomaru *et al.*, *Phys. Lett. A* **301** (2002) 215–219.
167. T. Uchiyama *et al.*, *Phys. Lett. A* **261** (1999) 5–11.
168. T. Uchiyama *et al.*, *Phys. Lett. A* **273** (2000) 310–315.
169. T. Tomaru *et al.*, *Jpn. J. Appl. Phys.* **47** (2008) 1771–1774.
170. S. Miyoki *et al.*, *Cryogenics* **41** (2001) 415–420.
171. T. Tomaru *et al.*, *Phys. Lett. A* **283** (2001) 80–84.
172. T. Tomaru *et al.*, *Class. Quantum Grav.* **19** (2002) 2045–2049.
173. S. Miyoki *et al.*, *Class. Quantum Grav.* **21** (2004) S1173–S1181.
174. K. Agatsuma *et al.*, *Phys. Rev. Lett.* **104** (2010) 040602.
175. Y. Ikushima *et al.*, *TEION KOGAKU* **42** (2007) 1–8.
176. K. Kuroda, *J. Cryo. Super. Soc. Jpn.* **46** (2011) 385–391 (in Japanese).
177. F. Kawazoe *et al.*, *Class. Quantum Grav.* **25** (2008) 195008.
178. http://www.et-gw.eu/descriptionelites.
179. T. Nakamura *et al.*, *Sensing Gravitational Waves* (Kyoto University Press, 1998) (in Japanese).
180. D. Hils and J. L. Hall, *Rev. Sci. Instrum.* **58** (1987) 1406–1412.
181. P. R. Saulson, *Phys. Rev. D* **30** (1984) 732–736.
182. J. C. Driggers *et al.*, *Phys. Rev. D* **86** (2012) 102001.

Chapter 12

Gravitational wave detection in space

Wei-Tou Ni

School of Optical-Electrical and Computer Engineering,
University of Shanghai for Science and Technology,
516, Jun Gong Rd., Shanghai 200093, P. R. China

Kavli Institute for Theoretical Physics China,
CAS, Beijing 100190, P. R. China
weitouni@163.com

weitou@gmail.com

Gravitational Wave (GW) detection in space is aimed at low frequency band (100 nHz–100 mHz) and middle frequency band (100 mHz–10 Hz). The science goals are the detection of GWs from (i) Supermassive Black Holes; (ii) Extreme-Mass-Ratio Black Hole Inspirals; (iii) Intermediate-Mass Black Holes; (iv) Galactic Compact Binaries and (v) Relic GW Background. In this paper, we present an overview on the sensitivity, orbit design, basic orbit configuration, angular resolution, orbit optimization, deployment, time-delay interferometry (TDI) and payload concept of the current proposed GW detectors in space under study. The detector proposals under study have arm length ranging from 1000 km to 1.3×10^9 km (8.6 AU) including (a) Solar orbiting detectors — (ASTROD Astrodynamical Space Test of Relativity using Optical Devices (ASTROD-GW) optimized for GW detection), Big Bang Observer (BBO), DECi-hertz Interferometer GW Observatory (DECIGO), evolved LISA (e-LISA), Laser Interferometer Space Antenna (LISA), other LISA-type detectors such as ALIA, TAIJI etc. (in Earthlike solar orbits), and Super-ASTROD (in Jupiterlike solar orbits); and (b) Earth orbiting detectors — ASTROD-EM/LAGRANGE, GADFLI/GEOGRAWI/g-LISA, OMEGA and TIANQIN.

Keywords: Gravitational waves; space gravitational wave detectors; dark energy; galaxy co-evolution with black holes; inflation; galactic compact binaries.

PACS Numbers(s): 04.80.Nn, 04.80.−y, 95.30.Sf, 95.55.Ym, 98.62.Ai, 98.80.Es

1. Introduction

Gravitational Wave (GW) detection has been a focused research subject for some time. With the announcement of LIGO direct GW detection,[1,2] we are fully ushered into the age of GW astronomy. Second-generation ground-based interferometers are being upgraded/completed for GW detection in the high-frequency band

(10–100 kHz; see Refs. 3–5 for a complete spectral classification of GWs).[6] Observational data from Pulsar Timing Arrays (PTAs) are being accumulated for the first GW detection in the very low frequency band (300 pHz–100 nHz).[7] Collaborations working on Cosmic Microwave Background (CMB) observations are actively pushing their sensitivities further for detecting imprints of primordial GWs in the Hubble frequency band (1 aHz–10 fHz) on B-mode polarizations.[8] LISA (Laser Interferometer Space Antenna)[9] Pathfinder[10] launched on 3 December 2015 has successfully demonstrated the drag-free technology[11] for space detection of GWs in the middle and low frequency band (0.1 Hz–10 Hz; 100 nHz–0.1 Hz). The activities are mounting in this centennial year (2015–2016) of the establishment of general relativity.

With the invention of lasers in 1960, the implementation of satellite laser ranging and lunar laser ranging in 1960s and the development of drag-free navigation for geodesy in 1970s, concept of laser interferometry in space for GW detection were developed in 1980s. The first public proposal on space interferometers for GW detection was presented at the Second International Conference on Precision Measurement and Fundamental Constants (PMFC-II), 8–12 June 1981, in Gaithersburg.[12,13] In this seminal proposal, Faller and Bender raised possible GW mission concepts in space using laser interferometry. Two basic ingredients were addressed — drag-free navigation for the reduction of perturbing forces on the spacecraft (S/C) and laser interferometry for the sensitivity of measurement. LISA-like S/C orbit formation was reached in 1985 in the proposal Laser Antenna for Gravitational-radiation Observation in Space (LAGOS).[14] A schematic of LISA-type orbit configuration is shown in Fig. 1. It is natural for people like Bender and Faller working in lunar laser ranging and measuring free-fall acceleration using interferometry to propose such an experiment. In fact, test mass free fall inside a falling shroud in vacuum in the interferometric measurement of the Earth's gravitational acceleration can be considered as a passive drag-free navigation device.[15] The discrepancy in the absolute gravimeter comparison at the Bureau International des Poids et Mesures (BIPM) is partially resolved using correction to interferometric measurements of absolute gravity arising from the finite speed of light.[16] In the S/C tracking, the finite velocity of light has always been incorporated. Both the test mass for GW missions and the test mass of interferometric gravimeter can be regarded as freely falling objects in the solar system and tracked using astrodynamical equation. Thus, we see the interplay among space geodesy, Galileo Equivalence Principle (Universality of Free Fall) experiments in space and GW detection missions. Recent development for a GRACE follow-on mission SAGM (Space Advanced Gravity Measurements),[17] TEPO[18] (testing the equivalence principle with optical readout in space) and TIANQIN[19] (a space-borne GW detector) can be considered as such an example.

A big step for the GW detection in space is the 1993 ESA M3 Assessment study of LISA and later recommendation as the third cornerstone of "Horizon 2000 Plus". After 2000, LISA became a joint ESA–NASA mission until the 2011 NASA

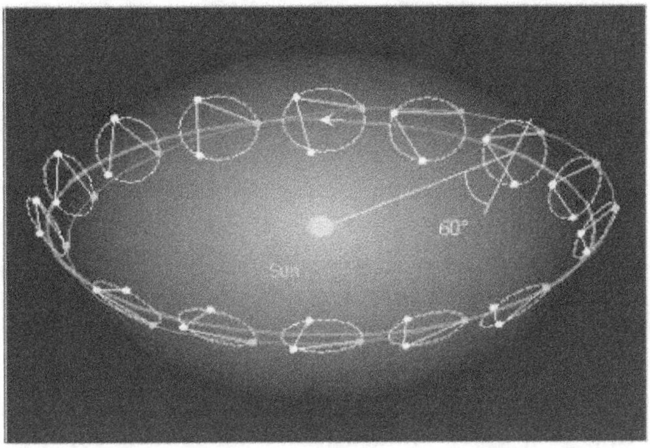

Fig. 1. Schematic of LISA-type orbit configuration in Earthlike solar orbit.[9] (For color version, see page I-CP14.)

withdrawal. In 1998, LISA Pathfinder was selected as the second of the European Space Agency's Small Missions for Advanced Research in Technology (SMART) to develop and to test the demanding drag-free technology. At this occasion of Centennial Celebration of General Relativity, ESA has successfully launched the LISA Pathfinder on a Vega rocket from Europe's spaceport in Kourou, French Guiana on 3 December 2015, and has successfully demonstrated the drag-free technology[11] for observing GWs from space. Based on the ongoing technological development for LISA Pathfinder, ESA has sponsored a technology reference study (completed in 2008) for the fundamental physics explorer as a common bus for fundamental physics missions.[20] New Gravitational-wave Observatory (NGO)/evolved LISA (eLISA),[21] down-scaled from 5 million km to 1 million km arm length, was proposed in 2011 to accommodate the budget change and received excellent evaluation. In November 2013, ESA announced the selection of the Science Themes for the L2 and L3 launch opportunities — the "Hot and Energetic Universe" for L2 and "The Gravitational Universe" for L3.[22] ESA L3 mission is likely to have a launch opportunity in 2034.[22] Since eLISA/NGO GW mission concept is the major candidate at this time and it takes one year to transfer to the science orbit, a starting time for science phase is likely in 2035. Since 2035 is still 20 years away, it is not yet the time to freeze the specific mission concept. At present a comparison of laser measurement technology and atom interferometry is underway in ESA.

The general concept of Astrodynamical Space Test of Relativity using Optical Devices (ASTROD) is to have a constellation of drag-free S/Cs navigate through the solar system and range with one another using optical devices to map the solar system gravitational field, to measure related solar system parameters, to test relativistic gravity, to observe solar g-mode oscillations and to detect GWs. A baseline implementation of ASTROD was proposed in 1993 and has been under concept and laboratory studies since then.[23–30] In 1996, ASTROD I (Mini-ASTROD) with one

S/C ranging with ground stations was proposed for testing relativistic gravity and mapping the solar system.[23] The mission study shows that the precision of testing relativistic gravity in the solar system is achievable to 10^{-9}–10^{-8} in terms of Eddington parameter γ, which is more than three orders of improvement over the present precision, with accompanying improvement in other aspects of relativistic gravity.[31–35] Early in 2009, responding to the call for GW mission studies of Chinese Academy of Sciences (CAS), a dedicated mission concept ASTROD optimized for Gravitational Wave detection (ASTROD-GW) for GW detection with 3 S/C (spacecraft) orbiting near Sun–Earth Lagrange points L3, L4 and L5 respectively with nominal arm length of 260 million km was proposed and studied.[3,36–40] A schematic of ASTROD-GW orbit configuration with inclination is shown in Fig. 2.[3,41] Before the ASTROD-GW proposal, Super-ASTROD which was proposed in 1996[23] with S/C's in Jupiterlike orbits was studied as a dual mission for GW measurement and for cosmological model/relativistic gravity test in 2008.[42] With the proposal of ASTROD-GW, the baseline GW configuration of Super-ASTROD makes 3 out of 4–5 S/C orbiting near Sun–Jupiter Lagrange points L3, L4 and L5, respectively. For the possibility of a down scaled version of ASTROD-GW mission, the ASTROD-EM with the orbits of 3 S/C near Earth–Moon Lagrange points L3, L4 and L5 respectively has been under study.[43]

DECi-hertz Interferometer GW Observatory (DECIGO)[44] was proposed in 2001 with the aim of detecting GWs from early universe in the middle frequency observation band between the terrestrial band and the low frequency band of other space GW detectors. It will use a Fabry–Perot method (instead of a delay line method) as in the ground interferometers but with a 1000 km arm length. As a LISA follow-on, Big Bang Observer (BBO)[45] with arm length 50,000 km was proposed in the United States with a similar goal. A likely version of DECIGO/BBO is to have 12 S/Cs with correlated detection. They will be used for the direct measurement of the stochastic GW background by correlation analysis.[46] 6S/C-ASTROD-GW with two sets of ASTROD-GW has also been considered to possibly explore the relic GWs in the lower part of the low frequency band.[39,40] ALIA[47] of arm length

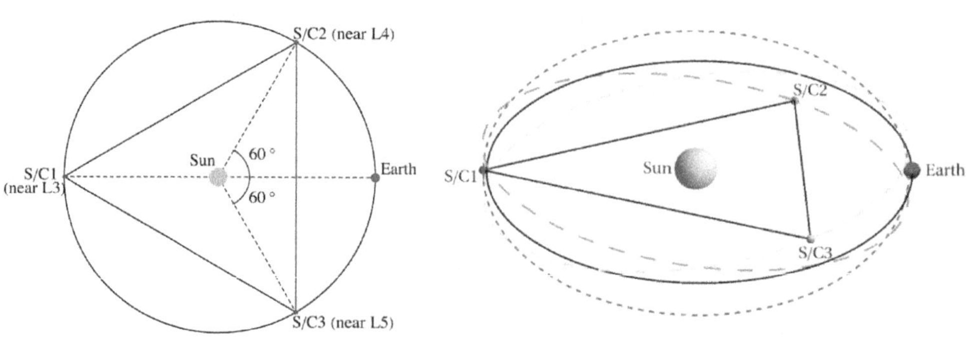

Fig. 2. Schematic of ASTROD-GW orbit configuration with inclination. Left, projection on the ecliptic plane; Right, 3D view with the scale of vertical axis multiplied tenfold.[3,41] (For color version, see page I-CP15.)

500,000 km was proposed as a less-ambitious LISA follow-on. TAIJI (also called ALIA descope)[48] of arm length 3 million km has also been proposed and under study with the main goal of detecting intermediate mass black hole binaries at high redshift.

After the end in 2011 of ESA-NASA partnership for flying LISA, NASA solicited "Concepts for the NASA Gravitational Wave Mission" proposals on 27 September 2011 for study of low cost GW missions (http://nspires.nasaprs.com/external/). geosynchronous LISA/GEOstationary GRAvitational Wave Interferometer (gLISA/GEOGRAWI),[49–51] Geostationary Antenna for Disturbance-Free Laser Interferometry (GADFLI),[52] and Laser Gravitational-wave Antenna at Geo-lunar Lagrange points (LAGRANGE)[53] was proposed and Orbiting Medium Explorer for Gravitational Astronomy (OMEGA)[54,55] re-emerged. OMEGA of arm length 1 million km was first proposed as a low-cost alternative to LISA in the 1990s. An artist's conception of the OMEGA mission configuration is shown in Fig. 3. In China, a GW mission in Earth orbit called TIANQIN[19] of arm length 110,000 km has been proposed and under study.

Table 1 lists the orbit configuration, arm length, orbit period, S/C number, acceleration noise and laser metrology noise of various GW space mission proposals. Figures 4–1 show respectively the strain Power Spectral Density (PSD) amplitude $[S_h(f)]^{1/2}$ versus frequency plot, the characteristic strain h_c versus frequency plot and the normalized GW spectral energy density $\Omega_{\rm gw}$ versus frequency plot for various GW detectors and sources in the low-frequency band and middle frequency band. The characteristic strain h_c, the strain PSD amplitude $[S_h(f)]^{1/2}$ and the normalized GW spectral energy density $\Omega_{\rm gw}$ are related as follows:

$$\begin{aligned} h_c(f) &= f^{1/2}[S_h(f)]^{1/2}; \\ \Omega_{\rm gw}(f) &= \left(\frac{2\pi^2}{3{H_0}^2}\right) f^3 S_h(f) = \left(\frac{2\pi^2}{3{H_0}^2}\right) f^2 {h_c}^2(f). \end{aligned} \tag{1}$$

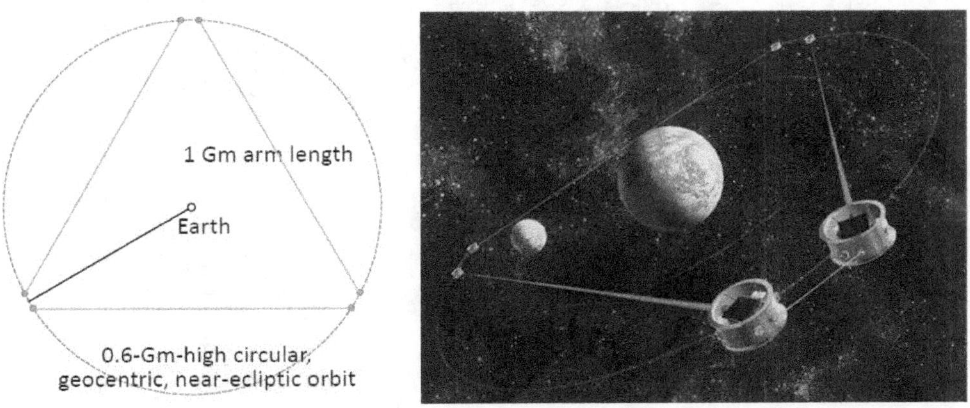

Fig. 3. Schematic (left) and artist's conception (right) of the OMEGA mission configuration.[55] (For color version, see page I-CP16.)

Table 1. A compilation of GW mission proposals.

Mission concept	S/C configuration	Arm length	Orbit period	S/C #	Acceleration noise [$fm/s^2/Hz^{1/2}$]	Laser metrology noise [$pm/Hz^{1/2}$]
	Solar-Orbit GW Mission Proposals					
LISA[9]	Earthlike solar orbits with 20° lag	5 Gm	1 year	3	3	20
eLISA[21]	Earthlike solar orbits with 10° lag	1 Gm	1 year	3	3	12(10)
ASTROD-GW[36–40]	Near Sun–Earth L3, L4, L5 points	260 Gm	1 year	3	3	1000
Big Bang Observer[45]	Earthlike solar orbits	0.05 Gm	1 year	12	0.03	1.4×10^{-5}
DECIGO[44]	Earthlike solar orbits	0.001 Gm	1 year	12	0.0004	2×10^{-6}
ALIA[47]	Earthlike solar orbits	0.5 Gm	1 year	3	0.3	0.6
TAIJI (ALIA-descope)[48]	Earthlike solar orbits	3 Gm	1 year	3	3	5–8
Super-ASTROD[42]	Near Sun–Jupiter L3, L4, L5 points (3 S/C), Jupiterlike solar orbit(s)(1–2 S/C)	1300 Gm	11 year	4 or 5	3	5000
	Earth-Orbit GW Mission Proposals					
OMEGA[54,55]	0.6 Gm height orbit	1 Gm	53.2 days	6	3	5
gLISA/GEOGRAWI[49–51]	Geostationary orbit	0.073 Gm	24 h	3	3, 30	0.3, 10
GADFLI[52]	Geostationary orbit	0.073 Gm	24 h	3	0.3, 3, 30	1
TIANQIN[19]	0.057 Gm height orbit	0.11 Gm	44 h	3	1	1
ASTROD-EM[43]	Near Earth–Moon L3, L4, L5 points	0.66 Gm	27.3 days	3	1	1
LAGRANGE[53]	Earth–Moon L3, L4, L5 points	0.66 Gm	27.3 days	3	3	5

Detailed accounts and explanations of Figs. 4–1 are given in Secs. 3–6 and in Ref. 5. A large part of these figures are taken from the corresponding low frequency band and middle frequency band of Figs. 2–4 in Ref. 5.

In the following section, we discuss the link of gravity (including GW) with orbit observations/experiments in the solar system. In Sec. 3, we review the methods and the most recent experimental results of radio Doppler spacecraft tracking. In Sec. 4, we explain the basic principle of laser-interferometric space mission for GW detection. In Sec. 5, we address the sensitivity spectra and review basic noises. In Sec. 6, we discuss the scientific goals of GW space missions. In Sec. 7, we address the basic orbit design using eLISA and ASTROD-GW as concrete examples. In

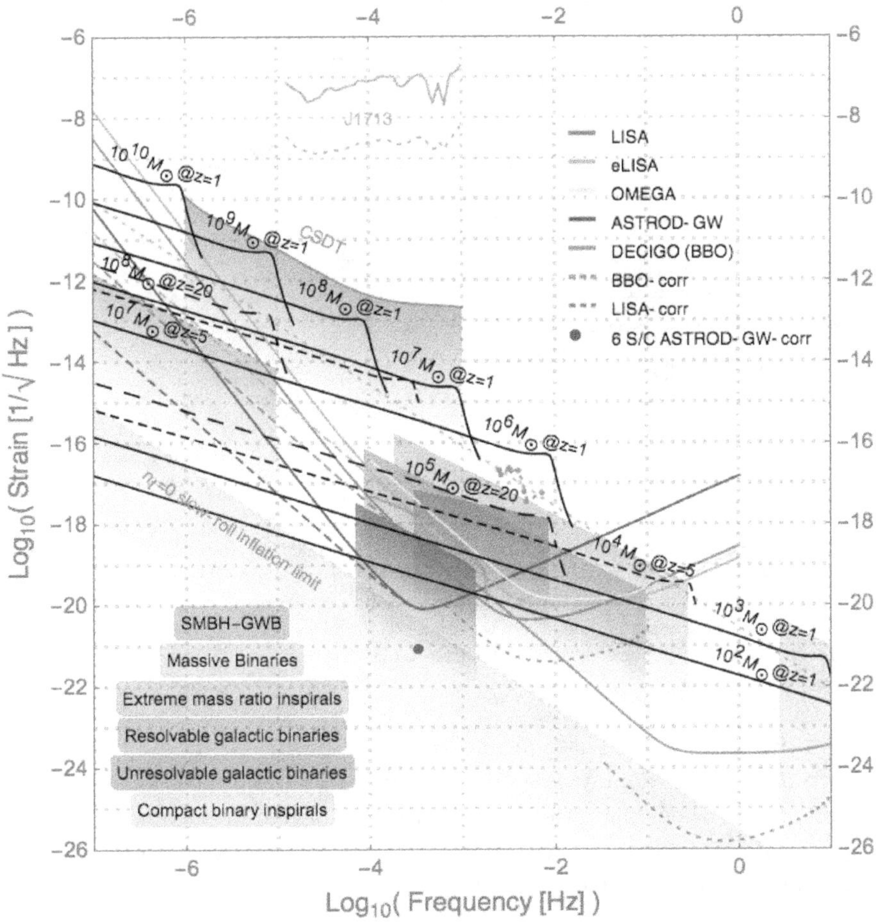

Fig. 4. Strain PSD amplitude versus frequency for various GW detectors and GW sources. The black lines show the inspiral, coalescence and oscillation phases of GW emission from various equal-mass black-hole binary mergers in circular orbits at various redshift: solid line, $z = 1$; dashed line, $z = 5$; long-dashed line $z = 20$. See text for more explanation. [Cassini Spacecraft Doppler Tracking (CSDT); Supermassive Black Hole-GW Background (SMBH-GWB).] (For color version, see page I-CP14.)

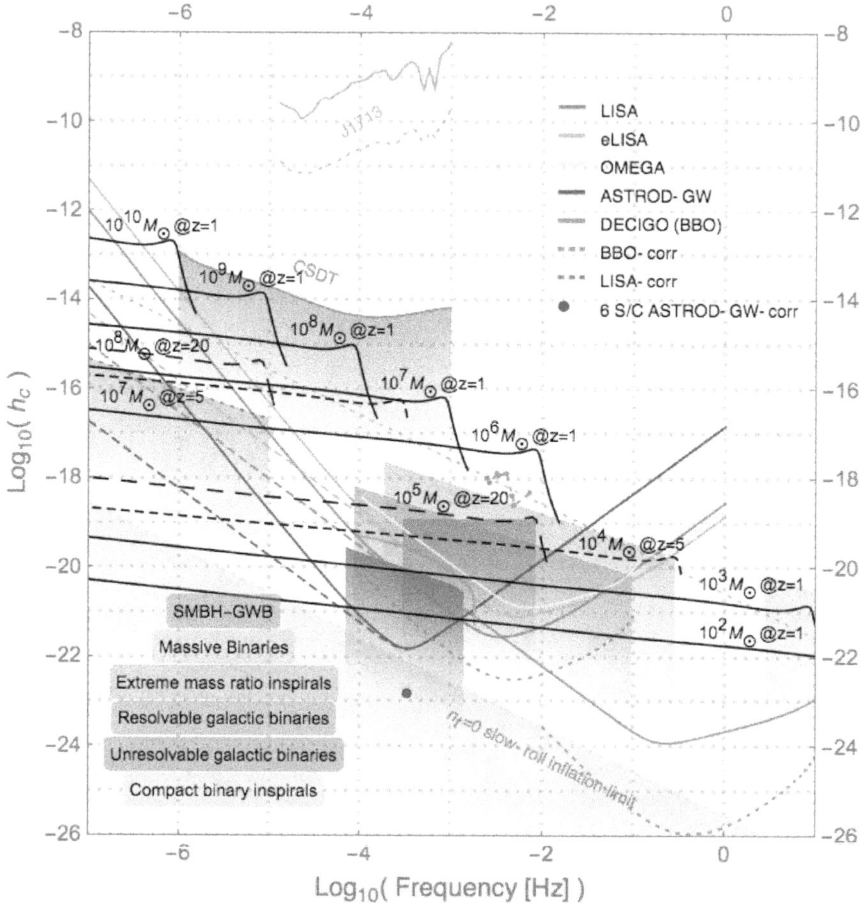

Fig. 5. Characteristic strain h_c versus frequency for various GW detectors and sources. The black lines show the inspiral, coalescence and oscillation phases of GW emission from various equal-mass black-hole binary mergers in circular orbits at various redshift: solid line, $z = 1$; dashed line, $z = 5$; long-dashed line $z = 20$. See text for more explanation. [Cassini Spacecraft Doppler Tracking (CSDT); Supermassive Black Hole-GW Background (SMBH-GWB).] (For color version, see page I-CP15.)

Sec. 8, we discuss the orbit design and orbit optimization using ephemerides. In Sec. 9, we discuss the deployment of spacecraft to various positions of Earthlike solar orbit, their propellant ratios and the total mass requirements. In Sec. 10, we discuss time delay interferometry (TDI). In Sec. 11, we discuss the payload. In Sec. 12, we summarize the paper and present an outlook.

2. Gravity and Orbit Observations/Experiments in the Solar System

Historically, the orbit and gravity observations/experiments in the solar system have been important resources for the development of fundamental physical laws as the

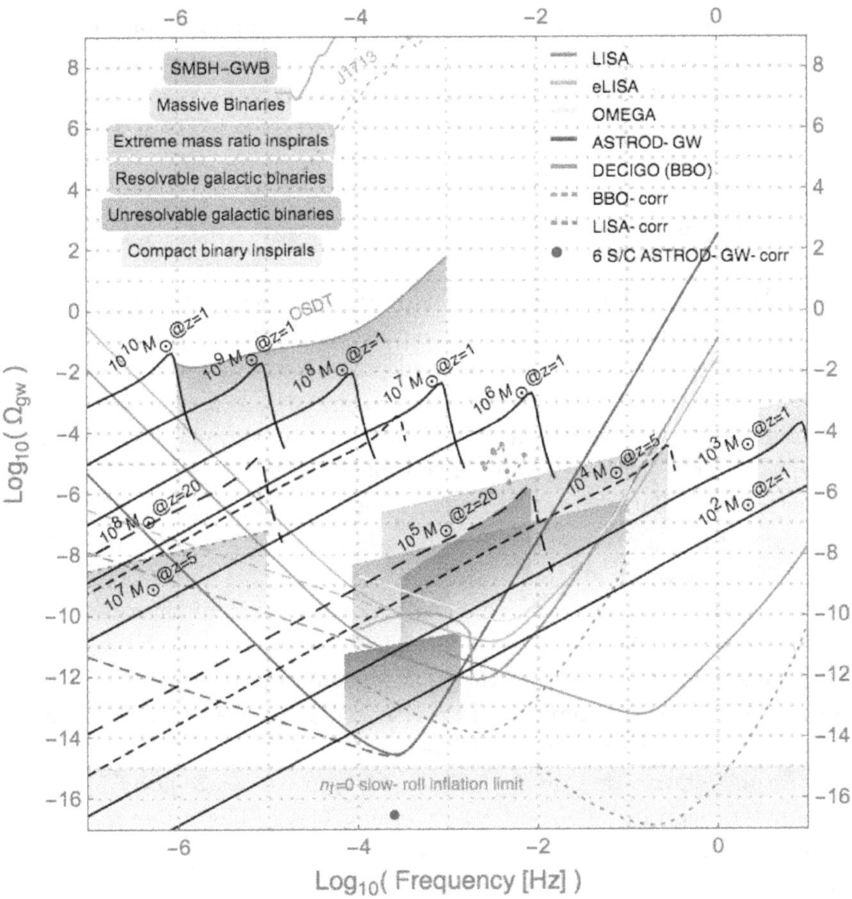

Normalized GW spectral energy density Ω_{gw} versus frequency for various GW detectors and GW sources. The black lines show the inspiral, coalescence and oscillation phases of GW emission from various equal-mass black-hole binary mergers in circular orbits at various redshift: solid line, $z = 1$; dashed line, $z = 5$; long-dashed line $z = 20$. See text for more explanation. [Cassini Spacecraft Doppler Tracking (CSDT); Supermassive Black Hole-GW Background (SMBH-GWB).] (For color version, see page I-CP16.)

precision and accuracy are improved. It is so for both the developments of Newtonian world system and Einstein's general relativity.[56–58] With the eminent improvement for orbit and gravity measurements pending, we are in a historical epoch for a great stride in the testing and development of fundamental laws. The gravitational field in the solar system is determined by three factors: the dynamic distribution of matter in the solar system; the dynamic distribution of matter outside the solar system (galactic, cosmological, etc.) and GWs propagating through the solar system. Different relativistic/cosmological theories of gravity make different predictions of the solar system gravitational field. Hence, precise measurements of the solar system gravitational field test these relativistic theories, in addition to enabling GW observations, determination of the matter distribution in the solar system and determination of the observable (testable) influence of our galaxy and cosmos. To measure the solar system gravitational field, we measure/monitor distance between different natural and/or artificial celestial bodies. In the solar system, the equation of motion of a celestial body or a spacecraft is given by the astrodynamical equation

$$\mathbf{a} = \mathbf{a}_{\rm N} + \mathbf{a}_{\rm 1PN} + \mathbf{a}_{\rm 2PN} + \mathbf{a}_{\rm Gal\text{-}Cosm} + \mathbf{a}_{\rm GW} + \mathbf{a}_{\rm nongrav}, \tag{2}$$

where $\mathbf{a}$ is the acceleration of the celestial body or spacecraft, $\mathbf{a}_{\rm N}$ is the acceleration due to Newtonian gravity, $\mathbf{a}_{\rm 1PN}$ the acceleration due to first post-Newtonian effects, $\mathbf{a}_{\rm 2PN}$ the acceleration due to second post-Newtonian effects, $\mathbf{a}_{\rm Gal\text{-}Cosm}$ the acceleration due to Galactic and cosmological gravity, $\mathbf{a}_{\rm GW}$ the acceleration due to GWs, and $\mathbf{a}_{\rm nongrav}$ the acceleration from all nongravitational origins.[3] Distances between spacecraft depend critically on the solar system gravity (including gravity induced by solar oscillations), underlying gravitational theory and incoming GWs. A precise measurement of these distances as a function of time will enable the cause of variation to be determined.

Ideally, it would be desirable to have a constellation of drag-free spacecraft navigate through the solar system and range with one another using optical devices (or other sensitive devices) to map the solar system gravitational field, to measure related solar system parameters, to test relativistic gravity, to observe solar g-mode oscillations, and to detect GWs.[3,59] Practically, certain orbit configurations are good for testing relativistic gravity; certain configurations are good for measuring solar parameters; certain are good for detecting GWs. These factors are integral part of mission designs for various purposes.[3,59]

To test relativistic gravity, the spacecraft needs to go into inner solar orbit where the solar gravity is stronger or to send signals passing near the solar limbs to get stronger influence from solar gravity. ASTROD I during the superior solar conjunctions to measure the Shapiro delay of light and with continuous laser ranging of 1 mm accuracy to improve the determination of relativistic parameters is such a mission proposal.[31–33] BepiColombo to be launched in 2017 is an ESA-JAXA mission under implementation.[60,61] One of its goals of radio science is to test relativistic gravity. In determining its orbit about Mercury, it will indirectly find the motion of the center of mass of Mercury with an accuracy several orders of magnitude better than what is possible by radar ranging to its surface. This is a good opportunity to measure Mercury's perihelion advance and the Shapiro time delay, and to improve on the other post-Newtonian parameters by a couple of orders of magnitude.[62]

To measure or to improve solar and planetary parameters, the spacecraft needs to go near the measured body or to have supreme sensitivity. NEAR (Near Earth Asteroid Rendezvous Mission: determined the mass $(6.687 \pm 0.003) \times 10^{18}$ gm and density $2.67 \pm 0.03\,\mathrm{gm/cm^3}$ of asteroid 433 Eros, its lower order gravitational-harmonics, and its rotation state using ground-based Doppler and range tracking of the NEAR spacecraft orbiting Eros together with images of the asteroid's surface landmarks), MESSENGER (MErcury Surface, Space ENvironment, GEochemistry, and Ranging: entered orbit around Mercury on March 18, 2011, deorbited as planned, and impacted the surface of Mercury on April 30, 2015. During this period, MESSENGER measured the gravity of Mercury and the state of the planetary core by utilizing the spacecraft's positioning data.) and ASTROD I (a mission proposal having a Venus swing-by for gravity assistance and for improved measurement of Venus gravity/multipole moments, with laser ranging of accuracy about 1 mm for improvement on the parameter determination of planets and asteroids) — are such examples.

For laser-interferometric GW detection without fast Doppler tracking (e.g. using optical combs), nearly equal arm lengths are required; LISA-like mission concepts and ASTROD-GW-like mission concepts are examples.

3. Doppler Tracking of Spacecraft

Radio Doppler tracking of spacecraft in a space mission can be used to constrain (or detect) the level of low-frequency GWs. The separated test masses of this GW detector are the Doppler tracking radio antenna on Earth and a distant S/C. Doppler tracking measures relative distance change. From these measurements, GWs can be detected or constrained. In 1967, Braginsky and Gertsenshtein[65] first proposed to use Doppler data of spacecraft tracking for GW searches. In 1971, Anderson[66] pursued this method of search with preexisting data. Davis[67] worked out the GW response of Doppler tracking for special cases in 1974; Estabrook and Walquist[68] analyzed the effect of GWs passing through the line-of-sight of S/C on the Doppler tracking frequency measurements in general in 1975 (see also Ref. 69).

In Doppler tracking of S/C, a highly stable master clock on Earth is used as a reference to control a monochromatic radio wave for transmitting to S/C (uplink). When S/C transponder receives the monochromatic radio wave, it phase-locks the local oscillator with or without a frequency offset and transponds the local oscillator signal back (to Earth station; downlink) coherently.

The one-way Doppler response $y(t)$ is defined as

$$y(t) \equiv \frac{\delta\nu}{\nu_0} \equiv \frac{(\nu_1(t) - \nu_0)}{\nu_0}, \tag{3}$$

where ν_0 is the frequency of emitted signal and ν_1 is the frequency of received signal. Far from the GW sources as it is in the present experimental/observational situations, the plane wave approximation is valid. For weak plane waves propagating in the z-direction in general relativity, we have the following spacetime metric:

$$ds^2 = dt^2 - (\delta_{ij} + h_{ij}(ct - z))dx^i dx^j, \quad |h_{ij}| \ll 1, \tag{4}$$

where Latin indices run from 1 to 3 and sum over repeated indices is assumed. Estabrook and Walquist[68,69] derived the one-way and two-way Doppler responses to plane GWs in weak field approximation (4) in the transverse traceless gauge in general relativity. Written in the notation of Armstrong *et al.*[70] the formula for one-way Doppler response on board S/C 2 received from S/C 1 is

$$y(t) = (1 - \underline{\boldsymbol{k}} \cdot \underline{\boldsymbol{n}})[\Psi(t - (1 + \underline{\boldsymbol{k}} \cdot \underline{\boldsymbol{n}})L) - \Psi(t)], \tag{5}$$

where $\underline{\boldsymbol{k}}[= (\underline{k}^i) = (\underline{k}^1, \underline{k}^2, \underline{k}^3)]$ is the unit vector in the GW propagation direction, $\underline{\boldsymbol{n}}[= (\underline{n}^i) = (\underline{n}^1, \underline{n}^2, \underline{n}^3)]$ the unit vector along the link from spacecraft 1 to spacecraft 2 and L is the path length of the Doppler link. The function $\Psi(t)$ is defined as

$$\Psi(t) \equiv \frac{\underline{n}^i h_{ij}(t) \underline{n}^j}{\{2[1 - (\underline{\boldsymbol{k}} \cdot \underline{\boldsymbol{n}})^2]\}}. \tag{6}$$

With one-way Doppler response known, two-way and multiple way response can easily be written down. As noticed and derived by Tinto and da Silva Alves,[71] for GW solutions in any metric theories of gravity of the form (4), the Doppler response formula (5) with the definition (6) is valid also.

Doppler tracking of the Viking S/C (S-band, 2.3 GHz),[72] the Voyager I S/C (S-band uplink + coherently transponded S-band and X-band (8.4 GHz) downlink),[73] Pioneer 10 (S band),[74] and Pioneer 11 (S band)[75] have been used for GW measurement and have given constraints on GW background in the low-frequency band.

The most recent measurements came from the CSDT. Armstrong *et al.*[76] used the Cassini multilink radio system during 2001–2002 solar opposition to derive improved observational limits on an isotropic background of low-frequency GWs. The Cassini multilink radio system consists of a sophisticated multilink radio system that simultaneously receives two uplink signals at frequencies of X and Ka bands and transmits three downlink signals with X-band coherent with the X-band uplink, Ka-band coherent with the X-band uplink, and Ka-band coherent with the Ka-band uplink. X band is a standard deep space communication frequency band about 8.4 GHz; Ka band is another deep space communication frequency band about 32 GHz. Armstrong *et al.*[76] used the Cassini multilink radio system with higher frequencies and an advanced tropospheric calibration system to remove the effects of leading noises — plasma and tropospheric scintillation to a level below the other noises. The resulting data were used to construct upper limits on the strength of an isotropic background in the 1 μHz–1 mHz band.[76] The characteristic strain upper limit curve labeled CSDT in Fig. 5 is a smoothed version of the curve in Fig. 4 of Ref. 76. The corresponding CSDT curves on the strain PSD amplitude in Fig. 4 and the normalized spectral energy density in Fig. 1 are calculated using Eq. (1) for conversion. The minimal points on these curves are

$$\begin{aligned} [S_h(f)]^{1/2} &< 8 \times 10^{-13}, \quad \text{at several frequencies in the 0.2–0.7 mHz band;} \\ h_c(f) &< 2 \times 10^{-15}, \quad \text{at frequency about 0.3 mHz;} \\ \Omega_{\rm gw}(f) &< 0.03, \quad \text{at frequency } 1.2\,\mu\text{Hz}. \end{aligned} \tag{7}$$

The GW sensitivity of spacecraft Doppler tracking could still be improved by 1–2 order of magnitude with a space borne optical clock on board.[77]

In the radio tracking of spacecraft, the received frequency of the signals is tracked. Its integral is the phase. In the radio ranging of spacecraft, the received phase of the signals is measured. The derivative of the phase is the frequency. For coherent transponding, the phase measured is basically a range up to an additive constant which needs to be determined.

Pulse laser ranging. Another way to measure the range is by using pulse timing. This is what being done in satellite laser ranging and lunar laser ranging. For ranging through the Earth's atmosphere, the best way to find the atmospheric delay is to use two colors (two wavelengths) to measure the atmospheric delay and subtract it. The distance determination of satellite laser ranging with two colors (two wavelengths) has reached millimeter accuracy. With the newer generation of lunar laser ranging,[78,79] the accuracy of lunar distance determination has also reached millimeter accuracy. On board timing accuracy of 3 ps (0.9 mm) has already achieved by the Time Transfer by Laser Link (T2L2) event timer onboard Jason 2 satellite.[80,81] Based on these developments, the one-way ranging technical capability over the whole solar system could have a millimeter accuracy. With this accuracy and extended ranges of 20 AU, the capability of probing the fundamental laws of spacetime and mapping the solar system gravity will be greatly enhanced.[32–35] For 1 mm out of 20 AU, the fractional uncertainty is 3×10^{-16}. It requires laser stability and clock accuracy to reach this level of fractional uncertainty; the accuracy is already achieved in the laboratory and will be available in space. ASTROD I[31–35] using a space borne precision clock has included as one of its goals GW sensitivity improvement of the CSDT by one order of magnitude. In fact, the fractional accuracies of optical clocks have already reached the 10^{-18} level. *When space optical clocks reach this level, pulse laser ranging together with drag-free technology will be an important alternative for detection of GWs in the lower part of low frequency band.*

The basic principle of spacecraft Doppler tracking, of spacecraft laser ranging, of space laser interferometers, and of Pulsar Timing Arrays (PTAs) for GW detection are similar. In the development of GW detection methods, spacecraft Doppler tracking method and pulse laser ranging method have stimulated significant inspirations. The methods using space laser interferometers and using PTAs are becoming two important methods of detecting GWs. The PTAs and their sensitivity are addressed in Refs. 5 and 7. Interferometric space missions and their sensitivities will be addressed in the following section.

4. Interferometric Space Missions

In a Michelson interferometer, the wave front is split into two parts to go in two different paths and then the two wave fronts are recombined to interfere. For white light, Michelson had to match the two optical path lengths very precisely in order to have interference fringes. After laser was invented, the coherence length became longer. One could build unequal arm Michelson interferometer. An alternative configuration of the Michelson interferometer is the Mach–Zehnder Interferometer.

Two-way Doppler tracking can be considered as an unequal arm Michelson interferometer; the local oscillator splits off a beam directing to the uplink spacecraft and the return beam from the spacecraft transponder interferes with the local oscillator. The phase (and frequency) of the beat is measured as a function of time. The Doppler response of a single link is given by (5). Using (5) the response of two-way Doppler tracking[68,69] is given by

$$y(t) = -(1 - \underline{\boldsymbol{k}} \cdot \underline{\boldsymbol{n}})\Psi(t) - 2(\underline{\boldsymbol{k}} \cdot \underline{\boldsymbol{n}})\Psi(t - (1 + \underline{\boldsymbol{k}} \cdot \underline{\boldsymbol{n}})L) + (1 + \underline{\boldsymbol{k}} \cdot \underline{\boldsymbol{n}})\Psi(t - 2L). \tag{8}$$

The three terms in (8) correspond, respectively, to the projected amplitude of the wave at the event of reception of the Doppler tracking signal at Earth, the transponding event at the spacecraft, and the emission event of the tracking signal from Earth.

Since the deviation of the speed of the electromagnetic wave from that of vacuum in plasma is inversely proportional to the square of the frequency, the time uncertainty due to solar wind or ionized gas in the microwave propagation is smaller in the Ka band (32 GHz) and X band (8.4 GHz) than S band (2.3 GHz). This is one of two motivations for Doppler tracking of Cassini spacecraft to use Ka band and X band for better noise performance. The other motivation is with shorter wavelength, the measurement precision increases. At optical frequency, the wavelength is more than fourth-order smaller and the plasma effect is eighth-order smaller. Therefore, when better sensitivities in the optical path length measurement was needed in GW detection, the GW community started to use optical method. When sensitivity is increased, we need to suppress spurious noise below the aimed sensitivity level. This requires that (i) we reduce the acceleration noise and implement the drag-free technology; (ii) we reduce the laser noise as much as possible. The basic drag-free technology is now demonstrated by LISA Pathfinder.[11] For reducing laser noise, we need laser stabilization. The best way is to implement absolute stabilization; e.g. to lock to an iodine molecular line. However, laser stabilization alone is not enough for the required strain sensitivity of the order of 10^{-21}. To lessen the laser noise requirement, TDI came to rescue.

For space laser-interferometric GW antenna, the arm lengths vary according to solar system orbit dynamics. In order to attain the requisite sensitivity, laser frequency noise must be suppressed below the secondary noises such as the optical path noise, acceleration noise etc. For suppressing laser frequency noise, it is necessary to use TDI in the analysis to match the optical path length of different beams closely. The better match of the optical path lengths are, the better cancellation of the laser frequency noise and the easier to achieve the requisite sensitivity. In case of exact match, the laser frequency noise is fully canceled, as in the original Michelson interferometer.

The TDI was first used in the study of ASTROD mission concept.[23,25,26] In the deep-space interferometry, long distances are invariably involved. Due to long distances, laser light is attenuated to a great extent at the receiving spacecraft.

To transfer the laser light back or to another spacecraft, amplification is needed. The procedure is to phase lock the local laser to the incoming weak laser light and to transmit the local laser light back or to another spacecraft. Liao *et al.*[29,30] have demonstrated the phase locking of a local oscillator with 2-pW laser light in laboratory. Dick *et al.*[82] have demonstrated phase locking to 40-fW incoming weak laser light. The power requirement feasibility for both e-LISA/NGO and ASTROD-GW is met with these developments. In the 1990s, Ni *et al.*[23,25,26] used the following two TDI configurations during the study of ASTROD interferometry and obtained numerically the path length differences using Newtonian dynamics.

These two TDI configurations are the unequal arm Michelson TDI configuration and the Sagnac TDI configuration for three spacecraft formation flight. The principle is to have two split laser beams to go to Paths 1 and 2 and interfere at their end path. For unequal arm Michelson TDI configuration, one laser beam starts from spacecraft 1 (S/C1) directed to and received by spacecraft 2 (S/C2), and optical phase locking the local laser in S/C2; the phase locked laser beam is then directed to and received by S/C1, and optical phase locking another local laser in S/C1; and so on following Path 1 to return to S/C1:

$$\text{Path 1: S/C1} \rightarrow \text{S/C2} \rightarrow \text{S/C1} \rightarrow \text{S/C3} \rightarrow \text{S/C1}. \tag{9}$$

The second laser beam starts from S/C1 also, but follows Path 2 route:

$$\text{Path 2: S/C1} \rightarrow \text{S/C3} \rightarrow \text{S/C1} \rightarrow \text{S/C2} \rightarrow \text{S/C1}, \tag{10}$$

to return to S/C1 and to interfere coherently with the first beam. If the two paths has exactly the same optical path length, the laser frequency noises cancel out; if the optical path length difference of the two paths are small, the laser frequency noises cancel to a large extent. In the Sagnac TDI configuration, the two paths are:

$$\begin{aligned} &\text{Path 1: S/C1} \rightarrow \text{S/C2} \rightarrow \text{S/C3} \rightarrow \text{S/C1}, \\ &\text{Path 2: S/C1} \rightarrow \text{S/C3} \rightarrow \text{S/C2} \rightarrow \text{S/C1}. \end{aligned} \tag{11}$$

Since then we have worked out the same things numerically for LISA,[83] eLISA/NGO,[84] LISA-type with 2×10^6 km arm length,[84] ASTROD-GW with no inclination,[85,86] and ASTROD-GW with inclination.[41]

TDI has been worked out for LISA much more thoroughly on various aspects since 1999.[87,88] First-generation and second-generation TDIs are proposed. In the first-generation TDIs, static situations are considered, while in the second-generation TDIs, motions are compensated to certain degrees. The two configurations considered above are first-generation TDI configurations in the sense of Armstrong *et al.*[87,88] We will discuss numerical TDI more in Sec. 10. For many other aspects of TDI, we refer the readers to the excellent review.[88]

In Table 1, we have compiled various interferometric space mission proposals for GW detection. Among the proposed science orbits, there are basically three categories — ASTROD-GW-like, LISA-like and OMEGA-like.

(i) LISA-like (LAGO-like)[14] science orbits: As in Fig. 1, the Earth-like solar orbits of the three spacecraft are appropriately inclined so that they form a nearly equilateral triangle formation having a tilt of $\pm 60^\circ$ (in the figure, the tilt is 60°) with respect to the ecliptic plane.[14] The formation rotates once per year clockwise or counterclockwise facing the Sun. Sections 7 and 8 give more detailed orbit analysis. LISA,[9] eLISA,[21] ALIA[47] and TAIJI (ALIA-descope)[48] have this kind of LISA-like science orbits. The ultimate configuration of Big Bang Observer[45] and DECIGO[44] has 12 spacecraft distributed in the Earth orbit in three groups separated by 120° in orbit; two groups has three spacecraft each in a LISA-like triangular formation and the third group has six spacecraft with two LISA-like triangles forming a star configuration (Fig. 6). An alternate configuration is that each group has four spacecraft forming a nearly square configuration (also has a tilt of 60° with respect to the ecliptic plane).

(ii) OMEGA-like science orbits: These orbits are Earth orbits away from (either inside or outside) Moon's orbit around the Earth. An example is the OMEGA mission orbit configuration. OMEGA mission proposed to NASA as a candidate MIDEX mission in 1998, and again as a mission-concept white paper in 2011. The OMEGA[54,55] mission consists of six identical spacecraft in a 600,000 km-high Earth orbit, two spacecraft at each vertex of a nearly equilateral triangle formation (Fig. 3). These orbits are stable, allowing for three years of planned science operations, as well as the possibility of an extended mission if desired. The arm length of the triangle formation is about 1 million km (1 Gm). The mission formation is outside of Moon's orbit.

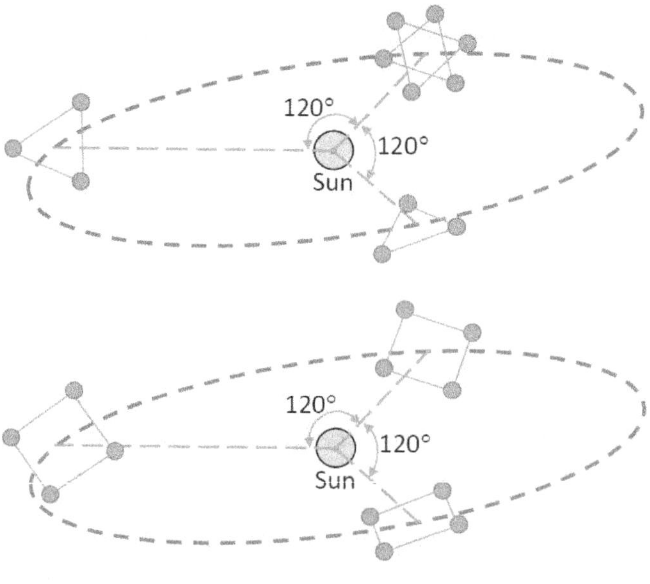

Fig. 6. Two schematic configurations of BBO and DECIGO in Earth-like solar orbits.

There are two mission proposals — GEOGRAWI[49]/gLISA[50,51] and GADFLI[52] using geostationary orbit formation. The three spacecraft of the formation are in the geostationary orbits forming a nearly equilateral triangle with arm length about 73,000 km.

TianQin is a GW mission proposal with 57,000 km-high orbit. The three spacecraft form a nearly equilateral triangle with arm length about 110,000 km with Earth-orbiting period 44 h.[19]

The orbits and spacecraft configuration of all these missions are near ecliptic plane. There are times the Sun light comes along the line-of-sight of telescope links. Sunlight shields are required when the line-of-sight cross the Sun. A solution has been proposed from the OMEGA mission proposal[55] which could be used for other missions in this category.

(iii) ASTROD-GW-like science orbits: The basic ASTROD-GW configuration consists of three spacecraft in the vicinity of the Sun–Earth Lagrange points L3, L4 and L5 respectively with near-circular orbits around the Sun, forming a nearly equilateral triangle as shown by Fig. 2 with the three arm lengths about 2.6×10^8 km (1.732 AU).[3,36–40] The dominant force on the spacecraft is from the Sun in the restricted three-body problem of Earth–Sun-spacecraft system. Since the Earth–Sun orbit is elliptical, the Lagrange points are not stationary in the Earth–Sun rotating frame. The motion of test particles at L3, L4 and L5 deviates from circular orbit by a fraction of $O(e)$ where e ($= 0.0167$) is the eccentricity of the Earth orbit around the Sun. However, the spacecraft can be in the halo orbit of the respective Lagrange points largely compensating the nonstationary motion of the Lagrange points to remain nearly circular orbits of the Sun. The circular orbits of spacecraft near the L3, L4 and L5 points are stable or virtually stable in 20 years (their orbits are also stable or quasi-stable with respect to their respective Lagrange points so that the deviations from circular orbit of their respective Lagrange point are of the order of $O(e^2)$ in AU) and the deviation of the spacecraft triangle from an equilateral triangle is of order of $O(e^2)$ in arm length. For a nonprecession planar formation, the angular resolution has antipodal ambiguity. To resolve this issue, we need to have precession orbit formation inclined with respect to the ecliptic. When the orbits of spacecraft have a small inclination λ (in radians) with respect to the ecliptic plane the arm length variation is of the order of $O(\lambda^2)$. Therefore, the added variation due to these two causes is of the order $O(e^2, \lambda^2)$. For these two causes to match (to $O(10^{-4})$), λ should be of the order of $O(1°)$. In Sec. 7, we review the inclined orbit analytically in the solar gravitational field and explain the angular resolution together with how to resolve the antipodal ambiguity. In Sec. 8, we will use solar system ephemeris to design and optimize the orbit configuration and will see that the perturbation from all planets except Earth is of the order of $O(10^{-4})$. The influence of Earth is already taken into consideration since the L3, L4 and L5 points are effectively stable in 20 years. Hence, suitable inclined circular orbits could be our basic orbits to start with and the deviations from actual optimized orbit should be on the order of $O(10^{-4})$.

For Super-ASTROD,[42] we could also place the three spacecraft with small inclination angle to Jovian solar orbit plane near Sun–Jupiter L3, L4 and L5 points with the other 1 or two spacecraft having large inclination(s).

For ASTROD-EM,[43] the three spacecraft will be placed in near Earth–Moon L3, L4 and L5 points. For the spacecraft dynamics, we have restricted 4-body (Earth, Moon, Sun, and the spacecraft whose gravitational filed can be neglected) problem to work out.

5. Frequency Sensitivity Spectrum

The space GW detectors are basically real-time free-mass detectors. As we have already discussed in Ref. 5 in general, there are two crucial issues in these proposed detectors: (i) to lower the disturbance effects and/or to model them for subtraction: drag-free to decrease the effects of surrounding disturbances, and appropriate modeling of the motion and the disturbances for subtraction to lower the residuals; (ii) to increase measurement sensitivity: microwave sensing, optical sensing, X-ray sensing, atom sensing, molecule sensing and timing. Associated with these two issues, there are two basic noises — the acceleration noise and the metrology noise. For laser-optic missions, the metrology noise is the laser metrology noise. The planned upper limits of these two kinds of basic noise for GW mission proposals are listed in the last two columns of Table 1. In space GW detection, the basic noise model is the LISA/eLISA noise model. Due to more stringent technological requirements, Big Bang Observer and DECIGO belong to second-generation space detector proposals. Super-ASTROD is also a second-generation space detector proposal due to its distance and power requirements. All others in Table 1 are first-generation space detector proposals. In Figs. 4–1, we plot the sensitivity curves of three typical first-generation space detectors (LISA/eLISA, ASTROD-GW and OMEGA) and two second-generation space detectors (Big Bang Observer and DECIGO). In the first generation category, for missions with arm length shorter than LISA, the planned strain upper limits are smaller than that of LISA in the higher frequency part; for missions with arm length longer than LISA, the planned strain upper limits are smaller than that of LISA in the lower frequency part.

As shown in Fig. 4, typical frequency sensitivity spectrum of strain PSD amplitude for space GW detection consists of three regions, the acceleration/vibration noise dominated region, the shot noise (flat for current space detector projects like LISA in strain PSD) dominated region, if any, and the antenna response restricted region. The lower frequency region for the detector sensitivity is dominated by vibration, acceleration noise or gravity-gradient noise. The higher frequency part of the detector sensitivity is restricted by antenna response (or storage time). In a power-limited design, sometimes there is a middle flat region in which the sensitivity is limited by the photon shot noise.[9,23,40]

The shot noise sensitivity limit in the strain for GW detection is inversely proportional to $P^{1/2}L$ with P the received power and L the distance or arm length.

Since P is inversely proportional to L^2 and $P^{1/2}L$ is constant, this sensitivity limit is independent of the distance. For 1–2 W emitting power, the limit is around $10^{-21}\,\text{Hz}^{-1/2}$. As noted in the LISA study,[9] making the arms longer shifts the time-integrated sensitivity curve to lower frequencies while leaving the bottom of the curve at the same level. Hence, ASTROD-GW with longer arm length has better sensitivity at lower frequency. e-LISA, ALIA, TAIJI (ALIA-descope), and GW interferometers in Earth orbit have shorter arms and therefore have better sensitivities at higher frequency.

In Figs. 4–1, we plot sensitivity curves for LISA, e-LISA and ASTROD-GW for the low-frequency GW band. In the Mock LISA Data Challenge (MLDC) program, the consensus goal for the LISA instrumental noise density amplitude ${}^{(\text{MLDC})}S_{\text{Ln}}{}^{1/2}(f)$ is

$$ {}^{(\text{MLDC})}S_{\text{Ln}}{}^{1/2}(f) = \left(\frac{1}{L_{\text{L}}}\right) \times \left\{\left[\left(1 + 0.5\left(\frac{f}{f_{\text{L}}}\right)^2\right)\right] \times S_{\text{Lp}} + \left[1 + \left(\frac{10^{-4}\,\text{Hz}}{f}\right)^2\right]\left(\frac{4S_{\text{a}}}{(2\pi f)^4}\right)\right\}^{1/2} \text{Hz}^{-1/2}, \quad (12\text{a}) $$

where $L_{\text{L}} = 5 \times 10^9\,\text{m}$ is the LISA arm length, $f_{\text{L}} = c/(2\pi L_{\text{L}})$ is the LISA arm transfer frequency, $S_{\text{Lp}} = 4 \times 10^{-22}\,\text{m}^2\,\text{Hz}^{-1}$ is the LISA (white) position noise (power) level due to photon shot noise, and $S_{\text{a}} = 9 \times 10^{-30}\,\text{m}^2\,\text{s}^{-4}\,\text{Hz}^{-1}$ is the LISA white acceleration noise (power) level.[89] Note that (12a) contains the "reddening" factor $[1 + (10^{-4}\,\text{Hz}/f)^2]$ in the acceleration noise term.

In 2003, Bender[90] looked into the possible LISA sensitivity below 100 μHz. From a careful analysis of noises of test mass and capacitive sensing, Bender suggested a specific sensitivity goal at frequencies down to 3 μHz which contained a milder (than MLDC) "reddening factor". For frequency between 10 μHz and 100 μHz, he suggested to put in the "reddening factor" $[(10^{-4}\,\text{Hz}/f)^{1/2}]$ and for frequency between 3 μHz and 10 μHz, the "reddening factor" $[3.16 \times (10^{-5}\,\text{Hz}/f)]$. To drop this "reddening factor" might be difficult. However, with monitoring the gap of capacitive sensing and the positions of major mass distribution, the factor may be alleviated to certain extent. To completely drop the factor or to go beyond, one may need to go to optical sensing and optical feedback control.[24,27,28,91–93] If we drop the "reddening factor", the enhanced LISA instrumental noise density amplitude ${}^{(\text{Enhanced})}S_{\text{Ln}}{}^{1/2}(f)$ becomes

$$ {}^{(\text{Enhanced})}S_{\text{Ln}}{}^{1/2}(f) = \left(\frac{1}{L_{\text{L}}}\right) \times \left\{\left[\left(1 + 0.5\left(\frac{f}{f_{\text{L}}}\right)^2\right)\right] \times S_{\text{Lp}} + \left[\frac{4S_{\text{a}}}{(2\pi f)^4}\right]\right\}^{1/2} \text{Hz}^{-1/2}. \quad (12\text{b}) $$

After NASA's withdrawal from ESA-NASA collaboration of LISA in 2011, the European eLISA/NGO for space detection of GWs emerged. The orbit configuration is the same as LISA, but with arm length shrunk five times to 1 million km, the orbits slowly drifting away from the Earth and the nominal mission duration two years (extendable to five years) to save weight, fuel and costs. The three spacecraft will consist of one "mother" and two simpler "daughters," with interferometric measurements along only two arms with the "mother" at the vertex.[21] The eLISA/NGO strain noise PSD goal is also shown in Fig. 4. For the lower frequency part of the power spectrum of eLISA/NGO, we choose to use the same acceleration noise with reddening factor (solid line) and without reddening factor (dashed line) as those of LISA to obtain the eLISA/NGO strain noise for easy comparison.

The eLISA arm length L_{eL} is five times shorter. Its instrumental noise density amplitude ${}^{(\mathrm{MLDC})}S_{\mathrm{eLn}}{}^{1/2}(f)$ is

$$
\begin{aligned}
{}^{(\mathrm{MLDC})}S_{\mathrm{eLn}}{}^{1/2}(f) = \left(\frac{1}{L_{\mathrm{eL}}}\right) \times \Bigg\{ &\left[\left(1+0.5\left(\frac{f}{f_{\mathrm{eL}}}\right)^2\right)\right] \times S_{\mathrm{eLp}} \\
&+ \left[1+\left(\frac{10^{-4}\,\mathrm{Hz}}{f}\right)^2\right]\left(\frac{4S_{\mathrm{a}}}{(2\pi f)^4}\right)\Bigg\}^{1/2} \mathrm{Hz}^{-1/2}, \quad (13\mathrm{a})
\end{aligned}
$$

where $L_{\mathrm{eL}} = 10^9\,\mathrm{m}$ is the eLISA arm length, $f_{\mathrm{eL}} = c/(2\pi L_{\mathrm{eL}})$ is the eLISA arm transfer frequency, $S_{\mathrm{eLp}} = 1 \times 10^{-22}\,\mathrm{m}^2\,\mathrm{Hz}^{-1}$ is the eLISA (white) position noise level due to photon shot noise assuming that the telescope diameter is 25 cm (compared with 40 cm for that of LISA) and that the laser power is the same as LISA. With these assumptions, the eLISA position noise amplitude would be $10\,\mathrm{pm/Hz}^{1/2}$ listed in parentheses in the eLISA entry, comparable to $12\,\mathrm{pm/Hz}^{1/2}$ used in Ref. 94. The corresponding enhanced eLISA instrumental noise density amplitude ${}^{(\mathrm{Enhanced})}S_{\mathrm{eLn}}{}^{1/2}(f)$ is

$$
\begin{aligned}
&{}^{(\mathrm{Enhanced})}S_{\mathrm{eLn}}{}^{1/2}(f) \\
&= \left(\frac{1}{L_{\mathrm{eL}}}\right) \times \left\{\left[\left(1+0.5\left(\frac{f}{f_{\mathrm{eL}}}\right)^2\right)\right] \times S_{\mathrm{eLp}} + \left(\frac{4S_{\mathrm{a}}}{(2\pi f)^4}\right)\right\}^{1/2} \mathrm{Hz}^{-1/2}. \quad (13\mathrm{b})
\end{aligned}
$$

For ASTROD-GW, our goal on the instrumental strain noise density amplitude is

$$
S_{\mathrm{An}}{}^{1/2}(f) = \left(\frac{1}{L_{\mathrm{A}}}\right) \times \left\{\left[\left(1+0.5\left(\frac{f}{f_{\mathrm{A}}}\right)^2\right)\right] \times S_{\mathrm{Ap}} + \left[\frac{4S_{\mathrm{a}}}{(2\pi f)^4}\right]\right\}^{1/2} \mathrm{Hz}^{-1/2}, \quad (14)
$$

over the frequency range of $100\,\mathrm{nHz} < f < 1\,\mathrm{Hz}$. Here $L_{\mathrm{A}} = 260 \times 10^9\,\mathrm{m}$ is the ASTROD-GW arm length, $f_{\mathrm{A}} = c/(2\pi L_{\mathrm{A}})$ is the ASTROD-GW arm transfer frequency, $S_{\mathrm{a}} = 9 \times 10^{-30}\,\mathrm{m}^2\,\mathrm{s}^{-4}\,\mathrm{Hz}^{-1}$ is the white acceleration noise level (the same as that for LISA), and $S_{\mathrm{Ap}} = 10{,}816 \times 10^{-22}\,\mathrm{m}^2\,\mathrm{Hz}^{-1}$ is the (white) position noise level due to laser shot noise which is 2704 ($= 52^2$) times that for LISA.[3,36–40] The corresponding noise curve for the ASTROD-GW instrumental noise density amplitude ${}^{(\mathrm{MLDC})}S_{\mathrm{An}}{}^{1/2}(f)$ with the same "reddening" factor as specified in MLDC program is

$$
{}^{(\mathrm{MLDC})}S_{\mathrm{An}}{}^{1/2}(f) = \left(\frac{1}{L_{\mathrm{A}}}\right) \times \left\{\left[\left(1 + 0.5\left(\frac{f}{f_{\mathrm{A}}}\right)^2\right)\right] \times S_{\mathrm{Ap}} + \left[1 + \left(\frac{10^{-4}}{f}\right)^2\right]\left(\frac{4S_{\mathrm{a}}}{(2\pi f)^4}\right)\right\}^{1/2} \mathrm{Hz}^{-1/2}, \quad (14a)
$$

over the frequency range of $100\,\mathrm{nHz} < f < 1\,\mathrm{Hz}$. The sensitivity curves from the six formulas (12a), (12b) to (14a) are shown in Fig. 4. The corresponding sensitivity curves in terms of $h_c(f)$ and $\Omega_{\mathrm{gw}}(f)$ are shown in Figs. 5 and 1, respectively. The ones with reddening factor are shown with dashed line in the lower frequency part.

With the same laser power as that of LISA, the ASTROD-GW sensitivity would be shifted to lower frequency by a factor up to 52 if other frequency-dependent requirements can be shifted and met. The sensitivity curve would then be shifted toward lower frequency as a whole. Since the main constraints on the lower frequency part of the sensitivity is from the accelerometer noise, this translational shift depends on whether the accelerometer noise requirement for ASTROD-GW could be lowered (more stringent) from that of LISA requirement at a particular frequency. Since ASTROD is in a time frame later than LISA, if the absolute metrological accelerometer/inertial sensor could be developed, there is a potential to go toward this requirement. However, *to be simple*, we have taken a conservative stand and assume that *the LISA accelerometer noise goal and all other local requirements are taken as they are in the above equations and in the plotting of sensitivity curves in Figs.* 4–1. Since the strain sensitivity is mainly the accelerometer noise divided by arm length at low frequency, at a particular low frequency limited by accelerometer noise, the strain sensitivity for ASTROD-GW is 52 times lower than LISA (or 260 times lower than eLISA) due to longer arm length whether we take (12a) (or (13a)) and (14a) to compare or (12b) (or (13b)) and (14) to compare. With better lower-frequency resolution, the confusion limit of Galactic compact binary background for ASTROD-GW would be somewhat lower than that for LISA. The confusion limit for eLISA would be somewhat higher than that for LISA. In Figs. 4–1, the confusion limit curves are for LISA. ASTROD-GW will complement LISA and PTAs in exploring single events and backgrounds of MBH–MBH binary GWs in the important frequency range 100 nHz–1 mHz to study black hole co-evolution with galaxies, dark energy and other issues (Sec. 6).

OMEGA has 1 million km arms just as eLISA. The sensitivity goal of OMEGA is: (i) The acceleration noise PSD is the same as LISA and eLISA; (ii) the (white) position noise amplitude is fourfold lower than LISA and twofold lower than eLISA. The sensitivity curve of OMEGA plotted on Fig. 4 is from Ref. 55 with corresponding curves shown on Figs. 5 and 1. The lower frequency part and the flat part are close to eLISA while the antenna-response-limited part is slightly better. The small difference as compared to that in Table 1 may be because of OMEGA has three pairs of S/C with one more link for interferometry or just because of different sources of drawing.

For GW mission proposals listed in Table 1 with formations inside the lunar orbit around the Earth, the acceleration noise requirements are about the same level or slightly more stringent than OMEGA and eLISA while the requirement on the position noise amplitude is lower because of more power received. The goal sensitivity curves in the higher frequency part is slightly better for the two mission proposals in geostationary orbits, gLISA/GEOGRAWI and GADFLI. TIANQIN with 0.11 Gm arm length aims at first and sure detection of a GW source in space, the required sensitivity on ${S_\mathrm{a}}^{1/2}$ and ${S_\mathrm{p}}^{1/2}$ are ${S_\mathrm{a}}^{1/2} = 1 \times 10^{-15}\,\mathrm{m\,s^{-2}\,Hz^{-1/2}}$ and ${S_\mathrm{p}}^{1/2} = 1\,\mathrm{pm\,Hz^{-1/2}}$ at 6 mHz.

ALIA in solar orbit as a LISA follow-on aims at better sensitivity at frequency above 1 mHz. It has arm length of 0.5 Gm (0.5 million km) — ten times shorter than LISA and two times shorter than eLISA. The acceleration noise requirement is tenfold more stringent than LISA, i.e. ${S_\mathrm{a}}^{1/2} = 0.3 \times 10^{-15}\,\mathrm{m\,s^{-2}\,Hz^{-1/2}}$. The position noise amplitude requirement is 30 times more stringent than LISA, i.e. ${S_\mathrm{p}}^{1/2} = 0.6 \times 10^{-15}\,\mathrm{pm\,Hz^{-1/2}}$. TAIJI (ALIA-descope) has arm length of 3 Gm and aims at a detection of intermediate black hole coalescence in addition to other scientific goals common to most space mission proposals. Its sensitivity is relaxed from ALIA to ${S_\mathrm{a}}^{1/2} = 3 \times 10^{-15}\,\mathrm{m\,s^{-2}\,Hz^{-1/2}}$ (the same as LISA) and ${S_\mathrm{p}}^{1/2} = 5$–$8\,\mathrm{pm\,Hz^{-1/2}}$.

The three spacecraft of ASTROD-EM and of LAGRANGE will be located near L3, L4 and L6 Lagrange points of Earth–Moon system, respectively. Due to the inclination of the Moon-Earth orbit plane to the ecliptic, the spacecraft formation plane will not intersect the Sun. Hence, unlike other missions in Earth orbit, the Sun light will not come along the line-of-sight of telescope links. Sunlight shields are not required. The spacecraft orbit dynamics is a restricted 4-body (Earth, Moon, Sun and the spacecraft) problem which we are still working on.[43] The acceleration noise and the laser metrology noise requirements are listed in Table 1.

BBO and DECIGO have similar goals of detecting primordial GWs. BBO has a delay line implementation. DECIGO uses a Fabry–Perot implementation. The acceleration noise ${S_\mathrm{a}}^{1/2}$ and the laser metrology noise ${S_\mathrm{p}}^{1/2}$ requirements of BBO are ${S_\mathrm{a}}^{1/2} = 3 \times 10^{-17}\,\mathrm{m\,s^{-2}\,Hz^{-1/2}}$ and ${S_\mathrm{p}}^{1/2} = 1.4 \times 10^{-5}\,\mathrm{pm\,Hz^{-1/2}}$, respectively; those of DECIGO are ${S_\mathrm{a}}^{1/2} = 4 \times 10^{-19}\,\mathrm{m\,s^{-2}\,Hz^{-1/2}}$ and ${S_\mathrm{p}}^{1/2} = 2 \times 10^{-6}\,\mathrm{pm\,Hz^{-1/2}}$. The strain sensitivity curve of a single DECIGO interferometer as shown in Fig. 5 is from Ref. 95. BBO has a similar single-interferometer sensitivity curve. One-sigma, power-law integrated sensitivity curve for BBO (BBO-corr) as shown in Fig. 5 is obtained by Thrane and Romano.[96] That of DECIGO is similar. We also put in the plot their LISA autocorrelation measurement sensitivity curve (LISA-corr) in a single detector assuming perfect subtraction of instrumental noise and/or any unwanted astrophysical foreground.[96] The minimum autocorrelation sensitivity using the same method for ASTROD-GW is also estimated and plotted in Fig. 5; this would also be the level that 6 S/C ASTROD-GW[40] (6 S/C ASTROD-GW-corr) could reach. All of the corresponding curves are plotted in Figs. 4 and 1. *Considering the sensitivity requirements or arm length involved, DECIGO, BBO and Super-ASTROD belongs to the second-generation space interferometers.* For the sensitivity of Super-ASTROD, we assume ${S_{\rm a}}^{1/2} = 3 \times 10^{-15}\,{\rm m\,s^{-2}\,Hz^{-1/2}}$ (the same as LISA) and ${S_{\rm p}}^{1/2} = 5000\,{\rm pm\,Hz^{-1/2}}$.

Atom Interferometry. The development in atom interferometry is fast and promising. It already contributes to precision measurement and fundamental physics. A proposal using atom interferometry to detect GWs has been raised at Stanford University as an alternate method to LISA on the LISA bandwidth.[97,98] Issues have arisen on its realization of LISA sensitivity for this proposal.[99,100] In Observatoire de Paris, SYRTE has started the first stage of its project — Matter-wave laser Interferometric Gravitation Antenna (MIGA)[101] of building a 300 m long optical cavity to interrogate atom interferometers at the underground laboratory Laboratoire Souterrain à Bas Bruit (LSBB) in Rustrel. In the second stage of the project (2018–2023), MIGA will be dedicated to science runs and data analyses in order to probe the spatio-temporal structure of the local field of the LSBB region. In the meantime, MIGA will assess future potential applications of atom interferometry to GW detection in the middle frequency band (0.1–10 Hz).

6. Scientific Goals

In this section, we review and summarize the scientific goals for space GW mission proposals and projects.[3,9,21,39,40,94] More studies on the scientific goals and data analysis in the next few years will be worthy for the preparation of space GW missions.

6.1. *Massive black holes and their co-evolution with galaxies*

Relations have been discovered between the MBH mass and the bulge mass of host galaxy, and between the MBH mass and the velocity-dispersion of host galaxy. These relations indicate that the central MBHs are linked to the evolution of galactic structure. Observational evidence indicate that MBHs reside in most local galaxies. Newly fueled quasar may come from the gas-rich major merger of two massive galaxies. GW observation in the low frequency band (100 nHz–100 mHz) by space interferometers and very low frequency band (300 pHz–100 nHz) by PTAs will be a major tool to study the co-evolution of galaxy with BHs.

The standard theory of MBH formation is the merger-tree theory with various Massive Black Hole Binary (MBHB) inspirals acting. The GWs from these MBHB inspirals can be detected and explored to cosmological distances using space GW detectors and PTAs depending on the masses of MBHBs. Although there are different merger-tree models and models with BH seeds, they all give significant detection rates for space GW detectors and PTAs,[7,102–104] NGO/eLISA[21] and

ASTROD-GW.[40] PTAs are most sensitive in the frequency range 300 pHz–100 nHz, NGO/eLISA space GW detector is most sensitive in the frequency range 2 mHz–0.1 Hz, while ASTROD-GW is most sensitive in the frequency range 100 nHz–2 mHz (Figs. 4–1). NGO/eLISA and ASTROD-GW will be able to directly observe how MBHs form, grow, and interact over the entire history of galaxy formation. ASTROD-GW will detect stochastic GW background from MBH binary mergers in the frequency range 100 nHz to 100 μHz. These observations are significant and important to the study of co-evolution of galaxies with MBHs. The expected rate of MBHB sources is 10–100 year^{-1} for NGO/eLISA and 10–1000 year^{-1} for LISA.[21] For ASTROD-GW, we are expecting similar number of sources but with better angular resolution (Sec. 7.3).[40]

A sample of MBHB merger sources are drawn on Figs. 4–1. The black lines show the inspiral, coalescence and oscillation phases of GW emission from various equal-mass black-hole binary mergers in circular orbits at various redshift: solid line, $z = 1$; dashed line, $z = 5$; long-dashed line $z = 20$. The $10^6\,M_\odot$–$10^6\,M_\odot$ MBHB merger at $z = 1$, $10^5\,M_\odot$–$10^5\,M_\odot$ MBHB merger at $z = 20$, and $10^4\,M_\odot$–$10^4\,M_\odot$ MBHB merger at $z = 5$ are from Schutz[105] for Fig. 4; others by scaling; the corresponding curves in Figs. 5 and 1 by transformation equation (1). MBHB merger events have large signal to noise ratio for space detectors. Some of these events with equal mass (from 10^2–$10^{10}\,M_\odot$) and circular orbit are shown in Figs. 4–1. They are all candidates for space-borne detectors. Some could be in earlier phases for future ground-based detectors.

With the detection of MBHB merger events and background, the properties and distribution of MBHs could be deduced and underlying population models could be tested.

PTAs have been collecting data for decades for detection of stochastic GW background from MBHB mergers. In modeling the MBHB stochastic GW background spectra, various authors obtained the following frequency dependence:

$$h_c(f) = A_{\text{year}} \left[\frac{f}{(1\,\text{year}^{-1})} \right]^{\alpha}, \tag{15}$$

with $\alpha = -(2/3)$.[7] PTAs have improved greatly on the sensitivity for GW detection recently.[106–108] They have put upper limits on the isotropic stochastic background assuming the frequency dependence (15) with $\alpha = -(2/3)$ as follows: from European PTA (EPTA), $A_{\text{year}} < 3 \times 10^{-15}$; from Parks PTA (PPTA), $A_{\text{year}} < 1 \times 10^{-15}$ and North American Nanohertz Observatory for Gravitational Waves (NANOGrav), $A_{\text{year}} < 1.5 \times 10^{-15}$. The three experiments form a robust upper limit of 1×10^{-15} on A_{year} at 95% confidence level ruling out most models of supermassive black hole formation. The limit is shown as constraint on the Supermassive Black Hole Binary GW Background (SBHB-GWB) in Figs. 2–4 of Ref. 5 as solid line in the frequency range 10^{-9}–10^{-7} Hz. The GW energy released from co-evolution with galaxies must go somewhere. More energy of GWs might be emitted with higher frequency in the hierarchy of supermassive black hole formation. Hence, we have extrapolated this constraint linearly (instead with a knee around $f \sim 100\,\mathrm{nHz}$ in most existed models) with dotted line to $10\,\mu\mathrm{Hz}$ with some confidence in our review.[5] We adopt the same thing in Figs. 4–1 here. Constraints with other α values have similar order of magnitudes.

6.2. *Extreme mass ratio inspirals*

EMRIs are GW sources for space GW detectors. The NGO/eLISA sensitive range for central MBH masses is 10^4–$10^7\,\mathrm{M}_\odot$. The expected number of NGO/eLISA detections over 2 year is 10–20 (Ref. [21]); for LISA, a few tens[21]; for ASTROD-GW, similar or more with sensitivity toward larger central BH's and with better angular resolution (Sec. 7.3).[40]

6.3. *Testing relativistic gravity*

An important scientific goal of LISA[9,21] and NGO/eLISA[21,94] is to test general relativity and to study black hole physics with precision in strong gravity. With better precision in 100 nHz–1 mHz frequency range, ASTROD is going to push this goal further in many aspects. These include testing strong-field gravity, precision probing of Kerr spacetime and measuring/constraining the mass of graviton. Some considerations have been given in Refs. 109 and 110. Lower frequency sensitivity is significant in improving the precision of various tests.[109,110] Further studies in these respects would be of great value.

6.4. *Dark energy and cosmology*

In the dark energy issue,[111] it is important to determine the value of w in the equation of state of dark energy,

$$w = \frac{p}{\rho}, \tag{16}$$

as a function of different epochs where p is the pressure and ρ the density of dark energy. For cosmological constant as dark energy, $w = -1$. From cosmological observations, our universe is close to being flat. In a flat Friedman–Lemaître–Robertson–Walker (FLRW) universe, the luminosity distance is given by

$$d_L(z) = (1+z)(H_0)^{-1} \int_0^z dz' [\Omega_m (1+z')^3 + \Omega_{\mathrm{DE}} (1+z')^{3(1+w)}]^{-\frac{1}{2}}, \tag{17}$$

where H_0 is Hubble constant, Ω_{DE} is the present dark energy density parameter, and the equation of state of the dark energy w is assumed to be constant. In the case of nonconstant w and nonflat FLRW universe, similar but more complicated expression can be derived. Here, we show (17) for illustrative purpose. From the observed relation of luminosity distance versus redshift z, the parameter w of the equation of state as a function of redshift z can be solved for and compared with various cosmological models. Dark energy cosmological models can be tested this way. Luminosity distance from supernova observations and from gamma ray burst observations versus redshift observations are the focus for the current dark energy probes.

Space GW detectors observing MBHB inspirals and EMRIs are good probes to determine the luminosity distances. With the redshift of the source determined by the electromagnetic observations of associated galaxies or cluster of galaxies, these space GW detectors are also dark energy probes. In the merging of MBHs during the galaxy co-evolution processes, gravitational waveforms generated give precise, gravitationally calibrated luminosity distances to high redshift. The inspiral signals of these binaries can serve as standard candles/sirens.[112,113] With better angular resolution (Sec. 7.3), ASTROD-GW will have better chance to identify the associated electromagnetic redshift and therefore will be better for the determination of the dark energy equation of state.[39,40]

6.5. *Compact binaries*

Space GW detectors are also sensitive to the GWs from Galactic compact binaries.[9,21] These detectors will be able to survey compact stellar-mass binaries and study the structure of the Galaxy. NGO/eLISA will detect about 3000 double white dwarf binaries individually with most in the GW frequency band 3–6 mHz (orbit period about 150–300 s); for LISA, about 10,000 double white dwarf binaries.[9,21] These sources constitute the population which has been proposed as progenitors of normal type Ia and peculiar supernovae. For a review on the electromagnetic counterparts of GW mergers of compact objects, see, e.g. Ref. 114. At the frequency band 3–6 mHz, NGO/eLISA is more sensitive than ASTROD-GW (Fig. 4). Since NGO/eLISA will be flying first these GW signals will serve as a calibration for ASTROD-GW in addition to the verification binaries. The eight verification binaries selected by NGO/eLISA are shown on Fig. 4 as red squares with two-year integration time (from Ref. 21, p. 14, Fig. 1).

At GW frequencies below a few mHz, millions of ultra-compact binaries will form a detectable foreground for NGO/eLISA and ASTROD-GW. At these frequencies, ASTROD-GW is more sensitive than NGO/eLISA (Fig. 4). More sources will be resolved individually and ASTROD-GW can improve on the observational results of NGO/eLISA.

6.6. *Relic GWs*

For direct detection of primordial (inflationary, relic) GWs in space, one may go to frequencies lower or higher than LISA bandwidth,[3,115] where there are potentially less foreground astrophysical sources[116] to mask detection. DECIGO[44] and Big Bang Observer[45] look for GWs in the higher frequency range while ASTROD-GW[3,115] looks for GWs in the lower frequency range. Their instrument sensitivity goals all reach 10^{-17} in terms of critical density. The main issue is the level of foreground and whether foreground could be separated.

The straight line in the bottom left corner of Fig. 4 corresponds to $\Omega_{\rm gw} = 10^{-15}$ Cosmic Polarization Background (CMB) upper limit (See, e.g. Ref. 5) of inflationary GW background. For ASTROD-GW, when a 6-S/C formation is used for correlated detection of stochastic GWs, the sensitivity can reach this region. However, the anticipated upper limit of MBH–MBH GW background is above the 3-S/C ASTROD-GW sensitivity. If this background is detected, then the detectability of inflationary GW of the strength $\Omega_{\rm gw} = 10^{-16}$–$10^{-17}$ from 6-S/C formation in the ASTROD-GW frequency region depends on whether this MBH–MBH GW 'foreground' could be separated due to different frequency dependence or other signatures.[40]

Other potentially possible GW sources in the relevant frequency band, e.g. cosmic strings, should also be studied. See Ref. 94 for cosmic strings and some other sources.

7. Basic Orbit Configuration, Angular Resolution and Multi-Formation Configurations

In this section, we review and summarize the basic LISA-like and ASTRO-GW configurations, their angular resolutions and multi-formation configurations. These basic configurations can be used for starting numerical design and numerical orbit optimization for missions in these two categories.

7.1. *Basic LISA-like orbit configuration*

As in Fig. 1, the center of mass of the basic LISA-like configuration[9,14,84,117–122] follows a circular orbit of radius R (= 1 AU) around the Sun. Since the distance (arm length) L between the spacecraft is much smaller than the circular orbit radius 1 AU, we could treat the spacecraft orbits as perturbed orbits from the circular orbit. The equations for the perturbed orbit are known as Euler–Hill equations, Hill equations, or Clohessy and Wiltshire equations. Hill used these equations for researches in the lunar theory in the 19th century.[123] Clohessy and Wiltshire[124] derived and used these equations for designing terminal guiding system for satellite rendezvous in 1960 after the space era began at 1957. Clohessy and Wiltshire used a frame — called $\mathcal{CW}$ frame with its origin on the circular reference (center of configuration) orbit and with the frame rotating with angular velocity Ω the same as that of reference orbit rotation. For the perturbed orbit to keep the same distance to the origin and to remain stationary in the $\mathcal{CW}$ frame, it is clear by calculating the difference of the perturbed orbit and fiducial orbit that the eccentricity e and the inclination i with respect to the ecliptic need to be

$$e = 3^{-1/2}\frac{L}{(2R)}; \quad i = \alpha \equiv \frac{L}{(2R)}; \tag{18}$$

to first-order in the perturbation or to $\mathrm{O}(L/(2R))$ $[= \mathrm{O}(\alpha)]$. One way to form a nearly triangular configuration with side or arm length $L(1+\mathrm{O}(\alpha))$ is to require the orbit nodes be separated by 120°, and to choose the true anomalies and arguments of perihelion such that each spacecraft at its aphelion is also at its maximum height above (north of) the ecliptic (first configuration); the other way is with aphelion at its minimum height below (south of) the ecliptic (second configuration).[9] With these choices, the mission configuration plane is at 60° from the ecliptic with the intersection to ecliptic tangential to the fiducial (center of configuration) circular orbit. For square configuration, just require the orbit nodes to be separated by 90°; one can similarly construct any regular polygon configuration or any planar configuration. The first configuration rotates clockwise; the second rotates counter-clockwise. Thus, one reaches the conclusion — *in the CW frame there are just two planes which make angles of* $\pm 60°$ *with the reference orbit plane, in which space-craft (test particles) obeying the CW equations and perform rigid rotations about the origin with angular velocity* $-\Omega$.

We follow Dhurandhar *et al.*,[122] and Wang and Ni[84] to write down the equation for the basic orbits of the three spacecraft for the LISA-like configurations. First, the equation of an elliptical orbit in the general X–Y plane is given by

$$X = R(\cos\psi + e), \quad Y = R(1 - e^2)^{1/2}\sin\psi, \tag{19}$$

where R is the semi-major axis of the ellipse, e the eccentricity and ψ the eccentric anomaly.

Define α to be the ratio of the planned arm length L of the orbit configuration to twice radius R (1 AU) of the mean Earth orbit around the Sun, i.e. $\alpha = L/(2R)$. Choose the initial time t_0 to be a specific epoch in the Julian calendar and work in the Heliocentric Coordinate System (X, Y, Z). X-axis is in the direction of vernal equinox. A set of elliptical S/C orbits can be defined as

$$\begin{gathered} X_{\mathrm{f}} = R(\cos\psi_{\mathrm{f}} + e)\cos\varepsilon, \quad Y_{\mathrm{f}} = R(1 - e^2)^{1/2}\sin\psi_{\mathrm{f}}, \\ Z_{\mathrm{f}} = R(\cos\psi_{\mathrm{f}} + e)\sin\varepsilon. \end{gathered} \tag{20}$$

Here, $R = 1$ AU; $e = 0.001925$; $\varepsilon = 0.00333$. The eccentric anomaly ψ_{f} is related to the mean anomaly $\Omega\ (t - t_0)$ by

$$\psi_{\mathrm{f}} + e\sin\psi_{\mathrm{f}} = \Omega(t - t_0). \tag{21}$$

Here, Ω is defined as 2π/(one sidereal year). The eccentric anomaly ψ_{f} can be solved by numerical iteration. Define ψ_k to be implicitly given by

$$\psi_k + e\sin\psi_k = \Omega(t - t_0) - 120°(k - 1), \quad \text{for } k = 1, 2, 3. \tag{22}$$

Define $X_{\mathrm{f}k}$, $Y_{\mathrm{f}k}$, $Z_{\mathrm{f}k}$, $(k = 1, 2, 3)$ to be

$$\begin{aligned} X_{\mathrm{f}k} &= R(\cos\psi_k + e)\cos\varepsilon, \\ Y_{\mathrm{f}k} &= R(1 - e^2)^{1/2}\sin\psi_k, \\ Z_{\mathrm{f}k} &= R(\cos\psi_k + e)\sin\varepsilon. \end{aligned} \tag{23}$$

Define $\varphi_0 \equiv \psi_E - 10°$ with ψ_E is the position angle of Earth with respect to the X-axis at t_0. Define $X_{f(k)}$, $Y_{f(k)}$, $Z_{f(k)}$, $(k = 1, 2, 3)$, i.e. $X_{f(1)}$, $Y_{f(1)}$, $Z_{f(1)}$; $X_{f(2)}$, $Y_{f(2)}$, $Z_{f(2)}$; $X_{f(3)}$, $Y_{f(3)}$, $Z_{f(3)}$ to be

$$\begin{aligned} X_{f(k)} &= X_{fk} \cos[120°(k-1) + \varphi_0] - Y_{fk} \sin[120°(k-1) + \varphi_0], \\ Y_{f(k)} &= X_{fk} \sin[120°(k-1) + \varphi_0] + Y_{fk} \cos[120°(k-1) + \varphi_0], \\ Z_{f(k)} &= Z_{fk}. \end{aligned} \tag{24}$$

The basic orbits of the three S/C (for one-body central problem) are

$$\begin{aligned} \mathbf{R}_{S/C1} &= (X_{f(1)}, Y_{f(1)}, Z_{f(1)}), \\ \mathbf{R}_{S/C2} &= (X_{f(2)}, Y_{f(2)}, Z_{f(2)}), \\ \mathbf{R}_{S/C3} &= (X_{f(3)}, Y_{f(3)}, Z_{f(3)}). \end{aligned} \tag{25}$$

The initial positions can be obtained by choosing $t = t_0$ and initial velocities by calculating the derivatives with respect to time at $t = t_0$. As an example, if we choose t_0 = JD2459215.5 (2021-Jan-1st 00:00:00), the initial conditions (states) of three spacecraft of eLISA/NGO in J2000.0 solar system barycentric Earth mean equator and equinox coordinates are calculated and tabulated in the third column of Table 2 (from Table 2 of Ref. 84). From these initial conditions, one could start to design and optimize the orbit configuration numerically using planetary and lunar ephemeris as in Sec. 8.2. For other choice at a different epoch (e.g. at an epoch in 2035 closer to eLISA/NGO planned arrival at science orbit), the procedure is the same.

7.2. *Basic ASTROD orbit configuration*

In the original proposal, the ASTROD-GW orbits are chosen in the ecliptic plane with inclination $\lambda = 0$. The angular resolution in the sky has antipodal ambiguity. Although over most of sky the resolution is good, near the ecliptic poles the resolution is poor. After 2010, we have designed the basic orbits of ASTROD-GW to have small inclinations in order to resolve these issues while keeping the variation of the arm lengths in the tolerable range.[39,40]

Following Refs. 39 and 40, the basic idea is that if the orbits of the ASTROD-GW spacecraft are inclined with a small angle λ, the interferometry plane with appropriate design is also inclined with similar angle and when the ASTROD-GW formation evolves, the interferometry plane can be designed to modulate in the ecliptic solar system barycentric frame. With this angular positions of GW sources both near the polar region and off the polar region are resolved without antipodal ambiguity (see also Sec. 7.3).

Let us first consider a circular orbit of a spacecraft in the Newtonian gravitational central problem (one-body central problem) in spherical coordinates (r, θ, φ):

$$r = a, \quad \theta = 90°, \quad \varphi = \omega t + \varphi_0, \tag{26}$$

Table 2. Initial states (conditions) of 3 S/C of eLISA/NGO at epoch JD2459215.5 (2021-Jan-1st 00:00:00) for our initial choice (third column), after first stage optimization (fourth column) and after all optimizations (fifth column) in J2000 equatorial (Earth mean equator and equinox coordinates) solar system barycentric coordinate system.

		Initial choice of S/C initial states	Initial states of S/C after first stage optimization	Initial states of S/C after final optimization
S/C1	X	$-1.53222193865 \times 10^{-2}$	$-1.53222193865 \times 10^{-2}$	$-1.53221933735 \times 10^{-2}$
Position	Y	$9.23347976632 \times 10^{-1}$	$9.23347976632 \times 10^{-1}$	$9.23345222988 \times 10^{-1}$
(AU)	Z	$4.04072005496 \times 10^{-1}$	$4.04072005496 \times 10^{-1}$	$4.04070800735 \times 10^{-1}$
S/C1	V_x	$-1.71752389145 \times 10^{-2}$	$-1.71926502995 \times 10^{-2}$	$-1.71928071373 \times 10^{-2}$
Velocity	V_y	$-1.41699055355 \times 10^{-4}$	$-1.41837311087 \times 10^{-4}$	$-1.41838556464 \times 10^{-4}$
(AU/day)	V_z	$-6.11987395198 \times 10^{-5}$	$-6.12586807155 \times 10^{-5}$	$-6.12592206525 \times 10^{-5}$
S/C2	X	$-1.86344993528 \times 10^{-2}$	$-1.86344993528 \times 10^{-2}$	$-1.86344993528 \times 10^{-2}$
Position	Y	$9.22658604804 \times 10^{-1}$	$9.22658604804 \times 10^{-1}$	$9.22658604804 \times 10^{-1}$
(AU)	Z	$3.98334135807 \times 10^{-1}$	$3.98334135807 \times 10^{-1}$	$3.98334135807 \times 10^{-1}$
S/C2	V_x	$-1.72244923440 \times 10^{-2}$	$-1.72419907995 \times 10^{-2}$	$-1.72419907995 \times 10^{-2}$
Velocity	V_y	$-1.88198725403 \times 10^{-4}$	$-1.88384533079 \times 10^{-4}$	$-1.88384533079 \times 10^{-4}$
(AU/day)	V_z	$-2.71845314386 \times 10^{-5}$	$-2.72100311132 \times 10^{-5}$	$-2.72100311132 \times 10^{-5}$
S/C3	X	$-1.19599845212 \times 10^{-2}$	$-1.19599845212 \times 10^{-2}$	$-1.19599845212 \times 10^{-2}$
Position	Y	$9.22711604030 \times 10^{-1}$	$9.22711604030 \times 10^{-1}$	$9.22711604030 \times 10^{-1}$
(AU)	Z	$3.98357113784 \times 10^{-1}$	$3.98357113784 \times 10^{-1}$	$3.98357113784 \times 10^{-1}$
S/C3	V_x	$-1.72249891952 \times 10^{-2}$	$-1.72424881557 \times 10^{-2}$	$-1.72424881557 \times 10^{-2}$
Velocity	V_y	$-9.59855278460 \times 10^{-4}$	$-9.60776184172 \times 10^{-4}$	$-9.60776184172 \times 10^{-4}$
(AU/day)	V_z	$-9.55537821052 \times 10^{-5}$	$-9.56487660660 \times 10^{-5}$	$-9.56487660660 \times 10^{-5}$

where a, ω, and φ_0 are constants. For spacecraft in this discussion, we have $a = 1$ AU, $\omega = 2\pi/T_0$ with $T_0 = 1$ sidereal year, and φ_0 is the initial phase in the coordinate considered. The spacecraft orbit at time t in Cartesian coordinates is

$$x = a\cos\varphi = a\cos(\omega t + \varphi_0); \quad y = a\sin\varphi = a\sin(\omega t + \varphi_0); \quad z = 0. \tag{27}$$

Let us transform this orbit actively into an orbit with inclination λ, and with the intersection of the orbit plane and xy-plane (the ecliptic) at the line $\varphi = \Phi_0$ in the xy-plane. The active transformation matrix is

$$\mathrm{R}(\lambda;\Phi_0) = \begin{pmatrix} \cos^2\Phi_0 + \sin^2\Phi_0\cos\lambda & \sin\Phi_0\cos\Phi_0(1-\cos\lambda) & \sin\Phi_0\sin\lambda \\ \sin\Phi_0\cos\Phi_0(1-\cos\lambda) & \sin^2\Phi_0 + \cos^2\Phi_0\cos\lambda & -\cos\Phi_0\sin\lambda \\ -\sin\Phi_0\sin\lambda & \cos\Phi_0\sin\lambda & \cos\lambda \end{pmatrix}. \tag{28}$$

The new spacecraft orbit is

$$\begin{pmatrix} x' \\ y' \\ z' \end{pmatrix} = \begin{pmatrix} a[1-\sin^2\Phi_0(1-\cos\lambda)]\cos\varphi + a\sin\Phi_0\cos\Phi_0(1-\cos\lambda)\sin\varphi \\ a\cos\Phi_0\sin\Phi_0(1-\cos\lambda)\cos\varphi + a[1-\cos^2\Phi_0(1-\cos\lambda)]\sin\varphi \\ -a\sin\Phi_0\sin\lambda\cos\varphi + a\cos\Phi_0\sin\lambda\sin\varphi \end{pmatrix}. \tag{29}$$

For the three orbits with inclination λ (in radian), we choose:

$$\begin{aligned} \text{S/C I: } \Phi_0(\text{I}) &= 270^\circ, \qquad \varphi_0(\text{I}) = 0^\circ; \\ \text{S/C II: } \Phi_0(\text{II}) &= 150^\circ, \qquad \varphi_0(\text{II}) = 120^\circ; \\ \text{S/C III: } \Phi_0(\text{III}) &= 30^\circ, \qquad \varphi_0(\text{III}) = 240^\circ. \end{aligned} \tag{30}$$

Defining

$$\xi \equiv 1 - \cos\lambda = 0.5\lambda^2 + \text{O}(\lambda^4), \tag{31}$$

from Eqs. (29) and (30), we have

(i) for the orbit of S/C I

$$\begin{pmatrix} x^{\text{I}} \\ y^{\text{I}} \\ z^{\text{I}} \end{pmatrix} = \begin{pmatrix} a\cos\omega t - \xi a\cos\omega t \\ a\sin\omega t \\ a\cos\omega t\sin\lambda \end{pmatrix}, \tag{32}$$

(ii) for the orbit of S/C II

$$\begin{pmatrix} x^{\text{II}} \\ y^{\text{II}} \\ z^{\text{II}} \end{pmatrix} = \begin{pmatrix} a\left[\left(-\frac{1}{2}\right)\cos\omega t - \left(\frac{3^{1/2}}{2}\right)\sin\omega t\right] \\ \quad + \left(\frac{a}{2}\right)\xi\left[\left(\frac{3^{1/2}}{2}\right)\sin\omega t - \frac{1}{2}\cos\omega t\right] \\ a\left[\left(-\frac{1}{2}\right)\sin\omega t + \left(\frac{3^{1/2}}{2}\right)\cos\omega t\right] \\ \quad + \left(\frac{3^{1/2}}{2}\right)a\xi\left[\left(\frac{3^{1/2}}{2}\right)\sin\omega t - \frac{1}{2}\cos\omega t\right] \\ a\sin\lambda\left[\left(\frac{3^{1/2}}{2}\right)\sin\omega t - \frac{1}{2}\cos\omega t\right] \end{pmatrix}, \tag{33}$$

(iii) for the orbit of S/C III

$$\begin{pmatrix} x^{\text{III}} \\ y^{\text{III}} \\ z^{\text{III}} \end{pmatrix} = \begin{pmatrix} a\left[\left(-\frac{1}{2}\right)\cos\omega t + \left(\frac{3^{1/2}}{2}\right)\sin\omega t\right] \\ \quad + \left(\frac{a}{2}\right)\xi\left[\left(\frac{3^{1/2}}{2}\right)\sin\omega t - \frac{1}{2}\cos\omega t\right] \\ a\left[\left(-\frac{1}{2}\right)\sin\omega t - \left(\frac{3^{1/2}}{2}\right)\cos\omega t\right] \\ \quad - \left(\frac{3^{1/2}}{2}\right)a\xi\left[\left(-\frac{3^{1/2}}{2}\right)\sin\omega t - \frac{1}{2}\cos\omega t\right] \\ a\sin\lambda\left[\left(-\frac{3^{1/2}}{2}\right)\sin\omega t - \frac{1}{2}\cos\omega t\right] \end{pmatrix}. \tag{34}$$

One can readily check that $[(x^{\rm I})^2+(y^{\rm I})^2+(z^{\rm I})^2]^{1/2}=[(x^{\rm II})^2+(y^{\rm II})^2+(z^{\rm II})^2]^{1/2}=[(x^{\rm III})^2+(y^{\rm III})^2+(z^{\rm III})^2]^{1/2}=a$ hold for consistency.

Calculate the arm vectors $\underline{V}_{\text{II-I}}=\underline{r}^{\rm II}-\underline{r}^{\rm I}$, $\underline{V}_{\text{III-II}}=\underline{r}^{\rm III}-\underline{r}^{\rm II}$ and $\underline{V}_{\text{I-III}}=\underline{r}^{\rm III}-\underline{r}^{\rm I}$:

$$\underline{V}_{\text{II-I}}=\begin{pmatrix} a\left[-\left(\frac{3}{2}\right)\cos\omega t-\left(\frac{3^{1/2}}{2}\right)\sin\omega t\right]+a\xi\left[\left(\frac{3^{1/2}}{4}\right)\sin\omega t+\left(\frac{3}{4}\right)\cos\omega t\right] \\ a\left[-\left(\frac{3}{2}\right)\sin\omega t+\left(\frac{3^{1/2}}{2}\right)\cos\omega t\right]+a\xi\left[\left(\frac{3}{4}\right)\sin\omega t-\left(\frac{3^{1/2}}{4}\right)\cos\omega t\right] \\ a\sin\lambda\left[\left(\frac{3^{1/2}}{2}\right)\sin\omega t-\left(\frac{3}{2}\right)\cos\omega t\right] \end{pmatrix}, \tag{35}$$

$$\underline{V}_{\text{III-II}}=\begin{pmatrix} 3^{1/2}a\sin\omega t-\left(\frac{3^{1/2}}{2}\right)a\xi\sin\omega t \\ -3^{1/2}a\cos\omega t+\left(\frac{3^{1/2}}{2}\right)a\xi\cos\omega t \\ -3^{1/2}a\sin\lambda\sin\omega t \end{pmatrix}, \tag{36}$$

$$\underline{V}_{\text{I-III}}=\begin{pmatrix} a\left[\left(\frac{3}{2}\right)\cos\omega t-\left(\frac{3^{1/2}}{2}\right)\sin\omega t\right]+a\xi\left[\left(\frac{3^{1/2}}{4}\right)\sin\omega t-\left(\frac{3}{4}\right)\cos\omega t\right] \\ a\left[\left(\frac{3}{2}\right)\sin\omega t+\left(\frac{3^{1/2}}{2}\right)\cos\omega t\right]+a\xi\left[-\left(\frac{3}{4}\right)\sin\omega t-\left(\frac{3^{1/2}}{4}\right)\cos\omega t\right] \\ a\sin\lambda\left[\left(\frac{3^{1/2}}{2}\right)\sin\omega t+\left(\frac{3}{2}\right)\cos\omega t\right] \end{pmatrix}. \tag{37}$$

The closure relation $\underline{V}_{\text{II-I}}+\underline{V}_{\text{III-II}}+\underline{V}_{\text{I-III}}=\underline{0}$ is checked for verifying calculations also. The arm lengths are calculated to be

$$\begin{aligned} |\underline{V}_{\text{II-I}}| &= 3^{1/2}a\left[(1-\xi/2)^2+\sin^2\lambda\sin^2(\omega t-60^\circ)\right]^{1/2}, \\ |\underline{V}_{\text{III-II}}| &= 3^{1/2}a\left[(1-\xi/2)^2+\sin^2\lambda\sin^2(\omega t)\right]^{1/2}, \\ |\underline{V}_{\text{I-III}}| &= 3^{1/2}a\left[(1-\xi/2)^2+\sin^2\lambda\sin^2(\omega t+60^\circ)\right]^{1/2}. \end{aligned} \tag{38}$$

The fractional arm length variation is within $(1/2)\sin^2\lambda$ which is about 10^{-4} for λ around 1°.

The cross-product vector $\underline{N}(t)\equiv\underline{V}_{\text{III-II}}\times\underline{V}_{\text{I-III}}$ is normal to the orbit configuration plane and has the following components:

$$\underline{N}=\left[\left(\frac{3^{3/2}}{2}\right)(1-\xi/2)a^2\right]\begin{pmatrix} -\sin\lambda\cos 2\omega t \\ -\sin\lambda\sin 2\omega t \\ 1-\xi/2 \end{pmatrix}. \tag{39}$$

The normalized unit normal vector $\underline{n}$ is then:

$$\underline{n} = \left[\sin^2\lambda + (1-\xi/2)\right]^{1/2} \begin{pmatrix} -\sin\lambda\cos 2\omega t \\ -\sin\lambda\sin 2\omega t \\ 1-\xi/2 \end{pmatrix}. \tag{40}$$

The geometric center $\underline{V}_c$ of the ASTROD-GW spacecraft configuration is

$$\underline{V}_c = \begin{pmatrix} -\left(\frac{1}{2}\right)\xi a\cos\omega t \\ \left(\frac{1}{2}\right)\xi a\sin\omega t \\ 0 \end{pmatrix}. \tag{41}$$

There are three interferometers with two arms in the ASTROD-GW configuration. The geometric center of each of these three interferometers is at a distance of about 0.25 AU from the Sun. Numerical simulation and optimization of orbit configuration for inclination of 0.5°, 1°, 1.5°, 2°, 2.5° and 3° have been worked out using planetary ephemeris to take into account of the planetary perturbations in Ref. 41. The case with inclinations of 1° is reviewed in Sec. 8.3 for illustration. When LISA configuration orbits are around the Sun, it is equivalent to multiple detector arrays distributed in 1 AU orbit. The extension of ASTROD-GW is already of 1 AU. When ASTROD-GW orbits are around the Sun, it is also equivalent to multiple detector arrays distributed in 1 AU orbit.

7.3. *Angular resolution*

Consider angular resolution of a coherent GW source. Consider first the LISA case as an example. The detector formation of LISA is modulated in its orbit around the Sun. The azimuth modulation amplitude is 2π rad with inclination 1.05 rad (60°) so that the antenna pattern sweeps around the sky in one year. The antenna response is not isotropic but the averaged linear angular resolution (in a year) of monochromatic GW sources for LISA differs by less than a factor of three among all directions.[9] This is also true for all LISA-like formations. *The angular resolution is basically proportional to the inverse of the strain signal to noise ratio.* If the inclination is of the order 0.017–0.052 rad (1–3°) for LISA, the polar resolution would be worsened by 30–10 times (approximately the ratio sin of 1.05 rad to sin of 0.03–0.1 rad); the steradian localization in the celestial sphere would be worsened by square of this factor. Away from the polar region ($\theta \gg 0.017$–0.052 rad), the steradian localization in the celestial sphere would be by $\sin^2\theta$. If the signal to noise ratio is downgraded by 5 (as in eLISA/NGO or in OMEGA in its low frequency part due to shorter arm length), the linear angular resolution is worsened by five times. ASTROD-GW has less sensitivity above 1 mHz compared with LISA, therefore the angular resolution will be worsened by both factors. In the 100 nHz–1 mHz region, ASTROD-GW has better sensitivity compared with LISA, in most part by 52 times. Hence, the angular resolution in the polar region is similar to that of LISA, while in other regions, the linear resolution is enhanced by roughly $52 \times \sin\theta$ (upgraded by 52 but downgraded by $\sin^2\theta$ in sterad [by sin θ in rad]). Although there is a mild dependence on the configuration inclination angle λ, within a factor of 3, the averaged antenna pattern for ASTROD-GW away from the polar region is better by a factor of $52 \times \sin\theta$ compared to that of LISA. Since the antenna pattern of ASTROD-GW sweeps over the whole sky in half year as can been seen from Eq. (40), the time of average needed is half a year instead of a year.[39,40]

For more complicated sources like chirping GW sources from binary black holes (BBHs), one needs to do fitting in order to obtain the accuracy of the parameters. However, the tendency of accuracy of parameters is the same: for similar situations, it is proportional to the inverse of the strain signal to noise ratio.

For Super-ASTROD, the strain signal to noise ratio would be even better than ASTROD-GW by five times toward the lower frequency region, therefore the angular resolution would be better by five times. For polar resolution, the ASTROD inclination strategy could be applied. However, since Super-ASTROD has 1 or 2 S/C in off-ecliptic orbit, this may not be needed. For ASTROD-EM, since the lunar orbit is inclined about 5° to the ecliptic and the node precession period is 18.61 tropical years, the Earth–Moon Lagrange points also precess together. Depending on the time and duration of mission, it might or might not be desirable to use slightly inclined orbit.[43]

Most of the Earth orbit GW missions have dipolar ambiguity and the resolution is poor in the polar region. However, this is not a big issue since we just need to look at both polarity for identification of electromagnetic counterparts and the polar region is only a small portion of the sky.

7.4. *Six/twelve spacecraft formation*

In order to detect relic GWs using correlated detection, Big Bang Observer[45] and DECIGO[44] proposals have 12 spacecraft distributed in the Earth orbit in three groups separated by 120° in orbit; two groups has three spacecraft each in a LISA-like triangular formation and the third group has six spacecraft with two LISA-like triangles forming a star configuration (Fig. 6). An alternate configuration is that each group has four spacecraft forming a nearly square configuration (also has a tilt of 60° with respect to the ecliptic plane).

For a more sensitive detection of background or relic GWs, correlated detection with two sets of triangular ASTROD-GW formation are required, i.e. a 6-S/C constellation. The second nearly triangular formation could be put again near L3, L4, L5 respectively, but separated from the first formation by 1×10^6 km to 5×10^6 km for the respective S/C.[40]

8. Orbit Design and Orbit Optimization Using Ephemerides

Although Sun is dominant in the solar system, there are other planets and celestial objects affecting the orbit, notably Jupiter, Venus and Earth. Ephemerides is a must for orbit design. At present, there are three complete fundamental ephemerides of the solar system — Development Ephemerides (DE),[125] Ephemerides of Planets and Moon (EPM)[126] and Intégrateur Numérique Planétaire de l'Observatoire de Paris (INPOP).[127] Any of these three ephemerides could be used in the orbit design and orbit optimization. For easier in numerical processing, we normally use Center for Gravitation and Cosmology (CGC) ephemeris framework together with initial conditions taken from DE ephemerides at a certain epoch for evolving with post-Newtonian approximation.

8.1. *CGC ephemeris*

In 1998, we started orbit simulation and parameter determination for ASTROD.[128,129] We worked out a post-Newtonian ephemeris of the solar system including the solar quadrupole moment, the eight planets, the Pluto, the moon and the three biggest asteroids. We term this working ephemeris CGC 1 (CGC: Center for Gravitation and Cosmology). Using this ephemeris as a deterministic model and adding stochastic terms to simulate noise, we generate simulated ranging data and use Kalman filtering to determine the accuracies of fitted relativistic and solar system parameters for 1050 days of the ASTROD mission.

For a better evaluation of the accuracy of $\dot{G}/G$, we need also to monitor the masses of other asteroids. For this, we considered all known 492 asteroids with diameter greater than 65 km to obtain an improved ephemeris framework — CGC 2, and calculated the perturbations due to these 492 asteroids on the ASTROD spacecraft.[130,131]

In building CGC ephemeris framework, we use the post-Newtonian barycentric metric and equations of motion as derived in Brumberg[132] with Parametrized Post-Newtonian (PPN) parameters β and γ for solar system bodies (with the gauge parameter α set to zero). These equations are used to build our computer-integrated ephemeris (with the PPN parameters $\gamma = \beta = 1$, $J_2 = 2 \times 10^{-7}$) for eight-planets, the Pluto, the Moon and the Sun. The initial positions and initial velocities at the epoch 2005-June-10 0:00 are taken from the DE403 ephemeris. The evolution is solved by using the fourth-order Runge–Kutta method with the step size $h = 0.01$ day. In Ref. 129, the 11-body evolution is extended to 14-body to include the three big asteroids — Ceres, Pallas and Vesta (CGC 1 ephemeris). Since the tilt of the axis of the solar quadrupole moment to the perpendicular of the elliptical plane is small (7°), in CGC 1 ephemeris, we have neglected this tilt. In CGC 2 ephemeris, we have added the perturbations of additional 489 asteroids.

In our first optimization of ASTROD-GW orbits,[133–135] we have used CGC 2.5 ephemeris in which only three biggest minor planets are taken into accounts, but the Earth's precession and nutation are added; the solar quadratic zonal harmonic and the Earth's quadratic to quartic zonal harmonic are considered. In later simulation, we add the perturbation of additional 349 asteroids and call it CGC 2.7 ephemeris.[84–86] The differences in orbit evolution compared with DE405 for Earth for 3700 days starting at JD2461944.0 (2028-Jun-21 12:00:00) are shown in Fig. 5 of Ref. 40. The differences in radial distances are less than about 200 m. The differences for other inner planets are smaller. The differences in latitude and longitude for Earth are less than 1 mas.

8.2. *Numerical orbit design and orbit optimization for eLISA/NGO*

The mission orbit configuration of eLISA/NGO is similar to that of LISA but with a shorter arm length and a closer distance to Earth. The distance of any two of three spacecraft must be maintained as close as possible during geodetic flight. LISA orbit configuration has been studied analytically and numerically in various previous works.[117–122,136,137] For eLISA/NGO, we followed the analytical procedure of Dhurandhar *et al.*[122] (see Sec. 7.1) in making our initial choice of the initial conditions in Ref. 84; these initial conditions are listed in column 3 of Table 2 in Sec 7.1. With this orbit choice, we started numerical orbit design and used the CGC ephemeris to numerically optimize the orbit configuration in Ref. 84 as we have done for ASTROD-GW orbit[85,86,133–135,138] design. In this section, we review and summarize the procedure following [84].

The goal of the eLISA/NGO mission orbit optimization is to equalize the three arm lengths of the eLISA/NGO formation and to reduce the relative line-of-sight velocities between three pairs of spacecraft as much as possible. In the solar system, the eLISA/NGO spacecraft orbits are perturbed by the planets. With the initial states of the three spacecraft as listed in column three of Table 2, we calculated the eLISA/NGO orbit configuration for 1000 days using CGC 2.7. The variations of arm lengths and velocities in the line-of-sight direction are drawn in Fig. 2 of [84]. The largest variations are caused by Earth, Jupiter and Venus. Our method of optimization is to modify the initial velocities and initial heliocentric distances so that (i) the perturbed orbital periods for 1000 day average remains close to one another, and (ii) the average major axes are adjusted to make arms nearly equal. We do this iteratively as follows. From Fig. 2 of Ref. 84, we noticed that the variation of Arm1 (between S/C2 and S/C3) is small. First, we adjust the initial conditions of S/C2 and S/C3 to make the variation of Arm1 satisfy the mission requirements that arm length variations are within 2% and Doppler velocities are within 10 m/s. Then, we adjust the initial conditions of S/C1 so that Arm2 and Arm3 satisfy the mission requirements. Adjustments are always performed in the ecliptic heliocentric coordinate system.

The actual adjustment procedure is described as follows. Firstly, the magnitudes of initial velocities of S/C2 and S/C3 were adjusted so that their average periods (367.474 days) in 3 year were a little bit longer than one sidereal year. Within a definite range, when the periods become longer, the variations of Arm1 become smaller. The initial velocities were adjusted so that the Arm1 satisfied the eLISA/NGO arm length and Doppler velocity requirements. After this, we adjusted the initial velocities of S/C1 to make its orbital period approach those of S/C2 and S/C3, and Arm2 and Arm3 nearly equal. If the results obtained from the above procedure did not satisfy the requirements or better results were expected, we could adjust the orbital periods of S/C2 and S/C3 a little bit longer again under the constraint that the eLISA/NGO requirements for Arm1 is satisfied. Up to this stage, only initial velocities have been adjusted. After we have completed this stage, the initial conditions of the 3 S/C are listed in column 4 of Table 2; the variations of arm lengths and velocities in the line-of-sight direction are drawn in Fig. 3 of Ref. 84.

After the first stage, we optimized the orbital period of S/C1 by adjusting the initial velocity and the semi-major axis until the eLISA/NGO requirements were satisfied. The initial conditions of the 3 S/C, after optimization, are listed in column 5 of Table 2; the variations of arm lengths (within 2%) and velocities in the line-of-sight direction (within 5.5 m/s) are drawn in Fig. 7. In Fig. 7, we also draw the angle between S/C and Earth subtended from Sun in 1000 days; it starts at 10° behind Earth (in solar orbit) and varies between 9° and 16° with a quasi-period of variation about one sidereal year mainly due to Earth's elliptic motion.

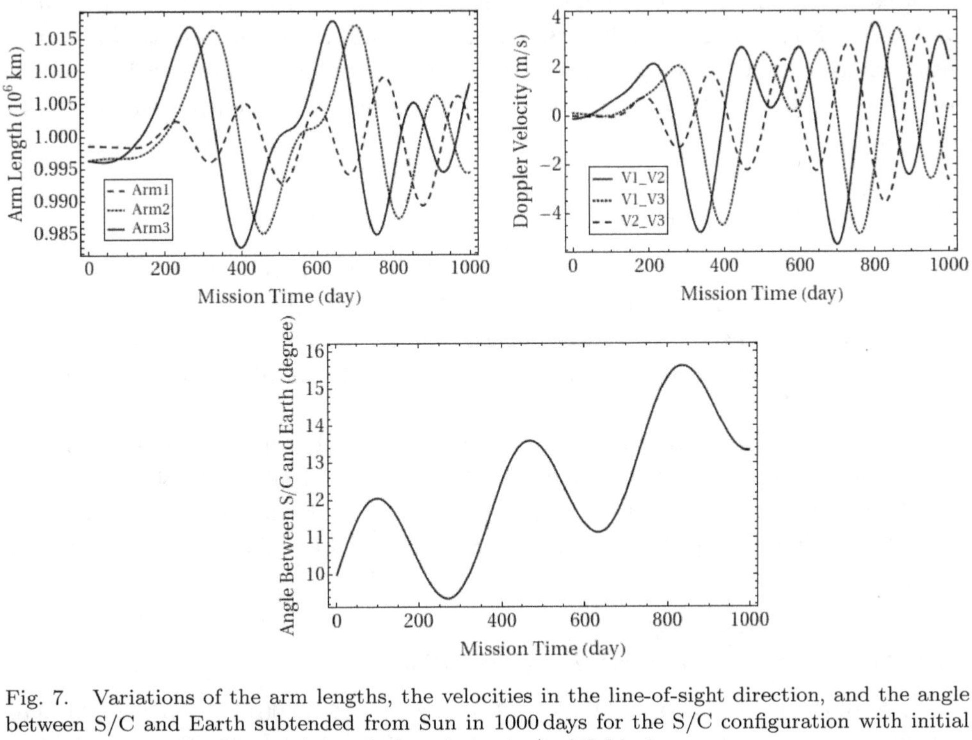

Fig. 7. Variations of the arm lengths, the velocities in the line-of-sight direction, and the angle between S/C and Earth subtended from Sun in 1000 days for the S/C configuration with initial conditions given in column 5 (after final optimization) of Table 2.

8.3. *Orbit optimization for ASTROD-GW*[41]

The goal of the ASTROD-GW mission orbit optimization is to equalize the three arm lengths of the ASTROD-GW formation and to reduce the relative line-of-sight velocities between three pairs of spacecraft as much as possible. In our first optimization, the time of start of the science part of the mission is chosen to be noon, June 21, 2025 (JD2460848.0) and the optimization is for a period of 3700 days using CGC 2.5 ephemeris.[133–135] Since the preparation of the mission may take longer time and there is a potential that the extended mission life time may be longer than 10 year, in later optimizations,[85,86,138] we started at noon, June 21, 2028 (JD2461944.0) and optimize for a period of 20 years using CGC 2.7 ephemeris including more asteroids than those of CGC 2.5. In both of these optimizations, the orbit configuration is set in the ecliptic plane and we have the inclination angle $\lambda = 0$. With the basic configuration of ASTROD-GW changed into an inclined precession orbit formation, we re-design and re-optimize our orbit configuration numerically starting at noon, June 21, 2035 (JD2464500.0) for 10 years for the inclination angle 0.5°, 1°, 1.5°, 2°, 2.5° and 3° using the CGC 2.7.1 ephemeris.[41]

In this section, we illustrate the design and optimization method with inclined precession orbit formation for the case having inclination angle 1° following Ref. 41 which uses CGC 2.7.1. The differences between CGC 2.7.1 and CGC 2.7 (summarized in Sec. 8.1) is detailed in Sec. 8.3.1. In Sec. 8.3.2, we review how to obtain the initial choice of S/C initial conditions as a starting point for numerical optimization. In Sec. 8.3.3, we discuss method of optimization and summarize the results of optimization.

8.3.1. *CGC* 2.7.1 *ephemeris*

In the CGC 2.7.1 ephemeris framework, we pick up 340 asteroids besides the Ceres, Pallas and Vesta from the Lowell database. The masses of 340 asteroids are given by Lowell data[139] instead of estimating the masses based on the classification in CGC 2.7.[83,84,86] The orbit elements of these asteroids are also updated from the Lowell database.

For a 10 year duration starting at June 21, 2035, the differences between the Earth's heliocentric distances calculated by CGC 2.7.1 and DE430 are within 150 m, and that the differences in longitudes and latitudes are within 1.4 mas and 0.45 mas, respectively. These differences do not affect the results of our TDI calculations.

8.3.2. *Initial choice of spacecraft initial conditions*

The R.A. of the Earth at JD2464500 (2035-June-21st 12:00:00) is $17^{h}57^{m}45.09^{s}$, i.e. 269.438° from DE 430 ephemeris. The initial positions of the 3 S/Cs are obtained by choosing the ωt as 89.44° for $\varphi = \omega t + \varphi_0$ in Eq. (29). The initial velocities are

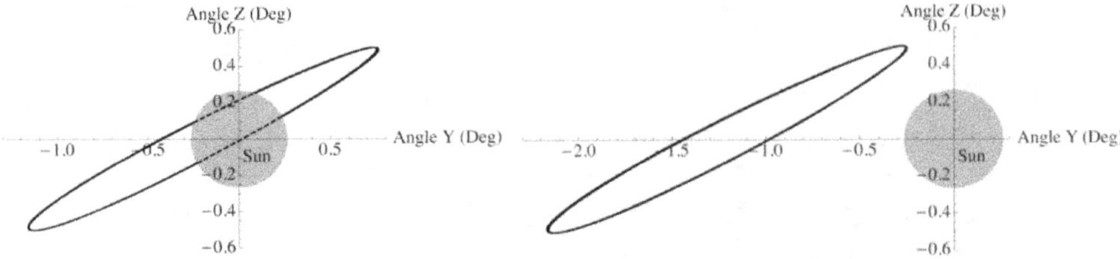

Fig. 8. S/C1 view from Earth before rotating the initial conditions by an angle (left diagram) and after rotating by an angle 2.0°(right diagram) for the case of inclination angle 1.0°.[41]

derived from Eq. (29) by calculating the derivatives with respect to t. The S/C1 orbit near the Lagrange point L3 is partly obscured by Sun from the line-of-sight of Earth (left diagram of Fig. 8). It would obstruct the communication with the Earth stations. To avoid the obscuration, we rotate the initial angle Φ_0 and φ_0 forward of by 2.0° for inclination angle 1.0°. The S/C1 orbit is shown on the right diagram of Fig. 8. The initial choice of initial states for the 3 S/Cs in this case is listed in column 3 of Table 3.

8.3.3. *Method of optimization*

Our optimization method is to modify the initial velocities and initial heliocentric distances to reach the aim of (i) equalizing the three arm lengths of the ASTROD-GW formation as much as possible and (ii) reducing the relative Doppler velocities between three pairs of spacecraft as much as possible.

During the actual optimization procedure, we use the following equation to modify the average period of the orbit:

$$\boldsymbol{V}_{\text{new}} = \boldsymbol{V}_{\text{prev}} + \Delta\boldsymbol{V} \approx \left(1 - \frac{1}{3}\frac{\Delta T}{T}\right)\boldsymbol{V}_{\text{prev}}. \tag{42}$$

For the case of inclination angle of 1°, we calculate the orbits of the 3 S/C with the initial choice of initial conditions listed in column 3 of Table 3 using the CGC 2.7.1 ephemeris. The average periods of the 3 S/Cs in 10 years are 365.256 days (S/C1), 365.267 days (S/C2) and 365.266 days (S/C3), respectively. We use Eq. (42) to change the initial velocities so that the average period of S/C1, S/C2 and S/C3 is adjusted to 365.255 days, 365.257 days and 365.257 days, respectively. The initial conditions after this step are listed in column 4 of Table 3. In the next step, we use the following equations to trim the S/C eccentricities to be nearly circular:

$$\begin{aligned} \boldsymbol{R}_{\text{new}} &= \boldsymbol{R}_{\text{prev}} + \Delta\boldsymbol{R} \approx \left(1 + \frac{\Delta R}{R}\right)\boldsymbol{R}_{\text{prev}}, \\ \boldsymbol{V}_{\text{new}} &= \boldsymbol{V}_{\text{prev}} + \Delta\boldsymbol{V} \approx \left(1 - \frac{\Delta R}{R}\right)\boldsymbol{V}_{\text{prev}}. \end{aligned} \tag{43}$$

Here, R is the initial heliocentric distance of spacecraft. The fractional adjustment $\pm(\Delta R/R)$ in $\boldsymbol{R}_{\text{prev}}$ and $\boldsymbol{V}_{\text{prev}}$ would adjust eccentricity without adjust the period

Table 3. Initial states of S/Cs for the configuration with the inclination angle 1° at epoch JD2464500.0 for initial choice, after period optimization, and after all optimizations in J2000 equatorial solar system barycentric coordinate system.[41]

$\lambda = 1.0°$		Initial choice of S/C initial states	Initial states of S/Cs after period optimization	Initial states of S/C after final optimization
S/C1	X	$-2.8842263289715 \times 10^{-2}$	$-2.8842263289715 \times 10^{-2}$	$-2.8842514605546 \times 10^{-2}$
Position	Y	$9.1157742309044 \times 10^{-1}$	$9.1157742309044 \times 10^{-1}$	$9.1158659433458 \times 10^{-1}$
(AU)	Z	$3.9552690922456 \times 10^{-1}$	$3.9552690922456 \times 10^{-1}$	$3.9553088730467 \times 10^{-1}$
S/C1	V_x	$-1.7188548244458 \times 10^{-2}$	$-1.7188535691176 \times 10^{-2}$	$-1.7188363750567 \times 10^{-2}$
Velocity	V_y	$-2.8220395391983 \times 10^{-4}$	$-2.8220375159556 \times 10^{-4}$	$-2.8220098038726 \times 10^{-4}$
(AU/day)	V_z	$-4.4970276654173 \times 10^{-4}$	$-4.4970243993363 \times 10^{-4}$	$-4.4969796642665 \times 10^{-4}$
S/C2	X	$8.7453598387569 \times 10^{-1}$	$8.7453598387569 \times 10^{-1}$	$8.7453598387569 \times 10^{-1}$
Position	Y	$-4.3802677355114 \times 10^{-1}$	$-4.3802677355114 \times 10^{-1}$	$-4.3802677355114 \times 10^{-1}$
(AU)	Z	$-2.0634980179207 \times 10^{-1}$	$-2.0634980179207 \times 10^{-1}$	$-2.0634980179207 \times 10^{-1}$
S/C2	V_x	$8.2301784322477 \times 10^{-3}$	$8.2301033726700 \times 10^{-3}$	$8.2301033726700 \times 10^{-3}$
Velocity	V_y	$1.3797379424198 \times 10^{-2}$	$1.3797253460590 \times 10^{-2}$	$1.3797253460590 \times 10^{-2}$
(AU/day)	V_z	$6.1425805519808 \times 10^{-3}$	$6.1425244722884 \times 10^{-3}$	$6.1425244722884 \times 10^{-3}$
S/C3	X	$-8.5683596527799 \times 10^{-1}$	$-8.5683596527799 \times 10^{-1}$	$-8.5679330969623 \times 10^{-1}$
Position	Y	$-4.8998222347472 \times 10^{-1}$	$-4.8998222347472 \times 10^{-1}$	$-4.8995800210059 \times 10^{-1}$
(AU)	Z	$-1.9592963105165 \times 10^{-1}$	$-1.9592963105165 \times 10^{-1}$	$-1.9591994878015 \times 10^{-1}$
S/C3	V_x	$8.9788714330506 \times 10^{-3}$	$8.9787977300014 \times 10^{-3}$	$8.9792464008067 \times 10^{-3}$
Velocity	V_y	$-1.3530263187520 \times 10^{-2}$	$-1.3530152097744 \times 10^{-2}$	$-1.3530828362023 \times 10^{-2}$
(AU/day)	V_z	$-5.6998631854817 \times 10^{-3}$	$-5.6998163886731 \times 10^{-3}$	$-5.7001012664635 \times 10^{-3}$

of the orbit. The initial conditions after all optimization are listed in column 5 of Table 3.

For the inclination angles 0.0°, 0.5°, 1.5°, 2°, 2.5° and 3°, the optimization processes are similar to the inclination 1.0° and the results can be found in Ref. 41.

9. Deployment of Formation in Earthlike Solar Orbit

The deployment to orbit around Earth, to halo orbit of Earth–Moon Lagrange points and to Sun–Earth L1 and L2 points are well studied. Here, we say a few words on the deployment of a spacecraft to different positions of an Earthlike solar orbit. A preliminary design of the transfer orbits of the spacecraft from the separations of the launch vehicles to the mission orbits near L3, L4 and L5 points has been given in Refs. 40 and 140. Let us review this preliminary design first.

In the mission study of ASTROD I, the ASTROD I S/C is given an appropriate delta-V before the last stage of launcher separation in the Low Earth Orbit (LEO) and is injected directly to the solar orbit going geodetic to Venus swing-by. We can use the same strategy to launch the ASTROD-GW S/C directly into the solar transfer orbits near the designated Hohmann transfer orbits or Venus swing-by orbit. This way, the only major delta V needed for each S/C to reach the destination occurs near the destination to boost the S/C to stay near the destined Lagrange point. In row 2–4 of Table 4, we list types of transfer orbits, transfer times, the values of solar transfer delta-V and propellant mass ratio for three ASTROD-GW S/C. These estimates are good for any other S/C deployed to the same positions. The propellant mass ratios are around 0.5–0.55, 0.280 and 0.47 for S/C 1, 2 and 3. The total masses in case of ASTROD-GW S/C correspond to a dry mass of 500 kg are 1111–1266, 723 and 1035 kg for 3 S/C respectively (including the propellant and the propulsion module with mass of 10% of the propellant).

For deployment to other location in the solar orbit, we made estimates and list them in row 5–7 of Table 4. The baseline is: (i) S/C is propelled by a high efficient propulsion module (including the propellant with specific impulse 320 s and the propulsion module with mass of 10% of the propellant) for large delta-V maneuvers and for delivery to the destination; (ii) This module is to be separated when the destination state is achieved.

Further studies on the optimizations of deployment from separation of launcher(s) for the orbit configurations with inclinations and for a period of 20 years are ongoing for both LISA-like missions and ASTROD-GW-like missions.[141]

10. Time Delay Interferometry

In Sec. 4, we start discussing TDI, now we continue. *To achieve required GW sensitivity, TDI to suppress laser frequency noise is required for space GW missions.*

Schematic orbit configuration of LISA-type mission design[9] and ASTROD-GW mission design[41] are shown in Figs. 1 and 2, respectively. For the numerical evaluation, we take a common receiving time epoch for both beams; the results would be

Table 4. Estimated delta-V and propellant mass ratio for solar transfer of S/C.

° ahead of Earth in solar orbit	Transfer orbit	Transfer time	Solar transfer delta-V after injection from LEO to solar transfer orbit	Solar transfer propellant mass ratio (Isp = 320 s)
180° (near L3)	Venus flyby transfer	1.3–1.5 year	2.2–2.5 km/s	0.50–0.55
60° (near L4)	Inner Hohmann, 2 Revolutions	1.833 year	1.028 km/s	0.280
300° (−60°) (near L5)	Outer Hohmann, 1 Revolutions	1.167 year	2 km/s	0.47
0–60°	Inner Hohmann, ≤ 2 Revolutions	Less than 1.833 year	Less than 1.028 km/s	Less than 0.280
60°–300°	Venus flyby transfer	1.3–1.5 year	2.2–2.5 km/s	0.50–0.55
300°–360°	Outer Hohmann, 1 Revolutions	Less than 1.167 year	Less than 2 km/s	Less than 0.47

very close to each other numerically if we take the same start time epoch and calculate the path differences. We refer to the path S/C1 → S/C2 → S/C1 as a (path) and the path S/C1 → S/C3 → S/C1 as b (path). Hence, the difference ΔL between Paths 1 and 2 for the unequal-arm Michelson can be denoted as $ab - ba \equiv [a, b]$. Here ab means a path followed by b path. The unequal-arm Michelson is now commonly called X-configuration.[87,88] The result of this TDI calculation for ASTROD-GW orbit with 1° inclination is shown in Fig. 9.

The first-generation and second-generation TDIs are proposed since 1999.[87,88] In the first-generation TDIs, static situations are considered. While in the second-generation TDIs, motions are compensated to a certain degree. The X-configurations considered above belong to the first-generation TDI configurations. We note that the numerical method has the advantage of taking care of all generations into a single calculation format. We shall not review more about these developments here, but refer the readers to the excellent review paper by Tinto and Dhurandhar[88] for a comprehensive treatment.

We compile for comparison the resulting differences for second-generation two-arm TDIs with $n = 1$ and $n = 2$ (n is the degree of polynomial in ab, ba, a^2 and b^2) due to arm length variations for various mission proposals — eLISA/NGO, an NGO-LISA-type mission with a nominal arm length of 2×10^6 km, LISA and ASTROD-GW in Table 5.

We note that:

(i) All the second-generation TDIs considered for the one-detector case for eLISA/NGO, for NGO-LISA-type with 2×10^6 km arm length, for LISA and for ASTROD-GW with 1° inclination basically satisfy the requirement. (Table 5)

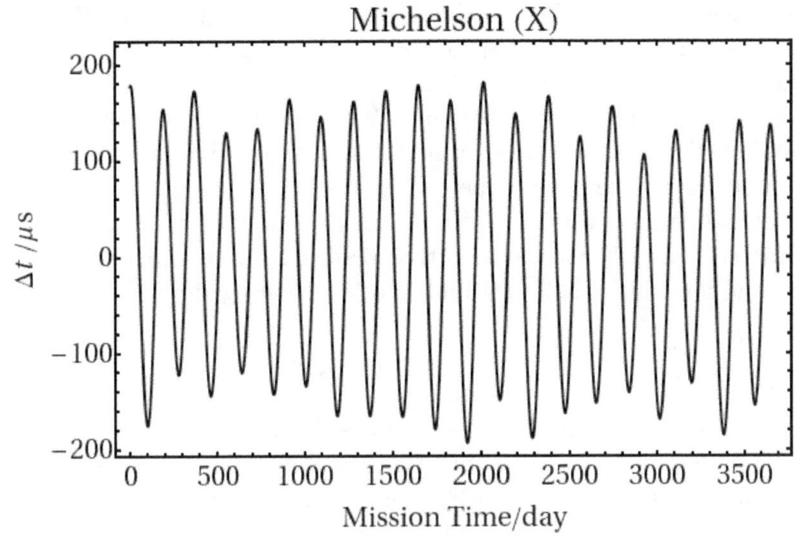

Fig. 9. Path length differences between two optical paths of the unequal-arm Michelson TDI configuration (X-configuration) for ASTROD-GW orbit formation with 1° inclination.

(ii) The requirement for unequal arm Michelson (X-configuration) TDI of ASTROD-GW needs to be relaxed by about two orders. (Fig. 9 and Table 5).
(iii) In view of the possibility of a GW mission in Earth orbit, numerical TDI study for GW missions in Earth orbit are desired.
(iv) Experimental demonstration of TDI in laboratory for LISA has been implemented in 2010.[142] eLISA and the original ASTROD-GW TDI requirement are based on LISA requirement, and hence also demonstrated. With the present

Table 5. Comparison of the resulting differences for second-generation TDIs ($n = 1$ and $n = 2$) due to arm length variations for various mission proposals — eLISA/NGO, an NGO-LISA-type mission with a nominal arm length of 2×10^6 km, LISA and ASTROD-GW.

TDI configuration		TDI path difference ΔL			
		eLISA/NGO[84]	NGO-LISA-type with 2×10^6 km arm length[84]	LISA[83]	ASTROD-GW (1° inclination)[41]
Duration		1000 days	1000 days	1000 days	10 years
$n = 1$	$[ab, ba]$	−1.5 to +1.5 ps	−11 to +12 ps	−70 to +80 ps	−228 to +228 μs
$n = 2$	$[a^2b^2, b^2a^2]$	−11 to +12 ps	−90 to +100 ps	−600 to +650 ps	−1813 to +1813 ns
	$[abab, baba]$	−6 to +6 ps	−45 to +50 ps	−300 to +340 ps	−907 to +907 ns
	$[ab^2a, ba^2b]$	−0.0032 to +0.0034 ps	−0.0036 to +0.004 ps	−0.015 to +0.013 ps	−0.66 to +0.66 ns
Nominal arm length		1 Gm (1 Mkm)	2 Gm	5 Gm	260 Gm
Requirement on ΔL		10 m (30 ns)	20 m (60 ns)	50 m (150 ns)	500 m (1500 ns)

pace of development in laser technology, the laser frequency noise requirement is expected to be able to compensate for 2–3 order of TDI requirement relaxation in 20 years.

(v) X-configuration TDI sensitivity for GW sources has been studied extensively for eLISA.[21] It satisfies the present technological requirements well. With enhanced laser technology expected, it would also be good for studying the ASTROD-GW and various GW missions in Earth orbit. The study for GW sensitivity and GW sources for other first-generation and second-generation TDIs and for other missions would also be encouraged.

11. Payload Concept

GW detection in space basically measures the distance change between two S/C (or celestial bodies) as GW comes by. The two S/C (or celestial bodies) must be in geodesic motion (or such motion can be deduced). The distance measurement must be ultra-sensitive as the GWs are weak. A typical implementation (mission) consists of three spacecraft in an almost equilateral triangle formation. The three spacecraft range interferometrically with one another. Each spacecraft carries a payload of two proof masses, two telescopes, two lasers, a weak light detection and handling system, a laser stabilization system, and a drag-free system. For lower part of space GW band or for possibly higher precision, a precision/optical clock, or an absolute laser stabilization system, and an absolute laser metrology system may be used.

Weak light phase locking and handling: For solar orbit missions, this is important. For ASTROD-GW with a distance of 260 Gm (1.73 AU), there is a need to phase lock a local laser to 100 fW incoming light to amplify and manipulate it. For 100 fW ($\lambda = 1064$ nm) weak light, there are 5×10^5 photons/s. This would be good for 100 kHz frequency tuning. For LISA, 85 pW weak light phase locking is required. In Tsing Hua University, 2 pW weak-light phase-locking with 0.2 mW local oscillator has been demonstrated.[29,30] In Jet Propulsion Laboratory (JPL), Dick *et al.*[82] have achieved offset phase locking to 40 fW incoming laser light. It would be good for future development focusing on frequency-tracking, modulation–demodulation and coding–decoding to make it a mature experimental technique. This is also important for the deep space optical communication.

Drag-free system design and development: Drag-free system consists of a high precision accelerometer/inertial sensor to detect non-drag-free motions and a micro-thruster system to do the feedback to keep the spacecraft drag-free. LISA Pathfinder successfully demonstrated and tested the drag-free technology in the frequency range above 100 μHz to satisfy not just the requirement of LISA Pathfinder, but also the requirement of LISA.[11] The success paved the road of knowledge for all the space mission proposals in Table 1. However, for lower part (100 nHz–100 μHz) of the space frequency band, there needs more work. We have discussed frequency sensitivity spectrum and reddening factors in Sec. 5. To suppress the reddening factors requires position sensing noise to be flat down to 100 nHz and gravity acceleration due to spacecraft to be small and modeled to the required level at low frequencies. The self-gravity-acceleration needs to be stable or subtracted in real-time. An absolute laser metrology system to monitor positions of major mass distribution in the S/C will be implemented to do this. To completely drop the factor or to go beyond, one may need to go to optical sensing and optical feedback control. As to the accelerometer/inertial sensor design of ASTROD, an absolute laser metrology system is proposed to push the noise down, in particular in the lower frequency region. In addition, ASTROD is proposed to monitor the positions of various parts of the spacecraft, to facilitate gravitational modeling.[27,28]

Micro-thruster system: For drag-free feedback control, micro-thrusters are needed. Field Emission Electric Propulsion (FEEP) system with its high specific thrust is a good candidate for the micro-thruster system. The sensitivity of FEEP system is good and is in the μN range. The main issue for FEEPs is lifetime. Due to technical problems during the development of the FEEP technology, the cold gas thrusters have become the alternative choice. The GAIA mission carries cold gas thrusters for the attitude and orbit control system (AOCS).[143] MICROSCOPE[144] and LISA Pathfinder are equipped with cold gas thrusters based on the GAIA thrusters. The main disadvantage of cold gas thrusters compared to FEEPs is the higher mass per delta-V. The total mission duration is limited by the amount of propellant stored in the tanks. Therefore, the FEEP technology would be preferred if it is available at a later time.

Laser system: Nd:YAG nonplanar ring oscillators pumped by laser diodes are available with output power of 2 W for use. The frequency noises must be suppressed to very low level. The strategy is like the one adopted by NGO/eLISA using pre-stabilization, arm locking[21] and TDI (Sec. 10).

Laser frequency standard/Clock: Space optical clocks and optical comb frequency synthesizer technologies are important in the realization and simplification of the GW mission target sensitivity at lower frequency. Another use of the optical clock and optical comb frequency synthesizer is to calibrate the optical metrology for ASTROD-GW-like missions. This is important for the laser metrology inertial sensor and for monitoring distances inside spacecraft, to correct local gravity changes due to, for example, thermal effects. All these measurements use lasers as standard rods. They need to be calibrated using optical frequency standards or absolutely stabilized laser frequency standard referenced to an atomic or molecular line. The advent of optical clocks and optical combs in space may possibly simplify the experimental design of ASTROD-GW-like mission.

At present, optical clocks in the laboratory[145] have reached a fractional inaccuracy at 10^{-18} level; and they are improving. Clocks of this accuracy level or better can be used for exquisitely sensitive measurements of gravity, motion, and inertial navigation. The use of this kind of clocks certainly will facilitate the detection of the lower frequency GWs and stimulate the needs of re-design the implementation schemes of the lower frequency space GW detection.

Absolute laser metrology system: With an ultraprecise laser frequency standard/clock, an absolute laser metrology system can be built to monitor the positions of various parts of the spacecraft to facilitate gravitational modeling.

Radiation monitor: A small radiation detector onboard the spacecraft will monitor test-mass charging of the inertial drag-free sensors. This radiation monitor can also be used for measuring Solar Energetic Particles (SEPs) and Galactic Cosmic Rays (GCRs) in the area of solar and galactic physics with corresponding applications to space weather.[146,147]

12. Outlook

White dwarf was discovered in 1910 with its density soon estimated. Now we understand that GWs from white dwarf binaries in our Galaxy form a stochastic GW background ("confusion limit")[148] for space GW detection in general relativity. The characteristic strain for confusion limit is about 10^{-20} in 0.1–1 mHz band. As to individual sources, some can have characteristic strain around this level for frequency 1–3 mHz in low-frequency band. One hundred year ago, the sensitivity of astrometric observation through the atmosphere around this band is about 1 arcsec. This means the strain sensitivity to GW detection is about 10^{-5}; 15 orders away from the required sensitivity.

The first artificial satellite Sputnik was launched in 1957. The technological demonstration mission LISA Pathfinder was launched on 3 December 2015. This mission successfully tested and demonstrated the drag-free technology to satisfy not just the requirement of LISA Pathfinder, but also basically the drag-free requirement of LISA GW space mission concept.[11] Thus, the major issue in the technological gap of 15 orders of magnitude is successfully abridged during last hundred years. The success paved the road for all the space mission proposals (Table 1). At present the space GW missions are expected to be launched in two decades. Weak-light phase locking is demonstrated in laboratories.[29,30,82] Weak-light technology still needs developments. And we do anticipate the possibility of an earlier launch date for eLISA (or a substitute mission) and possible earlier flight of other missions. With the first direct detection of GWs by LIGO and the success of LISA Pathfinder mission, the outlook of space detection of GWs is bright.

The science goals of space GW detectors are the detection of GWs from (i) Massive Black Holes; (ii) Extreme-Mass-Ratio Black Hole Inspirals; (iii) Intermediate-Mass Black Holes; (iv) Galactic Compact Binaries and (v) Relic GW Background. As we can readily see from Figs. 4–1, the signal-to-noise ratios (S/N) for GW detection of MBHB mergers are very high, and for the high S/N detection of more massive mergers the strain sensitivity at lower part (100 nHz–100 μHz) of the space detection band is important. For doing this, longer arms have advantages. Longer arm missions would be good to compliment PTAs in the exploration of black hole co-evolution with galaxies. Longer arm missions with its better angle resolution are also more effective in the determination of the equation of state of dark energy, testing relativistic gravity and, possibly, probing the inflationary physics. Efforts in minimizing the accelerometer/inertial sensor noise over the MLDC formula or beyond will strengthen these goals. Deployment of S/C to any position in the Earth-like solar orbit could be less than 1.8 years with propellant mass ratio less than 0.55. This is within the practical range of launcher implementation.

Now, we list important issues for further studies in order to realize and sharpen our expectations for GW detection in the frequency range 100 nHz–100 μHz:

(i) Manipulating weak light.
(ii) Improvement of low-frequency acceleration noise.
(iii) Fourier spectrum of perturbations due to celestial bodies in the solar system and the precision needed to know the positions of solar system bodies in order to separate this spectrum from GW spectrum.
(iv) Further studies in optimizing deployment delta-V and propellant ratio.
(v) Optimizing the inclination angle of the ASTROD-GW-like constellation.
(vi) Extraction of GW signals based on precise numerical orbits.
(vii) Further studies in the angular resolution of GW sources.
(viii) Separation of weak lensing effects from GW signals.

It is time to think seriously about second-generation space GW detectors — BBO, DECIGO, Super-ASTROD and the like. Optical clocks in the laboratory have reached a fractional inaccuracy at 10^{-18} level and their inaccuracy is still improving. Clocks of this accuracy level will be developed for space use. This development is good for laser pulse ranging scheme for Super-ASTROD. The laser pulse timing accuracy of 3 ps is already achieved in T2L2 on board JASON2 satellite.[149] 0.9 mm (3 ps) out of 1300 Gm (8.6 AU) is 7×10^{-16}. At 1 μHz, the characteristic strain would reach 7×10^{-19} precision. This is comparable to the best of the lower frequency strain acceleration noise level in Fig. 5. Pulse timing accuracy is still improving. It would be good to study this scheme in more detail.

Acknowledgments

I would like to thank Wei-Ping Pan for his help in drawing Figs. 4–1, and to thank An-Ming Wu and Gang Wang for helpful discussions during many collaboration years. I would also like to thank Science and Technology Commission of Shanghai Municipality (STCSM-14140502500) and Ministry of Science and Technology of China (MOST-2013YQ150829, MOST-2016YFF0101900) for supporting this work in part, and to thank the Kavli Institute for Theoretical Physics, China (KITPC) for funding the Next Detectors for Gravitational Astronomy Program (during the program, the writing of this review was started), and for their hospitality.

References

1. B. P. Abbott *et al.*, *Phys. Rev. Lett.* **116** (2016) 061102.
2. B. P. Abbott *et al.*, *Phys. Rev. Lett.* **116** (2016) 241103.

3. W.-T. Ni, *Mod. Phys. Lett. A* **25** (2010) 922.
4. The complete frequency classification of GWs, http://astrod.wikispaces.com/file/view/GW-classification.pdf.
5. K. Kuroda, W.-T. Ni and W.-P. Pan, Gravitational waves: Classification, methods of detection, sensitivities, and sources, in *One Hundred Years of General Relativity: From Genesis and Empirical Foundations to Gravitational Waves, Cosmology and Quantum Gravity*, Chap. 10, ed. W.-T. Ni (World Scientific, Singapore, 2016); *Int. J. Mod. Phys. D* **24** (2015) 1530031.
6. K. Kuroda, Ground based gravitational wave detectors, in *One Hundred Years of General Relativity: From Genesis and Empirical Foundations to Gravitational Waves, Cosmology and Quantum Gravity*, Chap. 11, ed. W.-T. Ni (World Scientific, Singapore, 2016); *Int. J. Mod. Phys. D* **24** (2015) 1530032.
7. R. N. Manchester, Pulsars and gravity, in *One Hundred Years of General Relativity: From Genesis and Empirical Foundations to Gravitational Waves, Cosmology and Quantum Gravity*, Chap. 9, ed. W.-T. Ni (World Scientific, Singapore, 2016); *Int. J. Mod. Phys. D* **24** (2015) 1530018.
8. M. Bucher, Physics of the cosmic microwave background anisotropy, in *One Hundred Years of General Relativity: From Genesis and Empirical Foundations to Gravitational Waves, Cosmology and Quantum Gravity*, Chap. 15, ed. W.-T. Ni (World Scientific, Singapore, 2016); *Int. J. Mod. Phys. D* **24** (2015) 1530004.
9. LISA Study Team, LISA (Laser Interferometer Space Antenna) — A Cornerstone Mission for the Observation of Gravitational Waves, ESA System and Technology Study Report, ESA-SCI 11 (2000).
10. P. W. McNamara, *Int. J. Mod. Phys. D* **22** (2013) 1341001.
11. M. Armano *et al.*, *Phys. Rev. Lett.* **116** (2016) 231101.
12. J. E. Faller and P. L. Bender, A possible laser gravitational wave experiment in space, in *Program and Abstracts of Second Int. Conf. Precision Measurement and Fundamental Constants (PMFC-II)*, 8–12 June 1981, National Bureau of Standards, Gaithersburg, Maryland, USA.
13. J. E. Faller and P. L. Bender, A possible laser gravitational wave experiment in space, in *Precision Measurement and Fundamental Constants II*, eds. B. N. Taylor and W. D. Phillips, Vol. 617, Special Publication (National Bureau of Standards, USA, 1984), pp. 689–690.
14. J. E. Faller, P. L. Bender, J. L. Hall, D. Hils and M. A. Vincent, Space antenna for gravitational wave astronomy, in *Proc. Colloquium on Kilometric Optical Arrays in Space* (ESA, 1985), SP-226.
15. B. Iyer and W.-T. Ni, *Int. J. Mod. Phys. D* **22** (2013) 1302001.
16. K. Kuroda and N. Mio, *Metrologia* **28** (1991) 75.
17. H.-C. Yeh, Q.-Z. Yan, Y.-R. Liang, Y. Wang and J. Luo, *Rev. Sci. Instrum.* **82** (2011) 044501.
18. F. Gao, Z. B. Zhou and J. Luo, *Chin. Phys. Lett.* **28** (2011) 080401.
19. J. Luo *et al.*, *Class. Quantum Grav.* **33** (2016) 035010; e-mail correspondence with Hsien-Chi Yeh on May 21, 2015.
20. D. A. Binns, N. Rando and L. Cacciapuoti, *Adv. Space Res.* **43** (2009) 1158.
21. O. Jennrich *et al.*, NGO (New Gravitational wave Observatory) assessment study Report, ESA/SRE (2011) 19.
22. http://www.esa.int/Our_Activities/Space_Science/ESA_s_new_vision_to_study_the_invisible_Universe.
23. W.-T. Ni, ASTROD and gravitational waves, in *Gravitational Wave Detection*, eds. K. Tsubono, M.-K. Fujimoto and K. Kuroda (Universal Academy Press, Tokyo, Japan, 1997), pp. 117–129.

24. W.-T. Ni, S.-s. Pan, G.-S. Peng, J.-T. Shy, S.-M. Tseng, S.-L. Tsao, S.-E. Wang, J. Wu, S.-A. Wu and H.-C. Yeh, *Class. Quantum Grav.* **13** (1996) A311.
25. W.-T. Ni *et al.*, *Proc. SPIE 3116: Small Spacecraft, Space Environments, and Instrumentation Technologies* (1997), p. 105.
26. W.-T. Ni *et al.*, *Adv. Space Res.* **32** (2003) 1437.
27. X. Xu and W.-T. Ni, Gravitational modeling of the proof-mass for the ASTROD mission, *Paper Presented* in *31st COSPAR Scientific Assembly*, Birmingham, 14–21, July, 1996; *National Tsing Hua University Preprint GP-076*, July, 1996; *Adv. Space Res.* **32** (2003) 1443.
28. W.-T. Ni, J.-T. Shy, S.-M. Tseng and H.-C. Yeh, *Adv. Space Res.* **32** (2003) 1283.
29. A.-C. Liao, W.-T. Ni and J.-T. Shy, *Publ. Yunnan Obs.* **3** (2002) 88 (in Chinese).
30. A.-C. Liao, W.-T. Ni and J.-T. Shy, *Int. J. Mod. Phys. D* **11** (2002) 1075.
31. H. Selig, C. Lämmerzahl and W.-T. Ni, *Int. J. Mod. Phys. D* **22** (2013) 1341003.
32. C. Braxmaier, H. Dittus and B. Foulon, *Exp. Astron.* **34** (2012) 181, arXiv:1104.0060.
33. T. Appouchaux *et al.*, *Exp. Astron.* **23** (2009) 491.
34. W.-T. Ni *et al.*, ASTROD I: Mission concept and Venus flybys, in *Proc. 5th IAA Int. Conf. Low-Cost Planetary Missions, ESTEC, Noordwijk*, The Netherlands, 24–26 September 2003, ESA SP-542, November 2003, pp. 79–86.
35. W.-T. Ni *et al.*, *Acta Astron.* **59** (2006) 598.
36. W.-T. Ni, ASTROD Optimized for Gravitational-wave Detection: ASTROD-GW — A pre-Phase A study proposal submitted to Chinese Academy of Sciences, February 26, 2009.
37. W.-T Ni *et al.*, ASTROD Optimized for Gravitational Wave Detection: ASTROD-GW, in *Proc. Sixth Deep Space Exploration Technology Symp.*, December 3–6, 2009, Sanya, Hainan, China, pp. 122–128 (in Chinese).
38. W.-T. Ni *et al.*, ASTROD Optimized for Gravitational Wave Detection: ASTROD-GW, *Paper (COSPAR Paper Number H05-0007-10) Presented in the 38th COSPAR Scientific Assembly*, 18–25 July 2010, Bremen, Germany.
39. W.-T. Ni, Dark energy, co-evolution of massive black holes with galaxies, and ASTROD-GW, *Paper (COSPAR Paper Number H05-0017-10) Presented in the 38th COSPAR Scientific Assembly*, 18–25 July 2010, Bremen, Germany (2010); *Adv. Space Res.* **51** (2013) 525–534, arXiv:1104.5049.
40. W.-T. Ni, *Int. J. Mod. Phys. D* **22** (2013) 1431004.
41. G. Wang and W.-T Ni, *Chin. Phys. B* **24** (2015) 059501.
42. W.-T. Ni, *Class. Quantum Grav.* **26** (2009) 075021, arXiv:0812.0887.
43. W.-T. Ni and A.-M. Wu, Orbit design of ASTROD-EM, paper in preparation.
44. S. Kawamura *et al.*, *Class. Quantum Grav.* **23** (2006) S125.
45. J. Crowder and N. J. Cornish, *Phys. Rev. D* **72** (2005) 083005.
46. N. Seto, *Phys. Rev. D* **73** (2006) 063001.
47. P. L. Bender, *Class. Quantum Grav.* **21** (2004) S1203.
48. X. Gong *et al.*, *J. Phys.: Conf. Ser.* **610** (2015) 012011, arXiv:1410.7296.
49. M. Tinto, J. C. N. de Araujo, O. D. Aguiar and M. E. S. Alves, arXiv:1111.2576.
50. M. Tinto, J. C. N. de Araujo, O. D. Aguiar and M. E. S. Alves, *Astropart. Phys.* **48** (2013) 50.
51. M. Tinto, D. Debra, S. Buchman and S. Tilley, *Rev. Sci. Instrum.* **86** (2015) 014501.
52. S. T. McWilliams, Geostationary Antenna for Disturbance-Free Laser Interferometry (GADFLI), arXiv:1111.3708v1.
53. J. W. Conklin *et al.*, LAGRANGE: LAser GRavitational-wave ANtenna at GEo-lunar Lagrange points, arXiv:1111.5264v2.
54. B. Hiscock and R. W. Hellings, *Bull. Am. Astron. Soc.* **29** (1997) 1312.

55. R. Hellings, S. L. Larson, S. Jensen, C. fish, M. Benacquista, N. Cornish and R. Lang, A low-cost, high-performance space gravitational astronomy mission: A mission-concept white paper submitted to NASA, 2011, http://pcos.gsfc.nasa.gov/studies/rfi/GWRFI-0007-Hellings.pdf.
56. W.-T. Ni, Solar-system tests of relativistic gravity, in *One Hundred Years of General Relativity: From Genesis and Empirical Foundations to Gravitational Waves, Cosmology and Quantum Gravity*, Chap. 8, ed. W.-T. Ni (World Scientific, Singapore, 2016); *Int. J. Mod. Phys. D* **25** (2016) 1630003.
57. C. M. Will, *Living Rev. Rel.* **17** (2014) 4, http://www.livingreviews.org/lrr-2014-4.
58. W.-T. Ni, *Int. J. Mod. Phys. D* **14** (2005) 901.
59. W.-T. Ni, *Int. J. Mod. Phys. D* **17** (2008) 921.
60. BepiColombo, https://en.wikipedia.org/wiki/BepiColombo.
61. BepiColombo, http://sci.esa.int/bepicolombo/.
62. A. Milani, D. Vokrouhlický, D. Villani, C. Bonanno and A. Rossi, *Phys. Rev. D* **66** (2002) 082001.
63. D. K. Yeomans *et al.*, *Science* **289** (2000) 2085.
64. NASA, "Press Release: NASA Completes MESSENGER Mission with Expected Impact on Mercury's Surface". April 30, 2015. Retrieved May 2, 2015.
65. V. B. Braginsky and M. E. Gertsenshtein, *Sov. Phys.-JETP Lett.* **5** (1967) 287.
66. A. J. Anderson, *Nature* **229** (1971) 547.
67. R. W. Davis, in *Colloque Internationalaux CNRS No.* 220, '*Ondes et Radiations Gravitationelles*', 1974, Institut Henri Poincare: Paris, p. 33.
68. F. B. Estabrook and H. D. Wahlquist, *Gen. Relativ. Gravit.* **6** (1975) 439.
69. H. D. Wahlquist, *Gen. Relativ. Gravit.* **19** (1987) 1101.
70. J. W. Armstrong, F. B. Estabrook and M. Tinto, *Astrophys. J.* **527** (1999) 814.
71. M. Tinto and M. E. da Silva Alves, *Phys. Rev. D* **82** (2010) 122003.
72. J. W. Armstrong, R. Woo and F. B. Estabrook, *Astrophys. J.* **230** (1979) 570.
73. R. W. Hellings, P. S. Callahan, J. D. Anderson and A. T. Moffett, *Phys. Rev. D* **23** (1981) 844.
74. J. D. Anderson, J. W. Armstrong, F. B. Estabrook, R. W. Hellings, E. K. Law and H. D. Wahlquist, *Nature* **308** (1984) 158.
75. J. W. Armstrong, F. B. Estabrook and H. D. Wahlquist, *Astrophys. J.* **318** (1987) 536.
76. J. W. Armstrong, L. Iess, P. Tortora and B. Bertotti, *Astrophys. J.* **599** (2003) 806.
77. M. Tinto, G. J. Dick, J. D. Prestage and J. W. Armstrong, *Phys. Rev. D* **79** (2009) 102003.
78. T. W. Murphy *et al.*, *Class. Quantum Grav.* **29** (2012) 184005.
79. T. W. Murphy *et al.*, *Publ. Astron. Soc. Pac.* **120** (2008) 20, arXiv:0710.0890.
80. P. Exertier, É. Samain, P. Bonnefond and P. Guillemot, *Adv. Space Res.* **46** (2010) 1559.
81. É. Samain, Clock comparisons based on laser ranging technologies, in *One Hundred Years of General Relativity: From Genesis and Empirical Foundations to Gravitational Waves, Cosmology and Quantum Gravity*, Chap. 7, ed. W.-T. Ni (World Scientific, Singapore, 2016); *Int. J. Mod. Phys. D* **24** (2015) 1530016.
82. G. J. Dick *et al.*, IPN Progress Report **42** (2008) 175.
83. S. V. Dhurandhar, W.-T. Ni and G. Wang, *Adv. Space Res.* **51** (2013) 198.
84. G. Wang and W.-T. Ni, *Class. Quantum Grav.* **30** (2013) 065011.
85. G. Wang and W.-T. Ni, *Chin. Astron. Astrophys.* **36** (2012) 211.
86. G. Wang and W.-T Ni, *Chin. Phys. B* **22** (2013) 049501.
87. J. W. Armstrong, F. B. Estabrook and M. Tinto, *Astrophys. J.* **527** (1999) 814.

88. M. Tinto and S. V. Dhurandhar, *Living Rev. Rel.* **17** (2014) 6.
89. K. G. Arun *et al.*, *Class. Quantum Grav.* **26** (2009) 094027.
90. P. L. Bender, *Class. Quantum Grav.* **20** (2003) S301.
91. C. C. Speake and S. M. Aston, *Class. Quanutum Grav.* **22** (2005) S269.
92. K.-X. Sun, G. Allen, S. Buchman, D. Debra and R. Byer, *Class. Quantum Grav.* **22** (2005) S287.
93. A. Pulido Patón, *Int. J. Mod. Phys. D* **17** (2008) 941.
94. P. Amaro-Seoane *et al.*, *Class. Quantum Grav.* **29** (2012) 124016.
95. M. Ando, presented original DECIGO target sensitivity in the form of numerical data.
96. E. Thrane and J. D. Romano, *Phys. Rev. D* **88** (2013) 124032.
97. J. M. Hogan *et al.*, *Gen. Relativ. Gravit.* **43** (2011) 1953.
98. J. M. Hogan and M. A. Kasevich, Atom interferometric gravitational wave detection using heterodyne laser links, arXiv:1501.06797.
99. P. L. Bender, *Phys. Rev. D* **84** (2011) 028101.
100. S. Dimopoulos *et al.*, *Phys. Rev. D* **84** (2011) 028102.
101. R. Geiger *et al.*, Matter-Wave Laser Interferometric Gravitation Antenna (MIGA): New perspectives for fundamental physics and geosciences, in *Proc. 50th Rencontres de Moriond "100 years after GR"*, La Thuile (Italy), 21–28 March 2015, arXiv:1505.07137.
102. A. Sesana, A. Vecchio and C. N. Colacino, *Mon. Not. R. Astron. Soc.* **390** (2008) 192.
103. A. Sesana, A. Vecchio and M. Volonteri, *Mon. Not. R. Astron. Soc.* **394** (2009) 2255.
104. P. Demorest *et al.*, Gravitational Wave Astronomy Using Pulsars: Massive Black Hole Mergers & The Early Universe, *White paper submitted to the Astro2010 Decadal Survey*, 2009, arXiv:0902.2968.
105. B. F. Schutz, Revealing a hidden universe, ppt presentation on April 02, 2012, slide no. 25.
106. L. Lentati *et al.*, *Mon. Not. R. Astron. Soc.* **453** (2015) 2576.
107. R. M. Shannon *et al.*, *Science* **349** (2015) 1522.
108. Z. Arzoumanian *et al.*, *Astrophy. J.* **821** (2016) 13.
109. K. G. Arun and A. Pai, *Int. J. Mod. Phys. D* **22** (2013) 1341012.
110. K. Yagi, *Int. J. Mod. Phys. D* **22** (2013) 1341013.
111. E. J. Copeland, M. Sami and S. Tsujikawa, *Int. J. Mod. Phys. D* **15** (2006) 1753.
112. B. F. Schutz, *Nature* **323** (1986) 310.
113. B. S. Sathyaprakash and B. F. Schutz, *Living Rev. Rel.* **12** (2009) 2, http://www.livingreviews.org/lrr-2009-2.
114. A. Kamble, *Int. J. Mod. Phys. D* **22** (2013) 1341011.
115. W.-T. Ni, *Int. J. Mod. Phys. A* **24** (2009) 3493, arXiv:0903.0756v1.
116. A. J. Farmer and E. S. Phinney, *Mon. Not. R. Astron. Soc.* **346** (2003) 1197.
117. M. A. Vincent and P. L. Bender, in *Proc. Astrodynamics Specialist Conf.* (Kalispell USA, Univelt, San Diego, 1987), p. 1346.
118. W. M. Folkner, F. Hechler, T. H. Sweetser, M. A. Vincent and P. L. Bender, *Class. Quantum Grav.* **14** (1997) 1405.
119. C. Cutler, *Phys. Rev. D* **57** (1998) 7089.
120. S. P. Hughes, Preliminary optimal orbit design for laser interferometer space antenna, *25th Ann. AAS Guidance and Control Conf.* (Breckenridge CO, Feb. 2002).
121. F. Hechler and W. M. Folkner, *Adv. Space Res.* **32** (2003) 1277.
122. S. V. Dhurandhar, K. Rajesh Nayak, S. Koshti and J.-Y. Vinet, *Class. Quantum Grav.* **22** (2005) 481.

123. G. W. Hill, *Am. J. Math.* **1** (1878) 5.
124. W. H. Clohessy and R. S. Wiltshire, *J. Aerosp. Sci.* **27** (1960) 653.
125. W. M. Folkner, J. G. Williams, D. H. Boggs, R. S. Park and P. Kuchynka, The Planetary and Lunar Ephemerides DE430 and DE431, Interplanetary Network Progress Report, 196:C1, (2014).
126. E. V. Pitjeva, *Solar Syst. Res.* **47** (2013) 386.
127. A. Fienga, H. Manche, J. Laskar and M. Gastineau, *Astron. Astrophys.* **477** (2008) 315.
128. D.-W. Chiou and W.-T. Ni, *Adv. Space Res.* **25** (2000) 1259.
129. D.-W. Chiou and W.-T. Ni, Orbit Simulation for the Determination of Relativistic and Solar-System Parameters for the ASTROD Space Mission, *Paper Presented in 33rd COSPAR Scientific Assembly*, Warsaw, 16–23 July 2000, arXiv:astro-ph/0407570.
130. C.-J. Tang and W.-T. Ni, Asteroid Perturbations and Mass Determination for the ASTROD Space Mission, *Paper Presented in the 33rd COSPAR Scientific Assembly*, Warsaw, 16–23 July, 2000, arXiv:astro-ph/0407606.
131. C.-J. Tang and W.-T. Ni, *Publ. Yunnan Obs.* **3** (2002) 21 (in Chinese).
132. V. A. Brumberg, *Essential Relativistic Celestial Mechanics* (Adam Hilger, Bristol, 1991), pp. 177–178.
133. J. R. Men, W.-T. Ni and G. Wang, ASTROD-GW mission orbit design, in *Proc. Sixth Deep Space Exploration Technology Symp.*, December 3–6, 2009 Sanya, China (2009), p. 47.
134. J. R. Men, W.-T. Ni and G. Wang, *Acta Astron. Sin.* **51** (2009) 198.
135. J. R. Men, W.-T. Ni and G. Wang, *Chin. Astron. Astrophys.* **34** (2010) 434.
136. Z. Yi *et al.*, *Int. J. Mod. Phys. D* **17** (2008) 1005.
137. G. Li *et al.*, *Int. J. Mod. Phys. D* **17** (2008) 1021.
138. G. Wang and W.-T. Ni, *Acta Astron. Sin.* **52** (2011) 427.
139. The Asteroid Orbital Elements Database, ftp://ftp.lowell.edu/pub/elgb/astorb.html.
140. A.-M. Wu and W.-T. Ni, *Int. J. Mod. Phys. D* **22** (2013) 1341005.
141. A.-M. Wu, W.-T. Ni and G. Wang, Formation Design for Various Gravitational Wave Missions, Paper No. IAC-16-A2.1.6, *67th Int. Astronautical Congress* (*IAC*), Guadalajara, Mexico, 26–30 September 2016.
142. G. de Vine, B. Ware, K. McKenzie, R. E. Spero, W. M. Klipstein and D. A. Shaddock, *Phys. Rev. Lett.* **104** (2010) 211103.
143. D. Risquez and R. Keil, GAIA Attitude Model, Micro-Propulsion Sub-System, Technical Report, Leiden Observatory, GAIA-C2-TN-LEI-DRO-003 (2010).
144. https://microscope.cnes.fr/en/MICROSCOPE/GP_mission.htm.
145. A. D. Ludlow, M. M. Boyd, J. Ye, E. Peik and P. O. Schmidt, *Rev. Mod. Phys.* **87** (2015) 637.
146. C. Grimani, *Int. J. Mod. Phys. D* **22** (2013) 1341006.
147. D. Shaul, K. Aplin, H. Araujo, R. Bingham, J. Blake, G. Branduardi-Raymont, S. Buchman, A. Fazakerley, L. Finn, L. Fletcher, A. Glover, C. Grimani, M. Hapgood, B. Kellet, S. Matthews, T. Mulligan, W.-T. Ni, P. Nieminen, A. Posner, J. Quenby, P. Roming, H. Spence, T. Sumner, H. Vocca, P. Wass and P. Young, *AIP Conf. Proc.* **873** (2006) 172.
148. P. L. Bender and D. Hils, *Class. Quantum Grav.* **14** (1997) 1439.
149. É. Samain, Clock comparison based on laser ranging technologies, in *One Hundred Years of General Relativity*: *From Genesis and Empirical Foundations to Gravitational Waves, Cosmology and Quantum Gravity*, Chap. 7, ed. W.-T. Ni (World Scientific, Singapore, 2016); *Int. J. Mod. Phys. D* **24** (2015) 1530021.

Subject Index

F

G

H

I

N

O

P

Q

T

U

V

W

Author Index

Authors of chapters and the first 3 authors of references

C

D

H

I

J

M

T

X

Y

Z

www.ingramcontent.com/pod-product-compliance
Lightning Source LLC
LaVergne TN
LVHW081353011026
849161LV00013BA/2807
* 9 7 8 9 8 1 4 6 7 8 4 8 3 *